# The Preparato Manual of Black Powder and Pyrotechnics Version 4.0

# Volume 1

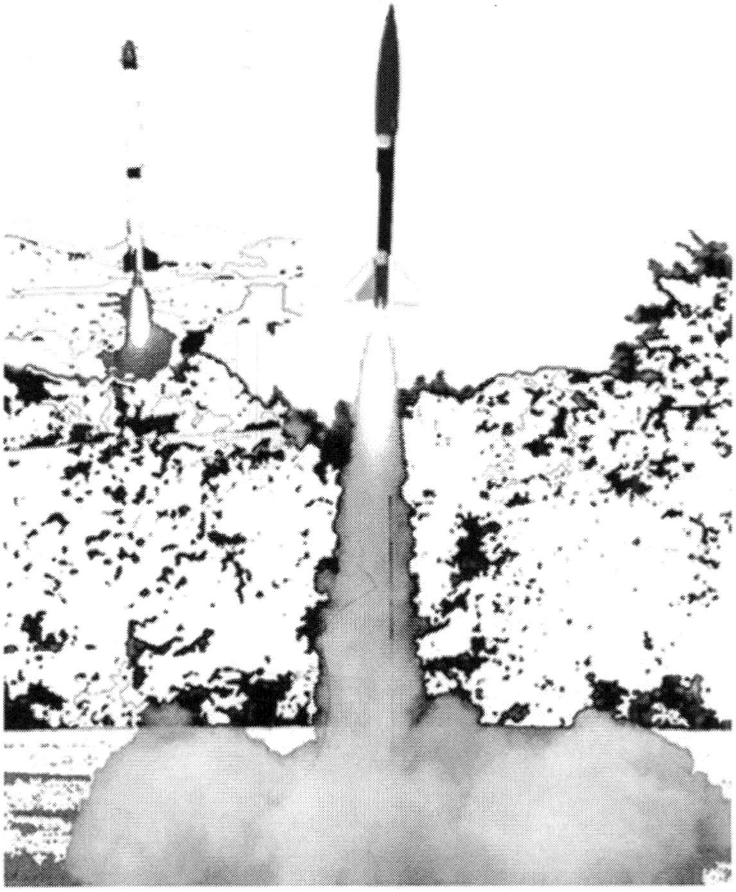

**The Preparatory Manual of Black Powder and Pyrotechnics Version 4.0 Volume 1®**





**The Preparatory Manual of Black Powder and Pyrotechnics Version 4.0 Volume 1**
**ISBN NO. 9781980538271**

# - Table of contents -

# DISCLAIMER WARNING!

Volumes 1 and 2 are a professional collection of over 1100+ pyrotechnic formulations, some of which are intended for professional pyrotechnicians only; however, despite the obvious, many of the formulations can be made by amateurs as long as the given procedure(s) are followed with exact detail. Nevertheless, some of the recipes and mixtures in this book can be dangerous under extra ordinary circumstances, and should not be prepared by inexperienced personnel unless other wise stated in the procedure(s); all safety and hazard info will be given in detail for each entry. Furthermore, all of the recipes and mixtures in this book have been thoroughly tested and work, without exceptions. This book is intended to catalogue and organize the various and countless number of pyrotechnic compositions and only the best of the thousands of possible pyrotechnic compositions have been selected for use in this book. Case specific hazards in this book mainly include the compounds of the following:

1. Nitrocellulose
2. Nitroglycerine
3. Azide compounds
4. Lead styphnates
5. Potassium chlorate
6. Antimony trisulfide (in combination with potassium chlorate and/or glass powder)
7. Radioactive compounds (used in experimental compositions)
8. Arsenic and mercury compounds
9. Picric acid

Nitrocellulose is a case specific hazard due its ability to self deflagrate, meaning it ignites and burns violently when heated, touched by a flame, or spark. Therefore only trained and qualified personnel should handle nitrocellulose, or attempt to prepare it unless other wise stated in the procedure(s).

Nitroglycerine is a case specific hazard mostly in part to the fact it produces severe headaches and migraines when ingested, absorbed through the skin, or inhaled. Nitroglycerine in the past has been widely criticized as an explosion hazard, buts its stability is greater then once preceded; however, only trained and qualified personnel should attempt to make nitroglycerine, or handle it in any way. Nitroglycerine is widely used in high performance gun propellants for military use, and hence, all nitroglycerine-containing compositions in this book have been thoroughly tested, and are generally regarded as stable, but amateurs should not attempt to prepare such mixtures.

Azide compounds are case specific hazards due to their ability to detonate, especially the heavy metal azides. Lead azide is one such example, which is easily detonated by heat, spark, friction, or flame. Other metal azides such as sodium azide are highly toxic compounds and should not be handled by untrained and unqualified personnel.

Heavy metal styphnates such as lead styphnate are case specific hazards due to the fact that they are easily detonated by heat, shock, flame, or spark, and should not be prepared or used by untrained unqualified personnel in anyway unless other wise stated in the procedure(s).

Potassium chlorate is a case specific hazard due to its ability to form friction sensitive compositions, many of which have a tendency to deflagrate without warning when rubbed, or by other friction means. Potassium chlorate mixtures containing sulfur listed in this book have been thoroughly tested and are generally regarded as safe. Potassium chlorate and its compositions can react and ignite and even explode spontaneously when exposed to concentrated sulfuric acid. Never mix potassium chlorate or its compositions with concentrated sulfuric acid; as well, never mix potassium chlorate or potassium perchlorate and their compositions with hydrochloric acid, as toxic fumes of chlorine gas will be evolved; note: this is actually a positive test for chlorates and perchlorates used in analytical chemistry, i.e. titration using conc. hydrochloric acid.

Antimony trisulfide in combination with potassium chlorate and glass powder is regarded as a case specific hazard due to its tendency to easily ignite when rubbed or by friction. All the compositions listed in this book that contain antimony trisulfide in combination with potassium chlorate and glass powder are safe, and have been thoroughly tested; however, untrained and unqualified personnel should not attempt to handle or deal with such compositions unless other wise stated in the procedure(s).

Radioactive compounds are obviously case specific hazards due to the radiation they emit. Any radioactive containing compositions listed in this book have been widely tested and are safe; however, obviously untrained and unqualified personnel should not attempt to work with radioactive materials in anyway; proper shielding is necessary, not to mention radioactive compounds are technically illegal to poses without proper licensing and proper storage and shielding facilities.

Arsenic and mercury containing compositions are case specific hazards as they are highly toxic. Arsenic and mercury compounds are highly toxic through ingestion, skin absorption, or inhalation, and should only be handled by trained and qualified professionals unless other wise stated in the procedure(s).

Picric acid is a case specific hazard due to the fact that reacts with bases and metal oxides forming primary explosives, all of which easily detonate or deflagrate when heated, or touched by a flame or spark. Metal picrates are known to self-deflagrate when touched by a flame or spark.

### A note to beginners, and amateurs

It should be known that the field of pyrotechnics is extremely vast, and encompasses multiple levels ranging from fireworks, to high performance rocket fuels, gun propellants, all the way down to simple concoctions, to specialty compositions. This book will discuss all categories of pyrotechnics, and as previously mentioned only the best of formulations have been selected. Now it needs to be understood that working with pyrotechnics does have its hazards; these hazards are more common sense, rather then written-in-stone "commandments" detailing every aspect of health and safety regarding the chemicals used, and the ways they are mixed; with this in mind, pyrotechnic formulations are simply just that, "mixed" chemicals to form a formulation other wise known as a "composition". As you will learn, many pyrotechnic formulations contain an "oxidizer", and a "fuel". Now, in the world of pyrotechnics, many of the oxidizers are selected from the group of "chlorates", "perchlorates", "nitrates", "dichromates", "permanganates", and "metal peroxides", and the fuel is usually a combustible material such as charcoal, sulfur, or any combustible agent including sugar to powdered metals. Now at first, these oxidizers may be scary to new comers, and mixing them with combustible materials may seem dangerous at first glance, but in fact, no pyrotechnic formulation would even exist unless the proper amount of various chemicals were thoroughly "mixed" at some time throughout the centuries to begin with; this mixing can make you nervous or scared at first. For example, "what's the best method of mixing the chemicals"?, "will the mixture explode"?, or the common: "should I really be doing this"?.........ect., ect.; yes it is important to understand the dangers involved in mixing potentially explosive materials, as such, all safety and hazard information will be given for each of the procedures in this book. As you will learn there are several main types of mixing apparatus available, mainly "ball mills" and "vertical mixers". In some cases, one method of mixing is desired over the other; with this in mind, I prefer vertical mixers to ball mills, but each work satisfactorily. It should be noted, that mixing certain chemicals by hand, such as using a mortar & pestle can be quite dangerous due to the friction involved during the mixing process. This is why I prefer vertical mixers because the blades rotate, minimizing friction and heat build up.

If you are afraid to mix chemicals then maybe pyrotechnics is not your thing. Many sites, mainly social media, will make you afraid of mixing certain chemicals, and they will act like it's instant death. However, in most cases, this is not true. The industry of pyrotechnics spans a galaxy of compounds and centuries of trial and error. With this in mind, at no point would I subject anyone to dangerous circumstances as a result of the formulas listed in this book; if you are afraid, then perhaps you should become an accountant. Yes, some mixtures are dangerous, but like I said, using common sense defeats this.

### Note on using granulated chemicals and/or chemicals in the form of "pellets", or "crystals" otherwise in non-powdered form

Many of the chemicals used in this book, including such materials as ammonium nitrate or sodium nitrate come delivered from the manufacturer or seller in the form of granulated pellets or otherwise a non-powdered product. For use in this book, any and all chemicals that come delivered from the manufacturer/seller in the form of pellets, grains, chips, flakes, cubes, or crystals of any size considered "non-powdered", need to be pulverized into a powder before use. This pulverization can easily be achieved by placing the granulated product or crystals, for example, ammonium or sodium nitrate into a standard kitchen style blender utilizing non-serrated knife-type steel blades, and dry blending the compound on moderate speed until a fine powder is obtained. Make sure to wash out the blender after each use, and make sure to use only one compound at a time.

# Chapter 1: Pyrotechnic compositions

### 1a. Introduction to pyrotechnics

The field of pyrotechnics is very old, and dates back to the ancient Chinese dynasties. Of coarse in those days, there where only a few known pyrotechnics compositions, one of which being black powder. Black powder is probably one of the oldest chemical concoctions known to man. This concoction, however very simple, has re-defined history, culture, leaders, governments, and kingdoms. Empires have risen and fallen on the back of black powder, and our modern civilization that we known of, was shaped and coexists because of black powder, and its simple pyrotechnic ability. Today, black powder is seldom even considered for use in anything other then fireworks, old muskets guns used by hobbyists, and cannons for old war re-enactments. Other then these entities, black powder has very little use outside the area of fireworks, and remains an ingredient for the hobbyist, mainly black powder time fuse, and personally owned cannons for home and recreational use, such as firing salutes during weekend celebrations, or during a touchdown by the home team. Today, the world of pyrotechnics has expanded far beyond the realm of simple black powder cannons and small arms weapons such as muzzle-loaded firearms. Today, pyrotechnics occupies a much broader field not just in fireworks, but also in the areas of gun propellants, smoke devices, signals and flares, high performance rocket propellants, and gas generating compositions for various military, industrial, and commercial applications. This book will discuss many of these applications, but will focus mainly on the preparation of such compositions and their ultimate importance in our modern civilization.

### 2b. What makes pyrotechnic compositions what they are?

For most of you, you are probably already aware of what makes pyrotechnic compositions work; however, some readers may not be aware to any significant extent, so I will provide a brief yet simple tutorial on pyrotechnics. First of all, 95% of all pyrotechnic compositions have a fundamental characteristic to one another, and that fundamental characteristic is oxidizers, and reducing agents. Oxidizing agents and reducing agents go hand in hand to make pyrotechnic compositions work. Oxidizers are a classification of chemical compounds that contain an abundance of oxygen within the molecules of the oxidizing substance. This abundance of oxygen gives oxidizing agents specific physical properties, and makes them reactive towards reducing agents. Reducing agents are a class of chemical compound that do not necessarily contain oxygen, but contain atoms that are excited when ignited in the presence of oxidizing compounds. The excited atoms within reducing agents allow the reducing agents to become "oxidized" by the oxidizer, meaning the excited atoms, when ignited in the presence of oxidizing compounds, become oxidized, meaning they take up oxygen from the oxidizing agent and hence become "oxidized". When reducing agents take up oxygen from an oxidizing substance, the oxidizer itself becomes reduced as it looses its oxygen, and the reducing agent becomes oxidized as it gains oxygen. This strange process is called oxidation/reduction and is the foundation to the chemistry of pyrotechnic compositions. Note: reducing agents are sometimes called "combustion" agents, as many of them tend to be flammable. Lets look at some examples of oxidizing and reducing agents:

| COMMON OXIDIZERS | COMMON REDUCERS |
|---|---|
| Sodium nitrate | Aluminum powder |
| Potassium nitrate | Magnesium powder |
| Potassium chlorate | Sulfur |
| Sodium perchlorate | Charcoal |
| Sodium peroxide | Sugar |
| Potassium permanganate | Antimony sulfide |
| Potassium dichromate | Zinc phosphide |

All of the oxidizing agents contain an abundance of oxygen, making these compounds extra reactive. Notice how the reducing agents listed in the table have no oxygen. Now, just because they don't contain oxygen does not make them reducing agents, but what does, is the fact that all of the substances contain atoms that are easily oxidized, meaning the atoms prefer to be in a higher state of energy. To understand this higher state of energy, you must understand the various energies in the atoms themselves. Lets look at the energies of sodium chlorate, and aluminum powder. The central atom in sodium chlorate is a chlorine atom. This chlorine atom has an oxidation state of +5, making it more reactive then most typical compounds. When we look at aluminum, the aluminum is in its ground state, and is easily oxidized, meaning it wants to be in a higher state of energy. To better explain this look at the following diagram.

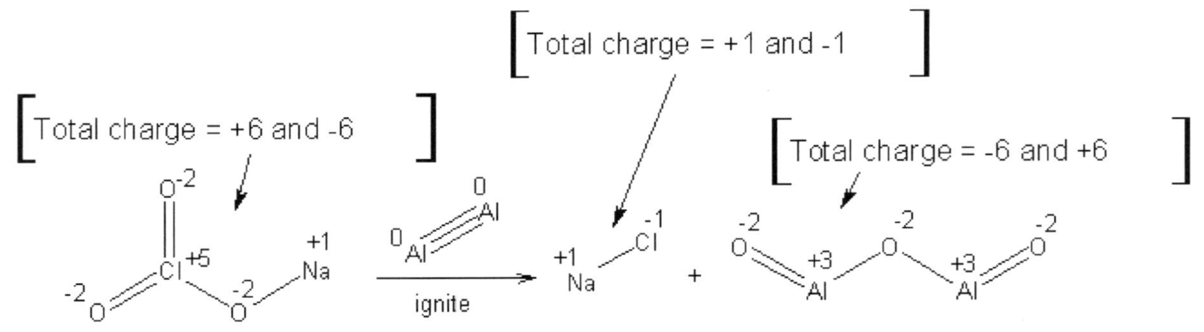


Notice in the previous diagram how the chlorine atom on the left is in the +5 oxidation state, and after the reaction (on the right side of the arrow), we see the chlorine atom is in the –1 oxidation state. So what does this mean? Well because the chlorine atom was in a +5 oxidation state, it wants to be in its –1 oxidation state, therefore, any compound that contains a +5 chlorine atom will be more reactive then most compounds, and hence makes the compound an oxidizing agent, i.e., sodium chlorate. Now during the reaction, the +5 chlorine atom gains electrons from the oxygen atoms forming chlorine and oxygen. This chlorine reacts directly with the sodium forming sodium chloride and the oxygen reacts with the aluminum, forming aluminum oxide. The reason why the oxygen reacts with the aluminum, and not with the sodium is because the oxygen has a higher affinity towards the aluminum then the chlorine, and so then the chlorine has only the sodium to bond to. The entire reaction can take only seconds in its entirety, and the energy released is given off as products, light, heat, and fire. The latter is what is used in pyrotechnics to achieve a specific effect, as seen in fireworks.

Most oxidizing agents give up oxygen when ignited or heated. Note: In most pyrotechnic compositions, the oxidizer is the source of oxygen, and this same oxygen is what "burns" the reducing agent producing light, heat, and fire. When this oxygen is liberated, it is in a highly reactive state, and bonds immediately and violently with the reducing agent. If we look at the burning of a piece of wood for example, the oxidizer, being the oxygen from the air, combines with products from the reducing agent, being wood. In all oxidation and reduction examples, matter is neither created nor destroyed, but simply converted from one form to the next. In the reaction of sodium chlorate with aluminum powder, the reactants are sodium chlorate and aluminum, and the products of the reaction are sodium chloride (NaCl) and aluminum oxide ($Al_2O_3$). In the reaction of oxygen with wood, we get carbon, carbon dioxide, carbon monoxide, water, nitrogen, and large cyclic hydrocarbons, which we see as "smoke". In the reaction of sodium chlorate with aluminum, we see a white smoke, which is composed of the aluminum oxide, and we see a residue left behind after the reaction, which is the sodium chloride. Also, a major product in combustion is light, heat, and of coarse, fire.

If your still having trouble understating oxidation and reduction, lets look at one more example. The following diagram shows potassium nitrate reacting with sulfur to yield the products of sulfur dioxide, potassium oxide, and nitrogen gas. In the reaction, the nitrogen within the potassium nitrate is in the +5 oxidation state; however it does not want to be at this state, so it is active and waits to react with a reducing agent. When the potassium nitrate reacts with the sulfur, the central nitrogen in the +5 state gains electrons from the oxygens', liberating nitrogen and oxygen. The oxygen then bonds with the sulfur and potassium forming sulfur dioxide and potassium oxide. Now, why doesn't the nitrogen bond with the potassium like the chlorine with the sodium in the previous example? The answer is because nitrogen is much more inert then chlorine and likes to bond with itself forming diatomic nitrogen. Before you get confused, remember that these characteristics are residing in general chemistry, which a whole new ball game in itself.

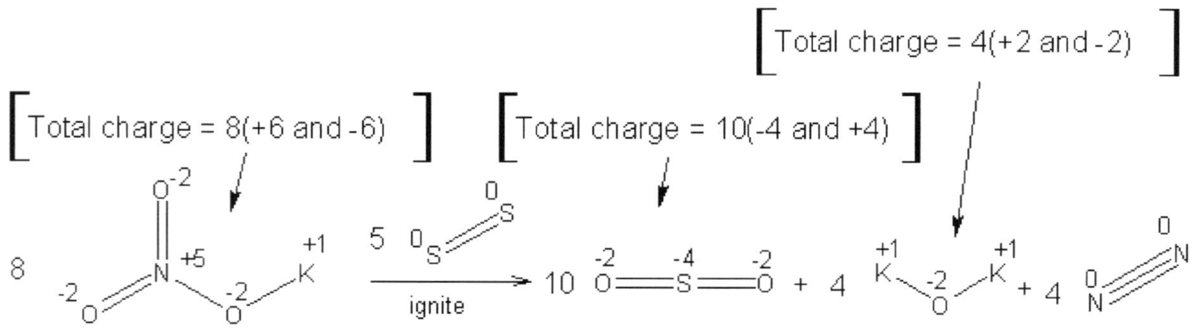


Lets now examine the oxidation states of the various atoms found in this book. These oxidation states are getting into general chemistry, which is something I want to avoid, but I feel that a quick lesson in the oxidation states of simple atoms will better explain the chemistry involved during combustion of the pyrotechnic compounds. Lets look at the following table:

| Oxidizing agent | Oxidation number of the central atom thereof |
|---|---|
| Sodium nitrate - $NaNO_3$ | Nitrogen = +5 (wants to be in the 0 state) |
| Potassium nitrate - $KNO_3$ | Nitrogen = +5 (wants to be in the 0 state) |
| Potassium chlorate - $KClO_3$ | Chlorine = +5 (wants to be in the –1 state) |
| Potassium perchlorate - $KClO_4$ | Chlorine = +7 (really wants to be in –1 state) |
| Potassium permanganate - $KMnO_4$ | Manganese = +7 (really wants to be in the +2, or +4 state) |
| Potassium dichromate - $K_2Cr_2O_7$ | Chromium = +6 (really wants to be in 0, or +2 state) |

Now, look at the following diagram to get a rough idea of the oxidation states in each central atom in the corresponding oxidizing agent.

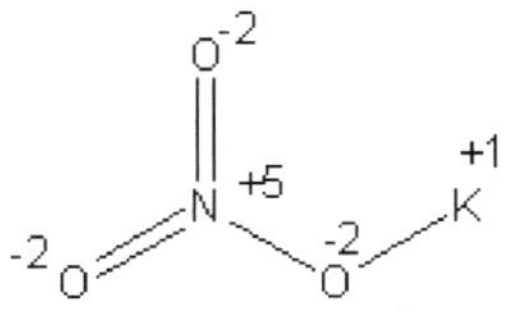

Potassium nitrate

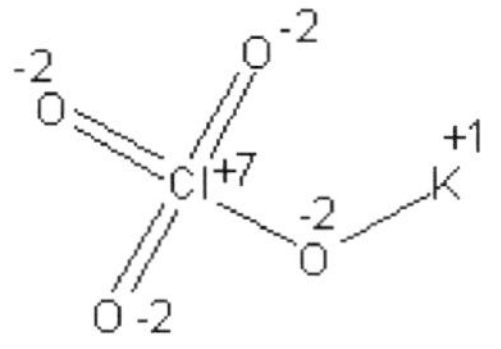

Potassium perchlorate

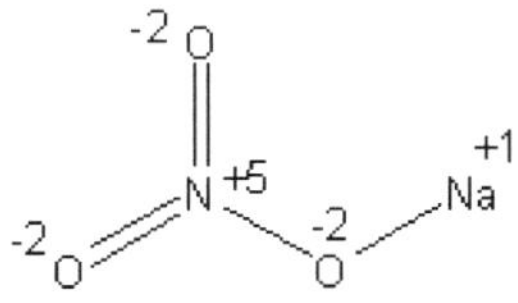

Sodium nitrate

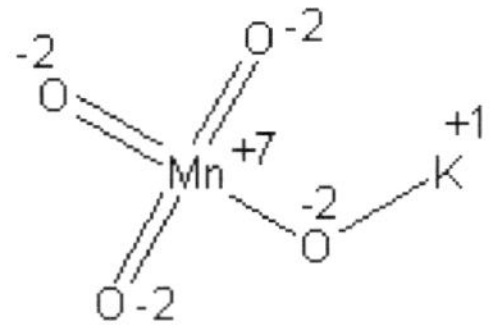

Potassium permanganate

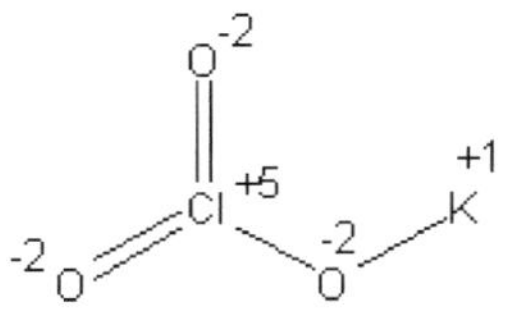

Potassium chlorate

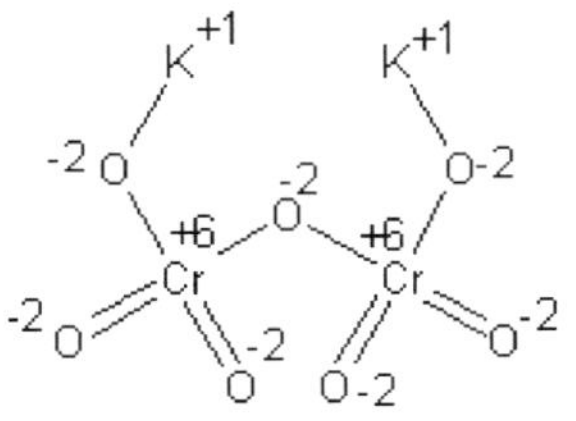

Potassium dichromate

If you notice, all molecules listed above have an equal number of positive charges, and an equal number of negative charges. These charges are perfectly balanced and in a state of equilibrium. Many of these compounds that are in equilibrium are called "inorganic" compounds, as they contain charged particles of positive and negative. All oxidizing agents contain central atoms in a high oxidation state; to get an idea of what lower oxidation states would be, look at the following diagram of stable, and ***non-oxidizing agents***.

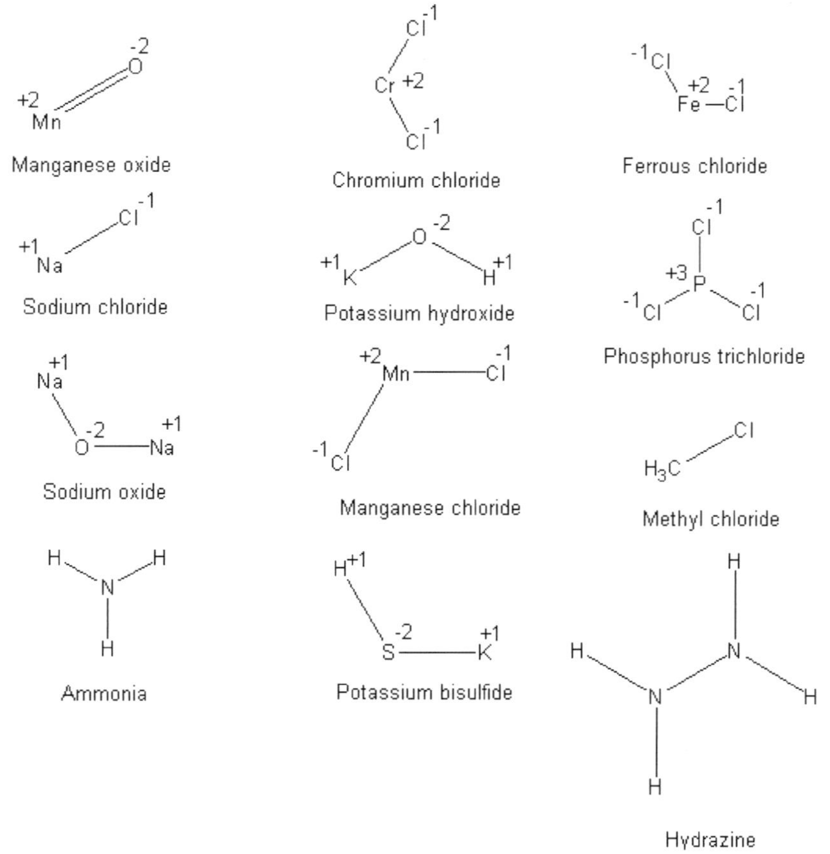


If you notice in the diagram, ammonia, hydrazine, and methyl chloride have no oxidation states written next to them. The reason is because these molecules are covalent. Now, what does this mean? Well, lets not worry about that as covalent molecules differ from inorganic molecules, and I'm trying to confuse you!! Remember that most oxidizing agents are inorganic molecules, containing positive and negative charges.

Now, as stated earlier, when an oxidizing agent "oxidizes" a reducing agent, the reducing agent becomes "oxidized", and then the oxidizing agent becomes "reduced". If this is confusing, look at the following diagram and try to understand this principle.

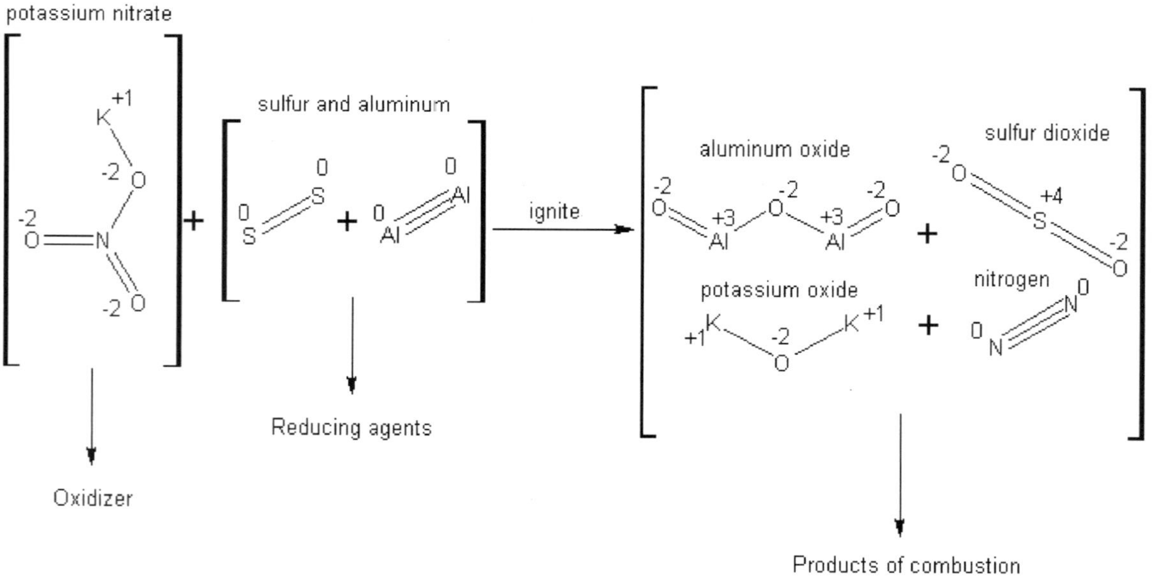


In the above diagram, potassium nitrate is the oxidizer, and sulfur and aluminum are the reducing agents. During the combustion process, the potassium nitrate oxidizes the sulfur and the aluminum forming products. During the oxidation process, the potassium nitrate is "reduced" to potassium oxide, nitrogen, and aluminum oxide. At the same time, the sulfur is "oxidized" to sulfur dioxide, and the aluminum is "oxidized" to the aluminum oxide. Remember that nitrogen tends to like being by itself so it evolves as nitrogen gas, or diatomic nitrogen seen as $N_2$.

This process of oxidation and reduction is the bases behind 99.9% of all pyrotechnics. As mentioned earlier, other products in the oxidation and reduction process include light and fire. These two elements are the bases for fireworks, and all the beautiful colors and effects produced by such pyrotechnic mixtures. The color of the fire and light produced during an oxidation/reduction process depends on the oxidizing agent and reducing agents involved in the reaction. For example, barium oxidizing salts produce green colors, and strontium oxidizing salts produce brilliant reddish colors during the combustion process.

### 3a. The nature of reducing agents verse oxidizing agents

So now lets discuss the nature of reducing agents, and their importance in pyrotechnic compositions. As previously discussed, reducing agents contain atoms that are easily oxidized. These atoms exist in oxidation states that are not preferred by the reducing agent. So why do these compounds even exist if they contain atoms that want to be in different states to begin with? This is a good question, and takes us to the fact of chemistry and its reactions. As we cannot discuss all these reactions, because this is not a chemistry book, we can talk a little about the reducing agents used in pyrotechnics. Lets look at the following diagram.

**Common reducing agents used in pyrotechnic compositions**

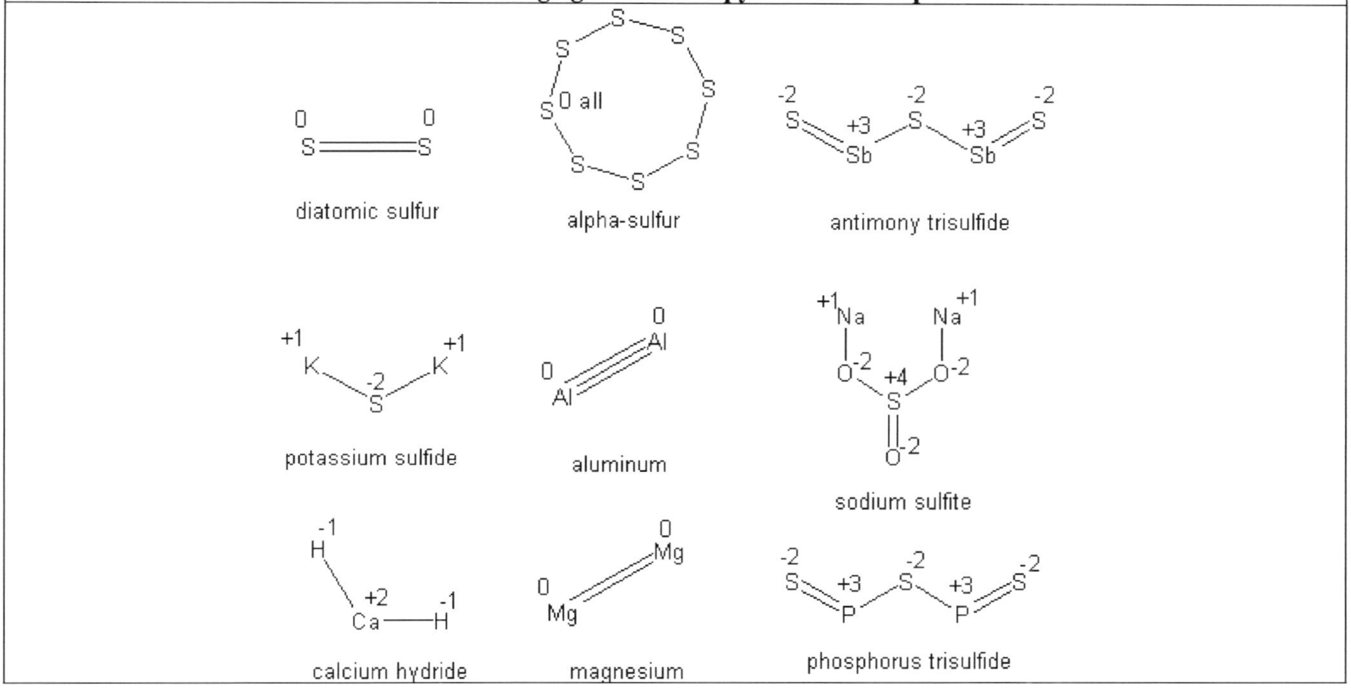


In the above illustration we see that antimony trisulfide and phosphorus trisulfide have antimony and phosphorus atoms in the +3 oxidations state. These two elements prefer to be in a +5 oxidation state where they are most comfortable. Therefore, antimony trisulfide and phosphorus trisulfide are reducing agents, as their central atoms, i.e., antimony and phosphorus want be in a higher state. Now, as previously mentioned, if antimony and phosphorus prefer to be in a higher state, why do they even exits in the states shown in antimony and phosphorus trisulfide? Well, to better explain this, is to understand the nature of chemical reactions. Not all elements end up in their desired states during a chemical reaction. In fact, each chemical reaction is unique in its own light. Some chemical reactions put the atoms in different states, but not necessarily the exact state they feel most comfortable. At the same time, other chemical reactions, specifically those of combustion, found in ignition of pyrotechnic mixtures, push these atoms into their desired states immediately without any interference. What oxidation states each atom has, it directly proportional to the reactants involved and the reaction conditions. In pyrotechnic reactions, all the atoms involved enter the exact states they feel most comfortable in, because during the ignition of the pyrotechnic composition, allot of energy is released in the form of heat. It's this heat that pushes the atoms into their desired states.

### 4a. Deflagration and the process of combustion

Now is the time to start talking about what the actual process of the combustion of a pyrotechnic mixture is called. This process is called "deflagration". Deflagration is the process of rapid and violent combustion. Many pyrotechnic mixtures undergo this type of combustion when they are ignited. As discussed earlier, when an oxidizer in a mixture reacts with a reducing agent, oxygen is liberated from the oxidizer and it then reacts with the reducing agent. The rate of reaction is very fast, and this rapid reaction rate is what causes the deflagration process. Deflagration is crucial in the pyrotechnics arena, and is ultimately the reason why many pyrotechnic mixtures form beautiful colors, sparks, flames, ect, upon ignition. Without the process of deflagration, rockets, gun propellants, stars, cones, fountains, and many other pyrotechnic devices would not function properly as their rapid combustion is what gives them the energy to perform as pyrotechnics. In all pyrotechnic mixtures/devices, it's the deflagration process that produces the pressures needed for the pyro concoctions to work properly. These pressures can be used to propel rockets, provide energy for flares and signals, allow military personnel to lay down smoke screens, and/or propel solid projectiles, such as bullets for security and protection. Now propulsion of bullets may seem like a negative use for the deflagration process, but all deflagration processes go hand in hand, and are equal. For example, when we take a sample of black powder, and ignite it in an open space, it flares up and burns out quite quickly. If we take this black powder, and instead of igniting it in the open, we insert it into an open end of a tube, such as a gun barrel, and then place a ball or similar projectile therein, and ignite the powder, a loud bang results, and the ball or projectile goes flying down range. In both cases the deflagration process is the result of the rapid combustion of the pyrotechnic mixture. When most pyrotechnic compositions are ignited in open spaces, they burn quickly, and when they are ignited within enclosed areas, they burn even quicker, but in both cases the deflagrations equal in energy, but the pressures can vary. Some mixtures however, may not burn as fast, or as long as other mixtures, but in any sense, the deflagration process is the driving force behind the energy and the process.

### 5a. Non-oxidizing pyrotechnics, and their nature

It should be noted, that not all pyrotechnic compositions contain straight oxidizers and reducing agents. Some mixtures contain compounds that undergo a process called "self deflagration". Self-deflagration is when a compound supplies oxygen from itself, to itself. Self-deflagrating compounds are nitrocellulose, nitro starch, picrates salts, picramates, ADN, and other similar compounds. These molecules are capable of self-combustion, and upon ignition, they will supply their own oxygen to the rest of the molecule to support combustion. Compounds like nitrocellulose can be ignited by themselves, and do not require an oxidizer or reducing agent to complete the deflagrating process. Self-deflagrating compounds have the advantage over conventional oxidizer/reducer mixtures, in that self-deflagrating compositions tend to burn cleaner and leave behind fewer residues. The reason is because self-deflagrating compositions tend to produce higher gas volumes with limited residue. Nitrocellulose and related self-deflagrating compositions decompose almost entirely into gas leaving very little residue. On the other end, potassium nitrate compositions, especially that of black powder, leave behind significant residue upon ignition; this creates problems especially when dealing with firearms.

### 6a. Methods of forming pyrotechnic compositions

In some cases, the desired pyrotechnic composition can be readily formed by simply blending the ingredients together using any desired means. However, some times pyrotechnic compositions can only be formed when they are mixed in special ways. Black powder for example, cannot simply be made by just blending the ingredients together. To form black powder and similar pyrotechnic mixtures, the ingredients of the mixtures need to be blended very thoroughly to form a uniform mix; solvents are also widely used to facilitate proper blending to maximize surface area.. To mix these ingredients very thoroughly requires special mixing devices. There are two main types of mixing apparatus available. The most common is the "ball mill", and "vertical mixer". A ball mill is a simple device utilizing a central container, similar to an empty coffee can laying on its side, and a small electric motor. The electric motor rotates the coffee can utilizing a fan belt or similar band. The rotating speed can very from 50 to 1500 RPM, depending on the type of mixing desired. In some cases, steel balls, or other materials are added to the mill, so they produce a high surface area of blending and pulverizing when the can rotates. Ball mills are one of the most common apparatus used to form pyrotechnic compositions, such as black powder. Also, as mentioned, in other cases, an inert solvent is added to facilitate proper pulverization and coheasion with maximum surface area. The second most common type of mixer is a vertical mixer. There are two types of "vertical mixers"; a) a common blender, as found in your kitchen, using stainless steel blades, and b) a large capacity "vertical mixer" utilizing plastic or stainless steel pallet blades that rotate widely and at a uniform speed; similar to a bread blender you might have in your home. All ingredients are added when necessary, and the mixture can be dry blended, or "wet" blended using an inert solvent to facilitate maximum surface area. Now, these two types of mixers may have different names, but their method of work is the same. You can find out more about these mixing devices by conducting a search on the Internet.

**Common workbench ball mills.**

**Common industrial grade vertical mixers.**

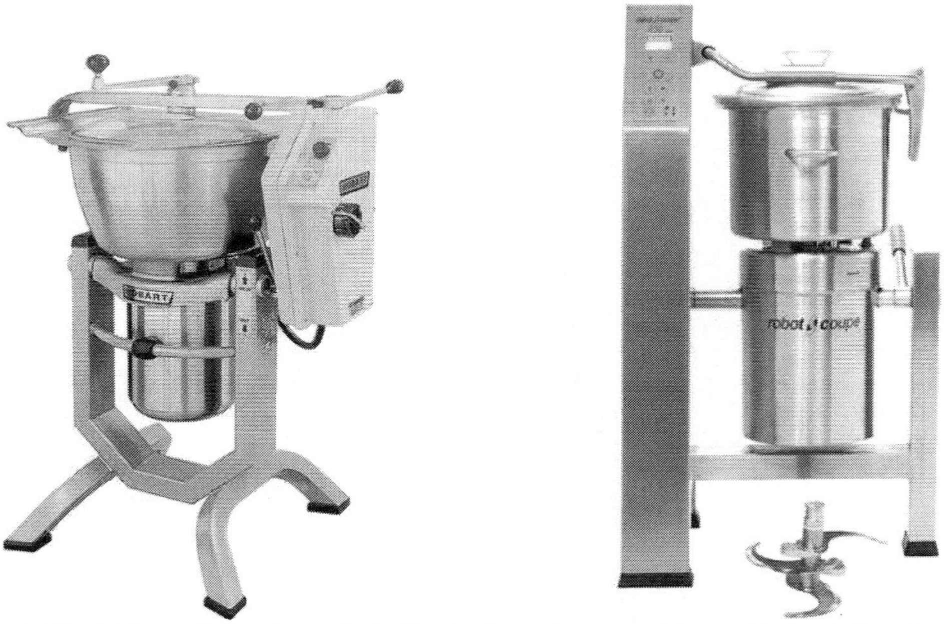


**Special Sealable vertical mixers. Both illustrations show special sealable heated vertical mixers.**

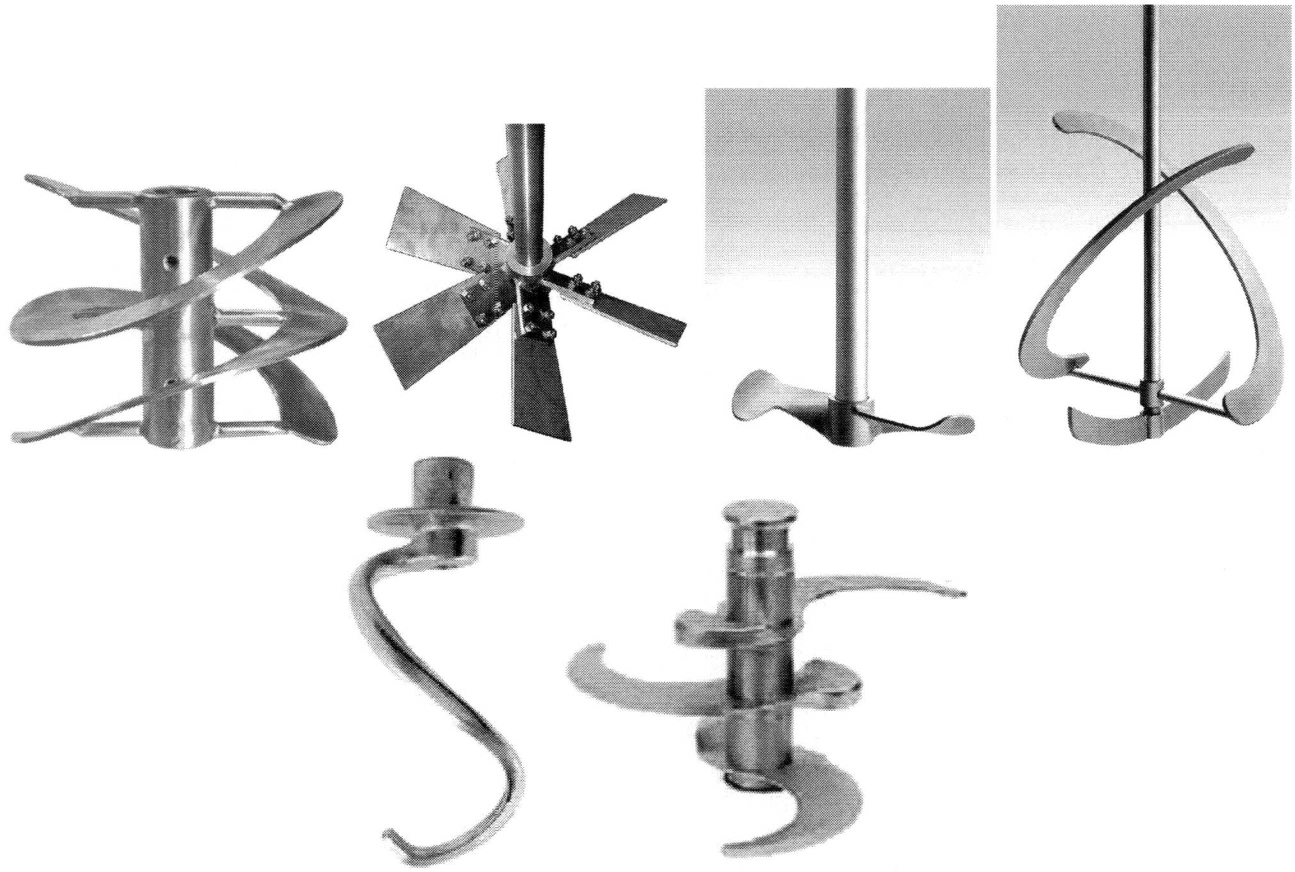

**Suitable vertical mixer stirring blades. Most are stainless steel, but some come in Teflon. Note: upper left, and bottom right are suitable blades for grinding and pulverizing (as many ingredients are in the forms of flakes/pellets and hence, must be ground into powder).**

## 7a. How to use this manual

In the beginning of each section, there is a brief examination of the various chemicals used in the preparation of the pyrotechnic compositions. Each section is then followed by the actual procedures, which have all been written to explain how the given mixture is obtained. You will notice as you read, that many of the procedures are very similar in nature. At the end of each procedure, a percentage is given to show the amount of each substance in the given mixture. At the end of each percentage entry, you will notice that it sometimes says, "mixed balance", "mixed residual balance", ect. What this means is the percentage has been cut-off at the hundredth mark of each decimal place to forgo extensive numbering. For example, in one given procedure it calls for 160 grams of animal glue, 450 grams of potassium chlorate, 210 grams of powdered glass, 40 grams of plaster of Paris powder, 40 grams of zinc oxide, 20 grams of" whiting", 30 grams flours of sulfur, and 100 grams of red phosphorus. The total amount of solids used is 1050 grams. Therefore the "exact" percentages would be written as 42.8571428571% potassium chlorate, 20% powdered glass, 15.2380952381% animal glue, ect., ect. However, to save time, we would write the percentages as 42.85% potassium chlorate, 20% powdered glass, 15.23% animal glue, 9.52% red phosphorus, 3.8% plaster of Paris, 3.8% zinc oxide, 2.85% sulfur, 1.9% whiting, and 0.05% residue, or "mixed balance"

# Chapter 2. Black Powder

## Section 1: Black Powder

### *Chemicals used in this section*

| 1. Potassium nitrate | | 2. Sulfur |
|---|---|---|
| 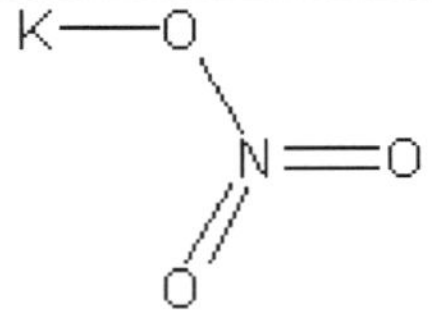 | | 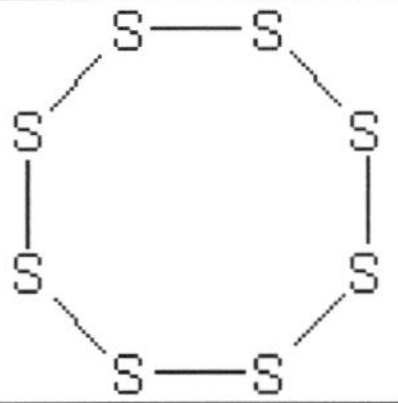 |
| Potassium nitrate forms colorless transparent or white crystals, or a crystalline powder. The salt has a melting point of 333 Celsius, and begins to decompose when heated to 400 Celsius. Its solubility in water is 1 gram in 2.8 milliliters of water at 25 Celsius. The salt dissolves in water with a decrease in temperature. | | Sulfur forms a yellow odorless powder. The dry solid is flammable and burns forming choking fumes of sulfur dioxide. The solid is insoluble in water and most solvents, but readily soluble in liquid ammonia. |
| **Process for preparation 1:**<br>1. Dissolve 100 grams of sodium nitrate into 110 milliliters of water at room temperature.<br>2. Dissolve 88 grams of potassium chloride into 252 milliliters of water at room temperature.<br>3. Pour the potassium chloride solution into the sodium nitrate solution, and then place the combined mixture into an ice bath, and chill it to 0 Celsius.<br>4. Allow the mixture to chill at 0 Celsius for about 1 hour. Thereafter, filter-off the precipitated potassium nitrate crystals, and then vacuum dry or air-dry the filtered-off potassium nitrate crystals.<br>**Process for preparation 2:**<br>1. Place 130 grams of 70% nitric acid into a flask, and then add in 250 milliliters of water at room temperature to dilute the acid.<br>2. Slowly add, in small portions at a time, 89 grams of potassium hydroxide, or 110 grams of potassium carbonate, over a period of 1 hour.<br>3. After the addition, boil the mixture to remove the water, and recover the potassium nitrate, or use any desired means. | | Sulfur is obtained in pure form from many different sources |
| **3. Charcoal** | | |
| Charcoal various in degree and quality, but normal charcoal usually exists of any desired soft wood, which is roasted in a kiln or suitable device. The roasting, otherwise called destructive distillation, produces charcoal, and by-products such as methanol, acetic acid, and various hydrocarbon gases including methane, ethane, and propylene. | | |

**- *Black Powder Compositions in this section* -**

| | |
|---|---|
| **1. 02-01-001A: Black Powder Composition 1 (standard black powder):** *75% potassium nitrate, 15% charcoal, 10% sulfur* | **2. 02-01-001B: Black Powder Composition 2 (Chinese Grade, 10th Century):** *67% potassium nitrate, 21.99% charcoal, 10.99% sulfur, 0.01% sulfides, 0.01% arsenic* |
| **3. 02-01-001C: Black Powder Composition 3 (Whitehorne black powder, 15th century Europe)**: *50% potassium nitrate, 33.33% charcoal, 16.61% sulfur, 0.06% impurities* | **4. 02-01-001D: Black Powder Composition 4 (English royal black powder, 17th century England)**: *50% potassium nitrate, 33.33% charcoal, 16.61% sulfur, 0.06% impurities* |
| **5. 02-01-001E: Black Powder Composition 4 (English royal black powder, 17th century England)**: *75% potassium nitrate, 12.5% charcoal, 12.5% sulfur* | **6. 02-01-001F: Black Powder Composition 5 (French specific black powder, 17th century)**: *78% potassium nitrate, 19% charcoal, 3% sulfur* |
| **7. 02-01-001G: Black Powder Composition 6 (American black powder, 19th century:** *72.9% potassium nitrate, 15.62% charcoal, 11.45% sulfur, 0.03% impurities* | **8. 02-01-001H: Standard black powder (precipitation method):** *75% potassium nitrate, 15% charcoal, 10% sulfur* |
| **9. 02-01-001I: Standard pyrotechnics Black Powder (PGI optimum, fireworks standard):** *74% potassium nitrate, 14% charcoal, 12% sulfur* | **10. 02-01-002A: 19th-century American black powder composition:** *74% potassium nitrate, 14% charcoal, 12% sulfur* |
| **11. 02-01-003A: "Reduced wastage" black powder composition (burns more evenly with ejection of fewer un-burned particles):** *74.8% potassium nitrate, 13.2% charcoal, 11.8% sulfur, 0.176% impurities, 0.024% paraffin* | **12. 02-01-004A: Standard black powder composition (precipitation method 2):** *75% potassium nitrate, 14.6% charcoal, 10.4% sulfur* |
| **13. 02-01-005A: Standard "sporting powder" black powder composition (basic tumbling method):** *74% potassium nitrate, 15.6% charcoal, 10.4% sulfur* | **14. 02-01-006A: Black powder composition:** *74% potassium nitrate, 15.6% charcoal, 10.4% sulfur* |
| **15. 02-01-007A: Black powder composition:** *75% potassium nitrate, 15% charcoal, 10% sulfur* | **16. 02-01-008A: Black powder composition:** *76.27% potassium nitrate, 13.55% charcoal, 10.16% sulfur, 0.02% mixed balance* |
| **17. 02-01-009A: Black powder composition:** *72.72% potassium nitrate, 17.27% charcoal, 10% sulfur, 0.01% mixed residue* | **18. 02-01-010A: Black powder composition:** *74.74% potassium nitrate, 15.15% charcoal, 10.10% sulfur, 0.01% mixed residual balance* |
| **19. 02-01-011A: Black powder composition:** *75% potassium nitrate, 15% charcoal, 10% sulfur* | **20. 02-01-012A: "Explosive" Black powder composition:** *75% potassium nitrate, 15% charcoal, 10% sulfur* |

**02-01-001A: Black Powder Composition 1 (standard black powder)**:
Into a standard ball mill, place ***225 grams of potassium nitrate*** of at least 98% purity, followed by ***45 grams of charcoal***, and then followed by ***30 grams of sulfur*** of a purity of at least 99%. Then add in 50 milliliters of 95% ethyl alcohol—denatured alcohol can be used. Then add in about 100 grams of steel shot. The steel shot should be of 4 to 5 millimeters in diameter, and should be of stainless steel—no rust, or dirty shot. Note: Teflon coated steel balls are preferred. Thereafter, turn on the tumbler machine, and allow the mixture to churn for several hours at 180 RPM. Note: the charcoal may vary, but usually consists of charcoal made by the pyrolysis of willow trees, grapevines, pines, or laurel trees. Although wood charcoal can vary in chemical composition and texture, either form of wood charcoal is satisfactory, but only from soft woods. After rotating the mixture for several hours, turn off the machine, and then remove the contents within the drum. Place these contents onto a shallow pan, and allow the mixture to thoroughly air-dry for 2 to 3 days. Note: After 2 or 3 days, the black powder should have no odor of solvent. After the black powder has thoroughly dried, what you do with it now is up to you. For example, if you wish to make grains for firearms, the black powder should be carefully broken up, and then passed through any selected sieves or screens to separate out the different grain sizes. A Black powder with grains of 0.50 to 1.5 millimeters in diameter is best suited for firearms. A Black powder with grains of 2 to 3.5 millimeters is best suited for cannons. If you wish use black powder in finely divided form, place your dried mixture into a clean ball mill, and then ball mill it just like before, with clean steel shot until it's thoroughly ground; a standard vertical mixer can be used as well. This finely divided black powder can be used for any suitable operation where powdered black powder is desired.
**Burn rate:** Depends on grain size.
**Water resistance:** Moderate.
**Stability:** Can be stored for decades.
**Sensitivity:** Stable towards normal friction; stable towards shock; stable to temperatures up to 300 Celsius.
**Flammability (1 to 10, un-contained):** 9 powdered; 7.6 small grains; 6 medium grains; 5 large grains.
**Ease of ignition (1 to 10 un-contained):** 9 powdered; grain size varies and results in different ignition rates.
**Tendency to cake:** May cake on prolonged exposure to moisture.
**Detonation ability:** Cannot be detonated under normal conditions.
**Explosive ability (contained; deflagration):** Excellent; has great "heaving" action.
**Percentage:** ***75% potassium nitrate, 15% charcoal, 10% sulfur***
**Classification:** Deflagrating explosive (classified as propellant).

**Danger of preparation:** Can be made by amateurs.

### 02-01-001B: Black Powder Composition 2 (Chinese Grade, 10th Century):

Into a clay or wood drum, place ***188 grams of Chinese saltpeter***, followed by ***66 grams of charcoal***, and then followed by ***33 grams of flours of sulfur***. Then add in 50 milliliters of a 25% saltpeter solution. Note: this saltpeter solution would have contained traces of sulfide minerals and arsenic. Then add in about 75 grams of limestone or marble balls of 10 to 15 millimeters in diameter. The limestone or marble balls should be pre-coated with charcoal, sulfide ore dust, or with a low volatile resin. Thereafter, rotate the drum vigorously back and forth, like a washing machine, for several hours to form a uniform pasty mixture. Note: the charcoal may vary, and the exact charcoal used by Chinese alchemists is not fully known, but probably consists of any soft wood from local areas in central China. After rotating the mixture for several hours, remove the contents from the drum, and place it onto a shallow clay pan or spread out over leaves, and allow the mixture to thoroughly air-dry for 4 to 5 days. Note: After 4 or 5 days, the black powder should be completely dry. After the black powder has thoroughly dried, it needs to be milled in the same manner as before but dry milled using clean pre-coated limestone or marble balls of similar diameter for several hours to form a uniform finely divided powder. This finely divided black powder can be used for any suitable operation where black powder is used. Note: this brand of black powder may still exist in some products of Chinese fireworks.
**Burn rate:** Depends on grain size.
**Water resistance:** Moderate.
**Stability:** Can be stored for decades.
**Sensitivity:** Stable towards normal friction; stable towards shock; stable to temperatures up to 300 Celsius.
**Flammability (1 to 10, un-contained):** 9 powdered; 7.6 small grains; 6 medium grains; 5 large grains.
**Ease of ignition (1 to 10 un-contained):** 9 powdered; grain size varies and results in different ignition rates.
**Tendency to cake:** May cake on prolonged exposure to moisture.
**Detonation ability:** Cannot be detonated under normal conditions.
**Explosive ability (contained; deflagration):** Excellent; has great "heaving" action.
**Percentage:** ***67% potassium nitrate, 21.99% charcoal, 10.99% sulfur, 0.01% sulfides, 0.01% arsenic***
***Note: will also contain traces of limestone, calcium carbonate, sodium chloride, and resin***
**Classification:** Deflagrating explosive (classified as propellant).

**Danger of preparation:** Can be made by amateurs.

### 02-01-001C: Black Powder Composition 3 (Whitehorne black powder, 15th century Europe):

Into a bronze or lead drum or barrel, place ***150 grams of potassium nitrate*** (recrystallized from boiling water), then add in ***99.9 grams of charcoal (from rye straw)***, followed by ***49.8 grams of flours of sulfur***. Thereafter, add in 15 milliliters of cold water, followed by about 75 grams of copper or lead shot of 6 to 9 millimeters in diameter. Thereafter, rotate the bronze or lead drum back and forth vigorously, like a washing machine, for several hours. After rotating the mixture for several hours, remove the contents within the drum. Then place these contents onto a shallow pan, and allow the mixture to thoroughly air-dry for several days. After the black powder has thoroughly dried, it needs to be milled for several hours in an identical manner using clean copper or lead shot in a clean bronze or lead drum or barrel. Thereafter, the milled black powder can be passed through various sizes of sieves to separate out various grain sizes. After this, the grains are ready for use. Black powder grains of 0.50 to 1.5 millimeter in diameter are best suited for firearms. Black powder grains of 2 to 3.5 millimeters are best suited for mortars, and grains of 3 to 6 millimeters are best for cannons. If you wish to use black powder in finely divided form, place your dried mixture into a clean rotating drum, and then ball mill it just like before, with clean steel shot until it's thoroughly ground. This finely divided black powder can be used for any suitable operation where powdered black powder is desired.
**Burn rate:** Depends on grain size.
**Water resistance:** Moderate.
**Stability:** Can be stored for decades.
**Sensitivity:** Stable towards normal friction; stable towards shock; stable to temperatures up to 300 Celsius.
**Flammability (1 to 10, un-contained):** 8 powdered; 7 small grains; 6 medium grains; 5 large grains.
**Ease of ignition (1 to 10 un-contained):** 8 powdered; grain size varies and results in different ignition rates.
**Tendency to cake:** May cake on prolonged exposure to moisture.
**Detonation ability:** Cannot be detonated under normal conditions.
**Explosive ability (contained; deflagration):** Excellent; has great "heaving" action.
**Percentage:** ***50% potassium nitrate, 33.33% charcoal, 16.61% sulfur, 0.06% impurities***
**Classification:** Deflagrating explosive (classified as propellant).

**Danger of preparation:** Can be made by amateurs.

## 02-01-001D: Black Powder Composition 4 (English royal black powder, 17th century England):

Into a steel drum or barrel, place ***300 grams of potassium nitrate*** (recrystallized from boiling water), followed by ***50 grams of soft wood charcoal or rye straw charcoal***, followed by ***50 grams of flours of sulfur***. Thereafter, add in 20 milliliters of mineral water, followed by about 100 grams of iron or lead shot of 6 to 8 millimeters in diameter. Thereafter, rotate the steel drum or barrel back and forth vigorously, like a washing machine, for several hours. After rotating the mixture for several hours, remove the contents within the drum. Then place these contents onto a shallow pan (minus the iron or lead shot), and allow the mixture to thoroughly air-dry for several days. After the black powder has thoroughly dried, it can be broken up into pieces and then passed through a sieve with 1.5 millimeter holes for firearms, or 3 millimeter holes for mortars, or 7 millimeter holes for cannons. For forming powder for time fuse and for blasting powders, the dry mixture needs to be milled for several hours in an identical manner using clean iron or lead shot in a clean steel drum or barrel. This finely divided black powder can be used for any suitable operation where powdered black powder is desired.
**Burn rate:** Depends on grain size.
**Water resistance:** Moderate.
**Stability:** Can be stored for decades.
**Sensitivity:** Stable towards normal friction; stable towards shock; stable to temperatures up to 300 Celsius.
**Flammability (1 to 10, un-contained):** 9 powdered; 7.6 small grains; 6 medium grains; 5 large grains.
**Ease of ignition (1 to 10 un-contained):** 9 powdered; grain size varies and results in different ignition rates.
**Tendency to cake:** May cake on prolonged exposure to moisture.
**Detonation ability:** Cannot be detonated under normal conditions.
**Explosive ability (contained; deflagration):** Excellent; has great "heaving" action.
**Percentage:** ***75% potassium nitrate, 12.5% charcoal, 12.5% sulfur***
**Classification:** Deflagrating explosive (classified as propellant).

**Danger of preparation:** Can be made by amateurs.

## 02-01-001E: Black Powder Composition 5 (French specific black powder, 17th century):

Into a hard wood barrel or drum, place ***312 grams of potassium nitrate***, followed by ***76 grams of rye straw charcoal***, followed by ***12 grams of flours of sulfur***. Thereafter, add in 20 milliliters of water, followed by about 100 grams of iron or lead shot of 4 to 7 millimeters in diameter. Thereafter, as usual, rotate the wood drum or barrel back and forth vigorously, like a washing machine, for several hours. After rotating the mixture for several hours, remove the contents within the drum. Then place these contents onto a shallow pan (minus the iron or lead shot), and allow the mixture to thoroughly air-dry for several days. After the black powder has thoroughly dried, it can be broken up into pieces and then passed through a sieve with 1.2 to 1.5 millimeter holes for firearms, or 2.5 to 4 millimeter holes for mortars, or 5.2 to 12 millimeter holes for cannons. For forming powder for time fuse and for blasting powders, the dry mixture needs to be milled for several hours in an identical manner using clean iron or lead shot in a clean hard wood drum or iron drum or barrel. This finely divided black powder can be used for any suitable operation where powdered black powder is desired.
**Burn rate:** Depends on grain size.
**Water resistance:** Moderate.
**Stability:** Can be stored for decades.
**Sensitivity:** Stable towards normal friction; stable towards shock; stable to temperatures up to 300 Celsius.
**Flammability (1 to 10, un-contained):** 9 powdered; 7.6 small grains; 6 medium grains; 5 large grains.
**Ease of ignition (1 to 10 un-contained):** 9 powdered; grain size varies and results in different ignition rates.
**Tendency to cake:** May cake on prolonged exposure to moisture.
**Detonation ability:** Cannot be detonated under normal conditions.
**Explosive ability (contained; deflagration):** Excellent; has great "heaving" action.
**Percentage:** ***78% potassium nitrate, 19% charcoal, 3% sulfur***
**Classification:** Deflagrating explosive (classified as propellant).

**Danger of preparation:** Can be made by amateurs.

## 02-01-001F: Black Powder Composition 6 (American black powder, 19th century):

Into a galvanized iron barrel or drum, place ***280 grams of precipitated potassium nitrate***, i.e., recrystallized from boiling water, followed by ***60 grams of soft wood charcoal***, followed by ***44 grams of flours of sulfur***. Thereafter, add in 15 milliliters of water, followed by about 75 grams of steel shot of 4 to 6 millimeters in diameter. Thereafter, rotate the steel drum or barrel for several hours at 150 RPM. After rotating the mixture for several hours, remove the contents within the drum. Then place these contents onto a shallow pan (minus the shot), and allow the mixture to thoroughly air-dry for several days. After the black powder has thoroughly dried, it can be broken up into pieces and then passed through a sieve with 1.0 to 1.6 millimeter holes for firearms, or 2.0 to 4.1 millimeter holes for mortars, or 5.0 to 12.6 millimeter holes for cannons. For forming powder for time fuse and for blasting powders, the dry mixture needs to be milled for several hours in an identical manner using clean steel or lead shot in a clean steel barrel or drum. This finely divided black powder can be used for any suitable operation where powdered black powder is desired.
**Burn rate:** Depends on grain size.
**Water resistance:** Moderate.
**Stability:** Can be stored for decades.
**Sensitivity:** Stable towards normal friction; stable towards shock; stable to temperatures up to 300 Celsius.
**Flammability (1 to 10, un-contained):** 9 powdered; 7.6 small grains; 6 medium grains; 5 large grains.
**Ease of ignition (1 to 10 un-contained):** 9 powdered; grain size varies and results in different ignition rates.
**Tendency to cake:** May cake on prolonged exposure to moisture.
**Detonation ability:** Cannot be detonated under normal conditions.
**Explosive ability (contained; deflagration):** Excellent; has great "heaving" action.
**Percentage:** ***72.9% potassium nitrate, 15.62% charcoal, 11.45% sulfur, 0.03% impurities***
**Classification:** Deflagrating explosive (classified as propellant).

**Danger of preparation:** Can be made by amateurs.

## 02-01-001G: Standard black powder (precipitation method):

Place ***45 grams of standard wood charcoal*** and ***30 grams of flours of sulfur*** into a ball mill, and then throw in about 25 grams of steel shot of 4 to 5 millimeters in diameter. Thereafter, tumble the contents for about 1 or 2 hours at 150 to 180 RPM. After the mixing process, remove the charcoal/sulfur mixture from the drum. Now, into a large suitable beaker, glass or stainless steel container, place 650 milliliters of water, followed by ***225 grams of potassium nitrate***. Note: your container should be big enough so that after adding in the water and potassium nitrate, your container is just under half full. Thereafter, boil the water and allow the potassium nitrate to dissolve. Once the potassium nitrate dissolves, add in the charcoal/sulfur mixture previously prepared, and then continue boiling the entire mixture while rapidly stirring for about 1 hour. Note: during the boiling process, water will obviously be lost due to evaporation; as a result, never mind this water loss. After 1 hour, remove your heat source, and allow the total mixture to cool to about 70 or 80 degrees Celsius. During this cool down time, rapidly stir the entire blackish, mucky mixture. When the temperature of the mixture reaches 70 to 80 Celsius, quickly add to it, 600 milliliters of ice-cold isopropyl alcohol, or 700 milliliters of ice-cold denatured alcohol. Now, quickly place this mixture (your container) into an ice bath, and chill the contents of your mixture to 0 Celsius. Allow the mixture to stand at this temperature for at least 30 minutes. During the whole time, rapidly stir the black mucky mixture. After allowing the mixture to chill for at least 30 minutes, filter the mixture to remove the insoluble materials. Now, you can use any of two methods of filtration, i.e., gravity filtration or vacuum filtration. If using gravity filtration, it will take some time to recover the black powder, so vacuum filtration is recommended. If using vacuum filtration, suck the black powder dry. If you used gravity filtration, once your black powder has been removed, recover the pasty black powder material from the filter paper, and place it onto a shallow pan and spread it out, and allow it to air-dry until completely dry. Either way, once your black powder has been recovered and dried, it can be crushed up to make various sizes of grains.
**Burn rate:** Depends on grain size.
**Water resistance:** Moderate.
**Stability:** Can be stored for decades.
**Sensitivity:** Stable towards normal friction; stable towards shock; stable to temperatures up to 300 Celsius.
**Flammability (1 to 10, un-contained):** 9 powdered; 7.6 small grains; 6 medium grains; 5 large grains.
**Ease of ignition (1 to 10 un-contained):** 9 powdered; grain size varies and results in different ignition rates.
**Tendency to cake:** May cake on prolonged exposure to moisture.
**Detonation ability:** Cannot be detonated under normal conditions.
**Explosive ability (contained; deflagration):** Excellent; has great "heaving" action.
**Percentage:** ***75% potassium nitrate, 15% charcoal, 10% sulfur***
**Classification:** Deflagrating explosive (classified as propellant).

**Danger of preparation:** Can be made by amateurs.

### 02-01-001H: Standard pyrotechnics Black Powder (PGI optimum, fireworks standard):

Into a specialty heated ball mill, place ***296 grams of potassium nitrate*** of at least 98% purity, followed by ***56 grams of willow tree, grapevine, pine, or laurel tree charcoal***, and then followed by ***48 grams of sulfur*** of a purity of at least 99%. Then add in 75 milliliters of 95% ethyl alcohol, or denatured alcohol. Thereafter, add in about 100 grams of steel shot of 4 to 5 millimeters in diameter. Then, turn on the ball mill, and allow the mixture to churn for several hours at 180 RPM and a temperature of about 40 Celsius. After rotating the mixture for several hours, turn off the machine, and then remove the contents within the drum. Place these contents onto a shallow pan, and allow the mixture to thoroughly air-dry for 2 to 3 days. Note: After 2 or 3 days, the black powder should have no odor of solvent. After the black powder has thoroughly dried, you can do several things with it. For example, if you wish to make grains for firing aerial display shells, the dried black powder mixture need to be broken up, and then shaken through sieves with 3.5 to 5 millimeter holes. For use in time fuse, or firecrackers, the dried black powder mixture should be placed in a clean rotating drum, and then ball milled with clean steel shot until it's thoroughly ground. This finely divided black powder can be used for any suitable operation where powdered black powder is desired.
**Burn rate:** Depends on grain size (excellent for firing aerial display shells in fireworks).
**Water resistance:** Moderate.
**Stability:** Can be stored for decades.
**Sensitivity:** Stable towards normal friction; stable towards shock; stable to temperatures up to 300 Celsius.
**Flammability (1 to 10, un-contained):** 9 powdered; 7.6 small grains; 6 medium grains; 5 large grains.
**Ease of ignition (1 to 10 un-contained):** 9 powdered; grain size varies and results in different ignition rates.
**Tendency to cake:** May cake on prolonged exposure to moisture.
**Detonation ability:** Cannot be detonated under normal conditions.
**Explosive ability (contained; deflagration):** Excellent; has great "heaving" action.
**Percentage:** ***74% potassium nitrate, 14% charcoal, 12% sulfur***
**Classification:** Deflagrating explosive (classified as propellant).

**Danger of preparation:** Can be made by amateurs.

### 02-01-002A: 19th-century American black powder composition:

Into a suitable iron, tin, steel, or bronze pot, place 2 liters of water followed by ***750 grams of potassium nitrate***, and the boil the solution using a wood fire, or steam bath. When the water begins to boil, gently stir in ***150 grams of charcoal***, followed by ***120 grams of flours of sulfur***, and then continue to boil the mixture. Note: the mixture should be continuously stirred using a wood spatula all through out the boiling process. Continue to boil the mixture until about 98% of the water has been evaporated, and only a pasty mass exists. When this point is achieved, remove the pot from the heat source, and allow its contents to cool to room temperature. Note: during this cool down period, continuously stir the pasty mixture with a wooden spatula. Once the mixture has cooled to room temperature, lay out the mixture out on a shallow pan and allow it to thoroughly dry. Once the mass is dry, it can then be pulverized into different grain sizes that can be separated using desired sieves or screens.
**Burn rate:** Depends on grain size (granules 1 to 12 millimeters are suitable for fire-arms, mortars, and cannons, and fine powder can be used as blasting agent).
**Water resistance:** Moderate.
**Stability:** Can be stored for decades.
**Sensitivity:** Stable towards normal friction; stable towards shock; stable to temperatures up to 300 Celsius.
**Flammability (1 to 10, un-contained):** 9 powdered; 7.6 small grains; 6 medium grains; 5 large grains.
**Ease of ignition (1 to 10 un-contained):** 9 powdered; grain size varies and results in different ignition rates.
**Tendency to cake:** May cake on prolonged exposure to moisture.
**Detonation ability:** Cannot be detonated under normal conditions.
**Explosive ability (contained; deflagration):** Excellent; has great "heaving" action.
**Percentage:** ***74% potassium nitrate, 14% charcoal, 12% sulfur***
**Classification:** Deflagrating explosive (classified as propellant).

**Danger of preparation:** Can be made by amateurs.

### 02-01-003A: "Reduced wastage" black powder composition (burns more evenly with ejection of fewer un-burned particles):

Into a standard heated ball mill containing 250 grams of steel shot of 10 millimeters in diameter, place ***500 grams of potassium nitrate*** of at least 98% purity, followed by ***88.8 grams of willow tree, grapevine, pine, or laurel tree charcoal***, followed by ***79.3 grams of flours of sulfur,*** followed by ***166 milligrams of paraffin,*** and finally followed by 175 milliliters of 95% ethyl alcohol, or denatured alcohol. Thereafter, tumble the mixture for several hours at 200 RPM and a temperature of about 40 Celsius. After rotating the mixture for several hours, turn off the machine, and then remove the contents within the drum. Place these contents onto a shallow pan, and allow the mixture to thoroughly air-dry for 2 to 3 days. Note: After 2 or 3 days, the black powder should have no odor of solvent. After the black powder has

thoroughly dried, you can do several things with it. You can pulverize it by breaking it up in a crucible and then separating the various grain sizes using the desired sieves, or you can pulverize it using a clean ball mill to form a uniform mixture.
**Burn rate:** Depends on grain size.
**Water resistance:** Moderate.
**Stability:** Can be stored for decades.
**Sensitivity:** Stable towards normal friction; stable towards shock; stable to temperatures up to 300 Celsius.
**Flammability (1 to 10, un-contained):** 9 powdered; 7.6 small grains; 6 medium grains; 5 large grains.
**Ease of ignition (1 to 10 un-contained):** 9 powdered; grain size varies and results in different ignition rates.
**Tendency to cake:** May cake on prolonged exposure to moisture.
**Detonation ability:** Cannot be detonated under normal conditions.
**Explosive ability (contained; deflagration):** Excellent; has great "heaving" action.
**Percentage:** ***74.8% potassium nitrate, 13.2% charcoal, 11.8% sulfur, 0.176% impurities, 0.024% paraffin,***
**Classification:** Deflagrating explosive (classified as propellant).

**Danger of preparation:** Can be made by amateurs.
**Use:** Most suitable for use in cannons using 7 to 12 millimeter grains.

### 02-01-004A: Standard black powder composition (precipitation method 2):

Into a suitable beaker or container, place ***225 grams of potassium nitrate,*** followed by 225 milliliters of water. Then, heat the mixture to 80 Celsius with good stirring to make sure the potassium nitrate dissolves. Once the nitrate dissolves, quickly prepare a second mixture by adding 675 milliliters of 95% ethyl alcohol or denatured alcohol, followed by adding in ***31.2 grams of powdered sulfur***, and then followed by ***43.8 grams of regular powdered wood charcoal***, and then rapidly stir the entire mixture for about 30 minutes. During this 30-minute period, maintain the hot potassium nitrate solution at 80 Celsius with good stirring. After 30 minutes, place the alcohol mixture into an ice bath, and chill to 15 Celsius. Thereafter, slowly and gently add in the hot potassium nitrate solution while rapidly stirring the alcohol mixture. After the hot potassium nitrate solution has been added, continue to stir the mucky mass rapidly for about 1 hour. After 1 hour, filter the mixture using gravity filtration or preferably vacuum filtration to recover the insoluble mass. After the filtration process, place the filtered mass onto a shallow pan and allow it to dry at room temperature for several days or so. After this, the composition is ready to be used. To use, it can be beaten or pulverized and the various grain sizes then separated using the desired sieves or screens, or the mass can be pulverized by hand, and then pressed under high pressure (10,000 psi) into pellets of any desired size or shape, or the mass can be ball milled using steel shot of 5 to 10 millimeters in diameter to form a uniform mixture.
**Burn rate:** Depends on grain size.
**Water resistance:** Moderate.
**Stability:** Can be stored for decades.
**Sensitivity:** Stable towards normal friction; stable towards shock; stable to temperatures up to 300 Celsius.
**Flammability (1 to 10, un-contained):** 9 powdered; 7.6 small grains; 6 medium grains; 5 large grains.
**Ease of ignition (1 to 10 un-contained):** 9 powdered; grain size varies and results in different ignition rates.
**Tendency to cake:** May cake on prolonged exposure to moisture.
**Detonation ability:** Cannot be detonated under normal conditions.
**Explosive ability (contained; deflagration):** Excellent; has great "heaving" action.
**Percentage:** ***75% potassium nitrate, 14.6% charcoal, 10.4% sulfur***
**Classification:** Deflagrating explosive (classified as propellant).

**Danger of preparation:** Can be made by amateurs.
**Use:** Most suitable for use in small arms weapons, and in fireworks.

### 02-01-005A: Standard "sporting powder" black powder composition (basic tumbling method):

Into a suitable ball mill, filled with 500 grams of heavy steel shot, place ***156 grams of soft wood charcoal***, followed by ***104 grams of flours of sulfur***. Thereafter, tumble the mixture a 400 RPM for about 3 hours. Thereafter, add in ***740 grams of potassium nitrate***, and then continue to tumble the mixture at 250 RPM for about 5 hours at room temperature. Thereafter, separate the tumbled mixture from the steel shot using any desired screen in the usual manner, and then place the tumbled separated mixture into a suitable mixing bowl, equipped with motorized stirrer, and then add in 250 milliliters of ether. Thereafter, blend the mixture on moderate speed for about 15 minutes to form a uniform paste. Thereafter, the mixture is ready for casting. To do so, the wet material should be pressed through any desired die cast machine, into spaghetti like strands of 2 to 3 millimeters in diameter by 150 to 300 millimeters in length. Thereafter, the spaghetti like strands should be cut into lengths of 3 to 5 millimeters. Finally, the grains should be cured in an oven at normal temperature. The cured grains can then be poured into shell casings for use in standard ammunition, or can be placed into baggies for use in canons or the like.
**Burn rate:** Depends on grain size.
**Water resistance:** Moderate.
**Stability:** Can be stored for decades.

**Sensitivity:** Stable towards normal friction; stable towards shock; stable to temperatures up to 300 Celsius.
**Flammability (1 to 10, un-contained):** 9 powdered; 7.6 small grains; 6 medium grains; 5 large grains.
**Ease of ignition (1 to 10 un-contained):** 9 powdered; grain size varies and results in different ignition rates.
**Tendency to cake:** May cake on prolonged exposure to moisture.
**Detonation ability:** Cannot be detonated under normal conditions.
**Explosive ability (contained; deflagration):** Excellent; has great "heaving" action.
**Percentage:** ***74% potassium nitrate, 15.6% charcoal, 10.4% sulfur***
**Classification:** Deflagrating explosive (classified as propellant).

**Danger of preparation:** Can be made by amateurs.
**Use:** Used for small arms weapons, cannons, and fireworks.

### 02-01-006A: Black powder composition:

Into a suitable ball mill, filled with 250 grams of lead shot, place ***78 grams of soft wood charcoal***, followed by ***52 grams of flours of sulfur,*** and then followed by ***370 grams of potassium nitrate***. Thereafter, tumble the mixture at 250 RPM for about 2 hours. Note: During the 2-hour mixing process, spray into the ball mill, in small portions at a time, 125 milliliters of cold water. After the 2-hour mixing period, the mixture is ready for extrusion. To do so, the material should be pressed through any desired die cast machine, into spaghetti like strands of 2 to 3 millimeters in diameter by 150 to 300 millimeters in length. Thereafter, the spaghetti like strands should be cut into lengths of 1.5 to 3 millimeters. Finally, the grains should be cured in an oven at normal temperature. The cured grains can then be poured into shell casings for use in standard ammunition, or can be placed into baggies for use in canons or the like. If desired, the final material can be thoroughly dried. And then pulverized in a separate mill to form a fine powder.
**Burn rate:** Depends on grain size.
**Water resistance:** Moderate.
**Stability:** Can be stored for decades.
**Sensitivity:** Stable towards normal friction; stable towards shock; stable to temperatures up to 300 Celsius.
**Flammability (1 to 10, un-contained):** 9 powdered; 7.6 small grains; 6 medium grains; 5 large grains.
**Ease of ignition (1 to 10 un-contained):** 9 powdered; grain size varies and results in different ignition rates.
**Tendency to cake:** May cake on prolonged exposure to moisture.
**Detonation ability:** Cannot be detonated under normal conditions.
**Explosive ability (contained; deflagration):** Excellent; has great "heaving" action.
**Percentage:** ***74% potassium nitrate, 15.6% charcoal, 10.4% sulfur***
**Classification:** Deflagrating explosive (classified as propellant).

**Danger of preparation:** Can be made by amateurs.
**Use:** Used for small arms weapons, cannons, and fireworks.

### 02-01-007A: Black powder composition:

Into a suitable ball mill, containing 200 grams of aluminum shot, place ***375 grams of pre-dried potassium nitrate***, followed by ***75 grams of soft wood charcoal,*** and then followed by ***50 grams of flours of sulfur***. Thereafter, add in 175 milliliters of alcohol spirits of camphor, and then tumble the combined mixture at 350 RPM for about 1 hour. After the 1 hour tumbling process, the mixture is ready for extrusion. To do so, the material should be pressed through any desired die cast machine, or formed into grains using any desired method in the usual manner. The grains in any case, need to be cured in an oven at ordinary temperatures in the usual manner. The cured grains can then be placed loosely into shell casings for use in standard ammunition, or can be placed into baggies for use in canons or the like.
**Burn rate:** Depends on grain size.
**Water resistance:** Moderate.
**Stability:** Can be stored for decades.
**Sensitivity:** Stable towards normal friction; stable towards shock; stable to temperatures up to 300 Celsius.
**Flammability (1 to 10, un-contained):** 9 powdered; 7.6 small grains; 6 medium grains; 5 large grains.
**Ease of ignition (1 to 10 un-contained):** 9 powdered; grain size varies and results in different ignition rates.
**Tendency to cake:** May cake on prolonged exposure to moisture.
**Detonation ability:** Cannot be detonated under normal conditions.
**Explosive ability (contained; deflagration):** Excellent; has great "heaving" action.
**Percentage:** ***75% potassium nitrate, 15% charcoal, 10% sulfur***
**Classification:** Deflagrating explosive (classified as propellant).

**Danger of preparation:** Can be made by amateurs.
**Use:** Used for small arms weapons, cannons, and fireworks.

### 02-01-008A: Black powder composition:

Into a suitable ball mill, containing 500 grams of steel shot, place ***450 grams of pre-dried potassium nitrate***, followed by ***80 grams of soft wood charcoal,*** and then followed by ***60 grams of flours of sulfur***. Thereafter, add in 125 milliliters of 95% ethyl alcohol, and then tumble the mixture at 500 RPM for about 2 hours. After 2 hours, place the mixture into a suitable vacuum apparatus, and remove the ethyl alcohol under mild vacuum. Thereafter, the mixture is ready for use. To do so, the material should be pressed through any desired screen or die cast machine, or formed into grains using any desired method in the usual manner. The grains in any case, need to be cured in an oven at ordinary temperature in the usual manner. The cured grains can then be placed loosely into shell casings for use in standard ammunition, or can be placed into baggies for use in canons or the like.
**Burn rate:** Depends on grain size.
**Water resistance:** Moderate.
**Stability:** Can be stored for decades.
**Sensitivity:** Stable towards normal friction; stable towards shock; stable to temperatures up to 300 Celsius.
**Flammability (1 to 10, un-contained):** 9 powdered; 7.6 small grains; 6 medium grains; 5 large grains.
**Ease of ignition (1 to 10 un-contained):** 9 powdered; grain size varies and results in different ignition rates.
**Tendency to cake:** May cake on prolonged exposure to moisture.
**Detonation ability:** Cannot be detonated under normal conditions.
**Explosive ability (contained; deflagration):** Excellent; has great "heaving" action.
**Percentage:** ***76.27% potassium nitrate, 13.55% charcoal, 10.16% sulfur, 0.02% mixed balance***
**Classification:** Deflagrating explosive (classified as propellant).

**Danger of preparation:** Can be made by amateurs.
**Use:** Used for small arms weapons, cannons, and fireworks.

### 02-01-009A: Black powder composition:

Into a vertical mixer, or any suitable mixing drum, bowl, ect, equipped with motorized stirrer, place 500 milliliters of hexane, followed by ***800 grams of potassium nitrate of average commercial mesh***, followed by ***190 grams of soft wood charcoal,*** and then followed by ***110 grams of double sublimed sulfur***. Thereafter, blend the entire mixture at moderate speed for about 30 minutes. Thereafter, filter-off the insoluble mass, and then place this mass onto a shallow tray or pan, and allow it to thoroughly air-dry. Once it has, place the dried mass into any desired ball mill, filled with 500 grams of lead shot, and then tumble the mixture at 250 RPM for about 2 hours. Thereafter, the mixture is ready for use. To use, the material can be moistened with alcohol or acetone, and then pressed through any desired screen or die cast machine, or formed into grains using any desired method. Thereafter, cure the grains in an oven at 60 to 90 Celsius until dry and hard. The cured grains can then be placed loosely into shell casings for use in standard ammunition, or the ball-milled product can be placed directly into baggies for use in canons, or in fireworks, ect., ect.
**Burn rate:** Depends on grain size.
**Water resistance:** Moderate.
**Stability:** Can be stored for decades.
**Sensitivity:** Stable towards normal friction; stable towards shock; stable to temperatures up to 300 Celsius.
**Flammability (1 to 10, un-contained):** 9 powdered; 7.6 small grains; 6 medium grains; 5 large grains.
**Ease of ignition (1 to 10 un-contained):** 9 powdered; grain size varies and results in different ignition rates.
**Tendency to cake:** May cake on prolonged exposure to moisture.
**Detonation ability:** Cannot be detonated under normal conditions.
**Explosive ability (contained; deflagration):** Excellent; has great "heaving" action.
**Percentage:** ***72.72% potassium nitrate, 17.27% charcoal, 10% sulfur, 0.01% mixed residue***
**Classification:** Deflagrating explosive (classified as propellant).

**Danger of preparation:** Can be made by amateurs.
**Use:** Used for small arms weapons, cannons, and fireworks.

**02-01-010A: Black powder composition:**
Into a vertical mixer, or any suitable mixing drum, bowl, ect, equipped with motorized stirrer, place 1500 milliliters of hot water, followed by ***740 grams of potassium nitrate of average commercial mesh***. Thereafter, blend the mixture on moderate speed for about 45 minutes. Thereafter, add in ***150 grams of soft wood charcoal,*** and then followed by ***100 grams of sulfur powder***. Thereafter, continue to blend the entire mixture at moderate speed for about 30 additional minutes. Thereafter, add in 750 milliliters of acetone, and then continue to blend the entire mixture for about 1 hour. Thereafter, filter-off the insoluble mass, and then place this mass onto a shallow tray or pan, and allow it to thoroughly air-dry. Note: the acetone can be recovered if desired by using any desired means. Once the dried mass has dried, it is ready for use. To use, the material can be moistened with alcohol or acetone, and then pressed through any desired screen or die cast machine, or formed into grains using any desired method. Thereafter, cure the grains in an oven at 60 to 90 Celsius until dry and hard. The cured grains can then be placed loosely into shell casings for use in standard ammunition, or the ball-milled product can be placed directly into baggies for use in canons, or in fireworks, ect., ect.
**Burn rate:** Depends on grain size.
**Water resistance:** Moderate.
**Stability:** Can be stored for decades.
**Sensitivity:** Stable towards normal friction; stable towards shock; stable to temperatures up to 300 Celsius.
**Flammability (1 to 10, un-contained):** 9 powdered; 7.6 small grains; 6 medium grains; 5 large grains.
**Ease of ignition (1 to 10 un-contained):** 9 powdered; grain size varies and results in different ignition rates.
**Tendency to cake:** May cake on prolonged exposure to moisture.
**Detonation ability:** Cannot be detonated under normal conditions.
**Explosive ability (contained; deflagration):** Excellent; has great "heaving" action.
**Percentage:** ***74.74% potassium nitrate, 15.15% charcoal, 10.10% sulfur, 0.01% mixed residual balance***
**Classification:** Deflagrating explosive (classified as propellant).

**Danger of preparation:** Can be made by amateurs.
**Use:** Used for small arms weapons, cannons, and fireworks.

**02-01-011A: Black powder composition:**
Into a suitable ball mill, filled with 250 grams of heavy steel shot, place ***150 grams of soft wood charcoal***, and then followed by ***100 grams of sulfur powder***. Thereafter, tumble the mixture at 500 RPM for about 1 hour. Thereafter, place this tumbled mixture into a suitable mixing bowl, equipped with motorized stirrer, and then add in ***750 grams of potassium nitrate***, and then followed by 400 milliliters of 95% ethyl alcohol. Thereafter, blend the mixture on moderate speed for about 2 hours. Thereafter, add in 400 milliliters of cold water, then continue to blend the mixture on moderate speed for about 1 hour. Thereafter, filter-off the insoluble mass, and then place the wet mass into a suitable ball mill, filled with 350 grams of steel shot, and equipped with vacuum. Thereafter, tumble the mixture at 150 RPM at room temperature and a suitable vacuum to remove the alcohol and water. Once the mixture is dried, remove the vacuum, and continue to tumble the mixture at 150 RPM for about 1 hour. Thereafter, the mixture is ready for use. To use, the mixture can be used directly as a loose powder, or moistened with a little alcohol, and then extruded through any die cast machine to form grains of any desired size. The "wet" grains then need to be cured in an oven under the usual means.
**Burn rate:** Depends on grain size.
**Water resistance:** Moderate.
**Stability:** Can be stored for decades.
**Sensitivity:** Stable towards normal friction; stable towards shock; stable to temperatures up to 300 Celsius.
**Flammability (1 to 10, un-contained):** 9 powdered; 7.6 small grains; 6 medium grains; 5 large grains.
**Ease of ignition (1 to 10 un-contained):** 9 powdered; grain size varies and results in different ignition rates.
**Tendency to cake:** May cake on prolonged exposure to moisture.
**Detonation ability:** Cannot be detonated under normal conditions.
**Explosive ability (contained; deflagration):** Excellent; has great "heaving" action.
**Percentage:** ***75% potassium nitrate, 15% charcoal, 10% sulfur***
**Classification:** Deflagrating explosive (classified as propellant).

**Danger of preparation:** Can be made by amateurs.
**Use:** Used for small arms weapons, cannons, and fireworks.

**02-01-012A: "Explosive" Black powder composition:**

Into a standard kitchen blender, add in ***750 grams of potassium nitrate***, ***100 grams of sulfur powder***, followed by ***150 grams of soft wood charcoal***, and then turn on the blender to medium speed and blend for 5 to 7 seconds, then stop; thereafter, tap and rotate the blender container, place back onto the blender machine, then turn on the blender to medium speed and blend for 5 to 7 seconds; thereafter, stop; tap and rotate the blender container, place back onto the blender machine, then turn on the blender to medium speed and blend for 5 to 7 seconds. Thereafter, stop blending and allow the ingredients to stand for 30 minutes. After 30 minutes, turn on the blender to medium speed and blend for 5 to 7 seconds, then stop; thereafter, tap and rotate the blender container, place back onto the blender machine, then turn on the blender to medium speed and blend for 5 to 7 seconds; thereafter, stop; tap and rotate the blender container, place back onto the blender machine, then turn on the blender to medium speed and blend for 5 to 7 seconds. Thereafter, stop blending and allow the ingredients to stand for 30 minutes. After 30 minutes, turn on the blender to medium speed and blend for 5 to 7 seconds, then stop; thereafter, tap and rotate the blender container, place back onto the blender machine, then turn on the blender to medium speed and blend for 5 to 7 seconds; thereafter, stop; tap and rotate the blender container, place back onto the blender machine, then turn on the blender to medium speed and blend for 5 to 7 seconds. Thereafter, stop blending and allow the ingredients to stand for 30 minutes. Now, allow the ingredients in the blender to stand for 2 hours; thereafter, repeat the aforementioned process 3 more times. Note: the idea here is to thoroughly blend the ingredients without build up of heat. Once finished, the process is complete and the "proper" explosive black powder has been prepared. Thereafter, open the blender container, and pour the powdered mixture into a "zip lock" bag, and seal air-tight. The "zip lock" bag can be stored at room temperature, below room temperature, or can be stored at a temperature below 130 Fahrenheit for as long as necessary. In any sense, storing the powder is usually carried out at room temperature; as long as the "Zip lock" bag is sealed air-tight, the powder will never cake, and will remain a fresh free flowing powder.

**Proper blender blades for mixing potassium nitrate, sulfur, and charcoal (do not use serrated blades; the stainless steel blades should be smooth).**

**Burn rate:** 0.2 seconds per inch.
**Water resistance:** Moderate.
**Stability:** Very stable; can be stored for decades.
**Sensitivity:** Stable towards normal friction; stable towards shock; stable to temperatures up to 300 Celsius.
**Flammability (1 to 10, un-contained):** 10 powdered.
**Ease of ignition (1 to 10 un-contained):** 10 powdered.
**Tendency to cake:** May cake on prolonged exposure to moisture.
**Detonation ability:** Can be detonated when confined and ignited.
**Explosive ability (contained; explosive deflagration):** detonates; 1,983 meters/per/second
**Percentage:** ***75% potassium nitrate, 15% charcoal, 10% sulfur***
**Classification:** Deflagrating explosive (classified as explosive).

**Danger of preparation:** Can be made by amateurs, as long as the procedure is followed exactly.
**Use:** Bursting charges; blasting agent; firecrackers, improvised explosive devices. Note: do not use in firearms, cannons, or large bore guns (explosion of the weapon will result).

# Chapter 3. Modified Black powder

## Section 1: Black Powder/Modified Black Powder

***Chemicals used in this section***

| | | |
|---|---|---|
| **1. Potassium nitrate (see Black Powder)** | | **2. Sulfur (see Black Powder)** |
| **3. Charcoal (see Black Powder)** | | **4. Sugar Carbon** |
| | | Sugar carbon is prepared by roasting sugar at 150 to 300 Celsius until nothing but a black residue remains. |
| **5. Barium Chromate** | | **6. Potassium Perchlorate** |
| 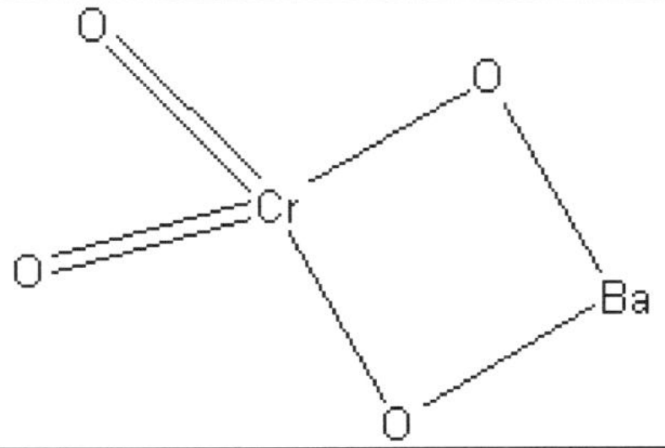 | | 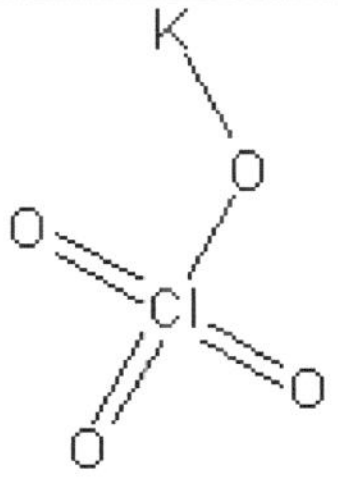 |
| Barium Chromate forms yellow monoclinic, orthorhombic crystals. The crystals are toxic, so users should wear gloves when handling. | | Potassium perchlorate forms colorless or white crystals, or crystalline powder. The salt decomposes when heated to 400 Celsius. |
| | | **Method of preparation 1:** 1. Dissolve 100 grams of sodium perchlorate into 100 milliliters of water at room temperature. 2. Dissolve 61 grams of potassium chloride into 179 milliliters of water room temperature. 3. Thoroughly blend both solutions at room temperature, and then chill the mixture in an ice bath at 0 Celsius for about 1 hour. After 1 hour, filter-off the precipitated potassium perchlorate, and then vacuum dry or air-dry the solid. |
| | | **Method of Preparation 2:** Into a suitable crucible, place 100 grams of potassium chlorate, followed by 50 grams of potassium chloride. Thereafter, briefly shake the crucible to mix the two salts. Then gently and slowly heat the mixture using a Bunsen burner to about 370 Celsius. Be sure to carefully monitor the temperature using a temperature probe or equivalent. Shortly thereafter, a gentle gas evolution of oxygen will take place. Keep heating the mixture at about 380 Celsius until the evolution of gas ceases. After the evolution of gas has ceased, immediately remove the heat source, and allow the mixture to cool to room temperature. Then, using a knife or other sharp and hard instrument, fracture and pulverize the solid mass in the crucible, and then place this pulverized mass into a suitable beaker, or container. Thereafter, add in 200 milliliters of water (at room temperature), and then rapidly stir the entire mixture for about 1 hour to dissolve the potassium chloride. After stirring for 1 hour, filter-off the insoluble potassium perchlorate, and then vacuum dry or air-dry the crystals. |
| | | |

**7. Potassium Dichromate**

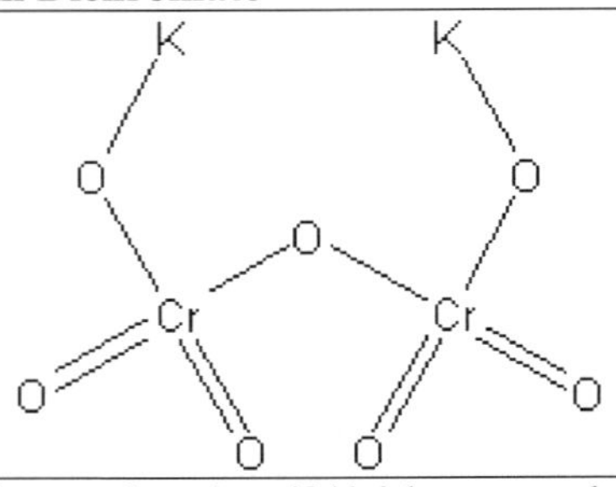


Potassium Dichromate forms beautiful bright orange-red crystals. The crystals are very stable in air, and can be stored for years.

**8. Ammonium Bisulfide**

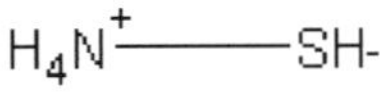


Forms white tetragonal orthorhombic crystals, which sublime quite readily. The commercial product is obtained in porcelain lumps to preserve it more from decomposition. The salt is readily soluble in water and other common solvents.

**9. Potassium chlorate**

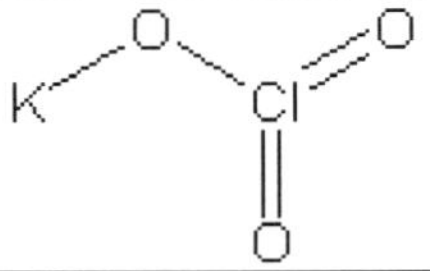


Potassium chlorate forms colorless to white crystalline solid. The salt has a melting point of 368 Celsius, and decomposes into oxygen at higher temperatures. Potassium chlorate is relatively insoluble in water and most common solvents.

**10. Carbon Disulfide**

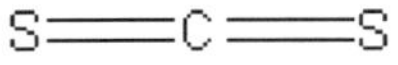


Carbon disulfide is a colorless refractive liquid with foul smelling odor. It has a boiling point of 46.5 Celsius. The liquid is toxic, and users should wear proper gloves and use proper ventilation when handling. The liquid is prepared by dry distillation of sulfur and charcoal at high temperature

**Method of preparation 1:**

1. Dissolve 100 grams of sodium chlorate, into 110 milliliters of water at room temperature.
2. Dissolve 70 grams of potassium chloride into 200 milliliters of water at room temperature.
3. Thoroughly mix both mixtures, and the chill the combined mixture at 0 Celsius for about 1 hour. Thereafter, filter-off the precipitated potassium chlorate, and then vacuum dry or air-dry the solid.

**11. Nitrocellulose; gun cotton**

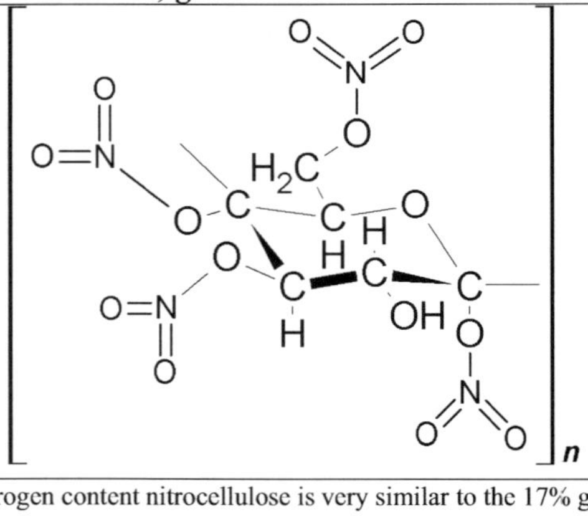


14% nitrogen content nitrocellulose is very similar to the 17% grade, but has lesser value for use in explosives then the 17% grade; nevertheless, it is still widely used in gelatin explosives when mixed with high energy liquid explosives; with potassium perchlorate, and aluminum powder as fillers for blasting operations for multiple commercial and military applications in the usual manner. The 14% grade is extensively used in military "smokeless" powders for firearms, mortar shell propellants, artillery shell propellants, and for tank fired gun propellants, and is the most preferred nitrogen percentage for use in making high performance gun propellants, hence its nickname "gun cotton". On that note, the 14% nitrogen grade nitrocellulose forms white, fluffy white linters, powder, granules, spheres, pellets, or beautiful white trimmed needles. This particular grade is somewhat soluble in acetone, but is insoluble in water and most organic solvents. As mentioned with the 17% grade, the variety of gun propellant and propulsion propellant compositions containing this 14% nitrogen grade ranges in the thousands of formulations, and it would be impossible to list even a fraction of the documented "recipes" and compositions that have been created. In the aforementioned gelatin explosive mixtures the filler such as potassium perchlorate can be replaced with potassium chlorate, sodium peroxide, ammonium nitrate, potassium nitrate, and similar oxidizers, and the aluminum powder can be replaced with magnesium powder, sodium hypophosphite, powdered sulfides, and even sulfites. **Danger of preparation:** can be prepared by non-scientists as long as all procedure(s) and safety info is strictly followed.

**Method of Preparation 1:** Into a suitable beaker, equipped with mechanical stirrer, thermometer, and sitting on top of hot plate, pour in 800 milliliters of 70% nitric acid, and then slowly and carefully add in 300 milliliters of 98% sulfuric acid. Once all the sulfuric acid has been added, turn on the mechanical stirrer to high speed, and then turn on the hot plate to 50 Celsius. When the nitrating acid reaction mixture reaches 50 Celsius, throw in 70 grams of regular cotton balls, and allow the cotton balls to slowly submerge themselves while continuing to heat the reaction mixture at 50 Celsius under rapid stirring. Note: at first the cotton balls will merely float on the surface of the nitrating acid reaction mixture, but in time, with proper heating and rapid and vigorous stirring, the nitration of the cotton balls will proceed, and the desired product will become homogenous in the reaction mixture. In any case, temperature needs to be carefully monitored so as to keep the sulfuric acid from carbonizing the cotton balls. Continue to heat the reaction mixture at 50 Celsius under vigorous stirring for 3 hours. After 3 hours, remove the heat source and then allow the reaction mixture to cool to room temperature. Once it has, filter-off the insoluble desired nitrated cellulose product using a clean Buchner funnel under moderate suction, wash with six 250-milliliter portions of 10% sodium carbonate solution, followed by washing with ten 500-milliliter portions of warm water. Finally, recover the suction dried product from the Buchner funnel, place onto a shallow pan or tray, and allow to thoroughly air-dry. Once dry, collect the dried product, and if desired to store for prolonged period of time, moisten the nitrated cotton with 10% isopropyl alcohol, and store in a clear plastic or glass bottle sealed air tight, and store in a cool dry place until use.

**Method of Preparation 3:** Into a beaker, place 192 grams of 60% nitric acid, and then carefully add 274 grams of 98% sulfuric acid. Thereafter, heat the acid mixture to about 50 Celsius. When the temperature reaches about 50 Celsius, rapidly add 100 grams of a cellulose mixture, prepared by mixing 50 grams of cellophane pieces (cut into pieces by hand), with 50 grams of ground-up cotton. After the addition, stir the acid mixture rapidly for 90 minutes at 50 Celsius. After heating and stirring for 90 minutes, remove the heat source, and cool the reaction mixture to room temperature. Then drown the reaction mixture into 3000 milliliters of ice water, and then stir the mixture for ten minutes. After which, filter-off the insoluble product, wash with 500 milliliters of cold water, four 150-milliliter portions of 10% baking soda solution, followed by 500 milliliters of water. Then, vacuum dry or air-dry the product. Note: The filter cake may contain some impurities.

**Method of Preparation 5:** Into a beaker, place 2 kilograms of 98% sulfuric acid, and then place this acid mixture into a ice bath and chill to 0 Celsius. Thereafter, place 2840 grams of newspaper into a separate beaker, and then moisten the newspaper by the addition of a small amount of water. Note: add just enough water to moisten the paper, but not to the point where the newspaper becomes fragile, and easily torn. Then, add this moistened newspaper, in portions, to the sulfuric acid at such a rate to keep the sulfuric acid at 0 Celsius at all times. During the addition, rapidly stir the sulfuric acid. After the addition, rapidly stir the acid mixture at 0 Celsius for about 1 hour. Then, filter-off the pulpy cellulose mass, wash with 1000 milliliters of warm water, and then dry by pressing the cellulosic mass to exclude as much of the liquids as possible. Thereafter, allow the cellulosic mass to dry at room temperature for several days. Thereafter, place the dry cellulosic material into a grinding machine, and grind into a fine particle size. After which, place 2 kilograms of 70% nitric acid into a beaker, and then carefully add 1000 grams of 98% sulfuric acid. Then heat this acid mixture to 40 Celsius with constant stirring. Thereafter, add in portions, the dry cellulosic material obtained earlier over a period of about 30 minutes. After the addition, rapidly stir the reaction mixture at 40 Celsius for 2 hours. Thereafter, remove the heat source, and allow the reaction mixture to cool to room temperature. Then add the reaction mixture to 10-liters of ice, and then stir the mixture for ten minutes. After which, filter-off the insoluble product, wash with 500 milliliters of cold water, then with four 500-milliliter portions of 10% baking soda solution, and then 1000 milliliters of water. After the washing, vacuum dry or air-dry the product.

**Method of Preparation 2:** Into a beaker, place 200 grams of wood pulp (paper making grade), and then add 140 milliliters of water. Thereafter, add 200 grams of 15% sodium hydroxide, and then heat the mixture for 1 hour at 80 to 100 Celsius. After the heating period, remove the heat source, allow the mixture to cool to room temperature, and then add 100 milliliters of carbon disulfide. Thereafter, stir the resulting gelatinous mass for 1 hour, and then filter-off the mass, wash with several hundred millimeters of warm water, and then dry by pressing it upon a block (to extrude as much of the liquid content as possible, and then air-dry the cellulosic material for several days. Thereafter, place the cellulosic material into a grinding machine, and grind into a fine particle size. After which, place 314 grams of 99% nitric acid into a beaker, and then slowly add 62 grams of 98% sulfuric acid. Afterwards, slowly add 12 grams of water. Then heat this acid mixture to 40 Celsius with constant stirring. Then, add in portions, the dry cellulosic material obtained earlier over a period of about 30 minutes. After the addition, rapidly stir the reaction mixture at 40 Celsius for 1 hour. Thereafter, remove the heat source, and allow the reaction mixture to cool to room temperature. Then, add the reaction mixture into 1000 milliliters of ice water, and then stir for ten minutes. After which, filter-off the insoluble product, wash with 500 milliliters of cold water, then with four 150-milliliter portions of 10% baking soda solution, and then 500 milliliters of water. After the washing, vacuum dry or air-dry the product.

**Method of Preparation 4:** Into a beaker, place 10 kilograms of 98% sulfuric acid, and then place this acid mixture into a ice bath and chill to 0 Celsius. Thereafter, slowly add in portions, 2 kilograms of standard cotton at a rate sufficient to keep the sulfuric acid at 0 Celsius at all times. During the addition, rapidly stir the sulfuric acid. Note: allowing the temperature to rise may result in carbonization of the cotton. After the addition, continue to stir until a pulpy-mass is obtained. When it is, gradually pour in 4 kilograms of 70% nitric acid at a rate sufficient to keep the reaction mixtures temperature at 0 Celsius. During the nitric acid addition, rapidly stir the reaction mixture. After the addition of the nitric acid, continue to stir at 0 Celsius for 3 hours. After 3 hours, remove the ice bath, and allow the reaction mixture to warm to room temperature. Thereafter, Add 10 liters of cold water to the reaction mixture, and then stir for ten minutes. Then filter-off the precipitated mass, wash with two 500-milliliter portions of cold water, then four 500-milliliter portions of 10% baking soda solution, and then with two 500-milliliter portions of warm water. Afterwards, vacuum dry or air-dry the product.

**Method of Preparation 6:** Into a beaker, place 314 grams of 99% nitric acid, and then 140 grams of 98% sulfuric acid. Thereafter, slowly add 64 grams of water. Then heat this nitrating acid mixture to about 40 Celsius with rapid stirring, and then add 106 grams of wood cellulose (94% dry bases; obtained from alkali lye extraction using a boiling 20% sodium hydroxide solution upon wood chips). After the addition, rapidly stir the acid mixture at 40 Celsius for 1 hour, and then cool the reaction mixture to room temperature. Then drown the reaction mixture into 1500 milliliters of ice water, and then stir the mixture for ten minutes. Then, filter-off the insoluble product, wash with 1000 milliliters of cold water, four 150-milliliter portions of 10% baking soda solution, followed by 1000 milliliters of water. Then, vacuum dry or air-dry the product. Note: The filter cake may contain some impurities.

## 12. Lead Tetraoxide

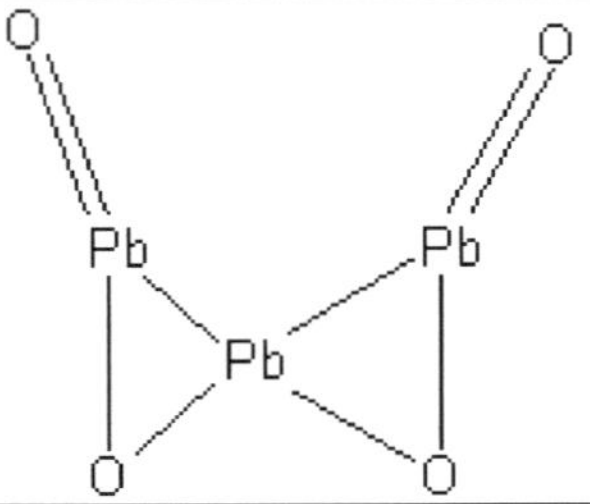


Forms a bright red, bulky powder. The powder is poisonous, and a strong oxidizer. Lead Tetraoxide is insoluble in water and alcohol, but soluble in hot acid. The powder is good oxidizer, which looses oxygen when heated to 500 Celsius.

## 13. Diphenylamine

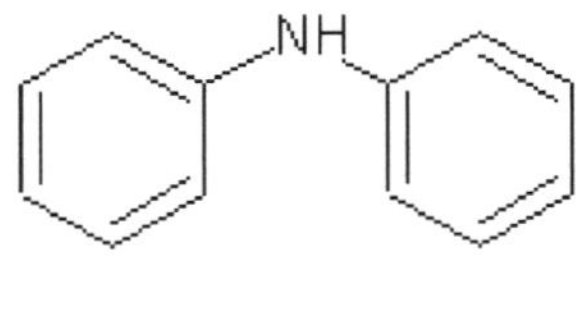


Diphenylamine forms crystals with a floral odor. The crystals have a melting point of 54 Celsius, and a boiling point of 302 Celsius. The crystals are insoluble in water, but soluble in alcohol, and other common solvents. The solid is prepared by heating a mixture of aniline with aniline hydrochloride.

## 14. Titanium Dioxide

O═Ti═O

Titanium Dioxide forms a white powder with a high melting point of 1855 Celsius. The powder is insoluble in water and most solvents. Titanium Dioxide exists in nature.

## 15. Manganese Dioxide

O═Mn═O

Black powder forming tetragonal crystals. The compound is insoluble in water in the usual organic solvents. Manganese Dioxide is a strong oxidizing agent.

## 16. Sodium Benzoate

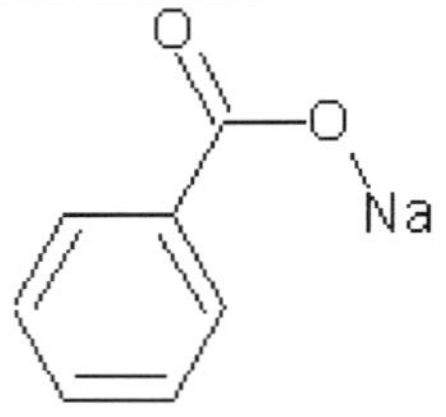


Sodium Benzoate forms a colorless to white crystalline granules, or powder. The crystals are soluble in water, but relatively insoluble in most solvents. Sodium Benzoate is non-toxic and is widely used as a preservative in foodstuffs.

## 17. Ammonium Nitrate

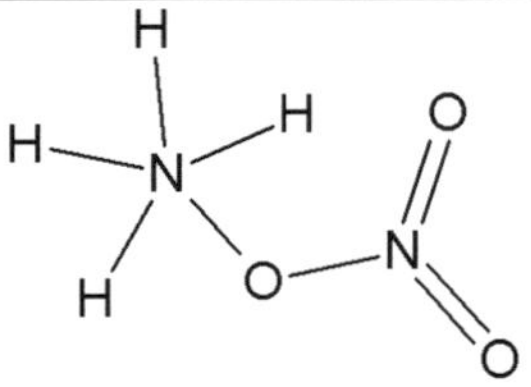


Ammonium nitrate forms odorless, transparent, hygroscopic, and deliquescent crystals or white granules. The crystals are highly stable, but decomposes when heated to about 210 Celsius forming water and nitrous oxide (laughing gas). Ammonium nitrate is very soluble in water, and alcohol, but insoluble in the usual organic solvents. Note: even though this book outlines the process for preparing ammonium nitrate, the compound is readily available and can be legally purchased from numerous on-line sites. In any sense ammonium nitrate is manufactured by the millions of tones per year, and much of it is used as a fertilizer, plant growth enhancer, and for nitrogen fixation of soil. On the other side, ammonium nitrate is widely used in manufacturing of explosives, such as RDX, but it is also widely used by itself in rock blasting for commercial mining operations. In this case, the ammonium nitrate is usually simply sprayed directly into the bore holes, blasting caps are then fitted, wired, and fired. Ammonium nitrate is also used when admixed with fuel oil to form very stable "prills" that are safe to handle and can be stored for many years; in the past prills were very common for use in make-shift military demolition operations for example, such as clearing trees and debris for creating landing zones for aircraft, however, ammonium nitrate prills are much less common these days. The compound is widely used in commercial and military dynamites when admixed with TNT/aluminum for use in general demolition operations and various similar applications; as well, ammonium nitrate is used to make "octol" as described under TNT, which is widely used in missile warheads for anti-tank purposes. Ammonium nitrate is widely used in making high performance smokeless rocket fuels, and used in pyrotechnic formulations for a variety of commercial and military applications. Ammonium nitrate can safely be stored in glass, plastic, or poly bags and stored in a cool dry place, and can be stored for many years. **Danger of preparation:** can be prepared by non-scientists as long as all procedure(s) and safety info is strictly followed.

**Method of Preparation 1:** Dissolve 206 grams of 70% nitric acid into 600 milliliters of water. Then cool the diluted nitric acid solution to 0 Celsius by means of an ice bath while stirring continuously. Next, bubble 40 grams of ammonia gas into the nitric acid mixture over a period of one hour while stirring and maintaining the reaction mixtures temperature at 0 Celsius. Or slowly add 136 grams of 28 – 30% ammonia solution, or 400 grams of a 10% ammonia solution into the diluted nitric acid solution while stirring and maintaining the diluted nitric acids temperature at 0 Celsius. Upon completion of the ammonia addition, recrystallize the ammonium nitrate from the reaction mixture, and then vacuum dry or air dry the product. If a rotary evaporator is available, place the filtered reaction mixture there into, and remove the water under high vacuum. Then, recrystallize the ammonium nitrate from methanol (2 grams ammonium nitrate dissolves in 16 milliliters methanol), and then vacuum dry or air-dry the product. The product will weigh about 180 to 188 grams and will be of 99% purity.

**Method of Preparation 2:** Dissolve 1000 grams of sodium nitrate into 2300 milliliters of water. Then add 740 grams of 28 – 30% ammonia solution, or 2000 grams of 10% ammonia solution, or pass 200 grams of anhydrous ammonia into the nitrate solution while stirring the nitrate solution. Thereafter, place this mixture into an ice bath, and chill to 0 Celsius. Then, pass 3530 grams of dry carbon dioxide into the reaction mixture at a steady rate while stirring the reaction mixture and maintaining its temperature at 0 Celsius, or add in pieces, 3000 grams of dry ice. After the carbon dioxide addition, continue to stir the reaction mixture for 30 minutes at 0 Celsius, and then filter-off the precipitated sodium bicarbonate. Thereafter, recrystallize the ammonium nitrate from the reaction mixture. If a rotary evaporator is available, place the filtered reaction mixture there into, and remove the water under high vacuum. After the recrytsallization process, vacuum dry or air-dry the crystals. Then, recrystallize the ammonium nitrate from methanol (2 grams ammonium nitrate dissolves in 16 milliliters methanol), and then vacuum dry or air-dry the product. Recrystallize the ammonium nitrate from methanol is not necessary if the ammonium nitrate is to be used in making prills. If the ammonium nitrate is to be alloyed with sensitive high explosives such as picric acid, picric acid salts, styphnic acid, or styphnic acid salts, the ammonium nitrate should be recrystallized from methanol.

## 18. Calcium Carbonate

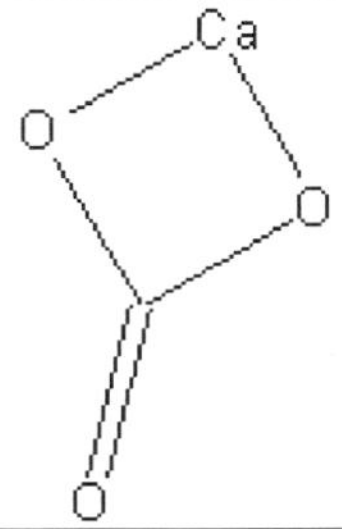


Calcium Carbonate is a white powder, granules, or solid material of various sizes. The product is very common and exists as chalk, and other natural materials. The dry solid breaks down to lime when heated to high temperatures.

## 19. Sodium Nitrate

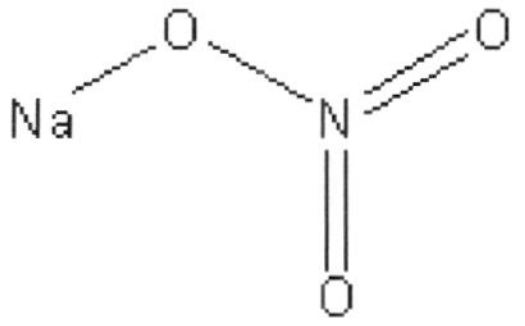


Sodium Nitrate forms colorless to white crystals, granules, or powder. The crystals have a melting point of 308 Celsius, but they begin to decompose at higher temperatures. The solid compound is very soluble in water, slightly in alcohol, but relatively insoluble in most solvents.

**Method of Preparation 1:**

1. Dissolve 130 grams of concentrated nitric acid into 200 milliliters of water at room temperature.
2. Slowly add, in small portions at a time to the nitric acid solution, 63 grams of sodium hydroxide, or 133 grams of baking soda.
3. After the addition, boil the solution to remove water, and recover the sodium nitrate.

## 20. Urea

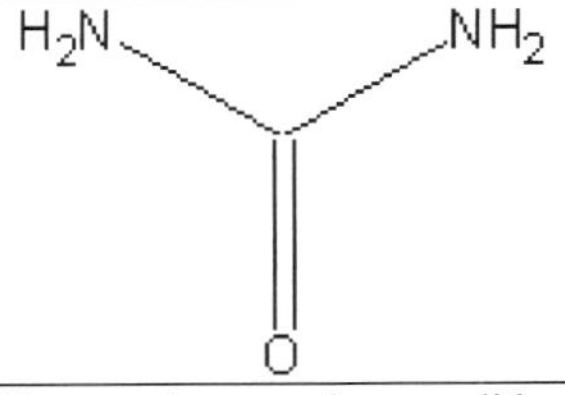


Urea forms a crystalline powder, granules, or solid material of tetragonal prisms. The crystals have a melting point of 132 Celsius, but tend to develop the odor of ammonia at room temperature. The solid is soluble in water, and alcohol, soluble in acids, but relatively insoluble in most common laboratory solvents.

## 21. Lead Nitrate

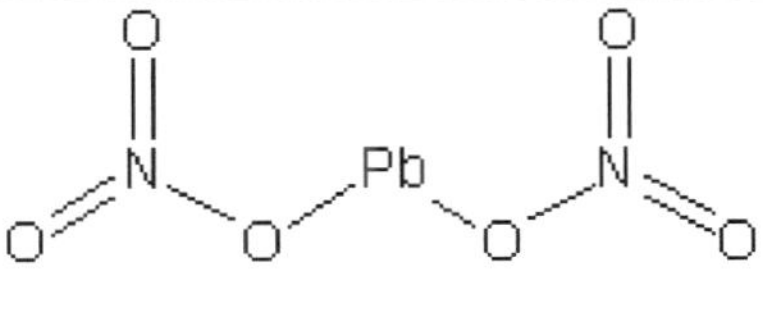


Lead nitrate forms colorless to white, translucent crystals. The crystals are poisonous so users should wear gloves when handling. Lead nitrate is soluble in water, but not very soluble in other solvents. It is prepared by dissolving lead oxides into nitric acid.

## 22. Nitro Starch

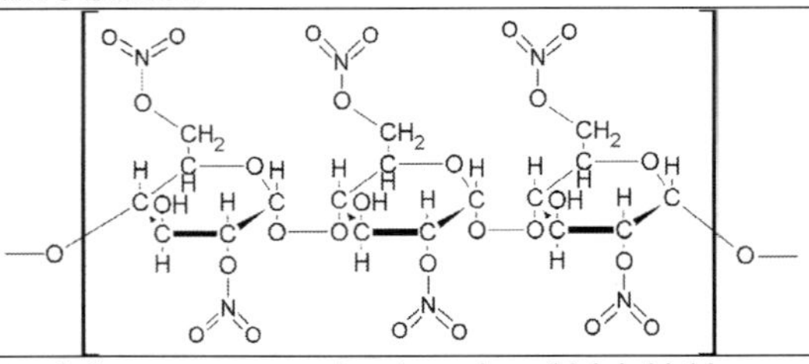

Nitro starch forms an orange colored powder, which is highly flammable. It has a average nitrogen content of 10 to 13%. The compound is insoluble in water and most organic solvents, but it has moderate solubility in a mixture of 95% ethanol and 99% diethyl ether (ratio of 1 to 1). Nitro starch is highly flammable, and like its great grandfather, nitrocellulose, it is used in making smokeless gun powders, rocket propellants, fireworks, and blasting compositions, the latter when mixed with sodium or potassium nitrate, TNT, RDX, ammonium nitrate, or perchlorates. When dry, it is easily ignited and explodes when mixed with reducing agents and ignited. Note: a small charge of aluminum flash powder confined, and ignited using a simple cannon fuse, is capable of detonating nitro starch. The dry compound should never be stored dry, but should be stored wet with 10% water or alcohol. Like nitrocellulose, nitro starch is very easy to make and very cheap, and hence can be a simple source of explosive power for use in guerrilla operations/insurgent operations for making improvised munitions. Danger of preparation: can be prepared by non-scientists as long as all procedure(s) and safety info is strictly followed.

**Method of Preparation 1:** Into a suitable sized 3-neck flask, equipped with mechanical stirrer, thermometer, and wide mouth funnel, and sitting inside a ice/salt bath, pour in 400 milliliters of 55% nitric acid and then chill this dilute nitric acid to -5 Celsius by means of the ice/salt bath. Once the dilute nitric acid has chilled to -5 Celsius, turn on the mechanical stirrer to high speed, and then add in, in small portions at a time, 22 grams of finely divided dry cornstarch while stirring and maintaining the 55% nitric acid at -5 Celsius. After the addition, into a separate clean beaker, prepare a nitrating acid mixture by carefully adding 890 milliliters of 98% sulfuric acid into 300 milliliters of anhydrous nitric acid/methylene chloride solution obtained from a stock solution prepared in "chapter 6e. Preparation of nitric acid". Thereafter, manually stir this nitrating acid mixture for about 5 minutes, and then slowly add this anhydrous nitric acid/methylene chloride mixture to the dilute nitric acid/cornstarch reaction mixture over a period of time so as to maintain the temperature of the reaction mixture below 10 Celsius under rapid stirring. Once all the anhydrous nitric acid/methylene chloride mixture has been added, remove the ice/salt bath, and allow the reaction mixture to warm up under rapid stirring for 3 hours. After 3 hours, stop stirring, disassemble the apparatus, slowly and carefully, pour the entire reaction mixture into 1500 milliliters of ice water contained in a suitable sized beaker, equipped with mechanical stirrer, and then stir this aqueous mixture on high speed for about 45 minutes. Thereafter, filter-off the precipitated product of nitrostarch using a clean Buchner funnel under rapid suction, wash with three 500-milliliter portions of 10% sodium carbonate solution, followed by washing with six 500-milliliter portions of warm water, and then suction dry as best as possible. Finally, recover the suction dried product from the Buchner funnel, place onto a shallow pan or tray, and allow to thoroughly air dry. Once dry, recover the dried product, grind it into a powder, and then store in a glass or plastic container and store in a cool dry place until use.

***- Modified Black Powder Compositions in this section -***

| | |
|---|---|
| **1. 03-01-001A: Standard "reduced burn rate" time fuse mixture:** *99.99% black powder, 0.01% oil* | **2. 03-01-001B: "Reduced burn rate" fuse powder directly from potassium nitrate:** *72% potassium nitrate, 21% charcoal, 6.5% sulfur, and 0.5% oil* |
| **3. 03-01-002A: "Reduced burn rate" fuse powder directly from potassium nitrate:** *72.8% potassium nitrate, 14.5% charcoal, 9.7% sulfur, and 3% rosin* | **4. 03-01-003A: Modified black powder for use in fuse (slow burning black powder composition):** *74% potassium nitrate, 12.4% sugar carbon, 10.4% sulfur, 3.1% charcoal, 0.10% impurities* |
| **5. 03-01-004A: Gasless pyrotechnic composition for use in gasless and smokeless time fuse:** *71.2% barium chromate, 16.8% zirconium/nickel alloy, 7.9% potassium perchlorate, 2.9% ethyl cellulose, 0.99% potassium dichromate, 0.21% residue* | **6. 03-01-004B: Gasless pyrotechnic composition for use in gasless and smokeless time fuse (modified burn rate):** *62.3% barium chromate, 21.8% zirconium/nickel alloy, 14.5% potassium perchlorate, 1.3% potassium dichromate, 0.10% residue* |
| **7. 03-01-004C: Gasless pyrotechnic composition for use in gasless and smokeless time fuse (modified burn rate 2):** *56.1% barium chromate, 27.5% zirconium/nickel alloy, 14.7% potassium perchlorate, 1.4% potassium dichromate, 0.30% residue* | **8. 03-01-005A: Chlorate slow burning fuse composition:** *41% Brazilian natural rubber, 27.3% charcoal, 13.6% glass, 13.6% potassium chlorate, 4.1% ammonium bisulfide, 0.40% residue* |
| **9. 03-01-006A: Pyrotechnic composition for time fuses:** *37.5% Indian rubber, 25% soft wood charcoal, 12.5% potassium chlorate, 12.5% glass, 12.5% carbon disulfide* | **10. 03-01-007A: Nitrocellulose containing pyrotechnic composition for fuses (gasless):** *25% nitrocellulose, 24% dibutyl phthalate, 20.5% lead tetraoxide, 16.5% potassium nitrate, 13% silicon, 1% diphenylamine* |
| **11. 03-01-008A: Modified "high reduced burn" black powder composition:** *62.74% potassium nitrate, 23.52 sulfur, 11.76% charcoal, 1.96% alpha-napthyl amine, 0.02% mixed residual balance* | **12. 03-01-008B: Modified "high reduced burn" black powder composition:** *67.96% potassium nitrate, 29.12% charcoal, 2.91% alpha-napthyl amine, 0.01% mixed residual balance* |
| **13. 03-01-009A: Modified black powder composition for use in time fuse and delays:** *51.36% potassium nitrate, 31.5% titanium dioxide, 10.27% charcoal, 6.84% sulfur, 0.03% balance* | **14. 03-01-009B: Modified black powder composition for use in time fuse and delays:** *52.81% potassium nitrate, 29.57% manganese dioxide, 10.56% charcoal, 7.04% sulfur, 0.02% balance* |
| **15. 03-01-009A: "Graphite" black powder composition:** *74% potassium nitrate, 14% graphite, 12% sulfur* | **16. 03-01-09B: "Graphite" black powder composition (modified):** *74% potassium nitrate, 14% graphite, 12% sulfur* |
| **17. 03-01-010A: Thermally stable modified black powder composition:** *60% sodium nitrate, 17.8% charcoal, 10% sulfur, 10% ammonium nitrate, 1.2% calcium carbonate, 1% urea* | **18. 03-01-011A: Reduced chamber pressure modified black powder suitable for small arms:** *44.5% potassium nitrate, 18.8% potassium perchlorate, 10.8% sodium benzoate, 8.9% hard wood charcoal, 5.9% of sulfur, 5.9% dicyandiamide, 3.9% dextrin, 0.49% wax, 0.49% graphite, 0.32% residue* |
| **19. 03-01-012A: Modified black powder composition utilizing cellulose fibers for added stability and energy output:** *60% potassium nitrate, 20% cellulose, 20% erythorbic acid* | **20. 03-01-012B: Modified black powder composition utilizing cellulose fibers for making consumable gun case cartridges:** *60% potassium nitrate, 20% cellulose, 20% erythorbic acid* |
| **21. 03-01-013A: Modified black powder composition utilizing coated sodium nitrate:** *71.6% sodium nitrate, 15.9% charcoal, 11.9% sulfur, 0.49% graphite, 0.11% mixed impurities* | **22. 03-01-014A: Basic modified black powder composition for use in canons:** *75% potassium nitrate, 25% gum Arabic* |
| **23. 03-01-015A: Modified black powder composition with increased burn rate:** *75% potassium nitrate, 15% sodium benzoate, 10% sulfur* | **24. 03-01-016A: Specialty modified black powder composition:** *72% lead nitrate, 18% sulfur, 10% nitro starch* |
| **25. 03-01-017A: Reduced burn black powder composition for various uses:** *72.8% sodium nitrate, 14.5% charcoal, 9.7% sulfur, 3% rosin* | **26. 03-01-017B: Reduced burn black powder composition for various uses:** *72.8% potassium nitrate, 14.5% charcoal, 9.7% sulfur, 3% gum Arabic* |
| **27. 03-01-018A: Modified "high reduced burn" black powder composition:** *73.52% potassium nitrate, 16.66% charcoal, 7.84% sulfur, 1.96% alpha-napthyl amine, 0.02% mixed residual balance* | **28. 03-01-019A: Modified "reduced burn" black powder composition:** *74% potassium nitrate, 16.5% carbon black, 9.5% sulfur* |
| | |

| | |
|---|---|
| **29. 03-01-020A: Modified "reduced burn" black powder composition:** *74% potassium nitrate, 10.4% sulfur, 9.36% soft wood charcoal, 6.24% sugar carbon* | **30. 03-01-021A: "Explosive" graphite based black powder composition**: *74% potassium nitrate, 14% graphite, 12% sulfur* |

### 03-01-001A: Standard "reduced burn rate" time fuse mixture:

Place ***200 grams of regular black powder*** (of any grain and brand), into a ball mill, and then add in ***2 grams of caster oil, linseed oil, olive oil, corn oil, cotton seed oil, or any lubricating oil***, followed by 50 grams of steel balls of 4 to 5 millimeters in diameter. Thereafter, rotate the mixture using the ball mill for 1 hour at nominal speed (about 180 to 200 RPM). After rotating for 1 hour, pour in 100 milliliters of *distilled water*, and continue to tumble the mixture for 1 hour at 180 to 200 RPM. Thereafter, remove the entire bulky mass, and spread it out onto a shallow pan, and allow it to dry fully. Once the mass has been dried, thoroughly pulverize the mass using either a crucible or clean ball mill. This mixture should then be packed into fuses using the normal techniques for fuse preparation.
**Burn rate:** 55 to 57 seconds per foot.
**Water resistance:** Above moderate.
**Stability:** Can be stored for decades.
**Sensitivity:** Stable towards friction; stable towards shock; stable to temperatures up to 400 Celsius.
**Flammability (1 to 10, un-contained):** 6 powdered.
**Ease of ignition (1 to 10 un-contained):** 7 powdered.
**Tendency to cake:** Will not cake.
**Detonation ability:** Cannot be detonated under normal conditions.
**Explosive ability (contained; deflagration):** Low.
**Percentage:** ***99.99% black powder, 0.01% oil***
**Classification:** Deflagrating explosive (classified as propellant).

**Danger of preparation:** Can be made by amateurs.
**Use:** Time fuse, delay fuzes for grenades and other fuzes, and in delay elements for a wide variety of munitions.

### 03-01-001B: "Reduced burn rate" fuse powder directly from potassium nitrate:

Into a standard ball mill, place ***144 grams of potassium nitrate*** of 99%+ purity, followed by ***13 grams of sulfur*** of 99% purity, followed by ***42 grams of standard charcoal***, followed finally by ***1 gram of castor oil, linseed oil, olive oil, corn oil, cotton seed oil, or any lubricating oil***. Thereafter, add in 50 grams of steel balls of 4 to 5 millimeters in diameter. Thereafter, rotate the mixture using the ball mill for 1 hour at 100 to 180 RPM. Afterwards, remove the mass, and them spread it out on a shallow pan and allow it to cure for several days. This mixture should then be packed into fuses using the normal techniques for fuse preparation.
Note: The potassium nitrate can be replaced with the same gram amount of *sodium nitrate, lithium nitrate*, or *any anhydrous nitrate* of 98% purity or better (with the exception of ammonium nitrate and copper nitrate).
Note: Nitrates containing magnesium, calcium, or metals might have burning rates from 45 to 60 seconds per foot. *Ammonium nitrate should not be used.*
**Burn rate:** 52 to 57 seconds per foot.
**Water resistance:** Above moderate.
**Stability:** Can be stored for decades.
**Sensitivity:** Stable towards friction; stable towards shock; stable to temperatures up to 400 Celsius.
**Flammability (1 to 10, un-contained):** 6 powdered.
**Ease of ignition (1 to 10 un-contained):** 7 powdered.
**Tendency to cake:** Will not cake.
**Detonation ability:** Cannot be detonated under normal conditions.
**Explosive ability (contained; deflagration):** Low.
**Percentage:** ***72% nitrate, 21% charcoal, 6.5% sulfur, and 0.5% oil***
**Classification:** Deflagrating explosive (classified as propellant).

**Danger of preparation:** Can be made by amateurs.
**Use:** Time fuse, delay fuzes for grenades and other fuzes, and in delay elements for a wide variety of munitions.

### 03-01-002A: "Reduced burn rate" fuse powder directly from potassium nitrate:

Into a standard ball mill, place ***145.6 grams of powdered potassium nitrate*** or ***sodium nitrate***, followed by ***19.4 grams of powdered sulfur***, ***29 grams of regular wood charcoal*** (any quality wood charcoal), and then ***6 grams of standard rosin***. Then quickly add in 25 grams of steel balls of 4 to 5 millimeters in diameter. Thereafter, tumble the entire mixture for 1 hour at 180 RPM. Thereafter, remove the entire mass, and then lay it onto a shallow pan or tray and allow it to air-dry fully. Thereafter, thoroughly pulverize the mixture using a crucible or clean ball mill. This mixture should then be packed into fuses using the normal techniques for fuse preparation.
**Burn rate:** 45 seconds per foot.
**Water resistance:** Above moderate.
**Stability:** Can be stored for decades.
**Sensitivity:** Stable towards friction; stable towards shock; stable to temperatures up to 400 Celsius.
**Flammability (1 to 10, un-contained):** 4 powdered.
**Ease of ignition (1 to 10 un-contained):** 6 powdered.
**Tendency to cake:** Will not cake.
**Detonation ability:** Cannot be detonated under normal conditions.
**Explosive ability (contained; deflagration):** Low.
**Percentage:** ***72.8% nitrate, 14.5% charcoal, 9.7% sulfur, and 3% rosin***
**Classification:** Deflagrating mixture (classified as flammable solid).

**Danger of preparation:** Can be made by amateurs.
**Use:** Time fuse, delay fuzes for grenades and other fuzes, and in delay elements for a wide variety of munitions.

### 03-01-003A: Modified black powder for use in fuse (slow burning black powder composition):

Into a suitable vertical mixer equipped with plastic stir blade, place 350 milliliters of 95% ethyl alcohol, and then add in ***52 grams of flours of sulfur***, followed by ***15.6 grams of standard wood charcoal,*** and then followed by ***62.4 grams of pulverized sugar carbon*** (prepared by thoroughly burning sugar until nothing but the carbon remains, and then pulverizing by hand). Then thoroughly blend the mixture on low speed for about 30 minutes. After 30 minutes, throw in ***370 grams of potassium nitrate***, and then continue blending the mixture on low speed for about 2 hours to form a good mix. After blending the mixture for 2 hours, filter the mixture using gravity or vacuum filtration to collect the insoluble mixture. After the filtration process, place the filtered mass onto a shallow pan, and allow it to air-dry for several days. After then, place the dried mass into a ball mill, filled with steel shot of 5 millimeters in diameter or so, and then tumble the mixture at 300 RPM for about 1 hour to form a uniform ground mixture. Thereafter, to use the mixture, it should be packed into the desired fuse tube under pressure.
**Burn rate:** about 66 seconds per foot
**Water resistance:** Good.
**Stability:** Can be stored for decades.
**Sensitivity:** Stable towards friction; stable towards shock; stable to temperatures up to 400 Celsius.
**Flammability (1 to 10, un-contained):** 4 powdered.
**Ease of ignition (1 to 10 un-contained):** 5.5 powdered.
**Tendency to cake:** Will not cake.
**Detonation ability:** Cannot be detonated under normal conditions.
**Explosive ability (contained; deflagration):** Low.
**Percentage:** ***74% potassium nitrate, 12.4% sugar carbon, 10.4% sulfur, 3.1% charcoal, 0.10% impurities***
**Classification:** Deflagrating mixture (classified as flammable solid).

**Danger of preparation:** Can be made by amateurs.
**Use:** Used in time fuse, and for delay trains in blasting caps and detonators.

### 03-01-004A: Gasless pyrotechnic composition for use in gasless and smokeless time fuse:

Into a suitable ball mill, filled with 150 grams of steel shot of 5 millimeters in diameter, place ***85 grams of finely powdered zirconium/nickel alloy*** (commercially available), and containing 50% zirconium and 50% nickel, and then add in ***5 grams of potassium dichromate***, and then tumble the mixture for about 15 to 20 minutes to coat the metal alloy with the dichromate. Thereafter, remove the coated zirconium/nickel alloy, and place it into a suitable mixing bowl, blender, or similar container, equipped with motorized stirrer utilizing plastic stir blade, and then add in ***40 grams of potassium perchlorate***, followed by ***15 grams of ethyl cellulose***, and then followed by ***360 grams of barium chromate***, and then blend the mixture on moderate speed for about 1 hour to form a uniform paste. Thereafter, place the pasty mass onto a shallow tray, and allow it to thoroughly air-dry. Once it has, place the dried mass into a suitable clean ball mill, filled the usual amount of steel shot of the usual diameter, and then tumble the mixture for about 45 minutes to 1 hour at 250 RPM to form a uniform powder. Now, to use the powder, it simply needs to be pressed into fuse bodies or tubes under pressure using the normal techniques.
**Burn rate:** 6 seconds per inch at 0.2 inch diameter fuse body.

**Water resistance:** Above moderate.
**Stability:** Can be stored for decades.
**Sensitivity:** Stable towards friction; stable towards shock; stable to temperatures up to 400 Celsius.
**Flammability (1 to 10, un-contained):** 6 powdered.
**Ease of ignition (1 to 10 un-contained):** 7.2 powdered.
**Tendency to cake:** Will not cake.
**Detonation ability:** Cannot be detonated under normal conditions.
**Explosive ability (contained; deflagration):** Low.
**Percentage:** ***71.2% barium chromate, 16.8% zirconium/nickel alloy, 7.9% potassium perchlorate, 2.9% ethyl cellulose, 0.99% potassium dichromate, 0.21% residue***
**Classification:** Deflagrating mixture (classified as flammable solid).

**Danger of preparation:** Can be made by amateurs.
**Use:** Used in smokeless and gasless time fuses for multiple uses.

### 03-01-004B: Gasless pyrotechnic composition for use in gasless and smokeless time fuse (modified burn rate):

Into a suitable ball mill, filled with 150 grams of steel shot of 5 millimeters in diameter, place ***105 grams of finely powdered zirconium/nickel alloy*** (commercially available), and containing 30% zirconium and 70% nickel, and then add in ***6.5 grams of potassium dichromate***, and then tumble the mixture for about 15 to 20 minutes to coat the metal alloy powder with the dichromate. Thereafter, remove the dichromate coated zirconium/nickel alloy, and place it into a suitable mixing bowl, blender, or similar container, equipped with motorized stirrer utilizing plastic stir blades, and then add in ***70 grams of potassium perchlorate***, followed by ***300 grams of barium chromate***, and then blend the mixture on moderate speed for about 1 hour to form a uniform paste. Thereafter, place the pasty mass onto a shallow tray, and allow it to thoroughly air-dry. Once it has, place the dried mass into a suitable clean ball mill, filled the usual amount of steel shot of the usual diameter, and then tumble the mixture for about 45 minutes to 1 hour at 250 RPM to form a uniform powder. Now, to use the powder, it simply needs to be pressed into fuse bodies or tubes under pressure using the normal techniques.
**Burn rate:** 12 seconds per inch at 0.20 inch diameter fuse body.
**Water resistance:** Above moderate.
**Stability:** Can be stored for decades.
**Sensitivity:** Stable towards friction; stable towards shock; stable to temperatures up to 400 Celsius.
**Flammability (1 to 10, un-contained):** 5.2 powdered.
**Ease of ignition (1 to 10 un-contained):** 6.1 powdered.
**Tendency to cake:** Will not cake.
**Detonation ability:** Cannot be detonated under normal conditions.
**Explosive ability (contained; deflagration):** Low.
**Percentage:** ***62.3% barium chromate, 21.8% zirconium/nickel alloy, 14.5% potassium perchlorate, 1.3% potassium dichromate, 0.10% residue***
**Classification:** Deflagrating mixture (classified as flammable solid).

**Danger of preparation:** Can be made by amateurs.
**Use:** Used in smokeless and gasless time fuses for multiple uses.

### 03-01-004C: Gasless pyrotechnic composition for use in gasless and smokeless time fuse (modified burn rate 2):

Into a suitable ball mill, filled with 150 grams of steel shot of 5 millimeters in diameter, place ***140 grams of finely powdered zirconium/nickel alloy*** (commercially available), and containing 30% zirconium and 70% nickel, and then add in ***7.5 grams of potassium dichromate***, and then tumble the mixture for about 15 to 20 minutes to coat the metal alloy powder with the dichromate. Thereafter, remove the dichromate coated zirconium/nickel alloy, and place it into a suitable mixing bowl, blender, or similar container, equipped with motorized stirrer utilizing plastic stir blade, and then add in ***75 grams of potassium perchlorate***, followed by ***285 grams of barium chromate***, and then blend the mixture on moderate speed for about 1 hour to form a uniform paste. Thereafter, place the pasty mass onto a shallow tray, and allow it to thoroughly air-dry. Once it has, place the dried mass into a suitable clean ball mill, filled the usual amount of steel shot of the usual diameter, and then tumble the mixture for about 45 minutes to 1 hour at 250 RPM to form a uniform powder. Now, to use the powder, it simply needs to be pressed into fuse bodies or tubes under pressure using the normal techniques.
**Burn rate:** 20 seconds per inch at 0.20 inch diameter fuse body.
**Water resistance:** Above moderate.
**Stability:** Can be stored for decades.
**Sensitivity:** Stable towards friction; stable towards shock; stable to temperatures up to 400 Celsius.
**Flammability (1 to 10, un-contained):** 4.5 powdered.
**Ease of ignition (1 to 10 un-contained):** 6.9 powdered.
**Tendency to cake:** Will not cake.

**Detonation ability:** Cannot be detonated under normal conditions.
**Explosive ability (contained; deflagration):** Low.
**Percentage:** ***56.1% barium chromate, 27.5% zirconium/nickel alloy, 14.7% potassium perchlorate, 1.4% potassium dichromate, 0.30% residue***
**Classification:** Deflagrating mixture (classified as flammable solid).

**Danger of preparation:** Can be made by amateurs.
**Use:** Used in smokeless and gasless time fuses for multiple uses.

### 03-01-005A: Chlorate slow burning fuse composition:

Into a suitable vertical mixing bowl, such a stainless steel wide mouth bowl, place ***100 grams of powdered soft wood charcoal,*** followed by ***50 grams of glass powder,*** followed by ***50 grams of potassium chlorate,*** followed by ***150 grams of natural Brazilian rubber,*** followed by 75 milliliters of hexane, and then thoroughly blend the mixture using a stainless steel blade or similar item for about 30 minutes to form a uniform fluidized pasty mass. After 30 minutes, add in ***15 grams of ammonium bisulfide***, and then continue to blend the mixture until the bulk of the hexane has been evaporated. When most of the hexane has evaporated, and a dough like mass remains, the fuse mixture is ready to be cured. To cure, simply place the dough-like material onto any shallow pan or tray and allow it to thoroughly air-dry. Thereafter, place the dried mass into a clean ball mill, filled with 150 grams of stainless steel shot of 10 millimeters in diameter, and then tumble the mixture for about 1 hour at 150 RPM to form a uniform powdered mixture. Thereafter, the fuse powder is ready to be used. To do so, it needs to be pressed into any desirable fuse casing using high pressure or similar means.
**Burn rate:** Very slow burn rate.
**Water resistance:** Above moderate.
**Stability:** Can be stored for decades.
**Sensitivity:** Stable towards friction; stable towards shock; stable to temperatures up to 400 Celsius.
**Flammability (1 to 10, un-contained):** 3.1 powdered.
**Ease of ignition (1 to 10 un-contained):** 4 powdered.
**Tendency to cake:** Will not cake.
**Detonation ability:** Cannot be detonated under normal conditions.
**Explosive ability (contained; deflagration):** Low.
**Percentage:** ***41% Brazilian natural rubber, 27.3% charcoal, 13.6% glass, 13.6% potassium chlorate, 4.1% ammonium bisulfide, 0.40% residue***
**Classification:** Deflagrating mixture (classified as flammable solid).

**Danger of preparation:** Can be made by amateurs.
**Use:** Used in making time fuses for a variety of applications.

### 03-01-006A: Pyrotechnic composition for time fuses:

Into a suitable vertical mixing bowl, equipped with motorized stirrer, place ***100 grams of powdered soft wood charcoal,*** followed by ***50 grams of glass powder,*** followed by ***50 grams of potassium chlorate,*** followed by ***150 grams of Indian rubber,*** followed by 175 milliliters of hexane, and then thoroughly blend the mixture until the bulk of the hexane evaporates. Thereafter, add in ***50 grams of carbon disulfide***, and then continue to blend the mixture until a tough, waxy paste remains. When a pasty-like mass remains, the fuse mixture is ready to be used. To use, simply press the paste into any desirably fuse casing using high pressure or similar means, and then cure in an oven at 70 Celsius for 24 hours.
**Burn rate:** Very slow burn rate.
**Water resistance:** Above moderate.
**Stability:** Can be stored for decades.
**Sensitivity:** Stable towards friction; stable towards shock; stable to temperatures up to 400 Celsius.
**Flammability (1 to 10, un-contained):** 3 powdered.
**Ease of ignition (1 to 10 un-contained):** 3.8 powdered.
**Tendency to cake:** Will not cake.
**Detonation ability:** Cannot be detonated under normal conditions.
**Explosive ability (contained; deflagration):** Low.
**Percentage:** ***37.5% Indian rubber, 25% soft wood charcoal, 12.5% potassium chlorate, 12.5% glass, 12.5% carbon disulfide***
**Classification:** Deflagrating mixture (classified as flammable solid).

**Danger of preparation:** Can be made by amateurs.
**Use:** Used in making short time fuses for detonators and other delay elements.

### 03-01-007A: Nitrocellulose containing pyrotechnic composition for fuses (gasless):

Into a ball mill, filled with 500 grams of light steel shot, place ***410 grams of lead tetraoxide,*** followed by ***330 grams of potassium nitrate,*** followed by ***260 grams of finely divided silicon,*** followed by ***500 grams of nitrocellulose of average nitrogen content,*** followed by ***480 grams of dibutyl phthalate***, and finally followed by ***20 grams of diphenylamine***. Thereafter, tumble the mixture at 350 RPM for about 3 hours at room temperature. Thereafter, place the mixture into a beaker or any similar container, and then heat the mixture at 100 Celsius for about 8 hours. Thereafter, blend the mixture at 100 Celsius for about 5 minutes. Thereafter, the mixture is ready for extruding. To do so, the mixture should be extruded through time fuse of any desired diameter, or the mixture can be pressed into pellets or rods for use in mechanical fuzes of the usual operation.
**Burn rate:** Slow; average burn rate: 67 seconds per foot at 2mm diameter fuse.
**Water resistance:** Above moderate.
**Stability:** Can be stored for decades.
**Sensitivity:** Stable towards friction; stable towards shock; stable to temperatures up to 400 Celsius.
**Flammability (1 to 10, un-contained):** 6.7 powdered.
**Ease of ignition (1 to 10 un-contained):** 8 powdered.
**Tendency to cake:** Will not cake.
**Detonation ability:** Cannot be detonated under normal conditions.
**Explosive ability (contained; deflagration):** Below moderate.
**Percentage:** ***25% nitrocellulose, 24% dibutyl phthalate, 20.5% lead tetraoxide, 16.5% potassium nitrate, 13% silicon, 1% diphenylamine,***
**Classification:** Deflagrating explosive (classified as propellant).

**Danger of preparation:** Can be made by amateurs.
**Use:** Used for making times fuses, and for use in mechanical fuzes.

### 03-01-008A: Modified "high reduced burn" black powder composition:

Into a suitable ball mill, filled with 250 grams of steel shot, preferably Teflon coated, place ***10 grams of alpha-napthyl amine***, followed by ***320 grams of potassium nitrate,*** followed by ***60 grams of alder wood charcoal (pre-pulverized),*** and then followed by ***120 grams of flours of sulfur***. Thereafter, add in 50 milliliters of acetone, and then followed by 75 milliliters of warm water, and then tumble the mixture at 250 RPM for about 90 minutes. Thereafter, the mixture is ready for use. To do so, the mixture needs to be extruded into time fuse cord, or pressed into tablets, pellets, or rods. The composition needs to be dried in the usual manner thereafter.
**Burn rate:** 400 seconds per yard.
**Water resistance:** Above moderate.
**Stability:** Can be stored for decades.
**Sensitivity:** Stable towards friction; stable towards shock; stable to temperatures up to 400 Celsius.
**Flammability (1 to 10, un-contained):** 5 powdered.
**Ease of ignition (1 to 10 un-contained):** 8 powdered.
**Tendency to cake:** Will not cake.
**Detonation ability:** Cannot be detonated under normal conditions.
**Explosive ability (contained; deflagration):** Moderate.
**Percentage:** ***62.74% potassium nitrate, 23.52 sulfur, 11.76% charcoal, 1.96% alpha-napthyl amine, 0.02% mixed residual balance***
**Classification:** Deflagrating explosive (classified as propellant).

**Danger of preparation:** Can be made by amateurs.
**Use:** Used in time fuse or as a delay element for multiple purposes.

### 03-01-008B: Modified "high reduced burn" black powder composition:

Into a suitable ball mill, filled with 150 grams of lead shot, place ***15 grams of alpha-napthyl amine***, followed by ***350 grams of potassium nitrate,*** and then followed by ***150 grams of alder wood charcoal (pre-pulverized)***. Thereafter, add in 50 milliliters of acetone, and then followed by 75 milliliters of warm water, and then tumble the mixture at 250 RPM for about 90 minutes. Thereafter, the mixture is ready for use. To do so, the mixture needs to be extruded into time fuse cord, or pressed into tablets, pellets, or rods. The composition needs to be dried in the usual manner thereafter.

**Burn rate:** 300 seconds per yard.
**Water resistance:** Above moderate.
**Stability:** Can be stored for decades.
**Sensitivity:** Stable towards friction; stable towards shock; stable to temperatures up to 400 Celsius.
**Flammability (1 to 10, un-contained):** 4.9 powdered.
**Ease of ignition (1 to 10 un-contained):** 7.8 powdered.
**Tendency to cake:** Will not cake.
**Detonation ability:** Cannot be detonated under normal conditions.
**Explosive ability (contained; deflagration):** Below moderate.
**Percentage:** ***67.96% potassium nitrate, 29.12% charcoal, 2.91% alpha-napthyl amine, 0.01% mixed residual balance***
**Classification:** Deflagrating explosive (classified as propellant).

**Danger of preparation:** Can be made by amateurs.
**Use:** Used in time fuse or as a delay element for multiple purposes.

### 03-01-009A: Modified black powder composition for use in time fuse and delays:

Into a suitable ball mill, containing 250 grams of lead shot, place ***375 grams of potassium nitrate***, followed by ***75 grams of soft wood charcoal,*** and then followed by ***50 grams of flours of sulfur***. Thereafter, add in 125 milliliters of 95% ethyl alcohol, and then tumble the combined mixture at 300 RPM for about 1 hour. Thereafter, add in ***230 grams of titanium dioxide***, and then continue to tumble the mixture at 300 RPM for about 1 hour. After the 1 hour tumbling process, the mixture is ready for use. To do so, the material should be pressed into any desirable time fuse cord, tube, ect., and then cure in an oven at ordinary temperatures in the usual manner.

**Burn rate:** 170 seconds per yard on average.
**Water resistance:** Above moderate.
**Stability:** Can be stored for decades.
**Sensitivity:** Stable towards friction; stable towards shock; stable to temperatures up to 400 Celsius.
**Flammability (1 to 10, un-contained):** 4.4 powdered.
**Ease of ignition (1 to 10 un-contained):** 6.1 powdered.
**Tendency to cake:** Will not cake.
**Detonation ability:** Cannot be detonated under normal conditions.
**Explosive ability (contained; deflagration):** Below moderate.
**Percentage:** ***51.36% potassium nitrate, 31.5% titanium dioxide, 10.27% charcoal, 6.84% sulfur, 0.03% balance***
**Classification:** Deflagrating mixture (classified as flammable solid).

**Danger of preparation:** Can be made by amateurs.
**Use:** Used in time fuse cord, and as a delay element for explosive munitions.

### 03-01-009B: Modified black powder composition for use in time fuse and delays:

As in the previous example, into a suitable ball mill, containing 300 grams of Teflon coated steel shot, place ***375 grams of potassium nitrate***, followed by ***75 grams of soft wood charcoal,*** and then followed by ***50 grams of flours of sulfur***. Thereafter, add in 125 milliliters of 95% ethyl alcohol, and then tumble the combined mixture at 300 RPM for about 1 hour. Thereafter, add in ***210 grams of manganese dioxide***, and then continue to tumble the mixture at 300 RPM for about 1 hour. After the 1 hour tumbling process, the mixture is ready for use. To do so, the material should be pressed into any desirable time fuse cord, tube, ect., and then cured in an oven at ordinary temperatures in the usual manner.

**Burn rate:** 115 seconds per yard on average.
**Water resistance:** Above moderate.
**Stability:** Can be stored for decades.
**Sensitivity:** Stable towards friction; stable towards shock; stable to temperatures up to 400 Celsius.
**Flammability (1 to 10, un-contained):** 4.4 powdered.
**Ease of ignition (1 to 10 un-contained):** 6.1 powdered.
**Tendency to cake:** Will not cake.
**Detonation ability:** Cannot be detonated under normal conditions.

**Explosive ability (contained; deflagration):** Below moderate.
**Percentage:** ***52.81% potassium nitrate, 29.57% manganese dioxide, 10.56% charcoal, 7.04% sulfur, 0.02% balance***
**Classification:** Deflagrating mixture (classified as flammable solid).

**Danger of preparation:** Can be made by amateurs.
**Use:** Used in time fuse cord, and as a delay element for explosive munitions.

### 03-01-009A: "Graphite" black powder composition:

Into a suitable iron, tin, steel, or bronze pot, place 2 liters of water followed by ***750 grams of potassium nitrate***, and the boil the solution using a wood fire, or steam bath. When the water begins to boil, gently stir in ***150 grams of finely divided graphite***, followed by ***120 grams of flours of sulfur***, and then continue to boil the mixture. Note: the mixture should be continuously stirred using a wood spatula all through out the boiling process. Continue to boil the mixture until about 98% of the water has been evaporated, and only a pasty thick mass exists. When this point is achieved, remove the pot from the heat source, and allow its contents to cool to room temperature. Note: during this cool down period, continuously stir the pasty mixture with a wooden spatula. Once the mixture has cooled to room temperature, lay out the mixture on a shallow pan and allow it to thoroughly dry, or press the pasty mass through any desired screen mesh to form grains of any desired size and then cure in the usual manner.
**Burn rate:** Depends on grain size (granules/grains 1 to 12 millimeters are suitable for firearms, mortars, and cannons; fine powder can be used as a blasting agent).
**Water resistance:** Moderate.
**Stability:** Can be stored for decades.
**Sensitivity:** Stable towards friction; stable towards shock; stable to temperatures up to 400 Celsius.
**Flammability (1 to 10, un-contained):** 9 powdered.
**Ease of ignition (1 to 10 un-contained):** 9 powdered.
**Tendency to cake:** Will not cake.
**Detonation ability:** Can be detonated under normal conditions.
**Explosive ability (contained; explosive deflagration):** High when in powder form.
**Percentage:** ***74% potassium nitrate, 14% graphite, 12% sulfur***
**Classification:** Deflagrating explosive (classified as propellant).

**Danger of preparation:** Can be made by amateurs as long as the procedure is followed exactly.
**Use:** Suitable for use in firearms, mortars, cannons, and in blasting agents.

### 03-01-009B: "Graphite" black powder composition (modified):

Into a suitable ball mill containing 300 grams of steel shot of 5 millimeters in diameter, followed by ***750 grams of potassium nitrate***, followed by ***150 grams of finely divided graphite***, followed by ***120 grams of flours of sulfur***, and then followed by 75 milliliters of 95% ethyl alcohol, or 75 milliliters of denatured alcohol, and then tumble the mixture at 500 RPM for about 2 hours. After 2 hours, remove the mixture, and place it onto a shallow pan and allow it to thoroughly air-dry, or press the pasty mass through any desired screen mesh to form grains of any desired size and then cure in the usual manner. If desired, you can place the dry mass into a ball mill filled with clean steel shot of 10 millimeters in diameter and tumble the mixture for 2 hours at 250 RPM to form a uniform finely divided powder.
**Burn rate:** Depends on grain size (granules/grains 1 to 12 millimeters are suitable for firearms, mortars, and cannons; fine powder can be used as a blasting agent).
**Water resistance:** Moderate.
**Stability:** Can be stored for decades.
**Sensitivity:** Stable towards friction; stable towards shock; stable to temperatures up to 400 Celsius.
**Flammability (1 to 10, un-contained):** 9 powdered.
**Ease of ignition (1 to 10 un-contained):** 9 powdered.
**Tendency to cake:** Will not cake.
**Detonation ability:** Can be detonated under normal conditions.
**Explosive ability (contained; explosive deflagration):** High when in powder form.
**Percentage:** ***74% potassium nitrate, 14% graphite, 12% sulfur***
**Classification:** Deflagrating explosive (classified as propellant).

**Danger of preparation:** Can be made by amateurs as long as the procedure is followed exactly.

**Use:** Suitable for use in firearms, mortars, cannons, and in blasting agents.

## 03-01-010A: Thermally stable modified black powder composition:

Into a suitable ball mill containing 175 grams of Teflon coated steel shot of 5 millimeters in diameter, add in ***300 grams of sodium nitrate***, followed by ***50 grams of flours of sulfur***, followed by ***50 grams of ammonium nitrate***, followed by ***89 grams of regular charcoal,*** followed by ***6 grams of calcium carbonate***, and then followed by ***5 grams of urea.*** Thereafter, throw in 75 milliliters of hexane, and then rotate the mixture at 250 RPM for about 2 hours at room temperature. After 2 hours, remove the mixture, and place it onto a shallow pan and allow it to thoroughly air-dry, or press the pasty mass through any desired screen mesh to form grains of any desired size and then cure in the usual manner. If desired, once the mass is completely dry, it can be powdered in the usual manner to form a uniform finely divided powder.
**Burn rate:** Depends on grain size (granules/grains 1 to 12 millimeters are suitable for firearms, mortars, and cannons; fine powder can be used as a blasting agent).
**Water resistance:** Moderate.
**Stability:** Can be stored for decades.
**Sensitivity:** Stable towards friction; stable towards shock; stable to temperatures up to 400 Celsius.
**Flammability (1 to 10, un-contained):** 8.8 powdered.
**Ease of ignition (1 to 10 un-contained):** 9 powdered.
**Tendency to cake:** Will not cake.
**Detonation ability:** Can be detonated under normal conditions.
**Explosive ability (contained; explosive deflagration):** High when in powder form.
**Percentage:** ***60% sodium nitrate, 17.8% charcoal, 10% sulfur, 10% ammonium nitrate, 1.2% calcium carbonate, 1% urea***
**Classification:** Deflagrating explosive (classified as propellant).

**Danger of preparation:** Can be made by amateurs as long as the procedure is followed exactly.
**Use:** Suitable for use in firearms, mortars, cannons, and in blasting agents.

## 03-01-011A: Reduced chamber pressure modified black powder suitable for small arms:

Into a standard vertical mixer, utilizing stainless steel mixing blades, place ***225 gram of potassium nitrate***, followed by ***45 grams of finely divided hard wood charcoal***, followed by ***30 grams of flours of sulfur***, followed by ***95 grams of potassium perchlorate***, followed by ***55 grams of sodium benzoate***, followed by ***30 grams of dicyandiamide***, and then followed by ***20 grams of dextrin***. Thereafter, turn on the vertical mixer, and dry blend the ingredients for about 30 minutes. After 30 minutes, add in 160 milliliters of water while continuing to blend the ingredients, and after addition of the water, throw in about ***2.5 grams of standard wax***, followed by ***2.5 grams of finely divided graphite***, and then continue to blend the mixture for about 1 hour at room temperature. After 1 hour, the mixture is ready to use; to do so, it should be removed from the mixer, and placed onto a shallow tray and allowed to air-dry for several days; or press the pasty mass through any desired screen mesh to form grains of any desired size and then cure the grains in the usual manner; or after the dried mass has been dried for several days, the dried mass can be screened through any desirable mesh screen to form various grain sizes; however, a more convenient method of forming granules is to utilize the solvent evaporation technique. To do so, place the entire dried mass into a clean vertical mixer, utilizing Teflon coated stir blades, and then add in 250 milliliters of 95% alcohol. Thereafter, blend the mixture on moderate speed until the alcohol has evaporated to the point where only a wet tacky paste remains. Now, once the paste has formed, continue mixing until the pasty mass begins to dry and form into tiny balls or granules—this granulation technique will proceed as the solvent evaporates to near dryness. The size of the granules is dependant on how long you continue to blend the mixture once the original bulky paste has formed, and the speed at which your blending takes place. The longer you continue to blend the mixture, the more solvent evaporates. The more solvent that evaporates, the smaller the granules will be when blending on high speed. The least amount of solvent that evaporates after the original formation of the paste, the larger the granules will be when blending on low speed. Moderate sized granules can be formed by continuing to blend the mixture after the original formation of the paste, on low speed until the mixture is near dryness; or stop blending, and press the pasty mass through any desired screen mesh to form grains of any desired size and then cure in the usual manner. Either way, once your desired granules/grains have formed, allow the granules/grains to cure in an oven at 60 Celsius until completely dry. To use the grains, simply place them (loosely) into any desirable shell casing, or they can be used directly in "black powder" muzzle loaded weapons.
**Burn rate:** Depends on grain size (granules/grains 1 to 12 millimeters are suitable for firearms, mortars, and cannons).
**Chamber pressure:** 7800 psi at 60 grains
**Water resistance:** Moderate.
**Stability:** Can be stored for decades.
**Sensitivity:** Stable towards friction; stable towards shock; stable to temperatures up to 400 Celsius.
**Flammability (1 to 10, un-contained):** 8 powdered.
**Ease of ignition (1 to 10 un-contained):** 8.1 powdered.
**Tendency to cake:** Will not cake.
**Detonation ability:** Can be detonated under extreme conditions only.
**Explosive ability (contained; explosive deflagration):** Moderately high when in powder form.
**Percentage:** ***44.5% potassium nitrate, 18.8% potassium perchlorate, 10.8% sodium benzoate, 8.9% hard wood charcoal, 5.9% of sulfur, 5.9% dicyandiamide, 3.9% dextrin, 0.49% wax, 0.49% graphite, 0.32% residue***
**Classification:** Deflagrating explosive (classified as propellant).

**Danger of preparation:** Can be made by amateurs as long as the procedure is followed exactly.
**Use:** Used in small arms weapons mainly for pistols and shotguns, and also in grenade launchers.

### 03-01-012A: Modified black powder composition utilizing cellulose fibers for added stability and energy output:

Into a suitable vertical mixer, equipped with a plastic stir blade, place ***300 grams of potassium nitrate***, followed by ***200 grams of cellulosic fibers*** (commercially available and sold under a wide variety of names; in this process the brand name is designated as KS1016, available from the James River Corporation), and then add in 175 to 200 milliliters of cold water, and then blend the mixture to form a slurry or paste. Thereafter, into a separate clean vertical mixer or equivalent, place ***300 grams of potassium nitrate***, followed by ***200 grams of erythorbic acid***, and then add in 2400 milliliters of warm water, and then stir the mixture to form a clear solution. Note: if all the solids fail to dissolve, add in some more water until they do. Thereafter, pour this clear solution into the vertical mixer containing the potassium nitrate/cellulose mixture, and then blend the mixture on high for about 1 hour. After 1 hour, pour the entire mixture onto a shallow pan, and then allow it to thoroughly air-dry for several days or so. Once a dry material is obtained, place the dried mass into a clean vertical mixer or equivalent, and then add in 500 milliliters of acetone, and then blend the mixture until most of the acetone evaporates, and a pasty-like mass is obtained. Now, like in previous granulation processes, the size of the granules is dependant on how long you continue to blend the mixture once the original bulky paste has formed, and the speed at which your blending takes place. The longer you continue to blend the mixture, the more solvent evaporates. The more solvent that evaporates, the smaller the granules will be when blending on high speed. The least amount of solvent that evaporates, after the original formation of the paste, the larger the granules will be when blending on low speed. Moderate sized granules can be formed by continuing to blend the mixture after the original formation of the paste on low speed until the mixture is near dryness, or press the pasty mass through any desired screen mesh to form grains of any desired size and then cure in the usual manner. Either way, once your desired granules have formed, allow the granules/grains to cure in an oven at 50 to 60 Celsius until completely dry. To use the grains, simply place them (loosely) into any desirable shell casing, or they can be used directly in "black powder" muzzle loaded weapons.
**Burn rate:** Depends on grain size (granules/grains 1 to 12 millimeters are suitable for firearms, mortars, and cannons).
**Chamber pressure:** 7600 psi at 50 grains
**Water resistance:** Moderate.
**Stability:** Can be stored for decades.
**Sensitivity:** Stable towards friction; stable towards shock; stable to temperatures up to 400 Celsius.
**Flammability (1 to 10, un-contained):** 8 powdered.
**Ease of ignition (1 to 10 un-contained):** 8 powdered.
**Tendency to cake:** Will not cake.
**Detonation ability:** Cannot be detonated under normal conditions.
**Explosive ability (contained; explosive deflagration):** Low.
**Percentage:** ***60% potassium nitrate, 20% cellulose, 20% erythorbic acid***
**Classification:** Deflagrating explosive (classified as propellant).

**Danger of preparation:** Can be made by amateurs.
**Use:** Used in small arms weapons of the usual type.

### 03-01-012B: Modified black powder composition utilizing cellulose fibers for making consumable gun case cartridges:

Into a suitable vertical mixer or equivalent, equipped with a plastic stir blade, place ***300 grams of potassium nitrate***, followed by ***200 grams of cellulosic fibers*** (commercially available and sold under a wide variety of names; in this process the brand name is designated as KS1016, available from the James River Corporation), and then add in 175 to 200 milliliters of cold water, and then blend the mixture to form a slurry or paste. Thereafter, into a separate mixing bowl, beaker, or equivalent, place ***300 grams of potassium nitrate***, followed by ***200 grams of erythorbic acid***, and then add in 2400 milliliters of warm water, and then stir the mixture manually to form a clear solution. Note: if all the solids fail to dissolve, add in some more water until they do. Thereafter, pour this clear solution into the vertical mixer containing the potassium nitrate/cellulose mixture, and then blend the mixture on high for about 1 hour. After 1 hour, pour the entire mixture onto a shallow pan, and then allow it to thoroughly air-dry for several days or so. Once a dry material is obtained, it needs to be pressed into pre fabricated gun cartridge molds under a pressure of 10,000 to 14,000 psi. Note: a heated mold, heated between 40 and 60 Celsius is preferred to aid in flow compaction during the pressing process. Once the gun cartridges are formed, they can be filled with any suitable gun propellant powder or granules, fitted with suitable and proper gun primer, and then capped with bullet or projectile. The primer and bullet or projectile bodies can be secured to the gun cartridge by using any average acetone based rubber cement composition or binder (modelers cement), followed by drying. Thereby the munition is ready for firing.
**Burn rate:** Nominal.
**Chamber pressure:** N/A
**Water resistance:** Above moderate.
**Stability:** Can be stored for decades.
**Sensitivity:** Stable towards friction; stable towards shock; stable to temperatures up to 400 Celsius.

**Flammability (1 to 10, un-contained):** 8
**Ease of ignition (1 to 10 un-contained):** 7
**Tendency to cake:** Will not cake.
**Detonation ability:** Cannot be detonated under normal conditions.
**Explosive ability (contained; explosive deflagration):** Low.
**Percentage:** ***60% potassium nitrate, 20% cellulose, 20% erythorbic acid***
**Classification:** Deflagrating explosive (classified as propellant).

**Danger of preparation:** Can be made by amateurs.
**Use:** Used in forming non-metal, totally consumable gun cartridge cases for handgun, rifle, and high caliber ammunition casings; thereby decreasing weight of overall ammunition without sacrificing chamber pressure and force.

### 03-01-013A: Modified black powder composition utilizing coated sodium nitrate:

Into a suitable empty ball mill, place ***360 grams of dry sodium nitrate***, and then add in ***2.5 grams of powdered graphite***, and then tumble the mixture for about 12 hours. Note: sealing the ball meal to protect it from moisture is recommended. After 12 hours, the sodium nitrate will be properly coated. Now, add in ***60 grams of finely powdered sulfur***, followed by ***80 grams of standard soft wood charcoal***, and then add in 150 grams of Teflon coated steel shot of 5 millimeters in diameter, and then tumble the mixture at 50 RPM for about 12 hours. After 12 hours, the powder is ready for loading. To do so, it simply needs to be poured into any desirable black powder rifle, cannon, ect, utilizing the normal techniques, or it can be loaded into shell casings, and then tapped with a bullet, utilizing the normal techniques.
**Burn rate:** Depends on grain size (granules/grains 1 to 12 millimeters are suitable for firearms, mortars, and cannons).
**Chamber pressure:** 8200 psi at 50 grains.
**Water resistance:** Above moderate.
**Stability:** Can be stored for decades.
**Sensitivity:** Stable towards friction; stable towards shock; stable to temperatures up to 400 Celsius.
**Flammability (1 to 10, un-contained):** 8 powdered.
**Ease of ignition (1 to 10 un-contained):** 8 powdered.
**Tendency to cake:** Will not cake.
**Detonation ability:** Cannot be detonated under normal conditions.
**Explosive ability (contained; explosive deflagration):** Moderate.
**Percentage:** ***71.6% sodium nitrate, 15.9% charcoal, 11.9% sulfur, 0.49% graphite, 0.11% mixed impurities***
**Classification:** Deflagrating explosive (classified as propellant).

**Danger of preparation:** Can be made by amateurs.
**Use:** Used in black powder rifles and cannons of the usual diameter.

### 03-01-014A: Basic modified black powder composition for use in canons:

Into a suitable beaker of similar container, place ***544 grams of gum Arabic***, and then add in 1080 milliliters of water, and then stir the entire mixture for about 10 minutes to dissolve the gum Arabic. Thereafter, prepare a second solution by placing into another suitable sized beaker or similar container, ***136 grams of potassium nitrate***, followed by 2700 milliliters of warm water, and then stir the mixture to dissolve all the nitrate. Thereafter, mix both solutions, and stir the combined mixture for about 30 minutes. Thereafter, slowly add in, drop wise, over a period of about 30 to 40 minutes, 550 milliliters of 95% ethyl alcohol, while stirring the gum Arabic/nitrate mixture. After the addition of the alcohol, stir the mixture for about 30 minutes, and then filter-off the precipitated mass, and then press it or extrude it from any desirable extruding machine in the usual manner. If desired, the filtered-off mass can be pressed directly into cakes, or molds of various sizes for forming single solid propellant cakes that can be loaded directly into any desirable cannon. Note: the excess gum Arabic and alcohol mixture can be recycled by distilling-off the alcohol, and then allowing the remaining solution to evaporate to recover the gum Arabic.
**Burn rate:** Depends on grain size (granules/grains 1 to 12 millimeters are suitable for firearms, mortars, and cannons).
**Chamber pressure:** Depends on grain size; small grains: 12,000 psi at 100 grain.
**Water resistance:** Above moderate.
**Stability:** Can be stored for decades.
**Sensitivity:** Stable towards friction; stable towards shock; stable to temperatures up to 400 Celsius.
**Flammability (1 to 10, un-contained):** 8.5 powdered.
**Ease of ignition (1 to 10 un-contained):** 8 powdered.
**Tendency to cake:** Will not cake.
**Detonation ability:** Cannot be detonated under normal conditions.
**Explosive ability (contained; explosive deflagration):** Moderate.
**Percentage:** ***75% potassium nitrate, 25% gum Arabic***
**Classification:** Deflagrating explosive (classified as propellant).

**Danger of preparation:** Can be made by amateurs.
**Use:** Used in black powder cannons. Not suitable for use in rifles.

### 03-01-015A: Modified black powder composition with increased burn rate:

Into a mortar and pestle, place ***750 grams of potassium nitrate***, followed by ***150 grams of sodium benzoate***, followed by ***100 grams of flours of sulfur***, and then gently blend the mixture with the pestle for about 1 hour or less to form a uniform powder. Thereafter, the mixture should be ball milled in a suitable ball mill, minus the steel shot at 150 RPM for about 30 minutes. Thereafter, the mixture is ready for use. To use, the mixture simply needs to be loaded into any desirable shell casing in the usual manner.
**Burn rate:** Depends on grain size (granules/grains 1 to 12 millimeters are suitable for firearms, mortars, and cannons).
**Chamber pressure:** Depends on grain size; small grains: 9,000 psi at 75 grain.
**Water resistance:** Above moderate.
**Stability:** Can be stored for decades.
**Sensitivity:** Stable towards friction; stable towards shock; stable to temperatures up to 400 Celsius.
**Flammability (1 to 10, un-contained):** 8.5 powdered; 7.6 small grains; 6 medium grains; 5 large grains.
**Ease of ignition (1 to 10 un-contained):** 8 powdered; grain size varies and results in different ignition rates.
**Tendency to cake:** Will not cake.
**Detonation ability:** Cannot be detonated under normal conditions.
**Explosive ability (contained; explosive deflagration):** Moderate.
**Percentage:** ***75% potassium nitrate, 15% sodium benzoate, 10% sulfur***
**Classification:** Deflagrating explosive (classified as propellant).

**Danger of preparation:** Can be made by amateurs.
**Use:** Used in handguns in general, and mainly in .44, and .45 caliber shells. Can also be used in fireworks.

### 03-01-016A: Specialty modified black powder composition:

Into a suitable ball mill, filled with 250 grams of lead shot, place ***360 grams of lead nitrate***, followed by ***90 grams of flours of sulfur,*** and then followed by ***50 grams of nitro starch***. Thereafter, tumble the mixture at 250 RPM for about 2 hours. Note: During the 2-hour mixing process, spray into the ball mill, in small portions at a time, 150 milliliters of butyl acetate. After the 2-hour mixing period, the mixture is ready for extrusion. To do so, the material should be pressed through any desired die cast machine, into spaghetti like strands of 2 to 3 millimeters in diameter by 150 to 300 millimeters in length; or press the pasty mass through any desired screen mesh to form grains of any desired size and then cure in the usual manner. Thereafter, the spaghetti like strands should be cut into lengths of 1.5 to 3 millimeters. Finally, the grains should be cured in an oven at normal temperature. The cured grains can then be poured into shell casings for use in standard ammunition, or can be placed into baggies for use in canons or the like. If desired, the final material can be thoroughly dried, and then pulverized in a separate mill to form a fine powder.
**Burn rate:** Depends on grain size (granules/grains 1 to 12 millimeters are suitable for firearms, mortars, and cannons).
**Chamber pressure:** Depends on grain size; small grains: 9980 psi at 25 grain.
**Water resistance:** Above moderate.
**Stability:** Can be stored for decades.
**Sensitivity:** Stable towards friction; stable towards shock; stable to temperatures up to 400 Celsius.
**Flammability (1 to 10, un-contained):** 9 powdered; 8 small grains; 8.1 medium grains; 7.3 large grains.
**Ease of ignition (1 to 10 un-contained):** 9 powdered; grain size varies and results in different ignition rates.
**Tendency to cake:** Will not cake.
**Detonation ability:** Can be detonated under proper conditions.
**Explosive ability (contained; explosive deflagration):** Above moderate.
**Percentage:** ***72% lead nitrate, 18% sulfur, 10% nitro starch***
**Classification:** Deflagrating explosive (classified as propellant).

**Danger of preparation:** Can be made by amateurs as long the procedure is followed exactly.
**Use:** Used in specialty sporting ammunition such as handguns and shotguns; can also be used as a blasting agent when properly primed.

### 03-01-017A: Reduced burn black powder composition for various uses:

Into a suitable ball mill filled with 200 grams of Teflon coated steel shot, place ***364 grams of pre-dried sodium nitrate***, followed by ***48.5 grams of flours of sulfur,*** followed by ***72.5 grams of soft wood charcoal (pre-pulverized),*** and then followed by ***15 grams of standard commercial rosin***. Thereafter, add in 100 milliliters of ether, and then tumble the mixture at 100 RPM for about 1 hour. Note: During the mixing process, the ball mill should be kept closed to prevent evaporation of the solvent. Thereafter, the mixture is ready for use. To do so, the mixture can be extruded through any desired die cast machine, formed into grains using any desired method, pressed into time fuse, tablets, pellets, or formed into any desired means if desired; or press the pasty mass through any desired screen mesh to form grains of any desired size and then cure in the usual manner. In any case, the mixture should be cured in an oven at normal temperature or dried under vacuum to recover the solvent.

**Burn rate:** Depends on grain size (granules/grains 1 to 12 millimeters are suitable for firearms, mortars, and cannons).
**Chamber pressure:** Depends on grain size; small grains: 9980 psi at 25 grain.
**Water resistance:** Above moderate.
**Stability:** Can be stored for decades.
**Sensitivity:** Stable towards friction; stable towards shock; stable to temperatures up to 400 Celsius.
**Flammability (1 to 10, un-contained):** 8.4 powdered; 7.5 small grains; 7 medium grains; 6.2 large grains.
**Ease of ignition (1 to 10 un-contained):** 8.9 powdered; grain size varies and results in different ignition rates.
**Tendency to cake:** Will not cake.
**Detonation ability:** Cannot be detonated under normal conditions.
**Explosive ability (contained; explosive deflagration):** Above moderate.
**Percentage:** ***72.8% sodium nitrate, 14.5% charcoal, 9.7% sulfur, 3% rosin***
**Classification:** Deflagrating explosive (classified as propellant).

**Danger of preparation:** Can be made by amateurs.
**Use:** Used in fireworks, for small arms ammunition, "fast burn" time fuse for aerial fireworks shells, or any desired means.

### 03-01-017B: Reduced burn black powder composition for various uses:

Into a suitable ball mill filled with 200 grams of Teflon coated steel shot of 5mm diameter, place ***364 grams of pre-dried potassium nitrate***, followed by ***48.5 grams of flours of sulfur,*** followed by ***72.5 grams of hard wood charcoal (pre-pulverized),*** and then followed by ***15 grams of gum Arabic***. Thereafter, add in 250 milliliters of 95% ethyl alcohol, and then tumble the mixture at 100 RPM for about 1 hour. Thereafter, the mixture is ready for use. To do so, the mixture can be extruded through any desired die cast machine, formed into grains using any desired method, pressed into time fuse, tablets, pellets, or formed into any desired means if desired. In any case, the mixture should be cured in an oven at normal temperature.

**Burn rate:** Depends on grain size (granules/grains 1 to 12 millimeters are suitable for firearms, mortars, and cannons).
**Chamber pressure:** Depends on grain size; small grains: 9000 psi at 70 grain.
**Water resistance:** Above moderate.
**Stability:** Can be stored for decades.
**Sensitivity:** Stable towards friction; stable towards shock; stable to temperatures up to 400 Celsius.
**Flammability (1 to 10, un-contained):** 8 powdered; 7.5 small grains; 7 medium grains; 6.2 large grains.
**Ease of ignition (1 to 10 un-contained):** 7.8 powdered; grain size varies and results in different ignition rates.
**Tendency to cake:** Will not cake.
**Detonation ability:** Cannot be detonated under normal conditions.
**Explosive ability (contained; explosive deflagration):** Above moderate.
**Percentage:** ***72.8% potassium nitrate, 14.5% charcoal, 9.7% sulfur, 3% gum Arabic***
**Classification:** Deflagrating explosive (classified as propellant).

**Danger of preparation:** Can be made by amateurs.
**Use:** Used in fireworks, for small arms ammunition, time fuse, or any other desired means.

### 03-01-018A: Modified "high reduced burn" black powder composition:

Into a suitable ball mill filled with 200 grams of Teflon coated steel shot of 10mm diameter, place ***10 grams of alpha-napthyl amine***, followed by ***375 grams of potassium nitrate,*** followed by ***85 grams of alder wood charcoal (pre-pulverized),*** and then followed by ***40 grams of flours of sulfur***. Thereafter, add in 150 milliliters of 95% ethyl alcohol, and then tumble the mixture at 100 RPM for about 1 hour. Thereafter, the mixture is ready for use. To do so, the mixture can be extruded through any desired die cast machine, formed into grains using any desired method, pressed into time fuse, tablets, pellets, or formed into any desired means if desired. In any case, the mixture should be cured in an oven at normal temperature.

**Burn rate:** Depends on grain size (granules/grains 1 to 12 millimeters are suitable for firearms, mortars, and cannons).
**Chamber pressure:** Depends on grain size; small grains: 9500 psi at 50 grain.

**Water resistance:** Above moderate.
**Stability:** Can be stored for decades.
**Sensitivity:** Stable towards friction; stable towards shock; stable to temperatures up to 400 Celsius.
**Flammability (1 to 10, un-contained):** 8 powdered; 7.5 small grains; 7 medium grains; 6.2 large grains.
**Ease of ignition (1 to 10 un-contained):** 7.8 powdered; grain size varies and results in different ignition rates.
**Tendency to cake:** Will not cake.
**Detonation ability:** Cannot be detonated under normal conditions.
**Explosive ability (contained; explosive deflagration):** Above moderate.
**Percentage:** ***73.52% potassium nitrate, 16.66% charcoal, 7.84% sulfur, 1.96% alpha-napthyl amine, 0.02% mixed residual balance***
**Classification:** Deflagrating explosive (classified as propellant).

**Danger of preparation:** Can be made by amateurs.
**Use:** Used in fireworks, for small arms ammunition, time fuse, or any other desired means.

### 03-01-019A: Modified "reduced burn" black powder composition:

Into a suitable ball mill filled with 200 grams of steel shot, place ***370 grams of potassium nitrate***, followed by ***47.5 grams of flours of sulfur,*** and then followed by ***82.5 grams of carbon black***. Thereafter, add in 175 milliliters of 95% ethyl alcohol, and then tumble the mixture at 200 RPM for about 1 hour. Thereafter, the mixture is ready for use. To do so, the mixture can be extruded through any desired die cast machine, formed into grains using any desired method, pressed into time fuse, tablets, pellets, or formed into any desired means if desired. In any case, the mixture should be cured in an oven at normal temperature.
**Burn rate:** 75 seconds per yard on average.
**Chamber pressure:** Depends on grain size; small grains: 11,200 psi at 100 grain.
**Water resistance:** Above moderate.
**Stability:** Can be stored for decades.
**Sensitivity:** Stable towards friction; stable towards shock; stable to temperatures up to 400 Celsius.
**Flammability (1 to 10, un-contained):** 8.4 powdered; 7.9 small grains; 7.2 medium grains; 6.8 large grains.
**Ease of ignition (1 to 10 un-contained):** 8 powdered; grain size varies and results in different ignition rates.
**Tendency to cake:** Will not cake.
**Detonation ability:** Cannot be detonated under normal conditions.
**Explosive ability (contained; explosive deflagration):** Above moderate.
**Percentage:** ***74% potassium nitrate, 16.5% carbon black, 9.5% sulfur***
**Classification:** Deflagrating explosive (classified as propellant).

**Danger of preparation:** Can be made by amateurs.
**Use:** Used in fireworks, for small arms ammunition, time fuse, or any other desired means.

### 03-01-020A: Modified "reduced burn" black powder composition:

Into a suitable ball mill filled with 350 grams of Teflon coated steel shot of 5mm diameter, place ***370 grams of potassium nitrate***, followed by ***52 grams of flours of sulfur,*** followed by ***46.8 grams of soft wood charcoal,*** and then followed by ***31.2 grams of finely ground "sugar" carbon***. Thereafter, add in 150 milliliters of 95% ethyl alcohol, and then tumble the mixture at 125 RPM for about 1 hour. Thereafter, the mixture is ready for use. To do so, the mixture can be extruded through any desired die cast machine, formed into grains using any desired method, pressed into time fuse, tablets, pellets, or formed into any desired means if desired. In any case, the mixture should be cured in an oven at normal temperature.
**Burn rate:** 35 seconds per yard on average.
**Chamber pressure:** Depends on grain size; small grains: 11,500 psi at 100 grain.
**Water resistance:** Above moderate.
**Stability:** Can be stored for decades.
**Sensitivity:** Stable towards friction; stable towards shock; stable to temperatures up to 400 Celsius.
**Flammability (1 to 10, un-contained):** 8.7 powdered; 8 small grains; 7.4 medium grains; 7 large grains.
**Ease of ignition (1 to 10 un-contained):** 8 powdered; grain size varies and results in different ignition rates.
**Tendency to cake:** Will not cake.
**Detonation ability:** Cannot be detonated under normal conditions.
**Explosive ability (contained; explosive deflagration):** Above moderate.
**Percentage:** ***74% potassium nitrate, 10.4% sulfur, 9.36% soft wood charcoal, 6.24% sugar carbon***
**Classification:** Deflagrating explosive (classified as propellant).

**Danger of preparation:** Can be made by amateurs.
**Use:** Used in fireworks, for small arms ammunition, "fast burn" time fuse for aerial fireworks shells, or any other desired means.

### 03-01-021A: "Explosive" graphite based black powder composition:

Into a standard kitchen blender, add in ***750 grams of potassium nitrate***, ***150 grams of finely divided graphite***, followed by ***120 grams of flours of sulfur***, and then turn on the blender to medium speed and blend for 5 to 7 seconds, then stop; thereafter, tap and rotate the blender container, place back onto the blender machine, then turn on the blender to medium speed and blend for 5 to 7 seconds; thereafter, stop; tap and rotate the blender container, place back onto the blender machine, then turn on the blender to medium speed and blend for 5 to 7 seconds. Thereafter, stop blending and allow the ingredients to stand for 30 minutes. After 30 minutes, turn on the blender to medium speed and blend for 5 to 7 seconds, then stop; thereafter, tap and rotate the blender container, place back onto the blender machine, then turn on the blender to medium speed and blend for 5 to 7 seconds; thereafter, stop; tap and rotate the blender container, place back onto the blender machine, then turn on the blender to medium speed and blend for 5 to 7 seconds. Thereafter, stop blending and allow the ingredients to stand for 30 minutes. After 30 minutes, turn on the blender to medium speed and blend for 5 to 7 seconds, then stop; thereafter, tap and rotate the blender container, place back onto the blender machine, then turn on the blender to medium speed and blend for 5 to 7 seconds; thereafter, stop; tap and rotate the blender container, place back onto the blender machine, then turn on the blender to medium speed and blend for 5 to 7 seconds. Thereafter, stop blending and allow the ingredients to stand for 30 minutes. Now, allow the ingredients in the blender to stand for 2 hours; thereafter, repeat the aforementioned process 3 more times. Note: the idea here is to thoroughly blend the ingredients without build up of heat. Once finished, the process is complete and the "proper" explosive black powder has been prepared. Thereafter, open the blender container, and pour the powdered mixture into a "zip lock" bag, and seal air-tight. The "zip lock" bag can be stored at room temperature, below room temperature, or can be stored at a temperature below 130 Fahrenheit for as long as necessary. In any sense, storing the powder is usually carried out at room temperature; as long as the "Zip lock" bag is sealed air-tight, the powder will never cake, and will remain a fresh free flowing powder.

**Proper blender blades for mixing potassium nitrate, sulfur, and graphite (do not use serrated blades; the stainless steel blades should be smooth).**

**Burn rate:** 0.091 seconds per inch.
**Water resistance:** Above moderate.
**Sensitivity:** Stable towards friction; stable towards shock; stable to temperatures up to 190 Celsius.
**Stability:** Very stable; can be stored for decades.
**Flammability (1 to 10, un-contained):** 10 powdered.
**Ease of ignition (1 to 10 un-contained):** 10 powdered.
**Tendency to cake:** May cake on prolonged exposure to moisture.
**Detonation ability:** Can be detonated under normal conditions.
**Explosive ability (contained; explosive deflagration):** detonates; 1,817 meters/per/second
**Percentage:** ***74% potassium nitrate, 14% graphite, 12% sulfur***
**Classification:** Deflagrating explosive (classified as explosive).

**Danger of preparation:** Can be made by amateurs, as long as the procedure is followed exactly.
**Use:** Bursting charges; blasting agent; firecrackers, improvised explosive devices. Note: do not use in firearms, cannons, or large bore guns (explosion of the weapon will result).

# Chapter 4. High Performance Rocket Propellants

## Section 1: Ammonium Perchlorate Rocket Propellants

***Chemicals used in this section (binders are not included)***

| | | |
|---|---|---|
| **1. Potassium nitrate (see Black Powder)** | | **2. Sulfur (see Black Powder)** |
| **3. Charcoal (see Black Powder)** | | **4. Sugar Carbon (see Modified Black Powder)** |
| **5. Barium Chromate (see Modified Black Powder)** | | **6. Potassium Perchlorate (see Modified Black Powder)** |
| **7. Potassium Dichromate (see Modified Black Powder)** | | **8. Ammonium Bisulfide (see Modified Black Powder)** |
| **9. Potassium chlorate (see Modified Black Powder)** | | **10. Carbon Disulfide (see Modified Black Powder)** |
| **11. Nitrocellulose (see Modified Black Powder)** | | **12. Lead Tetraoxide (see Modified Black Powder)** |
| **13. Diphenylamine (see Modified Black Powder)** | | **14. Titanium Dioxide (see Modified Black Powder)** |
| **15. Manganese Dioxide (see Modified Black Powder)** | | **16. Sodium Benzoate (see Modified Black Powder)** |
| **17. Ammonium Nitrate (see Modified Black Powder)** | | **18. Calcium Carbonate (see Modified Black Powder)** |
| **19. Sodium Nitrate (see Modified Black Powder)** | | **20. Urea (see Modified Black Powder)** |
| **21. Lead Nitrate (see Modified Black Powder)** | | **22. Nitro Starch (see Modified Black Powder)** |
| **23. Ammonium Perchlorate** | | **24. Aluminum powder** |
| 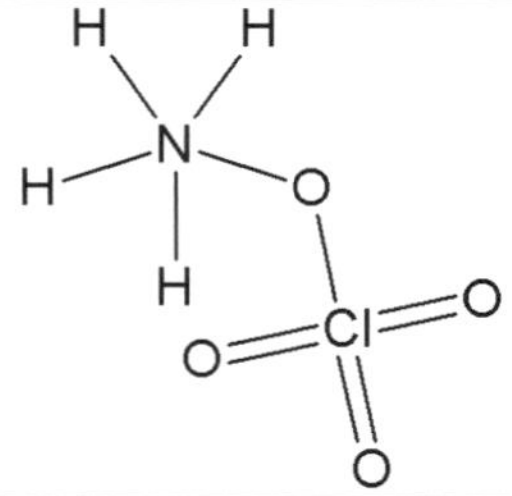 | | 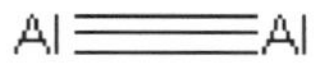 |
| Ammonium perchlorate forms colorless orthorhombic crystals or smooth silky white powder. The crystals decompose when heated into oxygen and ammonium chloride. The crystals are very soluble in water, but practically insoluble in most organic solvents. Ammonium perchlorate has very limited use in explosives and explosive compositions, but its use in propellant compositions is absolutely overwhelming. Practically 98% of the ammonium perchlorate manufactured around the world is used in making rocket propellants and pyrotechnics. The salt is heavily used in solid rocket propellants when mixed with aluminum or other metals, and reducing agents along with a polymer binder/epoxy binder. Note: NASA's retired space shuttle's used twin rocket boosters composed of ammonium perchlorate/aluminum/binder. As expected, when properly mixed in the correct proportions, mixtures of APC and any combustible agents ignite and burn rapidly and violently. Ammonium perchlorate can be stored in glass or plastic bottles sealed air-tight, or stored in poly bags, and stored in a cool dry place away from sunlight. **Danger of preparation:** can be prepared by non-scientists as long as all procedure(s) and safety info is strictly followed. | | Aluminum powder forms a tin-white free-flowing powder that is very stable; it often ignites explosively when subjected to air and a flame. Aluminum powder is readily available commercially, and is sold under a variety of particle sizes. |
| | | |

**Method of Preparation 1:** Into a beaker, equipped with mechanical stirrer add and dissolve 470 grams of sodium perchlorate into 1000 milliliters of water. Thereafter add in 320 grams of a 28 - 30% ammonia solution, or 900 grams of 10% ammonia solution, or pass 96 grams of anhydrous ammonia gas into the perchlorate solution while stirring. Thereafter, add in pieces, 3000 grams of crushed dry ice over a period of 30 minutes while continuing to stir the reaction mixture. After all the crushed dry ice has been added, continue to rapidly stir the reaction mixture until all the dry ice has evaporated; at this point the insoluble sodium bicarbonate would have full precipitated. Thereafter, stop stirring and then filter-off the insoluble by-product of sodium bicarbonate using a clean Buchner funnel under moderate suction, and then pour the filtered reaction mixture onto a shallow pan, and heat the pan at 80 Celsius while blowing air over the surface of the liquid (a small portable cooling fan can be used). Continue to heat and blow air over the shallow pan until dry solid remains. Once dry solid remains, immediately remove the shallow pan from the heat source, and stop blowing air-over the shallow pan, and then immediately allow the ammonium perchlorate to cool, and then place into a plastic container, sealed air-tight, and store in a cool dry place until use.

## 25. Magnesium Sulfide

**Mg=S**

Magnesium sulfide forms reddish-brown crystals.

## 26. Copper Chromite

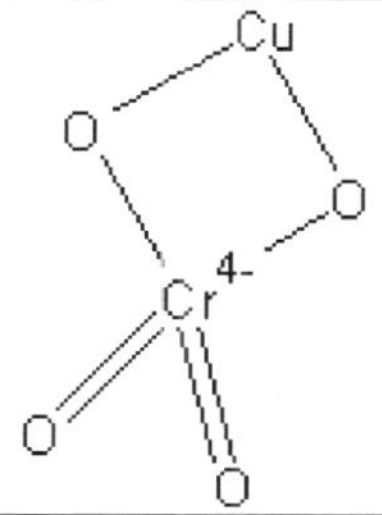


Forms grayish to black tetragonal crystals. The crystals are insoluble in water and the usual solvents, Upon heating to high temperatures, the crystals decompose into chromium dioxide and chromium trioxide.

## 27. Nitroguanidine

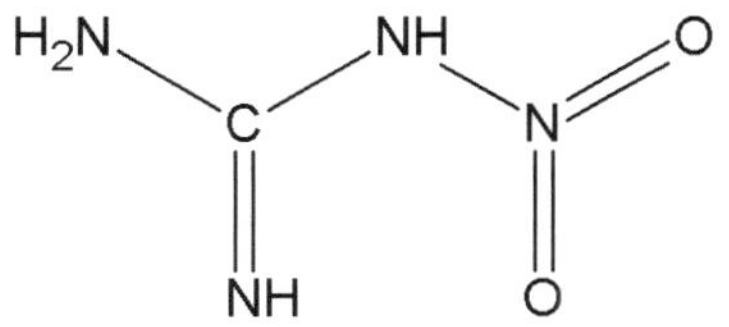


Nitroguanidine forms white or colorless needles or prisms, which begin to decompose when heated to 225 Celsius, evolving ammonia. The crystals are insoluble in water and alcohol, but soluble in concentrated acids from where the corresponding acid salt is precipitated by pouring into cold water. The nitrate and perchlorate salts are quite common and are used in fireworks, and a variety of pyrotechnic compositions. Nitroguanidine usually occurs as the alpha form, but the beta form can be prepared by treating guanidine sulfate with 90% nitric acid. The beta form is converted back to the alpha form by treatment with sulfuric acid, from which the sulfate salt is precipitated by pouring into cold water; the free base (nitroguanidine) is then liberated by treating the sulfate with sodium carbonate. Nitroguanidine is slowly decomposed in alkali solutions. Nitroguanidine is used primarily in the preparation of other explosives, but it is widely used in medium velocity explosives with ammonium nitrate for use in borehole rock blasting operations. Nitroguanidine is widely used in pyrotechnic compositions for various applications, and is widely used when alloyed with low melting nitrobenzenes and nitramines, to form explosive composites for various commercial and industrial applications. Nitroguanidine is also widely used in rocket propellants, either with a high-energy plasticizer, and/or when mixed with a nitrate, chlorate, perchlorate and powdered aluminum or magnesium and then heat casted with an epoxy binder. **Danger of preparation:** can be prepared by non-scientists as long as all procedure(s) and safety info is strictly followed.

## 28. Ammonium Sulfate

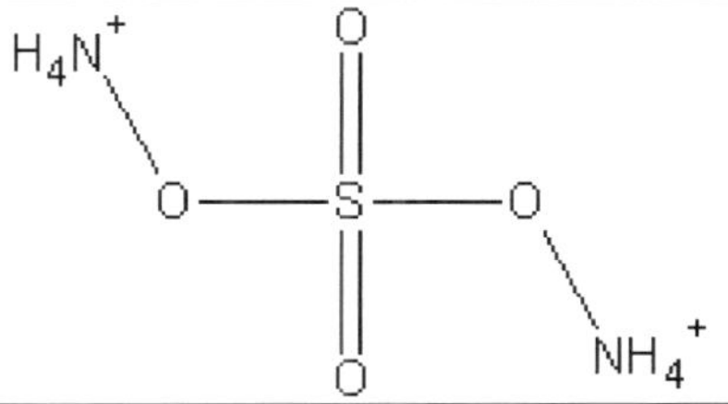


Forms odorless crystalline powder, granules, or masses. The salt decomposes when heated to 280 Celsius. The salt is soluble in water, but insoluble in the usual organic solvents. It can be made by dissolving ammonia into dilute sulfuric acid.

**Method of Preparation 1:** Into a suitable sized beaker, equipped with mechanical stirrer, and sitting inside an ice bath, pour in 900 milliliters of 98% sulfuric acid, and then allow the sulfuric acid to chill to 0 Celsius. When the temperature of the sulfuric acid reaches 0 Celsius, turn on the mechanical stirrer to high speed, and then slowly add in, in small portions at a time, 120 grams of guanidine while rapidly stirring the reaction mixture and maintaining its temperature below 5 Celsius. After the addition of the guanidine, continue to rapidly stir the reaction mixture at 0 to 5 Celsius for about 15 minutes, and thereafter, slowly add, in small portions at a time, 184 grams of 70% nitric acid over a period of 1 hour while keeping the reaction mixtures temperature at 0 Celsius under rapid stirring. After addition of the 70% nitric acid, remove the ice bath, and then continue to rapidly stir the reaction mixture for about 45 minutes. Thereafter, stop stirring, and then slowly and carefully pour the entire reaction mixture onto 1 kilogram of ice, contained in a large clean beaker, equipped with mechanical stirrer, allow the ice to melt, and then turn on the mechanical stirrer and blend the aqueous mixture for 45 minutes. After about 45 minutes, stop stirring, and then filter-off the precipitated product, wash with six 250-milliliter portions of 10% sodium carbonate solution, followed by washing with copious amounts of warm water, and then suction dry as best as possible. Finally, recover the suction dried product from the Buchner funnel, and place onto a shallow pan or tray and allow to thoroughly air-dry. Once dry, collect the nitroguanidine product and store in a plastic container in a cool dry place until use.

## 29. Nitroglycerine

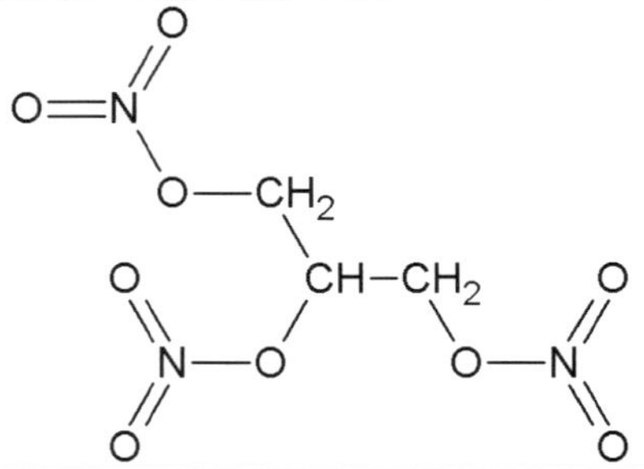

Nitroglycerine forms a pale yellow, oily liquid, that has a sweet acrid burning taste. Note: it should be noted, that tasting, skin absorption, and inhalation of vapors can cause severe migraines and headaches, so caution should be taken at all cost. In fact, nitroglycerine is more dangerous as a toxin then it is to handle as an explosive, and the liquid explosive has been given a marked name as being "dangerous" and "unstable", which is hardly the truth; in fact, nitroglycerine is far more stable then people have been led to believe. In any sense, nitroglycerine is insoluble in water, but soluble in conc. nitric and sulfuric acids, as well it has good solubility in many organic solvents, and is miscible with acetone, ether, benzene, ethyl acetate, and chloroform, from which prolonged storage of nitroglycerine it is advisable to dissolve the nitroglycerine in acetone up to 70% by weight; these mixtures are very stable, and can be stored for decades. On the lighter side, there is a given hazard when working with nitroglycerine, as the liquid, when subjected to a mild flame (such as from a common cigarette lighter), a sample will immediately begin to decompose evolving dark red fumes of nitrogen oxides, followed by detonation. Obviously, when rapidly and strongly heated, the explosive liquid detonates within seconds of heat contact. Nitroglycerine has been around since the 1400's, but use as a worthwhile secondary high explosive did not really take place until it was discovered that diatomaceous earth absorbs nitroglycerine forming a chunky material, which could be packed into paper tubes, and primed. Hence, this started a revolution of sorts in the explosives industry, as nitroglycerine factories popped up all over the world, and every so often, some factories would explode! Nitroglycerine is known to be sensitive to shock and percussion, which renders the liquid explosive practically useless to be used by itself as a liquid explosive (although it can be if desired, i.e,. in making improvised liquid explosives for terrorist uses); therefore, the liquid, which as described, is absorbed with diatomaceous earth, but also with aluminum oxide, and even silica gel forming stable, yet tedious explosive charges, that still require a blasting cap for proper initiation. In essence, nitroglycerine begins to decompose when heated to about 50 Celsius, however, the pure liquid becomes volatile when slowly and gently

## 30. Magnesium Oxide

**Mg=O**

Magnesium Oxide forms a nice fine white powder that exists in two forms: light, and heavy. Both forms absorb carbon dioxide from the air, and react with water to form magnesium hydroxide. Magnesium oxide is prepared by roasting magnesium carbonate.

| | | |
|---|---|---|
| heated to 100 Celsius under vacuum; this fact is amazing, and details why the liquid is not just dangerous to handle in the sense of toxicity, but safe to work with under the proper given conditions. Pure nitroglycerine can safely be stored in a refrigerator, and protected from light. For prolonged storage as previously described, it is best to store nitroglycerine in an inert solvent such as acetone—whereby the nitroglycerine can be separated as an oily insoluble layer upon the addition to cold water. The use of nitroglycerine in dynamite charges has greatly decreased in the past 70 years due to various reasons; such reasons include, 1) its difficult to handle in the since that it causes severe headaches and migraines, as factory workers were often exposed to the agent; 2) it's stability has been in question as the result of numerous accidents, i.e., explosions due to improper handling, poor manufacturing techniques, and individual error; and 3) its explosive power is high, and this results in tendency to cause cave-ins and unsatisfactory rock fracturing when used in mining operations; as a result, its use in mining operations and engineering projects has been significantly decreased, and in this case, nitroglycerine has been practically taken over by ammonium nitrate explosive charges for use in mining and rock blasting operations. On another note, even though nitroglycerine's use in dynamites and blasting charges has significantly decreased, it is still highly prized and manufactured primarily for making high performance propellant compositions for smokeless gun powders and high performance rocket compositions. In any regards, nitroglycerine is still a cheap and readily available material for use in military, or commercial explosives. Note: nitroglycerine is used in medicine as a vasodilator. Danger of preparation: can be prepared by non-scientists as long as all procedure(s) and safety info is strictly followed. | | |
| **Method of Preparation 1:** Into a suitable sized 3-neck flask, equipped with mechanical stirrer, thermometer, and addition funnel, and sitting inside a ice/salt bath, pour in 150 milliliters of 70% nitric acid, followed by slowly and carefully adding in 128 milliliters of 98% sulfuric acid; thereafter, turn on the mechanical stirrer to low speed, and then chill this nitrating acid mixture to 0 Celsius. In the meantime, pour into the addition funnel 40 grams of warm glycerol (heated to 30 Celsius), and when the nitrating acid mixture has cooled, begin to drip in, the warm glycerol into the nitrating acid reaction mixture over a period of time sufficient to keep the nitrating acid mixture below 5 Celsius at all times. After addition of the glycerol, continue to stir the reaction mixture on low speed at 5 Celsius for 1 hour. After 1 hour, remove the ice/salt bath, and then continue to stir the reaction mixture on low speed at room temperature for 2 hours. After 2 hour, stop stirring, disassemble the apparatus, and then slowly, and carefully pour the entire reaction mixture onto 1 kilogram of ice, contained in a suitable sized beaker, equipped with mechanical stirrer, allow the ice to melt, and then mechanically stir this diluted two-phase reaction mixture for 45 minutes. Now, after stirring for 45 minutes, stop stirring, and then pour the entire two-phase mixture into a large separatory funnel, let stand for 10 minutes, and then drain-off the bottom acid layer. Thereafter, pour into this separatory funnel, 1000 milliliters of pre-chilled 10% sodium carbonate solution (pre-chill to 5 Celsius), and then gently shake the separatory funnel manually for about 10 minutes. Thereafter, let the separatory funnel stand for 10 minutes, and then carefully drain-off the bottom nitroglycerine layer into a clean beaker. Once all the nitroglycerine has been collected, add to it, 35 grams of anhydrous magnesium sulfate, and then manually stir for about 15 minutes (to absorb moisture); thereafter, stop stirring, and then gently gravity filter-off the magnesium sulfate, collecting the dry nitroglycerine in a clean beaker. Once all the nitroglycerine has been collected, pour it into an amber glass bottle and store in a refrigerator until use. | | |
| **31. Iron-II-oxide** | | **32. Zirconium Hydride** |
|| ||
| Ferrous oxide forms a jet-black powder, which readily absorbs carbon dioxide from the air. The dry solid is insoluble in water and all organic solvents, but the dry powder readily dissolves in acids. | | Forms a grayish-black powder, which is stable towards water. It is formed by reacting calcium hydride with zirconium oxide in the presence of hydrogen gas at 600 Celsius. A powerful reducing agent. |
| | | |

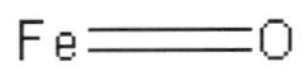

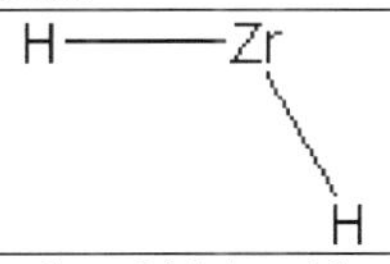

**33. Teflon**

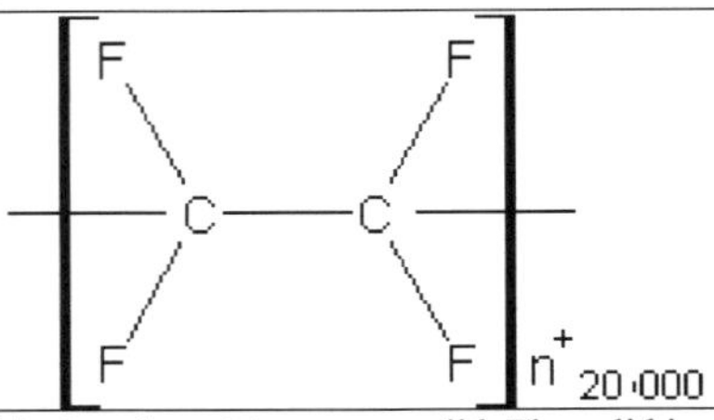


Teflon is a white translucent to opaque solid. The solid is almost resistant to all chemicals, and is widely used for industrial and commercial applications. The powered solid melts when heated to 327 Celsius forming a viscous gel, which volatizes when heated to 400 Celsius. Teflon is insoluble in all known solvents. Although Teflon is chemical resistant, in powdered form it is capable of acting as an energy source in pyrotechnic compositions, specifically high performance rocket propellants.

**34. Zinc Oxide**

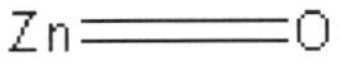


White or slightly yellowish to light brown powder. Zinc oxide is insoluble in water and most organic solvents, but does dissolve in acids and alkaline solutions. It can be made by roasting zinc carbonate or hydroxide.

**35. Ammonium Dichromate**

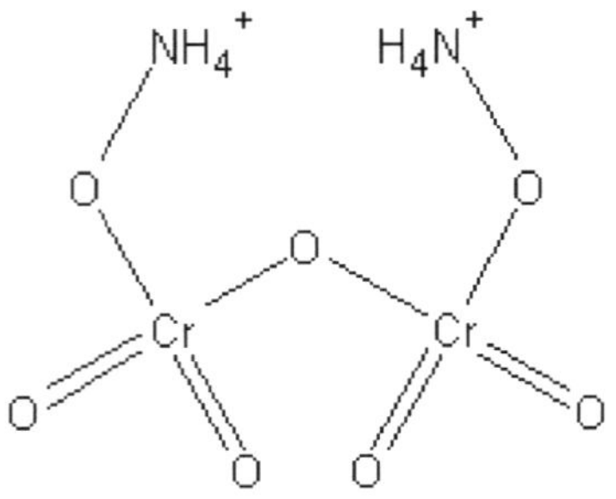


The dichromate forms beautiful bright orange-red crystals. The crystals are flammable. Ammonium dichromate is readily soluble in water, but not readily soluble in the usual solvents. The crystals decompose when heated to 180 Celsius, and the decomposition becomes self-sustained when heated to 225 Celsius. The product can be prepared by reacting ammonium sulfate solution with sodium dichromate, followed by recrytsallization.

**36. Lithium Perchlorate**

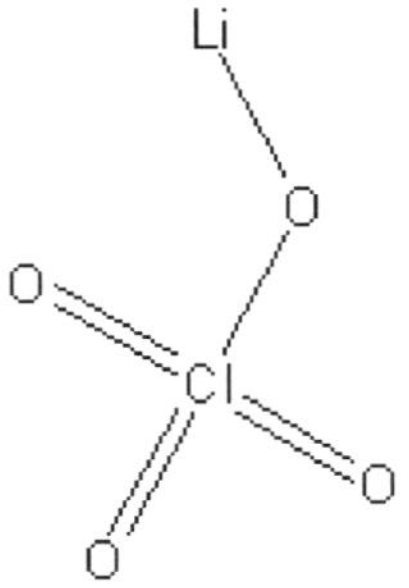


Lithium perchlorate forms colorless to white crystals, granules, or powder. The crystals begin to decompose when to 400+ Celsius, but the crystals will melt when gently heated to 236 Celsius. Lithium perchlorate is one of the most soluble salts in water, and is very soluble in ether, alcohol, and acetone.

**37. Magnesium powder**

**Mg=Mg**

Magnesium powder forms a white to silvery white powder with hexagonal system. The powder is readily flammable and burns producing intense light. The solid is insoluble in water and all organic solvents; however the powder will slowly react with water at ordinary temperatures. Magnesium powder readily oxidizes when heated or ignited.

**38. Lithium Aluminum Hydride**

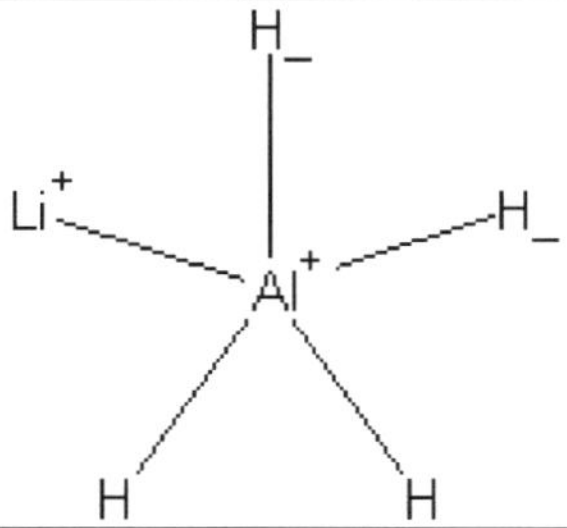


White to light gray solid. The dry solid is very reactive and reduces many organic compounds. Lithium Aluminum Hydride is prepared by reacting ether solutions of lithium hydride and anhydrous aluminum chloride.

| | |
|---|---|
| **39. Iron-III-oxide (red iron oxide)** | **40. PVC** |
| 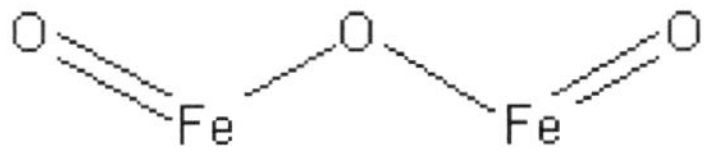 | 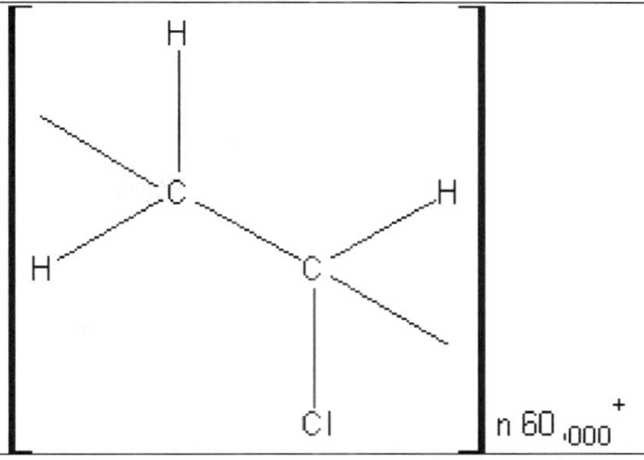 |
| Red Iron oxide is formed by roasting ferrous hydroxide, oxides, or hydrates at red heat. The compound is insoluble in water, and becomes insoluble in acids when roasted at high temperatures for prolonged periods of time. The oxide can act as a secondary source of oxygen in pyrotechnic mixtures. | PVC forms a white to translucent solid, which discolors in the presence of light, so commercial grades often contain stabilizers not desired for use in pyrotechnic compositions. The solid is obtained by the polymerization of vinyl chloride using various catalysts. The solid is readily soluble in cylcohexanone, tetrahydrofuran, and methylene chloride. |
| **41. Copper Sulfide** | **42. Sodium Hydride** |
| 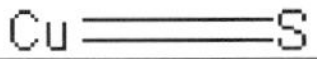 | 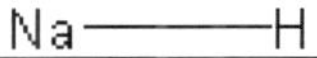 |
| Copper sulfide forms a black powder, which is stable in dry air, but is slowly oxidized to copper sulfate in moist air. The dry solid is insoluble in water, and all organic solvents. | Sodium hydride forms a grayish-white powder. It is very reactive and reacts vigorously with water, and many other chemicals. It is prepared by heating metallic sodium in a stream of hydrogen under inert atmosphere. |
| **43. Titanium Hydride** | **44. Silicon Nitride** |
| 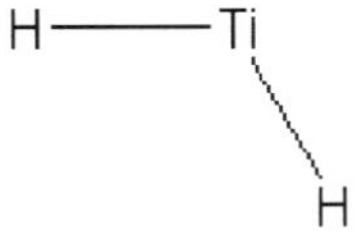 | 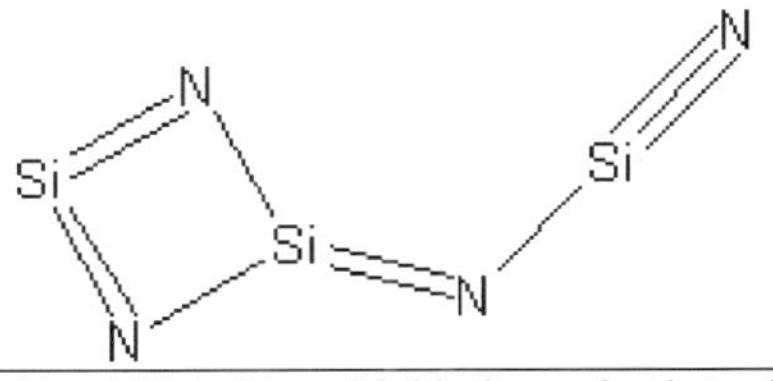 |
| Titanium Hydride forms a grayish-black powder, which is relatively stable towards moisture and air. Titanium hydride is prepared by heating titanium oxide with calcium hydride in the presence of hydrogen at 600 Celsius. | Silicon nitride is a white, to brownish-black powder that exists in two forms—differentiated by heating. Silicon nitride is very resistant to heat and forms a non-oxide ceramic material. |

***- High Performance Ammonium Perchlorate Propellants in this section -***

| | |
|---|---|
| **1. 04-01-001A: High Performance Military grade rocket propellant:** *68.13% ammonium perchlorate, 16.03% aluminum, 15.53% kraton polymer binder, 0.30% aziridine resin, 0.01% impurities* | **2. 04-01-002A: High Performance Military grade rocket propellant:** *74.77%% ammonium perchlorate, 15.95% dioctyl adipate binder, 8.47% RB-100 resin, 0.79% lecithin catalyst, 0.02% impurities* |
| **3. 04-01-003A: High Performance Military grade rocket propellant with PVC:** *78.43% ammonium perchlorate, 19.6% PVC polymer, 0.98% Epon 815 epoxy resin, 0.98% magnesium sulfide catalyst, 0.01% impurities* | **4. 04-01-004A: High Performance military rocket propellant (JPL X500 type):** *77.8% ammonium perchlorate, 11% dioctyl azelate, 6.6% hexane triol plasticizer, 1.6% 2,4-toluene diisocyanate, 1.5% poly propylene glycol plasticizer, 1.1% copper chromite burn rate accelerator, 0.05% ferric acetylacetonate curing catalyst, 0.35% impurities* |
| **5. 04-01-005A: High Performance military rocket propellant:** *70% ammonium perchlorate, 18.6% aluminum, 5.4% ploy-1,2-butylene glycol plasticizer, 4% hexamethylene diisocyante curative, 1.2% polyoxypropylene derivative of trimethylolpropane curative, 0.10% sulfur, 0.70% moisture* | **6. 04-01-006A: High Performance military rocket propellant (with negative pressure exponent):** *58.2% ammonium perchlorate, 18% ammonium sulfate, 14% poly propylene glycol plasticizer, 6% dioctyl adipate, 1.5% 2,4-toluene diisocyanate, 1.2% poly propylene glycol adduct of trimethylol propane, 0.75% triethylamine catalyst, 0.05% ferric acetylacetonate curing catalyst, 0.30% impurities* |
| **7. 04-01-007A: High Performance military rocket propellant (reduced flash):** *32% ammonium perchlorate, 26.4% nitroguanidine, 21.9% polyethylene glycol binder, 12.7% aluminum, 6.3% ammonium nitrate, 0.70% impurities* | **8. 04-01-008A: High Performance polymeric military rocket propellant:** *68% ammonium perchlorate, 16% aluminum, 11.8% polybutylene glycol diamine, 4.1% epoxy resin D.E.R. 332, 0.10%, residue* |

| | |
|---|---|
| **9. 04-01-009A: High Performance "plateau-burning" polymeric rocket propellant:** *75% ammonium perchlorate, 20.4% polybutyl acrylic acid binder, 3.6% epon 828 epoxy resin, 1% lithium fluoride stabilizer* | **10. 04-01-010A: High Performance polymeric rocket propellant:** *62.2% ammonium perchlorate, 17.2% aluminum, 12.6% diglycidyl ether polypropylene glycol binder, 6.4% diglycidyl ether bisphenol A binder, 1.1% monoethanolamine catalyst, 0.21% diethylentriamn curing catalyst, 0.29% residue* |
| **11. 04-01-010A: High Performance polymeric rocket propellant (modified with increased burn rate—aluminum free):** *75.2% ammonium perchlorate, 15.3% diglycidyl ether polypropylene glycol binder, 7.7% diglycidyl ether bisphenol A binder, 1.4% monoethanolamine catalyst, 0.25% diethylentriamn curing catalyst, 0.15% residue* | **12. 04-01-011A: High Performance polyurethane rocket propellant:** *73% ammonium perchlorate, 15% aluminum, 7.1% polybutadiene diol binder, 3.9% dioctylsebacate plasticizing agent, 0.4% 2,6,-toluene diisocyanate, 0.3% polyprpylene oxide triol binding agent, 0.15% nickel-II-hydroxide curing agent, 0.08% residue, 0.05% dihydroxypropyl bis(cyanoethylamine curing catalyst, 0.02% ferric acetylacetonate* |
| **13. 04-01-012A: High Performance CMDB rocket propellant (composite modified Double Base propellant):** *49% ammonium perchlorate, 26% nitroglycerine, 14% aluminum, 4.6% triacetin, 4% plastisol nitrocellulose, 1% magnesium oxide burn catalyst, 1% 2-nitrodiphenylamine, 0.32% 2,6-toluene diisocyanate curative, 08% residue* | **14. 04-01-012A: High Performance rocket propellant:** *72% ammonium perchlorate, 22% aluminum, 2.5% dioctyl adipate liquid plasticizer, 2.1% HTPB polymer binder, 1% iron-II-oxide burn catalyst, 0.3% di-issocyante curing agent, 0.09% impurities, 0.01% triphenyl bismuth catalyst* |
| **15. 04-01-013A: High Performance JPL X500 rocket propellant:** *68.6% ammonium perchlorate, 23.4% polypropylene glycol 2025 plasticizer, 4.7% toluene diisocyante, 1.9% copper chromite burn rate catalyst, 1.1% 1,2,6-hexanetriol, 0.21% mixed impurities, 0.09% ferric acetyl acetonate* | **16. 04-01-013B: High Performance JPL X500 rocket propellant (with decreased burn rate):** *70% ammonium perchlorate, 23.9% polypropylene glycol 2025 plasticizer, 4.8% toluene diisocyante, 1.2% 1,2,6-hexanetriol, 0.1% ferric acetyl acetonate* |
| **17. 04-01-014A: High Performance rocket propellant with reduced electromagnetic interference exhaust:** *69.6% ammonium perchlorate, 14% aluminum, 14% binder, 2% molybdenum, 0.35% iron oxide, 0.05% residue* | **18. 04-01-015A: High Performance impulse rocket propellant:** *50% zirconium hydride, 45% ammonium perchlorate, 5% Viton A copolymer* |
| **19. 04-01-016A: High Performance Teflon based rocket propellant:** *69.3% ammonium perchlorate, 18.8% aluminum powder, 8.9% butyl rubber, 2.9% Teflon, 0.10% residue* | **20. 04-01-016B: High Performance Teflon based rocket propellant with reduced burn rate:** *67.3% ammonium perchlorate, 18.2% aluminum powder, 8.6% butyl rubber, 2.8% Teflon, 2.8% triphenlyphosine burn reducer, 0.30% residue* |
| **21. 04-01-016C: High Performance Teflon based "smokeless" rocket propellant with reduced burn rate:** *77% ammonium perchlorate, 16% Teflon, 7% butyl rubber* | **22. 04-01-017A: High Performance smokeless polymer rocket propellant:** *81.8% ammonium perchlorate, 10.8% Butadiene copolymer, 2.1% dibutoxyethyoxyethyl formal plasticizer, 2.1% carbon black filler, 1.9% milori blue burning catalyst, 0.38% mixed residues, 0.32% zinc oxide catalyst, 0.32% flexamine antioxidant agent, 0.10% aerosol-OT wetting agent, 0.10% SA-113 vulcanization accelerator, 0.08% sulfur* |
| **23. 04-01-018A: High Performance smokeless polymer rocket propellant:** *75.4% ammonium perchlorate, 8.2% ammonium dichromate, 5.6% butadiene copolymer, 5.4% polybutadiene plasticizer, 2.7% milori blue burn catalyst, 2.1% philblack E furnace black, 0.42% magnesium oxide, 0.18% mixed residues* | **24. 04-01-019A: High Performance ammonium perchlorate rocket propellant:** *48.2% ammonium perchlorate, 34.4% magnesium, 17.2% aluminum, 0.20% mixed residues* |
| **25. 04-01-020A: High Performance ammonium perchlorate rocket propellant:** *63.4% boron, 31.7% ammonium perchlorate, 2.9% glyceryl monoricinoleate, 1.77% toluene diisocyanate, 0.08% Ferric acetyl acetonate, 0.08% diethanol oleamide, 0.07% balance* | **26. 04-01-021A: High Performance ammonium perchlorate rocket propellant:** *45.47% ammonium perchlorate, 23.1% urea, 11.99% lithium perchlorate, 7.98% epoxy binder, 7.33% polypropylene glycol, 2.49% tallow monoglyceride, 1.35% maleic anhydride, 0.1699% tin octoate curing catalyst, 0.1201% balance (exact decimal places where removed to reduce complications).* |
| | |

| | |
|---|---|
| **27. 04-01-022A: High Performance meltable-wax ammonium perchlorate rocket propellant:** *74.4% ammonium perchlorate, 19% urea, 6.6% wax* | **28. 04-01-023A: High Performance polymeric ammonium perchlorate rocket propellant:** *61.47% ammonium perchlorate, 17.6% epoxy binder, 12.29% urea, 8.19% aluminum, 0.409% tin octoate, and 0.111% balance* |
| **29. 04-01-024A: High Performance ammonium perchlorate rocket propellant:** *84% ammonium perchlorate, 14.8% diglycidyl ether, 1% triethylene tetramine, 0.2% balance* | **30. 04-01-025A: High Performance and flexible, ammonium perchlorate rocket propellant with semicarbazide perchlorate burn increaser:** *54.7% ammonium perchlorate, 29.85% Paralax P-10 binder, 14.9% semicarbazide perchlorate, 0.497% benzoyl peroxide, 0.053% rounded balance* |
| **31. 04-01-026A: High Performance ammonium perchlorate rocket propellant containing magnetic iron oxide burn catalyst:** *41.41% ammonium perchlorate, 29.29% nitroglycerine, 17.17% nitrocellulose, 8% triacetin, 2% magnetic iron-III-oxide catalyst, 1% 2-nitrodiphenylamine, 1% resorcinol, 0.13% balance* | **32. 04-01-027A: High Performance rubber-like ammonium perchlorate rocket propellant:** *42.9% polymerized binder, 42% ammonium perchlorate, 15% lithium aluminum hydride, 0.1% carbon black, 0.7% mixed residues* |
| **33. 04-01-027B: High Performance rubber-like ammonium perchlorate rocket propellant:** *57.89% ammonium perchlorate, 21% aluminum, 21% partially polymerized binder, 0.11% balance* | **34. 04-01-028A: High Performance ammonium perchlorate rocket propellant utilizing asphalt:** *75% ammonium perchlorate, 10.5% asphalt, 8% castor oil, 3.5% dibutyl sebacate, 3% cetyl acetamide* |
| **35. 04-01-029A: Specialty ammonium perchlorate semi-solid (gelled) rocket propellant:** *71.25% ammonium perchlorate, 22.35% unleaded gasoline, 3.75% carbon black, 2.65% aluminum naphthenate* | **36. 04-01-030A: Specialty high performance sterile ammonium perchlorate rocket propellant for use in space rockets:** *64% ammonium perchlorate, 18.8% polyurethane, 16% aluminum, 1.2% ethylene oxide addition agent* |
| **37. 04-01-031A: Classic high performance ammonium perchlorate rocket propellant:** 69% ammonium perchlorate, 16% aluminum, 12.98% polymer binder, 1% iron oxide burn catalyst, 0.89% IPDI compound, 0.13% AO-2246 curing catalyst | **38. 04-01-032A: High performance ammonium perchlorate rocket propellant:** *65% ammonium perchlorate, 16.8% polypropylene glycol, 15% aluminum powder, 2.14% toluene diisocyante polymerization process, 0.5% alosperse, 0.27% trimethylol propane, 0.25% phenyl beta-naphthylamine, 0.04% possible moisture* |
| **39. 04-01-033A: High performance ammonium perchlorate rocket propellant:** *79.59% ammonium perchlorate, 10.1% dibutyl sebacate, 10.1% polyvinyl chloride, 0.2% dibutyl tin dilaurate, 0.01% possible moisture* | **40. 04-01-034A: High performance smokeless ammonium perchlorate rocket propellant:** *67.9% ammonium perchlorate, 26.3% polybutadiene polymer, 2% aluminum, 1.5% diglycidyl ether, 1% butadiene polymer, 1% dioctyl adipate, 0.036% chromium compound catalyst, 0.264% balanced* |
| **41. 04-01-034B: High performance smokeless ammonium perchlorate rocket propellant (modified):** *67.9% ammonium perchlorate, 26.3% polybutadiene polymer, 2.1% polymer butadiene, 2% aluminum, 1.47% diglycidyl ether, 0.06% chromium compound catalyst, 0.17% balanced* | **42. 04-01-034C: High performance smokeless ammonium perchlorate rocket propellant (modified):** *67.9% ammonium perchlorate, 26.3% polybutadiene polymer, 2.1% polymer butadiene, 2% aluminum, 1.47% epoxy compound, 0.06% chromium compound catalyst, 0.17% balanced* |
| **43. 04-01-035A: High performance smokeless ammonium perchlorate rocket propellant with copper sulfide burn accelerator:** *83.3% ammonium perchlorate, 14.44% polybutadiene resin, 2.22% copper-II-sulfide, 0.04% residual balance* | **44. 04-01-036A: High performance ammonium perchlorate rocket propellant with copper sulfide burn accelerator (aluminum based):** *76.33% ammonium perchlorate, 13.23% aluminum powder, 7.63% epoxy resin, 2.79% copper-II-sulfide, 0.02% impurities* |
| **45. 04-01-037A: High performance ammonium perchlorate rocket propellant (with metal hydride accelerator component):** *43.01% ammonium perchlorate, 26.88% ammonium nitrate, 12.54% sodium hydride, 8.96% epoxy resin, 4.2% magnesium powder, 2.5% titanium hydride, 1.79% copper-II-chromite burn rate catalyst, 0.12% residual impurities* | **46. 04-01-038A: High performance ammonium perchlorate rocket propellant (with silicon monoxide accelerator component):** *84.53% ammonium perchlorate, 12.25% butadiene-1,3 binder, 2.42% methacrylic acid catalyst, 0.64% magnesium oxide filler, 0.146% silicon monoxide burn rate accelerator, 0.014% residual balance* |
| | |

| | |
|---|---|
| **47. 04-01-038B: High performance ammonium perchlorate rocket propellant (with silicon monoxide and silicon nitride accelerator component):** *76.39% ammonium perchlorate, 11.07% butadiene-1,3, 3.97% sulfur, 2.98% potassium dichromate, 2.25% silicon nitride, 2.18% stearic acid, 0.79% silicon monoxide, 0.26% magnesium oxide filler, 0.11% residual balance* | **48. 04-01-039A: High performance castable ammonium perchlorate rocket propellant:** *83.65% ammonium perchlorate, 12.92% PVC, 3.42% dioctyl sebacate, 0.01% mixed impurities* |
| **49. 04-01-039B: High performance castable ammonium perchlorate rocket propellant (with silicon nitride and calcium carbide burn enhancer)**: *78.56% ammonium perchlorate, 11.94% PVC, 3.52% dioctyl sebacate, 3.14% silicon nitride, 2.83% calcium carbide, 0.01% impurities* | **50. 04-01-040A: High Performance ammonium perchlorate rocket propellant with nitrocellulose base:** *47.79% ammonium perchlorate, 25.97% nitrocellulose, 20.77% triacetin, 5.19% tri-butoxyethyl phosphate, 0.25% lecithin, 0.03% balance* |
| **51. 04-01-040B: High Performance ammonium perchlorate rocket propellant with nitrocellulose base:** 48.32% ammonium perchlorate, 21.59% nitrocellulose, 15.42% triacetin, 10.28% methyl phthalyl ethyl glycolate, 3.85% nitroglycerine, 0.25% lecithin, 0.25% diphenyl amine, 0.04% residual balance | **52. 04-01-040C: High Performance ammonium perchlorate rocket propellant with nitrocellulose base:** *63.15% ammonium perchlorate, 18.04% nitrocellulose, 10.82% diethyl phthalate, 7.21% triacetin, 0.75% lecithin, 0.03% residual balance* |
| **53. 04-01-040D: High Performance ammonium perchlorate rocket propellant with nitrocellulose base:** 49.38% ammonium perchlorate, 24.69% nitrocellulose, 12.34% diethyl phthalate, 12.34% bis-(3,5,5-trimethoxyhexyl) adipitate, 1.23% lecithin, 0.02% residual impurities | **54. 04-01-041A: Specialty high Performance ammonium perchlorate rocket propellant:** *67.9% ammonium perchlorate, 15.99% aluminum powder, 15.72% hydroxy terminated liquid resin binder, 0.2999% polymerized tartaric acid compound, 0.0901% residue* |
| **55. 04-01-042A: High Performance ammonium perchlorate rocket propellant:** 49.6% ammonium perchlorate, 49.2% polychlorotrifluoroethylene copolymer, 1.2% paraffin wax binder | **56. 04-01-043A: High performance low temperature burning smokeless propellant:** *85% ammonium perchlorate, 9.33% n-butyl acrylate binder, 3.58% polyester binder compound, 1.44% methyl acrylate binder, 0.4% methyl amyl ketone peroxide cure catalyst, 0.25% lecithin wetting agent.* |
| **57. 04-01-043B: High performance moderate temperature burning smokeless propellant:** 75% ammonium perchlorate, 12.35% polyester binder, 12.35% styrene, 0.25% cumene hydroperoxide cure catalyst, 0.05% lecithin wetting agent | **58. 04-01-044A: High performance smokeless propellant:** *75% ammonium perchlorate, 12.25% PVC, 12.25% dibutyl sebacate, 0.49% barium ricinoleate stabilizer, 0.01% mixed balance* |
| **59. 04-01-045A: High performance smokeless propellant:** *65% ammonium perchlorate, 16.8% polypropylen glycol plasticizer, 15% aluminum powder, 2.14% toluene diisocyante polymerization catalyst, 0.5% "alosperse" polymerization agent, 0.27% trimethylol propane binder aid, 0.25% phenyl beta-naphthylamine binder catalyst, 0.04% ferric acetyl acetonate catalyst* | **60. 04-01-046A: Specialty high Performance polymer-based rocket propellant:** *64% ammonium perchlorate, 18.8% polyurethane, 16% aluminum powder, 1.2% ethylene oxide polymerization agent* |
| **61. 04-01-047A: Specialty high Performance "hypersonic" rocket propellant:** *68.84% ammonium perchlorate, 28.38% aluminum powder, 2.77% graphite, 0.01% residue* | |

**04-01-001A: High Performance Military grade rocket propellant:**
Add and dissolve ***93 grams of a plastic elastomer*** (commercially available and sold under a variety of trade names; the trade name used in this procedure is Kraton 1107) into 150 milliliters of toluene. The polymer will readily dissolve in a few minutes at room temperature to form a solution. Next, add in ***1.8 grams of a standard aziridine BA114 compound resin*** (commercially available), followed by ***96 grams of aluminum powder***. After the additions, vigorously stir the toluene mixture for about 15 minutes. Thereafter, add in ***408 grams of finely divided ammonium perchlorate*** while maintaining strong stirring of the toluene mixture. After adding in the ammonium perchlorate, stir the mixture for 2 hours at room temperature. Afterwards, pour the entire toluene mixture into a shallow pan with a high surface area and allow the toluene to evaporate (may take up to 3 days). After the odor of toluene is gone, remove the dry solid propellant and then chop it into tiny pieces. Now, to use the propellant, place the your chopped-up pieces of dry solid propellant into a beaker or suitable chemical resistant container, and then gently heat to 150 Celsius until the propellant becomes a viscous liquid. When the propellant has become a viscous liquid, you may then pour this liquid into your rocket motor or rocket mold, which you should already have designed and built (make sure your rocket motor or rocket motor mold can withstand a temperature of 150 Celsius). Once your rocket motor or mold has been poured, gently but firmly shake (for several minutes) your rocket motor or rocket mold to remove any air bubbles that might be trapped. Note: removing air bubbles is essential to prevent possible rocket failure (explosions). Thereafter, cure your rocket motor or mold for several days in a dark place

at room temperature. After several days, your rocket motor is ready for test firing, or deployment. If you used a mold, your solid rocket propellant can then be removed from the mold or container in one piece. **Note:** Do not use a blasting cap or detonator for initiation.
**Burn rate:** 2.5 seconds per inch on average; tends to vary based on diameter and density of rocket grain.
**Water resistance:** Above moderate.
**Stability:** Can be stored for decades.
**Sensitivity:** Stable towards friction; stable towards shock; stable to temperatures up to 400 Celsius.
**Flammability (1 to 10, un-contained):** 9.2 powdered; 8 small grains; 7.4 medium grains; 7 large grains.
**Ease of ignition (1 to 10 un-contained):** 7 powdered; grain size varies and results in different ignition rates.
**Tendency to cake:** Will not cake.
**Detonation ability:** Can be detonated but only severe conditions—requires TNT, RDX, or HMX booster.
**Explosive ability (contained; explosive deflagration):** Above moderate.
**Percentage:** ***68.13% ammonium perchlorate, 16.03% aluminum, 15.53% kraton polymer binder, 0.30% aziridine resin, 0.01% impurities***
**Classification:** Deflagrating explosive (classified as propellant).

**Danger of preparation:** See note below.
**Use:** Standard military H.P. rocket fuel for rockets and missiles; has excellent thrust capability, and is used for propelling high explosive warheads.

**Note:** Inexperienced personnel should not attempt this procedure, as there are hazards involved. During the heating process caution should be taken to avoid excessive heat or friction. Make sure to thoroughly stir the mixture during the heating process to avoid lumps or clumps, which can lead to excessive heat build-up. This procedure is perfectly safe for experienced persons. To gain experience, preparers should use only small quantities to begin with.

### 04-01-002A: High Performance Military grade rocket propellant:

Add ***68 grams of RB-810 resin*** (commercially available), followed by ***128 grams of dioctyl adipate*** (commercially available), and then followed by ***6.4 grams of lecithin*** to any suitable beaker or heat resistant container, and then heat the contents in the beaker or container to 71 Celsius. Thereafter, rapidly blend the ingredients in the beaker or container for 15 to 30 minutes at 71 Celsius. Afterwards, slowly and carefully add in ***600 grams of finely divided ammonium perchlorate***. During the addition of the ammonium perchlorate continue heating and blending. After adding in the ammonium perchlorate, continue heating and stirring the mixture for about 30 minutes. Afterwards, pour the propellant mixture into your rocket motor or rocket mold, which you should already have designed and built (make sure your rocket motor or rocket motor mold can withstand a temperature of 71 Celsius). Once your rocket motor or mold has been poured, gently but firmly shake (for several minutes) your rocket motor or rocket mold to remove any air bubbles that might be trapped. Note: removing air bubbles is essential to prevent possible rocket failure (explosions). Thereafter, cure your rocket motor or mold for several days in a dark place at room temperature. After several days, your rocket motor is ready for test firing, or deployment. If you used a mold, your solid rocket propellant can then be removed from the mold or container in one piece. **Note:** Do not use a blasting cap or detonator for initiation.
**Burn rate:** 3.1 seconds per inch on average; tends to vary based on diameter and density of rocket grain.
**Water resistance:** Above moderate.
**Stability:** Can be stored for decades.
**Sensitivity:** Stable towards friction; stable towards shock; stable to temperatures up to 475 Celsius.
**Flammability (1 to 10, un-contained):** 8.2 powdered; 7.4 small grains; 7.4 medium grains; 7 large grains.
**Ease of ignition (1 to 10 un-contained):** 7 powdered; grain size varies and results in different ignition rates.
**Tendency to cake:** Will not cake.
**Detonation ability:** Can be detonated but only severe conditions—requires TNT, RDX, or HMX booster.
**Explosive ability (contained; explosive deflagration):** Above moderate.
**Percentage:** ***74.77%% ammonium perchlorate, 15.95% dioctyl adipate binder, 8.47% RB-100 resin, 0.79% lecithin catalyst, 0.02% impurities***
**Classification:** Deflagrating explosive (classified as propellant).

**Danger of preparation:** See note below.
**Use:** Standard military H.P. rocket fuel for rockets and missiles. Has excellent thrust capability, and is used for propelling high explosive warheads.

**Note:** Inexperienced personnel should not attempt this procedure, as there are hazards involved. During the heating process caution should be taken to avoid excessive heat or friction. Make sure to thoroughly stir the mixture during the heating process to avoid lumps or clumps, which can lead to excessive heat build-up. This procedure is perfectly safe for experienced persons. To gain experience, preparers should use only small quantities to begin with.

## 04-01-003A: High Performance Military grade rocket propellant with PVC:

Into a suitable beaker or container, add 600 milliliters of tetrahydrofuran solvent. Thereafter, add in ***80 grams of PVC*** and then stir the mixture to dissolve the PVC. Note: PVC pipes from the hardware store can be used. Once the PVC has dissolved, filter the solution, to remove any insoluble materials, and then pass the filtered solvent mixture (three times) through a layer of aluminum oxide contained in a silica gel column. Thereafter, add to your filtered solvent mixture containing the dissolved PVC, ***4 grams of Epon 815 epoxy agent***, followed by ***4 grams of magnesium sulfide***, and then followed by ***320 grams of finely divided ammonium perchlorate***. Thereafter, rapidly stir the entire mixture for about 15 to 30 minutes at a temperature around 40 Celsius. Afterwards, remove the heat source, and then quickly add in, before the mixture cools, 600 milliliters of warm water. Thereafter, allow the entire mixture to cool to room temperature, and rapidly stir it during this cool down period. Once the mixture has cooled, continue to rapidly stir the mixture at room temperature for about 30 minutes. After 30 minutes, filter the mixture to collect the insoluble propellant mixture. You can use gravity filtration or vacuum filtration. Gravity filtration takes the longest amount time, so vacuum filtration is recommended. If using, vacuum filtration, suck the propellant mixture dry. If you used gravity filtration, once your propellant mixture has been removed, recover the pasty material from the filter paper, and place it onto a shallow pan and spread it out, and allow it to air-dry for several days or until completely dry. Either way, once your propellant is completely dry, you can move onto the packing process. Now, place your propellant into a suitable beaker or container, and then add in 150 milliliters of 99% isopropyl alcohol or 95% ethyl alcohol, and then vigorously stir the entire mixture to form a paste of similar mixture. Thereafter, press this mixture into your rocket motor or rocket mold under a pressure of 10,000 psi. Once your rocket motor or mold has been pressed and filled, place your rocket motor or mold into a warm dry place and allow it to cure for up to 7 days.

**Burn rate:** 3 seconds per inch on average; tends to vary based on diameter and density of rocket grain.
**Water resistance:** Above moderate.
**Stability:** Can be stored for decades.
**Sensitivity:** Stable towards friction; stable towards shock; stable to temperatures up to 300 Celsius.
**Flammability (1 to 10, un-contained):** 9 powdered; 7 small grains; 7 medium grains; 6.8 large grains.
**Ease of ignition (1 to 10 un-contained):** 7 powdered; grain size varies and results in different ignition rates.
**Tendency to cake:** Will not cake.
**Detonation ability:** Can be detonated but only severe conditions—requires TNT, RDX, or HMX booster.
**Explosive ability (contained; explosive deflagration):** Above moderate.
**Percentage:** ***78.43% ammonium perchlorate, 19.6% PVC polymer, 0.98% Epon 815 epoxy resin, 0.98% magnesium sulfide catalyst, 0.01% impurities***
**Classification:** Deflagrating explosive (classified as propellant).

**Danger of preparation:** See note below.
**Use:** Standard military H.P. rocket fuel for rockets and missiles. Has excellent thrust capability, and is used for propelling high explosive warheads.

**Note:** Inexperienced personnel should not attempt this procedure, as there are hazards involved. During the heating process caution should be taken to avoid excessive heat or friction. Make sure to thoroughly stir the mixture during the heating process to avoid lumps or clumps, which can lead to excessive heat build-up. This procedure is perfectly safe for experienced persons. To gain experience, preparers should use only small quantities to begin with.

## 04-01-004A: High Performance military rocket propellant (JPL X500 type):

Into a standard ball mill, place ***196 grams of ammonium perchlorate*** of 100 mesh, followed by ***84 grams of ammonium perchlorate*** of 30 microns, followed by ***40 grams of dioctyl azelate***, followed by ***4 grams of copper chromite***, followed by ***200 milligrams of ferric acetylacetonate***, and then 100 grams of Teflon coated steel shot of 4 to 5 millimeters in diameter. Thereafter, thoroughly tumble the mixture for 1 hour at room temperature. At the same time, into a standard vertical mixer, equipped with plastic stirring blades rather then sharp steel ones, place ***5.4 grams of poly propylene glycol***, followed by ***24 grams of hexane triol***. Thereafter, thoroughly blend this mixture for 1 hour at room temperature on high speed. After preparing these two mixtures, place both of them into a specially designed heated blender with vacuum pump and gauge. Thereafter, thoroughly blend the two mixtures on high speed at 30 Celsius under a vacuum of 10 millimeters of mercury for about 35 minutes. Immediately after 35 minutes, equalize the reduced pressure, open the machine, and toss in ***6 grams of 2,4-toluene diisocyante***, and then continue to heat and blend the mixture on high speed and under a vacuum of about 20 millimeters of mercury at 23 to 25 Celsius for about 10 minutes. Afterwards, the propellant mixture is ready to cast. To do so, simply pour and press the propellant into any desired rocket motor or mold under a pressure of 10,000 psi. Finally, let your motor or mold cure for about 1 to 2 weeks at room temperature.

**Burn rate:** 3.8 seconds per inch on average; tends to vary based on diameter and density of rocket grain.
**Water resistance:** Above moderate.
**Stability:** Can be stored for decades.
**Sensitivity:** Stable towards friction; stable towards shock; stable to temperatures up to 300 Celsius.
**Flammability (1 to 10, un-contained):** 9 powdered; 7 small grains; 7 medium grains; 6.7 large grains.
**Ease of ignition (1 to 10 un-contained):** 7.9 powdered; grain size varies and results in different ignition rates.
**Tendency to cake:** Will not cake.
**Detonation ability:** Can be detonated but only severe conditions—requires TNT, RDX, or HMX booster.
**Explosive ability (contained; explosive deflagration):** Above moderate.

**Percentage:** ***77.8% ammonium perchlorate, 11% dioctyl azelate, 6.6% hexane triol plasticizer, 1.6% 2,4-toluene diisocyanate, 1.5% poly propylene glycol plasticizer, 1.1% copper chromite burn rate accelerator, 0.05% ferric acetylacetonate curing catalyst, 0.35% impurities***
**Classification:** Deflagrating explosive (classified as propellant).

**Danger of preparation:** Can be made by amateurs, as long as the procedure is followed exactly.
**Use:** Standard military rocket fuel for rockets and missiles.

### 04-01-005A: High Performance military rocket propellant:

Into a special heated vertical mixer with plastic stir blades rather then sharp steel, place ***21 grams of poly-1,2-butylene glycol***, followed by ***4.8 grams of a polyoxypropylene derivative of trimethylolpropane***. Thereafter, blend on high for about 30 minutes at 45 Celsius. After 30 minutes, continue heating and stirring the mixture, and add in ***400 milligrams of flours of sulfur***, and then followed by ***72 grams of powdered aluminum***. Thereafter, continue to thoroughly mix the mixture on high at 45 Celsius for about 2 hours. After 2 hours, add in ***272 grams of powdered ammonium perchlorate***, followed immediately by ***16 grams of hexamethylene diisocyanate***, and then followed by 15 milliliters of cold water. Then continue to rapidly blend the mixture on high at 45 Celsius for a bout 1 hour until a fluidized mixture is obtained. After about 1 hour, the propellant is ready to be poured and pressed. To do so, pour the fluidized propellant into your rocket motor or mold, and then press at 5,000 psi using a standard Arbor press machine. Finally, let your motor or mold cure for about 1 to 2 weeks at room temperature.
**Burn rate:** 3 seconds per inch at 5,000 psi; tends to vary based on diameter and density of rocket grain.
**Water resistance:** Above moderate.
**Stability:** Can be stored for decades.
**Sensitivity:** Stable towards friction; stable towards shock; stable to temperatures up to 300 Celsius.
**Flammability (1 to 10, un-contained):** 8.9 powdered; 7.3 small grains; 7 medium grains; 6.7 large grains.
**Ease of ignition (1 to 10 un-contained):** 8 powdered; grain size varies and results in different ignition rates.
**Tendency to cake:** Will not cake.
**Detonation ability:** Can be detonated but only severe conditions—requires TNT, RDX, or HMX booster.
**Explosive ability (contained; explosive deflagration):** Above moderate.
**Percentage:** ***70% ammonium perchlorate, 18.6% aluminum, 5.4% ploy-1,2-butylene glycol plasticizer, 4% hexamethylene diisocyante curative, 1.2% polyoxypropylene derivative of trimethylolpropane curative, 0.10% sulfur, 0.70% moisture***
**Classification:** Deflagrating explosive (classified as propellant).

**Danger of preparation:** Can be made by amateurs, as long as the procedure is followed exactly.
**Use:** Standard military rocket fuel for rockets and missiles.

**Machines for pressing rocket propellants**

**Standard "arbor" presses. The presses are used to "jam" and "press" powdered or varying grain sizes of rocket fuels under high pressure into a rocket motor or rocket mold; the result is a compacted virtual solid grain of varying density. The varying density is based on the pressure adhered to the propellant grain; in some cases, higher density grains have more efficient thrust-to-weight ratios, but not always. As well, burn rates can fluctuate depending on diameter of the solid rocket grain and its density.**

**Die-castings for forming solid rocket grains in rocket motors and/or rocket molds**

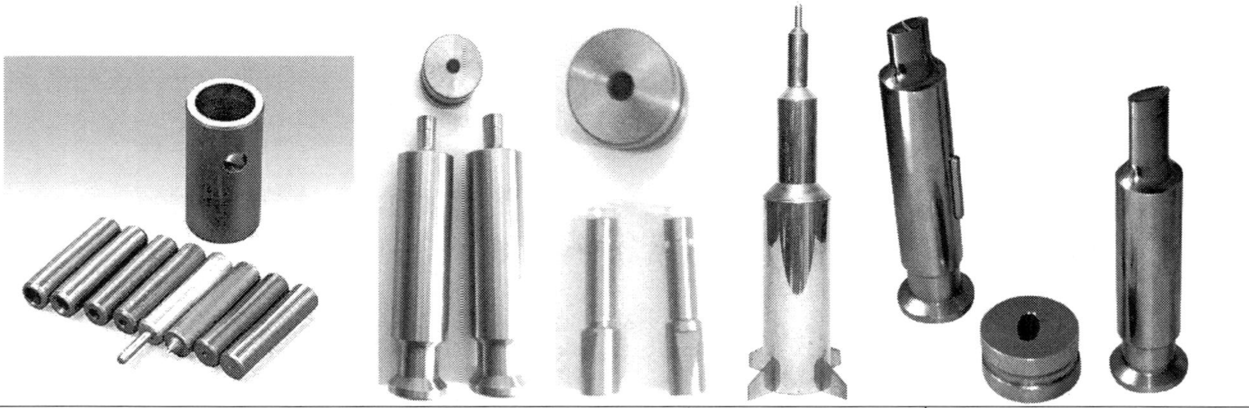

| **Classic rocket design** | **Details** |
| --- | --- |
| 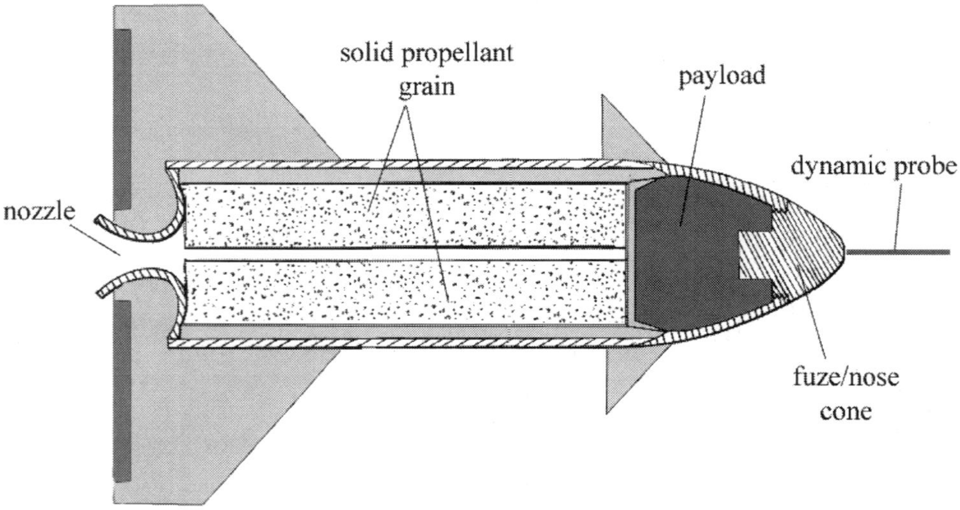  | Rocket design is virtually the same in any field and exists of a solid core of rocket grain, a nozzle, and fins for stabilization. The rocket grain can be fired using any ignition means in the usual manner. The adjacent rocket design portraits the rocket grain, and omits the fuze and payload, which will be included in later illustrations. |

**04-01-006A: High Performance military rocket propellant (with negative pressure exponent):**
Into a special heated vertical mixer, equipped with plastic blades, place ***232 grams of finely divided ammonium perchlorate***, followed by ***72 grams of ammonium sulfate***, followed by ***56 grams of poly propylene glycol***, followed by ***5 grams of poly propylene glycol adduct of trimethylol propane***, followed by ***3 grams of triethylamine***, followed by ***200 milligrams of ferric acetylacetonate***, followed by ***6.3 grams of 2,4-toluene diisocyanate***, and then finally followed by ***24 grams of dioctyl adipate***. Thereafter, thoroughly blend this mixture for 2 hours at 45 Celsius under high speed. After blending the mixture for 2 hours, its ready to use. To use, simply pour and press the propellant into any desired rocket motor or mold under a pressure of 9,000 psi. Finally, let your motor or mold cure for about 1 to 2 weeks at room temperature.
**Burn rate:** 6.2 seconds per inch at 9,000 psi; tends to vary based on diameter and density of rocket grain.
**Water resistance:** Above moderate.
**Stability:** Can be stored for decades.
**Sensitivity:** Stable towards friction; stable towards shock; stable to temperatures up to 300 Celsius.
**Flammability (1 to 10, un-contained):** 8 powdered; 7 small grains; 7 medium grains; 6 large grains.
**Ease of ignition (1 to 10 un-contained):** 8 powdered; grain size varies and results in different ignition rates.
**Tendency to cake:** Will not cake.
**Detonation ability:** Can be detonated but only severe conditions—requires TNT, RDX, or HMX booster.
**Explosive ability (contained; explosive deflagration):** Above moderate.
**Percentage:** ***58.2% ammonium perchlorate, 18% ammonium sulfate, 14% poly propylene glycol plasticizer, 6% dioctyl adipate, 1.5% 2,4-toluene diisocyanate, 1.2% poly propylene glycol adduct of trimethylol propane, 0.75% triethylamine catalsyst, 0.05% ferric acetylacetonate curing catalyst, 0.30% impurities***
**Classification:** Deflagrating explosive (classified as propellant).

**Danger of preparation:** Can be made by amateurs, as long as the procedure is followed exactly.
**Use:** Specialty military rocket fuel for missiles.

**04-01-007A: High Performance military rocket propellant (reduced flash):**
Into a suitable vertical mixer equipped with plastic stirring blade, place ***344 grams of polyethylene glycol***, followed by ***416 grams of nitroguanidine***, and then thoroughly blend this mixture for 15 to 30 minutes at room temperature on high. Thereafter, add in ***200 grams of aluminum powder***, and then thoroughly blend for 15 to 30 minutes at room temperature on high. Finally, add in ***510 grams of powdered ammonium perchlorate***, and ***100 grams of powdered ammonium nitrate***, and then thoroughly blend the mixture on high speed for 2 hours at room temperature. After the blending, the propellant mixture needs to be pressed into your rocket motor or mold under a pressure ranging from 4,000 to 6,000 psi; thereafter the rocket motor should be cured for several days at 30 Celsius.
**Burn rate:** 2.2 seconds per inch at 4,000 psi; tends to vary based on diameter and density of rocket grain.
**Water resistance:** Above moderate.
**Stability:** Can be stored for decades.
**Sensitivity:** Stable towards friction; stable towards shock; stable to temperatures up to 300 Celsius.
**Flammability (1 to 10, un-contained):** 9 powdered.
**Ease of ignition (1 to 10 un-contained):** 8.8 powdered; grain size varies and results in different ignition rates.
**Tendency to cake:** Will not cake.
**Detonation ability:** Can be detonated but only severe conditions—requires TNT, RDX, or HMX booster.
**Explosive ability (contained; explosive deflagration):** Above moderate.
**Percentage:** ***32% ammonium perchlorate, 26.4% nitroguanidine, 21.9% polyethylene glycol binder, 12.7% aluminum, 6.3% ammonium nitrate, 0.70% impurities***
**Classification:** Deflagrating explosive (classified as propellant).

**Danger of preparation:** Can be made by amateurs, as long as the procedure is followed exactly.
**Use:** Specialty military rocket fuel for missiles.

**04-01-008A: High Performance polymeric military rocket propellant:**
Into a suitable specialty heated vertical mixer, or any suitable mixing container, equipped with motorized stirrer, place ***20.8 grams of epoxy resin D.E.R. 332***, followed by ***59.2 grams of polybutylene glycol diamine***, followed by ***80 grams of finely ground aluminum powder***, and then heat the mixture to about 60 Celsius with moderate stirring for about 20 minutes. After 20 minutes, add in ***340 grams of ammonium perchlorate*** of standard mesh, and then continue to blend the mixture at 60 Celsius for another 20 minutes or so. Thereafter, the mixture is ready to be cast. To do so, it needs to be poured and pressed into any desirable rocket motor or mold at 7,000 psi, and then cured in an oven at 60 Celsius for about 2 hours; followed by curing in the same oven at 80 Celsius for about 17 hours. After 17 hours, the rocket motor will be entirely cured, and should then be removed from the oven and cooled. The solid propellant grain will have a rubbery like consistency.
**Burn rate:** 2.2 seconds per inch at 7,000 psi; tends to vary based on diameter and density of rocket grain.
**Water resistance:** Above moderate.

**Stability:** Can be stored for decades.
**Sensitivity:** Stable towards friction; stable towards shock; stable to temperatures up to 300 Celsius.
**Flammability (1 to 10, un-contained):** 9 powdered.
**Ease of ignition (1 to 10 un-contained):** 8 powdered; grain size varies and results in different ignition rates.
**Tendency to cake:** Will not cake.
**Detonation ability:** Can be detonated but only severe conditions—requires TNT, RDX, or HMX booster.
**Explosive ability (contained; explosive deflagration):** Above moderate.
**Percentage:** ***68% ammonium perchlorate, 16% aluminum, 11.8% polybutylene glycol diamine, 4.1% epoxy resin D.E.R. 332, 0.10%, residue***
**Classification:** Deflagrating explosive (classified as propellant).

**Danger of preparation:** Can be made by amateurs, as long as the procedure is followed exactly.
**Use:** Used as a solid fuel for military rockets, missiles, and in rockets for space exploration.

### 04-01-009A: High Performance "plateau-burning" polymeric rocket propellant:

Into a suitable vertical mixer or similar container, place ***102 grams of polybutyl acrylic acid terpolymer binder***, followed by ***18 grams of Epon 828 epoxy resin***, followed by ***5 grams of lithium fluoride powder***, and then followed by ***375 grams of finely divided ammonium perchlorate*** of average mesh, and then blend the mixture with plastic stir blades on moderate speed for about 15 to 20 minutes. Thereafter, pour and press the propellant mixture into any desirable rocket motor, engine, or mold under a pressure of 4,000 psi; and then cure the rocket motor, engine, or mold in an oven at 55 Celsius for about 3 days.
**Burn rate:** 2.9 seconds per inch at 4,000 psi; tends to vary based on diameter and density of rocket grain.
**Water resistance:** Above moderate.
**Stability:** Can be stored for decades.
**Sensitivity:** Stable towards friction; stable towards shock; stable to temperatures up to 300 Celsius.
**Flammability (1 to 10, un-contained):** 9 powdered.
**Ease of ignition (1 to 10 un-contained):** 8 powdered; grain size varies and results in different ignition rates.
**Tendency to cake:** Will not cake.
**Detonation ability:** Can be detonated but only severe conditions—requires TNT, RDX, or HMX booster.
**Explosive ability (contained; explosive deflagration):** Above moderate.
**Percentage:** ***75% ammonium perchlorate, 20.4% polybutyl acrylic acid binder, 3.6% epon 828 epoxy resin, 1% lithium fluoride stabilizer***
**Classification:** Deflagrating explosive (classified as propellant).

**Danger of preparation:** Can be made by amateurs, as long as the procedure is followed exactly.
**Use:** Used in high performance rockets for military and commercial use.

### 04-01-010A: High Performance polymeric rocket propellant:

Into a suitable specialty heated vertical mixer, place ***32.8 grams of diglycidyl ether of bisphenol A*** (having an epoxide equivalent weight from 475 to 575; commercially available), followed by ***65 grams of diglycidyl ether of polypropylene glycol*** (having an epoxide equivalent weight of 330 and a viscosity of 42 cps.; commercially available), followed by ***88.5 grams of aluminum powder of 325 mesh***, followed by ***95 grams of ammonium perchlorate of 250 mesh,*** followed by ***224 grams of standard ammonium perchlorate*** of average mesh, and then blend the mixture at 30 Celsius on moderate speed for about 10 to 15 minutes. Immediately thereafter, add in ***6 grams of monoethanolamine,*** followed by ***1.1 grams of diethylentriamine,*** and then continue to blend the mixture at 30 Celsius on moderate speed for about 15 minutes. After 15 minutes, pour and press the mixture into any desirable rocket motor, engine, or mold under a pressure ranging from 4,000 to 5,000 psi, and then cure the rocket motor, engine, or mold in an oven at 108 Celsius for about 2 hours.
**Burn rate:** 3 seconds per inch at 4,000 psi; tends to vary based on diameter and density of rocket grain.
**Water resistance:** Above moderate.
**Stability:** Can be stored for decades.
**Sensitivity:** Stable towards friction; stable towards shock; stable to temperatures up to 300 Celsius.
**Flammability (1 to 10, un-contained):** 9 powdered.
**Ease of ignition (1 to 10 un-contained):** 8 powdered; grain size varies and results in different ignition rates.
**Tendency to cake:** Will not cake.
**Detonation ability:** Can be detonated but only severe conditions—requires TNT, RDX, or HMX booster.
**Explosive ability (contained; explosive deflagration):** Above moderate.
**Percentage:** ***62.2% ammonium perchlorate, 17.2% aluminum, 12.6% diglycidyl ether polypropylene glycol binder, 6.4% diglycidyl ether bisphenol A binder, 1.1% monoethanolamine catalyst, 0.21% diethylentriamine curing catalyst, 0.29% residue***
**Classification:** Deflagrating explosive (classified as propellant).

**Danger of preparation:** Can be made by amateurs, as long as the procedure is followed exactly.
**Use:** Used in high performance rockets for military and commercial use.

| Rocket assisted artillery shell | Details: |
|---|---|
| 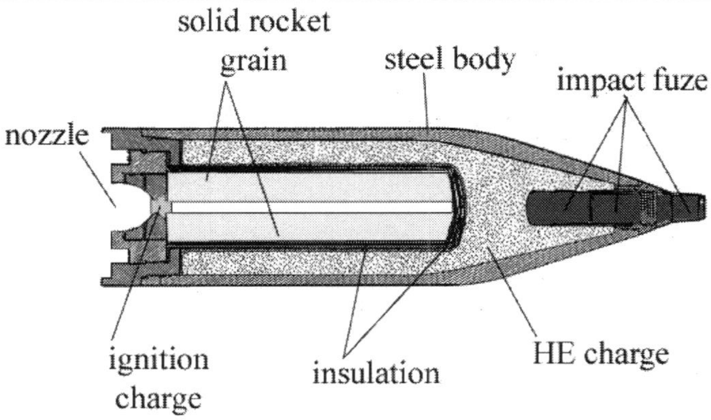  | The adjacent illustration outlines the design of a simple rocket assisted projectile; more commonly seen in 155 and 280-millimeter shells. The rocket assistance can increase the overall range of the shell by about 2 to 3 kilometers; however, the payload weight, being an explosive charge in most cases, is decreased as would be expected due to weight limitations. |

### 04-01-010A: High Performance polymeric rocket propellant (modified with increased burn rate—aluminum free):

Into a suitable specialty heated vertical mixer, place ***32.8 grams of diglycidyl ether of bisphenol A*** (having an epoxide equivalent weight from 475 to 575, commercially available), followed by ***65 grams of diglycidyl ether of polypropylene glycol*** (having an epoxide equivalent weight of 330 and a viscosity of 42 cps., commercially available), followed by ***95 grams of ammonium perchlorate of 250 mesh,*** followed by ***224 grams of standard ammonium perchlorate*** of average mesh, and then blend the mixture at 30 Celsius on moderate speed for about 10 to 15 minutes. Immediately thereafter, add in ***6 grams of monoethanolamine,*** followed by ***1.1 grams of diethylentriamine,*** and then continue to blend the mixture at 30 Celsius on moderate speed for about 15 minutes. After 15 minutes, pour and press the mixture into any desirable rocket motor, engine, or mold at 5,000 psi, and then cure the rocket motor, engine, or mold in an oven at 108 Celsius for about 2 hours.
**Burn rate:** 3.1 seconds per inch at 5,000 psi; tends to vary based on diameter and density of rocket grain.
**Water resistance:** Above moderate.
**Stability:** Can be stored for decades.
**Sensitivity:** Stable towards friction; stable towards shock; stable to temperatures up to 300 Celsius.
**Flammability (1 to 10, un-contained):** 9 powdered.
**Ease of ignition (1 to 10 un-contained):** 8 powdered; grain size varies and results in different ignition rates.
**Tendency to cake:** Will not cake.
**Detonation ability:** Can be detonated but only severe conditions—requires TNT, RDX, or HMX booster.
**Explosive ability (contained; explosive deflagration):** Above moderate.
**Percentage:** ***75.2% ammonium perchlorate, 15.3% diglycidyl ether polypropylene glycol binder, 7.7% diglycidyl ether bisphenol A binder, 1.4% monoethanolamine catalyst, 0.25% diethylentriamine curing catalyst, 0.15% residue***
**Classification:** Deflagrating explosive (classified as propellant).

**Danger of preparation:** Can be made by amateurs, as long as the procedure is followed exactly.
**Use:** Used in high performance rockets for military and commercial use.

### 04-01-011A: High Performance polyurethane rocket propellant:

Into a suitable specialty heated vertical mixer, place ***75 grams of aluminum powder of standard mesh***, followed by ***19.5 grams of dioctylsebacate plasticizer***, followed by ***35.8 grams of polybutadiene diol*** (of 6000 molecular weight, available from Phillips Petroleum Company), followed by ***1.5 grams of polypropylene oxide triol*** (of 700 molecular weight), followed by ***250 milligrams of dihydroxypropyl bis(cyanoethylamine),*** followed by ***2 grams of 2,6-toluene diisocyante***, and then followed by ***100 milligrams of ferric acetylacetonate,*** and then blend the mixture on moderate speed for about 10 minutes. Thereafter, add in ***365 grams of standard ammonium perchlorate*** of average mesh, and then continue to blend the mixture on moderate speed for about 15 to 20 minutes to form a uniform mix. Thereafter, add in ***750 milligrams of finely divided nickel-II-hydroxide***, and then continue to blend the mixture on moderate speed for about 5 minutes. Finally, the mixture is ready to cure. To do so, it should be pressed into any desirable rocket motor, engine, or mold at 7,000 psi, and then cured in an oven at 57 Celsius for 10 days. The excessive amount of curing time is to allow proper polymerization. Requires proper ignition composition.
**Burn rate:** 3.4 seconds per inch at 7,000 psi; tends to vary based on diameter and density of rocket grain.
**Water resistance:** Above moderate.

**Stability:** Can be stored for decades.
**Sensitivity:** Stable towards friction; stable towards shock; stable to temperatures up to 300 Celsius.
**Flammability (1 to 10, un-contained):** 9 powdered.
**Ease of ignition (1 to 10 un-contained):** 8 powdered; grain size varies and results in different ignition rates.
**Tendency to cake:** Will not cake.
**Detonation ability:** Can be detonated but only severe conditions—requires TNT, RDX, or HMX booster.
**Explosive ability (contained; explosive deflagration):** Above moderate.
**Percentage:** ***73% ammonium perchlorate, 15% aluminum, 7.1% polybutadiene diol binder, 3.9% dioctylsebacate plasticizing agent, 0.4% 2,6,-toluene diisocyanate, 0.3% polyprpylene oxide triol binding agent, 0.15% nickel-II-hydroxide curing agent, 0.08% residue, 0.05% dihydroxypropyl bis(cyanoethylamine curing catalyst, 0.02% ferric acetylacetonate,***
**Classification:** Deflagrating explosive (classified as propellant).

**Danger of preparation:** Can be made by amateurs, as long as the procedure is followed exactly.
**Use:** Used in high performance rockets for military and commercial use.

**Laser guided missile system**

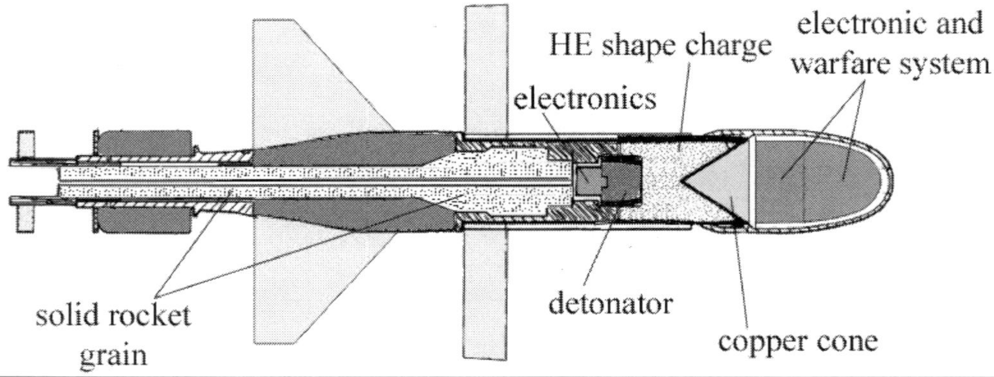


**Description/Details**

The above illustration shows a laser guided or wire guided rocket assisted anti-tank projectile.

### 04-01-012A: High Performance CMDB rocket propellant (composite modified Double Base propellant):

Into a suitable vertical mixer, place ***23 grams of triacetin***, followed by ***70 grams of finely divided aluminum powder***, followed by ***5 grams of finely divided magnesium oxide,*** followed by ***5 grams of 2-nitrodiphenylamine***, followed by ***20 grams of plastisol nitrocellulose,*** followed by ***245 grams of ammonium perchlorate***, and then blend the mixture on low speed at room temperature for about 10 minutes. After 10 minutes, slowly add in ***130 grams of nitroglycerine,*** followed by ***1.6 grams of 2,6-toluene diisocyanate***, and then continue to blend the mixture on low speed 10 to 15 minutes to form a uniform mixture. Thereafter, the mixture is ready to be used. To use, the propellant needs to be pressed into any desirable rocket motor or mold at 4,000 to 6,000 psi, and then allowed to cure at room temperature for about several days. Note: moderate heat may be used to speed up the curing process, but caution should be taken. Requires proper ignition composition.
**Burn rate:** 3.1 seconds per inch at 6,000 psi; tends to vary based on diameter and density of rocket grain.
**Water resistance:** Above moderate.
**Stability:** Can be stored for decades.
**Sensitivity:** Stable towards friction; stable towards shock; stable to temperatures up to 100 Celsius.
**Flammability (1 to 10, un-contained):** 9 solid grain.
**Ease of ignition (1 to 10 un-contained):** 7 solid grain; grain size varies and results in different ignition rates.
**Tendency to cake:** Will not cake.
**Detonation ability:** Can be detonated but only severe conditions—requires TNT, RDX, or HMX booster.
**Explosive ability (contained; explosive deflagration):** Medium high.
**Percentage:** ***49% ammonium perchlorate, 26% nitroglycerine, 14% aluminum, 4.6% triacetin, 4% plastisol nitrocellulose, 1% magnesium oxide burn catalyst, 1% 2-nitrodiphenylamine, 0.32% 2,6-toluene diisocyanate curative, 08% residue***
**Classification:** Deflagrating explosive (classified as propellant).

**Danger of preparation:** Recommended that only experienced pyrotechnicians carryout the procedure.
**Use:** Widely used in high performance rockets for military and commercial use. Also used as a high performance gun propellant, and can be granulated using various techniques listed in the gun propellants section.

**Note:** Inexperienced personnel should not attempt this procedure, as there are hazards involved. Make sure to thoroughly stir the mixture to avoid lumps or clumps, which can lead to excessive heat build-up. This procedure is perfectly safe for experienced persons. To gain experience, preparers should use only small quantities to begin with. Keep in mind the dangers of working with nitroglycerine.

### 04-01-012A: High Performance rocket propellant:

Into a suitable vertical mixer, equipped with stainless steel stir blades, place 200 milliliters of acetone, followed by ***110 grams of standard powdered aluminum***, followed by ***5 grams of finely dived red iron-III-oxide***, followed by ***10.5 grams of HTPB polymer binder*** (hydroxy terminated polybutadiene), followed by ***12.5 grams of dioctyl adipate liquid plasticizer***, followed by ***50 milligrams of triphenyl bismuth***, and then blend the mixture until nearly all the acetone has evaporated. Thereafter, add in ***360 grams of ammonium perchlorate***, followed by ***1.5 grams of di-isocyanate curing agent***, and then continue to blend the mixture on moderate speed for about 10 to 15 minutes. After the 10 to 15 minute blending period, the mixture is ready to be casted and cured. To do so, pour and press the mixture into any suitable rocket motor or mold at 5,000 psi, and then cure the rocket motor or mold in an oven at 40 to 50 Celsius for several hours. Use a standard ignition composition for proper burn.

**Burn rate:** 2.3 seconds per inch at 5,000 psi; tends to vary based on diameter and density of rocket grain.
**Water resistance:** Above moderate.
**Stability:** Can be stored for decades.
**Sensitivity:** Stable towards friction; stable towards shock; stable to temperatures up to 300 Celsius.
**Flammability (1 to 10, un-contained):** 9.1 solid grain.
**Ease of ignition (1 to 10 un-contained):** 9 powdered; grain size varies and results in different ignition rates.
**Tendency to cake:** Will not cake.
**Detonation ability:** Can be detonated but only severe conditions—requires TNT, RDX, or HMX booster.
**Explosive ability (contained; explosive deflagration):** Above moderate.
**Percentage:** ***72% ammonium perchlorate, 22% aluminum, 2.5% dioctyl adipate liquid plasticizer, 2.1% HTPB polymer binder, 1% iron-II-oxide burn catalyst, 0.3% di-issocyante curing agent, 0.09% impurities, 0.01% triphenyl bismuth catalyst***
**Classification:** Deflagrating explosive (classified as propellant).

**Danger of preparation:** Can be made by amateurs, as long as the procedure is followed exactly.
**Use:** Widely used in high performance rockets for military and commercial use.

**Rocket Propelled Grenade (RPG)**

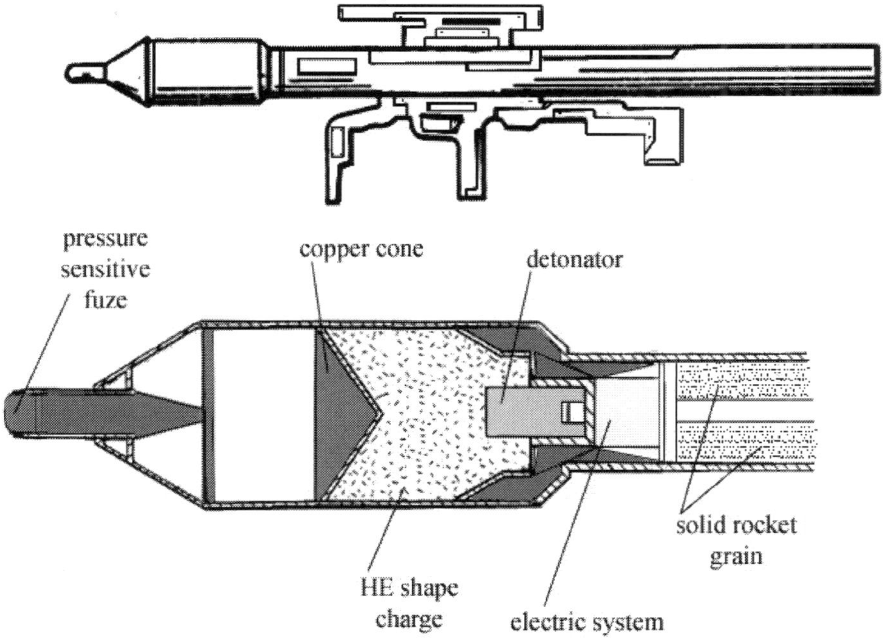


**Details:**

This illustration demonstrates an anti-tank warhead fired from a re-loadable rocket system called an RPG. Upon firing, the rockets grain ignites propelling the warhead towards the target. Upon impact, a special electric fuze, generates an electric current upon impact. This current initiates the explosive train, which detonates the shape charge. The shape charge produces a blast of tremendous force, which blasts a hole through any know substance, especially metal with ease.

### 04-01-013A: High Performance JPL X500 rocket propellant:

Into a suitable ball mill, filled with Teflon coated steel shot of the usual diameter and size, place ***105 grams of ground ammonium perchlorate***, followed by ***245 grams of ungrounded ammonium perchlorate***, followed by ***10 grams of copper chromite***, followed by ***500 milligrams of ferric acetyl acetonate***, and then tumble the mixture for about 45 minutes at 150 RPM to form a uniform mixture. Afterwards, into a suitable special heated and sealable vertical mixer, and equipped with vacuum gauge, place the dry tumbled mixture previously prepared, followed by ***119.5 grams of polypropylene glycol 2025***, followed by ***6 grams of 1,2,6-hexanetriol***, and then blend the mixture on moderate speed for about 30 minutes under a vacuum of 10 millimeters of mercury at a temperature of 23 Celsius. Thereafter, quickly remove the vacuum, open the mixing container, and then throw in ***24 grams of toluene diisocyante,*** and then reseal the container, apply a vacuum of 10 millimeters of mercury, and then continue to blend the mixture on moderate speed for about 10 minutes at 23 Celsius. Thereafter, the propellant mixture is ready to be casted. To do so, simply pour it into any desirable rocket motor, engine, or mold under a pressure of 7,000 psi, and then cure the rocket motor, engine, or mold in an oven at 60 Celsius for about 16 hours. Requires standard rocket propellant ignition composition.

**Burn rate:** 3.1 seconds per inch at 7,000 psi; tends to vary based on diameter and density of rocket grain.
**Water resistance:** Above moderate.
**Stability:** Can be stored for decades.
**Sensitivity:** Stable towards friction; stable towards shock; stable to temperatures up to 300 Celsius.
**Flammability (1 to 10, un-contained):** 8.9 solid grain.
**Ease of ignition (1 to 10 un-contained):** 9 powdered; grain size varies and results in different ignition rates.
**Tendency to cake:** Will not cake.
**Detonation ability:** Can be detonated but only severe conditions—requires TNT, RDX, or HMX booster.
**Explosive ability (contained; explosive deflagration):** Above moderate.
**Percentage:** ***68.6% ammonium perchlorate, 23.4% polypropylene glycol 2025 plasticizer, 4.7% toluene diisocyante, 1.9% copper chromite burn rate catalyst, 1.1% 1,2,6-hexanetriol, 0.21% mixed impurities, 0.09% ferric acetyl acetonate***
**Classification:** Deflagrating explosive (classified as propellant).

**Danger of preparation:** Can be made by amateurs, as long as the procedure is followed exactly.
**Use:** Widely used in high performance rockets for military and commercial use.

### 04-01-013B: High Performance JPL X500 rocket propellant (with decreased burn rate):

This procedure is identical to 03-01-013A. Into a suitable ball mill, filled with Teflon coated steel shot of the usual diameter and size, place ***105 grams of ground ammonium perchlorate***, followed by ***245 grams of ungrounded ammonium perchlorate***, followed by ***500 milligrams of ferric acetyl acetonate***, and then tumble the mixture for about 45 minutes at 150 RPM to form a uniform mixture. Afterwards, into a suitable special heated and sealable vertical mixer, and equipped with vacuum gauge, place the dry tumbled mixture previously prepared, followed by ***119.5 grams of polypropylene glycol 2025***, followed by ***6 grams of 1,2,6-hexanetriol***, and then blend the mixture on moderate speed for about 30 minutes under a vacuum of 10 millimeters of mercury at a temperature of 23 Celsius. Thereafter, quickly remove the vacuum, open the mixing container, and then throw in ***24 grams of toluene diisocyante,*** and then reseal the container, apply a vacuum of 10 millimeters of mercury, and then continue to blend the mixture on moderate speed for about 10 minutes at 23 Celsius. Thereafter, the propellant mixture is ready to be casted. To do so, simply pour it into any desirable rocket motor, engine, or mold under a pressure of 7,000 psi, and then cure the rocket motor, engine, or mold in an oven at 60 Celsius for about 16 hours. Requires standard rocket propellant ignition composition.

**Burn rate:** 3.3 seconds per inch at 7,000 psi; tends to vary based on diameter and density of rocket grain.
**Water resistance:** Above moderate.
**Stability:** Can be stored for decades.
**Sensitivity:** Stable towards friction; stable towards shock; stable to temperatures up to 300 Celsius.
**Flammability (1 to 10, un-contained):** 8.7 solid grain.
**Ease of ignition (1 to 10 un-contained):** 9 powdered; grain size varies and results in different ignition rates.
**Tendency to cake:** Will not cake.
**Detonation ability:** Can be detonated but only severe conditions—requires TNT, RDX, or HMX booster.
**Explosive ability (contained; explosive deflagration):** Above moderate.
**Percentage:** ***70% ammonium perchlorate, 23.9% polypropylene glycol 2025 plasticizer, 4.8% toluene diisocyante, 1.2% 1,2,6-hexanetriol, 0.1% ferric acetyl acetonate***
**Classification:** Deflagrating explosive (classified as propellant).

**Danger of preparation:** Can be made by amateurs, as long as the procedure is followed exactly.
**Use:** Widely used in high performance rockets for military and commercial use.

**Anti-tank projectile**

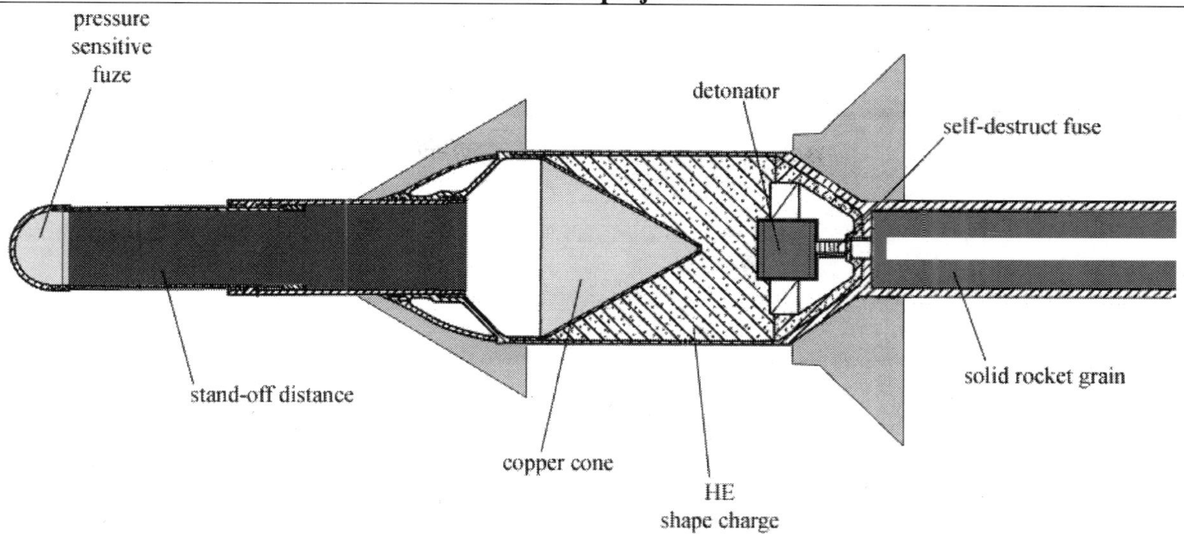


**Details:**

Another classic design of an anti-tank shape charge. A copper cone is placed inversed to the inverted cone shape of high explosive. Upon detonating of the shape charge, its energy is condensed into a fine stream composed of super heated gas and particles. This "jet" moves at incredible speed, topping nearly 29,000 miles per hour, and is capable of penetrating almost any substance.

### 04-01-014A: High Performance rocket propellant with reduced electromagnetic interference exhaust:

Into a suitable vertical mixer, place ***70 grams of finely powdered aluminum***, followed by ***1.75 grams of red iron-III-oxide***, followed by ***10 grams of finely divided molybdenum*** of 6 microns, followed by ***348 grams of ammonium perchlorate*** of average mesh, and then followed by ***70 grams of a binder*** represented as a carboxy terminated polybutadiene compound containing a polyfunctional aziridinyl compound and a trifunctional epoxy resin with cross linking agents and dioctyl adipate plasticizer sold as Grade 390-325 by the Sylvania Electronic Products, inc.; thereafter blend the mixture at room temperature on high speed for about 45 minutes to form a uniform mix. Thereafter, cast the mixture into any desirable rocket motor or engine at 4,000 to 9,000 psi, using the usual techniques, and then cure the rocket motor or engine in an oven at 71 Celsius for several hours.

**Burn rate:** 2.9 seconds per inch at 4,000 psi; tends to vary based on diameter and density of rocket grain.
**Water resistance:** Above moderate.
**Stability:** Can be stored for decades.
**Sensitivity:** Stable towards friction; stable towards shock; stable to temperatures up to 300 Celsius.
**Flammability (1 to 10, un-contained):** 9 solid grain.
**Ease of ignition (1 to 10 un-contained):** 9 powdered; grain size varies and results in different ignition rates.
**Tendency to cake:** Will not cake.
**Detonation ability:** Can be detonated but only severe conditions—requires TNT, RDX, or HMX booster.
**Explosive ability (contained; explosive deflagration):** Above moderate.
**Percentage:** ***69.6% ammonium perchlorate, 14% aluminum, 14% binder, 2% molybdenum, 0.35% iron oxide, 0.05% residue***
**Classification:** Deflagrating explosive (classified as propellant).

**Danger of preparation:** Can be made by amateurs, as long as the procedure is followed exactly.
**Use:** Used in specialty rockets and missiles, which utilize sensitive radar or radio-guided guidance systems, or sensitive electronic equipment where electromagnetic interferences may decrease said guidance/electrical systems effectiveness, such as in spacecraft or specialty military weapons.

### 04-01-015A: High Performance impulse rocket propellant:

Into a suitable sealable vertical mixer, place ***25 grams of Viton A*** (vinylidene fluoride), followed by ***250 grams of finely divided zirconium hydride***, followed by ***225 grams of ammonium perchlorate***, and then followed by 625 milliliters of acetone; thereafter close the mixing container to prevent evaporation of the acetone, and then blend the mixture at room temperature on high speed for about 15 minutes to form a uniform mixture. Thereafter, add in 1800 milliliters of hexane, and then continue to blend the mixture for about 15 minutes on moderate speed. After 15 minutes, stop the stirring process, and then allow the mixture to stand at room temperature for an hour or so. During which time, the propellant mixture will settle to the bottom of the mixing container. Thereafter, filter-off the insoluble propellant mixture, and then vacuum dry it, or air-dry it. Note: the acetone and hexane solvents can be recycled by distilling the solvent mixture to separate the acetone

and hexane solvents. Once the filtered-off propellant composition has been collected and dried, it is ready for pressing. To do so, the dried propellant should be pressed into any desirable rocket motor, engine, or mold under 5,000 psi.
**Burn rate:** 2.9 seconds per inch at 4,000 psi; tends to vary based on diameter and density of rocket grain.
**Water resistance:** Above moderate.
**Stability:** Can be stored for decades.
**Sensitivity:** Stable towards friction; stable towards shock; stable to temperatures up to 300 Celsius.
**Flammability (1 to 10, un-contained):** 9 solid grain.
**Ease of ignition (1 to 10 un-contained):** 9 powdered; grain size varies and results in different ignition rates.
**Tendency to cake:** Will not cake.
**Detonation ability:** Can be detonated but only severe conditions—requires TNT, RDX, or HMX booster.
**Explosive ability (contained; explosive deflagration):** Above moderate.
**Percentage:** ***50% zirconium hydride, 45% ammonium perchlorate, 5% Viton A copolymer***
**Classification:** Deflagrating explosive (classified as propellant).

**Danger of preparation:** Can be made by amateurs, as long as the procedure is followed exactly.
**Use:** Used in specialty rockets and missiles for various applications.

### 04-01-016A: High Performance Teflon based rocket propellant:

Into a suitable vertical mixer, place ***45 grams of standard butyl rubber***, followed by ***15 grams of finely divided Teflon powder***, followed by ***95 grams of finely divided aluminum powder of standard mexh,*** followed by ***350 ammonium perchlorate***, and then followed by 150 milliliters of acetone; thereafter, blend the mixture at room temperature until the bulk of the acetone evaporates. Thereafter, add in 250 milliliters of cold water, and then continue to blend the mixture for about 10 to 15 minutes on moderate speed. After 10 to 15 minutes, stop the stirring process, and then allow the mixture to stand at room temperature for an hour or so. During which time, the propellant mixture will settle to the bottom of the mixing container. Thereafter, filter-off the insoluble propellant mixture, and then dry it in a desiccator filled with metallic sodium. Once the filtered-off propellant composition has been collected and thoroughly dried, it is ready for pressing. To do so, the dried propellant should be pressed into any desirable rocket motor, engine, or mold under a pressure of 3,000 to 4,000 psi.
**Burn rate:** 2.8 seconds per inch at 4,000 psi; tends to vary based on diameter and density of rocket grain.
**Water resistance:** Above moderate.
**Stability:** Can be stored for decades.
**Sensitivity:** Stable towards friction; stable towards shock; stable to temperatures up to 300 Celsius.
**Flammability (1 to 10, un-contained):** 9 solid grain.
**Ease of ignition (1 to 10 un-contained):** 8 solid grain; grain size varies and results in different ignition rates.
**Tendency to cake:** Will not cake.
**Detonation ability:** Can be detonated but only severe conditions—requires TNT, RDX, or HMX booster.
**Explosive ability (contained; explosive deflagration):** Above moderate.
**Percentage:** ***69.3% ammonium perchlorate, 18.8% aluminum powder, 8.9% butyl rubber, 2.9% Teflon, 0.10% residue***
**Classification:** Deflagrating explosive (classified as propellant).

**Danger of preparation:** Can be made by amateurs, as long as the procedure is followed exactly.
**Use:** Used in high performance rockets for military and commercial use.

### 04-01-016B: High Performance Teflon based rocket propellant with reduced burn rate:

Into a suitable vertical mixer, place ***45 grams of standard butyl rubber***, followed by ***15 grams of finely divided Teflon powder***, followed by ***95 grams of finely divided aluminum powder of standard mesh,*** followed by ***350 ammonium perchlorate***, followed by ***15 grams of triphenylphosphine,*** and then followed by 150 milliliters of hexane; thereafter blend the mixture at room temperature until the bulk of the hexane evaporates. Thereafter, place the semi-pasty mass onto a shallow tray, and allow it to air-dry but in a moist free environment. Afterwards, place the dried mass into a clean ball mill, filled with the usual Teflon coated steel shot, and then tumble the mixture at 100 RPM for about 45 minutes. Thereafter, it is ready for pressing. To do so, the dried propellant should be pressed into any desirable rocket motor, engine, or mold under a pressure of 9,000 psi.
**Burn rate:** 2.9 seconds per inch at 9,000 psi; tends to vary based on diameter and density of rocket grain.
**Water resistance:** Above moderate.
**Stability:** Can be stored for decades.
**Sensitivity:** Stable towards friction; stable towards shock; stable to temperatures up to 300 Celsius.
**Flammability (1 to 10, un-contained):** 9.1 solid grain.
**Ease of ignition (1 to 10 un-contained):** 7.8 solid grain; grain size varies and results in different ignition rates.
**Tendency to cake:** Will not cake.
**Detonation ability:** Can be detonated but only severe conditions—requires TNT, RDX, or HMX booster.
**Explosive ability (contained; explosive deflagration):** Above moderate.

**Percentage:** ***67.3% ammonium perchlorate, 18.2% aluminum powder, 8.6% butyl rubber, 2.8% Teflon, 2.8% triphenlyphosine burn reducer, 0.30% residue***
**Classification:** Deflagrating explosive (classified as propellant).

**Danger of preparation:** Can be made by amateurs, as long as the procedure is followed exactly.
**Use:** Used in high performance rockets for military and commercial use.

| **Standard anti-tank missile design** |
|---|
| 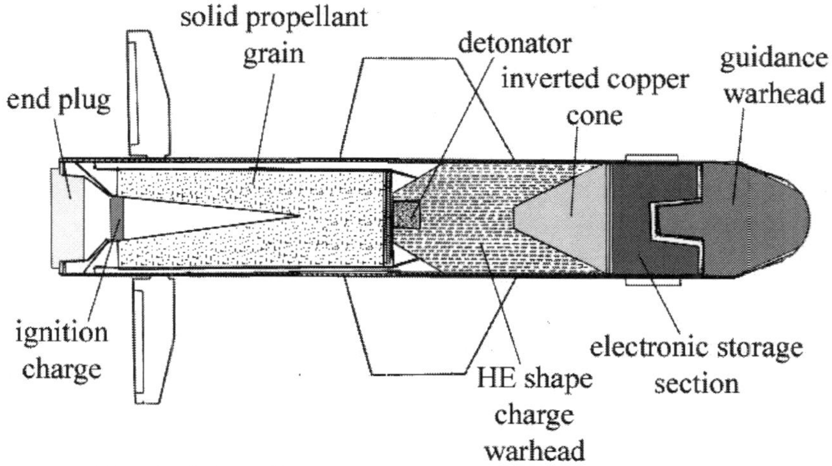 |
| **Description/Details:** |
| A typical wire guided missile for anti-tank purposes. |



**04-01-016C: High Performance Teflon based "smokeless" rocket propellant with reduced burn rate:**
As in the previous procedure, place into a suitable vertical mixer, place ***35 grams of standard butyl rubber***, followed by ***80 grams of finely divided Teflon powder***, followed by ***385 ammonium perchlorate***, and then followed by 150 milliliters of hexane; thereafter, blend the mixture at room temperature until the bulk of the hexane evaporates. Thereafter, place the semi-pasty mass onto a shallow tray, and allow it to air-dry. Afterwards, place the dried mass into a clean ball mill, filled with the usual Teflon coated steel shot, and then tumble the mixture at 100 RPM for about 45 minutes. Thereafter, it is ready for pressing. To do so, the dried propellant should be pressed into any desirable rocket motor, engine, or mold under a pressure of 4,000 to 5,000 psi.
**Burn rate:** 3.1 seconds per inch at 5,000 psi; tends to vary based on diameter and density of rocket grain.
**Water resistance:** Above moderate.
**Stability:** Can be stored for decades.
**Sensitivity:** Stable towards friction; stable towards shock; stable to temperatures up to 300 Celsius.
**Flammability (1 to 10, un-contained):** 9 solid grain.
**Ease of ignition (1 to 10 un-contained):** 7 solid grain; grain size varies and results in different ignition rates.
**Tendency to cake:** Will not cake.
**Detonation ability:** Can be detonated but only severe conditions—requires TNT, RDX, or HMX booster.
**Explosive ability (contained; explosive deflagration):** Above moderate.
**Percentage:** ***77% ammonium perchlorate, 16% Teflon, 7% butyl rubber***
**Classification:** Deflagrating explosive (classified as propellant).

**Danger of preparation:** Can be made by amateurs, as long as the procedure is followed exactly.
**Use:** Used in high performance rockets for military and commercial use.

### 04-01-017A: High Performance smokeless polymer rocket propellant:

Into a suitable vertical mixer, place ***375 milligrams of finely powdered sulfur***, followed by ***1.5 grams of zinc oxide***, followed by ***10 grams of carbon black***, followed by ***9 grams of Milori blue agent***, followed by ***376.5 grams of ammonium perchlorate***, followed by ***500 milligrams of Aerosol-OT wetting agent*** (dioctyl ester of sodium sulfosuccinic acid), and then followed by ***1.5 grams of Flexamine antioxidant*** (25% diarylamine-ketone complex, and complex mixture of amines and phenols, commercially available); thereafter blend the mixture for about 15 minutes. Thereafter, add in ***50 grams of a copolymer*** composed of butadiene of 2-methyl-5-vinylpyridine, followed by ***10 grams of dibutoxyethoxyethyl formal plasticizer***, followed by ***500 milligrams of SA-113***, which is N,N-dimethyl-tertiary butyl sulfenyl dithiocarbamate vulcanization accelerator, and then continue to blend the mixture for about 15 minutes. After 15 minutes, the mixture is ready to be casted and pressed. To do so, pour and press the mixture into any desirable rocket motor or mold under 6,000 psi, and then cure the munitions in an oven at 76 Celsius for 14 days. Can be ignited using any standard high performance rocket ignition composition.
**Burn rate:** 3.4 seconds per inch at 6,000 psi; tends to vary based on diameter and density of rocket grain.
**Water resistance:** Above moderate.
**Stability:** Can be stored for decades.
**Sensitivity:** Stable towards friction; stable towards shock; stable to temperatures up to 300 Celsius.
**Flammability (1 to 10, un-contained):** 8.9 solid grain.
**Ease of ignition (1 to 10 un-contained):** 6.9 solid grain; grain size varies and results in different ignition rates.
**Tendency to cake:** Will not cake.
**Detonation ability:** Can be detonated but only severe conditions—requires TNT, RDX, or HMX booster.
**Explosive ability (contained; explosive deflagration):** Above moderate.
**Percentage:** ***81.8% ammonium perchlorate, 10.8% Butadiene copolymer, 2.1% dibutoxyethyoxyethyl formal plasticizer, 2.1% carbon black filler, 1.9% milori blue burning catalyst, 0.38% mixed residues, 0.32% zinc oxide catalyst, 0.32% flexamine antioxidant agent, 0.10% aerosol-OT wetting agent, 0.10% SA-113 vulcanization accelerator, 0.08% sulfur***
**Classification:** Deflagrating explosive (classified as propellant).

**Danger of preparation:** Can be made by amateurs, as long as the procedure is followed exactly.
**Use:** Used for high performance rockets, missiles, and take-off-assisted rocket packs (JATO).

### 04-01-018A: High Performance smokeless polymer rocket propellant:

Into a suitable ball mill, filled with 200 grams of steel shot of 5mm diameter, place ***2.1 grams of magnesium oxide***, followed by ***41 grams of ammonium dichromate***, followed by ***375 grams of ammonium perchlorate***, followed by ***13.5 grams of Milori blue***, and then tumble the mixture at 100 RPM for about 30 minutes. Thereafter, or in the mean time, into a separate clean beaker, equipped with motorized stirrer, place ***28 grams of a copolymer 1,3-butadiene-2-methyl-5-vinylpyridine***, and then begin to blend this rubbery polymer. Shortly thereafter, add in ***10.5 grams of Philblack E*** (a furnace carbon black sold by the Phillips Petroleum Company), followed by ***27 grams of polybutadiene*** compound sold as "Butarez 15 plasticizer", and then continue to blend the mixture for about 5 minutes. After 5 minutes, immediately add the dry ingredients from the ball mill, and then continue to blend the mixture for about 20 minutes. Thereafter, the mixture is ready to be casted. To do so, it simply needs to be poured and pressed into any desirable rocket motor, engine, or mold under a pressure of 3,000 psi, and then cure in an oven at 120 Celsius for about 3 hours. Must be ignited using a high temperature ignition composition.
**Burn rate:** 6.1 seconds per inch at 3,000 psi; tends to vary based on diameter and density of rocket grain.
**Water resistance:** Above moderate.
**Stability:** Can be stored for decades.
**Sensitivity:** Stable towards friction; stable towards shock; stable to temperatures up to 300 Celsius.
**Flammability (1 to 10, un-contained):** 8.2 solid grain.
**Ease of ignition (1 to 10 un-contained):** 5 solid grain; grain size varies and results in different ignition rates.
**Tendency to cake:** Will not cake.
**Detonation ability:** Can be detonated but only severe conditions—requires TNT, RDX, or HMX booster.
**Explosive ability (contained; explosive deflagration):** Above moderate.
**Percentage:** ***75.4% ammonium perchlorate, 8.2% ammonium dichromate, 5.6% butadiene coplymer, 5.4% polybutadiene plasticizer, 2.7% milori blue burn catalyst, 2.1% philblack E furnace black, 0.42% magnesium oxide, 0.18% mixed residues***
**Classification:** Deflagrating explosive (classified as propellant).

**Danger of preparation:** Can be made by amateurs, as long as the procedure is followed exactly.
**Use:** Used in military and commercial rockets and missiles for multiple uses.

**04-01-019A: High Performance ammonium perchlorate rocket propellant:**
Into a suitable vertical mixer, equipped with plastic stir blades, place ***50 grams of finely powdered aluminum*** of average mesh, followed by ***100 grams of finely powdered magnesium*** of the usual mesh, and then followed by ***140 grams of ammonium perchlorate*** of average mesh. Immediately thereafter, add in 150 milliliters of diethyl ether, and then blend the mixture on high speed until enough ether has evaporated to form a dough-like material. Thereafter, the dough-like material is ready for use. To use, the dough simply needs to be pressed into any desirable rocket motor, engine, or mold under a pressure of 6,000 psi; the resulting engine or mold should be cured at ambient temperatures for several days. Should be ignited using a black powder charge or equivalent ignition composition.
**Burn rate:** 4 seconds per inch at 6,000 psi; tends to vary based on diameter and density of rocket grain.
**Water resistance:** Moderate.
**Stability:** Can be stored for decades.
**Sensitivity:** Stable towards friction; stable towards shock; stable to temperatures up to 300 Celsius.
**Flammability (1 to 10, un-contained):** 8 solid grain.
**Ease of ignition (1 to 10 un-contained):** 6.1 solid grain; grain size varies and results in different ignition rates.
**Tendency to cake:** Will not cake.
**Detonation ability:** Can be detonated but only severe conditions—requires TNT, RDX, or HMX booster.
**Explosive ability (contained; explosive deflagration):** Above moderate.
**Percentage:** ***48.2% ammonium perchlorate, 34.4% magnesium, 17.2% aluminum, 0.20% mixed residues***
**Classification:** Deflagrating explosive (classified as propellant).

**Danger of preparation:** Can be made by amateurs, as long as the procedure is followed exactly.
**Use:** Used in military and commercial rockets and missiles for multiple uses.

**04-01-020A: High Performance ammonium perchlorate rocket propellant:**
Into a suitable vertical mixer, equipped with stainless steel blades, place ***27.6 grams of glyceryl monoricinoleate (GMRO)***, followed by ***766 milligrams of diethanol oleamide (DEO)***, followed by ***16.8 grams of toluene diisocyanate (TDI)***, followed by 700 milliliters of acetone. Thereafter, blend the mixture for about 5 minutes at room temperature. After 5 minutes, add in ***600 grams of finely powdered boron***, followed by ***300 grams of ammonium perchlorate***, and then followed by ***766 milligrams of ferric acetyl acetonate***, and then continue to blend the mixture for about 30 to 40 minutes to form a homogenous mixture where no solid material settles out after stirring has stopped. Thereafter, press the mixture into desirable rocket motor or mold under a pressure of 5,000 psi in the usual manner, and then cure the rocket motor or mold in an oven at 60 Celsius for 3 hours.
**Burn rate:** 6.1 seconds per inch at 5,000 psi; tends to vary based on diameter and density of rocket grain.
**Water resistance:** Moderate.
**Stability:** Can be stored for decades.
**Sensitivity:** Stable towards friction; stable towards shock; stable to temperatures up to 300 Celsius.
**Flammability (1 to 10, un-contained):** 7.6 solid grain.
**Ease of ignition (1 to 10 un-contained):** 6 solid grain; grain size varies and results in different ignition rates.
**Tendency to cake:** Will not cake.
**Detonation ability:** Cannot be detonated.
**Explosive ability (contained; explosive deflagration):** Low.
**Percentage:** ***63.4% boron, 31.7% ammonium perchlorate, 2.9% glyceryl monoricinoleate, 1.77% toluene diisocyanate, 0.08% Ferric acetyl acetonate, 0.08% diethanol oleamide, 0.07% balance***
**Classification:** Deflagrating explosive (classified as propellant).

**Danger of preparation:** Can be made by amateurs, as long as the procedure is followed exactly.
**Use:** Used in medium velocity rocket motors for propelling artillery and mortar rounds to increase range.

**04-01-021A: High Performance ammonium perchlorate rocket propellant:**
Into a suitable specialty heated vertical mixer, place ***116 grams of urea***, followed by ***227.5 grams of ammonium perchlorate***, followed by ***60 grams of lithium perchlorate***, and then heat the mixture to 80 to 85 Celsius; immediately thereafter, begin blending the mixture, and once the temperature of the mixture reaches 80 to 85 Celsius, continue to blend the viscous material for 30 minutes. Thereafter, quickly add in ***39.95 grams of an epoxy binder*** called Unox 201 Epoxy resin, sold by Union carbide, followed by ***36.67 grams of polypropylene glycol***, followed by ***12.5 grams of tallow monoglyceride***, followed by ***6.8 grams of maleic anhydride***, and then followed by ***850 milligrams of tin octoate*** (containing 28% tin by weight), and then continue to blend the viscous mixture at 80 to 85 Celsius on moderate speed for about 15 minutes to form a uniform mix. Thereafter, pour and press the viscous mixture into any desirable rocket motor, engine, or mold, under a pressure of 2,000 psi, and then cure the rocket motor, engine, or mold at 80 Celsius for 20 hours; after 20 hours, remove the heat source, and then allow the rocket motor, engine, or mold to stand at room temperature for 7 days.
**Burn rate:** 4.1 seconds per inch at 2,000 psi; tends to vary based on diameter and density of rocket grain.

**Water resistance:** Moderate.
**Stability:** Can be stored for decades.
**Sensitivity:** Stable towards friction; stable towards shock; stable to temperatures up to 300 Celsius.
**Flammability (1 to 10, un-contained):** 7.9 solid grain; 8.2 powdered grain.
**Ease of ignition (1 to 10 un-contained):** 5 solid grain; grain size varies and results in different ignition rates.
**Tendency to cake:** Will not cake.
**Detonation ability:** Can be detonated but only severe conditions—requires TNT, RDX, or HMX booster.
**Explosive ability (contained; explosive deflagration):** Moderate.
**Percentage:** ***45.47% ammonium perchlorate, 23.1% urea, 11.99% lithium perchlorate, 7.98% epoxy binder, 7.33% polypropylene glycol, 2.49% tallow monoglyceride, 1.35% maleic anhydride, 0.1699% tin octoate curing catalyst, 0.1201% balance (exact decimal places where removed to reduce complications).***
**Classification:** Deflagrating explosive (classified as propellant).

**Danger of preparation:** Can be made by amateurs, as long as the procedure is followed exactly.
**Use:** Used in military and commercial rockets and missiles for multiple uses.

### 04-01-022A: High Performance meltable-wax ammonium perchlorate rocket propellant:

Into a suitable specialty heated vertical mixer, place ***95 grams of urea***, followed by ***372 grams of ammonium perchlorate***, and then followed by ***33 grams of polyethylene wax***; thereafter, heat the mixture to 120 to 125 Celsius under rapid stirring; Once the temperature has reached 125 Celsius, continue to rapidly stir the viscous mixture 15 minutes. Thereafter, the mixture is ready for use. To use, pour and press the hot viscous mixture into any desirable rocket motor or engine under a pressure of 3,000 psi, and then allow the rocket motor or engine to cool and cure at room temperature for 48 hours.
**Burn rate:** 2.3 seconds per inch at 3,000 psi; tends to vary based on diameter and density of rocket grain.
**Water resistance:** Good.
**Stability:** Can be stored for decades.
**Sensitivity:** Stable towards friction; stable towards shock; stable to temperatures up to 300 Celsius.
**Flammability (1 to 10, un-contained):** 9 solid grain; 8.5 powdered grain.
**Ease of ignition (1 to 10 un-contained):** 7.6 solid grain; grain size varies and results in different ignition rates.
**Tendency to cake:** Will not cake.
**Detonation ability:** Can be detonated but only severe conditions—requires TNT, RDX, or HMX booster.
**Explosive ability (contained; explosive deflagration):** Moderate.
**Percentage:** ***74.4% ammonium perchlorate, 19% urea, 6.6% wax***
**Classification:** Deflagrating explosive (classified as propellant).

**Danger of preparation:** Can be made by amateurs, as long as the procedure is followed exactly.
**Use:** Used in military and commercial rockets and missiles for multiple uses.

### 04-01-023A: High Performance polymeric ammonium perchlorate rocket propellant:

Into a suitable specialty heated vertical mixer, place ***60 grams of urea***, followed by ***300 grams of ammonium perchlorate***, followed by ***86 grams of an epoxy binder*** called Unox 201 Epoxy resin, sold by Union carbide, and then followed by ***40 grams of aluminum powder*** of average mesh; immediately thereafter begin blending the mixture on high speed and heat the mixture to 90 Celsius; once the temperature of the mixture reaches 90 Celsius, continue to rapidly blend the viscous mixture for 15 minutes, then add in ***2 grams of tin octoate*** (containing 28% tin by weight), and then continue to blend the viscous mixture on high speed for about 10 minutes. Thereafter, immediately pour and press the hot viscous mixture into any desirable rocket motor, engine, or mold under a pressure of 5,000 psi; then cure the rocket motor, engine, or mold at 90 Celsius for 30 minutes, and thereafter, remove the heat source and allow the rocket motor, engine, or mold to cure at room temperature for 3 days. Note: polymerization will begin shortly after adding in the tin octoate.
**Burn rate:** 2.7 seconds per inch at 5,000 psi; tends to vary based on diameter and density of rocket grain.
**Water resistance:** Good.
**Stability:** Can be stored for decades.
**Sensitivity:** Stable towards friction; stable towards shock; stable to temperatures up to 300 Celsius.
**Flammability (1 to 10, un-contained):** 9 solid grain; 9.2 powdered grain.
**Ease of ignition (1 to 10 un-contained):** 7.9 solid grain; grain size varies and results in different ignition rates.
**Tendency to cake:** Will not cake.
**Detonation ability:** Can be detonated but only severe conditions—requires TNT, RDX, or HMX booster.
**Explosive ability (contained; explosive deflagration):** Above moderate.
**Percentage:** ***61.47% ammonium perchlorate, 17.6% epoxy binder, 12.29% urea, 8.19% aluminum, 0.409% tin octoate, and 0.111% balance***
**Classification:** Deflagrating explosive (classified as propellant).

**Danger of preparation:** Can be made by amateurs, as long as the procedure is followed exactly.
**Use:** Used in military and commercial rockets and missiles for multiple uses.

### 04-01-024A: High Performance ammonium perchlorate rocket propellant:

Into a suitable vertical mixer, place ***180 grams of diglycidyl ether***, followed by 360 milliliters of methylene chloride, and then blend the mixture for about 10 minutes to dissolve the diglycidyl ether. Thereafter, add in ***1020 grams of ammonium perchlorate*** of 250 to 300 mesh, followed by ***13.2 grams of triethylene tetramine***, and then continue to blend the mixture for about 60 to 70 minutes at room temperature; thereafter the mixture should be a tacky mass. Thereafter, the mixture is ready for casting. To use, the tacky brittle mass needs to be pressed into any desirable rocket motor, engine, or mold under a pressure of 2,000 to 3,000 psi. Thereafter, the motor, engine, or mold needs to be placed into a vacuum machine, and a vacuum of 400 millimeters of mercury applied for 20 minutes. After 20 minutes, remove the vacuum, take out the motor, engine, or mold and cure in an oven at 100 Celsius for about 4 hours.
**Burn rate:** 3.1 seconds per inch at 2,000 psi; tends to vary based on diameter and density of rocket grain.
**Water resistance:** Good.
**Stability:** Can be stored for decades.
**Sensitivity:** Stable towards friction; stable towards shock; stable to temperatures up to 300 Celsius.
**Flammability (1 to 10, un-contained):** 8.8 solid grain; 9 powdered grain.
**Ease of ignition (1 to 10 un-contained):** 7.9 solid grain; grain size varies and results in different ignition rates.
**Tendency to cake:** Will not cake.
**Detonation ability:** Can be detonated but only severe conditions—requires TNT, RDX, or HMX booster.
**Explosive ability (contained; explosive deflagration):** Above moderate.
**Percentage:** ***84% ammonium perchlorate, 14.8% diglycidyl ether, 1% triethylene tetramine, 0.2% balance***
**Classification:** Deflagrating explosive (classified as propellant).

**Danger of preparation:** Can be made by amateurs, as long as the procedure is followed exactly.
**Use:** Used in military and commercial rockets and missiles for multiple uses.

### 04-01-025A: High Performance and flexible, ammonium perchlorate rocket propellant with semicarbazide perchlorate burn increaser:

**Step 1:** Preparation of semicarbazide perchlorate

Into a suitable beaker or similar container, place 133.8 grams of semicarbazide hydrochloride, and then add in 1200 milliliters of water. Then stir the mixture to dissolve all the semicarbazide hydrochloride. Then, into a separate beaker or similar container, place 248.4 grams of silver perchlorate, followed by 600 milliliters of water. Thereafter, stir the mixture to dissolve all of the perchlorate. Thereafter, mix both solutions, and then stir the combine mixture for several minutes. Thereafter, filter-off the precipitated silver chloride, and then recrystallize the semicarbazide perchlorate from the filtered liquid (the semicarbazide will remain dissolved in the water). Note: removing the water under vacuum can be used to speed up the crystallization process. Once the crystalline semicarbazide perchlorate has been collected, it should be dissolved in hot methanol, and then precipitated by adding in 6 times the volume of water or methylene chloride. The collected precipitate should them vacuum dried or air-dried. Note: these crystals should then be ground in a ball mill to pulverize the crystals. The dried pulverized crystals are then suitable for use in step 2.

**Step 2:** Preparation of propellant composition

Into a suitable vertical mixer, place ***90 grams of Paraplex P-10 binder***, followed by ***45 grams of semicarbazide perchlorate***, and then followed by ***165 grams of ammonium perchlorate***; thereafter, rapidly and thoroughly blend the mixture for about 15 minutes to form a uniform mixture. Note: the Paralax P-10 may even begin to polymerize at room temperature so take note. After the 15-minute blending period, add in ***1.5 grams of benzoyl peroxide***, and then briefly, yet thoroughly, blend the mixture on rapid speed for about 30 seconds to 1 minute. Thereafter, immediately pour and press the mixture into any desirable rocket motor or engine under a pressure of 4,000 psi, and then allow the mixture to cure at room temperature for several days. The rocket propellant can be fired using any desirable ignition composition.
**Burn rate:** 3 seconds per inch at 4,000 psi; tends to vary based on diameter and density of rocket grain.
**Water resistance:** Good.
**Stability:** Can be stored for decades.
**Sensitivity:** Stable towards friction; stable towards shock; stable to temperatures up to 300 Celsius.
**Flammability (1 to 10, un-contained):** 9 solid grain; 9.3 powdered grain.
**Ease of ignition (1 to 10 un-contained):** 8 solid grain; grain size varies and results in different ignition rates.
**Tendency to cake:** Will not cake.

**Detonation ability:** Can be detonated but only severe conditions—requires TNT, RDX, or HMX booster.
**Explosive ability (contained; explosive deflagration):** Above moderate.
**Percentage:** ***54.7% ammonium perchlorate, 29.85% Paralax P-10 binder, 14.9% semicarbazide perchlorate, 0.497% benzoyl peroxide, 0.053% rounded balance***
**Classification:** Deflagrating explosive (classified as propellant).

**Danger of preparation:** Can be made by amateurs, as long as the procedure is followed exactly.
**Use:** Used in military and commercial rockets and missiles for multiple uses.

### 04-01-026A: High Performance ammonium perchlorate rocket propellant containing magnetic iron oxide burn catalyst:

Into a suitable vertical mixer, place ***85 grams of nitrocellulose***, followed by ***205 grams of ammonium perchlorate***, followed by ***145 grams of nitroglycerine***, followed by ***40 grams of triacetin,*** followed by ***5 grams of 2-nitrodiphenylamine,*** followed by ***5 grams of resorcinol,*** and then followed by ***10 grams of magnetic iron-III-oxide***; thereafter, gently blend the mixture for about 1 hour at room temperature. Thereafter, the mixture is ready for use. To use, the mass needs to be pressed into any desirable rocket motor, engine, or mold under a pressure of 2,400 psi. Thereafter, the rocket motor, engine, or mold needs to be placed into an oven and cured at 50 Celsius for several hours. The cured propellant can be ignited using any standard means.
**Burn rate:** 3.2 seconds per inch at 2,400 psi; tends to vary based on diameter and density of rocket grain.
**Water resistance:** Good.
**Stability:** Can be stored for decades.
**Sensitivity:** Stable towards friction; stable towards shock; stable to temperatures up to 150 Celsius.
**Flammability (1 to 10, un-contained):** 8.9 solid grain; 9 powdered grain.
**Ease of ignition (1 to 10 un-contained):** 7.7 solid grain; grain size varies and results in different ignition rates.
**Tendency to cake:** Will not cake.
**Detonation ability:** Can be detonated but only severe conditions—requires TNT, RDX, or HMX booster.
**Explosive ability (contained; explosive deflagration):** Above moderate.
**Percentage:** ***41.41% ammonium perchlorate, 29.29% nitroglycerine, 17.17% nitrocellulose, 8% triacetin, 2% magnetic iron-III-oxide catalyst, 1% 2-nitrodiphenylamine, 1% resorcinol, 0.13% balance***
**Classification:** Deflagrating explosive (classified as Class A propellant).

**Danger of preparation:** Should only be prepared by experienced personnel.
**Use:** Used in military and commercial rockets and missiles for multiple uses.

**Note:** Inexperienced personnel should not attempt this procedure, as there are hazards involved. Make sure to thoroughly stir the mixture to avoid lumps or clumps, which can lead to excessive heat build-up. This procedure is perfectly safe for experienced persons. To gain experience, preparers should use only small quantities to begin with. Keep in mind the dangers of working with nitroglycerine.

### 04-01-027A: High Performance rubber-like ammonium perchlorate rocket propellant:

Into a suitable vertical mixer, place ***75 grams of lithium aluminum hydride***, followed by ***210 grams of ammonium perchlorate*** of average mesh, followed by ***500 milligrams of carbon black***, followed by ***214.5 grams of a partially polymerized binder*** (composed of 99.85% fluoroalkyl methacrylate monomer (*n* is 4 and R is $CH_3$), and 0.15% benzoyl peroxide). Thereafter, blend the mixture for about 20 minutes at room temperature. Thereafter, the mixture is ready for use. To use, press the mixture into any desirable rocket motor, engine, or mold under a pressure of 5,000 psi in the usual manner, and then place the rocket motor, engine, or mold into an oven and cure at 35 Celsius for 24 hours. After 24 hours, raise the heat to 45 Celsius and cure at said temperature for 24 additional hours. After 24 hours, raise the heat to 55 Celsius, and cure at said temperature for 120 hours. After the curing process, the propellant will be in the form a rubber-like mass. The cured propellant can be ignited using any standard ignition composition.
**Burn rate:** 6.7 seconds per inch at 5,000 psi; tends to vary based on diameter and density of rocket grain.
**Water resistance:** Good.
**Stability:** Can be stored for decades.
**Sensitivity:** Stable towards friction; stable towards shock; stable to temperatures up to 150 Celsius.
**Flammability (1 to 10, un-contained):** 8 solid grain.
**Ease of ignition (1 to 10 un-contained):** 7 solid grain.
**Tendency to cake:** Will not cake.
**Detonation ability:** Cannot be detonated.
**Explosive ability (contained; explosive deflagration):** Low.
**Percentage:** ***42.9% polymerized binder, 42% ammonium perchlorate, 15% lithium aluminum hydride, 0.1% carbon black, 0.7% mixed residues***
**Classification:** Deflagrating explosive (classified as propellant).

**Danger of preparation:** Can be made by amateurs, as long as the procedure is followed exactly.
**Use:** Used in military and commercial rockets and missiles for multiple uses.

### 04-01-027B: High Performance rubber-like ammonium perchlorate rocket propellant:

Into a suitable vertical mixer, place ***100 grams of finely pulverized aluminum***, followed by ***275 grams of ammonium perchlorate*** of average mesh, followed by ***100 grams of a partially polymerized binder*** (composed of 59.91% fluoroalkyl methacrylate monomer (*n* is 3 and R is $CH_3$), and 39.94% fluoroalkyl methacrylate monomer (*n* is 4 and R is $CH_3$), and 0.15% benzoyl peroxide. Thereafter, blend the mixture for 30 minutes at room temperature. Thereafter, the mixture is ready for use. To use, the mass needs to be pressed into any desirable rocket motor, engine, or mold under a pressure of 4,000 psi, and then the rocket motor, engine, or mold needs to be placed into an oven and cured at 50 Celsius for 72 hours. After 72 hours, raise the heat to 70 Celsius and cure for 24 hours at said temperature. After 24 hours, raise the heat to 85 Celsius and cure for 24 hours at said temperature. After the curing process, the propellant will be in the form a rubber-like mass. The cured propellant can be ignited using any standard ignition composition.
**Burn rate:** 4 seconds per inch at 4,000 psi; tends to vary based on diameter and density of rocket grain.
**Water resistance:** Good.
**Stability:** Can be stored for decades.
**Sensitivity:** Stable towards friction; stable towards shock; stable to temperatures up to 150 Celsius.
**Flammability (1 to 10, un-contained):** 8.2 solid grain.
**Ease of ignition (1 to 10 un-contained):** 7 solid grain.
**Tendency to cake:** Will not cake.
**Detonation ability:** Cannot be detonated.
**Explosive ability (contained; explosive deflagration):** Below moderate.
**Percentage:** ***57.89% ammonium perchlorate, 21% aluminum, 21% partially polymerized binder, 0.11% balance***
**Classification:** Deflagrating explosive (classified as propellant).

**Danger of preparation:** Can be made by amateurs, as long as the procedure is followed exactly.
**Use:** Used in military and commercial rockets and missiles for multiple uses.

### 04-01-028A: High Performance ammonium perchlorate rocket propellant utilizing asphalt:

Into a special heated vertical mixer, place ***52.5 grams of asphalt*** (with penetration values ranging from 7 to 9 millimeters per 5 seconds at 100 grams at a temperature of 720 Fahrenheit), and then heat the asphalt to 177 Celsius under rapid stirring, and allow the asphalt to soften. Thereafter, add in ***15 grams of cetyl acetamide***, followed by ***40 grams of castor oil***, followed by ***17.5 grams of dibutyl sebacate***, and then blend the mixture for 15 minutes. Thereafter, add in, in small portions at a time while stirring rapidly, ***375 grams of ammonium perchlorate***, and then continue to blend the mixture for 15 minutes at 177 Celsius. Thereafter, the softened mixture is ready for casting. To do so, simply press the hot mixture into any desirable rocket motor, engine, or mold under a pressure of 6,000 psi, and then allow the rocket motor, engine, or mold to cool and harden for 48 hours.
**Burn rate:** 3.9 seconds per inch at 6,000 psi; tends to vary based on diameter and density of rocket grain.
**Water resistance:** Good.
**Stability:** Can be stored for decades.
**Sensitivity:** Stable towards friction; stable towards shock; stable to temperatures up to 150 Celsius.
**Flammability (1 to 10, un-contained):** 8 solid grain; 9 powdered grain.
**Ease of ignition (1 to 10 un-contained):** 7 solid grain; grain size varies and results in different ignition rates.
**Tendency to cake:** Will not cake.
**Detonation ability:** Can be detonated but only severe conditions—requires TNT, RDX, or HMX booster.
**Explosive ability (contained; explosive deflagration):** Moderate.
**Percentage:** ***75% ammonium perchlorate, 10.5% asphalt, 8% castor oil, 3.5% dibutyl sebacate, 3% cetyl acetamide***
**Classification:** Deflagrating explosive (classified as propellant).

**Danger of preparation:** Can be made by amateurs, as long as the procedure is followed exactly.
**Use:** Used in military and commercial rockets and missiles for multiple uses.

**04-01-029A: Specialty ammonium perchlorate semi-solid (gelled) rocket propellant:**
Into a suitable vertical mixer, place ***447 grams of regular unleaded gasoline,*** and then add in ***53 grams of aluminum naphthenate*** (containing 7.55% aluminum by weight). Thereafter, stir the mixture for about 10 to 15 minutes to form a gel. Thereafter, add in ***75 grams of carbon black,*** and then followed by ***1425 grams of ammonium perchlorate***, and then blend the mixture rapidly for 1 hour at room temperature. Thereafter, the tacky gel is ready for use. To do so, it simply needs to be pressed into any desirable rocket motor, engine, or mold under a pressure of 10,000 psi, in the usual fashion. The gelled propellant mixture can be ignited using any suitable ignition composition. Note: test the bun rate using a small sample before committing to a rocket munition for safety—stories of rocket motors exploding upon launch have been reported, so use caution.
**Burn rate:** 3.2 to 4,1 seconds per inch at 10,000 psi.
**Water resistance:** Good.
**Stability:** Can be stored for decades.
**Sensitivity:** Stable towards friction; stable towards shock; stable to temperatures up to 100 Celsius.
**Flammability (1 to 10, un-contained):** 8.5
**Ease of ignition (1 to 10 un-contained):** 8.7
**Tendency to cake:** N/A.
**Detonation ability:** Cannot be detonated.
**Explosive ability (contained; explosive deflagration):** N/A.
**Percentage:** ***71.25% ammonium perchlorate, 22.35% unleaded gasoline, 3.75% carbon black, 2.65% aluminum naphthenate***
**Classification:** Deflagrating explosive (classified as propellant).

**Danger of preparation:** Should be prepared by experienced pyrotechnicians only.
**Use:** Used in specialty rocket propellants, as an incendiary agent, or used as a gas generator.

**04-01-030A: Specialty high performance sterile ammonium perchlorate rocket propellant for use in space rockets:**
Into a suitable vertical mixer, cooled in an ice bath chilled to 0 Celsius, place ***6 grams of liquid ethylene oxide,*** followed by ***94 grams of polyurethane,*** followed by ***80 grams of finely powdered aluminum,*** and then followed by ***320 grams of ammonium perchlorate***. Thereafter, blend the mixture for about 20 minutes at a temperature below 5 Celsius. Thereafter, the mixture is ready for use. To do so, it simply needs to be pressed into any desirable rocket motor or engine under a pressure of 4,000 to 6,000 psi, and then cure the rocket motor or engine in an oven at 60 Celsius for 72 hours.
**Burn rate:** 3 seconds per inch at 6,000 psi; tends to vary based on diameter and density of rocket grain.
**Water resistance:** Good.
**Stability:** Can be stored for decades.
**Sensitivity:** Stable towards friction; stable towards shock; stable to temperatures up to 300 Celsius.
**Flammability (1 to 10, un-contained):** 8.5 solid grain; 9 powdered grain.
**Ease of ignition (1 to 10 un-contained):** 7.9 solid grain; grain size varies and results in different ignition rates.
**Tendency to cake:** Will not cake.
**Detonation ability:** Can be detonated but only severe conditions—requires TNT, RDX, or HMX booster.
**Explosive ability (contained; explosive deflagration):** Moderate.
**Percentage:** ***64% ammonium perchlorate, 18.8% polyurethane, 16% aluminum, 1.2% ethylene oxide addition agent***
**Classification:** Deflagrating explosive (classified as propellant).

**Danger of preparation:** Can be made by amateurs, as long as the procedure is followed exactly.
**Use:** Used in military and commercial rockets and missiles for space operations.

**04-01-031A: Classic high performance ammonium perchlorate rocket propellant:**
Into a special heated vertical mixer, place ***345 grams of ammonium perchlorate,*** followed by ***80 grams of finely powdered aluminum,*** followed by ***5 grams of red iron-III-oxide,*** followed by ***64.9 grams of Poly B-D R-45M),*** followed by ***4.45 grams of IPDI*** (isophorone diisocyanate compound), and then followed by ***650 milligrams of AO-2246*** (2,2-methylenebis-4-methyl-6-tert-butylphenoland cure catalyst), and then heat and blend the mixture at 63 Celsius for 1 hour. Immediately thereafter, the viscous mixture is ready for casting. To do so, it simply needs to be poured and pressed into any desirable rocket motor, engine, or mold under a pressure of 5,000 psi, and then allowed to stand and cure for several days.
**Burn rate:** 3.2 seconds per inch at 5,000 psi; tends to vary based on diameter and density of rocket grain.
**Water resistance:** Good.
**Stability:** Can be stored for decades.
**Sensitivity:** Stable towards friction; stable towards shock; stable to temperatures up to 300 Celsius.
**Flammability (1 to 10, un-contained):** 8.7 solid grain; 9.1 powdered grain.
**Ease of ignition (1 to 10 un-contained):** 7.5 solid grain; grain size varies and results in different ignition rates.

**Tendency to cake:** Will not cake.
**Detonation ability:** Can be detonated but only severe conditions—requires TNT, RDX, or HMX booster.
**Explosive ability (contained; explosive deflagration):** Moderate.
**Percentage:** ***69% ammonium perchlorate, 16% aluminum, 12.98% polymer binder, 1% iron oxide burn catalyst, 0.89% IPDI compound, 0.13% AO-2246 curing catalyst***
**Classification:** Deflagrating explosive (classified as propellant).

**Danger of preparation:** Can be made by amateurs, as long as the procedure is followed exactly.
**Use:** Used in military and commercial rockets and missiles for multiple uses.

| Rocket assisted 40mm grenade projectile | Details |
|---|---|
| 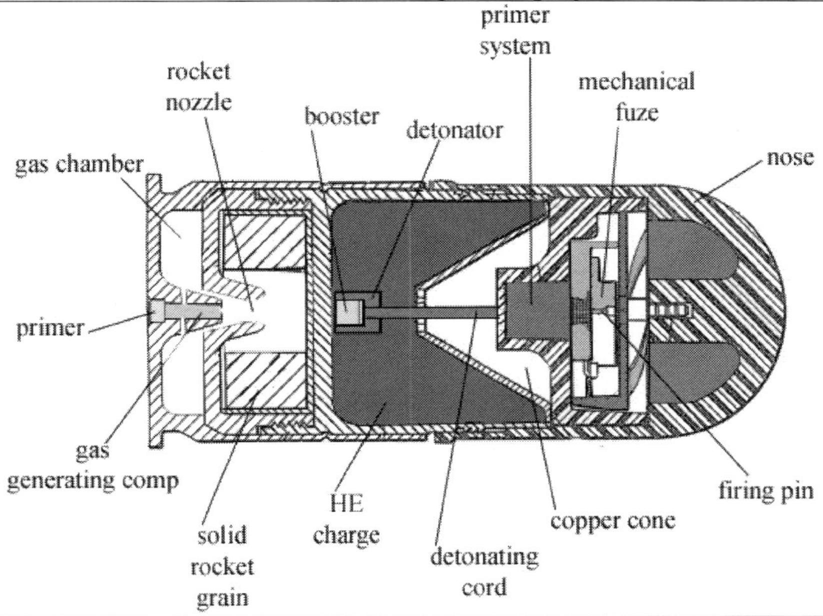 | The adjacent illustration outlines the design of a 40mm projectile to extend its range by a short distance. The projectile consists of a small shape charge capable of penetrating about 2 inches of steel. The projectile is fired using a low-pressure low recoil system, where by gas fills the gas chamber producing pressure. The pressure forces the grenade down range. Centrifugal force arms the grenade after about 6 to 9 revolutions. |

### 04-01-032A: High performance ammonium perchlorate rocket propellant:

Into a special sealed and heated vertical mixer equipped with vacuum port, place ***84 grams of polypropylene glycol,*** followed by ***1.35 grams of trimethylol propane***, followed by ***2.5 grams of "Alosperse"*** (which is a alkyl naphthalene sulfonate dispersion agent), followed by ***1.25 grams of phenyl beta-naphthylamine***. Thereafter, blend the mixture at 60 Celsius for about 20 minutes under a vacuum of 150mm of mercury. Note: vacuum must be applied to de-gas the mixture. Thereafter, reduce the temperature to 43 Celsius, temporarily equalize the pressure, open the vertical mixer, and then add in ***10.7 grams of toluene diisocyante***, seal the mixer, reapply the vacuum, and then continue to blend the mixture at 43 Celsius for 20 minutes. Thereafter, stop the blending and then allow the mixture to stand at 35 Celsius for about 2 hours under a vacuum of 150mm of mercury to allow the mixture to polymerize. After about 2 hours, equalize the pressure, open the mixer, and then pour the mixture into a suitable sized glass beaker. Thereafter, chill this glass beaker in a dry ice bath for 2 hours. After 2 hours, place the cold mixture into a ball mill, filled with 500 grams of Teflon coated heavy steel shot, and then tumble the mixture at 500 RPM for several hours to form a uniform pulverized mass. After several hours, remove the pulverized mass from the ball mill and place into a suitable glass beaker sitting on top of a hot plate, and equipped with portable mechanical stirrer using stainless steel blades, turn on said stirrer to high speed, turn on the hot plate to 60 Celsius, and then add in ***325 grams of ammonium perchlorate***, followed by ***200 milligrams of ferric acetyl acetonate***, and then followed by ***75 grams of aluminum powder*** of average mesh; thereafter, blend the mixture on high speed at 60 Celsius for 30 minutes. Afterwards, the mixture is ready to use. To use, the tacky mass needs to be poured and pressed into any desirable rocket motor, engine, or mold under a pressure of 4,000 psi in the usual fashion, and then cure the rocket motor, engine, or mold at room temperature for about 48 hours. The rocket motor, engine, or mold can be fired using any suitable ignition composition.
**Burn rate:** 3 seconds per inch at 4,000 psi; tends to vary based on diameter and density of rocket grain.
**Water resistance:** Good.
**Stability:** Can be stored for decades.
**Sensitivity:** Stable towards friction; stable towards shock; stable to temperatures up to 300 Celsius.
**Flammability (1 to 10, un-contained):** 8.6 solid grain; 9 powdered grain.
**Ease of ignition (1 to 10 un-contained):** 7.7 solid grain; grain size varies and results in different ignition rates.
**Tendency to cake:** Will not cake.
**Detonation ability:** Can be detonated but only severe conditions—requires TNT, RDX, or HMX booster.
**Explosive ability (contained; explosive deflagration):** Moderate.

**Percentage:** ***65% ammonium perchlorate, 16.8% polypropylene glycol, 15% aluminum powder, 2.14% toluene diisocyante polymerization process, 0.5% alosperse, 0.27% trimethylol propane, 0.25% phenyl beta-naphthylamine, 0.04% possible moisture***
**Classification:** Deflagrating explosive (classified as propellant).

**Danger of preparation:** Can be made by amateurs, as long as the procedure is followed exactly.
**Use:** Used in military and commercial rockets and missiles for multiple uses.

### 04-01-033A: High performance ammonium perchlorate rocket propellant:

Into a suitable vertical mixer, place ***50 grams of polyvinyl chloride*** (Geon Paste Resin 121), followed by ***50 grams of dibutyl sebacate***. Thereafter, blend the mixture at room temperature for about 20 minutes. After 20 minutes, place the mixture into a suitable beaker or similar container, and then chill this mixture in a dry ice bath for several hours. Thereafter, remove the dry ice bath, and then add in ***394 grams of ammonium perchlorate***, followed by ***1 gram of dibutyl tin dilaurate***, and then blend the mixture for about 10 minutes. Now, the mixture is ready for use. To use, the mixture needs to be pressed into a rocket engine mold under a pressure of 10,000 psi at room temperature (note: do not press the mixture directly into a rocket motor, as the mixture will expand and then de-expand during the heated curing process so this needs to be calculated for the dimensions of the rocket engine before attempting the curing process using the mold); after the pressing process, the temporary rocket engine mold needs to be heated to 90 Celsius. When the mixture has reached a temperature of 90 Celsius, it will begin to expand; therefore, once the mixture begins to expand, apply a pressure of 4,000 psi and maintain the temperature of the propellant mixture at 90 Celsius for 30 minutes. After 30 minutes, remove the heat source, and maintain a pressure of 4,000 psi during the cool down period. Once the propellant mixture has cooled, increase the pressure to 6,000 psi; thereafter, the solid propellant grain has cured properly and is then ready to be removed from its temporary mold, and then charged into your desired rocket engine.
**Burn rate:** 2.9 seconds per inch at 6,000 psi; tends to vary based on diameter and density of rocket grain.
**Water resistance:** Good.
**Stability:** Can be stored for decades.
**Sensitivity:** Stable towards friction; stable towards shock; stable to temperatures up to 300 Celsius.
**Flammability (1 to 10, un-contained):** 8.9 solid grain; 9.1 powdered grain.
**Ease of ignition (1 to 10 un-contained):** 7.5 solid grain; grain size varies and results in different ignition rates.
**Tendency to cake:** Will not cake.
**Detonation ability:** Can be detonated but only severe conditions—requires TNT, RDX, or HMX booster.
**Explosive ability (contained; explosive deflagration):** Moderate.
**Percentage:** ***79.59% ammonium perchlorate, 10.1% dibutyl sebacate, 10.1% polyvinyl chloride, 0.2% dibutyl tin dilaurate, 0.01% possible moisture***
**Classification:** Deflagrating explosive (classified as propellant).

**Danger of preparation:** Can be made by amateurs, as long as the procedure is followed exactly.
**Use:** Used in military and commercial rockets and missiles for multiple uses.

### 04-01-034A: High performance smokeless ammonium perchlorate rocket propellant:

Into a suitable vertical mixer, place ***450 milligrams of chromium 2-ethylhexanoate***, followed ***by 322 grams of a carboxy terminated polybutadiene*** with a molecular weight of 5,000 and a specific gravity of 0.91 at 60 degrees Fahrenheit containing 1.0 to 1.5 percent 2,2'-methylene-bis(4-methyl-6-tertbutyl) phenol as antioxidant, followed by ***18.5 grams of a diglycidyl ether of bisphenol A*** (with an epoxy equivalent of 174 and 98 percent pure) with N,N-ortho-tris(epoxypropyl) p-aminophenol (with an epoxy equivalent of 97.4 and 96 percent pure), followed by ***13 grams of dioctyl adipate***, followed by ***13 grams of a polymer of butadiene*** (containing a small amount of phenyl-beta-naphthylamine as stabilizer), thereafter followed by ***830 grams of ammonium perchlorate*** of average mesh, and then followed by ***25 grams of aluminum powder*** of average mesh. Thereafter, blend the mixture for about 1 hour to form a uniform mixture. Thereafter, the mixture is ready for use. To use, the mixture needs to be pressed into any desired rocket motor or engine under a pressure of 10,000 psi in the usual fashion.
**Burn rate:** 2.98 seconds per inch at 10,000 psi; tends to vary based on diameter and density of rocket grain.
**Water resistance:** Good.
**Stability:** Can be stored for decades.
**Sensitivity:** Stable towards friction; stable towards shock; stable to temperatures up to 300 Celsius.
**Flammability (1 to 10, un-contained):** 8.9 solid grain; 9.1 powdered grain.
**Ease of ignition (1 to 10 un-contained):** 7.4 solid grain; grain size varies and results in different ignition rates.
**Tendency to cake:** Will not cake.
**Detonation ability:** Can be detonated but only severe conditions—requires TNT, RDX, or HMX booster.
**Explosive ability (contained; explosive deflagration):** Moderate.
**Percentage:** ***67.9% ammonium perchlorate, 26.3% polybutadiene polymer, 2% aluminum, 1.5% diglycidyl ether, 1% butadiene polymer, 1% dioctyl adipate, 0.036% chromium compound catalyst, 0.264% balanced***
**Classification:** Deflagrating explosive (classified as propellant).

**Danger of preparation:** Can be made by amateurs, as long as the procedure is followed exactly.
**Use:** Used in military and commercial rockets and missiles for multiple uses where a visible smoke trail is not desired.

### 04-01-034B: High performance smokeless ammonium perchlorate rocket propellant (modified):

Into a suitable vertical mixer, place ***750 milligrams of chromium 2-ethylhexanoate***, followed by ***322 grams of a carboxy terminated polybutadiene*** with a molecular weight of 5,000 and a specific gravity of 0.91 at 60 degrees Fahrenheit containing 1.0 to 1.5 percent 2,2'-methylene-bis(4-methyl-6-tertbutyl) phenol as antioxidant, followed by ***18 grams of a diglycidyl ether of bisphenol A*** (with an epoxy equivalent of 174 and 98 percent pure) with N,N-ortho-tris(epoxypropyl) p-aminophenol (with an epoxy equivalent of 97.4 and 96 percent pure), followed by ***26.5 grams of a polymer of butadiene*** (containing a small amount of phenyl-beta-naphthylamine as stabilizer), thereafter followed by ***830 grams of ammonium perchlorate*** of average mesh, and then followed by ***25 grams of aluminum powder*** of average mesh. Thereafter, blend the mixture for about 1 hour to form a uniform mixture. Thereafter, the mixture is ready for use. To use, the mixture needs to be pressed into any desired rocket motor or engine under a pressure of 10,000 psi in the usual fashion.
**Burn rate:** 3.1 seconds per inch at 10,000 psi; tends to vary based on diameter and density of rocket grain.
**Water resistance:** Good.
**Stability:** Can be stored for decades.
**Sensitivity:** Stable towards friction; stable towards shock; stable to temperatures up to 300 Celsius.
**Flammability (1 to 10, un-contained):** 8.9 solid grain; 9.2 powdered grain.
**Ease of ignition (1 to 10 un-contained):** 7.3 solid grain; grain size varies and results in different ignition rates.
**Tendency to cake:** Will not cake.
**Detonation ability:** Cannot be detonated.
**Explosive ability (contained; explosive deflagration):** Below moderate.
**Percentage:** ***67.9% ammonium perchlorate, 26.3% polybutadiene polymer, 2.1% polymer butadiene, 2% aluminum, 1.47% diglycidyl ether, 0.06% chromium compound catalyst, 0.17% balanced***
**Classification:** Deflagrating explosive (classified as propellant).

**Danger of preparation:** Can be made by amateurs, as long as the procedure is followed exactly.
**Use:** Used in military and commercial rockets and missiles for multiple uses where a visible smoke trail is not desired.

### 04-01-034C: High performance smokeless ammonium perchlorate rocket propellant (modified):

Into a suitable vertical mixer, place ***750 milligrams of chromium 2-ethylhexanoate***, followed by ***322 grams of a carboxy terminated polybutadiene*** with a molecular weight of 5,000 and a specific gravity of 0.91 at 60 degrees Fahrenheit containing 1.0 to 1.5 percent 2,2'-methylene-bis(4-methyl-6-tertbutyl) phenol as antioxidant, followed by ***18 grams of 3,4-epoxycyclohexylmethyl carboxyl-3,4-epoxycyclohexane*** (with an epoxy equivalent of 130.6 and 96% pure) and the N,N-ortho-tris(epoxypropyl) p-aminophenol, followed by ***26.5 grams of a polymer of butadiene*** (containing a small amount of phenyl-beta-naphthylamine as stabilizer), thereafter followed by ***830 grams of ammonium perchlorate*** of average mesh, and then followed by ***25 grams of aluminum powder*** of average mesh. Thereafter, blend the mixture for about 1 hour to form a uniform mixture. Thereafter, the mixture is ready for use. To use, the mixture needs to be pressed into any desired rocket motor or engine under a pressure of 10,000 psi in the usual fashion.
**Burn rate:** 3 seconds per inch at 10,000 psi; tends to vary based on diameter and density of rocket grain.
**Water resistance:** Good.
**Stability:** Can be stored for decades.
**Sensitivity:** Stable towards friction; stable towards shock; stable to temperatures up to 300 Celsius.
**Flammability (1 to 10, un-contained):** 8.9 solid grain; 9.2 powdered grain.
**Ease of ignition (1 to 10 un-contained):** 7 solid grain; grain size varies and results in different ignition rates.
**Tendency to cake:** Will not cake.
**Detonation ability:** Cannot be detonated.
**Explosive ability (contained; explosive deflagration):** Below moderate.
**Percentage:** ***67.9% ammonium perchlorate, 26.3% polybutadiene polymer, 2.1% polymer butadiene, 2% aluminum, 1.47% epoxy compound, 0.06% chromium compound catalyst, 0.17% balanced***
**Classification:** Deflagrating explosive (classified as propellant).

**Danger of preparation:** Can be made by amateurs, as long as the procedure is followed exactly.
**Use:** Used in military and commercial rockets and missiles for multiple uses where a visible smoke trail is not desired.

**04-01-035A: High performance smokeless ammonium perchlorate rocket propellant with copper sulfide burn accelerator:**

Into a suitable vertical mixer, place ***1500 grams of ammonium perchlorate***, followed by ***260 grams of a hydroxyl terminated polybutadiene resin (Arco R45M),*** containing isophorone diisocyanate curing agent, and then add in ***40 grams of fine-grained copper-II-sulfide***. Thereafter, blend the mixture on high speed for 1 hour. Thereafter, the mixture is ready for use. To use, press the mixture into any desired rocket mold under a pressure of 6,000 psi in the usual manner, and then cure the rocket mold in an oven at 60 Celsius for 1 hour. After 1 hour, remove the heat source, and allow the solid rocket grain to cool to room temperature; once it has, remove the solid grain from the rocket mold, simply coat the solid grain with a thin layer of carbon black, and then insert this coated solid rocket grain into your pre-fabricated rocket motor.

**Burn rate:** 2.49 seconds per inch at 6,000 psi; tends to vary based on diameter and density of rocket grain.

**Water resistance:** Good.

**Stability:** Can be stored for decades.

**Sensitivity:** Stable towards friction; stable towards shock; stable to temperatures up to 300 Celsius.

**Flammability (1 to 10, un-contained):** 8.9 solid grain; 9.1 powdered grain.

**Ease of ignition (1 to 10 un-contained):** 7.8 solid grain; grain size varies and results in different ignition rates.

**Tendency to cake:** Will not cake.

**Detonation ability:** Can be detonated but only severe conditions—requires TNT, RDX, or HMX booster.

**Explosive ability (contained; explosive deflagration):** Moderate.

**Percentage:** ***83.3% ammonium perchlorate, 14.44% polybutadiene resin, 2.22% copper-II-sulfide, 0.04% residual balance***

**Classification:** Deflagrating explosive (classified as propellant).

**Danger of preparation:** Can be made by amateurs, as long as the procedure is followed exactly.

**Use:** Used in military and commercial rockets and missiles for multiple uses where a visible smoke trail is not desired.

| Rocket propelled projectile | Details |
| --- | --- |
| 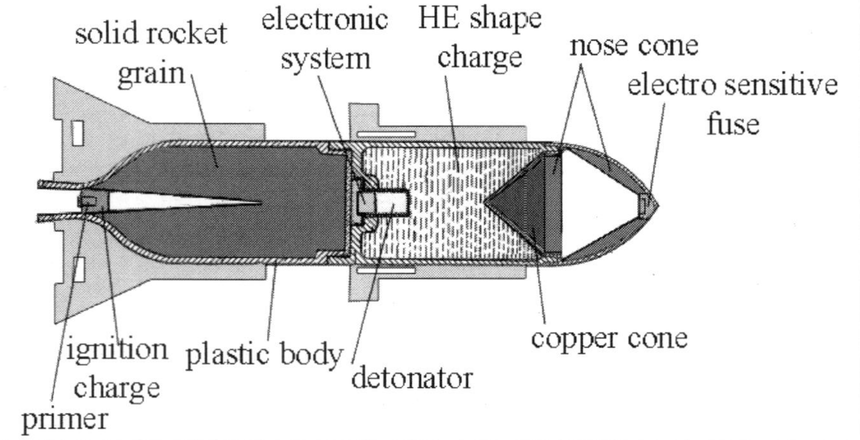  | Another example of a rocket assisted anti-tank munition. The electro sensitive fuse works by generating an electric current upon impact. |

**04-01-036A: High performance ammonium perchlorate rocket propellant with copper sulfide burn accelerator (aluminum based):**

Into a suitable vertical mixer, place 500 milliliters of hexane, and then add in ***1500 grams of ammonium perchlorate*** (20 microns), and then blend the mixture to form a pasty mass. Thereafter add in ***260 grams of powdered aluminum***, followed by ***55 grams of copper sulfide***, and then blend the mixture on moderate speed for about 40 minutes. Thereafter, place the mixture onto a shallow pan or tray, and then allow the mixture to thoroughly air-dry. Thereafter, place the dried mass into a suitable ball mill, filled with 500 grams of Teflon coated heavy steel shot, and then tumble the mixture for about 3 hours to form a uniform powder. Thereafter, place this pulverized mass into a special sealable clean vertical mixer, and then add in ***150 grams of Epon 815 epoxy resin***, or any other desired liquid epoxy compound, and then blend the entire mixture in the absence of air for about 30 minutes. Thereafter, the mixture is ready for use. To use, simply press the mixture into any desired rocket engine or mold under a pressure of 5,000 psi in the usual manner. Thereafter, allow the rocket engine or mold to cure at room temperature or in an oven at ordinary temperatures in the usual manner. Use typical ignition composition.

**Burn rate:** 2.8 seconds per inch at 5,000 psi; tends to vary based on diameter and density of rocket grain.

**Water resistance:** Good.

**Stability:** Can be stored for decades.

**Sensitivity:** Stable towards friction; stable towards shock; stable to temperatures up to 300 Celsius.

**Flammability (1 to 10, un-contained):** 9 solid grain; 9.3 powdered grain.

**Ease of ignition (1 to 10 un-contained):** 8 solid grain; grain size varies and results in different ignition rates.

**Tendency to cake:** Will not cake.

**Detonation ability:** Can be detonated but only severe conditions—requires TNT, RDX, or HMX booster.

**Explosive ability (contained; explosive deflagration):** Moderate.
**Percentage:** ***76.33% ammonium perchlorate, 13.23% aluminum powder, 7.63% epoxy resin, 2.79% copper-II-sulfide, 0.02% impurities***
**Classification:** Deflagrating explosive (classified as propellant).

**Danger of preparation:** Can be made by amateurs, as long as the procedure is followed exactly.
**Use:** Used in military and commercial rockets and missiles for multiple uses.

### 04-01-037A: High performance ammonium perchlorate rocket propellant (with metal hydride accelerator component):

Into a suitable sealable vertical mixer, equipped with vacuum port, place 500 milliliters of acetone, followed by adding in ***1200 grams of ammonium perchlorate***, followed by ***120 grams of magnesium powder***, followed by ***50 grams of copper-II-chromite***, and then followed by ***70 grams of titanium hydride***. Thereafter, dry blend the mixture on high speed for 45 minutes. Thereafter, add in ***750 grams of powdered ammonium nitrate,*** followed by ***350 grams of sodium hydride***, seal the vertical mixer, and then blend the mixture on moderate speed at STP for about 20 minutes to form a rough dough-like material. Thereafter, continue to blend the mixture on moderate speed, apply a vacuum of 50mm of mercury, and remove the solvent using the vacuum. Once the solvent has been evaporated, remove the vacuum, equalize the pressure, and then open the sealable vertical mixer; thereafter, quickly toss in ***250 grams of Epon 815 epoxy resin***, or any other desired liquid epoxy compound, re-seal the vertical mixer and blend the entire mixture in the absence of air for about 30 minutes. Thereafter, the mixture is ready for use. To use, simply press the mixture into any desired rocket engine or mold under a pressure of 9,000 psi in the usual manner, and then cure the rocket engine or mold in an oven at 60 Celsius for 1 hour. After 1 hour, the curing process will be complete so turn off the oven, and allow the rocket motor or mold to cool to room temperature.
**Burn rate:** 2.9 seconds per inch at 9,000 psi; tends to vary based on diameter and density of rocket grain.
**Water resistance:** Good.
**Stability:** Can be stored for decades.
**Sensitivity:** Stable towards friction; stable towards shock; stable to temperatures up to 300 Celsius.
**Flammability (1 to 10, un-contained):** 9.1 solid grain; 9.3 powdered grain.
**Ease of ignition (1 to 10 un-contained):** 8.5 solid grain; grain size varies and results in different ignition rates.
**Tendency to cake:** Will not cake.
**Detonation ability:** Can be detonated but only severe conditions—requires TNT, RDX, or HMX booster.
**Explosive ability (contained; explosive deflagration):** Moderate.
**Percentage:** ***43.01% ammonium perchlorate, 26.88% ammonium nitrate, 12.54% sodium hydride, 8.96% epoxy resin, 4.2% magnesium powder, 2.5% titanium hydride, 1.79% copper-II-chromite burn rate catalyst, 0.12% residual impurities***
**Classification:** Deflagrating explosive (classified as propellant).

**Danger of preparation:** Can be made by amateurs, as long as the procedure is followed exactly.
**Use:** Used in military and commercial rockets and missiles for multiple uses.

### 04-01-038A: High performance ammonium perchlorate rocket propellant (with silicon monoxide accelerator component):

Into a suitable sealable vertical mixer, place ***167 grams of butadiene-1,3,*** and then add in ***33 grams of methacrylic acid***. Thereafter, seal the vertical mixer, and then blend the mixture on high speed for 30 minutes to form a copolymer. Note: heat may be used to accelerate the process. After 30 minutes, quickly open the vertical mixer, and add in ***1152 grams of powdered ammonium perchlorate***, followed by ***8.8 grams of magnesium oxide***, and then followed by ***2 grams of silicon monoxide***. Thereafter, re-seal the vertical mixer and then continue to blend the mixture on high speed for about 1 hour. After 1 hour, the mixture is ready for use. To use, press the mixture into any rocket motor or mold under a pressure of 8,000 psi and then cure in an oven at 74 Celsius for 6 hours. After 6 hours, remove the heat source, and allow the rocket motor or mold to cool to room temperature.
**Burn rate:** 3 seconds per inch at 8,000 psi; tends to vary based on diameter and density of rocket grain.
**Water resistance:** Good.
**Stability:** Can be stored for decades.
**Sensitivity:** Stable towards friction; stable towards shock; stable to temperatures up to 300 Celsius.
**Flammability (1 to 10, un-contained):** 9 solid grain; 9.2 powdered grain.
**Ease of ignition (1 to 10 un-contained):** 8 solid grain; grain size varies and results in different ignition rates.
**Tendency to cake:** Will not cake.
**Detonation ability:** Can be detonated but only severe conditions—requires TNT, RDX, or HMX booster.
**Explosive ability (contained; explosive deflagration):** Moderate.
**Percentage:** ***84.53% ammonium perchlorate, 12.25% butadiene-1,3 binder, 2.42% methacrylic acid catalyst, 0.64% magnesium oxide filler, 0.146% silicon monoxide burn rate accelerator, 0.014% residual balance***
**Classification:** Deflagrating explosive (classified as propellant).

**Danger of preparation:** Can be made by amateurs, as long as the procedure is followed exactly.
**Use:** Used in military and commercial rockets and missiles for multiple uses.

### 04-01-038B: High performance ammonium perchlorate rocket propellant (with silicon monoxide and silicon nitride accelerator component):

Into a suitable sealable vertical mixer, place ***167 grams of butadiene-1,3,*** and then add in ***33 grams of stearic acid***. Immediately thereafter, seal the mixer, and blend the mixture for 15 minutes to form a copolymer compound. Note: heat may be used to accelerate the process. After 15 minutes, open the mixer, and then add in ***1152 grams of powdered ammonium perchlorate***, followed by ***4 grams of magnesium oxide***, and then followed by ***12 grams of silicon monoxide***. Thereafter, re-seal the mixer and continue to blend the mixture at high speed for about 1 hour. After 1 hour, quickly open the mixer, and then quickly add in ***60 grams of sulfur***, followed by ***45 grams of potassium dichromate***, and then followed by ***35 grams of silicon nitride***. Then re-seal the mixer, and continue to blend the mixture for 45 minutes. Thereafter, the mixture is ready for use. To use, press the mixture into any rocket motor or mold under a pressure of 8,000 psi, and then cure the rocket motor or mold in an oven at 74 Celsius for 1 hour. After 1 hour, remove the rocket motor or mold from the oven and allow it to cool.
**Burn rate:** 3 seconds per inch at 8,000 psi; tends to vary based on diameter and density of rocket grain.
**Water resistance:** Good.
**Stability:** Can be stored for decades.
**Sensitivity:** Stable towards friction; stable towards shock; stable to temperatures up to 300 Celsius.
**Flammability (1 to 10, un-contained):** 9 solid grain; 9.1 powdered grain.
**Ease of ignition (1 to 10 un-contained):** 8 solid grain; grain size varies and results in different ignition rates.
**Tendency to cake:** Will not cake.
**Detonation ability:** Can be detonated but only severe conditions—requires TNT, RDX, or HMX booster.
**Explosive ability (contained; explosive deflagration):** Moderate.
**Percentage:** ***76.39% ammonium perchlorate, 11.07% butadiene-1,3, 3.97% sulfur, 2.98% potassium dichromate, 2.25% silicon nitride, 2.18% stearic acid, 0.79% silicon monoxide, 0.26% magnesium oxide filler, 0.11% residual balance***
**Classification:** Deflagrating explosive (classified as propellant).

**Danger of preparation:** Can be made by amateurs, as long as the procedure is followed exactly.
**Use:** Used in military and commercial rockets and missiles for multiple uses.

| **Rocket assisted artillery shell** |
| --- |
| 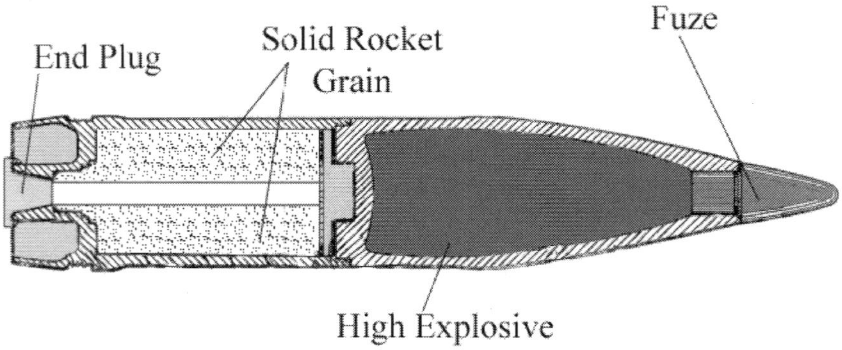  |
| **Description/Details:** |
| The above illustration shows the rocket grain, and explosive payload. The rocket is merely fired upon firing from the cannon and the resulting gas pressure produced by the propellant grain ignites the rocket grain. As previously stated, rocket assisted artillery shells have ranges extended by 2 to 3 kilometers on average. |

### 04-01-039A: High performance castable ammonium perchlorate rocket propellant:

Into a suitable heated vertical mixer, place ***170 grams of finely powdered PVC,*** followed by ***45 grams of dioctyl sebacate,*** and then followed by ***1100 grams of powdered ammonium perchlorate***. Thereafter, blend the mixture at moderate speed for about 2 hours to form a uniform mix. Thereafter, heat the mixture to 177 Celsius and blend the mixture until the mixture fuses. Once the mixture fuses, continue to blend the mixture for about 10 minutes at 177 Celsius. Immediately thereafter, the viscous free flowing mixture is ready for use. To use, pour and press fused mixture into any mold or rocket motor at 1,000 psi, and then allow the mold or rocket motor to cool and set for 48 hours.
**Burn rate:** 2.9 seconds per inch at 1,000 psi; tends to vary based on diameter and density of rocket grain.
**Water resistance:** Good.

**Stability:** Can be stored for decades.
**Sensitivity:** Stable towards friction; stable towards shock; stable to temperatures up to 300 Celsius.
**Flammability (1 to 10, un-contained):** 9.1 solid grain; 9.3 powdered grain.
**Ease of ignition (1 to 10 un-contained):** 8 solid grain; grain size varies and results in different ignition rates.
**Tendency to cake:** Will not cake.
**Detonation ability:** Can be detonated but only severe conditions—requires TNT, RDX, or HMX booster.
**Explosive ability (contained; explosive deflagration):** Moderate.
**Percentage:** ***83.65% ammonium perchlorate, 12.92% PVC, 3.42% dioctyl sebacate, 0.01% mixed impurities***
**Classification:** Deflagrating explosive (classified as propellant).

**Danger of preparation:** Can be made by amateurs, as long as the procedure is followed exactly; however, it is suggested that only experienced personnel attempt this procedure due to the heating process.
**Use:** Used in military and commercial rockets and missiles for multiple uses.

### 04-01-039B: High performance castable ammonium perchlorate rocket propellant (with silicon nitride and calcium carbide burn enhancer):

Into a suitable heated vertical mixer, place ***190 grams of finely powdered PVC,*** followed by ***56 grams of dioctyl sebacate,*** and then followed by ***1250 grams of powdered ammonium perchlorate***. Thereafter, blend the mixture for 30 minutes to form a uniform mix. Thereafter, add in ***50 grams of silicon nitride***, and then followed by ***45 grams of calcium carbide***, and then continue to blend the mixture for about 15 minutes. Thereafter, heat the mixture to 177 Celsius and blend the mixture until the mixture fuses. Once the mixture fuses, continue to blend the viscous mixture for about 10 minutes. Immediately thereafter, the mixture is ready for use. To use, pour and press the fused mixture into any mold or rocket motor at 1,000 psi, and then allow the mold or rocket motor to cool and set for 48 hours.
**Burn rate:** 2.9 seconds per inch at 1,000 psi; tends to vary based on diameter and density of rocket grain.
**Water resistance:** Good.
**Stability:** Can be stored for decades.
**Sensitivity:** Stable towards friction; stable towards shock; stable to temperatures up to 300 Celsius.
**Flammability (1 to 10, un-contained):** 9.1 solid grain; 9.3 powdered grain.
**Ease of ignition (1 to 10 un-contained):** 8 solid grain; grain size varies and results in different ignition rates.
**Tendency to cake:** Will not cake.
**Detonation ability:** Can be detonated but only severe conditions—requires TNT, RDX, or HMX booster.
**Explosive ability (contained; explosive deflagration):** Moderate.
**Percentage:** ***78.56% ammonium perchlorate, 11.94% PVC, 3.52% dioctyl sebacate, 3.14% silicon nitride, 2.83% calcium carbide, 0.01% impurities***
**Classification:** Deflagrating explosive (classified as propellant).

**Danger of preparation:** Can be made by amateurs, as long as the procedure is followed exactly; however, it is suggested that only experienced personnel attempt this procedure due to the heating process.
**Use:** Used in military and commercial rockets and missiles for multiple uses.

### 04-01-040A: High Performance ammonium perchlorate rocket propellant with nitrocellulose base:

Into a special sealable vertical mixer equipped with vacuum port, place ***200 grams of triacetin***, followed by ***50 grams of tri-butoxyethyl phosphate***, followed by ***250 grams of nitrocellulose*** (14% nitrogen content), and then followed by ***2.5 grams of lecithin***. Thereafter, blend the mixture on moderate speed for about 15 minutes to form a nice slurry. Thereafter, add in ***460 grams of powdered ammonium perchlorate***, and then blend the mixture on moderate speed for about 30 minutes. After blending for 30 minutes, seal the vertical mixer, and then de-gas the mixture by applying a vacuum of 100mm of mercury for 45 minutes; during the de-gassing process, continue to rapidly blend the mixture. After 45 minutes, remove the vacuum, equalize the pressure, and then open the vertical mixer. Now the mixture is ready for use. To use, the mixture needs to be poured and pressed into any desired rocket motor or engine in the usual manner under a pressure of 1,000 psi, and then the rocket motor or engine needs to be cured in an oven at 70 Celsius for about 2 days. A standard ignition composition can be used for proper burn.
**Burn rate:** 2.7 seconds per inch at 1,000 psi; tends to vary based on diameter and density of rocket grain.
**Water resistance:** Good.
**Stability:** Can be stored for decades.
**Sensitivity:** Stable towards friction; stable towards shock; stable to temperatures up to 300 Celsius.
**Flammability (1 to 10, un-contained):** 9.3 solid grain; 9.8 powdered grain.
**Ease of ignition (1 to 10 un-contained):** 9 solid grain; grain size varies and results in different ignition rates.
**Tendency to cake:** Will not cake.
**Detonation ability:** Can be detonated but only severe conditions—requires TNT, RDX, or HMX booster.

**Explosive ability (contained; explosive deflagration):** Above moderate.
**Percentage:** ***47.79% ammonium perchlorate, 25.97% nitrocellulose, 20.77% triacetin, 5.19% tri-butoxyethyl phosphate, 0.25% lecithin, 0.03% balance***
**Classification:** Deflagrating explosive (classified as propellant).

**Danger of preparation:** Can be made by amateurs, as long as the procedure is followed exactly; however, it is suggested that only experienced personnel attempt this procedure due to the flammability of nitrocellulose.
**Use:** Used in military and commercial rockets and missiles for multiple uses.

### 04-01-040B: High Performance ammonium perchlorate rocket propellant with nitrocellulose base:

First, into a suitable vertical mixer, place ***210 grams of nitrocellulose***, followed by ***37.5 grams of nitroglycerine***, and then followed by ***2.5 grams of diphenyl amine***. Thereafter, blend the mixture on moderate speed for about 15 minutes. Thereafter, into a separate sealable vertical mixer, equipped with vacuum port, place ***150 grams of triacetin***, followed by ***100 grams of methyl phthalyl ethyl glycolate***, followed by the 250 grams of nitrocellulose mixture (just prepared first), and then followed by ***2.5 grams of lecithin***. Thereafter, blend the mixture on moderate speed for about 15 minutes to form a nice slurry. After 15 minutes, add in ***470 grams of powdered ammonium perchlorate***, and then blend the mixture on moderate speed for about 30 minutes. Thereafter, seal the vertical mixer, and then de-gas the mixture by applying a vacuum of 100mm of mercury for 45 minutes; during the de-gassing process, continue to rapidly blend the mixture. After 45 minutes, remove the vacuum, equalize the pressure, and then open the vertical mixer. Now the mixture is ready for use. To use, the mixture needs to be poured and pressed into any desired rocket motor or engine under a pressure of 1,000 psi in the usual manner, and then cured in an oven at 70 Celsius for 7 days. A standard ignition composition can be used for proper burn.
**Burn rate:** 2.6 seconds per inch at 1,000 psi; tends to vary based on diameter and density of rocket grain.
**Water resistance:** Good.
**Stability:** Can be stored for decades.
**Sensitivity:** Stable towards friction; stable towards shock; stable to temperatures up to 150 Celsius.
**Flammability (1 to 10, un-contained):** 9.4 solid grain; 9.89 powdered grain.
**Ease of ignition (1 to 10 un-contained):** 9 solid grain; grain size varies and results in different ignition rates.
**Tendency to cake:** Will not cake.
**Detonation ability:** Can be detonated but only severe conditions—requires TNT, RDX, or HMX booster.
**Explosive ability (contained; explosive deflagration):** Above moderate.
**Percentage:** ***48.32% ammonium perchlorate, 21.59% nitrocellulose, 15.42% triacetin, 10.28% methyl phthalyl ethyl glycolate, 3.85% nitroglycerine, 0.25% lecithin, 0.25% diphenyl amine, 0.04% residual balance***
**Classification:** Deflagrating explosive (classified as propellant).

**Danger of preparation:** This procedure should only be carried out by experienced personnel due to the flammability of nitrocellulose, and the use of nitroglycerine. All safety and hazard info regarding nitroglycerine should be exercised.
**Use:** Used in military and commercial rockets and missiles for multiple uses.

### 04-01-040C: High Performance ammonium perchlorate rocket propellant with nitrocellulose base:

Into a sealable vertical mixer, equipped with vacuum port, place ***192 grams of triacetin***, followed by ***288 grams of diethyl phthalate***, followed by ***480 grams of nitrocellulose (average nitrogen content),*** followed by ***20 grams of lecithin***, and then followed by ***1680 grams of powdered ammonium perchlorate***; thereafter blend the mixture on high speed for about 45 minutes. Thereafter, seal the vertical mixer, and then de-gas the mixture by applying a vacuum of 100mm of mercury for 45 minutes; during the de-gassing process, continue to rapidly blend the mixture. After 45 minutes, remove the vacuum, equalize the pressure, and then open the vertical mixer. Now the mixture is ready for use. To use, the mixture needs to be poured and pressed into any desired rocket motor or engine under 1,000 psi in the usual manner, and then cured in an oven at 75 Celsius for 2 days. A standard ignition composition can be used for proper burn.
**Burn rate:** 2.7 seconds per inch at 1,000 psi; tends to vary based on diameter and density of rocket grain.
**Water resistance:** Good.
**Stability:** Can be stored for decades.
**Sensitivity:** Stable towards friction; stable towards shock; stable to temperatures up to 150 Celsius.
**Flammability (1 to 10, un-contained):** 9.3 solid grain; 9.8 powdered grain.
**Ease of ignition (1 to 10 un-contained):** 8.8 solid grain; grain size varies and results in different ignition rates.
**Tendency to cake:** Will not cake.
**Detonation ability:** Can be detonated but only severe conditions—requires TNT, RDX, or HMX booster.
**Explosive ability (contained; explosive deflagration):** Above moderate.
**Percentage:** ***63.15% ammonium perchlorate, 18.04% nitrocellulose, 10.82% diethyl phthalate, 7.21% triacetin, 0.75% lecithin, 0.03% residual balance***
**Classification:** Deflagrating explosive (classified as propellant).

**Danger of preparation:** Can be made by amateurs, as long as the procedure is followed exactly; however, it is suggested that only experienced personnel attempt this procedure due to the flammability of nitrocellulose.
**Use:** Used in military and commercial rockets and missiles for multiple uses.

### 04-01-040D: High Performance ammonium perchlorate rocket propellant with nitrocellulose base:

Into a sealable vertical mixer, equipped with vacuum port, place ***250 grams of bis-(3,5,5-trimethoxyhexyl) adipitate***, followed by ***250 grams of diethyl phthalate***, followed by ***500 grams of nitrocellulose (average nitrogen content),*** followed by ***25 grams of lecithin,*** and then followed by ***1000 grams of powdered ammonium perchlorate***, and then blend the mixture on moderate speed for about 45 minutes. Thereafter, seal the vertical mixer, and then de-gas the mixture by applying a vacuum of 100mm of mercury for 45 minutes; during the de-gassing process, continue to rapidly blend the mixture. After 45 minutes, remove the vacuum, equalize the pressure, and then open the vertical mixer. Now the mixture is ready for use. To use, the mixture needs to be poured and pressed into any desired rocket motor or engine under a pressure of 1,000 psi in the usual manner, and then cured in an oven at 75 Celsius for 2 days. A standard ignition composition can be used for proper burn.
**Burn rate:** 2.8 seconds per inch at 1,000 psi; tends to vary based on diameter and density of rocket grain.
**Water resistance:** Good.
**Stability:** Can be stored for decades.
**Sensitivity:** Stable towards friction; stable towards shock; stable to temperatures up to 150 Celsius.
**Flammability (1 to 10, un-contained):** 9.2 solid grain; 9.8 powdered grain.
**Ease of ignition (1 to 10 un-contained):** 8.9 solid grain; grain size varies and results in different ignition rates.
**Tendency to cake:** Will not cake.
**Detonation ability:** Can be detonated but only severe conditions—requires TNT, RDX, or HMX booster.
**Explosive ability (contained; explosive deflagration):** Above moderate.
**Percentage:** ***49.38% ammonium perchlorate, 24.69% nitrocellulose, 12.34% diethyl phthalate, 12.34% bis-(3,5,5-trimethoxyhexyl) adipitate, 1.23% lecithin, 0.02% residual impurities***
**Classification:** Deflagrating explosive (classified as propellant).

**Danger of preparation:** Can be made by amateurs, as long as the procedure is followed exactly; however, it is suggested that only experienced personnel attempt this procedure due to the flammability of nitrocellulose.
**Use:** Used in military and commercial rockets and missiles for multiple uses.

### 04-01-041A: Specialty high Performance ammonium perchlorate rocket propellant:

Into a suitable blender, equipped with stainless blades, place 400 milliliters of methanol, and then add and dissolve 43 grams of tris-1-(2-methyl aziridinyl) phosphine oxide, and then followed by 15 grams of tartaric acid. Thereafter, blend the mixture on low speed at room temperature for 48 hours. Thereafter, place this mixture onto a shallow pan or tray, and allow the methanol to evaporate; note: for commercial use, the mixture can be placed into a suitable vacuum apparatus, and the methanol recovered by removal under vacuum. Now, into a suitable vertical mixer, place ***272 grams of powdered ammonium perchlorate***, followed by ***64 grams of aluminum powder***, and then add in ***62.9 grams of a hydroxy-terminated liquid polybutadiene diisocyanate compound containing dioctyl adipate***, and then finally add in ***1.2 grams of the stock tartaric acid mixture*** previously prepared. Thereafter, blend the entire mixture on moderate speed for about 1 hour. After 1 hour, the mixture is ready for use. To use, the mixture needs to be pressed into any desired rocket motor, mold, or engine under a pressure of 4,000 psi in the usual manner, and then cure the rocket motor, mold, or engine in an oven at 60 Celsius for 24 hours.
**Burn rate:** 2.9 seconds per inch at 4,000 psi; tends to vary based on diameter and density of rocket grain.
**Water resistance:** Good.
**Stability:** Can be stored for decades.
**Sensitivity:** Stable towards friction; stable towards shock; stable to temperatures up to 150 Celsius.
**Flammability (1 to 10, un-contained):** 9.3 solid grain; 9.88 powdered grain.
**Ease of ignition (1 to 10 un-contained):** 8 solid grain; grain size varies and results in different ignition rates.
**Tendency to cake:** Will not cake.
**Detonation ability:** Can be detonated but only severe conditions—requires TNT, RDX, or HMX booster.
**Explosive ability (contained; explosive deflagration):** Moderate.
**Percentage:** ***67.9% ammonium perchlorate, 15.99% aluminum powder, 15.72% hydroxy terminated liquid resin binder, 0.2999% polymerized tartaric acid compound, 0.0901% residue***
**Classification:** Deflagrating explosive (classified as propellant).

**Danger of preparation:** Can be made by amateurs, as long as the procedure is followed exactly.
**Use:** Used in military and commercial rockets and missiles for multiple uses.

### 04-01-042A: High Performance ammonium perchlorate rocket propellant:

Into a heated vertical mixer, place ***248 grams of powdered ammonium perchlorate***, followed by ***246 grams of polychlorotrifluoroethylene***, and then followed by ***6 grams of paraffin wax***. Thereafter, heat the mixture to about 100 Celsius, and blend the mixture thoroughly throughout the heating process. Continue to heat the mixture at 100 Celsius for 20 minutes. Thereafter, the mixture is ready for casting. To do so, it needs to be poured into any desired rocket motor or mold at 1,000 psi, and then allow the rocket motor or mold to cool at room temperature for 24 hours. The propellant can be fired using any suitable ignition composition.
**Burn rate:** 3.9 seconds per inch at 1,000 psi; tends to vary based on diameter and density of rocket grain.
**Water resistance:** Good.
**Stability:** Can be stored for decades.
**Sensitivity:** Stable towards friction; stable towards shock; stable to temperatures up to 250 Celsius.
**Flammability (1 to 10, un-contained):** 8.5 solid grain; 9 powdered grain.
**Ease of ignition (1 to 10 un-contained):** 7.2 solid grain; grain size varies and results in different ignition rates.
**Tendency to cake:** Will not cake.
**Detonation ability:** Cannot be detonated.
**Explosive ability (contained; explosive deflagration):** Very low.
**Percentage:** ***49.6% ammonium perchlorate, 49.2% polychlorotrifluoroethylene copolymer, 1.2% paraffin wax binder***
**Classification:** Deflagrating explosive (classified as propellant).

**Danger of preparation:** Can be made by amateurs.
**Use:** Used in military and commercial rockets and missiles for multiple uses.

### 04-01-043A: High performance low temperature burning smokeless propellant:

Into a suitable blender, equipped with stainless steel blades, place 88.5 grams of adipic acid, followed by 86 grams of diethylene glycol, and then followed by 3.5 grams of maleic anhydride. Thereafter, blend the mixture for about 10 minutes. Thereafter, into a separate vertical mixer, place ***35.8 grams of this adipic acid/diethylene glycol/maleic anhydride mixture***, followed by adding in ***850 grams of powdered ammonium perchlorate***, followed by ***93.3 grams of n-butyl acrylate,*** followed by ***14.4 grams of methyl acrylate***, and then blend this mixture for about 15 minutes on high speed. Finally, add in ***4 grams of methyl amyl ketone peroxide,*** and then followed by ***2.5 grams of lecithin***. Thereafter, continue to blend the mixture on high speed for about 30 minutes at room temperature. Finally, the mixture is ready for casting. To do so, the mixture should be poured and pressed into any desirable rocket motor or mold under a pressure of 2,000 psi, and the following rocket motor or mold should then be cured in an oven at 105 Celsius for 24.
**Burn rate:** 4 seconds per inch at 2,000 psi; tends to vary based on diameter and density of rocket grain.
**Water resistance:** Good.
**Stability:** Can be stored for decades.
**Sensitivity:** Stable towards friction; stable towards shock; stable to temperatures up to 250 Celsius.
**Flammability (1 to 10, un-contained):** 8.4 solid grain; 8.7 powdered grain.
**Ease of ignition (1 to 10 un-contained):** 7 solid grain; grain size varies and results in different ignition rates.
**Tendency to cake:** Will not cake.
**Detonation ability:** Can be detonated but only severe conditions—requires TNT, RDX, or HMX booster.
**Explosive ability (contained; explosive deflagration):** Low.
**Percentage:** ***85% ammonium perchlorate, 9.33% n-butyl acrylate binder, 3.58% polyester binder compound, 1.44% methyl acrylate binder, 0.4% methyl amyl ketone peroxide cure catalyst, 0.25% lecithin wetting agent***
**Classification:** Deflagrating explosive (classified as propellant).

**Danger of preparation:** Can be made by amateurs, as long as the procedure is followed exactly.
**Use:** Used in military and commercial rockets and missiles for multiple uses where a visible smoke trail is not desired. Widely used in wire guided missiles.

**04-01-043B: High performance moderate temperature burning smokeless propellant:**
Into a suitable blender, equipped with stainless steel blades, place 74 grams of adipic acid, followed by 70.5 grams of diethylene glycol, and then followed by 5.3 grams of maleic anhydride. Thereafter, blend the mixture for about 10 minutes. Thereafter, into a separate vertical mixer, place ***123.5 grams of this adipic acid/diethylene glycol/maleic anhydride mixture***, followed by adding in ***750 grams of powdered ammonium perchlorate***, followed by ***123.5 grams of styrene,*** and then blend this mixture for about 10 minutes on moderate speed. Finally, add in ***2.5 grams of cumene hydroperoxide,*** and then followed by ***500 milligrams of lecithin***. Thereafter, continue to blend the mixture on moderate speed for 45 minutes at room temperature. Finally, the mixture is ready for casting. To do so, the mixture should be poured and pressed into any desirable rocket motor or mold under a pressure of 3,000 psi, and the following rocket motor or mold should then be cured in an oven at 105 Celsius for 24 hours.
**Burn rate:** 3.1 seconds per inch at 3,000 psi; tends to vary based on diameter and density of rocket grain.
**Water resistance:** Good.
**Stability:** Can be stored for decades.
**Sensitivity:** Stable towards friction; stable towards shock; stable to temperatures up to 250 Celsius.
**Flammability (1 to 10, un-contained):** 8.99 solid grain; 9.2 powdered grain.
**Ease of ignition (1 to 10 un-contained):** 7.4 solid grain; grain size varies and results in different ignition rates.
**Tendency to cake:** Will not cake.
**Detonation ability:** Can be detonated but only severe conditions—requires TNT, RDX, or HMX booster.
**Explosive ability (contained; explosive deflagration):** Moderate.
**Percentage:** ***75% ammonium perchlorate, 12.35% polyester binder, 12.35% styrene, 0.25% cumene hydroperoxide cure catalyst, 0.05% lecithin wetting agent***
**Classification:** Deflagrating explosive (classified as propellant).

**Danger of preparation:** Can be made by amateurs, as long as the procedure is followed exactly.
**Use:** Used in military and commercial rockets and missiles for multiple uses where a visible smoke trail is not desired. Widely used in wire guided missiles.

**04-01-044A: High performance smokeless propellant:**
Into a suitable vertical mixer, place ***250 grams of finely divided PVC*** (Geon 121), followed by ***10 grams of barium ricinoleate***, and then followed by ***250 grams of dibutyl sebacate***; thereafter, blend the mixture for about 15 minutes to form a uniform mixture. Thereafter, add in ***1530 grams of powdered ammonium perchlorate***, and then continue to blend the mixture for about 45 minutes at high speed. Thereafter, the viscous fluidized mixture is ready for casting. To do so, the mixture should be poured and pressed into any desirable rocket motor or mold under a pressure of 3,000 psi, and then the following rocket motor or mold should then be cured in an oven at 175 Celsius for 4 hours.
**Burn rate:** 3.2 seconds per inch at 3,000 psi, tends to vary based on diameter and density of rocket grain.
**Water resistance:** Good.
**Stability:** Can be stored for decades.
**Sensitivity:** Stable towards friction; stable towards shock; stable to temperatures up to 250 Celsius.
**Flammability (1 to 10, un-contained):** 8.89 solid grain; 9.1 powdered grain.
**Ease of ignition (1 to 10 un-contained):** 7.3 solid grain; grain size varies and results in different ignition rates.
**Tendency to cake:** Will not cake.
**Detonation ability:** Can be detonated but only severe conditions—requires TNT, RDX, or HMX booster.
**Explosive ability (contained; explosive deflagration):** Moderate.
**Percentage:** ***75% ammonium perchlorate, 12.25% PVC, 12.25% dibutyl sebacate, 0.49% barium ricinoleate stabilizer, 0.01% mixed balance***
**Classification:** Deflagrating explosive (classified as propellant).

**Danger of preparation:** Can be made by amateurs, as long as the procedure is followed exactly.
**Use:** Used in military and commercial rockets and missiles for multiple uses where a visible smoke trail is not desired. Widely used in wire guided missiles.

**04-01-045A: High performance smokeless propellant:**
Into a sealable heated vertical mixer, equipped with vacuum port, place ***2.7 grams of trimethylol propane***, followed by ***168 grams of polypropylene glycol***, followed by ***2.5 grams of phenyl beta-naphthylamine***, followed by ***5 grams of "alrosperse"*** (alkyl naphthalene sulfonate dispersing agent), and then followed by ***400 milligrams of ferric acetyl acetonate***. Thereafter, heat the mixture to 60 Celsius, and then blend the mixture at this temperature for 50 minutes. After 50 minutes, continue to blend the mixture at 60 Celsius, but seal the vertical mixer, and then de-gas the mixture by applying a vacuum of 100mm of mercury for 45 minutes; during the de-gassing process, continue to rapidly blend the mixture at 60 Celsius. After 45 minutes, reduce the temperature to 45 Celsius, quickly remove the vacuum, equalize the pressure, quickly open the vertical mixer, and then add in ***21.4 grams of toluene diisocyante***, followed by ***650 grams of powdered***

***ammonium perchlorate***, and then followed by ***150 grams of aluminum powder***. Immediately thereafter, re-seal the vertical mixer, and then continue to blend the mixture at 45 Celsius under vacuum for 45 minutes. After 45 minutes, remove the vacuum, equalize the pressure, and then open the vertical mixer. Now the mixture is ready for use. To do so, the mixture should immediately be poured and pressed into any desirable rocket motor or mold under a pressure of 2,000 psi, and the following rocket motor or mold should then be cured in an oven at 35 Celsius for several hours to allow for proper polymerization.
**Burn rate:** 3.4 seconds per inch at 2,000 psi; tends to vary based on diameter and density of rocket grain.
**Water resistance:** Good.
**Stability:** Can be stored for decades.
**Sensitivity:** Stable towards friction; stable towards shock; stable to temperatures up to 250 Celsius.
**Flammability (1 to 10, un-contained):** 8.8 solid grain; 9 powdered grain.
**Ease of ignition (1 to 10 un-contained):** 7.4 solid grain; grain size varies and results in different ignition rates.
**Tendency to cake:** Will not cake.
**Detonation ability:** Cannot be detonated.
**Explosive ability (contained; explosive deflagration):** Low.
**Percentage:** ***65% ammonium perchlorate, 16.8% polypropylen glycol plasticizer, 15% aluminum powder, 2.14% toluene diisocyante polymerization catalyst, 0.5% "alosperse" polymerization agent, 0.27% trimethylol propane binder aid, 0.25% phenyl beta-naphthylamine binder catalyst, 0.04% ferric acetyl acetonate catalyst***
**Classification:** Deflagrating explosive (classified as propellant).

**Danger of preparation:** Can be made by amateurs, as long as the procedure is followed exactly.
**Use:** Used in military and commercial rockets and missiles for multiple uses where a visible smoke trail is not desired. Widely used in wire guided missiles.

### 04-01-046A: Specialty high Performance polymer-based rocket propellant:

Into a suitable sealable vertical mixer, place ***94 grams of polyurethane,*** followed by ***80 grams of aluminum powder***, and then followed by ***320 grams of powdered ammonium perchlorate***. Thereafter, dry blend the mixture on high speed for 1 hour. Thereafter, add in ***6 grams of ethylene oxide***, seal the vertical mixer, and then blend the mixture on high speed for about 45 minutes. Thereafter, the mixture is ready for use. To use, the mixture simply needs to be pressed into any desirable rocket motor or mold under a pressure of 7,000 psi, and the resulting rocket motor or mold needs to cure in an oven at 60 Celsius for 72 hours.
**Burn rate:** 3.5 seconds per inch at 7,000 psi; tends to vary based on diameter and density of rocket grain.
**Water resistance:** Good.
**Stability:** Can be stored for decades.
**Sensitivity:** Stable towards friction; stable towards shock; stable to temperatures up to 250 Celsius.
**Flammability (1 to 10, un-contained):** 8.6 solid grain; 8.9 powdered grain.
**Ease of ignition (1 to 10 un-contained):** 7.9 solid grain; grain size varies and results in different ignition rates.
**Tendency to cake:** Will not cake.
**Detonation ability:** Cannot be detonated.
**Explosive ability (contained; explosive deflagration):** Low.
**Percentage:** ***64% ammonium perchlorate, 18.8% polyurethane, 16% aluminum powder, 1.2% ethylene oxide polymerization agent***
**Classification:** Deflagrating explosive (classified as propellant).

**Danger of preparation:** Can be made by amateurs, as long as the procedure is followed exactly.
**Use:** Used in military and commercial rockets and missiles for multiple uses.

| Multiple launch system (commonly found on helicopters) |
| --- |
| 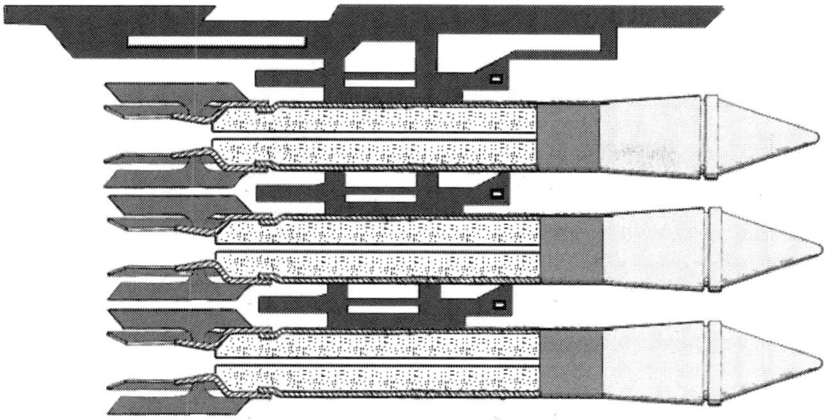 |
| **Details:** |
| The above illustration outlines a multiple rocket system using 120mm rockets, commonly found on helicopters or similar aircraft. These types of rockets are unguided, meaning they have no influence by laser or wire guidance. Un-guided rockets can be potent, but are less accurate. |

### 04-01-047A: Specialty high Performance "hypersonic" rocket propellant:

Into a suitable kitchen blender, equipped with razor sharp stainless steel blades, place ***174 grams of aluminum powder***, followed by ***422 grams of ammonium perchlorate***, and then followed by ***17 grams of finely divided graphite***. Thereafter, turn on the blender to medium speed and blend for 5 to 7 seconds, then stop; thereafter, tap and rotate the blender container, place back onto the blender machine, then turn on the blender to medium speed and blend for 5 to 7 seconds; thereafter, stop; tap and rotate the blender container, place back onto the blender machine, then turn on the blender to medium speed and blend for 5 to 7 seconds. Thereafter, stop blending and allow the ingredients to stand for 30 minutes. After 30 minutes, turn on the blender to medium speed and blend for 5 to 7 seconds, then stop; thereafter, tap and rotate the blender container, place back onto the blender machine, then turn on the blender to medium speed and blend for 5 to 7 seconds; thereafter, stop; tap and rotate the blender container, place back onto the blender machine, then turn on the blender to medium speed and blend for 5 to 7 seconds. Thereafter, stop blending and allow the ingredients to stand for 30 minutes. After 30 minutes, turn on the blender to medium speed and blend for 5 to 7 seconds, then stop; thereafter, tap and rotate the blender container, place back onto the blender machine, then turn on the blender to medium speed and blend for 5 to 7 seconds; thereafter, stop; tap and rotate the blender container, place back onto the blender machine, then turn on the blender to medium speed and blend for 5 to 7 seconds. Thereafter, stop blending and allow the ingredients to stand for 30 minutes. Now, allow the ingredients in the blender to stand for 2 hours; thereafter, repeat the aforementioned process 3 more times. Note: the idea here is to thoroughly blend the ingredients without build up of heat. Once finished, the process is complete and the "proper" rocket powder has been prepared. Thereafter, open the blender container, and pour the powdered mixture into a "zip lock" bag, and seal air-tight (for storage). Note: The "zip lock" bag can be stored at room temperature, below room temperature, or can be stored at a temperature below 130 Fahrenheit for as long as necessary. In any sense, storing the powder is usually carried out at room temperature; as long as the "Zip lock" bag is sealed air-tight (this is usually double bagged for prolonged storage), the powder will never cake, and will remain a fresh free flowing powder. Now, to prepare the rocket grain, pour the powdered mixture directly into a pre-fabricated rocket motor under a pressure of 12,000 psi.

**Proper blender blades for mixing ammonium perchlorate, powdered aluminum, and graphite (do not use serrated blades; the stainless steel blades should be smooth).**

**Burn rate:** 0.2445 seconds per inch at 12,000 psi; tends to vary based on diameter and density of rocket grain.
**Water resistance:** Good.
**Stability:** Can be stored for decades.
**Sensitivity:** Stable towards friction; stable towards shock; stable to temperatures up to 200 Celsius (pressed grain at 12,000 psi); powdered mixture will flash when subjected to external heat of 160 Celsius.
**Flammability (1 to 10, un-contained):** 10 solid grain; 10 powdered grain.
**Ease of ignition (1 to 10 un-contained):** 10 solid grain.
**Tendency to cake:** Will not cake.
**Detonation ability:** Can be detonated under normal conditions.
**Explosive ability (contained; explosive deflagration):** Detonates: 1,989 meters per second.
**Percentage:** ***68.84% ammonium perchlorate, 28.38% aluminum powder, 2.77% graphite, 0.01% residue***
**Classification:** Deflagrating explosive (classified as explosive).

**Danger of preparation:** Can be made by amateurs, as long as the procedure is followed exactly; however, it advisable that person or persons exercise extreme caution.
**Use:** Used in experimental hypersonic military missiles for shooting down "hypersonic" jet aircraft.

**Hypersonic anti-aircraft guided missile**

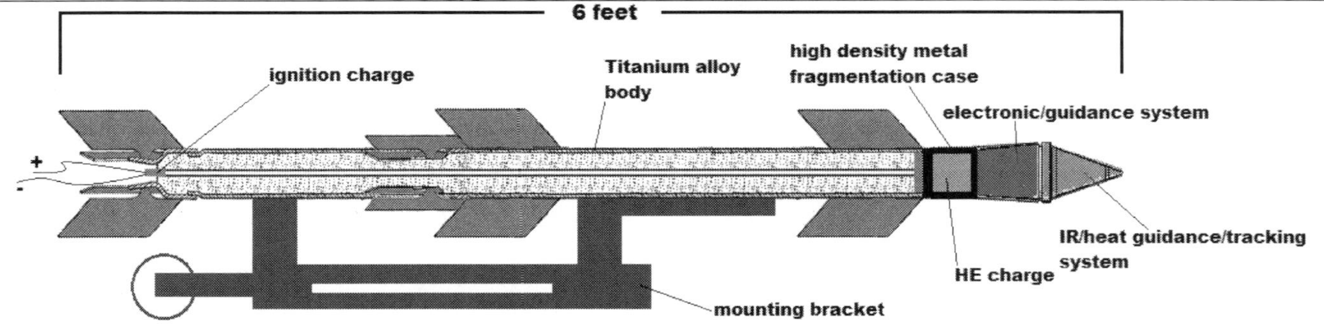


**Details:**

This illustration outlines a hypersonic experimental missile that can travel at speeds in excess of Mach 5.2. The length of the missile ranges from 6 feet to 9 feet, and is fired from a mounted platform attached to a vehicle. The range of the missile fluctuates from 3.4 to 6.9 kilometers, with a maximum slant angle of 10,000 feet.

# Section 2: Ammonium Dinitramide (ADN) Rocket Propellants

## *Chemicals used in this section (binders are not included)*

| | | |
|---|---|---|
| **1. Potassium nitrate (see Black Powder)** | | **2. Sulfur (see Black Powder)** |
| **3. Charcoal (see Black Powder)** | | **4. Sugar Carbon (see Modified Black Powder)** |
| **5. Barium Chromate (see Modified Black Powder)** | | **6. Potassium Perchlorate (see Modified Black Powder)** |
| **7. Potassium Dichromate (see Modified Black Powder)** | | **8. Ammonium Bisulfide (see Modified Black Powder)** |
| **9. Potassium chlorate (see Modified Black Powder)** | | **10. Carbon Disulfide (see Modified Black Powder)** |
| **11. Nitrocellulose (see Modified Black Powder)** | | **12. Lead Tetraoxide (see Modified Black Powder)** |
| **13. Diphenylamine (see Modified Black Powder)** | | **14. Titanium Dioxide (see Modified Black Powder)** |
| **15. Manganese Dioxide (see Modified Black Powder)** | | **16. Sodium Benzoate (see Modified Black Powder)** |
| **17. Ammonium Nitrate (see Modified Black Powder)** | | **18. Calcium Carbonate (see Modified Black Powder)** |
| **19. Sodium Nitrate (see Modified Black Powder)** | | **20. Urea (see Modified Black Powder)** |
| **21. Lead Nitrate (see Modified Black Powder)** | | **22. Nitro Starch (see Modified Black Powder)** |
| **23. Ammonium Perchlorate (see Ammonium Perchlorate Rocket Propellants)** | | **24. Aluminum powder (see Ammonium Perchlorate Rocket Propellants)** |
| **25. Magnesium Sulfide (see Ammonium Perchlorate Rocket Propellants)** | | **26. Copper Chromite (see Ammonium Perchlorate Rocket Propellants)** |
| **27. Nitroguanidine (see Ammonium Perchlorate Rocket Propellants)** | | **28. Ammonium Sulfate (see Ammonium Perchlorate Rocket Propellants)** |
| | | |

| | | |
|---|---|---|
| **29. Nitroglycerine (see Ammonium Perchlorate Rocket Propellants)** | | **30. Magnesium Oxide (see Ammonium Perchlorate Rocket Propellants)** |
| **31. Iron-II-oxide (see Ammonium Perchlorate Rocket Propellants)** | | **32. Zirconium Hydride (see Ammonium Perchlorate Rocket Propellants)** |
| **33. Teflon (see Ammonium Perchlorate Rocket Propellants)** | | **34. Zinc Oxide (see Ammonium Perchlorate Rocket Propellants)** |
| **35. Ammonium Dichromate (see Ammonium Perchlorate Rocket Propellants)** | | **36. Lithium Perchlorate (see Ammonium Perchlorate Rocket Propellants)** |
| **37. Magnesium powder (see Ammonium Perchlorate Rocket Propellants)** | | **38. Lithium Aluminum Hydride (see Ammonium Perchlorate Rocket Propellants)** |
| **39. Iron-III-oxide (red iron oxide) (see Ammonium Perchlorate Rocket Propellants)** | | **40. PVC (see Ammonium Perchlorate Rocket Propellants)** |
| **41. Copper Sulfide (see Ammonium Perchlorate Rocket Propellants)** | | **42. Sodium Hydride (see Ammonium Perchlorate Rocket Propellants)** |
| **43. Titanium Hydride (see Ammonium Perchlorate Rocket Propellants)** | | **44. Silicon Nitride (see Ammonium Perchlorate Rocket Propellants)** |
| **45. ADN** | | |
| 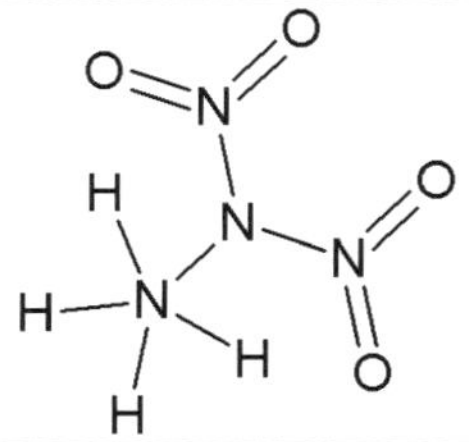 | | |
| ADN is a remarkable low velocity explosive which was first synthesized in 1971 by American chemists, but with little or no attention towards its explosive potential; however, German scientists declared its syntheses as early as 1956; in any case, since then, ADN has blossomed into a highly valuable secondary explosive primarily used as a substitute for ammonium nitrate explosives as ADN has a higher detonating velocity, has greater stability, shock and percussion insensitivity, great environmental persistence (where as ammonium nitrate has horrible environmental persistence; readily cakes, cracks, and easily dissolves in water), also has great brisance and has a greater solid-to-gas ratio with excellent shattering power. One interesting property ADN has, is that it is stable at ordinary and elevated environmental conditions, but does not require RDX/nitramine based blasting caps or detonators for proper initiation; whereas in mining and blasting operations, ammonium nitrate requires RDX or other nitramine base charges wired with detonating cord for proper ignition. In the past 7 years, the manufacture of ADN has remarkable increased, but its production in the US is limited; however, other countries like China, Russia, Germany, and France are producing larger quantities (estimated total annual manufacture: 180,000 metric tones). ADN, like ammonium nitrate can be alloyed with nitrobenzenes, such as TNT to form borehole blasting charges for mining, make-shift cratering, and for multi purpose demolition operations for both commercial and military importance. Another remarkable use for ADN is in the pyrotechnic field, where it is used in making high performance and smokeless gun propellants and rocket fuels; the latter has superior ballistic energy over ammonium nitrate based rocket fuels. Many other salts of ADN have been prepared, and will be discussed. ADN forms colorless to white to lightly yellowish crystals, off-white powder, or granules; no melting point has been achieved as upon heating the salt decomposes. The salt is only slightly soluble in water (solubility in water is increased by addition of nitric acid), with little solubility in the usual organic solvents, but has moderate solubility in absolute alcohol. In a non-explosive use, ADN is used in limited quantities (as of 2013) in making fireworks and other pyrotechnic compositions. Note: mixtures of ADN and aluminum powder form a powerful explosive with a detonating velocity of 3,980 meters per/second; this composition can be used as a detonator for other secondary explosives. **Danger of preparation:** can be prepared by non-scientists as long as all procedure(s) and safety info is strictly followed. | | |
| | | |

## Method of Preparation 1:

**Step 1:** Preparation of urea nitrate

Into a medium sized beaker, equipped with mechanical stirrer, add 100 milliliters of 33% diluted nitric acid, and then add in, in small portions, 28 grams of urea while stirring the 33% nitric acid vigorously. During the addition of the urea, the reaction temperature does not show any practical rise, so this reaction can be carried out at room temperature without cooling means. Immediately upon addition, urea nitrate forms as crystals. After the addition of the urea, continue to rapidly stir the reaction mixture for 15 minutes; thereafter, remove the mechanical stirrer, and then place the beaker into a freezer and let chill for 4 hours. After 4 hours, remove the beaker from the freezer, and then filter-off the precipitated crystals using a clean Buchner funnel, under moderate suction, and then suction dry as best as possible. Finally, remove the crystals from the Buchner funnel, and place them on a shallow pan or tray and allow them to air dry. Once dry they are ready for use in step 2.

**Step 2:** Preparation of nitrourea

Into a 3-neck flask, equipped with mechanical stirrer and thermometer, pour in 168 milliliters of 98% sulfuric acid, and chill the 3-neck flask in an ice/salt bath. Once the acid has chilled to -5 to 0 Celsius, add and dissolve in small portions at a time, 48 grams of urea nitrate (prepared in step 2) while rapidly stirring the 98% sulfuric acid mixture and maintaining its temperature below 20 Celsius. After the addition of the urea nitrate, remove the cooling bath, and then continue to stir the reaction mixture for an additional 1 hour until it warms to room temperature. Thereafter, stop stirring, and then pour the entire reaction mixture onto 750 grams of crushed ice contained in a suitable sized beaker, allow the ice to melt, then mechanically stir this diluted reaction mixture for 30 minutes. After 30 minutes, filter-off the precipitated crystals using a clean Buchner funnel under moderate suction, wash with 500 milliliters of pre-chilled ice cold water, followed by three 500-milliliter portions of 10% baking soda solution, followed by six 500-milliliter portions of warm water, and then suction dry as best as possible. Finally, remove the crystals from the Buchner funnel, and place them on a shallow pan or tray and allow them to air dry. Once dry they are ready for use in step 3.

**Step 3:** Preparation of ADN

In to a suitable sized beaker, equipped with mechanical stirrer and thermometer, add in the nitrourea (prepared in step 2), and then add in 200 milliliters of acetonitrile, turn on the mechanical stirrer to high speed, and then place the beaker into an ice bath. Once the ingredients have cooled, gradually add in 41 grams of nitronium tetrafluoroborate over a period of about 45 to 60 minutes, while continuing to rapidly stir the contents in the beaker (reaction mixture), while maintaining its temperature below 20 Celsius at all times. After addition of the nitronium tetrafluoroborate, continue to rapidly stir the reaction mixture while maintaining its temperature below 25 Celsius for about 1 hour. After 1 hour, remove the cooling bath, and then saturate the reaction mixture by bubbling into it, 25 to 30 grams of pure ammonia gas while continuing to rapidly blend the reaction mixture. Note: stop the addition of the ammonia gas when the odor of ammonia can be detected. After the addition of ammonia gas, continue to rapidly stir the reaction mixture for 45 minutes at room temperature, and then filter-off the by-product solids, and discard them. Now, pour the entire reaction mixture into a rotary evaporator, and evaporate-off the liquids at 50 Celsius under a high vacuum of 56mm of mercury, but only until 70% of the fluids have been removed by evaporation. When this point is achieved, stop the evaporation, equalize the pressure in the usual manner, and then recover the left over cloudy messy concentrated reaction mixture and pour into a clean suitable sized beaker, equipped with mechanical stirrer, and then add in 200 milliliters of ethyl acetate and the stir on moderate speed for about 10 minutes. After 10 minutes, stop stirring, and allow the mixture to stand for 10 minutes at room temperature; thereafter, immediately filter-off the insoluble by-products using a clean Buchner funnel under moderate suction. Now, pour the entire

## Method of Preparation 2:

Into a 3-neck flask, fitted with mechanical stirrer, thermometer, and wide mouth funnel, pour in 250 milliliters of acetonitrile, and then add 30 grams of nitronium tetrafluoroborate; thereafter, place the 3-neck flask into a dry ice/acetone bath, turn on the mechanical stirrer to high speed, and then allow the reaction mixture to chill to -20 Celsius by means of the cooling bath. Once all the nitronium tetrafluoroborate has dissolved into the acetonitrile, and the temperature of this reaction mixture has chilled down to about -20 Celsius, slowly add in, in small portions at a time, 8 grams of ammonium carbamate over a period of time sufficient to keep the temperature of the nitronium tetrafluoroborate reaction mixture at -20 to -10 Celsius at all times under rapid stirring. After the addition of the ammonium carbamate, continue to rapidly stir the reaction mixture and maintain its temperature around -20 to -5 Celsius for 1 hour. In the meantime, into a suitable large sized beaker, equipped with mechanical stirrer, prepare a special solution by adding and dissolving 26 grams of anhydrous ammonia into 480 milliliters of 99% isopropyl alcohol (simply bubble dry ammonia gas into the alcohol; do not use liquid ammonia), and then add to this ammonia/alcohol solution, 2000 milliliters of fresh diethyl ether and then turn on the mechanical stirrer to high speed and rapidly stir this ammonia/alcohol/ether solution at room temperature. Now, after the reaction mixture has been stirred for 1 hour, stop stirring, remove the dry ice/acetone bath, and then immediately, yet gradually pour this chilled reaction mixture into the ammonia/alcohol/ether solution while rapidly stirring the ammonia/alcohol/ether solution and maintaining its temperature below 30 Celsius at all times. After the addition, continue to stir this now combined reaction mixture on high speed for about 1 hour at room temperature. After stirring for 1 hour, stop stirring, and then place the mixture into a rotary evaporator, and evaporate-off the solvents at 30 Celsius under high vacuum of 65mm of mercury until dry solid remains. When no more liquids pass over or are collected, and a flocculent mass is left behind, stop the rotary evaporator, equalize the pressure in the usual manner, and then recover the left over residue from the rotary evaporator, and place into a suitable sized beaker, equipped with mechanical stirrer. Thereafter, pour in a solvent mixture pre-prepared by adding and dissolving 50 milliliters of acetone with 50 milliliters of ethyl acetate, and stir this new solvent mixture on high speed for about 30 minutes at room temperature. After 30 minutes of stirring, filter-off the insoluble impurities using a clean Buchner funnel under strong suction, and then pour the filtered solvent mixture into a clean rotary evaporator and evaporate-off the solvents at 50 Celsius under a vacuum of 170mm of mercury until dry solid remains When all the liquid solvents have evaporated and passed over, and a clean solid remains, stop the rotary evaporator, equalize the pressure in the usual manner, and then recover the left over product from the rotary evaporator, and then recrystallize it from 20 milliliters of n-butanol, collecting each crop of crystals in the same Buchner funnel, and once the volume of the n-butanol solvent mixture has been reduced down to 20% its total remaining volume (mother liquor), stop the recrystallization process, and then suction dry all the collected crystals of the desired product as best as possible; thereafter, place this suction dried desired product onto a shallow pan or tray, and allow it to air dry thoroughly until completely dry. Thereafter, collect the dry crystals and store in a suitable container in a cool dry place until use.

## Method of Preparation 3:

Into two different suitable sized beakers, dissolve 10 grams of KDN into 20 milliliters of water in the first beaker, and then in the second beaker dissolve 10 grams of ammonium sulfate into 20 milliliters of water; stir each solution by any means. Thereafter, simply mix both solutions with manual stirring. A white precipitate of potassium sulfate will immediately form. After mixing both solutions, stir the combined mixture for several minutes at room temperature, and then add in 200 milliliters of 99% isopropyl alcohol, and continue to stir the mixture for 30 minutes at room temperature. Thereafter, filter-off the precipitated potassium sulfate using a clean Buchner funnel under rapid suction, and then place the filtered reaction mixture into a beaker, and then place this beaker in a freezer, and chill at 0 Celsius for 12 hours. After 12 hours, remove the beaker from the freezer, and then add to it, 300 milliliters of petroleum ether and then stir the cold mixture for 15 minutes. Thereafter, filter-off the precipitated product using a clean Buchner funnel under moderate suction, wash with three 150-milliliter portions of fresh petroleum ether, reaction mixture into a clean rotary evaporator, and evaporate-off the liquids at 40 Celsius under a high vacuum of 77mm of mercury, but only until 70% of the fluids have been removed by evaporation. When his point is achieved, stop the evaporation, equalize the pressure in the usual manner, and then recover the left over cloudy messy concentrated mixture and pour into a clean suitable sized beaker, equipped with mechanical stirrer. Thereafter add in 200 milliliters of dry chloroform, and then turn on the mechanical stirrer to high speed, and blend the solvent mixture on high speed for 30 minutes at room temperature. Note: after addition of the chloroform, the desired ADN product will precipitate. After stirring the solvent mixture for about 30 minutes, stop stirring, and then filter-off the precipitated crystals of ADN using a clean Buchner funnel, under moderate suction, wash with three 250-milliliter portions of dry chloroform, and then suction dry as best as possible. Finally, remove the crystals from the Buchner funnel, and place them on a shallow pan or tray and allow them to air dry. Once dry, store the ADN product in glass containers and store in a cool dry place until use.

Note: Many other salts of ADN can be prepared by replacing dry ammonia gas with another base such as metal hydroxides or carbonates—preferably carbonates rather then hydroxides as strong hydroxides may decrease yields as the result of hydrolysis. For more info on ADN, see The Preparatory Manual of Chemical Warfare Agents Fourth Edition.

and then suction dry as best as possible. Thereafter, place this suction dried desired product onto a shallow pan or tray, and allow it to air dry thoroughly until completely dry. Thereafter, collect the dry crystals and store in a suitable container in a cool dry place until use.

## 46. Urea Nitrate

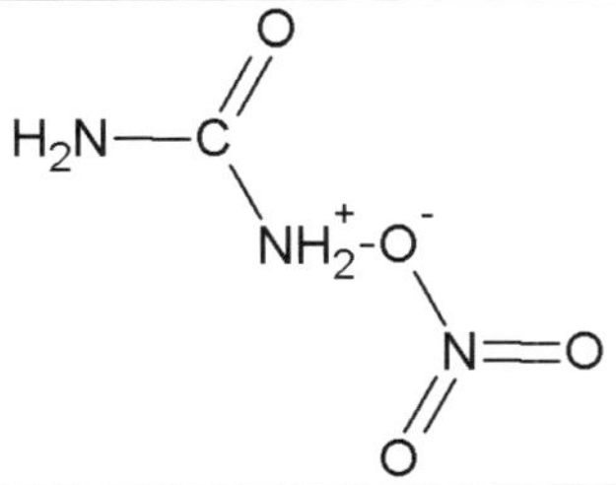


Urea Nitrate forms colorless, odorless leaflets with a melting point of 152 Celsius. The crystals are insoluble in cold water, but soluble in hot. The nitrate has been known to explode when struck or rubbed while heating.

**Method of preparation 1:** Into a medium sized beaker, equipped with mechanical stirrer, add 100 milliliters of 33% diluted nitric acid, and then add in, in small portions, 28 grams of urea while stirring the 33% nitric acid vigorously. During the addition of the urea, the reaction temperature does not show any practical rise, so this reaction can be carried out at room temperature without cooling means. Immediately upon addition, urea nitrate forms as crystals. After the addition of the urea, continue to rapidly stir the reaction mixture for 15 minutes; thereafter, remove the mechanical stirrer, and then place the beaker into a freezer and let chill for 4 hours. After 4 hours, remove the beaker from the freezer, and then filter-off the precipitated crystals using a clean Buchner funnel, under moderate suction, and then suction dry as best as possible. Finally, remove the crystals from the Buchner funnel, and place them on a shallow pan or tray and allow them to air dry. Once dry they are ready for use in step 2.

## 47. Silicon Powder

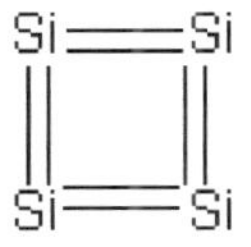


Silicon powder is a black to grayish material. The powder is stable at room temperature and is much more resistant to chemical attack then other powders. It is insoluble in water and the usual solvents.

## 48. Hexamine

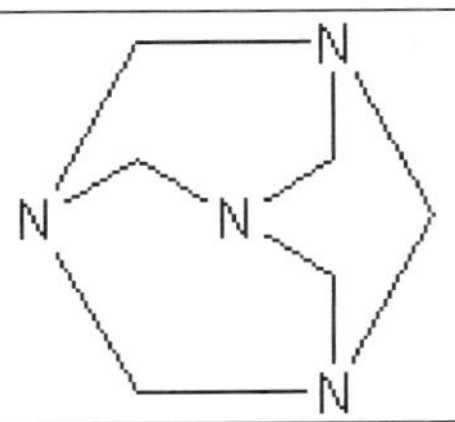


Hexamine, also known as methenamine, forms colorless to white crystals, granules, or powder. The melting point cannot be calculated at the crystals sublime upon heating. Hexamine is soluble in water, and slightly soluble in alcohol, but insoluble in most other solvents under the usual means.

## 49. Boron powder

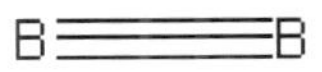


Boron powder is black to grayish-black material with a beautiful metallic luster. The powder is insoluble in water and the usual organic solvents. Boron powder is much less reactive then other powdered metals, but is capable of reacting with acids and bases.

## 50. Sodium hypophosphite

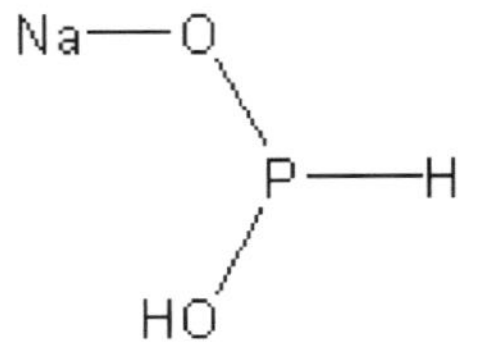

Sodium hypophosphite forms a monohydrate, which is a white, odorless granules or crystalline solid. The salt decomposes when strongly heated into phosphine, which bursts into flames. The salt is soluble in water, alcohol, and glycerol, but relatively insoluble in most organic solvents. Sodium hypophosphite can explode when mixed with a strong oxidizing agent and ignited.

## 51. KDN

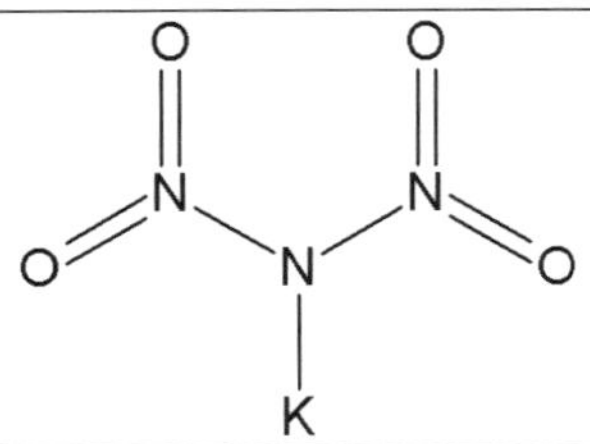

KDN forms beautiful colorless transparent rhombic bypyramidal crystals, or white powder or granules. Impure KDN may have a slight yellowish tint. The crystals are amazingly somewhat soluble in water, but insoluble in most common organic solvents, with moderate solubility in liquid ammonia and liquid sulfur dioxide. KDN is primarily used in formulating high performance gun and rocket propellants, but it is used in small amounts as an explosive for small time rock blasting and mining operations; KDN forms highly explosive mixtures with powdered metals; these compositions burn extremely fast, 0.009 inches per/second, and they detonate when confined and ignited; average detonation velocity equals around 3,800 meters per/second. One very useful trait that KDN has, is its ability to self deflagrate, this provides added ballistic energy when used in making high performance rocket fuels when admixed with nitrocellulose, wood meal powder, corn starch, sugar, and even flour. Note: mixtures of powdered metals with KDN cannot be used in gun or rocket propellants, as detonations will often occur destroying the firearm and/or rocket casing. **Danger of preparation:** can be prepared by non-scientists as long as all procedure(s) and safety info is strictly followed.

**Method of Preparation 1:** Into a 3-neck flask, fitted with mechanical stirrer, thermometer, and wide mouth funnel, pour in 250 milliliters of acetonitrile, and then add 30 grams of nitronium tetrafluoroborate; thereafter, place the 3-neck flask into a dry ice/acetone bath, turn on the mechanical stirrer to high speed, and then allow the reaction mixture to chill to -20 Celsius by means of the cooling bath. Once all the nitronium tetrafluoroborate has dissolved into the acetonitrile, and the temperature of this reaction mixture has chilled down to about -20 Celsius, slowly add in, in small portions at a time, 8 grams of ammonium carbamate over a period of time sufficient to keep the temperature of the nitronium tetrafluoroborate reaction mixture at -20 to -10 Celsius at all times under rapid stirring. After the addition of the ammonium carbamate, continue to rapidly stir the reaction mixture and maintain its temperature around -20 to -5 Celsius for 1 hour. In the meantime, into a suitable large sized beaker, equipped with mechanical stirrer, prepare a special solution by adding and dissolving 208 grams of potassium carbonate into 420 milliliters of warm water, and then add to this potassium carbonate solution, 2000 milliliters of fresh diethyl ether and then turn on the mechanical stirrer to high speed and rapidly stir this potassium carbonate/water/ether solution at room temperature. Now, after the reaction mixture has been stirred for 1 hour, stop stirring, remove the dry ice/acetone bath, and then immediately, yet gradually pour this chilled reaction mixture into the potassium carbonate/water/ether solution while rapidly stirring the potassium carbonate/water/ether solution and maintaining its temperature below 30 Celsius at all times. After the addition, continue to stir this now combined reaction mixture on high speed for about 1 hour at room temperature. After stirring for 1 hour, stop stirring, and then place the mixture into a rotary evaporator, and evaporate-off the solvents at 70 Celsius under high vacuum of 60mm of mercury until dry solid remains. When no more liquids pass over or are collected, and a flocculent mass is left behind, stop the rotary evaporator, equalize the pressure in the usual manner, and then recover the left over residue from the rotary evaporator, and place into a suitable sized beaker, equipped with mechanical stirrer. Thereafter, pour in 50 milliliters of distilled water, stir for 10 minutes, and then filter-off the insoluble impurities using a clean Buchner funnel under moderate suction. Thereafter, add to this filtered aqueous mixture a solvent mixture pre-prepared by adding and dissolving 50 milliliters of acetone with 50 milliliters of ethyl acetate, and stir this new solvent mixture on high speed for about 30 minutes at room temperature. After 30 minutes of stirring, filter-off the precipitated desired product using a clean Buchner funnel under moderate suction, and then suction dry as best as possible. Finally, place this suction dried desired product onto a shallow pan or tray, and allow it to air dry thoroughly until completely dry. Thereafter, collect the dry crystals and store in a suitable container in a cool dry place until use.

## Method of Preparation 2:

**Step 1:** Preparation of the potassium salt of Sulfamic Acid (potassium sulfamate)

Into a suitable sized beaker, equipped with mechanical stirrer (motorized stirrer), pour in 100 milliliters of water, and then add in 141 grams of sulfamic acid and then turn on the mechanical stirrer and begin to blend the suspension at room temperature. Thereafter, into a second clean beaker, prepare a second mixture by adding and dissolving 88 grams of potassium hydroxide into 100 milliliters of water. Note: dissolving the potassium hydroxide in water generates allot of heat, so allow this alkaline solution to cool to room temperature before using. Once it has cooled to room temperature, add the potassium hydroxide solution to the sulfamic acid suspension while continuing to rapidly stir the sulfamic acid mixture and maintaining its temperature at R.T. After the addition, continue to stir the combined mixture for 15 minutes at room temperature, and then pour the entire mixture into 428 milliliters of 95% ethanol, contained in a suitable sized beaker and equipped with mechanical stirrer, and stir the ethyl alcohol during the addition. After the addition, continue to rapidly stir the reaction mixture at room temperature for 30 minutes, and then filter-off the precipitated potassium salt of sulfamic acid using a Buchner funnel under moderate suction, wash with 100 milliliters of 95% ethanol, several times, using the same washing portion, and then suction dry as best as possible (vacuum dry). Finally, recover the suction dried product from the Buchner funnel, and place it onto a shallow pan or tray and allow it to thoroughly air dry; once dry, recover the dried product, grind it up into a fine powder, and proceed with step 2.

**Step 2:** Preparation of KDN

Into a suitable sized beaker equipped with mechanical stirrer, and sitting inside a dry ice/acetone bath at -30 Celsius, pour in 90 milliliters of 90% nitric acid, and then slowly and carefully add in 32 milliliters of 98% sulfuric acid. Thereafter, allow this nitrating acid solution to chill to around -30 Celsius, and once this cold temperature is achieved, slowly add in, small portions at a time, 34 grams of the potassium salt of sulfamic acid (prepared in step 1), in 500 milligram portions, while rapidly stirring the nitrating reaction mixture and maintaining its temperature below -20 Celsius at all times. After the addition of the potassium salt to the nitrating acid mixture, remove the beaker from the dry ice/acetone bath, and then immediately pour the entire reaction mixture into a beaker filled with 750 grams of ice, allow the ice to melt, and then mechanically stir this diluted mixture on high speed for about 45 minutes. Immediately thereafter, place this beaker containing the diluted reaction mixture back into the same dry ice/acetone bath, and chill to - 20 Celsius under rapid stirring. Thereafter, immediately pour in, a pre-prepared solution of potassium hydroxide, prepared by adding and dissolving 18 grams of potassium hydroxide into 40 milliliters of ice-cold water. During the addition rapidly stir the reaction mixture and maintain its temperature below -5 to 0 Celsius at all times.. After the addition of the potassium hydroxide solution, continue to stir the reaction mixture on high speed for about 30 minutes at 0 Celsius. Thereafter, remove the dry ice/acetone bath, and allow the reaction mixture to warm to room temperature; during this warm-up period, continue to rapidly stir the reaction mixture. Once the reaction mixture has warmed to room temperature, add in 200 milliliters of acetone, followed by 250 milliliters of 95% ethyl alcohol, and then continue to stir for about 45 minutes at room temperature. After stirring for about 45 minutes, filter-off the insoluble materials, using a clean Buchner funnel under moderate suction, wash with three 150-milliliter portions of 5% potassium carbonate solution, followed by washing with three 150-milliliter portions of fresh 99% isopropyl alcohol, and then suction dry as best as possible. Thereafter, recover the suction dried product from the Buchner funnel, and dissolve this solid material into 150 milliliters of cold water, stir for 5 minutes then quickly filter this aqueous solution using a clean Buchner funnel under rapid suction (to remove insoluble by-products), and then suction dry as best as possible. Thereafter, recrystallize the desired product from this filtered aqueous mixture, collecting each crop of crystals in the same Buchner funnel, and once the

| | | |
|---|---|---|
| | | volume of the aqueous mixture has been reduced down to 20% its total remaining volume (mother liquor), stop the recrystallization process, and then suction dry the collected crystals as best as possible. Finally, recover the suction dried crystals from the Buchner funnel, and place onto a shallow pan or tray and allow to air dry thoroughly. Once thoroughly dry, recover the dried crystals and store in a glass or plastic bottle in a cool dry place until use. |

***- High Performance Ammonium Dinitramide Propellants in this section -***

| | |
|---|---|
| **1. 04-02-001A High Performance Military grade rocket propellant of ADN:** *59% ADN, 24% nitrocellulose, 13% aluminum, 3.6% polyisocyante curative catalyst, 0.40% impurities* | **2. 04-02-002A: High Performance Military grade rocket propellant utilizing ADN:** *68.32% ADN, 18.63% epoxy binder, 12.42% aluminum powder, 0.62% copper chromite burn catalyst, 0.01% residual balance* |
| **3. 04-02-003A: High Performance Military grade rocket propellant utilizing ADN:** *57.14% ADN, 14.28% nitrocellulose, 12.69% aluminum powder, 10.31% epoxy binder, 5.55% urea nitrate, 0.03% balance* | **4. 04-02-004A: High Performance Military grade rocket propellant utilizing ADN:** *63.15% ADN, 12.63% aluminum powder, 9.47% nitrocellulose, 7.36% potassium nitrate, 7.36% PVC, 0.03% mixed balance* |
| **5. 04-02-005A: High Performance castable ADN rocket propellant:** *48.85% ADN, 25.44% dimethoxyethyl phthalate, 25.44% nitrocellulose, 0.25% lecithin, 0.,02% impurities* | **6. 04-02-006A: High Performance ADN rocket propellant:** *54.39% ADN, 16.73% urea nitrate, 12.55% PVC, 7.53% silicon, 7.53% hexamine, 1.25% iron-III-oxide, 0.02% balance* |
| **7. 04-02-007A: High Performance ADN rocket propellant:** *60.86% ADN, 10.43% magnesium powder, 8.69% ammonium perchlorate, 7.82% epoxy resin, 6.95% PVC, 5.21% nitrocellulose, 0.04 balance* | **8. 04-02-008A: High Performance AND/KDN rocket propellant:** *55.29% ADN or KDN, 13.82% potassium nitrate, 12.9% Teflon, 9.21% aluminum powder, 6.45% boron powder, 2.3% iron-III-oxide, 0.03% mixed balance* |
| **9. 04-02-009A: High Performance ADN rocket propellant:** *49% ADN, 26% ammonium perchlorate, 14% aluminum, 4.6% triacetin, 4% plastisol nitrocellulose, 1% magnesium oxide burn catalyst, 1% 2-nitrodiphenylamine, 0.32% 2,6-toluene diisocyanate curative, 08% residue* | **10. 04-02-010A High Performance Teflon based ADN rocket propellant:** *69.3% ADN, 18.8% aluminum powder, 8.9% butyl rubber, 2.9% Teflon, 0.10% residue* |
| **11. 04-02-011A High Performance ADN rocket propellant with boron modifier:** *63.4% boron, 31.7% ADN, 2.9% glyceryl monoricinoleate, 1.77% toluene diisocyanate, 0.08% Ferric acetyl acetonate, 0.08% diethanol oleamide, 0.07% impurities* | **12. 04-02-011B High Performance ADN rocket propellant with boron modifier:** *63.4% ADN, 31.7% boron powder, 2.9% glyceryl monoricinoleate, 1.77% toluene diisocyanate, 0.08% Ferric acetyl acetonate, 0.08% diethanol oleamide, 0.07% impurities* |
| **13. 04-02-012A High Performance ADN rocket propellant:** *67.58% ammonium perchlorate, 16.48% nitrocellulose, 14.83% diglycidyl ether, 1.08% triethylene tetramine, 0.03% residual balance* | **14. 04-02-013A High performance ADN rocket propellant:** *67.9% ADN, 26.3% polybutadiene polymer, 2.1% polymer butadiene, 2% aluminum, 1.47% epoxy compound, 0.06% chromium compound catalyst, 0.17% residual balance* |
| **15. 04-02-014A: High performance ADN rocket propellant (with metal hydride accelerator):** *43.01% ADN, 26.88% ammonium nitrate, 12.54% sodium hydride, 8.96% epoxy resin, 4.2% magnesium powder, 2.5% sodium hypophosphite, 1.79% copper-II-chromite burn rate catalyst, 0.12% residual impurities* | **16. 04-02-015A: High Performance JPL X360-ADN (new) rocket propellant:** *48.03% ADN, 23.43% polypropylene glycol plasticizer, 20.58% ammonium nitrate, 4.7% toluene diisocyante, 1.96% copper chromite burn rate catalyst, 1.17% 1,2,6-hexanetriol, 0.098% ferric acetyl acetonate, 0.032% mixed residual balance* |
| **17. 04-02-016A: High performance ADN, JATO rocket propellant:** *51.04% ADN, 21.89% ammonium nitrate, 8% 2,4,-dinitrodiphenyl ether, 8% ethylene glycol diglycolate, 4.74% cellulose acetate, 3% Prussian blue pigment catalyst, 2.3% acetonylacetone dioxime, 1% carbon black, 0.03% residue* | **18. 04-02-017A: High Performance ADN rocket propellant with nitroguanidine base:** *47.24% ADN, 26.24% nitroguanidine, 20.99% triacetin, 5.24% diethyl phthalate, 0.26% lecithin, 0.03% mixed balance* |
| | |

| | |
|---|---|
| **19. 04-02-018A: High Performance ADN rocket propellant:** *75.52% ADN, 8.25% ammonium dichromate, 5.63% copolymer binder, 5.43% polybutadiene, 2.71% Milori blue catalyst, 2% carbon black filler, 0.40% magnesium oxide catalyst, 0.06% mixed residual balance* | **20. 04-02-019A: High Performance ADN rocket propellant:** *50% ADN, 21.42% magnesium sulfite, 14.28% nitrocellulose, 7.14% aluminum powder, 5.71% magnesium, 1.42% aluminum stearate catalyst, 0.03% balance* |
| **21. 04-02-020A: High Performance ADN military rocket propellant:** *49.7% ADN, 24% nitroguanidine, 19.9% ammonium perchlorate, 4.7% ethyl acrylate-acrylic acid copolymer, 1% UNOX 221 linking agent, 0.49% carbon black catalyst, 0.210% moisture* | |

**04-02-001A: High Performance Military grade rocket propellant of ADN:**
Into a suitable vertical mixer, utilizing a plastic blade, not sharp steel, place ***14.4 grams of a polyisocyanate curative*** called Desmodur RTM N-100, and then followed by ***52 grams of powdered aluminum*** of fine grain. Thereafter, turn on the vertical mixer to low speed, and then blend the two ingredients for about 30 minutes at room temperature. Thereafter, simply add in ***97 grams of nitrocellulose*** (of average nitrogen content, i.e., 14%), and then followed by ***236 grams of ADN***. Immediately thereafter, continue to blend the mixture on low speed for about 30 minutes. Now, after blending for 30 minutes, the propellant is ready for use. To use, simply pour and press the mixture into any rocket motor or mold under a pressure of 4,000 psi, and allow the rocket motor or mold to cure for several days at room temperature.
**Burn rate:** 2 seconds per inch at 4,000 psi; tends to vary based on diameter and density of rocket grain.
**Water resistance:** Good.
**Stability:** Can be stored for decades.
**Sensitivity:** Stable towards friction; stable towards normal shock; stable to temperatures up to 150 Celsius.
**Flammability (1 to 10, un-contained):** 9.2 solid grain; 9.5 powdered grain.
**Ease of ignition (1 to 10 un-contained):** 9 solid grain; grain size varies and results in different ignition rates.
**Tendency to cake:** Will not cake.
**Detonation ability:** Can be detonated when subjected to molecular shock, i.e., from a blasting cap or detonator.
**Explosive ability (contained; explosive deflagration):** High.
**Percentage:** ***59% ADN, 24% nitrocellulose, 13% aluminum, 3.6% polyisocyante curative catalyst, 0.40% impurities***
**Classification:** Deflagrating explosive (classified as propellant).

**Danger of preparation:** it is advisable that only experienced pyrotechnicians attempt to carryout this procedure.
**Use:** Used in military and commercial rockets and missiles for multiple uses.

| Anti-tank rocket | Details |
|---|---|
| 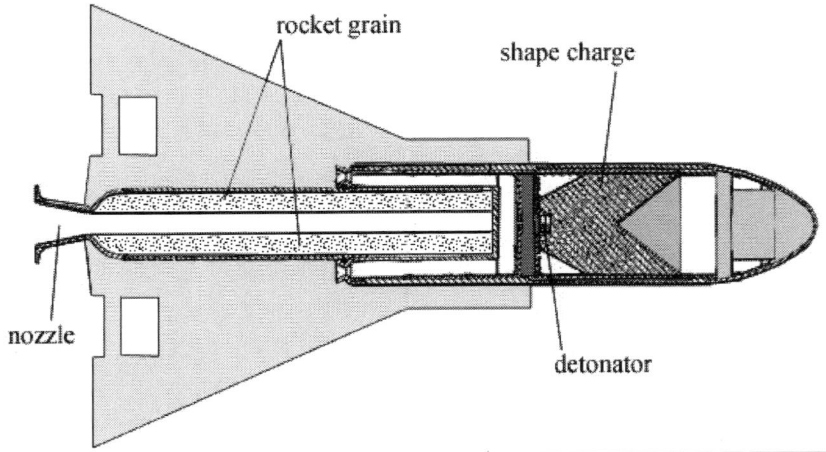  | Another illustration of an anti-tank rocket. |

### 04-02-002A: High Performance Military grade rocket propellant utilizing ADN:

Into a suitable vertical mixer, place ***550 grams of ADN,*** followed by ***100 grams of aluminum powder*** of average mesh, followed by ***5 grams of copper-II-chromite***, and then followed by 150 milliliters of acetone. Thereafter, blend the mixture on moderate speed for about 15 minutes. Thereafter, place the mixture onto a shallow pan or tray, and allow the mixture to thoroughly dry. Note: a vacuum apparatus can be used to recover the solvent if desired. Once the solvent has completely evaporated, and a dry solid remains, place the dry mixture into any a clean sealable vertical mixer, then add in ***150 grams of epon 815*** (any other liquid based epoxy resin will work), seal the vertical mixer, and then blend the mixture for about 30 minutes on high speed. Thereafter, the mixture is ready to use. To use, the mixture simply needs to be pressed into any desired rocket motor or mold under a pressure of 4,000 psi, and then cure the rocket motor or mold in an oven at 60 Celsius for 12 hours. A standard ignition composition can be used for proper burn ignition.
**Burn rate:** 2.1 seconds per inch at 4,000 psi; tends to vary based on diameter and density of rocket grain.
**Water resistance:** Good.
**Stability:** Can be stored for decades.
**Sensitivity:** Stable towards friction; stable towards normal shock; stable to temperatures up to 150 Celsius.
**Flammability (1 to 10, un-contained):** 9.13 solid grain; 9.4 powdered grain.
**Ease of ignition (1 to 10 un-contained):** 9 solid grain; grain size varies and results in different ignition rates.
**Tendency to cake:** Will not cake.
**Detonation ability:** Can be detonated but only severe conditions—requires TNT, RDX, or HMX booster.
**Explosive ability (contained; explosive deflagration):** Above moderate.
**Percentage:** ***68.32% ADN, 18.63% epoxy binder, 12.42% aluminum powder, 0.62% copper chromite burn catalyst, 0.01% residual balance***
**Classification:** Deflagrating explosive (classified as propellant).

**Danger of preparation:** it is advisable that only experienced pyrotechnicians attempt to carryout this procedure.
**Use:** Used in military and commercial rockets and missiles for multiple uses.

### 04-02-003A: High Performance Military grade rocket propellant utilizing ADN:

Into a suitable sealable vertical mixer, place ***360 grams of ADN,*** followed by ***80 grams of aluminum powder*** of average mesh, followed by ***35 grams of urea nitrate***, followed by ***90 grams of nitrocellulose***, and then followed ***65 grams of epon 815*** (any other liquid based epoxy resin will work); thereafter, seal the vertical mixer, and then blend the mixture on high speed for about 30 minutes. Thereafter, the mixture is ready to use. To use, the mixture simply needs to be pressed into any desired rocket motor or engine under a pressure of 5,000 psi, and then cure the rocket motor or engine in an oven at 60 Celsius for 12 hours. A standard ignition composition can be used for proper burn ignition.
**Burn rate:** 2 seconds per inch at 5,000 psi; tends to vary based on diameter and density of rocket grain.
**Water resistance:** Good.
**Stability:** Can be stored for decades.
**Sensitivity:** Stable towards friction; stable towards normal shock; stable to temperatures up to 150 Celsius.
**Flammability (1 to 10, un-contained):** 9.1 solid grain; 9.4 powdered grain.
**Ease of ignition (1 to 10 un-contained):** 9 solid grain; grain size varies and results in different ignition rates.
**Tendency to cake:** Will not cake.
**Detonation ability:** Can be detonated when subjected to molecular shock, i.e., from a blasting cap or detonator.
**Explosive ability (contained; explosive deflagration):** High.
**Percentage:** ***57.14% ADN, 14.28% nitrocellulose, 12.69% aluminum powder, 10.31% epoxy binder, 5.55% urea nitrate, 0.03% balance***
**Classification:** Deflagrating explosive (classified as propellant).

**Danger of preparation:** it is advisable that only experienced pyrotechnicians attempt to carryout this procedure.
**Use:** Used in military and commercial rockets and missiles for multiple uses.

### 04-02-004A: High Performance Military grade rocket propellant utilizing ADN:

Into a suitable vertical mixer, place ***600 grams of ADN,*** followed by ***70 grams of potassium nitrate***, followed by ***120 grams of aluminum powder***, followed by ***90 grams of nitrocellulose***, followed by ***70 grams of finely ground PVC***, and then add in 450 milliliters of tetrahydrofuran; thereafter, blend the mixture for 30 minutes on high speed. Thereafter, filter-off the insoluble mass, using a large Buchner funnel under suction, and then suction dry the mixture as best as possible. Now, recover the suction dried mixture from the Buchner funnel, and then press it into any desired rocket motor or mold under a pressure of 7,000 psi, and then cure the rocket motor or mold in an oven at 60 Celsius for 12 hours.
**Burn rate:** 2 seconds per inch at 7,000 psi; tends to vary based on diameter and density of rocket grain.
**Water resistance:** Good.
**Stability:** Can be stored for decades.

**Sensitivity:** Stable towards friction; stable towards normal shock; stable to temperatures up to 150 Celsius.
**Flammability (1 to 10, un-contained):** 9.1 solid grain; 9.4 powdered grain.
**Ease of ignition (1 to 10 un-contained):** 9 solid grain; grain size varies and results in different ignition rates.
**Tendency to cake:** Will not cake.
**Detonation ability:** Can be detonated when subjected to molecular shock, i.e., from a blasting cap or detonator.
**Explosive ability (contained; explosive deflagration):** High.
**Percentage:** ***63.15% ADN, 12.63% aluminum powder, 9.47% nitrocellulose, 7.36% potassium nitrate, 7.36% PVC, 0.03% mixed balance***
**Classification:** Deflagrating explosive (classified as propellant).

**Danger of preparation:** it is advisable that only experienced pyrotechnicians attempt to carryout this procedure.
**Use:** Used in military and commercial rockets and missiles for multiple uses.

### 04-02-005A: High Performance castable ADN rocket propellant:

Into a suitable sealable vertical mixer, equipped with vacuum port, place ***500 grams of dimethoxyethyl phthalate***, followed by ***500 grams of nitrocellulose (average nitrogen content)***, and then followed by ***5 grams of lecithin***. Thereafter, blend the mixture on moderate speed for about 15 minutes to form a viscous slurry. Thereafter, add in ***960 grams of ADN***, and then blend the mixture on moderate speed for about 30 minutes. After 30 minutes, seal the vertical mixer, and then de-gas the mixture by applying a vacuum of 120mm of mercury at room temperature for 12 hours. After 12 hours, remove the vacuum, equalize the pressure, and then open the vertical mixer. Thereafter, the mixture is ready for use. To use, the mixture needs to be pressed into any desired rocket motor or mold under a pressure of 3,000 psi, and then cure the rocket motor or mold in an oven at 70 Celsius for 24 hours. A standard ignition composition can be used for proper burn.
**Burn rate:** 2.5 seconds per inch at 3,000 psi; tends to vary based on diameter and density of rocket grain.
**Water resistance:** Good.
**Stability:** Can be stored for decades.
**Sensitivity:** Stable towards friction; stable towards normal shock; stable to temperatures up to 150 Celsius.
**Flammability (1 to 10, un-contained):** 9 solid grain; 9.2 powdered grain.
**Ease of ignition (1 to 10 un-contained):** 8.9 solid grain; grain size varies and results in different ignition rates.
**Tendency to cake:** Will not cake.
**Detonation ability:** Can be detonated but only severe conditions—requires TNT, RDX, or HMX booster.
**Explosive ability (contained; explosive deflagration):** Above moderate.
**Percentage:** ***48.85% ADN, 25.44% dimethoxyethyl phthalate, 25.44% nitrocellulose, 0.25% lecithin, 0.,02% impurities***
**Classification:** Deflagrating explosive (classified as propellant).

**Danger of preparation:** it is advisable that only experienced pyrotechnicians attempt to carryout this procedure.
**Use:** Used in military and commercial rockets and missiles for multiple uses.

### 04-02-006A: High Performance ADN rocket propellant:

Into a suitable vertical mixer, place ***650 grams of ADN***, followed by ***150 grams of fine-grained PVC***, and then followed by ***90 grams of finely divided silicon***. Thereafter, add in 250 milliliters of hexane, and then blend the mixture on moderate speed for about 15 minutes to form a pasty mass. Thereafter, add in ***200 grams of urea nitrate,*** followed by ***15 grams of iron-III-oxide***, and then followed by ***90 grams of hexamine***, and then continue to blend the mixture on moderate speed for about 30 minutes. Thereafter, the mixture is ready for use. To use, the mixture needs to be pressed into any desired rocket motor or mold under a pressure of 5,000 psi in the usual manner, and then cured in an oven at 60 Celsius for 12 hours. A standard ignition composition can be used for proper burn.
**Burn rate:** 2.6 seconds per inch at 5,000 psi; tends to vary based on diameter and density of rocket grain.
**Water resistance:** Good.
**Stability:** Can be stored for decades.
**Sensitivity:** Stable towards friction; stable towards normal shock; stable to temperatures up to 150 Celsius.
**Flammability (1 to 10, un-contained):** 9 solid grain; 9.2 powdered grain.
**Ease of ignition (1 to 10 un-contained):** 8.7 solid grain; grain size varies and results in different ignition rates.
**Tendency to cake:** Will not cake.
**Detonation ability:** Can be detonated but only severe conditions—requires TNT, RDX, or HMX booster.
**Explosive ability (contained; explosive deflagration):** Above moderate.
**Percentage:** ***54.39% ADN, 16.73% urea nitrate, 12.55% PVC, 7.53% silicon, 7.53% hexamine, 1.25% iron-III-oxide, 0.02% balance***
**Classification:** Deflagrating explosive (classified as propellant).

**Danger of preparation:** Can be made by amateurs, but it is advisable that only experienced pyrotechnicians attempt to carryout this procedure.
**Use:** Used in military and commercial rockets and missiles for multiple uses.

### 04-02-007A: High Performance ADN rocket propellant:

Into a suitable vertical mixer, place ***700 grams of ADN***, followed by ***100 grams of ammonium perchlorate***, followed by ***120 grams of magnesium powder*** (note: the magnesium powder should be coated with linseed oil utilizing a ball mill before adding), followed by ***60 grams of nitrocellulose***, followed by ***80 grams of fine grained PVC***, and then followed by ***90 grams of any desired epoxy liquid resin***. Thereafter, seal the vertical mixer, and then blend the mixture on moderate speed for about 45 minutes. Thereafter, the mixture is ready for use. To use, the mixture needs to be pressed into any desired rocket motor or mold, under a pressure of 7,000 psi in the usual manner, and then cure the rocket motor or mold in an oven at 60 Celsius for 18 hours. A standard ignition composition can be used for proper burn.
**Burn rate:** 2.7 seconds per inch at 7,000 psi; tends to vary based on diameter and density of rocket grain.
**Water resistance:** Good.
**Stability:** Can be stored for decades.
**Sensitivity:** Stable towards friction; stable towards normal shock; stable to temperatures up to 150 Celsius.
**Flammability (1 to 10, un-contained):** 9.1 solid grain; 9.2 powdered grain.
**Ease of ignition (1 to 10 un-contained):** 8.9 solid grain; grain size varies and results in different ignition rates.
**Tendency to cake:** Will not cake.
**Detonation ability:** Can be detonated but only severe conditions—requires TNT, RDX, or HMX booster.
**Explosive ability (contained; explosive deflagration):** Above moderate.
**Percentage:** ***60.86% ADN, 10.43% magnesium powder, 8.69% ammonium perchlorate, 7.82% epoxy resin, 6.95% PVC, 5.21% nitrocellulose, 0.04 balance***
**Classification:** Deflagrating explosive (classified as propellant).

**Danger of preparation:** Can be made by amateurs, but it is advisable that person or persons should have some knowledge of chemistry and pyrotechnics before carrying out this operation.
**Use:** Used in military and commercial rockets and missiles for multiple uses.

### 04-02-008A: High Performance ADN/KDN rocket propellant:

Into a standard vertical mixer, utilizing plastic blades/paddles, place ***600 grams of ADN or KDN***, followed by ***140 grams of finely powdered Teflon***, and then followed by ***70 grams of finely divided boron powder***. Thereafter, add in 250 milliliters of hexane, and then blend the mixture on low speed for 45 minutes to form a pasty mass. Thereafter, add in ***150 grams of potassium nitrate,*** followed by ***25 grams of red iron-III-oxide***, and then followed by ***100 grams of aluminum powder***, and then continue to blend the mixture on low speed for about 45 minutes. Thereafter, the mixture is ready for use. To use, the mixture needs to be pressed into any desired rocket motor or mold under a pressure of 5,000 psi, and then cured in an oven at 60 Celsius for 12 hours in the usual manner. A standard ignition composition can be used for proper burn.
**Burn rate:** 3 seconds per inch at 5,000 psi; tends to vary based on diameter and density of rocket grain.
**Water resistance:** Good.
**Stability:** Can be stored for decades.
**Sensitivity:** Stable towards friction; stable towards normal shock; stable to temperatures up to 150 Celsius.
**Flammability (1 to 10, un-contained):** 9 solid grain; 9.2 powdered grain.
**Ease of ignition (1 to 10 un-contained):** 8.7 solid grain; grain size varies and results in different ignition rates.
**Tendency to cake:** Will not cake.
**Detonation ability:** Can be detonated but only severe conditions—requires TNT, RDX, or HMX booster.
**Explosive ability (contained; explosive deflagration):** Above moderate.
**Percentage:** ***55.29% ADN or KDN, 13.82% potassium nitrate, 12.9% Teflon, 9.21% aluminum powder, 6.45% boron powder, 2.3% iron-III-oxide, 0.03% mixed balance***
**Classification:** Deflagrating explosive (classified as propellant).

**Danger of preparation:** Can be made by amateurs, but it is advisable that person or persons should have some knowledge of chemistry and pyrotechnics before carrying out this operation.
**Use:** Used in military and commercial rockets and missiles for multiple uses.

### 04-02-009A: High Performance ADN rocket propellant:

Into a suitable vertical mixer, place ***23 grams of triacetin***, followed by ***70 grams of finely divided aluminum powder***, followed by ***5 grams of finely divided magnesium oxide,*** followed by ***5 grams of 2-nitrodiphenylamine***, followed by ***20 grams of plastisol nitrocellulose,*** and then followed by ***245 grams of AND***; thereafter,blend the mixture on low speed at room temperature for about 15 minutes. After 15 minutes, slowly add in ***130 grams of ammonium nitrate,*** followed by ***1.6 grams of 2,6-toluene diisocyanate***, and then continue to blend the mixture on moderate speed for about 15 minutes to form a uniform mixture. Thereafter, the mixture is ready to be used. To use, the propellant needs to be pressed into any desirable rocket motor or mold under a pressure of 4,000 psi, and then allowed to cure at room temperature for several days.

**Burn rate:** 2.9 seconds per inch at 4,000 psi; tends to vary based on diameter and density of rocket grain.
**Water resistance:** Good.
**Stability:** Can be stored for decades.
**Sensitivity:** Stable towards friction; stable towards normal shock; stable to temperatures up to 150 Celsius.
**Flammability (1 to 10, un-contained):** 9 solid grain; 9.3 powdered grain.
**Ease of ignition (1 to 10 un-contained):** 8.6 solid grain; grain size varies and results in different ignition rates.
**Tendency to cake:** Will not cake.
**Detonation ability:** Can be detonated but only severe conditions—requires TNT, RDX, or HMX booster.
**Explosive ability (contained; explosive deflagration):** Above moderate.
**Percentage:** ***49% ADN, 26% ammonium nitrate, 14% aluminum, 4.6% triacetin, 4% plastisol nitrocellulose, 1% magnesium oxide burn catalyst, 1% 2-nitrodiphenylamine, 0.32% 2,6-toluene diisocyanate curative, 08% residue***
**Classification:** Deflagrating explosive (classified as propellant).

**Danger of preparation:** Can be made by amateurs, but it is advisable that person or persons should have some knowledge of chemistry and pyrotechnics before carrying out this operation.
**Use:** Used in military and commercial rockets and missiles for multiple uses.

### 04-02-010A: High Performance Teflon based ADN rocket propellant:

Into a suitable vertical mixer, place ***45 grams of standard butyl rubber***, followed by ***15 grams of finely divided Teflon powder***, followed by ***95 grams of finely divided aluminum powder of standard mesh,*** followed by ***350 grams of ADN***, and then followed by 150 milliliters of acetone, and then blend the mixture at room temperature until the bulk of the acetone evaporates (several hours). Thereafter, it is ready for pressing. To do so, press the propellant into any desirable rocket motor, engine, or mold, under a pressure of 6,000 psi, and the cure the rocket motor, engine, or mold in an oven at 60 Celsius for 24 hours.

**Burn rate:** 2.7 seconds per inch at 6,000 psi; tends to vary based on diameter and density of rocket grain.
**Water resistance:** Good.
**Stability:** Can be stored for decades.
**Sensitivity:** Stable towards friction; stable towards normal shock; stable to temperatures up to 150 Celsius.
**Flammability (1 to 10, un-contained):** 9.1 solid grain; 9.4 powdered grain.
**Ease of ignition (1 to 10 un-contained):** 8.9 solid grain; grain size varies and results in different ignition rates.
**Tendency to cake:** Will not cake.
**Detonation ability:** Can be detonated but only severe conditions—requires TNT, RDX, or HMX booster.
**Explosive ability (contained; explosive deflagration):** Above moderate.
**Percentage:** ***69.3% ADN, 18.8% aluminum powder, 8.9% butyl rubber, 2.9% Teflon, 0.10% residue***
**Classification:** Deflagrating explosive (classified as propellant).

**Danger of preparation:** Can be made by amateurs, but it is advisable that person or persons should have some knowledge of chemistry and pyrotechnics before carrying out this operation.
**Use:** Used in military and commercial rockets and missiles for multiple uses.

### 04-02-011A: High Performance ADN rocket propellant with boron modifier:

Into a suitable vertical mixer, place ***27.6 grams of glyceryl monoricinoleate (GMRO)***, followed by ***766 milligrams of diethanol oleamide (DEO)***, followed ***by 16.8 grams of toluene diisocyanate (TDI)***, followed by adding in ***600 grams of finely powdered boron***, followed by ***300 grams of ADN***, and then followed by ***766 milligrams of ferric acetyl acetonate***, and then blend the mixture on high for 45 minutes. Thereafter, the mixture is ready for use. To use, press the mixture into any desirable rocket motor or mold, under a pressure of 7,000 psi, and then cure the rocket motor or mold in an oven at 60 Celsius for 18 hours.

**Burn rate:** 6 seconds per inch at 7,000 psi; tends to vary based on diameter and density of rocket grain.
**Water resistance:** Good.
**Stability:** Can be stored for decades.
**Sensitivity:** Stable towards friction; stable towards normal shock; stable to temperatures up to 150 Celsius.

**Flammability (1 to 10, un-contained):** 8.6 solid grain; 8.9 powdered grain.
**Ease of ignition (1 to 10 un-contained):** 7.1 solid grain; grain size varies and results in different ignition rates.
**Tendency to cake:** Will not cake.
**Detonation ability:** Cannot be detonated.
**Explosive ability (contained; explosive deflagration):** Low.
**Percentage:** ***63.4% boron, 31.7% ADN, 2.9% glyceryl monoricinoleate, 1.77% toluene diisocyanate, 0.08% Ferric acetyl acetonate, 0.08% diethanol oleamide, 0.07% impurities***
**Classification:** Deflagrating explosive (classified as propellant).

**Danger of preparation:** Can be made by amateurs.
**Use:** Used in military and commercial rockets and missiles for multiple uses.

### 04-02-011B: High Performance ADN rocket propellant with boron modifier:

Into a suitable vertical mixer, place ***27.6 grams of glyceryl monoricinoleate (GMRO)***, followed by ***766 milligrams of diethanol oleamide (DEO)***, followed by ***16.8 grams of toluene diisocyanate (TDI)***, followed by ***600 grams of ADN***, followed by ***300 grams of finely ground boron powder***, and then followed by ***766 milligrams of ferric acetyl acetonate***; thereafter, blend the mixture on high speed for 45 minutes. Thereafter, the mixture is ready for use. To use, press the mixture into any desirable rocket motor or mold, under a pressure of 8,000 psi, and then cure the rocket motor or mold in an oven at 60 Celsius for 12 hours.
**Burn rate:** 3 seconds per inch at 8,000 psi; tends to vary based on diameter and density of rocket grain.
**Water resistance:** Good.
**Stability:** Can be stored for decades.
**Sensitivity:** Stable towards friction; stable towards normal shock; stable to temperatures up to 150 Celsius.
**Flammability (1 to 10, un-contained):** 9 solid grain; 9.3 powdered grain.
**Ease of ignition (1 to 10 un-contained):** 9 solid grain; grain size varies and results in different ignition rates.
**Tendency to cake:** Will not cake.
**Detonation ability:** Can be detonated but only severe conditions—requires TNT, RDX, or HMX booster.
**Explosive ability (contained; explosive deflagration):** Above moderate.
**Percentage:** ***63.4% ADN, 31.7% boron powder, 2.9% glyceryl monoricinoleate, 1.77% toluene diisocyanate, 0.08% Ferric acetyl acetonate, 0.08% diethanol oleamide, 0.07% impurities***
**Classification:** Deflagrating explosive (classified as propellant).

**Danger of preparation:** Can be made by amateurs, but it is advisable that person or persons should have some knowledge of chemistry and pyrotechnics before carrying out this operation.
**Use:** Used in military and commercial rockets and missiles for multiple uses.

### 04-02-012A: High Performance ADN rocket propellant:

Into a suitable vertical mixer, place ***180 grams of diglycidyl ether***, followed by ***820 grams of ADN***, followed by ***200 grams of nitrocellulose***, and then followed by ***13.2 grams of triethylene tetramine***; thereafter, blend the mixture for 60 minutes at room temperature. Thereafter, the mixture is ready for use. To use, the mixture should be pressed into any desirable rocket motor or mold, under a pressure of 5,000 psi. Thereafter, the rocket motor or mold should be cured in an oven at 100 Celsius for about 4 hours. The cured propellant can be ignited using any standard ignition composition.
**Burn rate:** 2.99 seconds per inch at 5,000 psi; tends to vary based on diameter and density of rocket grain.
**Water resistance:** Good.
**Stability:** Can be stored for decades.
**Sensitivity:** Stable towards friction; stable towards normal shock; stable to temperatures up to 150 Celsius.
**Flammability (1 to 10, un-contained):** 9.1 solid grain; 9.3 powdered grain.
**Ease of ignition (1 to 10 un-contained):** 9.1 solid grain; grain size varies and results in different ignition rates.
**Tendency to cake:** Will not cake.
**Detonation ability:** Can be detonated but only severe conditions—requires TNT, RDX, or HMX booster.
**Explosive ability (contained; explosive deflagration):** Above moderate.
**Percentage:** ***67.58% ADN, 16.48% nitrocellulose, 14.83% diglycidyl ether, 1.08% triethylene tetramine, 0.03% residual balance***
**Classification:** Deflagrating explosive (classified as propellant).

**Danger of preparation:** Can be made by amateurs, but it is advisable that person or persons should have some knowledge of chemistry and pyrotechnics before carrying out this operation.
**Use:** Used in military and commercial rockets and missiles for multiple uses.

### 04-02-013A: High performance ADN rocket propellant:

Into a suitable vertical mixer, place ***750 milligrams of chromium 2-ethylhexanoate***, followed by ***322 grams of a carboxy terminated polybutadiene*** with a molecular weight of 5,000 and a specific gravity of 0.91 at 60 degrees Fahrenheit containing 1.0 to 1.5 percent 2,2'-methylene-bis(4-methyl-6-tertbutyl) phenol as antioxidant, followed by ***18 grams of 3,4-epoxycyclohexylmethyl carboxyl-3,4-epoxycyclohexane*** (with an epoxy equivalent of 130.6 and 96% pure) and the N,N-ortho-tris(epoxypropyl) p-aminophenol, followed by ***26.5 grams of a polymer of butadiene*** (containing a small amount of phenyl-beta-naphthylamine as stabilizer), thereafter followed by ***830 grams of ADN*** of average mesh, and then followed by ***25 grams of aluminum powder*** of average mesh. Thereafter, blend the mixture for about 1 hour to form a uniform mixture. Thereafter, the mixture is ready for use. To use, the mixture needs to be pressed into any desired rocket motor or mold under a pressure of 6,000 psi in the usual fashion. Thereafter, the rocket motor or mold should be cured at room temperature for 6 days.
**Burn rate:** 3.1 seconds per inch at 6,000 psi; tends to vary based on diameter and density of rocket grain.
**Water resistance:** Good.
**Stability:** Can be stored for decades.
**Sensitivity:** Stable towards friction; stable towards normal shock; stable to temperatures up to 150 Celsius.
**Flammability (1 to 10, un-contained):** 9 solid grain; 9.1 powdered grain.
**Ease of ignition (1 to 10 un-contained):** 8.9 solid grain; grain size varies and results in different ignition rates.
**Tendency to cake:** Will not cake.
**Detonation ability:** Can be detonated but only severe conditions—requires TNT, RDX, or HMX booster.
**Explosive ability (contained; explosive deflagration):** Moderate.
**Percentage:** ***67.9% ADN, 26.3% polybutadiene polymer, 2.1% polymer butadiene, 2% aluminum, 1.47% epoxy compound, 0.06% chromium compound catalyst, 0.17% residual balance***
**Classification:** Deflagrating explosive (classified as propellant).

**Danger of preparation:** Can be made by amateurs, but it is advisable that person or persons should have some knowledge of chemistry and pyrotechnics before carrying out this operation.
**Use:** Used in military and commercial rockets and missiles for multiple uses.

### 04-02-014A: High performance ADN rocket propellant (with metal hydride accelerator):

Into a suitable sealable vertical mixer, place ***1200 grams of ADN***, followed by ***120 grams of magnesium powder***, followed by ***50 grams of copper-II-chromite***, followed by 250 milliliters of dry hexane, and then followed by ***70 grams of sodium hypophosphite***. Thereafter, seal the vertical mixer, and then blend the mixture on moderate speed for about 2 hours. After 2 hours, quickly open the vertical mixer, add in ***750 grams of ammonium nitrate,*** followed by ***350 grams of sodium hydride***, re-seal the vertical mixer, and then continue to blend the mixture on moderate speed for 45 minutes to form a rough dough. Thereafter, open the vertical mixer, add in ***250 grams of Epon 815 epoxy resin*** (any other desired liquid epoxy compound will work), re-seal the vertical mixer, and then blend the entire mixture for about 30 minutes. Thereafter, the mixture is ready for use. To use, simply press the mixture into any desired rocket engine or mold, under a pressure of 7,000 psi in the usual manner. Thereafter, allow the rocket engine or mold to cure in an oven at 60 Celsius for 12 hours. After 12 hours, remove the heat, and then allow the rocket engine or mold to cure at room temperature for 3 days.
**Burn rate:** 3.2 seconds per inch at 7,000 psi; tends to vary based on diameter and density of rocket grain.
**Water resistance:** Good.
**Stability:** Can be stored for decades when kept dry and free from moisture.
**Sensitivity:** Stable towards friction; stable towards normal shock; stable to temperatures up to 150 Celsius.
**Flammability (1 to 10, un-contained):** 9 solid grain; 9.1 powdered grain.
**Ease of ignition (1 to 10 un-contained):** 8.9 solid grain; grain size varies and results in different ignition rates.
**Tendency to cake:** Will not cake.
**Detonation ability:** Can be detonated but only severe conditions—requires TNT, RDX, or HMX booster.
**Explosive ability (contained; explosive deflagration):** Moderate.
**Percentage:** ***43.01% ADN, 26.88% ammonium nitrate, 12.54% sodium hydride, 8.96% epoxy resin, 4.2% magnesium powder, 2.5% sodium hypophosphite, 1.79% copper-II-chromite burn rate catalyst, 0.12% residual impurities***
**Classification:** Deflagrating explosive (classified as propellant).

**Danger of preparation:** Can be made by amateurs, but it is advisable that person or persons should have some knowledge of chemistry and pyrotechnics before carrying out this operation.
**Use:** Used in military and commercial rockets and missiles for multiple uses.

**High Performance High Altitude Model Rocket**

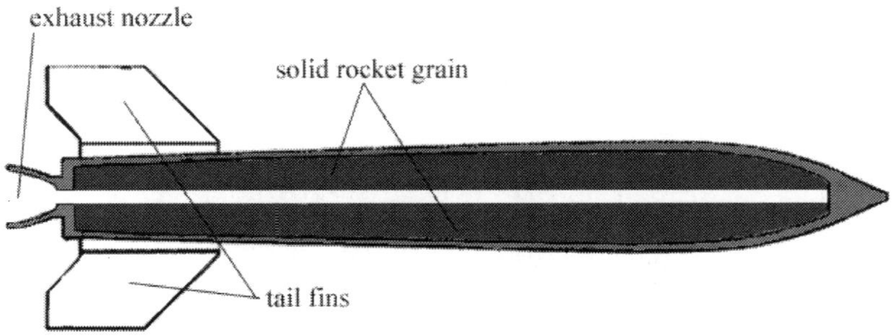


### 04-02-015A: High Performance JPL X360-ADN rocket propellant:

Into a special sealable heated vertical mixer, equipped with vacuum port, place ***105 grams of ground ammonium nitrate***, followed by ***245 grams of ADN***, followed by ***10 grams of copper chromite***, and then followed by ***119.5 grams of polypropylene glycol 2025***; thereafter, seal the vertical mixer and then blend the mixture on moderate speed for about 30 minutes under a vacuum of 10 millimeters of mercury at a temperature of 21 Celsius. After 30 minutes, quickly remove tee vacuum, equalize the pressure, open the vertical mixer, and then quickly add in ***500 milligrams of ferric acetyl acetonate***, followed by adding in ***24 grams of toluene diisocyante,*** and then re-seal the vertical mixer, re-apply the vacuum of 10 millimeters of mercury, and then continue to blend the mixture on moderate speed for about 15 minutes under said vacuum at 21 Celsius. Thereafter, again remove the vacuum, equalize the pressure, re-open the vertical mixer, quickly add in ***6 grams of 1,2,6-hexanetriol***, and then re-seal the vertical mixer, re-apply the vacuum at 10mm of mercury, and then continue to blend the mixture on moderate speed for about 15 additional minutes at 25 Celsius. After 15 minutes, remove the heat source, remove the vacuum, equalize the pressure, and then re-open the vertical mixer; at this point the mixture is ready. To use, press the mixture it into any desirable rocket motor or mold under a pressure of 9,000 psi, and then cure the rocket motor or mold in an oven at 60 Celsius for about 16 hours. Requires standard rocket propellant ignition composition for proper burn.
**Burn rate:** 3.3 seconds per inch at 9,000 psi; tends to vary based on diameter and density of rocket grain.
**Water resistance:** Good.
**Stability:** Can be stored for decades.
**Sensitivity:** Stable towards friction; stable towards normal shock; stable to temperatures up to 150 Celsius.
**Flammability (1 to 10, un-contained):** 9 solid grain; 9.1 powdered grain.
**Ease of ignition (1 to 10 un-contained):** 9 solid grain; grain size varies and results in different ignition rates.
**Tendency to cake:** Will not cake.
**Detonation ability:** Can be detonated but only severe conditions—requires TNT, RDX, or HMX booster.
**Explosive ability (contained; explosive deflagration):** Moderate.
**Percentage:** ***48.03% ADN, 23.43% polypropylene glycol plasticizer, 20.58% ammonium nitrate, 4.7% toluene diisocyante, 1.96% copper chromite burn rate catalyst, 1.17% 1,2,6-hexanetriol, 0.098% ferric acetyl acetonate, 0.032% mixed residual balance***
**Classification:** Deflagrating explosive (classified as propellant).

**Danger of preparation:** Can be made by amateurs, but it is advisable that person or persons should have some knowledge of chemistry and pyrotechnics before carrying out this operation.
**Use:** Used in military and commercial rockets and missiles for multiple uses.

### 04-02-016A: High performance ADN, JATO rocket propellant:

Into a special sealable heated vertical mixer, equipped with vacuum port, place ***40 grams of 2,4,-dinitrodiphenyl ether,*** followed by ***40 grams of ethylene glycol diglycolate***, and then heat the mixture to 140 Celsius and blend the mixture at this temperature for 15 minutes to form a homogenous mixture. Thereafter add in ***23.7 grams of cellulose acetate***, in small portions at a time, while thoroughly blending the mixture. After the addition of all the cellulose acetate, lower the temperature to 120 Celsius, and when this new temperature is achieved, add in ***11.5 grams acetonylacetone dioxime***, and then continue to blend the mixture for about 10 minute at 120 Celsius. Thereafter, add in ***255 grams of ADN***, and then continue to blend the mixture at 120 Celsius for about 10 minutes on high speed. After 10 minutes, quickly add in ***109.4 grams of ammonium nitrate*** of 325 mesh, followed by ***5 grams of finely divided carbon black***, and then followed by ***15 grams of Prussian blue pigment catalyst***, and then continue to blend the mixture on high speed at 120 Celsius for 30 minutes. After 30 minutes, seal the vertical mixer, and then de-gas the mixture using a vacuum of 15 millimeters of mercury at 120 Celsius for 30 minutes; during the de-gassing process, continue to rapidly blend the mixture. After 30 minutes, remove the heat, remove the vacuum, equalize the pressure, and then open the vertical mixer while the mixture is still hot. Thereafter, immediately pour and press the hot mixture into any desirable rocket motor or mold under a pressure of 3,000 psi, and then allow the rocket motor or mold to cure at room temperature for 3 days.
**Burn rate:** 3 seconds per inch at 3,000 psi; tends to vary based on diameter and density of rocket grain.
**Water resistance:** Good.
**Stability:** Can be stored for decades.
**Sensitivity:** Stable towards friction; stable towards normal shock; stable to temperatures up to 150 Celsius.
**Flammability (1 to 10, un-contained):** 9 solid grain; 9.1 powdered grain.

**Ease of ignition (1 to 10 un-contained):** 9 solid grain; grain size varies and results in different ignition rates.
**Tendency to cake:** Will not cake.
**Detonation ability:** Can be detonated but only severe conditions—requires TNT, RDX, or HMX booster.
**Explosive ability (contained; explosive deflagration):** Moderate.
**Percentage:** ***51.04% ADN, 21.89% ammonium nitrate, 8% 2,4,-dinitrodiphenyl ether, 8% ethylene glycol diglycolate, 4.74% cellulose acetate, 3% Prussian blue pigment catalyst, 2.3% acetonylacetone dioxime, 1% carbon black, 0.03% residue***
**Classification:** Deflagrating explosive (classified as propellant).

**Danger of preparation:** Can be made by amateurs, but it is advisable that person or persons should have some knowledge of chemistry and pyrotechnics before carrying out this operation.
**Use:** Used in military and commercial rockets and missiles for multiple uses.

### 04-02-017A: High Performance ADN rocket propellant with nitroguanidine base:

Into a special sealable vertical mixer, equipped with vacuum port, place ***200 grams of triacetin***, followed by ***50 grams of diethyl phthalate***, followed by ***250 grams of nitroguanidine***, and then followed by ***2.5 grams of lecithin***. Thereafter, blend the mixture on moderate speed for about 15 minutes to form a uniform mass. Thereafter, add in ***450 grams of ADN***, and then blend the mixture on moderate speed for about 30 minutes. After 30 minutes, seal the vertical mixer, and then de-gas the mixture using a vacuum of 50 millimeters of mercury for 30 minutes at room temperature; during the de-gassing process, continue to rapidly blend the mixture. After 30 minutes, remove the vacuum, equalize the pressure, and then open the vertical mixer. Thereafter, the mixture is ready for use. To use, the mixture needs to be poured and pressed into any desired rocket motor or mold under a pressure of 5,000 psi in the usual manner, and then cure the rocket motor or mold in an oven at 70 Celsius for 2 days. A standard ignition composition can be used for proper burn.
**Burn rate:** 2.9 seconds per inch at 5,000 psi; tends to vary based on diameter and density of rocket grain.
**Water resistance:** Good.
**Stability:** Can be stored for decades.
**Sensitivity:** Stable towards friction; stable towards normal shock; stable to temperatures up to 150 Celsius.
**Flammability (1 to 10, un-contained):** 9.1 solid grain; 9.3 powdered grain.
**Ease of ignition (1 to 10 un-contained):** 9 solid grain; grain size varies and results in different ignition rates.
**Tendency to cake:** Will not cake.
**Detonation ability:** Can be detonated but only severe conditions—requires TNT, RDX, or HMX booster.
**Explosive ability (contained; explosive deflagration):** Above moderate.
**Percentage:** ***47.24% ADN, 26.24% nitroguanidine, 20.99% triacetin, 5.24% diethyl phthalate, 0.26% lecithin, 0.03% mixed balance***
**Classification:** Deflagrating explosive (classified as propellant).

**Danger of preparation:** Can be made by amateurs, but it is advisable that person or persons should have some knowledge of chemistry and pyrotechnics before carrying out this operation.
**Use:** Used in military and commercial rockets and missiles for multiple uses.

### 04-02-018A: High Performance ADN rocket propellant:

Into a suitable vertical mixer, place ***375 grams of ADN,*** followed by ***13.5 grams of Milori blue pigment agent***, followed by ***2 grams of magnesium oxide (light),*** followed by ***10 grams of carbon black,*** followed by ***41 grams of ammonium dichromate***, followed by ***27 grams of polybutadiene,*** and then followed by ***28 grams of any desired epoxy resin or copolymer binder***. Thereafter, blend the mixture on high speed for about 45 minutes at room temperature. Thereafter, the mixture is ready for use. To use, the mixture needs to be pressed into any desired rocket motor or mold under a pressure of 9,000 psi, and the resulting rocket motor or mold needs to be cured in an oven at 90 Celsius for 6 days. A standard ignition composition can be used for proper burn.
**Burn rate:** 3.1 seconds per inch at 9,000 psi; tends to vary based on diameter and density of rocket grain.
**Water resistance:** Good.
**Stability:** Can be stored for decades.
**Sensitivity:** Stable towards friction; stable towards normal shock; stable to temperatures up to 150 Celsius.
**Flammability (1 to 10, un-contained):** 9.1 solid grain; 9.3 powdered grain.
**Ease of ignition (1 to 10 un-contained):** 9 solid grain; grain size varies and results in different ignition rates.
**Tendency to cake:** Will not cake.
**Detonation ability:** Can be detonated but only severe conditions—requires TNT, RDX, or HMX booster.
**Explosive ability (contained; explosive deflagration):** Above moderate.
**Percentage:** ***75.52% ADN, 8.25% ammonium dichromate, 5.63% copolymer binder, 5.43% polybutadiene, 2.71% Milori blue catalyst, 2% carbon black filler, 0.40% magnesium oxide catalyst, 0.06% mixed residual balance***
**Classification:** Deflagrating explosive (classified as propellant).

**Danger of preparation:** Can be made by amateurs, but it is advisable that person or persons should have some knowledge of chemistry and pyrotechnics before carrying out this operation.
**Use:** Used in military and commercial rockets and missiles for multiple uses.

### 04-02-019A: High Performance ADN rocket propellant:

Into a suitable vertical mixer, place ***700 grams of ADN,*** followed by ***300 grams of magnesium sulfite***, followed by ***200 grams of nitrocellulose,*** followed by ***20 grams of aluminum stearate,*** followed by ***100 grams of aluminum powder***, followed by ***80 grams of magnesium powder***, and then followed by 200 milliliters of dry hexane; thereafter, blend the mixture on high speed for 1 hour. Thereafter, the mixture is ready for use. To use, the mixture needs to be pressed into any desirable rocket motor or mold under a pressure of 10,000 psi, and then cure the rocket motor or mold in an oven at 60 Celsius for 12 hours.
**Burn rate:** 3.4 seconds per inch at 10,000 psi; tends to vary based on diameter and density of rocket grain.
**Water resistance:** Good.
**Stability:** Can be stored for decades.
**Sensitivity:** Stable towards friction; stable towards normal shock; stable to temperatures up to 150 Celsius.
**Flammability (1 to 10, un-contained):** 9 solid grain; 9.1 powdered grain.
**Ease of ignition (1 to 10 un-contained):** 8.7 solid grain; grain size varies and results in different ignition rates.
**Tendency to cake:** Will not cake.
**Detonation ability:** Can be detonated but only severe conditions—requires TNT, RDX, or HMX booster.
**Explosive ability (contained; explosive deflagration):** Above moderate.
**Percentage:** ***50% ADN, 21.42% magnesium sulfite, 14.28% nitrocellulose, 7.14% aluminum powder, 5.71% magnesium, 1.42% aluminum stearate catalyst, 0.03% balance***
**Classification:** Deflagrating explosive (classified as propellant).

**Danger of preparation:** Can be made by amateurs, but it is advisable that person or persons should have some knowledge of chemistry and pyrotechnics before carrying out this operation.
**Use:** Used in military and commercial rockets and missiles for multiple uses.

### 04-02-020A: High Performance ADN military rocket propellant:

Into a standard vertical mixture, place ***24 grams of ethyl acrylate-acrylic acid (95-5, copolymer binder),*** followed by ***5.2 grams of "UNOX 221" (cross linking agent),*** and then blend the mixture on high for about 30 minutes at room temperature. After 30 minutes, add in ***2.5 grams of carbon black***, followed by ***252.5 grams of ADN***, and then continue to thoroughly blend the mixture on high at room temperature for 30 minutes. After 30 minutes, add in ***101 grams of ammonium perchlorate*** and then continue to blend the mixture for about 30 minutes at room temperature. Once again, after 30 minutes, carefully add in ***122 grams of nitroguanidine***, and then continue to blend, but reduce speed to slow, and then gently continue blending for 45 minutes at room temperature. Afterwards, the propellant is ready to be pressed. To do so, it should be pressed into any desirable rocket motor or mold under a pressure of 9,000 psi using the usual techniques, and the resulting rocket motor or mold should then be cured in an oven at 60 Celsius for 3 days.
**Burn rate:** 3.67 seconds per inch at 9,000 psi; tends to vary based on diameter and density of rocket grain.
**Water resistance:** Good.
**Stability:** Can be stored for decades.
**Sensitivity:** Stable towards friction; stable towards normal shock; stable to temperatures up to 150 Celsius.
**Flammability (1 to 10, un-contained):** 9 solid grain; 9.1 powdered grain.
**Ease of ignition (1 to 10 un-contained):** 8 solid grain; grain size varies and results in different ignition rates.
**Tendency to cake:** Will not cake.
**Detonation ability:** Can be detonated but only severe conditions—requires TNT, RDX, or HMX booster.
**Explosive ability (contained; explosive deflagration):** Above moderate.
**Percentage:** ***49.7% ADN, 24% nitroguanidine, 19.9% ammonium perchlorate, 4.7% ethyl acrylate-acrylic acid copolymer, 1% UNOX 221 linking agent, 0.49% carbon black catalyst, 0.210% moisture***
**Classification:** Deflagrating explosive (classified as propellant).

**Danger of preparation:** Can be made by amateurs, but it is advisable that person or persons should have some knowledge of chemistry and pyrotechnics before carrying out this operation.
**Use:** Used in military and commercial rockets and missiles for multiple uses.

# Section 3: Ammonium Nitrate Containing rocket propellants

***Chemicals used in this section (binders are not included)***

| | | |
|---|---|---|
| **1. Potassium nitrate (see Black Powder)** | | **2. Sulfur (see Black Powder)** |
| **3. Charcoal (see Black Powder)** | | **4. Sugar Carbon (see Modified Black Powder)** |
| **5. Barium Chromate (see Modified Black Powder)** | | **6. Potassium Perchlorate (see Modified Black Powder)** |
| **7. Potassium Dichromate (see Modified Black Powder)** | | **8. Ammonium Bisulfide (see Modified Black Powder)** |
| **9. Potassium chlorate (see Modified Black Powder)** | | **10. Carbon Disulfide (see Modified Black Powder)** |
| **11. Nitrocellulose (see Modified Black Powder)** | | **12. Lead Tetraoxide (see Modified Black Powder)** |
| **13. Diphenylamine (see Modified Black Powder)** | | **14. Titanium Dioxide (see Modified Black Powder)** |
| **15. Manganese Dioxide (see Modified Black Powder)** | | **16. Sodium Benzoate (see Modified Black Powder)** |
| **17. Ammonium Nitrate (see Modified Black Powder)** | | **18. Calcium Carbonate (see Modified Black Powder)** |
| **19. Sodium Nitrate (see Modified Black Powder)** | | **20. Urea (see Modified Black Powder)** |
| **21. Lead Nitrate (see Modified Black Powder)** | | **22. Nitro Starch (see Modified Black Powder)** |
| **23. Ammonium Perchlorate (see Ammonium Perchlorate Rocket Propellants)** | | **24. Aluminum powder (see Ammonium Perchlorate Rocket Propellants)** |
| **25. Magnesium Sulfide (see Ammonium Perchlorate Rocket Propellants)** | | **26. Copper Chromite (see Ammonium Perchlorate Rocket Propellants)** |
| **27. Nitroguanidine (see Ammonium Perchlorate Rocket Propellants)** | | **28. Ammonium Sulfate (see Ammonium Perchlorate Rocket Propellants)** |
| **29. Nitroglycerine (see Ammonium Perchlorate Rocket Propellants)** | | **30. Magnesium Oxide (see Ammonium Perchlorate Rocket Propellants)** |
| **31. Iron-II-oxide (see Ammonium Perchlorate Rocket Propellants)** | | **32. Zirconium Hydride (see Ammonium Perchlorate Rocket Propellants)** |
| **33. Teflon (see Ammonium Perchlorate Rocket Propellants)** | | **34. Zinc Oxide (see Ammonium Perchlorate Rocket Propellants)** |
| **35. Ammonium Dichromate (see Ammonium Perchlorate Rocket Propellants)** | | **36. Lithium Perchlorate (see Ammonium Perchlorate Rocket Propellants)** |
| **37. Magnesium powder (see Ammonium Perchlorate Rocket Propellants)** | | **38. Lithium Aluminum Hydride (see Ammonium Perchlorate Rocket Propellants)** |
| **39. Iron-III-oxide (red iron oxide) (see Ammonium Perchlorate Rocket Propellants)** | | **40. PVC (see Ammonium Perchlorate Rocket Propellants)** |
| **41. Copper Sulfide (see Ammonium Perchlorate Rocket Propellants)** | | **42. Sodium Hydride (see Ammonium Perchlorate Rocket Propellants)** |
| **43. Titanium Hydride (see Ammonium Perchlorate Rocket Propellants)** | | **44. Silicon Nitride (see Ammonium Perchlorate Rocket Propellants)** |
| **45. ADN (see ADN Rocket Propellants)** | | **46. Urea Nitrate (see ADN Rocket Propellants)** |
| **47. Silicon Powder (see ADN Rocket Propellants)** | | **48. Hexamine (see ADN Rocket Propellants)** |
| **49. Boron powder (see ADN Rocket Propellants)** | | **50. Sodium hypophosphite (see ADN Rocket Propellants)** |
| **51. KDN (see ADN Rocket Propellants)** | | **52. Nickel Chloride** |
| | |   |
| | | Nickel chloride forms beautiful greenish crystals or golden-yellow white granules or powder. The solid is readily prepared by dissolving nickel metal into hot dilute hydrochloric acid, followed by recrystallization. |
| | | |

## 53. TNT

TNT forms slightly yellow to yellowish crystals, yellowish-white powder, or slight yellow granules with a melting point of 80 Celsius. It is insoluble in water, but shows moderate solubility in acetone, glacial acetic acid, THF, and benzene. TNT is one of the highest weight-to-weight produced high explosives in the world, and during WWII, so much TNT was made that its preparation was ranked higher by weight then any other chemical except conc. nitric and sulfuric acids. TNT has been used widely in grenades, explosive projectiles, bombs, torpedoes, rockets, demolition charges, artillery and mortar shells, and in almost any explosive munition imaginable. The use of TNT has been extensively used for over 100 years, and to this day, TNT remains one the most valuable and widely manufactured secondary high explosives (~500,000,000 metric tones annually). TNT is widely and extensively used in the explosives field either by itself (in powder, granular, or melt cast), and extensively used in melt cast composites with other high explosives of the nitramine, nitrobenzene, nitrophenyl, and amino nitrobenzene explosives, and practically any meltable explosive to form blocks/sheets for demolition purposes, linear shape charges for cutting in demolition operations, main charge for mortar/artillery shells, anti-tank weapons, drop bombs, grenades, land mines, rocket/missile warheads, anti-ship weapons, underwater mines, and practically any explosive munition. TNT is also commonly used when mixed with ammonium nitrate for use in making military dynamites. Due to TNT's low melting point, it cannot be used in rocket fuels, and its tendency to undergo incomplete combustion drastically decreases its desirability in making gun propellants. **Danger of preparation:** can be prepared by non-scientists as long as all procedure(s) and safety info is strictly followed.

### Method of Preparation 1:

**Step 1:** Preparation of 99% nitric acid/methylene chloride solution (batch process)

| | | |
|---|---|---|
| **Step 2:** Preparation of TNT<br><br>Into a 3-neck flask, equipped with mechanical stirrer, thermometer, and wide mouth funnel, and sitting inside an ice/salt bath, pour in the entire 99% nitric acid/methylene chloride solution (prepared in step 1), and then add gradually through the wide mouth funnel 8400 milliliters of 98% sulfuric acid over a period of 15 minutes while rapidly stirring the contents in the 3-neck flask. Thereafter, cool this acidic reaction mixture down to around 5 to 0 Celsius under constant stirring. When the acidic reaction mixture reaches around 0 Celsius, replace the wide mouth funnel with an addition funnel, and then pour into this addition funnel 920 grams of toluene, and then slowly drip in this toluene into the reaction mixture, over a period of time sufficient to keep the nitrating reaction mixture at a temperature around 5 to 0 Celsius at all times. Now, once all the toluene has been added, continue to rapidly stir the reaction mixture at 0 to 5 Celsius for 1 hour. After 1 hour, remove the salt/ice bath and allow the reaction mixture to warm to room temperature. Once this point is achieved, remove the wide mouth funnel from the 3-neck flask, and replace with a bulb reflux condenser, and then place said 3-neck flask into a round bottom heating mantle, and then turn on the heating mantle, and heat/reflux the reaction mixture at 70 Celsius under rapid stirring for 2 hours. After 2 hours, raise the temperature of the reaction mixture to 80 Celsius and hold this temperature for 2 hours while rapidly stirring the reaction mixture. Thereafter, remove the heat source, and allow the reaction mixture to cool to room temperature; during the cool down period continue to rapidly stir the reaction mixture. Once the reaction mixture has cooled to room temperature, disassemble the apparatus, carefully pour the entire reaction mixture onto 4 kilograms of ice, contained in a very large clean beaker, equipped with mechanical stirrer, allow the ice to melt, and then rapidly blend this aqueous mixture under high speed for 45 minutes. After stirring for 45 minutes, stop stirring, and then filter the aqueous reaction mixture using a clean large Buchner funnel, under strong suction, wash with ten 750-milliliter portions of 10% sodium carbonate solution, followed by washing with ten 1000-milliliter portions of warm water, and then suction dry as best as possible. Note: the filtered aqueous reaction mixture after filtration, will form a two-phase mixture, with the upper layer containing the methylene chloride; this methylene chloride can be recycled back to step 1 for another run of TNT manufacture. Now, to purify the crude suction dried TNT, which will contain small amounts of 2,3,4-trinitrotoluene, 3,4,5-trinitrotoluene, and tetranitromethane, place 2000 milliliters of 70% sulfuric acid into a large beaker equipped with mechanical stirrer, and then add the suction dried product from the Buchner funnel, and then turn on the mechanical stirrer and blend at room temperature on high speed for 3 hours. After 3 hours, pour in 2000 milliliters of diethyl ether, and continue to stir the mixture at room temperature for 1 hour. After1 hour, stop stirring, and then filter-off the TNT product using a clean Buchner funnel under rapid suction, wash with six 750-milliliter portions of 10% sodium carbonate solution, followed by six 750-milliliter portions of warm water, and then followed by washing with three 1500-milliliter portions of 99% isopropyl alcohol, and then suction dry as best as possible. Finally, place the suction dried desired product from the Buchner funnel and place onto a shallow pan or tray, and bake in an oven at 78 Celsius for 12 hours. After 12 hours, remove the dried desired product of TNT from the oven, and recover the dried TNT (once it has cooled), and then place into a glass container sealed air-tight and store in a cool dry place until use. | | |

*- High Performance Ammonium Nitrate Propellants in this section -*

| | |
|---|---|
| **1. 04-03-001A High Performance military rocket propellant (JPL X350 type):** *77.8% ammonium nitrate, 11% dioctyl azelate, 6.6% hexane triol plasticizer, 1.6% 2,4-toluene diisocyanate, 1.5% poly propylene glycol plasticizer, 1.1% copper chromite burn rate accelerator, 0.05% ferric acetylacetonate curing catalyst, 0.35% impurities* | **2. 04-03-002A: Nitrate based military rocket propellant (smokeless):** *66.9% ammonium nitrate, 25% nitrocellulose, 8% cornstarch, 0.10% residue* |
| **3. 04-03-003A: Ammonium nitrate based military rocket propellant with ADN increased power addictive:** *64% Ammonium nitrate, 15.8% thermoplastic phenoxy resin, 11.9% ADN, 3% carbon black filler, 2.2% acetyl triethylcitrate catalyst, 2% sodium barbiturate stabilizer, 1% 2,4-toluene diamine addictive, 0.10% residue* | **4. 04-03-003B: Ammonium nitrate based military rocket propellant with TNT power addictive:** *70% Ammonium nitrate, 13.3% thermoplastic phenoxy resin, 6.6% acetyl triethylcitrate catalyst, 4% TNT, 3% carbon black filler, 1.5% sodium barbiturate stabilizer, 0.8% 2,4-toluene diamine addictive, 0.4% ammonium oxalate burn rate catalyst, 0.2% N-phenylmorpholine inert filler, 0.20% residue* |
| **5. 04-03-004A: Inexpensive ammonium nitrate high performance rocket propellant:** *54.9% ammonium nitrate, 29.9% magnesium, 11.8% PPG polymer binder, 1.9% dioctyl adipate plasticizer, 1% isophorone diisocyante curative agent, 0.491% mixed residues, 0.009% dibutyltin diacetate curing catalyst* | **6. 04-03-05A High Performance JPL X360 rocket propellant:** *68.6% ammonium nitrate, 23.4% polypropylene glycol 2025 plasticizer, 4.7% toluene diisocyante, 1.9% copper chromite burn rate catalyst, 1.1% 1,2,6-hexanetriol, 0.21% mixed impurities, 0.09% ferric acetyl acetonate* |
| **7. 04-03-06A High Performance polymer rocket propellant:** *79.9% ammonium nitrate, 10.8% butadiene copolymer binder, 2.3% carbon black filler, 2.1% dibutoxyethoxyethyl formal plasticizer, 1.97% naphthenic acid catalyst, 1.9% Milori blue pigment burn rate catalsyt, 0.47% magnesium oxide filler, 0.31% binder catalyst, 0.25% wetting agent* | **8. 04-03-07A High Performance ammonium nitrate rocket propellant:** *83.2% ammonium nitrate, 9.1% polybutadiene polymer binder, 2% milori blue pigment curing agent, 1% 2-methyl-5-vinylpyridine copolymers, 1% carbon black catalyst filler, 0.81% benzophenone burn rate catalyst, 0.81% Pentary A copolymer catalyst, 0.74% epichlorohydrin polymerization catalyst, 0.30% zinc oxide filler, 0.30% flexamine binder addictive, 0.24% Butyl-eight rubber accelerator, 0.23% mixed residues, 0.17% sulfur burn modifier, 0.1% Aersol OT catalyst* |
| **9. 04-03-07B High Performance polymer ammonium nitrate rocket propellant (modified):** *78.7% ammonium nitrate, 9.7% polybutadiene polymer, 3.9% ammonium dichromate, 2.3% carbon black, 2.1% ZP-211 polyether binder, 1% 2-methyl-5-vinylpyridine, 0.97% nickel chloride curing catalyst, 0.45% mixed residues, 0.35% magnesium oxide filler, 0.32% flexamine, 0.21% milori blue pigment* | **10. 04-03-08A High Performance smokeless rubber-like rocket propellant:** *78.9% ammonium nitrate, 13.7% circosol oil, 4.1% rubber binder, 0.83% carbon black, 0.75% diphenyl guanidine, 0.5% Captax accelerator, 0.28% mixed residue, 0.25% tetramethyl thiua disulfide accelerator, 0.25% sulfur burn modifier, 0.20% RPA No. 3, 0.20% zinc oxide, 0.04% stearic acid catalyst* |
| **11. 04-03-09A High Performance smokeless polymer rocket propellant:** *81.8% ammonium nitrate, 10.8% Butadiene copolymer, 2.1% dibutoxyethyoxyethyl formal plasticizer, 2.1% carbon black filler, 1.9% milori blue burning catalyst, 0.38% mixed residues, 0.32% zinc oxide catalyst, 0.32% flexamine antioxidant agent, 0.10% aerosol-OT wetting agent, 0.10% SA-113 vulcanization accelerator, 0.08% sulfur* | **12. 04-03-010A High Performance smokeless polymer rocket propellant:** *75.4% ammonium nitrate, 8.2% ammonium dichromate, 5.6% butadiene coplymer, 5.4% polybutadiene plasticizer, 2.7% milori blue burn catalyst, 2.1% philblack E furnace black, 0.42% magnesium oxide, 0.18% mixed residues* |
| **13. 04-03-010B High Performance smokeless polymer rocket propellant:** *80.8% ammonium nitrate, 12.4% 1,3-butadiene copolymer, 2.7% carbon black, 1.9% milori burn rate catalyst, 1.3% 2-methyl-5-vinylpyridine, 0.53% formal plasticizer, 0.30% mixed residues, 0.07% zinc oxide curing catalyst* | **14. 04-03-011A High Performance rocket propellant:** *68.7% ammonium nitrate, 11.7% dinitrotoluene plasticizer, 7.8% PVC binder, 3.9% Prussian blue burn rate catalyst, 3.9% carbon black filler, 3.9% surfactant, 0.10% mixed residues* |
| **15. 04-03-011B High Performance rocket propellant (modified):** *76% ammonium nitrate, 7.8% dinitrotoluene plasticizer, 7.8% 2,4-dintrodiphenyl ether plasticizer, 5.2% PVC binder, 2% ammonium dichromate catalyst, 1% Prussian blue burn rate catalyst, 0.20% mixed residues* | **16. 04-03-012A: A high performance JATO (jet assisted take-off) rocket propellant:** *72.9% ammonium nitrate, 8% 2,4,-dinitrodiphenyl ether, 8% ethylene glycol diglycolate, 4.7% cellulose acetate, 3% Prussian blue pigment catalyst, 2.3% acetonylacetone dioxime, 1% carbon black, 0.10% residue* |
| | |

| | |
|---|---|
| **17. 04-03-013A High Performance ammonium nitrate "plastic-like" propellant:** *70% ammonium nitrate, 15% aluminum, 15% aluminum stearate* | **18. 04-03-014A High Performance ammonium nitrate propellant:** *80.34% ammonium nitrate, 8.54% ammonium dichromate, 6.83% binder, 4.27% ferrous ferrocyanide burn catalyst, 0.02% balance*<br>**Percentage 2 (binder—of total composition):** *2.28% rubbery polymer, 2.28% plasticizer, 1.14% reinforcing agent, 0.56% quaternizing agent, 0.23% wetting agent, 0.12% vulcanization accelerator, 0.12% metal oxide, 0.06% antioxidant, 0.04% sulfur* |
| **19. 04-03-015A High Performance ammonium nitrate rocket propellant utilizing asphalt:** *76.81% ammonium nitrate, 22.94% N,N-dimethyl acrylamide, 025% methyl amyl ketone peroxide curing catalyst* | **20. 04-03-016A High Performance ammonium nitrate rocket propellant utilizing ammonium dichromate burn rate catalyst:** *73.58% ammonium nitrate, 12.41% methyl acrylate, 4.37% adipic acid, 4.24% diethylene glycol, 2.68% styrene, 2% ammonium dichromate, 0.396% benzoyl peroxide, 0.17% maleic anhydride, 0.099% lecithin, 0.065% balance* |
| **21. 04-03-016B High Performance ammonium nitrate rocket propellant utilizing ammonium dichromate burn rate catalyst:** *76% ammonium nitrate, 12.4% methyl acrylate, 2.8% diethylene glycol, 2.8% styrene, 2.8% adipic acid, 2% ammonium dichromate, 0.4% lecithin, 0.4% methyl ethyl ketone peroxide, 0.2%, 0.1% cobalt octoate, 0.1% balance* | **22. 04-03-017A Specialty ammonium nitrate semi-solid (gelled) rocket propellant (with increase moldability):** *71.25% ammonium nitrate, 22.35% unleaded gasoline, 3.75% carbon black, 2.65% aluminum naphthenate* |
| **23. 04-03-018A Specialty ammonium nitrate rocket propellant with moderate acceleration for use in spacecraft:**<br>*50.89% ammonium nitrate, 21.55% hydroxyl terminated polypropylene oxide copolymer, 14.97% aluminum metal fuel, 5.98% 2,6-tolyl-diisocyante curing compound, 2.99% 1-decanol addictive, 2.99% trimethylol propane antioxidant, 0.59% ferric acetyl acetonate curing catalyst, 0.04% impurities* | **24. 04-03-019A: High Performance ammonium nitrate rocket propellant with nitrocellulose base:** *47.24% ammonium nitrate, 26.24% nitrocellulose, 20.99% triacetin, 5.24% diethyl phthalate, 0.26% lecithin, 0.03% mixed balance* |
| **25. 04-03-019B: High Performance ammonium nitrate rocket propellant with nitrocellulose base:** *48.85% ammonium nitrate, 25.44% dimethoxyethyl phthalate, 25.44% nitrocellulose, 0.25% lecithin, 0.,02% impurities* | **26. 04-03-020A: High Performance ammonium nitrate rocket propellant:** *62.18% ammonium nitrate, 11.83% acetyl triethyl citrate binder, 10.49% cellulose acetate binder, 10.44% dinitrophenoxyethanol, 3.04% carbon black filler, 0.99% disodium barbiturate catalyst, 0.99% toluene diamine cure catalyst, 0.04% mixed balance* |
| **27. 04-03-021A: High Performance ammonium nitrate rocket propellant:** *75.52% ammonium nitrate, 8.25% ammonium dichromate, 5.63% copolymer binder, 5.43% polybutadiene, 2.71% Milori blue catalyst, 2% carbon black filler, 0.40% magnesium oxide catalyst, 0.06% mixed residual balance* | **28. 04-03-022A: High Performance ammonium nitrate rocket propellant with ferric nitrate burn accelerator:** *77.27% ammonium nitrate, 18.18% polybutadiene, 4.54% ferric nitrate monohydrate, 0.01% mixed balance* |
| **29. 04-03-023A: High Performance ammonium nitrate rocket propellant:** *51.85% ammonium nitrate, 25.92% magnesium bicarbonate, 14.81% nitrocellulose, 7.4% aluminum powder, 0.02% balance* | **30. 04-03-024A: Specialty high Performance ammonium nitrate rocket propellant:** *78.88% ammonium nitrate, 12.36% ammonium tricatechol chromate, 8.76% potassium nitrate* |
| **31. 04-03-024B: Specialty high Performance ammonium nitrate rocket propellant:** *80% ammonium nitrate, 10.86% pyridine tricatechol chromate, 8.2% potassium nitrate, 0.94% impurities* | **32. 04-03-024C: Specialty high Performance ammonium nitrate rocket propellant:** *79.52% ammonium nitrate, 11.64% chromium acetylacetonate, 8.84% potassium nitrate* |
| **33. 04-03-025A: High Performance ammonium nitrate rocket propellant:** *50% ammonium nitrate, 21.42% magnesium bicarbonate, 14.28% nitrocellulose, 7.14% aluminum powder, 5.71% magnesium, 1.42% aluminum stearate catalyst, 0.03% balance* | **34. 04-03-026A: Specialty high Performance polymer-based ammonium nitrate rocket propellant:** *64% ammonium nitrate, 18.8% polyurethane, 16% aluminum powder, 1.2% ethylene oxide polymerization agent* |
| | |

| | |
|---|---|
| **35. 04-03-027A High Performance smokeless polymer rocket propellant:** *82.29% ammonium nitrate, 17.45% copolymer binder, 0.249%% milori blue catalyst, 0.011% residual balance* | |

### 04-03-001A: High Performance military rocket propellant (JPL X350 type):

Into a special sealable heated vertical mixer with vacuum port, place ***196 grams of ammonium nitrate*** of 100 mesh, followed by ***84 grams of ammonium nitrate*** of 30 microns, followed by ***40 grams of dioctyl azelate***, followed by ***4 grams of copper chromite***, followed by ***200 milligrams of ferric acetylacetonate***, and then thoroughly blend the mixture for 45 minutes at room temperature. After 45 minutes, add in ***5.4 grams of poly propylene glycol***, followed by ***24 grams of hexane triol***. Thereafter, thoroughly blend this mixture for 30 minutes at room temperature on high speed. Thereafter, heat the mixture to 30 Celsius, and thoroughly blend the mixture on high speed at 30 Celsius under a vacuum of 10 millimeters of mercury for 35 minutes. Immediately after 35 minutes, equalize the pressure, open the vertical mixer, and toss in ***6 grams of 2,4-toluene diisocyante***, re-seal the vertical mixer, and then continue to heat and blend the mixture at 30 Celsius under a vacuum of 20 millimeters of mercury for 15 minutes. Afterwards, the propellant mixture is ready to cast. To do so, remove the heat source, equalize the pressure, and then open the vertical mixer; thereafter, pour and press the propellant mixture into any desired rocket motor or mold under a pressure of 4,000 psi, and then allow the rocket motor or mold to cure for 1 week at room temperature.
**Burn rate:** 3.98 seconds per inch at 4,000 psi; tends to vary based on diameter and density of rocket grain.
**Water resistance:** Moderate.
**Stability:** Can be stored for decades when kept dry and free from moisture.
**Sensitivity:** Stable towards friction; stable towards normal shock; stable to temperatures up to 150 Celsius.
**Flammability (1 to 10, un-contained):** 8.7 solid grain; 9 powdered grain.
**Ease of ignition (1 to 10 un-contained):** 7.9 solid grain; grain size varies and results in different ignition rates.
**Tendency to cake:** Will not cake.
**Detonation ability:** Can be detonated but only severe conditions—requires TNT, RDX, or HMX booster.
**Explosive ability (contained; explosive deflagration):** Below moderate.
**Percentage:** ***77.8% ammonium nitrate, 11% dioctyl azelate, 6.6% hexane triol plasticizer, 1.6% 2,4-toluene diisocyanate, 1.5% poly propylene glycol plasticizer, 1.1% copper chromite burn rate accelerator, 0.05% ferric acetylacetonate curing catalyst, 0.35% impurities***
**Classification:** Deflagrating explosive (classified as propellant).

**Danger of preparation:** Can be made by amateurs.
**Use:** Used in military and commercial rockets and missiles for multiple uses.

**Rocket motor**

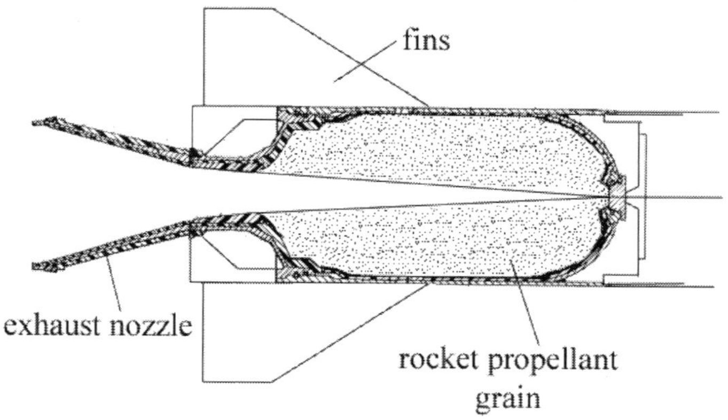


### 04-03-002A: Nitrate based military rocket propellant (smokeless):

Into a suitable vertical mixer, place 250 milliliters of acetone, followed by ***100 grams of nitrocellulose,*** followed by ***267 grams of powdered ammonium nitrate***, and then followed by ***32 grams of corn starch***; thereafter, blend the mixture until the bulk of the acetone evaporates (time may vary; do not use heat). Once the bulk of the acetone has evaporated, the mixture is ready to be used. To use, it simply needs to be pressed into any desirable rocket motor or mold under a pressure of 9,000 psi, and the resulting rocket motor or mold needs to be cured at 60 Celsius in an oven for 24 hours.
**Burn rate:** 3.2 seconds per inch at 9,000 psi; tends to vary based on diameter and density of rocket grain.
**Water resistance:** Moderate.
**Stability:** Can be stored for decades when kept dry and free from moisture.
**Sensitivity:** Stable towards friction; stable towards normal shock; stable to temperatures up to 150 Celsius.
**Flammability (1 to 10, un-contained):** 9 solid grain; 9.2 powdered grain.
**Ease of ignition (1 to 10 un-contained):** 8.8 solid grain; grain size varies and results in different ignition rates.

**Tendency to cake:** Will not cake.
**Detonation ability:** Can be detonated but only severe conditions—requires TNT, RDX, or HMX booster.
**Explosive ability (contained; explosive deflagration):** Above moderate.
**Percentage:** ***66.9% ammonium nitrate, 25% nitrocellulose, 8% cornstarch, 0.10% residue***
**Classification:** Deflagrating explosive (classified as propellant).

**Danger of preparation:** Can be made by amateurs, but it is advisable that person or persons should have some knowledge of chemistry and pyrotechnics before carrying out this operation.
**Use:** Used in military and commercial rockets and missiles for multiple uses.

### 04-03-003A: Ammonium nitrate based military rocket propellant with ADN increased power addictive:

Into a suitable vertical mixer, place ***79 grams of a thermoplastic phenoxy resin*** (commercially available as Bakelite PKDA-8500 resin), followed by ***59.5 grams of ADN***, followed by ***10 grams of sodium barbiturate***, followed by ***15 grams of finely divided standard carbon black***, and then followed by ***320 grams of powdered ammonium nitrate***; thereafter, blend the mixture for 15 minutes. After 15 minutes, add in ***11.2 grams of acetyl triethylcitrate***, and then immediately thereafter add in ***5 grams of 2,4-toluene diamine***, and then continue to blend the mixture on high speed for 10 minutes. Immediately thereafter, cast the mixture into any desirable rocket motor or mold, under a pressure of 6,000 psi, and then allow the rocket motor or mold to cure for 48 hours.
**Burn rate:** 3.1 seconds per inch at 6,000 psi; tends to vary based on diameter and density of rocket grain.
**Water resistance:** Moderate.
**Stability:** Can be stored for decades when kept dry and free from moisture.
**Sensitivity:** Stable towards friction; stable towards normal shock; stable to temperatures up to 150 Celsius.
**Flammability (1 to 10, un-contained):** 9 solid grain; 9.2 powdered grain.
**Ease of ignition (1 to 10 un-contained):** 8.8 solid grain; grain size varies and results in different ignition rates.
**Tendency to cake:** Will not cake.
**Detonation ability:** Can be detonated but only severe conditions—requires TNT, RDX, or HMX booster.
**Explosive ability (contained; explosive deflagration):** Above moderate.
**Percentage:** ***64% Ammonium nitrate, 15.8% thermoplastic phenoxy resin, 11.9% ADN, 3% carbon black filler, 2.2% acetyl triethylcitrate catalyst, 2% sodium barbiturate stabilizer, 1% 2,4-toluene diamine addictive, 0.10% residue***
**Classification:** Deflagrating explosive (classified as propellant).

**Danger of preparation:** Can be made by amateurs, but it is advisable that person or persons should have some knowledge of chemistry and pyrotechnics before carrying out this operation.
**Use:** Used in military and commercial rockets and missiles for multiple uses.

### 04-03-003B: Ammonium nitrate based military rocket propellant with TNT power addictive:

Into a suitable vertical mixer, place ***66.5 grams of a thermoplastic phenoxy resin*** (commercially available as Bakelite PKDA-8500 resin), followed by ***20 grams of TNT***, followed by ***7.5 grams of sodium barbiturate***, followed by ***15 grams of finely divided standard carbon black***, followed by ***1 gram of N-phenylmorpholine***, followed by ***2 grams of ammonium oxalate***, and then finally followed by ***350 grams of powdered ammonium nitrate***; thereafter, blend the mixture for about 30 minutes at high speed. After 30 minutes, add in ***33.3 grams of acetyl triethylcitrate***, and then immediately thereafter add in ***4 grams of 2,4-toluene diamine***, and then continue to blend the mixture for 15 minutes. After 15 minutes, press the mixture into any desirable rocket motor or mold under a pressure of 7,000 psi, and then allow the rocket motor or mold to cure for 48 hours.
**Burn rate:** 3.7 seconds per inch at 7,000 psi; tends to vary based on diameter and density of rocket grain.
**Water resistance:** Moderate.
**Stability:** Can be stored for decades when kept dry and free from moisture.
**Sensitivity:** Stable towards friction; stable towards normal shock; stable to temperatures up to 150 Celsius.
**Flammability (1 to 10, un-contained):** 8.9 solid grain; 9.2 powdered grain.
**Ease of ignition (1 to 10 un-contained):** 8 solid grain; grain size varies and results in different ignition rates.
**Tendency to cake:** Will not cake.
**Detonation ability:** Can be detonated but only severe conditions—requires TNT, RDX, or HMX booster.
**Explosive ability (contained; explosive deflagration):** Moderate.
**Percentage:** ***70% Ammonium nitrate, 13.3% thermoplastic phenoxy resin, 6.6% acetyl triethylcitrate catalyst, 4% TNT, 3% carbon black filler, 1.5% sodium barbiturate stabilizer, 0.8% 2,4-toluene diamine addictive, 0.4% ammonium oxalate burn rate catalyst, 0.2% N-phenylmorpholine inert filler, 0.20% residue***
**Classification:** Deflagrating explosive (classified as propellant).

**Danger of preparation:** Can be made by amateurs.
**Use:** Used in military and commercial rockets and missiles for multiple uses.

### 04-03-004A: Inexpensive ammonium nitrate high performance rocket propellant:

Into a suitable vertical mixer, place ***59.4 grams of PPG binder*** (polyosypropylene glycol, commercially available), followed by ***10 grams of commercially available dioctyl adipate plasticizer***, followed by ***150 grams of finely powdered magnesium*** of the usual mesh, and then followed by ***275 grams of powdered ammonium nitrate***, and then blend the mixture on high speed for 30 minutes. Thereafter, add in ***5.5 grams of isophorone diisocyanate curative agent***, followed immediately by ***50 milligrams of dibutyltin diacetate*** catalyst, and then continue to blend the mixer on high speed for 30 minutes. After 30 minutes, pour and press the mixture into any desirable rocket motor or mold under a pressure of 10,000 psi, and then cure the rocket motor or mold for several days at room temperature.
**Burn rate:** 4.1 seconds per inch at 10,000 psi; tends to vary based on diameter and density of rocket grain.
**Water resistance:** Moderate.
**Stability:** Can be stored for decades when kept dry and free from moisture.
**Sensitivity:** Stable towards friction; stable towards normal shock; stable to temperatures up to 150 Celsius.
**Flammability (1 to 10, un-contained):** 8.8 solid grain; 9.1 powdered grain.
**Ease of ignition (1 to 10 un-contained):** 7 solid grain; grain size varies and results in different ignition rates.
**Tendency to cake:** Will cake over time if exposed to moisture.
**Detonation ability:** Can be detonated but only severe conditions—requires TNT, RDX, or HMX booster.
**Explosive ability (contained; explosive deflagration):** Moderate.
**Percentage:** ***54.9% ammonium nitrate, 29.9% magnesium, 11.8% PPG polymer binder, 1.9% dioctyl adipate plasticizer, 1% isophorone diisocyante curative agent, 0.491% mixed residues, 0.009% dibutyltin diacetate curing catalyst***
**Classification:** Deflagrating explosive (classified as propellant).

**Danger of preparation:** Can be made by amateurs.
**Use:** Used in military and commercial rockets and missiles for multiple uses.

### 04-03-05A: High Performance JPL X360 rocket propellant:

Into a special sealable heated vertical mixer with vacuum port, place ***350 grams of powdered ammonium nitrate***, followed by ***10 grams of copper chromite***, and then blend the mixture for about 30 minutes to form a uniform mix. Thereafter, add in ***119.5 grams of polypropylene glycol 2025***, and then blend the mixture on moderate speed for about 30 minutes under a vacuum of 10 millimeters of mercury at a temperature of 21 Celsius. Thereafter, into a separate clean beaker, equipped with mechanical stirrer, prepare a curing mixture by adding and mixing, ***500 milligrams of ferric acetyl acetonate*** with ***24 grams of toluene diisocyante***; now, quickly equalize the pressure in the vertical mixer, then open the vertical mixer, and toss in this said curing mixture, and then re-seal the vertical mixer, and re-apply a vacuum of 10 millimeters of mercury, and then continue to blend the mixture on high speed for about 15 minutes at 21 Celsius. Thereafter, quickly equalize the pressure in the vertical mixer, open the vertical mixer, quickly add in ***6 grams of 1,2,6-hexanetriol***, and then re-seal the vertical mixer, apply the same vacuum as before (10mm of mercury), and then continue to blend the mixture on high speed for 10 additional minutes at 21 Celsius. After 10 minutes, the propellant mixture is ready to be casted. To do so, equalize the pressure, open the vertical mixer, and then pour and press the mixture into any desirable rocket motor or mold under a pressure of 9,000 psi, and then cure the rocket motor or mold in an oven at 60 Celsius for 16 hours.
**Burn rate:** 4.23 seconds per inch at 9,000 psi; tends to vary based on diameter and density of rocket grain.
**Water resistance:** Moderate.
**Stability:** Can be stored for decades when kept dry and free from moisture.
**Sensitivity:** Stable towards friction; stable towards normal shock; stable to temperatures up to 150 Celsius.
**Flammability (1 to 10, un-contained):** 8.6 solid grain; 9 powdered grain.
**Ease of ignition (1 to 10 un-contained):** 7 solid grain; grain size varies and results in different ignition rates.
**Tendency to cake:** Will cake over time if exposed to moisture.
**Detonation ability:** Cannot be detonated.
**Explosive ability (contained; explosive deflagration):** Below moderate.
**Percentage:** ***68.6% ammonium nitrate, 23.4% polypropylene glycol 2025 plasticizer, 4.7% toluene diisocyante, 1.9% copper chromite burn rate catalyst, 1.1% 1,2,6-hexanetriol, 0.21% mixed impurities, 0.09% ferric acetyl acetonate***
**Classification:** Deflagrating explosive (classified as propellant).

**Danger of preparation:** Can be made by amateurs.
**Use:** Used in military and commercial rockets and missiles for multiple uses.

### 04-03-06A: High Performance polymer rocket propellant:

Into a suitable heated vertical mixer, place ***55 grams of a butadiene methylvinylpyridine copolymer*** of a 90/10 mixture (commercially available), followed by ***12 grams of furnace quality carbon black***, and then blend the mixture on moderate speed for about 5 minutes. Thereafter, into an appropriate sized separate vertical mixer, place ***11 grams of dibutoxyethoxyethyl formal***, followed by ***9.7 grams of Milori blue pigment***, followed by ***10 grams of naphthenic acid***, followed by 50 grams of a ethylene oxide substituted alkly amines wetting agent (marketed by the Armour and Co), and then blend the mixture on high speed for about 10 minutes. Thereafter, allow this second mixture to stand for several minutes, and then decant-off the wetting agent, or use a Buchner funnel under suction to collect the solid material; either way, after removing the wetting agent, place the collected blended mixture back into the vertical mixer, and then add in the butadiene methylvinylpyridine copolymer/carbon black mixture (prepared first), and then continue to blend the combined mixture on high speed for about 15 minutes. After 15 minutes, add in ***2.4 grams of magnesium oxide***, followed by ***1.6 grams of a binder catalyst*** mixture (composed of 65% diarylaminoketone and 35% N,N'-diphenyl-p-phenylenediamine reaction products, and sold as Flexamine), and then turn on the heating element, and continue to blend the mixture on high speed until the temperature of the mixture reaches 68 Celsius. Once the mixtures temperature has reached 68 Celsius, slowly add in, in small portions at a time, ***405 grams of dry powdered ammonium nitrate***, and after all the nitrate has been added, continue to blend the mixture on high speed for 15 minutes at 68 Celsius. Thereafter, the mixture is ready to be pressed. To do so, simply press the propellant mixture into any desirable rocket motor or mold under a pressure of 5,000 psi, and then cure the rocket motor or mold in an oven at 76 Celsius for 18 hours.
**Burn rate:** 4.1 seconds per inch at 5,000 psi; tends to vary based on diameter and density of rocket grain.
**Water resistance:** Moderate.
**Stability:** Can be stored for decades when kept dry and free from moisture.
**Sensitivity:** Stable towards friction; stable towards normal shock; stable to temperatures up to 150 Celsius.
**Flammability (1 to 10, un-contained):** 8.5 solid grain; 9 powdered grain.
**Ease of ignition (1 to 10 un-contained):** 7 solid grain; grain size varies and results in different ignition rates.
**Tendency to cake:** Will cake over time if exposed to moisture.
**Detonation ability:** Cannot be detonated.
**Explosive ability (contained; explosive deflagration):** Below moderate.
**Percentage:** ***79.9% ammonium nitrate, 10.8% butadiene copolymer binder, 2.3% carbon black filler, 2.1% dibutoxyethoxyethyl formal plasticizer, 1.97% naphthenic acid catalyst, 1.9% Milori blue pigment burn rate catalsyt, 0.47% magnesium oxide filler, 0.31% binder catalyst, 0.25% wetting agent***
**Classification:** Deflagrating explosive (classified as propellant).

**Danger of preparation:** Can be made by amateurs.
**Use:** Used in military and commercial rockets and missiles for multiple uses.

### 04-03-07A: High Performance ammonium nitrate rocket propellant:

Into a special sealable heated vertical mixer with vacuum port (such as a "Baker-Perkins mixer"), place ***450 grams of polybutadiene***, followed by ***50 grams of 2-methyl-5-vinylpyridine***, and then blend the mixture for about 5 minutes at room temperature. Thereafter, add in ***50 grams of carbon black***, and then continue to blend the mixture for another 5 minutes. After 5 minutes, add in ***8.7 grams of finely powdered sulfur***, followed by ***15 grams of zinc oxide***, followed by ***5 grams of "Aerosol OT"*** (which is a dioctyl ester of sodium sulfosuccinic acid), followed by ***15 grams of "Felxamine***" (a commercially available binder composed of 65% complex diarylamineketone, and 35% N,N'-diphenyl-p-phenylenediamine), and continue to blend the mixture for about 5 minutes. After 5 minutes, add in ***40 grams of benzophenone***, followed by ***40 grams of Pentaryl A***" (which is a amylbiphenyl), and then continue to blend the mixture for about 60 minutes. After 60 minutes, seal the vertical mixer, apply a vacuum of 700 millimeters of mercury, and then blend the mixture on high speed for 5 minutes. After 5 minutes, remove the mild vacuum, equalize the pressure, open the vertical mixer, and then add in ***36.6 grams of epichlorohydrin***; thereafter, continue to blend the mixture without vacuum, for about 10 minutes. After 10 minutes, slowly add in, in small portions at a time, ***4083 grams of ammonium nitrate*** of 40 microns on average, and blend the mixture on moderate speed during the addition. After the addition of the nitrate, continue to blend the mixture for about 60 minutes to form a uniform mixture. After 60 minutes, add in ***99 grams of milori blue pigment agent***, and then blend the mixture for about 5 minutes. After 5 minutes, add in ***12.2 grams of "Butyl-Eight", which is a dithiocarbamate rubber accelerator***, and then blend the mixture on high speed for about 15 minutes. At this point, the temperature of the propellant mixture may have increased above room temperature. Either way, if the temperature of the propellant mixture is bellow 30 Celsius, heat the propellant mixture to 67 Celsius, and then blend the propellant mixture at this temperature for 15 minutes. After 15 minutes, remove the heat source, and then press the mixture into any desirable rocket motor or mold under a pressure of 7,000 psi, and then cure the rocket motor or mold in an oven at 80 Celsius for 18 hours.
**Burn rate:** 4.2 seconds per inch at 7,000 psi; tends to vary based on diameter and density of rocket grain.
**Water resistance:** Moderate.
**Stability:** Can be stored for decades when kept dry and free from moisture.
**Sensitivity:** Stable towards friction; stable towards normal shock; stable to temperatures up to 150 Celsius.
**Flammability (1 to 10, un-contained):** 8.6 solid grain; 9 powdered grain.
**Ease of ignition (1 to 10 un-contained):** 7 solid grain; grain size varies and results in different ignition rates.
**Tendency to cake:** Will cake over time if exposed to moisture.

**Detonation ability:** Can be detonated but only severe conditions—requires TNT, RDX, or HMX booster.
**Explosive ability (contained; explosive deflagration):** Below moderate.
**Percentage:** ***83.2% ammonium nitrate, 9.1% polybutadiene polymer binder, 2% milori blue pigment curing agent, 1% 2-methyl-5-vinylpyridine copolymers, 1% carbon black catalyst filler, 0.81% benzophenone burn rate catalyst, 0.81% Pentary A copolymer catalyst, 0.74% epichlorohydrin polymerization catalyst, 0.30% zinc oxide filler, 0.30% flexamine binder addictive, 0.24% Butyl-eight rubber accelerator, 0.23% mixed residues, 0.17% sulfur burn modifier, 0.1% Aersol OT catalyst,***
**Classification:** Deflagrating explosive (classified as propellant).

**Danger of preparation:** Can be made by amateurs.
**Use:** Used in military and commercial rockets and missiles for multiple uses.

### 04-03-07B: High Performance polymer ammonium nitrate rocket propellant (modified):

Into a special sealable heated vertical mixer with vacuum port (such as a "Baker-Perkins mixer"), place ***180 grams of polybutadiene***, followed by ***20 grams of 2-methyl-5-vinylpyridine***, and then blend them mixture for 5 minutes at room temperature. Thereafter, add in ***44 grams of carbon black***, and then continue to blend the mixture for another 5 minutes. After 5 minutes, add in ***40 grams of ZP-211*** (a high molecular weight polyether sold by the Thiokol corporation), followed by ***6 grams of "flexamine"***, followed by ***6.6 grams of magnesium oxide,*** and then followed by ***4 grams of milori blue pigment agent,*** and then blend the mixture on moderate speed for 15 minutes. After 15 minutes, add in ***72.8 grams of ammonium dichromate,*** followed by ***1447 grams of powdered ammonium nitrate,*** and then blend the mixture on moderate speed during the addition. After the addition of the nitrate, continue to blend the mixture for about 60 minutes to form a uniform mix. After 60 minutes, quickly add in ***18 grams of anhydrous nickel-II-chloride,*** and then quickly blend the mixture for about 5 minutes; thereafter, the mixture is ready for use. To use, pour and press the mixture into any desirable rocket motor or mold under a pressure 7,500 psi, and then cure the rocket motor or mold at room temperature for 48 hours.
**Burn rate:** 4 seconds per inch at 7,500 psi; tends to vary based on diameter and density of rocket grain.
**Water resistance:** Moderate.
**Stability:** Can be stored for decades when kept dry and free from moisture.
**Sensitivity:** Stable towards friction; stable towards normal shock; stable to temperatures up to 150 Celsius.
**Flammability (1 to 10, un-contained):** 8.3 solid grain; 9 powdered grain.
**Ease of ignition (1 to 10 un-contained):** 7 solid grain; grain size varies and results in different ignition rates.
**Tendency to cake:** Will cake over time if exposed to moisture.
**Detonation ability:** Can be detonated but only severe conditions—requires TNT, RDX, or HMX booster.
**Explosive ability (contained; explosive deflagration):** Below moderate.
**Percentage:** ***78.7% ammonium nitrate, 9.7% polybutadiene polymer, 3.9% ammonium dichromate, 2.3% carbon black, 2.1% ZP-211 polyether binder, 1% 2-methyl-5-vinylpyridine, 0.97% nickel chloride curing catalyst, 0.45% mixed residues, 0.35% magnesium oxide filler, 0.32% flexamine, 0.21% milori blue pigment***
**Classification:** Deflagrating explosive (classified as propellant).

**Danger of preparation:** Can be made by amateurs.
**Use:** Used in military and commercial rockets and missiles for multiple uses.

### 04-03-08A: High Performance smokeless rubber-like rocket propellant:

Into a suitable vertical mixer, place ***200 grams of GR-3 butadiene polymer rubber***, followed by ***60 grams of oil, sold as Cricosol 2XH***, followed by ***10 grams of RPA No. 3*** (a xylyl mercaptan solution in inert hydrocarbon solvent), followed by ***10 grams of zinc oxide***, and then followed by ***2 grams of stearic acid***. Thereafter, blend the mixture on high speed 30 minutes. Thereafter, add in ***600 grams of additional circosol 2XH oil***, and then continue to blend the mixture on high speed for another 30 minutes to form a uniform homogeneous mixture. Thereafter, add in ***24 grams of "Captax" (2-mercaptobenolthiazole)***, followed by ***12 grams of "Tuads" (tetramethyl thiuram disulfide),*** followed by ***36 grams of "DPG" (diphenyl guanidine)***, followed by ***40 grams of carbon black***, and then followed by ***12 grams of finely divided sulfur***; thereafter, continue to blend mixture for 15 minutes. Thereafter, add in ***3784 grams of ammonium nitrate*** of 100 mesh, and then continue to blend the mixture on high speed for 30 minutes. After 30 minutes, the mixture needs to be degassed, therefore, pour the mixture into a sealable clean vertical mixer, quipped with vacuum port, and blend this mixture on high speed under a vacuum of 100mm of mercury at room temperature for 1 hour. After 1 hour, the mixture is ready for use, so remove the vacuum, equalize the pressure, open the vertical mixer, and then pour and cast the mixture into any desirable rocket motor or mold under a pressure of 6,000 psi, and then cure the rocket motor or mold in an oven at 104 Celsius for 24 hours. The result will be a rubbery like material that needs to be ignited using a special high temperature ignition composition.
**Burn rate:** 4.1 seconds per inch at 6,000 psi; tends to vary based on diameter and density of rocket grain.
**Water resistance:** Very good.
**Stability:** Can be stored for decades.
**Sensitivity:** Stable towards friction; stable towards normal shock; stable to temperatures up to 150 Celsius.
**Flammability (1 to 10, un-contained):** 8.2 solid grain; 9 fine grain.

**Ease of ignition (1 to 10 un-contained):** 6.3 solid grain; grain size varies and results in different ignition rates.
**Tendency to cake:** Will not cake.
**Detonation ability:** Can be detonated but only severe conditions—requires TNT, RDX, or HMX booster.
**Explosive ability (contained; explosive deflagration):** Low.
**Percentage:** ***78.9% ammonium nitrate, 13.7% circosol oil, 4.1% rubber binder, 0.83% carbon black, 0.75% diphenyl guanidine, 0.5% Captax accelerator, 0.28% mixed residue,0.25% tetramethyl thiuram disulfide accelerator, 0.25% sulfur burn modifier, 0.20% RPA No. 3, 0.20% zinc oxide, 0.04% stearic acid catalyst,***
**Classification:** Deflagrating explosive (classified as propellant).

**Danger of preparation:** Can be made by amateurs.
**Use:** Used in military and commercial rockets and missiles for multiple uses.

### 04-03-09A: High Performance smokeless polymer rocket propellant:

Into a suitable vertical mixer, place ***375 milligrams of finely powdered sulfur***, followed by ***1.5 grams of zinc oxide***, followed by ***10 grams of carbon black***, followed by ***9 grams of Milori blue agent***, followed by ***376.5 grams of powdered ammonium nitrate***, followed by ***500 milligrams of Aerosol-OT wetting agent*** (dioctyl ester of sodium sulfosuccinic acid), and then followed by ***1.5 grams of Flexamine antioxidant*** (25% diarylamine-ketone complex, and complex mixture of amines and phenols; commercially available); thereafter, blend the mixture for 30 minutes. Thereafter, add in ***50 grams of a copolymer*** (composed of butadiene of 2-methyl-5-vinylpyridine), followed by ***10 grams of dibutoxyethoxyethyl formal plasticizer***, and then followed by ***500 milligrams of SA-113*** (which is N,N-dimethyl-tertiary butyl sulfenyl dithiocarbamate vulcanization accelerator); thereafter, continue to blend the mixture for 20 minutes on high speed. After 20 minutes, the mixture is ready to be pressed. To do so, pour and press the mixture into any desirable rocket engine or mold under a pressure of 9,000 psi, and then cure the rocket engine or mold in an oven at 76 Celsius for 7 days.
**Burn rate:** 4.4 seconds per inch at 9,000 psi; tends to vary based on diameter and density of rocket grain.
**Water resistance:** Good.
**Stability:** Can be stored for decades when kept from moisture.
**Sensitivity:** Stable towards friction; stable towards normal shock; stable to temperatures up to 150 Celsius.
**Flammability (1 to 10, un-contained):** 8.1 solid grain; 8.8 powdered grain.
**Ease of ignition (1 to 10 un-contained):** 6.7 solid grain; grain size varies and results in different ignition rates.
**Tendency to cake:** Will not cake.
**Detonation ability:** Can be detonated but only severe conditions—requires TNT, RDX, or HMX booster.
**Explosive ability (contained; explosive deflagration):** Low.
**Percentage:** ***81.8% ammonium nitrate, 10.8% Butadiene copolymer, 2.1% dibutoxyethyoxyethyl formal plasticizer, 2.1% carbon black filler, 1.9% milori blue burning catalyst, 0.38% mixed residues, 0.32% zinc oxide catalyst, 0.32% flexamine antioxidant agent, 0.10% aerosol-OT wetting agent, 0.10% SA-113 vulcanization accelerator, 0.08% sulfur***
**Classification:** Deflagrating explosive (classified as propellant).

**Danger of preparation:** Can be made by amateurs.
**Use:** Used in military and commercial rockets and missiles for multiple uses.

### 04-03-010A: High Performance smokeless polymer rocket propellant:

Into a suitable vertical mixer, place ***2.1 grams of magnesium oxide***, followed by ***41 grams of ammonium dichromate***, followed by ***375 grams of powdered ammonium nitrate***, followed by ***13.5 grams of Milori blue***, and then blend the mixture on high speed for 45 minutes. Thereafter, quickly add in ***28 grams of a copolymer 1,3-butadiene-2-methyl-5-vinylpyridine***, followed by ***10.5 grams of Philblack E*** (a furnace black sold by the Phillips Petroleum Company), followed by ***27 grams of polybutadiene*** compound (sold as "Butarez 15 plasticizer), and then continue to blend the mixture on high speed for 15 minutes. After 15 minutes, the mixture is ready to be pressed. To do so, pour and press the mixture into any desirable rocket motor or mold under a pressure of 7,000 psi, and then cure the rocket motor or mold in an oven at 120 Celsius for 3 hours.
**Burn rate:** 4.22 seconds per inch at 7,000 psi; tends to vary based on diameter and density of rocket grain.
**Water resistance:** Very good.
**Stability:** Can be stored for decades.
**Sensitivity:** Stable towards friction; stable towards normal shock; stable to temperatures up to 150 Celsius.
**Flammability (1 to 10, un-contained):** 8.21 solid grain; 9 fine grain.
**Ease of ignition (1 to 10 un-contained):** 6.4 solid grain; grain size varies and results in different ignition rates.
**Tendency to cake:** Will not cake.
**Detonation ability:** Can be detonated but only severe conditions—requires TNT, RDX, or HMX booster.
**Explosive ability (contained; explosive deflagration):** Low.
**Percentage:** ***75.4% ammonium nitrate, 8.2% ammonium dichromate, 5.6% butadiene coplymer, 5.4% polybutadiene plasticizer, 2.7% milori blue burn catalyst, 2.1% philblack E furnace black, 0.42% magnesium oxide, 0.18% mixed residues***

**Classification:** Deflagrating explosive (classified as propellant).

**Danger of preparation:** Can be made by amateurs.
**Use:** Used in military and commercial rockets and missiles for multiple uses.

### 04-03-010B: High Performance smokeless polymer rocket propellant:

Into a standard vertical mixer, place ***62.5 grams of 1,3-butadiene copolymer***, followed by ***6.9 grams of 2-methyl-5-vinylpyridine***, followed by ***13.8 grams of carbon black***, and then blend the mixture on high speed for about 15 minutes. After 15 minutes, add in ***407.5 grams of powdered ammonium nitrate***, followed by ***2.7 grams of dibutoxyethoxyethyl formal plasticizer***, followed by ***360 milligrams of zinc oxide*** curing catalyst, and then finally followed by ***10 grams of milori blue burning rate catalysts***; thereafter, continue to blend the mixture on high speed at room temperature for 40 minutes. Thereafter, the mixture is ready for pressing. To do so, the mixture simply needs to be pressed into any desirable rocket motor or mold under a pressure for 4,000 psi, and then the rocket motor or mold needs to be cured at 120 Celsius for 4 hours.
**Burn rate:** 4.22 seconds per inch at 4,000 psi; tends to vary based on diameter and density of rocket grain.
**Water resistance:** Good.
**Stability:** Can be stored for decades when kept from moisture.
**Sensitivity:** Stable towards friction; stable towards normal shock; stable to temperatures up to 150 Celsius.
**Flammability (1 to 10, un-contained):** 8.2 solid grain; 9 fine grain.
**Ease of ignition (1 to 10 un-contained):** 6.7 solid grain; grain size varies and results in different ignition rates.
**Tendency to cake:** Will not cake.
**Detonation ability:** Can be detonated but only severe conditions—requires TNT, RDX, or HMX booster.
**Explosive ability (contained; explosive deflagration):** Low.
**Percentage:** ***80.8% ammonium nitrate, 12.4% 1,3-butadinen copolymer, 2.7% carbon black, 1.9% milori burn rate catalyst, 1.3% 2-methyl-5-vinylpyridine, 0.53% formal plasticizer, 0.30% mixed residues, 0.07% zinc oxide curing catalyst***
**Classification:** Deflagrating explosive (classified as propellant).

**Danger of preparation:** Can be made by amateurs.
**Use:** Used in military and commercial rockets and missiles for multiple uses.

### 04-03-011A: High Performance rocket propellant:

Into a special heated vertical mixer, place ***75 grams of dinitrotoluene***, and then gradually heat the nitro compound to 120 Celsius under rapid stirring. When the nitro compound reaches a temperature of about 120 Celsius, add in ***50 grams of finely divided PVC***, and then continue to blend the mixture at 120 Celsius for about 5 minutes. After 5 minutes, add in ***25 grams of Prussian blue pigment***, followed by ***25 grams of finely powdered carbon black***, followed by ***25 oleyl dimethyl ethyl ammonium bromide surfactant***, followed by ***440 grams of powdered ammonium nitrate***, and then continue to blend the mixture at 120 Celsius for 1 hour. Thereafter, the propellant mixture is ready to be poured. To do so, pour and press the hot mixture into any desirable rocket motor or mold under a pressure of 3,000 psi, and then allow the rocket motor or mold to cool and cure for 3 days.
**Burn rate:** 3 seconds per inch at 3,000 psi; tends to vary based on diameter and density of rocket grain.
**Water resistance:** Good.
**Stability:** Can be stored for decades when kept from moisture.
**Sensitivity:** Stable towards friction; stable towards normal shock; stable to temperatures up to 150 Celsius.
**Flammability (1 to 10, un-contained):** 8.8 solid grain; 9 fine grain.
**Ease of ignition (1 to 10 un-contained):** 6.9 solid grain; grain size varies and results in different ignition rates.
**Tendency to cake:** Will not cake.
**Detonation ability:** Can be detonated but only severe conditions—requires TNT, RDX, or HMX booster.
**Explosive ability (contained; explosive deflagration):** Low.
**Percentage:** ***68.7% ammonium nitrate, 11.7% dinitrotoluene plasticizer, 7.8% PVC binder, 3.9% Prussian blue burn rate catalyst, 3.9% carbon black filler, 3.9% surfactant, 0.10% mixed residues***
**Classification:** Deflagrating explosive (classified as propellant).

**Danger of preparation:** Can be made by amateurs, but caution should be excersized as dinitrotoluene is highly toxic.
**Use:** Used as a propellant in anti-ship missiles, and long-range rockets and missiles.

### 04-03-011B: High Performance rocket propellant (modified):

Into a special heated vertical mixer, place ***39 grams of dinitrotoluene***, followed by ***39 grams of 2,4-dinitrodiphenyl ether***, and then gradually heat the nitro compounds to 120 Celsius under rapid stirring. When the nitro compounds reach a temperature of about 120 Celsius, add in ***26 grams of finely divided PVC***, and then continue to blend the mixture at 120 Celsius for about 5 minutes. After 5 minutes, add in ***5 grams of Prussian blue pigment***, followed by ***10 grams of ammonium dichromate***, and then followed by ***380 grams of powdered ammonium nitrate***, and then continue to blend the mixture at 120 Celsius for 1 hour. Thereafter, the propellant mixture is ready to be poured. To do so, pour and press the hot mixture into any desirable rocket motor or mold under a pressure of 3,000 psi, and then allow the rocket motor or mold to cool and cure for 3 days.

**Burn rate:** 3 seconds per inch at 3,000 psi; tends to vary based on diameter and density of rocket grain.
**Water resistance:** Good.
**Stability:** Can be stored for decades when kept from moisture.
**Sensitivity:** Stable towards friction; stable towards normal shock; stable to temperatures up to 150 Celsius.
**Flammability (1 to 10, un-contained):** 8.8 solid grain; 9 fine grain.
**Ease of ignition (1 to 10 un-contained):** 6.9 solid grain; grain size varies and results in different ignition rates.
**Tendency to cake:** Will not cake.
**Detonation ability:** Can be detonated but only severe conditions—requires TNT, RDX, or HMX booster.
**Explosive ability (contained; explosive deflagration):** Low.
**Percentage:** ***76% ammonium nitrate, 7.8% dinitrotoluene plasticizer, 7.8% 2,4-dintrodiphenyl ether plasticizer, 5.2% PVC binder, 2% ammonium dichromate catalyst, 1% Prussian blue burn rate catalyst, 0.20% mixed residues***
**Classification:** Deflagrating explosive (classified as propellant).

**Danger of preparation:** Can be made by amateurs, but caution should be excersized as dinitrotoluene is highly toxic.
**Use:** Used as a propellant in anti-ship missiles, and long-range rockets and missiles.

### 04-03-012A: A high performance JATO (jet assisted take-off) rocket propellant:

Into a special sealable heated vertical mixer equipped with vacuum port, place ***40 grams of 2,4,-dinitrodiphenyl ether,*** followed by ***40 grams of ethylene glycol diglycolate***, then heat the mixture to 140 Celsius, and then blend the mixture at this temperature for 15 minutes to form a homogenous mixture. Thereafter add in ***23.7 grams of cellulose acetate***, in small portions at a time, while thoroughly blending the mixture. After the addition of all the cellulose acetate, lower the temperature to 120 Celsius, and when this new temperature is achieved, add in ***11.5 grams acetonylacetone dioxime***, and then continue to blend the mixture for 15 minute at 120 Celsius. Thereafter, add in ***255 grams of ammonium nitrate*** of 325 mesh, and then continue to blend the mixture at 120 Celsius for 10 minutes on high speed. Thereafter, add in ***109.4 grams of ammonium nitrate*** of 325 mesh, followed by ***5 grams of finely divided carbon black***, and then followed by ***15 grams of Prussian blue pigment catalyst***, and then continue to blend the mixture for 15 minutes at 120 Celsius. Now, after 15 minutes, seal the vertical mixer, and then apply a vacuum of 15 millimeters of mercury (de-gassing) for 15 minutes at 120 Celsius. After 15 minutes of de-gassing, the mixture is ready for use, so remove the vacuum, equalize the pressure, and then open the vertical mixer; thereafter, pour and press the mixture, while still hot, into any desirable rocket motor or mold under a pressure of 4,000 psi, and then allow the rocket motor or mold to cure at room temperature for 4 days.

**Burn rate:** 3 seconds per inch at 4,000 psi; tends to vary based on diameter and density of rocket grain.
**Water resistance:** Good.
**Stability:** Can be stored for decades when kept from moisture.
**Sensitivity:** Stable towards friction; stable towards normal shock; stable to temperatures up to 150 Celsius.
**Flammability (1 to 10, un-contained):** 8.8 solid grain; 9 fine grain.
**Ease of ignition (1 to 10 un-contained):** 6.9 solid grain; grain size varies and results in different ignition rates.
**Tendency to cake:** Will not cake.
**Detonation ability:** Can be detonated but only severe conditions—requires TNT, RDX, or HMX booster.
**Explosive ability (contained; explosive deflagration):** Low.
**Percentage:** ***72.9% ammonium nitrate, 8% 2,4,-dinitrodiphenyl ether, 8% ethylene glycol diglycolate, 4.7% cellulose acetate, 3% Prussian blue pigment catalyst, 2.3% acetonylacetone dioxime, 1% carbon black, 0.10% residue***
**Classification:** Deflagrating explosive (classified as propellant).

**Danger of preparation:** Can be made by amateurs, but caution should be exercised during the heating process.
**Use:** Used in JATO rocket assisted munitions.

### 04-03-013A: High Performance ammonium nitrate "plastic-like" propellant:

Into a special sealable heated vertical mixer, place ***210 grams of powdered ammonium nitrate***, followed by ***45 grams of finely powdered aluminum***, and then followed by ***45 grams of aluminum stearate***. Thereafter, gently heat the mixture to about 120 Celsius, and blend the mixture during the heating process. When the temperature reaches 120 Celsius, the aluminum stearate should melt, forming a plastic mass. When the temperature of the mixture reaches 120 Celsius, rapidly blend the mixture for 10 minutes to form a gooey-mass. Note: monitor the temperature and the mixing time as the aluminum stearate may polymerize and then solidify. If solidification results, the mixture is worthless and the ammonium nitrate will have to be discarded or recovered by drowning into water. After the brief mixing time, the hot gooey mixture is ready to be cast. To do so, pour and press the mixture into any desirable rocket motor or mold under a pressure of 4,000 psi, and then allow the rocket motor or mold to cure at room temperature for 48 hours. Can be initiated using a black powder charge or standard ignition composition.

**Burn rate:** 2.9 seconds per inch at 4,000 psi; tends to vary based on diameter and density of rocket grain.
**Water resistance:** Good.
**Stability:** Can be stored for decades when kept from moisture.
**Sensitivity:** Stable towards friction; stable towards normal shock; stable to temperatures up to 150 Celsius.
**Flammability (1 to 10, un-contained):** 9 solid grain; 9.2 fine grain.
**Ease of ignition (1 to 10 un-contained):** 7.1 solid grain; grain size varies and results in different ignition rates.
**Tendency to cake:** Will not cake.
**Detonation ability:** Can be detonated but only severe conditions—requires TNT, RDX, or HMX booster.
**Explosive ability (contained; explosive deflagration):** Low.
**Percentage:** ***70% ammonium nitrate, 15% aluminum, 15% aluminum stearate***
**Classification:** Deflagrating explosive (classified as propellant).

**Danger of preparation:** Can be made by amateurs.
**Use:** Used in military and commercial rockets and missiles for multiple uses.

### 04-03-014A: High Performance ammonium nitrate propellant:

Into a suitable vertical mixer, place ***120 grams of a rubbery polymer*** (see binder ingredients below), followed by ***60 grams of a reinforcing agent*** (see binder ingredients below), followed by ***120 grams of plasticizer*** (see binder ingredients below), followed by ***12 grams of wetting agent*** (see binder ingredients below), followed by ***3.6 grams of an antioxidant*** (see binder ingredients below), followed by ***6 grams of a vulcanization accelerator*** (see binder ingredients below), followed by ***2.4 grams of sulfur,*** followed by ***30 grams of a quaternizing agent*** (see binder ingredients below), and then followed by ***6 grams of a metal oxide*** (see binder ingredients below). Thereafter, blend this mixture of chemicals for 15 minutes to form a uniform binder mixture. Thereafter, into clean separate vertical mixer, place ***470 grams of powdered ammonium nitrate,*** followed by ***50 grams of ammonium dichromate,*** followed by ***25 grams of ferrous ferrocyanide,*** and then followed by ***40 grams of the binder mixture prepared previously.*** Thereafter, blend this combined mixture for 40 minutes on high speed to form a uniform mixture. Thereafter, the mixture is read for use. To use, the mixture simply needs to be pressed into any desirable rocket motor or mold under a pressure of 10,000 psi in the usual manner. Can be ignited using the typical ignition compositions.

**Burn rate:** 4 seconds per inch at 10,000 psi; tends to vary based on diameter and density of rocket grain.
**Water resistance:** Good.
**Stability:** Can be stored for decades when kept from moisture.
**Sensitivity:** Stable towards friction; stable towards normal shock; stable to temperatures up to 150 Celsius.
**Flammability (1 to 10, un-contained):** 8.3 solid grain; 9 fine grain.
**Ease of ignition (1 to 10 un-contained):** 7 solid grain; grain size varies and results in different ignition rates.
**Tendency to cake:** Will not cake.
**Detonation ability:** Can be detonated but only severe conditions—requires TNT, RDX, or HMX booster.
**Explosive ability (contained; explosive deflagration):** Low.
**Percentage:** ***80.34% ammonium nitrate, 8.54% ammonium dichromate, 6.83% binder, 4.27% ferrous ferrocyanide burn catalyst, 0.02% balance***
**Percentage 2 (binder—of total composition):** ***2.28% rubbery polymer, 2.28% plasticizer, 1.14% reinforcing agent, 0.56% quaternizing agent, 0.23% wetting agent, 0.12% vulcanization accelerator, 0.12% metal oxide, 0.06% antioxidant, 0.04% sulfur***

**Binder ingredients:**

**A. Rubbery polymer:** Can be any polymer with a "Mooney" (ML-4) value of 10 to 40.
**B. Plasticizer:** Pentaryl A, Paraflux, Circosol-2XH, or dioctyl phthalate **.**
**C. Reinforcing agent:** carbon black, wood flour, lignin, styrene-divinylbenzene, or methyl acrylatedivinylbenzene.
**D. Quaternizing agent:** Methyl iodide, methyl bromide, methylene iodide, ethylene bromide, or methyl sulfate.
**E. Wetting agent (helps deflocculates or disperse the oxidizer):** Aersol OT, lecithin, or Duomeen C diacetate.
**F. Vulcanization accelerator:** N,N-dimethyl—S-tert-butyl sulfenyl dithiocarbamate, or butyl-eight (dithiocarbamate rubber accelerator).
**G. Metal oxide:** Iron oxide, zinc oxide, magnesium oxide, or calcium oxide.
**H. antioxidant:** Flexamine, phenyl-beta-naphthylamine, 2,2-methylene-bis-(4-methyl-6-tert-butylphenol).
**Classification:** Deflagrating explosive (classified as propellant).

**Danger of preparation:** Can be made by amateurs.
**Use:** Used in military and commercial rockets and missiles for multiple uses.

### 04-03-015A: High Performance ammonium nitrate rocket propellant utilizing asphalt:

Into a suitable vertical mixture, place ***114.7 grams of N,N-dimethyl acrylamide***, followed by ***384.05 grams of powdered ammonium nitrate***, and then blend the mixture for 15 minutes on high speed. Thereafter, add in ***1.25 grams of methyl amyl ketone peroxide***, and then continue to blend the mixture for 15 minutes to form a homogenous mixture. Thereafter, the mixture is ready for use. To use, the mixture simply needs to be pressed into any desirable rocket motor or mold under a pressure of 6,000 psi, in the usual fashion, and then place the rocket motor or mold in an oven and heat to 90 Celsius for 3 hours. The propellant material can be ignited using any suitable ignition composition.
**Burn rate:** 3.1 seconds per inch at 6,000 psi; tends to vary based on diameter and density of rocket grain.
**Water resistance:** Good.
**Stability:** Can be stored for decades when kept from moisture.
**Sensitivity:** Stable towards friction; stable towards normal shock; stable to temperatures up to 150 Celsius.
**Flammability (1 to 10, un-contained):** 8.7 solid grain; 9 fine grain.
**Ease of ignition (1 to 10 un-contained):** 7 solid grain; grain size varies and results in different ignition rates.
**Tendency to cake:** Will not cake.
**Detonation ability:** Can be detonated but only severe conditions—requires TNT, RDX, or HMX booster.
**Explosive ability (contained; explosive deflagration):** Low.
**Percentage:** ***76.81% ammonium nitrate, 22.94% N,N-dimethyl acrylamide, 025% methyl amyl ketone peroxide curing catalyst***
**Classification:** Deflagrating explosive (classified as propellant).

**Danger of preparation:** Can be made by amateurs.
**Use:** Used in military and commercial rockets and missiles for multiple uses.

### 04-03-016A: High Performance ammonium nitrate rocket propellant utilizing ammonium dichromate burn rate catalyst:

Into a suitable sealable vertical mixer, place ***21 grams of diethylene glycol***, followed by ***21.66 grams of adipic acid***, followed by ***856 milligrams of maleic anhydride,*** followed by ***13.3 grams of styrene,*** followed by ***61.4 grams of methyl acrylate,*** and then blend the mixture for 15 minutes on high speed. Thereafter, add in ***9.95 grams of ammonium dichromate,*** followed by ***363.95 grams of powdered ammonium nitrate,*** and then continue to blend the mixture for 15 minutes on high speed to form a homogenous mixture. Thereafter, add in ***490 milligrams of lecithin***, followed by ***1.96 grams of benzoyl peroxide***, seal the vertical mixer, and then continue to blend the mixture for 5 minutes on very rapid speed. Note: after adding the benzoyl peroxide polymerization catalyst, the mixture may begin to immediately set, so monitor the blending process carefully; immediately after 5 minutes, the mixture is ready for casting. To do so, it simply needs to be pressed into any desirable rocket motor or mold under a pressure of 5,000 psi, and then place the rocket motor or mold in an oven and heat to 90 Celsius for 3. The propellant material can be ignited using any suitable ignition composition.
**Burn rate:** 3.19 seconds per inch at 5,000 psi; tends to vary based on diameter and density of rocket grain.
**Water resistance:** Good.
**Stability:** Can be stored for decades when kept from moisture.
**Sensitivity:** Stable towards friction; stable towards normal shock; stable to temperatures up to 150 Celsius.
**Flammability (1 to 10, un-contained):** 8.7 solid grain; 9 fine grain.
**Ease of ignition (1 to 10 un-contained):** 7 solid grain; grain size varies and results in different ignition rates.
**Tendency to cake:** Will not cake.
**Detonation ability:** Can be detonated but only severe conditions—requires TNT, RDX, or HMX booster.
**Explosive ability (contained; explosive deflagration):** Low.
**Percentage:** ***73.58% ammonium nitrate, 12.41% methyl acrylate, 4.37% adipic acid, 4.24% diethylene glycol, 2.68% styrene, 2% ammonium dichromate, 0.396% benzoyl peroxide, 0.17% maleic anhydride, 0.099% lecithin, 0.065% balance***
**Classification:** Deflagrating explosive (classified as propellant).

**Danger of preparation:** Can be made by amateurs.
**Use:** Used in military and commercial rockets and missiles for multiple uses.

### 04-03-016B: High Performance ammonium nitrate rocket propellant utilizing ammonium dichromate burn rate catalyst:

Into a suitable sealable vertical mixer, place ***14 grams of diethylene glycol***, followed by ***14 grams of adipic acid***, followed by ***1 gram of maleic anhydride,*** followed by ***14 grams of styrene,*** followed by ***62 grams of methyl acrylate,*** and then blend the mixture for 10 minutes on high speed. Thereafter, add in ***10 grams of ammonium dichromate,*** followed by ***380 grams of powdered ammonium nitrate,*** and then continue to blend the mixture for 15 minutes on high speed to form a homogenous mixture. Thereafter, quickly add in ***2 grams of lecithin***, followed by ***2 grams of methyl ethyl ketone peroxide***, followed by ***500 milligrams of cobalt octoate,*** seal the vertical mixer, and then rapidly blend the mixture for about 5 minutes. Note: after adding the peroxide, lecithin, and cobalt compound polymerization catalysts, the mixture may begin to immediately set; therefore monitor the mixture closely. Immediately after, the mixture is ready for casting. To do so, open the vertical mixer, and then quickly press the mixture into any desirable rocket motor or mold under a pressure of 7,000 psi in the usual fashion, and then place the rocket motor or mold in an oven and cure to 60 Celsius for 3 hours. The propellant material can be ignited using any suitable ignition composition.

**Burn rate:** 3 seconds per inch at 7,000 psi; tends to vary based on diameter and density of rocket grain.
**Water resistance:** Good.
**Stability:** Can be stored for decades when kept from moisture.
**Sensitivity:** Stable towards friction; stable towards normal shock; stable to temperatures up to 150 Celsius.
**Flammability (1 to 10, un-contained):** 8.8 solid grain; 9 fine grain.
**Ease of ignition (1 to 10 un-contained):** 7 solid grain; grain size varies and results in different ignition rates.
**Tendency to cake:** Will not cake.
**Detonation ability:** Can be detonated but only severe conditions—requires TNT, RDX, or HMX booster.
**Explosive ability (contained; explosive deflagration):** Low.
**Percentage:** ***76% ammonium nitrate, 12.4% methyl acrylate, 2.8% diethylene glycol, 2.8% styrene, 2.8% adipic acid, 2% ammonium dichromate, 0.4% lecithin, 0.4% methyl ethyl ketone peroxide, 0.2%, 0.1% cobalt octoate, 0.1% balance***
**Classification:** Deflagrating explosive (classified as propellant).

**Danger of preparation:** Can be made by amateurs.
**Use:** Used in military and commercial rockets and missiles for multiple uses.

### 04-03-017A: Specialty ammonium nitrate semi-solid (gelled) rocket propellant (with increase moldability):

Into a suitable vertical mixer, place ***223.5 grams of regular unleaded gasoline,*** and then add in ***26.5 grams of aluminum naphthenate*** (containing 7.55% aluminum by weight). Thereafter, stir the mixture for about 15 minutes to form a gel. Thereafter, add in ***37.5 grams of carbon black,*** and then followed by ***712.5 grams of powdered ammonium nitrate***, and then continue to blend the mixture for 30 minutes on moderate speed. Thereafter, the mass is ready for use. To do so, it simply needs to be pressed into any desirable rocket motor under a pressure of 2,000 psi. The gelled propellant mixture can be ignited using any suitable ignition composition. Note: test the bun rate using a small sample before committing to a rocket motor for safety—stories of rocket motors exploding without warning have been reported, so use caution.

**Burn rate:** 3.4 seconds per inch at 2,000 psi.
**Water resistance:** Good.
**Stability:** Can be stored for decades.
**Sensitivity:** Stable towards friction; stable towards normal shock; stable to temperatures up to 200 Celsius.
**Flammability (1 to 10, un-contained):** 8.6
**Ease of ignition (1 to 10 un-contained):** 7
**Tendency to cake:** Will not cake.
**Detonation ability:** Cannot be detonated.
**Explosive ability (contained; explosive deflagration):** Low.
**Percentage:** ***71.25% ammonium nitrate, 22.35% unleaded gasoline, 3.75% carbon black, 2.65% aluminum naphthenate***
**Classification:** Deflagrating explosive (classified as propellant).

**Danger of preparation:** Can be made by amateurs.
**Use:** Used in military and commercial rockets and missiles for multiple uses.

**04-03-018A: Specialty ammonium nitrate rocket propellant with moderate acceleration for use in spacecraft:**
Into a suitable vertical mixer, place ***425 grams of powdered ammonium nitrate***, and then followed by ***125 grams of finely divided aluminum powder***. Thereafter, dry blend the mixture for 1 hour to form a uniform mix. After 1 hour, add in ***180 grams of a hydroxyl terminated polypropylene oxide*** (PPG; obtainable from Union Carbide Corporation and sold as PPG 2025), followed by ***25 grams of trimethylol propane*** (TMP), followed by ***25 grams of 1-decanol***, followed by ***50 grams of 2,6-tolyl-diisocyante compound***, and then continue to blend the mixture for 30 minutes on high speed. Thereafter, add in ***5 grams of ferric acetyl acetonate curing catalyst***, and then blend the mixture for 10 minutes. After 10 minutes, press the propellant mixture into any desirable rocket motor or mold under a pressure 4,000 psi, and then cure the rocket motor or mold in an oven at 60 Celsius for 3 days. Thereafter, the propellant can be fired using any desirable ignition composition.
**Burn rate:** 4.2 seconds per inch at 4,000 psi; tends to vary based on diameter and density of rocket grain.
**Water resistance:** Good.
**Stability:** Can be stored for decades when kept from moisture.
**Sensitivity:** Stable towards friction; stable towards normal shock; stable to temperatures up to 150 Celsius.
**Flammability (1 to 10, un-contained):** 8 solid grain; 8.9 fine grain.
**Ease of ignition (1 to 10 un-contained):** 6 solid grain; grain size varies and results in different ignition rates.
**Tendency to cake:** Will not cake.
**Detonation ability:** Cannot be detonated.
**Explosive ability (contained; explosive deflagration):** Very low.
**Percentage:** ***50.89% ammonium nitrate, 21.55% hydroxyl terminated polypropylene oxide copolymer, 14.97% aluminum metal fuel, 5.98% 2,6-tolyl-diisocyante curing compound, 2.99% 1-decanol addictive, 2.99% trimethylol propane antioxidant, 0.59% ferric acetyl acetonate curing catalyst, 0.04% impurities***
**Classification:** Deflagrating explosive (classified as propellant).

**Danger of preparation:** Can be made by amateurs.
**Use:** Used in military and commercial rockets and missiles for multiple uses; used to propel spacecraft; as well, used to propel wire guided missiles.

**04-03-019A: High Performance ammonium nitrate rocket propellant with nitrocellulose base:**
Into a special sealable vertical mixer equipped with vacuum port, place ***200 grams of triacetin***, followed by ***50 grams of diethyl phthalate***, followed by ***250 grams of nitrocellulose (average nitrogen content)***, and then followed by ***2.5 grams of lecithin***. Thereafter, blend the mixture on high speed for 15 minutes. After 15 minutes, add in ***450 grams of powdered ammonium nitrate***, and then continue to blend the mixture on moderate speed for 45 minutes. After 45 minutes, seal the vertical mixer, and then de-gas the mixture by applying a vacuum of 76mm of mercury, and hold this vacuum for 45 minutes at room temperature; during the de-gassing process, continue to rapidly blend the mixture. After 45 minutes, remove the vacuum, equalize the pressure, open the vertical mixer, and then pour and press the mixture into any desired rocket motor or mold under a pressure of 3,500 psi, and then cure the rocket motor or mold in an oven at 70 Celsius for 48 hours. A standard ignition composition can be used for proper burn.
**Burn rate:** 3.2 seconds per inch at 3,500 psi; tends to vary based on diameter and density of rocket grain.
**Water resistance:** Good.
**Stability:** Can be stored for decades when kept from moisture.
**Sensitivity:** Stable towards friction; stable towards normal shock; stable to temperatures up to 150 Celsius.
**Flammability (1 to 10, un-contained):** 9.1 solid grain; 9.2 fine grain.
**Ease of ignition (1 to 10 un-contained):** 7.8 solid grain; grain size varies and results in different ignition rates.
**Tendency to cake:** Will not cake.
**Detonation ability:** Can be detonated but only severe conditions—requires TNT, RDX, or HMX booster.
**Explosive ability (contained; explosive deflagration):** Above moderate.
**Percentage:** ***47.24% ammonium nitrate, 26.24% nitrocellulose, 20.99% triacetin, 5.24% diethyl phthalate, 0.26% lecithin, 0.03% mixed balance***
**Classification:** Deflagrating explosive (classified as propellant).

**Danger of preparation:** Can be made by amateurs, but it is advisable that experienced personnel carryout the process.
**Use:** Used in military and commercial rockets and missiles for multiple uses.

**Rocket with warhead assembly**

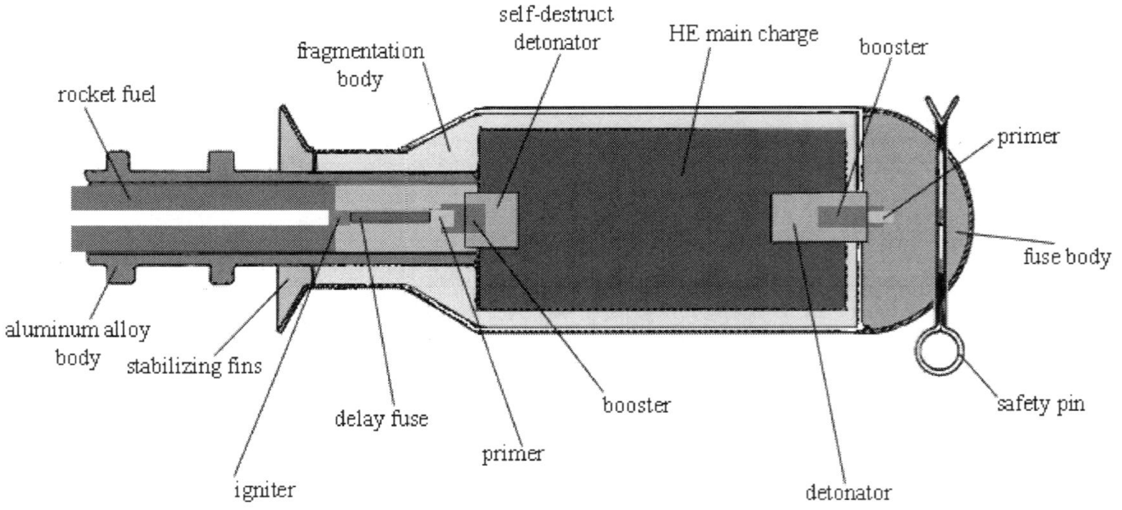


**Details:**

Illustration shows the rocket portion and the warhead in detail. The warhead consists of a standard high explosive charge, initiated by a forward detonator, or base detonator. This design shows a self-destruct mechanism. The self-destruct mechanism works by the burning of the rocket grain to an igniter. This igniter setoff a delay element, which burns to a primer, the primer sets-off the booster, which detonates the base detonator.

### 04-03-019B: High Performance ammonium nitrate rocket propellant with nitrocellulose base:

Into a special sealable vertical mixer equipped with vacuum port, place ***500 grams of dimethoxyethyl phthalate***, followed by ***500 grams of nitrocellulose (average nitrogen content)***, and then followed by ***5 grams of lecithin***. Thereafter, blend the mixture on high speed for 15 minutes. Thereafter, add in ***960 grams of powdered ammonium nitrate***, and then continue to blend the mixture on high speed for about 30 minutes. After 30 minutes, seal the vertical mixer, and then de-gas the mixture by applying a vacuum of 90mm of mercury, and hold this vacuum for 45 minutes at room temperature; during the de-gassing process, continue to rapidly blend the mixture. After 45 minutes, remove the vacuum, equalize the pressure, open the vertical mixer, and then pour and press the mixture into any desired rocket motor or mold under a pressure of 4,500 psi, and then cure the rocket motor or mold in an oven at 70 Celsius for 48 hours. A standard ignition composition can be used for proper burn.

**Burn rate:** 3.4 seconds per inch at 4,500 psi; tends to vary based on diameter and density of rocket grain.
**Water resistance:** Good.
**Stability:** Can be stored for decades when kept from moisture.
**Sensitivity:** Stable towards friction; stable towards normal shock; stable to temperatures up to 150 Celsius.
**Flammability (1 to 10, un-contained):** 9 solid grain; 9.3 fine grain.
**Ease of ignition (1 to 10 un-contained):** 7 solid grain; grain size varies and results in different ignition rates.
**Tendency to cake:** Will not cake.
**Detonation ability:** Cannot be detonated.
**Explosive ability (contained; explosive deflagration):** Moderate.
**Percentage:** ***48.85% ammonium nitrate, 25.44% dimethoxyethyl phthalate, 25.44% nitrocellulose, 0.25% lecithin, 0.,02% impurities***
**Classification:** Deflagrating explosive (classified as propellant).

**Danger of preparation:** Can be made by amateurs, but it is advisable that experienced personnel carryout the process.
**Use:** Used in military and commercial rockets and missiles for multiple uses.

### 04-03-020A: High Performance ammonium nitrate rocket propellant:

Into a special sealable vertical mixer equipped with vacuum port, place ***105 grams of cellulose acetate,*** followed by ***30.5 grams of finely divided carbon black***, followed by ***10 grams of disodium barbiturate,*** and then followed by ***104.5 grams of dinitrophenoxyethanol***. Thereafter, blend the mixture on high speed for 15 minutes. Thereafter, add in ***622.4 grams of powdered ammonium nitrate***, followed by ***118.5 grams of acetyl triethyl citrate***, and then followed by ***10 grams of toluene diamine***. Thereafter, continue to blend the mixture on high speed for 45 minutes at room temperature. After 45 minutes, seal the vertical mixer, and then de-gas the mixture by applying a vacuum of 88mm of mercury, and hold this vacuum for 45 minutes at room temperature; during the de-gassing process, continue to rapidly blend the mixture. After 45 minutes, remove the vacuum, equalize the pressure, open the vertical mixer, and then pour and press the mixture into any desired rocket motor or mold under a pressure of 6,000 psi, and then cure the rocket motor or mold in an oven at 70 Celsius for 24 hours. A standard ignition composition can be used for proper burn.
**Burn rate:** 3 seconds per inch at 6,000 psi; tends to vary based on diameter and density of rocket grain.

**Water resistance:** Good.
**Stability:** Can be stored for decades when kept from moisture.
**Sensitivity:** Stable towards friction; stable towards normal shock; stable to temperatures up to 150 Celsius.
**Flammability (1 to 10, un-contained):** 8.8 solid grain; 9 fine grain.
**Ease of ignition (1 to 10 un-contained):** 7 solid grain; grain size varies and results in different ignition rates.
**Tendency to cake:** Will not cake.
**Detonation ability:** Can be detonated but only severe conditions—requires TNT, RDX, or HMX booster.
**Explosive ability (contained; explosive deflagration):** Moderate.
**Percentage:** ***62.18% ammonium nitrate, 11.83% acetyl triethyl citrate binder, 10.49% cellulose acetate binder, 10.44% dinitrophenoxyethanol, 3.04% carbon black filler, 0.99% disodium barbiturate catalyst, 0.99% toluene diamine cure catalyst, 0.04% mixed balance***
**Classification:** Deflagrating explosive (classified as propellant).

**Danger of preparation:** Can be made by amateurs.
**Use:** Used in military and commercial rockets and missiles for multiple uses.

### 04-03-021A: High Performance ammonium nitrate rocket propellant:

Into a suitable vertical mixer, place ***375 grams of powdered ammonium nitrate,*** followed by ***13.5 grams of Milori blue pigment agent***, followed by ***2 grams of magnesium oxide (light),*** followed by ***10 grams of carbon black,*** followed by ***41 grams of ammonium dichromate***, followed by ***27 grams of polybutadiene,*** and then followed by ***28 grams of any desired epoxy resin or copolymer binder***. Thereafter, blend the mixture on high speed for 45 minutes at room temperature. Thereafter, the mixture is ready for use. To use, the mixture needs to be pressed into any desired rocket motor or mold under a pressure of 5,000 psi in the usual manner, and then cure the rocket motor or mold in an oven at 90 Celsius for 6 days. A standard ignition composition can be used for proper burn.
**Burn rate:** 2.6 seconds per inch at 5,000 psi; tends to vary based on diameter and density of rocket grain.
**Water resistance:** Good.
**Stability:** Can be stored for decades when kept from moisture.
**Sensitivity:** Stable towards friction; stable towards normal shock; stable to temperatures up to 150 Celsius.
**Flammability (1 to 10, un-contained):** 9 solid grain; 9.3 fine grain.
**Ease of ignition (1 to 10 un-contained):** 7 solid grain; grain size varies and results in different ignition rates.
**Tendency to cake:** Will not cake.
**Detonation ability:** Can be detonated but only severe conditions—requires TNT, RDX, or HMX booster.
**Explosive ability (contained; explosive deflagration):** Moderate.
**Percentage:** ***75.52% ammonium nitrate, 8.25% ammonium dichromate, 5.63% copolymer binder, 5.43% polybutadiene, 2.71% Milori blue catalyst, 2% carbon black filler, 0.40% magnesium oxide catalyst, 0.06% mixed residual balance***
**Classification:** Deflagrating explosive (classified as propellant).

**Danger of preparation:** Can be made by amateurs.
**Use:** Used in military and commercial rockets and missiles for multiple uses.

### 04-03-022A: High Performance ammonium nitrate rocket propellant with ferric nitrate burn accelerator:

Into a suitable vertical mixer, place ***425 grams of powdered ammonium nitrate*** of average mesh, and the followed by ***25 grams of ferric nitrate monohydrate***. Thereafter, blend the mixture on high speed for about 40 minutes. After 40 minutes, add in ***100 grams of polybutadiene*** (viscosity of 1500), and then continue to blend the mixture for about 1 hour at room temperature. Thereafter, the mixture is ready for use. To use, the mixture simply needs to be pressed into any desirable rocket motor or mold under a pressure of 7,000 psi in the usual manner, and the resulting rocket motor or mold needs to be cured in an oven at 82 Celsius for about 24 hours to form a rigid and hard propellant grain.
**Burn rate:** 2.7 seconds per inch at 7,000 psi; tends to vary based on diameter and density of rocket grain.
**Water resistance:** Good.
**Stability:** Can be stored for decades when kept from moisture.
**Sensitivity:** Stable towards friction; stable towards normal shock; stable to temperatures up to 150 Celsius.
**Flammability (1 to 10, un-contained):** 9 solid grain; 9.3 fine grain.
**Ease of ignition (1 to 10 un-contained):** 7.4 solid grain; grain size varies and results in different ignition rates.
**Tendency to cake:** Will not cake.
**Detonation ability:** Can be detonated but only severe conditions—requires TNT, RDX, or HMX booster.
**Explosive ability (contained; explosive deflagration):** Moderate.
**Percentage:** ***77.27% ammonium nitrate, 18.18% polybutadiene, 4.54% ferric nitrate monohydrate, 0.01% mixed balance***
**Classification:** Deflagrating explosive (classified as propellant).

**Danger of preparation:** Can be made by amateurs.
**Use:** Used in military and commercial rockets and missiles for multiple uses.

### 04-03-023A: High Performance ammonium nitrate rocket propellant:

Into a suitable vertical mixer, place ***700 grams of powdered ammonium nitrate,*** followed by ***350 grams of magnesium bicarbonate***, followed by ***200 grams of nitrocellulose,*** followed by ***100 grams of aluminum powder***, and then followed by 250 milliliters of hexane. Thereafter, blend the mixture on high speed for about 45 minutes at room temperature. Thereafter, the mixture is ready for use. To use, the mixture simply needs to be pressed into any desirable rocket motor or mold under a pressure of 10,000 psi, and then cure the rocket motor or mold in an oven at 60 Celsius for 24 hours.
**Burn rate:** 3.1 seconds per inch at 10,000 psi; tends to vary based on diameter and density of rocket grain.
**Water resistance:** Good.
**Stability:** Can be stored for decades when kept from moisture.
**Sensitivity:** Stable towards friction; stable towards normal shock; stable to temperatures up to 150 Celsius.
**Flammability (1 to 10, un-contained):** 8.9 solid grain; 9.2 fine grain.
**Ease of ignition (1 to 10 un-contained):** 7 solid grain; grain size varies and results in different ignition rates.
**Tendency to cake:** Will not cake.
**Detonation ability:** Cannot be detonated.
**Explosive ability (contained; explosive deflagration):** Below moderate.
**Percentage:** ***51.85% ammonium nitrate, 25.92% magnesium bicarbonate, 14.81% nitrocellulose, 7.4% aluminum powder, 0.02% balance***
**Classification:** Deflagrating explosive (classified as propellant).

**Danger of preparation:** Can be made by amateurs, but caution should be exercised my amateurs due to the use of nitrocellulose.
**Use:** Used in military and commercial rockets and missiles for multiple uses.

### 04-03-024A: Specialty high Performance ammonium nitrate rocket propellant:

Into a suitable vertical mixer, place ***394.4 grams of powdered ammonium nitrate***, followed by ***43.8 grams of potassium nitrate***, and then blend the mixture on high speed at room temperature for about 45 minutes. Thereafter, add in ***61.8 grams of ammonium tricatechol chromate***, followed by adding in 150 milliliters of 95% ethyl alcohol, and then continue to blend the mixture on high speed for about 45 minutes. Thereafter, the mixture is ready for pressing. To do so, the mixture needs to be pressed into any desired rocket motor or mold under a pressure of 7,000 psi, and then cure the rocket motor or mold in an oven at 60 Celsius for 12 hours.
**Burn rate:** 4 seconds per inch at 7,000 psi; tends to vary based on diameter and density of rocket grain.
**Water resistance:** Good.
**Stability:** Can be stored for decades when kept from moisture.
**Sensitivity:** Stable towards friction; stable towards normal shock; stable to temperatures up to 150 Celsius.
**Flammability (1 to 10, un-contained):** 8.7 solid grain; 9 fine grain.
**Ease of ignition (1 to 10 un-contained):** 6.2 solid grain; grain size varies and results in different ignition rates.
**Tendency to cake:** Will not cake.
**Detonation ability:** Can be detonated but only severe conditions—requires TNT, RDX, or HMX booster.
**Explosive ability (contained; explosive deflagration):** Moderate.
**Percentage:** ***78.88% ammonium nitrate, 12.36% ammonium tricatechol chromate, 8.76% potassium nitrate***
**Classification:** Deflagrating explosive (classified as propellant).

**Danger of preparation:** Can be made by amateurs.
**Use:** Used in military and commercial rockets and missiles for multiple uses.

**04-03-024B: Specialty high Performance ammonium nitrate rocket propellant:**
Into a suitable vertical mixer, place ***368 grams of powdered ammonium nitrate***, followed by ***41 grams of potassium nitrate***, and then blend the mixture on high speed for about 40 minutes. Thereafter, add in ***51 grams of pyridine tricatechol chromate***, and then add in 150 milliliters of 95% ethyl alcohol; thereafter, blend the mixture on high speed for about 45 minutes. After 45 minutes, the mixture is ready for pressing. To do so, the mixture needs to be pressed into any desired rocket motor or mold under a pressure of 6,500 psi, and then cure the rocket motor or mold in an oven at 60 Celsius for 24 hours.
**Burn rate:** 4.2 seconds per inch at 6,500 psi; tends to vary based on diameter and density of rocket grain.
**Water resistance:** Good.
**Stability:** Can be stored for decades when kept from moisture.
**Sensitivity:** Stable towards friction; stable towards normal shock; stable to temperatures up to 150 Celsius.
**Flammability (1 to 10, un-contained):** 8 solid grain; 9 fine grain.
**Ease of ignition (1 to 10 un-contained):** 6.2 solid grain; grain size varies and results in different ignition rates.
**Tendency to cake:** Will not cake.
**Detonation ability:** Can be detonated but only severe conditions—requires TNT, RDX, or HMX booster.
**Explosive ability (contained; explosive deflagration):** Below moderate.
**Percentage:** ***80% ammonium nitrate, 10.86% pyridine tricatechol chromate, 8.2% potassium nitrate, 0.94% impurities***
**Classification:** Deflagrating explosive (classified as propellant).

**Danger of preparation:** Can be made by amateurs.
**Use:** Used in military and commercial rockets and missiles for multiple uses.

**04-03-024C: Specialty high Performance ammonium nitrate rocket propellant:**
Into a suitable vertical mixer, place ***397.6 grams of powdered ammonium nitrate***, followed by ***44.2 grams of potassium nitrate***, and then blend the mixture on high speed for about 45 minutes. Thereafter, add in ***58.2 grams of chromium acetylacetonate***, followed by adding in 150 milliliters of 95% ethyl alcohol, and then continue to blend the mixture on high speed for about 45 minutes. Thereafter, the mixture is ready for pressing. To do so, the mixture needs to be pressed into any desired rocket motor or mold under a pressure for 8,000 psi, and then cure the rocket motor or mold in an oven at 60 Celsius for about 24 hours.
**Burn rate:** 4.3 seconds per inch at 8,000 psi; tends to vary based on diameter and density of rocket grain.
**Water resistance:** Good.
**Stability:** Can be stored for decades when kept from moisture.
**Sensitivity:** Stable towards friction; stable towards normal shock; stable to temperatures up to 150 Celsius.
**Flammability (1 to 10, un-contained):** 7.9 solid grain; 8.9 fine grain.
**Ease of ignition (1 to 10 un-contained):** 6 solid grain; grain size varies and results in different ignition rates.
**Tendency to cake:** Will not cake.
**Detonation ability:** Can be detonated but only severe conditions—requires TNT, RDX, or HMX booster.
**Explosive ability (contained; explosive deflagration):** Below moderate.
**Percentage:** ***79.52% ammonium nitrate, 11.64% chromium acetylacetonate, 8.84% potassium nitrate***
**Classification:** Deflagrating explosive (classified as propellant).

**Danger of preparation:** Can be made by amateurs.
**Use:** Used in military and commercial rockets and missiles for multiple uses.

**04-03-025A: High Performance ammonium nitrate rocket propellant:**
Into a suitable vertical mixer, place ***700 grams of powdered ammonium nitrate,*** followed by ***300 grams of magnesium bicarbonate***, followed by ***200 grams of nitrocellulose,*** followed by ***20 grams of aluminum stearate,*** followed by ***100 grams of aluminum powder***, and then followed by ***80 grams of magnesium powder***. Thereafter, blend the mixture on high speed for about 45 minutes. Thereafter, the mixture is ready for use. To use, the mixture simply needs to be pressed into any desirable rocket motor or mold under a pressure of 7,000 psi, and then cure the rocket motor or mold in an oven at 60 Celsius for 12 hours.
**Burn rate:** 4.4 seconds per inch at 7,000 psi; tends to vary based on diameter and density of rocket grain.
**Water resistance:** Good.
**Stability:** Can be stored for decades when kept from moisture.
**Sensitivity:** Stable towards friction; stable towards normal shock; stable to temperatures up to 150 Celsius.
**Flammability (1 to 10, un-contained):** 7.8 solid grain; 8.8 fine grain.
**Ease of ignition (1 to 10 un-contained):** 6 solid grain; grain size varies and results in different ignition rates.
**Tendency to cake:** Will not cake.
**Detonation ability:** Can be detonated but only severe conditions—requires TNT, RDX, or HMX booster.
**Explosive ability (contained; explosive deflagration):** Below moderate.

**Percentage:** ***50% ammonium nitrate, 21.42% magnesium bicarbonate, 14.28% nitrocellulose, 7.14% aluminum powder, 5.71% magnesium, 1.42% aluminum stearate catalyst, 0.03% balance***
**Classification:** Deflagrating explosive (classified as propellant).

**Danger of preparation:** Can be made by amateurs, but caution should be exercised when working with nitrocellulose.
**Use:** Used in military and commercial rockets and missiles for multiple uses.

### 04-03-026A: Specialty high Performance polymer-based ammonium nitrate rocket propellant:

Into a suitable sealable vertical mixer, place ***94 grams of polyurethane,*** followed by ***80 grams of aluminum powder***, and then followed by ***320 grams of powdered ammonium nitrate***. Thereafter, blend the mixture on high speed for about 45 minutes. After 45 minutes, add in ***6 grams of ethylene oxide***, seal the vertical mixer, and then continue to blend the mixture on high speed for about 15 minutes in the absence of air. Thereafter, the mixture is ready for use. To use, the mixture simply needs to be pressed into any desirable rocket motor or mold under a pressure of 5,500 psi, and then cure the rocket motor or mold in an oven at 60 Celsius for about 72 hours.
**Burn rate:** 2.98 seconds per inch at 5,500 psi; tends to vary based on diameter and density of rocket grain.
**Water resistance:** Good.
**Stability:** Can be stored for decades when kept from moisture.
**Sensitivity:** Stable towards friction; stable towards normal shock; stable to temperatures up to 150 Celsius.
**Flammability (1 to 10, un-contained):** 9 solid grain; 9.1 fine grain.
**Ease of ignition (1 to 10 un-contained):** 8.9 solid grain; grain size varies and results in different ignition rates.
**Tendency to cake:** Will not cake.
**Detonation ability:** Can be detonated but only severe conditions—requires TNT, RDX, or HMX booster.
**Explosive ability (contained; explosive deflagration):** Below moderate.
**Percentage:** ***64% ammonium nitrate, 18.8% polyurethane, 16% aluminum powder, 1.2% ethylene oxide polymerization agent***
**Classification:** Deflagrating explosive (classified as propellant).

**Danger of preparation:** Can be made by amateurs.
**Use:** Used in military and commercial rockets and missiles for multiple uses.

### 04-03-027A: High Performance smokeless polymer rocket propellant:

Into a suitable heated vertical mixer, place 500 grams of Butadiene/2-methyl-5-vinylpyridine copolymer, and then followed by 100 grams of carbon black. Thereafter, blend the mixture on high speed at 55 Celsius for about 1 hour. Thereafter, add in 100 grams of TP-90B (dibutoxy-ethoxy-ethyl-formal), followed by 5 grams of SA-113 (N,N-dimethyl-tert-butylsulfenyl-dithiocarbomate), followed by 3.75 grams of powdered sulfur, followed by 15 grams of zinc oxide, followed by 15 grams of "flexamine", and then followed by 5 grams of Aersol-OT (dioctyl ester of sodium sulfosuccinic acid). Thereafter, continue to blend the entire mixture for about 2 hours at 55 Celsius. After 2 hours, place into a clean separate heated vertical mixer, ***525 grams of the above binder mixture***, and then add in ***2475 grams powdered ammonium nitrate***, and then followed by ***7.5 grams of milori blue***. Thereafter, blend the combined mixture on high speed for about 6 hours at 70 Celsius. Afterwards, the mixture is ready to be casted. To do so, pour and press the mixture into any desirable rocket motor or mold under a pressure of 8,000 psi, and then cure the rocket motor or mold in an oven at 82 Celsius for 24 hours. Can be ignited using any standard high performance rocket ignition composition.
**Burn rate:** 2.9 seconds per inch at 8,000 psi; tends to vary based on diameter and density of rocket grain.
**Water resistance:** Good.
**Stability:** Can be stored for decades when kept from moisture.
**Sensitivity:** Stable towards friction; stable towards normal shock; stable to temperatures up to 150 Celsius.
**Flammability (1 to 10, un-contained):** 9 solid grain; 9 fine grain.
**Ease of ignition (1 to 10 un-contained):** 9 solid grain; grain size varies and results in different ignition rates.
**Tendency to cake:** Will not cake.
**Detonation ability:** Can be detonated but only severe conditions—requires TNT, RDX, or HMX booster.
**Explosive ability (contained; explosive deflagration):** Below moderate.
**Percentage:** ***82.29% ammonium nitrate, 17.45% copolymer binder, 0.249%% milori blue catalyst, 0.011% residual balance***
**Classification:** Deflagrating explosive (classified as propellant).

**Danger of preparation:** Can be made by amateurs.
**Use:** Used in military and commercial rockets and missiles for multiple uses.

# Section 4: Miscellaneous containing

***Chemicals used in this section (binders are not included)***

| | |
|---|---|
| **1. Potassium nitrate (see Black Powder)** | **2. Sulfur (see Black Powder)** |
| **3. Charcoal (see Black Powder)** | **4. Sugar Carbon (see Modified Black Powder)** |
| **5. Barium Chromate (see Modified Black Powder)** | **6. Potassium Perchlorate (see Modified Black Powder)** |
| **7. Potassium Dichromate (see Modified Black Powder)** | **8. Ammonium Bisulfide (see Modified Black Powder)** |
| **9. Potassium chlorate (see Modified Black Powder)** | **10. Carbon Disulfide (see Modified Black Powder)** |
| **11. Nitrocellulose (see Modified Black Powder)** | **12. Lead Tetraoxide (see Modified Black Powder)** |
| **13. Diphenylamine (see Modified Black Powder)** | **14. Titanium Dioxide (see Modified Black Powder)** |
| **15. Manganese Dioxide (see Modified Black Powder)** | **16. Sodium Benzoate (see Modified Black Powder)** |
| **17. Ammonium Nitrate (see Modified Black Powder)** | **18. Calcium Carbonate (see Modified Black Powder)** |
| **19. Sodium Nitrate (see Modified Black Powder)** | **20. Urea (see Modified Black Powder)** |
| **21. Lead Nitrate (see Modified Black Powder)** | **22. Nitro Starch (see Modified Black Powder)** |
| **23. Ammonium Perchlorate (see Ammonium Perchlorate Rocket Propellants)** | **24. Aluminum powder (see Ammonium Perchlorate Rocket Propellants)** |
| **25. Magnesium Sulfide (see Ammonium Perchlorate Rocket Propellants)** | **26. Copper Chromite (see Ammonium Perchlorate Rocket Propellants)** |
| **27. Nitroguanidine (see Ammonium Perchlorate Rocket Propellants)** | **28. Ammonium Sulfate (see Ammonium Perchlorate Rocket Propellants)** |
| **29. Nitroglycerine (see Ammonium Perchlorate Rocket Propellants)** | **30. Magnesium Oxide (see Ammonium Perchlorate Rocket Propellants)** |
| **31. Iron-II-oxide (see Ammonium Perchlorate Rocket Propellants)** | **32. Zirconium Hydride (see Ammonium Perchlorate Rocket Propellants)** |
| **33. Teflon (see Ammonium Perchlorate Rocket Propellants)** | **34. Zinc Oxide (see Ammonium Perchlorate Rocket Propellants)** |
| **35. Ammonium Dichromate (see Ammonium Perchlorate Rocket Propellants)** | **36. Lithium Perchlorate (see Ammonium Perchlorate Rocket Propellants)** |
| **37. Magnesium powder (see Ammonium Perchlorate Rocket Propellants)** | **38. Lithium Aluminum Hydride (see Ammonium Perchlorate Rocket Propellants)** |
| **39. Iron-III-oxide (red iron oxide) (see Ammonium Perchlorate Rocket Propellants)** | **40. PVC (see Ammonium Perchlorate Rocket Propellants)** |
| **41. Copper Sulfide (see Ammonium Perchlorate Rocket Propellants)** | **42. Sodium Hydride (see Ammonium Perchlorate Rocket Propellants)** |
| **43. Titanium Hydride (see Ammonium Perchlorate Rocket Propellants)** | **44. Silicon Nitride (see Ammonium Perchlorate Rocket Propellants)** |
| **45. ADN (see ADN Rocket Propellants)** | **46. Urea Nitrate (see ADN Rocket Propellants)** |
| **47. Silicon Powder (see ADN Rocket Propellants)** | **48. Hexamine (see ADN Rocket Propellants)** |
| **49. Boron powder (see ADN Rocket Propellants)** | **50. Sodium hypophosphite (see ADN Rocket Propellants)** |
| **51. KDN (see ADN Rocket Propellants)** | **52. Nickel Chloride (see Ammonium Nitrate Rocket Propellants)** |
| **53. TNT (see Ammonium Nitrate Rocket Propellants)** | **54. Acrylamide** |
| | 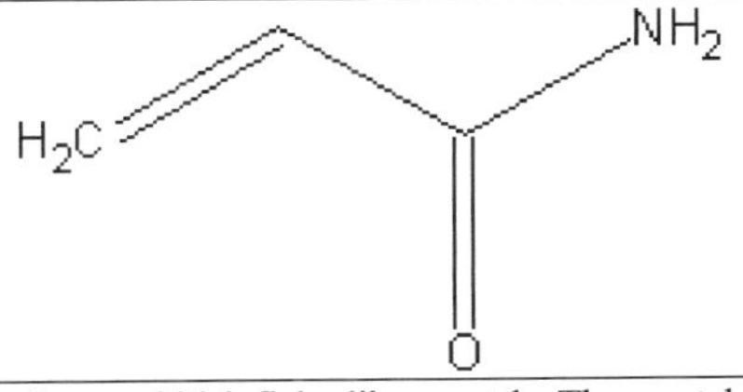  |
| | Acrylamide forms whitish flake-like crystals. The crystals are unstable and polymerize when heated or upon standing at room temperature. |
| | |

### 55. Ethylene Glycol

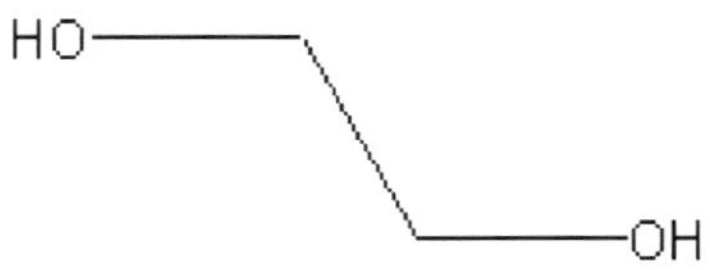

Ethylene Glycol is a colorless viscous toxic liquid with a boiling point of 197 Celsius, and a melting point of –13 Celsius. Ethylene Glycol is very soluble in water and many organic solvents, and is found in anti-freeze.

### 56. Magnesium Perchlorate

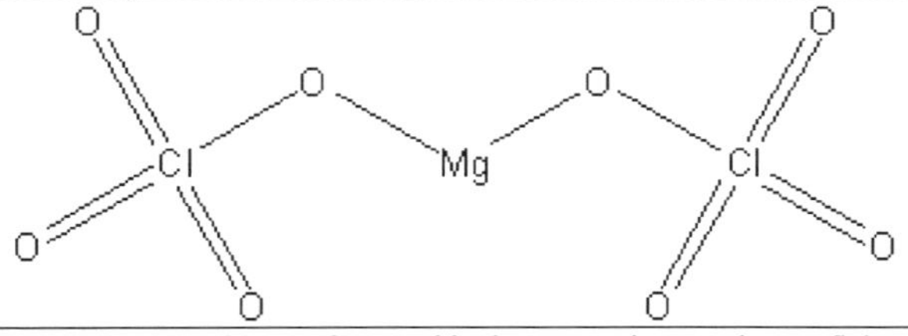

Magnesium Perchlorate forms white hygroscopic granules, or flaky mass. The salt decomposes when heated to 250 Celsius. It dissolves in water with the formation of considerable heat.

### 57. Metallic Lithium

Lithium forms a silvery-white metal, which rapidly tarnishes in moist air. Its melting point is 180 Celsius. The metal reacts violently with water and many organic chemicals. The commercial brand is sold stored under kerosene. Lithium is obtained by heating metallic sodium with lithium chloride.

### 58. Beryllium Hydride

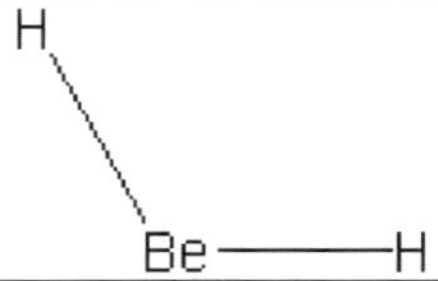

Beryllium Hydride forms a white solid, which begins to decompose when heated to 220 Celsius. The hydride is much more stable then other metal hydrides, and only slowly reacts with water. The salt is relatively insoluble in water and most organic solvents.

### 59. Red Phosphorus

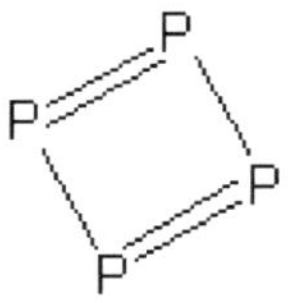

Red Phosphorus forms a red to violet powder, which is stable at room temperature. The solid sublimes when heated to 416 Celsius, but reverts to the white form when distilled at 290 Celsius. Red phosphorus is insoluble in all common solvents, but is capable of dissolving in phosphorus tribromide.

### 60. Sodium Borohydride

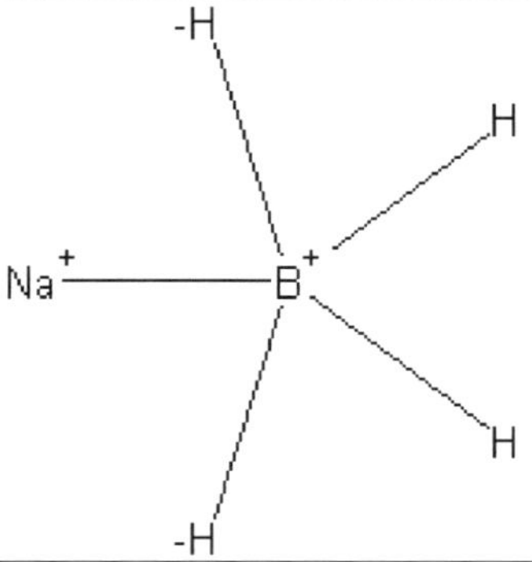

Sodium borohydride forms hygroscopic crystals that are capable of forming a dihydrate. The salt is soluble in water, and can be stored in aqueous solution for several days. Aqueous solutions decompose rapidly when boiled. The salt is very stable in dry air, as compared to other hydrides.

### 61. Sodium chlorate

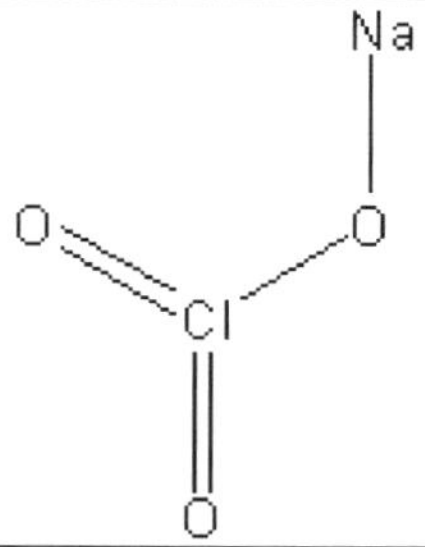

Sodium chlorate forms colorless, odorless crystals or white granules. The melting point of the crystals is 248 Celsius, but it begins to decompose into oxygen and sodium perchlorate when heated to 300 Celsius. The crystals are highly soluble in water, but relatively insoluble in alcohol, and most common organic solvents. Sodium chlorate is widely used in pyrotechnic compositions, and in the preparation of ammonium chlorate, and perchlorate, which are used in powerful solid rocket fuels.

## Method of Preparation 1:

**Step 1:** Setting-up a standard diaphragm "salt bridge cell"

***A. Introduction***

A standard diaphragm cell can be made from simple and inexpensive store bought materials. Anyone can easily assemble their own cell, and carryout interesting and useful procedures to make a variety of interesting products. First of all, you need to learn how to assemble your own homemade cell. The first thing you will need to do is go to the store and purchase the following materials:

1. 1 box of Epsom salt (available in any grocery store)
2. 1 standard clay pot (just under 1 quart size; 500 milliliters; try to find one that does not have a hole in the bottom)
3. 1 plastic Tupperware container w/lid (just under ½ gallon size; about 1800 milliliters)
4. 2 lead electrodes (about 6 inches long each).
5. 1 car battery charger; with power output with at least 12 volt 6 amp (available in most stores for $20.00 to $40.00)

**Note:** The lead electrodes should be rectangular with dimensions of 5 to 6 inches long, by ½ inch to ¼ inch in diameter (they don't have to be perfectly round, and can be square or rectangle). The lead electrodes may be circular if desired with dimensions of 6 inches long by 0.30 to 0.50 inch radius (9 millimeter to 12.9 millimeter). The lead electrodes can be made my melting pieces of lead into a make shift mold, made by forming a rectangular shaped mold using aluminum foil. After shaping the mold with the aluminum foil, place the pieces of lead there into, and then use a standard propane torch to melt the lead. After allowing the lead to cool, remove the aluminum foil. Lead can be found in car batteries, and in lead solder for welding copper and in electronics.

**Note:** The clay pot is a standard clay flowerpot, which can be found at any hardware store or pottery store. The clay pot should not have a hole in the bottom, as most of them do. *If a clay pot with no hole in the bottom is not available, use plumbers putty, or some similar water insoluble product to seal the hole on both sides.*

***B. Assemble your cell***

Now that you have your materials, assemble your cell. Use the following illustration to aid you. Note: For the electrodes, suitable sized holes should be cut into the plastic lid. One hole on the left or right side of the plastic lid, and the other directly in the center of the plastic lid. The holes if properly cut, can secure the electrodes firmly (with a tight fit). The clay pot is then placed into the plastic Tupperware container as shown in the illustration. Note: The plastic lid should be able to fit securely onto the plastic Tupperware container without disrupting the clay pot. A small hole should also be cut into the plastic lid to allow for gas venting. Note: This cell is not airtight. To make the cell airtight, the holes in which the electrodes protrude need to be sealed airtight. To do this, a glue gun may be used, or plumbers putty. The plastic lid should also be sealed gas tight. To do this, wrap Teflon tape (available at any hardware store) around the outer lip of the plastic Tupperware container many times, so that when the lid is attached, it forms a firm seal. Note: Other methods may be used to create an airtight cell. Use your own imagination, and see what you can come up with. An airtight seal of your cell is not necessary for most procedures. An airtight seal is only desired if a producing gas is the desired product, as in the production of chlorine gas.

5. Then attach each electrode to their respective electric wire from the battery charger. Note: The battery charger comes with a black clamp, and a red clamp. The black clamp is the "negative", and the red clamp is the "positive". Note: The battery charger only works when plugged in! Before plugging the battery charger in, secure both clamps to the cell. The black clamp goes on the negative (-) lead electrode, and the red clamp goes on the positive (+) lead electrode.
6. When everything is in place, plug the battery charger in, and allow it to run for about 6 hours. Note: During the process, hydrogen gas will be steadily evolved. Carryout the process in a ventilated area, such as a garage, bathroom, or shed.
7. After 6 hours, unplug the battery charger, and remove the clamps from each electrode. Then open the plastic lid, and pullout the clay pot. Note: The clay pot will be filled with a white precipitate. This white precipitate is *magnesium hydroxide*. The cathode liquid will contain a *dilute sulfuric acid solution*. This dilute sulfuric acid solution can be saved, if desired, or mixed with the contents in the clay pot. Mixing both solutions will reform magnesium sulfate (Epsom salt). If you wish to recycle the Epsom salt for future cell charge, mix both liquids. Note: To verify, for your own amusement, drop in a little baking soda to the anode liquid.
8. Finally, rinse out the clay pot with tap water, and do the same for the plastic Tupperware container. Your clay pot is now charged, so proceed with step 2.

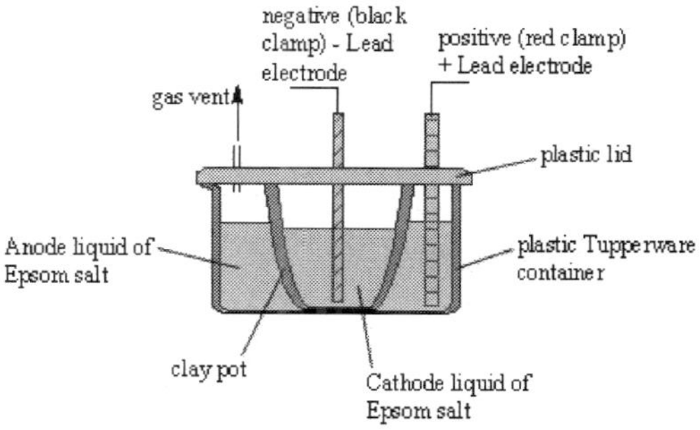


**Step 2:** Preparation of sodium chlorate

Now, dissolve 100 grams (3.5 oz.) of pickling salt into 350 milliliters (11.8 fluid oz.) of hot tap water. Thereafter, place this salt solution into the cathode compartment (-). Thereafter, dissolve 25 grams (1.0 oz.) of pickling salt into 350 milliliters (11.8 fluid oz.) of hot tap water, and then add in 5 drops of concentrated sulfuric acid. Then place this acidified salt solution into the anode compartment (+). Now, assemble the cell as illustrated below, using two lead electrodes, and then begin the operation by turning on your power supply, or by plugging in your battery charger. Allow the cell to electrolysis for about 15 hours at 6 Volt/6 Amp or equivalent—be sure to monitor the temperature so as it does not fall below 50 Celsius. The optimal temperature for maximizing chlorate formation in the cell is around 80 to 90 Celsius.

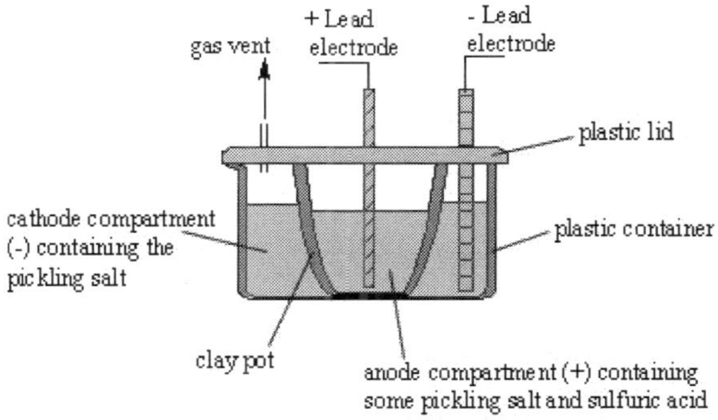


**Set-up for the preparation of sodium chlorate from pickling salt. The lead electrodes can be replaced with titanium, chromium, or graphite.**

After the 15-hour electrolysis process, unplug your power supply,

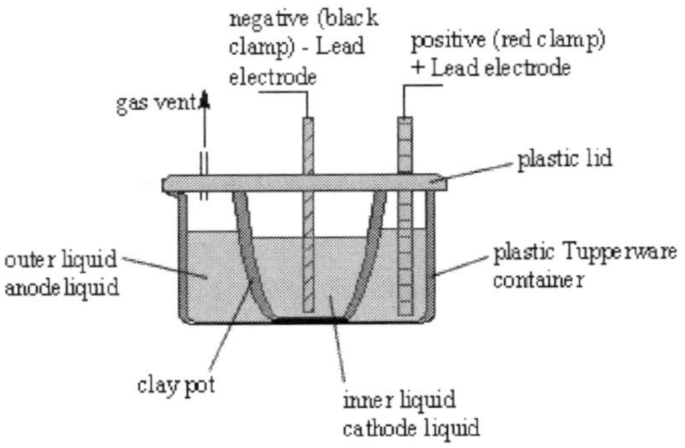


**Setup for the assembly of the cell. Note: The lead electrodes can be replaced with graphite, but lead is better to reduce corrosion of the electrodes.**

**C. Charge your cell**

Now that you have setup your cell, the next task is to "charge the cell". What does this mean? Well, to put it into simple terms, the clay pot acts as a salt bridge. In order for this salt bridge to work properly, it must be "Charged" with ions, so that during a particular process, the electrochemical reaction works properly. To charge the cell, carryout the following:

1. Into a beaker or other container, place 1500 milliliters (50.7 fluid oz.) of water. Then add 100 grams (3.6 oz.) of Epsom salt, and stir the water to completely dissolve the Epsom salt.
2. Place 500 milliliters (17 fluid oz.) of this Epsom salt solution into the clay pot. This will be called the "cathode" liquid, designating the "negative" side.
3. Place the remaining Epsom salt solution into the plastic Tupperware container (the outer compartment). This will be called the "anode" liquid, designating the "positive" side.
4. Then secure the plastic lid to the plastic Tupperware container, and then secure both electrodes.

and then open the cell. Carefully remove the clay pot, and then dump its contents into a clean beaker or similar glass container—the anode compartment will contain the bulk of the sodium chlorate, as well as some bleach, and small amounts of sodium perchlorate and sodium hydroxide. The cathode compartment will contain sodium chloride, and some sodium hydroxide, bleach, and sodium chlorate. You can discard the cathode liquid if desired, or you can recycle it for another crop of chlorate—to do this, simply fortify it with about 50 to 60 additional grams of salt, and then place it back into the cathode compartment. Your sodium chlorate solution (anode liquid) should be quickly filtered, to remove any insoluble materials, and then gently heated to a boil for about 10 minutes to drive-off any dissolved gasses, and to break down any bleach. Thereafter, for best results, pour the entire anode liquid onto a large shallow pan, and then allow to air dry. Blowing air over the surface of the pan using a cooling fan can help speed-up the process. Once all the water has evaporated, there will be left behind crystals of sodium chlorate, sodium hydroxide, and small amounts of sodium chloride, and sodium perchlorate. Now, scrape-up all the crystals, and then place them into a beaker, and then add to this beaker, 250 milliliters (8.5 fluid oz.) of a rubbing alcohol solution prepared by adding and mixing 100 milliliters (3.4 fluid oz.) of rubbing alcohol into 150 milliliters (5 fluid oz.) of warm tap water. Then gently swirl the beaker or similar container for several minutes, and then filter-off the insoluble crystals of the sodium chlorate. Finally, recrystallize these filtered-off crystals from 150 milliliters of hot tap water (see laboratory techniques guide on how to carryout a recrystallization). The final product after recrystallization will be about 95 to 96% sodium chlorate, well suitable for use in making fireworks, or pyrotechnic compositions. Further recrystallizations from boiling water can purify the sodium chlorate further.

## 62. Lead Dioxide

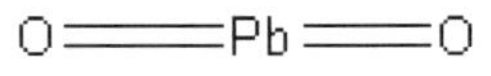

Lead Dioxide is a brownish, dark brown powder, which evolves oxygen gas when heated. Lead dioxide is a mild oxidizing agent. It is insoluble in water and the usual organic solvents.

## 63. Benzoyl Peroxide

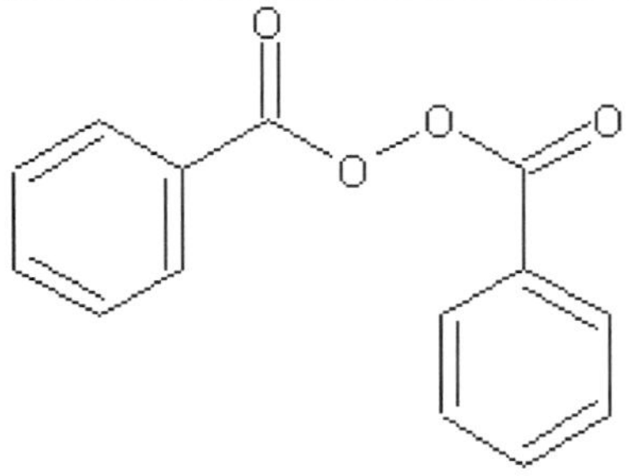

Benzoyl peroxide forms colorless crystals or crystalline powder. The pure material may explode when rapidly heated. The crystals have a melting point of about 106 Celsius. It is only slightly soluble in water and alcohol, but is soluble in most common solvents.

## 64. Barium Nitrate

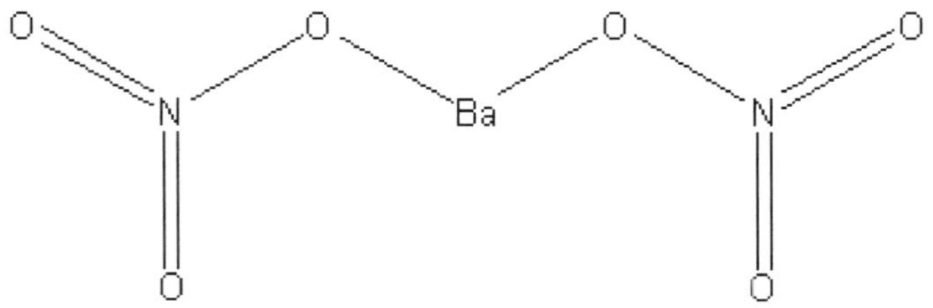

Colorless to white crystalline powder, granules, or solid mass. The crystals have a melting point of 500+ Celsius, with decomposition shortly thereafter. Barium nitrate is freely soluble in water, but insoluble in the usual organic solvents. Barium nitrate is toxic so users should wear gloves.

## 65. Cyanuric Acid

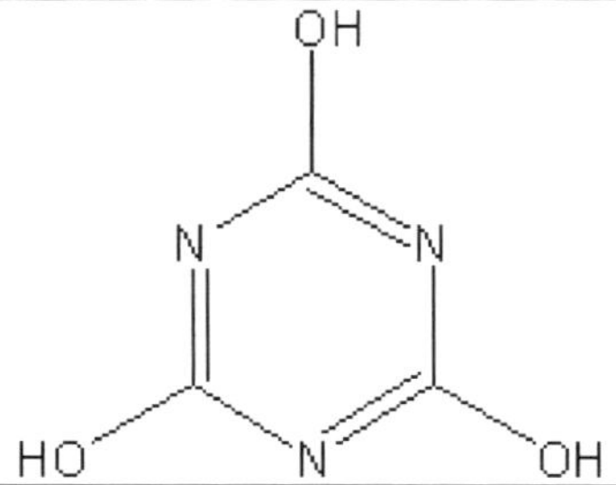

White crystalline solid, which decomposes when heated above 330 Celsius. The free crystals are only slightly soluble in water, and the usual solvents.

***- High Performance Miscellaneous Propellants in this section -***

| | |
|---|---|
| **1. 04-04-001A: High Performance military rocket propellant (M7 propellant):** *54.6% nitrocellulose, 35.4% nitroglycerine, 7.8% potassium perchlorate, 1.2% carbolac I, 0.90% ethyl centralite, 0.020% graphite, 0.080% moisture* | **2. 04-04-002A: High Performance military rocket propellant:** *49.7% HMX, 24% nitroglycerine, 19.9% ammonium perchlorate, 4.7% ethyl acrylate-acrylic acid copolymer, 1% UNOX 221 linking agent, 0.49% carbon black catalyst, 0.210% moisture* |
| **3. 04-04-003A: High Performance CMDB rocket propellants 2 (composite modified Double Base propellants):** *34.6% nitroglycerine, 21.9% ammonium perchlorate, 19.9% aluminum, 11% triacetin, 10.9% nitrocellulose, 1% 2-nitrodiphenylamine, 0.35% impurities, 0.29% 2,6-tolylene diisocynate curative, 0.06% lead 2-ethylhexoate catalyst* | **4. 04-04-003B: High Performance CMDB rocket propellants 2 (composite modified Double Base propellants):** *32.2% nitroglycerine, 20% ammonium perchlorate, 20% aluminum, 15% nitrocellulose, 10.3% di-n-propyl adipate, 1% 2-nitrodiphenylamine, 1% resorcinol diacetate, 0.38% 2,6-tolylene diisocyante curative, 0.11% residue, 0.01% ferric acetylacetonate* |
| **5. 04-04-004A: High Performance nitrocellulose/nitroglycerine based rocket propellant:** *49.11% nitrocellulose, 35.36% nitroglycerine, 15.02% triacetin, 0.49% nitrodiphenylamine, 0.02% impurities* | **6. 04-04-005A: High Performance "hard, yet flexible" propellant utilizing hydroxylamine perchlorate:** *80% hydroxylamine perchlorate, 10% D.E.R. 32 binder, 10% Epon 812 epoxy resin* |
| **7. 04-04-005B: High Performance "hard, yet flexible" propellant utilizing hydroxylamine perchlorate (modified):** *60% hydroxylamine perchlorate, 20% aluminum, 10% D.E.R. 32 binder, 10% Epon 812 epoxy resin* | **8. 04-04-006A: High Performance polymeric propellant utilizing guanidine perchlorate:** *38.5% guanidine perchlorate, 28.4% lithium perchlorate, 20% aluminum, 7% acrylamide, 4.6% ethylene glycol, 0.66% succinic anhydride polymerization catalyst, 0.53% N-methyl morpholine, 0.31% moisture and residue* |
| **9. 04-04-007A: Moderate performance "rubber-like" propellant utilizing potassium perchlorate and polysulfide binder:** *59.9% potassium perchlorate, 19.4% polysulfide binder, 9.9% ammonium perchlorate, 7.1% furfuryl alcohol plasticizer, 1.4% lead dioxide vulcanizing agent, 1.4% dibutyl phthalate dispersion medium, 0.56% mixed residues, 0.34% stearic acid catalyst,* | **10. 04-04-007B: Moderate performance "rubber-like" propellant utilizing potassium perchlorate and polysulfide binder (modified):** *50% potassium perchlorate, 23% polysulfide binder, 17% ammonium perchlorate, 7.5% furfuryl alcohol plasticizer, 1.3% zinc oxide, 0.68% stearic acid catalyst, 0.50% paraquinone dioxime, 0.02% mixed resides* |
| **11. 04-04-008A: High performance lithium perchlorate based propellant for various applications:** *71.4% lithium perchlorate, 28.5% copolymer of hexamethylene adipamide, 0.1% residue* | **12. 04-04-009A: High performance polymer rocket propellant based on potassium perchlorate:** *65.5% potassium perchlorate, 21.8% beta-caprolcatam pre-polymer, 12.3% N-acetyl caprolactam pre-polymer, 0.21% polymerization catalyst, 0.19% mixed impurities* |
| **13. 04-04-010A: High Performance potassium perchlorate rocket propellant:** *73.2% potassium perchlorate, 12.6% polyurethane compound binder, 12.1% aluminum, 2% copper chromite burn rate catalyst, 0.10% balanced* | **14. 04-04-011A: High Performance potassium perchlorate rocket propellant:** *76% potassium perchlorate, 14.5% Epoxy resin binder, 9.5% aluminum* |
| **15. 04-04-012A: Super high Performance potassium perchlorate rocket propellant containing hydroxylamine perchlorate:** *62% potassium perchlorate, 30.3% Thiokol LP-2 polysulfide binder, 4.77% hydroxylamine perchlorate, 1.9% p-quinone dioxime, 0.95% diphenyl guanidine, 0.08% mixed residues* | **16. 04-04-013A: Super high Performance flexible, hydroxylamine perchlorate rocket propellant:** *41.56% hydroxylamine perchlorate, 33.25% potassium perchlorate, 24.93% Paralax P-10 binder, 0.24% benzoyl peroxide, 0.02% balance* |
| **17. 04-04-014A: High Performance rocket propellant utilizing hydrazine nitrate:** *60% hydrazine nitrate, 25.6% methyl acrylate, 10.4% methyl methacrylate, 3.6% allyl diglycol carbonate, 0.4% tert-butyl peroxide* | **18. 04-04-015A: Specialty magnesium perchlorate semi-solid (gelled) rocket propellant:** *71.46% magnesium perchlorate, 22.99% unleaded gasoline, 3.72% lamp black, 1.81% aluminum naphthenate, 0.02% rounded balance* |
| **19. 04-04-016A: High Performance potassium perchlorate rocket propellant:** *73.2% potassium perchlorate, 12.6% polyurethane compound binder, 12.1% aluminum, 2% copper chromite burn rate catalyst, 0.10% balanced* | **20. 04-04-017A: High Performance nitronium perchlorate specialty rocket propellant:** *60% nitronium perchlorate, 30% polybutadiene, 10% metallic lithium* |
| **21. 04-04-018A: High Performance nitronium perchlorate specialty rocket propellant (modified cure):** *50% ammonium perchlorate, 25% polybutadiene, 25% metallic lithium* | **22. 04-04-019A: High Performance hydrazine nitrate rocket propellant:** *48% hydrazine nitrate, 32% beryllium hydride, 12% carboxy terminated polybutadiene, 4% dioctyl azelate, 4% mineral oil* |

| | |
|---|---|
| **23. 04-04-020A: High Performance hydrazine nitrate rocket propellant:** *40.46% hydrazine nitrate, 17.34% sodium borohydride, 17.34% epon 815 epoxy binder, 14.45% potassium nitrate, 10.4% red phosphorus, 0.01% residual balance* | **24. 04-04-021A: High Performance potassium perchlorate rocket propellant with nitrocellulose base:** *48.85% potassium perchlorate, 25.44% nitrocellulose, 20.35% triacetin, 5.08% butyl phthalyl butyl glycolate, 0.25% mixed balance* |
| **25. 04-04-022A: High Performance aluminum hydride propellant with nitrocellulose base:** *50% nitrocellulose, 29% aluminum hydride, 14% nitronium perchlorate, 7% polymeric dichlorostyrene* | **26. 04-04-023A: High Performance potassium perchlorate rocket propellant:** *74.55% potassium perchlorate, 24.85% silicone elastomer, 0.21% benzoyl peroxide, 0.39% mixed impurities* |
| **27. 04-04-023B: High Performance sodium perchlorate rocket propellant:** *69.58% sodium perchlorate, 29.82% silicone elastomer, 0.59% benzoyl peroxide, 0.01% mixed impurities* | **28. 04-04-023C: High Performance cesium perchlorate rocket propellant with infrared illumination:** *74.53% cesium perchlorate, 17.39% silicone gum, 7.45% aluminum perchlorate, 0.62% benzoyl peroxide, 0.01% balance* |
| **29. 04-04-024A: High Performance potassium perchlorate rocket propellant:** *59.76% potassium perchlorate, 24.9% alpha-caprolactam, 14.94% N-acetyl caprolactam, 0.39% sodium hydride catalyst, 0.01% residue* | **30. 04-04-025A: High performance thermally stable rocket propellant:** *38.51% barium nitrate, 26.45% cyanuric acid, 26.08% magnesium dust, 11% Teflon, 10.95% Viton A copolymer binder, 0.05% copper chromite burn catalyst* |
| **31. 04-04-026A: Moderate performance potassium nitrate based rocket propellant:** *75% potassium nitrate, 25% asphalt black* | |

**04-04-001A: High Performance military rocket propellant (M7 propellant):**
Into a vertical mixer, add in ***273 grams of nitrocellulose*** (14% nitrogen content), followed by 400 milliliters of ethyl acetate, and then followed by ***177 grams of nitroglycerine***. Thereafter, begin to moderately stir the mixture for about 10 minutes, and shortly thereafter, add in ***6 grams of Carbolac I***, and then continue to stir the entire mixture for about 30 minutes at room temperature. After 30 minutes, add in ***39 grams of potassium perchlorate,*** followed by ***4.5 grams of ethyl centralite***, and then continue stirring the mixture for 30 minutes at room temperature. After 30 minutes, add in 300 milliliters of warm water, and then continue blend the mixture on low speed for 1 hour. Thereafter, pour the entire mixture onto a shallow pan with a high surface area and allow it to thoroughly dry until the smell of solvent (ethyl acetate) is gone. When the smell of ethyl acetate is gone, place the remaining contents of the pan, including any water, into a large clean glass beaker, and then add in 2000 milliliters of water, and then gently heat the mixture with mild stirring to 50 Celsius, and then gently blend the mixture for 15 minutes. After 15 minutes, stop stirring and allow the mixture to stand at room temperature for 3 days, to allow the insoluble mixture to settle. Thereafter, use gravity filtration (no vacuum filtration) to recover the insoluble solids. Afterwards, place the dry solids into a clean vertical mixer, equipped with plastic stir blades, and then add in ***100 milligrams of finely divided graphite***; thereafter, gently blend the mixture on low speed for about 45 minutes at room temperature. After 45 minutes, the mixture is ready for pressing. To do so, press the mixture into any desirable rocket motor or mold under a pressure of 4,000 psi, and then cure the rocket motor or mold in an oven at 30 Celsius for 24 hours.
**Note:** Currently there are numerous formulations for M7 rocket propellant; however, they all share the same formulation containing nitrocellulose and nitroglycerine.
**Burn rate:** 2.8 seconds per inch at 4,000 psi; tends to vary based on diameter and density of rocket grain.
**Water resistance:** Good.
**Stability:** Can be stored for decades.
**Sensitivity:** Stable towards friction; stable towards normal shock; stable to temperatures up to 100 Celsius.
**Flammability (1 to 10, un-contained):** 9.2 solid grain; 9 fine grain.
**Ease of ignition (1 to 10 un-contained):** 9 solid grain; grain size varies and results in different ignition rates.
**Tendency to cake:** Will not cake.
**Detonation ability:** Can be detonated but only severe conditions—requires TNT, RDX, or HMX booster.
**Explosive ability (contained; explosive deflagration):** High.
**Percentage:** ***54.6% nitrocellulose, 35.4% nitroglycerine, 7.8% potassium perchlorate, 1.2% carbolac I, 0.90% ethyl centralite, 0.020% graphite, 0.080% moisture***
**Classification:** Deflagrating explosive (classified as Class A explosive).

**Danger of preparation:** Should only be prepared by experienced personnel.
**Use:** Used to propel anti-tank munitions; for example, US/NATO "TOW" missile system; Dragon, and Javelin anti-tank munitions.

**Note:** Inexperienced personnel should not attempt this procedure, as there are hazards involved. During the heating process caution should be taken to avoid excessive heat or friction. Make sure to thoroughly stir the mixture during the heating process to avoid lumps or clumps, which can lead to excessive heat build-up. This procedure should be preformed behind proper blast shielding. Keep in mind the dangers of working with nitroglycerine.

### 04-04-002A: High Performance military rocket propellant:

Into a standard vertical mixer, place ***24 grams of ethyl acrylate-acrylic acid*** (95-5, copolymer binder)**,** followed by ***5.2 grams of "UNOX 221"*** (cross linking agent), and then blend the mixture on high for about 30 minutes at room temperature. After 30 minutes, add in ***2.5 grams of carbon black***, followed by ***252.5 grams of powdered HMX*** (see The Preparatory Manual of Explosive Fourth Edition for preparation), and then continue to thoroughly blend the mixture on high at room temperature for 30 minutes. After 30 minutes, add in ***101 grams of powdered ammonium perchlorate*** and then continue to blend the mixture for about 30 minutes at room temperature. After 30 minutes, carefully add in ***122 grams of nitroglycerine***, and then continue to blend the mixture, but reduce blending speed to low, and continue blending for 45 minutes at room temperature. After 45 minutes, the propellant is ready to be casted. To do so, it should be gently pressed into any desirable rocket motor or mold under a pressure of 6,500 psi, using the usual techniques, and the resulting rocket motor or mold should then be cured for several days at ambient temperature.
**Burn rate:** 3.1 seconds per inch at 6,500 psi; tends to vary based on diameter and density of rocket grain.
**Water resistance:** Good.
**Stability:** Can be stored for decades.
**Sensitivity:** Stable towards friction; stable towards normal shock; stable to temperatures up to 100 Celsius.
**Flammability (1 to 10, un-contained):** 9 solid grain; 9.1 fine grain.
**Ease of ignition (1 to 10 un-contained):** 7.9 solid grain; grain size varies and results in different ignition rates.
**Tendency to cake:** Will not cake.
**Detonation ability:** Can be detonated but only certain conditions—requires blasting cap or detonator.
**Explosive ability (contained; explosive deflagration):** Below moderate.
**Percentage:** ***49.7% HMX, 24% nitroglycerine, 19.9% ammonium perchlorate, 4.7% ethyl acrylate-acrylic acid copolymer, 1% UNOX 221 linking agent, 0.49% carbon black catalyst, 0.210% moisture***
**Classification:** Deflagrating explosive (classified as Class A explosive).

**Danger of preparation:** Should only be prepared by experienced personnel.
**Use:** Used to propel anti-tank munitions; for example, US/NATO "TOW" missile system; Dragon, Javelin, Carl Gustov 84mm, Panzerfest IV anti-tank munitions.

**Note:** Inexperienced personnel should not attempt this procedure, as there are hazards involved. This procedure should be preformed behind proper blast shielding. Keep in mind the dangers of working with nitroglycerine.

### 04-04-003A: High Performance CMDB rocket propellant (composite modified Double Base propellant):

Into a suitable vertical mixer, place 290 milliliters of methylene chloride, followed by ***55.5 grams of triacetin***, followed by ***100 grams of finely divided aluminum powder***, followed by ***5.1 grams of 2-nitrodiphenylamine***, followed by ***55 grams of nitrocellulose*** (14% nitrogen content)**,** followed by ***110 grams of powdered ammonium perchlorate***, followed by slowly adding in ***173.5 grams of nitroglycerine,*** followed by ***350 milligrams of lead 2-ethylhexoate***, and then blend the mixture on low speed until only about 10% of the methylene chloride remains (about 40 to 60 minutes). Thereafter, add in ***1.5 grams of 2,6-tolylene diisocyanate***, and then continue to blend the mixture on low speed for 15 minutes to form a uniform mass. Thereafter, the mixture is ready to be casted and cured. To do so, the propellant mixture should be pressed into any desirable rocket motor or mold under a pressure of 7,000 psi, and then the rocket motor or mold should be cured in an oven at 40 Celsius for 16 hours.
**Burn rate:** 3 seconds per inch at 7,000 psi; tends to vary based on diameter and density of rocket grain.
**Water resistance:** Good.
**Stability:** Can be stored for decades.
**Sensitivity:** Stable towards friction; stable towards normal shock; stable to temperatures up to 100 Celsius.
**Flammability (1 to 10, un-contained):** 9.1 solid grain; 9.2 fine grain.
**Ease of ignition (1 to 10 un-contained):** 7.5 solid grain; grain size varies and results in different ignition rates.
**Tendency to cake:** Will not cake.
**Detonation ability:** Can be detonated but only certain conditions—requires blasting cap or detonator.
**Explosive ability (contained; explosive deflagration):** Moderate.
**Percentage:** ***34.6% nitroglycerine, 21.9% ammonium perchlorate, 19.9% aluminum, 11% triacetin, 10.9% nitrocellulose, 1% 2-nitrodiphenylamine, 0.35% impurities, 0.29% 2,6-tolylene diisocynate curative, 0.06% lead 2-ethylhexoate catalyst,***
**Classification:** Deflagrating explosive (classified as Class A explosive).

**Danger of preparation:** Should only be prepared by experienced personnel.
**Use:** Used to propel anti-tank munitions; for example, US/NATO "TOW" missile system; Dragon, Javelin, Carl Gustov 84mm, Panzerfest IV, "Hell fire missile" anti-tank munitions.

**Note:** Inexperienced personnel should not attempt this procedure, as there are hazards involved. This procedure should be preformed behind proper blast shielding. Keep in mind the dangers of working with nitroglycerine.

**Anti-tank rocket propelled munition (bazooka fired HEAT round)**

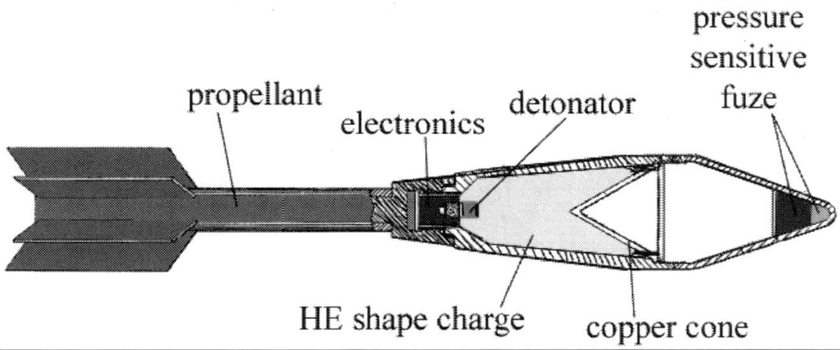


**04-04-003B: High Performance CMDB rocket propellant (composite modified Double Base propellants):**
Into a suitable vertical mixer, place 290 milliliters of methylene chloride, followed by ***51.5 grams of di-n-propyl adipate***, followed by ***100 grams of finely divided aluminum powder***, followed by ***5 grams of 2-nitrodiphenylamine***, followed by ***75 grams of nitrocellulose*** (14% nitrogen content), followed by ***100 grams of powdered ammonium perchlorate***, followed by slowly adding in ***161 grams of nitroglycerine,*** followed by ***5 grams of resorcinol diacetate***, followed by ***50 milligrams of ferric acetylacetonate***, and then blend the mixture on low speed until only about 10% of the methylene chloride remains (40 to 60 minutes). Thereafter, add in ***1.9 grams of 2,6-tolylene diisocyanate***, and then continue to blend the mixture on low speed for 15 minutes to form a uniform mass. Thereafter, the mixture is ready to be casted and cured. To do so, the propellant mixture should be pressed into any desirable rocket motor or mold under a pressure of 7,000 psi, and then cure the rocket motor or mold in an oven at 48 Celsius for 4 days.
**Burn rate:** 2.9 seconds per inch at 7,000 psi; tends to vary based on diameter and density of rocket grain.
**Water resistance:** Good.
**Stability:** Can be stored for decades.
**Sensitivity:** Stable towards friction; stable towards normal shock; stable to temperatures up to 100 Celsius.
**Flammability (1 to 10, un-contained):** 9.2 solid grain; 9.4 fine grain.
**Ease of ignition (1 to 10 un-contained):** 7.6 solid grain; grain size varies and results in different ignition rates.
**Tendency to cake:** Will not cake.
**Detonation ability:** Can be detonated but only certain conditions—requires blasting cap or detonator.
**Explosive ability (contained; explosive deflagration):** Moderate.
**Percentage:** ***32.2% nitroglycerine, 20% ammonium perchlorate, 20% aluminum, 15% nitrocellulose, 10.3% di-n-propyl adipate, 1% 2-nitrodiphenylamine, 1% resorcinol diacetate, 0.38% 2,6-tolylene diisocyante curative, 0.11% residue, 0.01% ferric acetylacetonate***
**Classification:** Deflagrating explosive (classified as Class A explosive).

**Danger of preparation:** Should only be prepared by experienced personnel.
**Use:** Used to propel anti-tank munitions; for example, US/NATO "TOW" missile system; Dragon, Javelin, Carl Gustov 84mm, Panzerfest IV, "Hell fire missile" anti-tank munitions.

**Note:** Inexperienced personnel should not attempt this procedure, as there are hazards involved. This procedure should be preformed behind proper blast shielding. Keep in mind the dangers of working with nitroglycerine.

**04-04-004A: High Performance nitrocellulose/nitroglycerine based rocket propellant:**
Into a suitable vertical mixer, place ***250 grams of nitrocellulose*** (14% nitrogen content), followed by ***180 grams of nitroglycerin***, followed by ***76.5 grams of triacetin***, and then followed by ***2.5 grams of nitrodiphenylamine.*** Thereafter, blend the mixture on moderate speed for about 45 minutes. Thereafter, the mixture is ready for use. To use, press the mixture into any desired rocket motor or mold under a pressure of 6,500 psi, and then cure the rocket motor or mold in an oven at 60 Celsius for 5 days.
**Burn rate:** 2.98 seconds per inch at 6,500 psi; tends to vary based on diameter and density of rocket grain.
**Water resistance:** Good.
**Stability:** Can be stored for decades.
**Sensitivity:** Stable towards friction; stable towards normal shock; stable to temperatures up to 100 Celsius.
**Flammability (1 to 10, un-contained):** 9.29 solid grain; 9.4 fine grain.

**Ease of ignition (1 to 10 un-contained):** 8 solid grain; grain size varies and results in different ignition rates.
**Tendency to cake:** Will not cake.
**Detonation ability:** Can be detonated but only certain conditions—requires blasting cap or detonator.
**Explosive ability (contained; explosive deflagration):** Above moderate.
**Percentage:** ***49.11% nitrocellulose, 35.36% nitroglycerine, 15.02% triacetin, 0.49% nitrodiphenylamine, 0.02% impurities***
**Classification:** Deflagrating explosive (classified as Class A explosive).

**Danger of preparation:** Should only be prepared by experienced personnel.
**Use:** Used to propel anti-tank munitions; for example, US/NATO "TOW" missile system; Dragon, Javelin, Carl Gustov 84mm, Panzerfest IV, "Hell fire missile" anti-tank munitions.

**Note:** Inexperienced personnel should not attempt this procedure, as there are hazards involved. This procedure should be preformed behind proper blast shielding. Keep in mind the dangers of working with nitroglycerine.

### 04-04-005A: High Performance "hard, yet flexible" propellant utilizing hydroxylamine perchlorate:

Into a standard sealable vertical mixer, place ***50 grams of a binder compound D.E.R. 332,*** followed by ***50 grams of Epon 812 resin,*** followed by ***400 grams of powdered hydroxylamine perchlorate,*** and then seal the vertical mixer; thereafter, blend the mixture on high speed for 15 minutes at room temperature to form a uniform mixture. After blending, the fluidized mixture is then ready for use. To use, it should be poured directly into your rocket motor or mold under a pressure of 2,000 psi, and the resulting rocket motor or mold needs to be cured in an oven at 50 Celsius for 24 hours.
**Burn rate:** 3 seconds per inch at 2,000 psi; tends to vary based on diameter and density of rocket grain.
**Water resistance:** Good.
**Stability:** Can be stored for decades.
**Sensitivity:** Stable towards friction; stable towards normal shock; stable to temperatures up to 180 Celsius.
**Flammability (1 to 10, un-contained):** 9.1 solid grain; 9.3 fine grain.
**Ease of ignition (1 to 10 un-contained):** 8 solid grain; grain size varies and results in different ignition rates.
**Tendency to cake:** Will not cake.
**Detonation ability:** Can be detonated but only severe conditions—requires TNT, RDX, or HMX booster.
**Explosive ability (contained; explosive deflagration):** High.
**Percentage:** ***80% hydroxylamine perchlorate, 10% D.E.R. 32 binder, 10% Epon 812 epoxy resin***
**Classification:** Deflagrating explosive (classified as Class A explosive).

**Danger of preparation:** Can be prepared by amateurs, but caution should be exercised.
**Use:** Used to propel specialty rockets and missiles for military and commercial applications.

### 04-04-005B: High Performance "hard, yet flexible" propellant utilizing hydroxylamine perchlorate (modified):

Into a standard sealable vertical mixer, place ***50 grams of a binder compound D.E.R. 332,*** followed by ***50 grams of Epon 812 resin,*** followed by ***100 grams of aluminum powder of 325 mesh,*** seal the vertical mixer, and then blend the mixture on high speed for about 30 minutes. Thereafter, add in ***300 grams of hydroxylamine perchlorate,*** and then continue to blend the mixture on high for about 20 minutes at room temperature to form a uniform mix. After blending for 20 minutes, the fluidized mixture is ready for use. To use, it should be poured directly into your rocket motor under a pressure of 2,000 psi, and the resulting rocket motor should then be cured at room temperature for about 2 days.
**Burn rate:** 2.5 seconds per inch at 2,000 psi; tends to vary based on diameter and density of rocket grain.
**Water resistance:** Good.
**Stability:** Can be stored for decades.
**Sensitivity:** Stable towards friction; stable towards normal shock; stable to temperatures up to 180 Celsius.
**Flammability (1 to 10, un-contained):** 9.3 solid grain; 9.5 fine grain.
**Ease of ignition (1 to 10 un-contained):** 8.1 solid grain; grain size varies and results in different ignition rates.
**Tendency to cake:** Will not cake.
**Detonation ability:** Can be detonated but only severe conditions—requires TNT, RDX, or HMX booster.
**Explosive ability (contained; explosive deflagration):** High.
**Percentage:** ***60% hydroxylamine perchlorate, 20% aluminum, 10% D.E.R. 32 binder, 10% Epon 812 epoxy resin***
**Classification:** Deflagrating explosive (classified as Class A explosive).

**Danger of preparation:** Can be prepared by amateurs, but caution should be exercised.
**Use:** Used to propel specialty rockets and missiles for military and commercial applications.

### 04-04-006A: High Performance polymeric propellant utilizing guanidine perchlorate:

Into a special heated sealable vertical mixer, place ***575 grams of guanidine perchlorate***, followed by ***425 grams of anhydrous lithium perchlorate***, and then gently and slowly heat the mixture to 140 Celsius with vigorous stirring. When the temperature reaches 140 Celsius, the mixture will melt and fuse into a molten mass. Once the mixture has melted, continue to rapidly blend the fused mixture for 10 minutes at 140 Celsius. After 10 minutes, reduce the temperature of the mixture to 85 Celsius, and once this temperature is achieved, quickly open the vertical mixer, and quickly add in ***105 grams of acrylamide***, followed by ***70 grams of ethylene glycol***, and then followed by ***8 grams of N-methyl morpholine***; immediately thereafter, re-seal the vertical mixer and then continue to blend the mixture at 85 Celsius for 30 minutes. After 30 minutes, add in ***300 grams of standard mesh powdered aluminum***, and then continue to blend the mixture for about 30 minutes at 85 Celsius. After 30 minutes, quickly add in ***10 grams of succinic anhydride***, and continue to heat and stir the mixture for 5 minutes at 85 Celsius. Note: once the succinic anhydride is added, polymerization will begin, so immediately after blending for 5 minutes, the mixture needs to be poured at 85 Celsius into the rocket motor (pre-fabricated before use) under a pressure of 2,000 psi. Note: in order for proper polymerization to proceed, your rocket motor must be heated in an oven at 85 Celsius prior to pouring in the hot propellant mixture. Simply pre-heat your rocket motor to 85 Celsius, and then proceed with the casting process as discussed; thereafter, cure the rocket motor in an oven at 85 Celsius for 3 hours.
**Burn rate:** 2.9 seconds per inch at 2,000 psi; tends to vary based on diameter and density of rocket grain.
**Water resistance:** Good.
**Stability:** Can be stored for decades.
**Sensitivity:** Stable towards friction; stable towards normal shock; stable to temperatures up to 180 Celsius.
**Flammability (1 to 10, un-contained):** 9.1 solid grain; 9.4 fine grain.
**Ease of ignition (1 to 10 un-contained):** 8 solid grain; grain size varies and results in different ignition rates.
**Tendency to cake:** Will not cake.
**Detonation ability:** Can be detonated but only severe conditions—requires TNT, RDX, or HMX booster.
**Explosive ability (contained; explosive deflagration):** Above moderate.
**Percentage:** ***38.5% guanidine perchlorate, 28.4% lithium perchlorate, 20% aluminum, 7% acrylamide, 4.6% ethylene glycol, 0.66% succinic anhydride plymerization catalyst, 0.53% N-methyl morpholine, 0.31% moisture and residue***
**Classification:** Deflagrating explosive (classified as explosive propellant).

**Danger of preparation:** Can be prepared by amateurs, but caution should be exercised.
**Use:** Used to propel specialty rockets and missiles for military and commercial applications.

### 04-04-007A: Moderate performance "rubber-like" propellant utilizing potassium perchlorate and polysulfide binder:

Into a standard vertical mixer, place ***97.5 grams of a polysulfide binder compound*** (sold as Thiokol LP-2), followed by ***7.5 grams of lead dioxide,*** followed by ***7.5 grams of dibutyl phthalate,*** followed by ***1.75 grams of stearic acid,*** and then blend the mixture on high speed for about 10 minutes. After 10 minutes, add in ***300 grams of potassium perchlorate,*** followed by ***50 grams of powdered ammonium perchlorate,*** followed by ***36 grams of furfuryl alcohol,*** and then continue to blend the mixture on high speed for about 15 minutes to form a uniform mix. After blending, the fluidized mixture is then ready for use. To use, it should be poured and pressed into your rocket motor under a pressure of 3,000 psi, and then cure the rocket motor at room temperature for 2 days.
**Burn rate:** 2.97 seconds per inch at 3,000 psi; tends to vary based on diameter and density of rocket grain.
**Water resistance:** Good.
**Stability:** Can be stored for decades.
**Sensitivity:** Stable towards friction; stable towards normal shock; stable to temperatures up to 180 Celsius.
**Flammability (1 to 10, un-contained):** 9.1 solid grain; 9.4 fine grain.
**Ease of ignition (1 to 10 un-contained):** 8 solid grain; grain size varies and results in different ignition rates.
**Tendency to cake:** Will not cake.
**Detonation ability:** Can be detonated but only severe conditions—requires TNT, RDX, or HMX booster.
**Explosive ability (contained; explosive deflagration):** Moderate.
**Percentage:** ***59.9% potassium perchlorate, 19.4% polysulfide binder, 9.9% ammonium perchlorate, 7.1% furfuryl alcohol plasticizer, 1.4% lead dioxide vulcanizing agent, 1.4% dibutyl phthalate dispersion medium, 0.56% mixed residues, 0.34% stearic acid catalyst***
**Classification:** Deflagrating explosive (classified as propellant).

**Danger of preparation:** Can be prepared by amateurs, but caution should be exercised.
**Use:** Used to propel specialty rockets and missiles for military and commercial applications; used to propel wire-guided missiles.

### 04-04-007B: Moderate performance "rubber-like" propellant utilizing potassium perchlorate and polysulfide binder (modified):

Into a standard vertical mixer, place ***115 grams of a polysulfide binder compound*** (sold as Thiokol LP-2), followed by ***2.5 grams of paraquinone dioxime,*** followed by ***3.4 grams of stearic acid,*** followed by ***6.5 grams of zinc oxide,*** and then blend the mixture on high speed for about 15 minutes. After 15 minutes, add in ***250 grams of potassium perchlorate,*** followed by ***85 grams of powdered ammonium perchlorate,*** followed by ***37.5 grams of furfuryl alcohol,*** and then continue to blend the mixture on high speed for 15 minutes to form a uniform mix. After blending, the fluidized mixture is then ready for use. To use, it should be poured and pressed directly into your rocket motor under a pressure of 2,500 psi, and then cure the rocket motor in an oven at 65 Celsius for 4 hours.
**Burn rate:** 2.99 seconds per inch at 2,500 psi; tends to vary based on diameter and density of rocket grain.
**Water resistance:** Good.
**Stability:** Can be stored for decades.
**Sensitivity:** Stable towards friction; stable towards normal shock; stable to temperatures up to 180 Celsius.
**Flammability (1 to 10, un-contained):** 9.1 solid grain; 9.4 fine grain.
**Ease of ignition (1 to 10 un-contained):** 8 solid grain; grain size varies and results in different ignition rates.
**Tendency to cake:** Will not cake.
**Detonation ability:** Can be detonated but only severe conditions—requires TNT, RDX, or HMX booster.
**Explosive ability (contained; explosive deflagration):** Moderate.
**Percentage:** ***50% potassium perchlorate, 23% polysulfide binder, 17% ammonium perchlorate, 7.5% furfuryl alcohol plasticizer, 1.3% zinc oxide, 0.68% stearic acid catalyst, 0.50% paraquinone dioxime, 0.02% mixed resides***
**Classification:** Deflagrating explosive (classified as propellant).

**Danger of preparation:** Can be prepared by amateurs, but caution should be exercised.
**Use:** Used to propel specialty rockets and missiles for military and commercial applications; used to propel wire-guided missiles.

### 04-04-008A: High performance lithium perchlorate based propellant for various applications:

Into a special heated sealable vertical mixer, equipped with nitrogen gas purge valve (for excluding air and moisture), quickly place ***500 grams of anhydrous lithium perchlorate***, followed by ***200 grams of a copolymer of hexamethylene adipamide and beta-aminocaproic acid***, seal the vertical mixer, begin a nitrogen gas purge, and then slowly bring the temperature of the mixture to 225 Celsius under rapid stirring; continue to heat and rapidly stir the mixture at 225 Celsius for 30 minutes. Note: during the heating process, maintain a dry nitrogen atmosphere by way of the nitrogen gas purge. After heating and blending at 225 Celsius for 30 minutes, remove the nitrogen gas purge, open the vertical mixer and then cast the hot mixture directly into your rocket motor or mold under a pressure of 2,000 psi, and then allow the rocket motor or mold to cool at room temperature for 7 days.
**Burn rate:** 2.8 seconds per inch at 2,000 psi; tends to vary based on diameter and density of rocket grain.
**Water resistance:** Fair when kept from prolonged exposure to moisture.
**Stability:** Can be stored for decades when kept from prolonged exposure to moisture.
**Sensitivity:** Stable towards friction; stable towards normal shock; stable to temperatures up to 180 Celsius.
**Flammability (1 to 10, un-contained):** 9.2 solid grain; 9.3 fine grain.
**Ease of ignition (1 to 10 un-contained):** 8.1 solid grain; grain size varies and results in different ignition rates.
**Tendency to cake:** Will not cake.
**Detonation ability:** Cannot be detonated.
**Explosive ability (contained; explosive deflagration):** Low.
**Percentage:** ***71.4% lithium perchlorate, 28.5% copolymer of hexamethylene adipamide, 0.1% residue***
**Classification:** Deflagrating explosive (classified as propellant).

**Danger of preparation:** Can be prepared by amateurs, but caution should be exercised.
**Use:** Used in rocket assisted take-off munitions for aircraft, and in rocket-assisted projectiles for artillery shells and the like.

**04-04-009A: High performance polymer rocket propellant based on potassium perchlorate:**
Into a sealable heated vertical mixer, place ***100 grams of beta-caprolactam***, followed by ***1 gram of sodium hydride***, followed by ***56.5 grams of N-acetyl caprolactam***, followed by ***300 grams of potassium perchlorate***, seal the vertical mixer, and then heat the contents to about 160 Celsius under a rapid stirring. When the temperature of the mixture reaches about 160 Celsius, it will become fluidized, so continue to rapidly blend the mixture at 160 Celsius for 10 minutes. Immediately thereafter, the fluidized mixture needs to be poured and pressed into any desirable rocket motor or mold, while still hot, under a pressure of 2,000 psi. Thereafter, allow the rocket motor or mold to cool for 48 hours.
**Burn rate:** 3.1 seconds per inch at 2,000 psi; tends to vary based on diameter and density of rocket grain.
**Water resistance:** Excellent.
**Stability:** Can be stored for decades.
**Sensitivity:** Stable towards friction; stable towards normal shock; stable to temperatures up to 210 Celsius.
**Flammability (1 to 10, un-contained):** 9 solid grain; 9.3 fine grain.
**Ease of ignition (1 to 10 un-contained):** 8 solid grain; grain size varies and results in different ignition rates.
**Tendency to cake:** Will not cake.
**Detonation ability:** Cannot be detonated.
**Explosive ability (contained; explosive deflagration):** Moderate.
**Percentage:** ***65.5% potassium perchlorate, 21.8% beta-caprolcatam pre-polymer, 12.3% N-acetyl caprolactam pre-polymer, 0.21% polymerization catalyst, 0.19% mixed impurities***
**Classification:** Deflagrating explosive (classified as propellant).

**Danger of preparation:** Can be prepared by amateurs, but caution should be exercised.
**Use:** Used in rocket assisted take-off munitions for aircraft, and in rocket-assisted projectiles for artillery shells and the like.

**04-04-010A: High Performance potassium perchlorate rocket propellant:**
Into a suitable vertical mixer, place ***190 grams of potassium perchlorate***, followed by ***50 grams of finely powdered aluminum*** of the usual mesh, and then followed by ***10 grams of copper chromite***. Immediately thereafter, add in 300 milliliters of diethyl ether, and then blend the mixture on high speed for 30 minutes. Thereafter, add in ***62.5 grams of a unhardened polyurethane compound*** (commercially available), followed by ***172.5 grams of potassium perchlorate***, followed by ***10 grams of finely powdered aluminum***, and then continue to blend the mixture on high speed for 20 minutes. Thereafter, the mixture is ready for use. To use, the mixture needs to be pressed into any desirable rocket motor or mold under a pressure of 5,000 psi, and the resulting rocket motor or mold should be cured at 50 Celsius for 3 hours. Should be ignited using a black powder charge or equivalent ignition composition.
**Burn rate:** 2.9 seconds per inch at 5,000 psi; tends to vary based on diameter and density of rocket grain.
**Water resistance:** Excellent.
**Stability:** Can be stored for decades.
**Sensitivity:** Stable towards friction; stable towards normal shock; stable to temperatures up to 210 Celsius.
**Flammability (1 to 10, un-contained):** 9.2 solid grain; 9.5 fine grain.
**Ease of ignition (1 to 10 un-contained):** 8.9 solid grain; grain size varies and results in different ignition rates.
**Tendency to cake:** Will not cake.
**Detonation ability:** Cannot be detonated.
**Explosive ability (contained; explosive deflagration):** Above moderate.
**Percentage:** ***73.2% potassium perchlorate, 12.6% polyurethane compound binder, 12.1% aluminum, 2% copper chromite burn rate catalyst, 0.10% balanced***
**Classification:** Deflagrating explosive (classified as propellant).

**Danger of preparation:** Can be prepared by amateurs, but caution should be exercised.
**Use:** Used to propel specialty rockets and missiles for military and commercial applications.

**04-04-011A: High Performance potassium perchlorate rocket propellant:**
Into a suitable sealable vertical mixer, place ***380 grams of potassium perchlorate***, followed by ***47.5 grams of powdered aluminum***, and then followed by ***72.5 grams of any epoxy resin or Armstrong Resin E-301-14***, seal the vertical mixer, and then blend the mixture on high speed for about 15 minutes. Thereafter, the mixture is ready for use. To use, press the mixture into any desirable rocket motor or mold under a pressure of 7,000 psi, and then cure the rocket motor or mold in an oven at 170 Celsius for 3 hours. The rocket propellant can be fired using any desirable ignition composition.
**Burn rate:** 2.9 seconds per inch at 7,000 psi; tends to vary based on diameter and density of rocket grain.
**Water resistance:** Excellent.
**Stability:** Can be stored for decades.
**Sensitivity:** Stable towards friction; stable towards normal shock; stable to temperatures up to 210 Celsius.
**Flammability (1 to 10, un-contained):** 9.2 solid grain; 9.5 fine grain.

**Ease of ignition (1 to 10 un-contained):** 8.9 solid grain; grain size varies and results in different ignition rates.
**Tendency to cake:** Will not cake.
**Detonation ability:** Cannot be detonated.
**Explosive ability (contained; explosive deflagration):** Above moderate.
**Percentage:** ***76% potassium perchlorate, 14.5% Epoxy resin binder, 9.5% aluminum***
**Classification:** Deflagrating explosive (classified as propellant).

**Danger of preparation:** Can be prepared by amateurs, but caution should be exercised.
**Use:** Used to propel specialty rockets and missiles for military and commercial applications.

## 04-04-012A: Super high Performance potassium perchlorate rocket propellant containing hydroxylamine perchlorate:

Into a suitable sealable vertical mixer, place ***159 grams of Thiokol LP-2 polysulfide binder***, followed by ***227.5 grams of potassium perchlorate of 100 mesh***, followed by ***97.5 grams of potassium perchlorate of 325 mesh,*** followed by ***10 grams of para-quinone dioxime,*** followed by ***5 grams of diphenyl guanidine,*** and then followed by ***25 grams of hydroxylamine perchlorate***, seal the vertical mixer, and then blend the mixture on high speed for 20 minutes. Thereafter, the mixture is ready for use. To use, press the tacky mass into any desirable rocket motor or mold under pressure of 5,000 psi, in the usual manner, and then cure the rocket motor or mold in an oven at 85 Celsius for 18 hours. The rocket propellant can be fired using any desirable ignition composition.
**Burn rate:** 3.1 seconds per inch at 5,000 psi; tends to vary based on diameter and density of rocket grain.
**Water resistance:** Excellent.
**Stability:** Can be stored for decades.
**Sensitivity:** Stable towards friction; stable towards normal shock; stable to temperatures up to 210 Celsius.
**Flammability (1 to 10, un-contained):** 9 solid grain; 9.2 fine grain.
**Ease of ignition (1 to 10 un-contained):** 8 solid grain; grain size varies and results in different ignition rates.
**Tendency to cake:** Will not cake.
**Detonation ability:** Cannot be detonated.
**Explosive ability (contained; explosive deflagration):** Moderate.
**Percentage:** ***62% potassium perchlorate, 30.3% Thiokol LP-2 polysulfide binder, 4.77% hydroxylamine perchlorate, 1.9% p-quinone dioxime, 0.95% diphenyl guanidine, 0.08% mixed residues***
**Classification:** Deflagrating explosive (classified as propellant).

**Danger of preparation:** Can be prepared by amateurs, but caution should be exercised.
**Use:** Used to propel specialty rockets and missiles for military and commercial applications.

## 04-04-013A: Super high Performance flexible, hydroxylamine perchlorate rocket propellant:

Into a suitable sealable vertical mixer, place ***150 grams of Paraplex P-10 binder***, followed by ***250 grams of hydroxylamine perchlorate of 200 mesh***, and then followed by ***200 grams of potassium perchlorate of average mesh,*** seal the vertical mixer, and then rapidly and thoroughly blend the mixture for 15 minutes to form a uniform mix. Note: the Paralax P-10 may even begin to polymerize at room temperature so take note. After the 15-minute blending period, quickly add in ***1.5 grams of benzoyl peroxide***, re-seal the vertical mixer and then briefly, yet rapidly blend the mixture for 1 minute. Thereafter, immediately press the mixture into any desirable rocket motor under a pressure of 6,000 psi, and then allow the mixture to cure at room temperature for 3 days. The rocket propellant can be fired using any desirable ignition composition.
**Burn rate:** 3 seconds per inch at 6,000 psi; tends to vary based on diameter and density of rocket grain.
**Water resistance:** Excellent.
**Stability:** Can be stored for decades.
**Sensitivity:** Stable towards friction; stable towards normal shock; stable to temperatures up to 210 Celsius.
**Flammability (1 to 10, un-contained):** 9.1 solid grain; 9.2 fine grain.
**Ease of ignition (1 to 10 un-contained):** 8.3 solid grain; grain size varies and results in different ignition rates.
**Tendency to cake:** Will not cake.
**Detonation ability:** Cannot be detonated.
**Explosive ability (contained; explosive deflagration):** Above moderate.
**Percentage:** ***41.56% hydroxylamine perchlorate, 33.25% potassium perchlorate, 24.93% Paralax P-10 binder, 0.24% benzoyl peroxide, 0.02% balance***
**Classification:** Deflagrating explosive (classified as propellant).

**Danger of preparation:** Can be prepared by amateurs, but caution should be exercised.
**Use:** Used to propel specialty rockets and missiles for military and commercial applications.

### 04-04-014A: High Performance rocket propellant utilizing hydrazine nitrate:

Into a suitable sealable vertical mixer, place ***128 grams of methyl acrylate,*** followed by ***18 grams of allyl diglycol carbonate,*** followed by ***52 grams of methyl methacrylate,*** followed by ***300 grams of hydrazine nitrate,*** seal the vertical mixer, and then blend the mixture on high speed for 20 minutes to form a uniform mix. Thereafter, add in ***2 grams of tert-butyl peroxide,*** re-seal the vertical mixer, and then blend the mixture for 15 minutes under rapid speed at room temperature. Thereafter the mixture is ready for use. To use, press the mixture into any desirable rocket motor under a pressure of 7,000 psi, and then allow the rocket motor to cure at room temperature for 7 days. The rocket propellant can be fired using any desirable ignition composition.
**Burn rate:** 3.2 seconds per inch at 7,000 psi; tends to vary based on diameter and density of rocket grain.
**Water resistance:** Excellent.
**Stability:** Can be stored for decades.
**Sensitivity:** Stable towards friction; stable towards normal shock; stable to temperatures up to 210 Celsius.
**Flammability (1 to 10, un-contained):** 9.2 solid grain; 9.1 fine grain.
**Ease of ignition (1 to 10 un-contained):** 8 solid grain; grain size varies and results in different ignition rates.
**Tendency to cake:** Will not cake.
**Detonation ability:** Cannot be detonated.
**Explosive ability (contained; explosive deflagration):** Moderate.
**Percentage:** ***60% hydrazine nitrate, 25.6% methyl acrylate, 10.4% methyl methacrylate, 3.6% allyl diglycol carbonate, 0.4% tert-butyl peroxide***
**Classification:** Deflagrating explosive (classified as propellant).

**Danger of preparation:** Can be prepared by amateurs, but caution should be exercised.
**Use:** Used to propel specialty rockets and missiles for military and commercial applications.

### 04-04-015A: Specialty magnesium perchlorate semi-solid (gelled) rocket propellant:

Into a suitable vertical mixer, place ***22 grams of aluminum naphthenate*** (containing 7.55% aluminum by weight), followed by ***278 grams of regular unleaded gasoline***. Thereafter, stir the mixture for about 15 minutes to form a gel. Thereafter, add in ***45 grams of lamp black,*** and then followed by ***864 grams of anhydrous magnesium perchlorate***, and then continue to blend the mixture for about 30 minutes on high speed. Thereafter, the tacky gel is ready for use. To do so, it simply needs to be pressed into any desirable rocket motor under a pressure of 2,000 psi, in the usual fashion. The gelled propellant mixture can be ignited using any suitable ignition composition. Note: test the bun rate using a small sample before committing to a rocket munition for safety—stories of rocket motors exploding without warning have been reported, so use caution.
**Burn rate:** 4.1 seconds per inch at 2,000 psi; tends to vary based on diameter and density of rocket grain.
**Water resistance:** Excellent.
**Stability:** Can be stored for decades.
**Sensitivity:** Stable towards friction; stable towards normal shock; stable to temperatures up to 100 Celsius.
**Flammability (1 to 10, un-contained):** 8.7
**Ease of ignition (1 to 10 un-contained):** 7.2
**Tendency to cake:** Will not cake.
**Detonation ability:** Can be detonated, but normally requires a blasting cap or detonator.
**Explosive ability (contained; explosive deflagration):** Low.
**Percentage:** ***71.46% magnesium perchlorate, 22.99% unleaded gasoline, 3.72% lampblack, 1.81% aluminum naphthenate, 0.02% rounded balance***
**Classification:** Deflagrating explosive (classified as propellant).

**Danger of preparation:** Can be prepared by amateurs, but caution should be exercised.
**Use:** Used to propel specialty rockets and missiles for military and commercial applications.

### 04-04-016A: High Performance potassium perchlorate rocket propellant:

Into a suitable vertical mixer, place ***190 grams of potassium perchlorate***, followed by ***50 grams of finely powdered aluminum*** of the usual mesh, and then followed by ***10 grams of copper chromite***. Immediately thereafter, add in 300 milliliters of dry hexane, and then blend the mixture on high speed to for 15 minutes. Thereafter, add in ***62.5 grams of a unhardened polyurethane compound*** (commercially available), followed by ***172.5 grams of powdered ammonium perchlorate***, followed by ***10 grams of finely powdered aluminum***, and then continue to blend the mixture on high speed for 30 minutes to form a uniform mass. Thereafter, the mixture is ready for use. To use, the mass needs to be pressed into any desirable rocket motor or mold under a pressure of 8,000 psi, under the usual conditions, and the resulting rocket motor or mold should be cured at 80 Celsius for 3 hours. Should be ignited using a black powder charge or equivalent ignition composition.
**Burn rate:** 2.99 seconds per inch at 8,000 psi; tends to vary based on diameter and density of rocket grain.
**Water resistance:** Excellent.
**Stability:** Can be stored for decades.
**Sensitivity:** Stable towards friction; stable towards normal shock; stable to temperatures up to 100 Celsius.
**Flammability (1 to 10, un-contained):** 9.2 solid grain; 9.3 fine grain.
**Ease of ignition (1 to 10 un-contained):** 8.4 solid grain; grain size varies and results in different ignition rates.
**Tendency to cake:** Will not cake.
**Detonation ability:** Can be detonated, but normally requires a blasting cap or detonator.
**Explosive ability (contained; explosive deflagration):** Above moderate.
**Percentage:** ***73.2% potassium perchlorate, 12.6% polyurethane compound binder, 12.1% aluminum, 2% copper chromite burn rate catalyst, 0.10% balanced***
**Classification:** Deflagrating explosive (classified as propellant).

**Danger of preparation:** Can be prepared by amateurs, but caution should be exercised.
**Use:** Used to propel specialty rockets and missiles for military and commercial applications.

### 04-04-017A: High Performance nitronium perchlorate specialty rocket propellant:

Into a special sealable vertical mixer equipped with nitrogen gas purge valve, place ***150 grams of polybutadiene*** (having a molecular weight of 3,000 and a viscosity of 300 poises), followed by ***50 grams of metallic lithium*** of 200 microns, seal the vertical mixer, begin a nitrogen gas purge, and then blend the mixture on high speed for about 45 minutes. Note: the mixing should be kept under inert atmosphere to exclude moisture and air. After blending for 45 minutes, quickly open the vertical mixer, add in ***300 grams of nitronium perchlorate*** of 100 microns, re-seal the vertical mixer, continue the nitrogen gas purge, and then continue to blend the mixture on high speed for 1 hour. Thereafter, the mixture is ready. To use, press the mixture into any desirable rocket motor or mold under a pressure of 10,000 psi, and then cure the rocket motor or mold at room temperature for 48 hours.
**Burn rate:** 2.5 seconds per inch at 10,000 psi; tends to vary based on diameter and density of rocket grain.
**Water resistance:** Poor.
**Stability:** Can be stored for up to 10 years when kept sealed in airtight and watertight containers.
**Sensitivity:** Stable towards friction; stable towards shock; stable to temperatures up to 100 Celsius.
**Flammability (1 to 10, un-contained):** 9.5 solid grain; 9.7 fine grain.
**Ease of ignition (1 to 10 un-contained):** 9.1 solid grain; grain size varies and results in different ignition rates.
**Tendency to cake:** Will not cake.
**Detonation ability:** Can be detonated, but normally requires a blasting cap or detonator.
**Explosive ability (contained; explosive deflagration):** Above moderate.
**Percentage:** ***60% nitronium perchlorate, 30% polybutadiene, 10% metallic lithium***
**Classification:** Deflagrating explosive (classified as propellant).

**Danger of preparation:** Should only be prepared by experienced personnel.
**Use:** Used to propel specialty rockets and missiles for military and commercial applications.

### 04-04-018A: High Performance nitronium perchlorate specialty rocket propellant (modified cure):

Into a special sealable vertical mixer equipped with nitrogen gas purge valve, place ***125 grams of polybutadiene*** (having a minimum of 0.4 milliequivalent of hydroxyl groups), followed by ***125 grams of metallic lithium*** of 200 microns, seal the vertical mixer, begin the nitrogen gas purge, and then blend the mixture on high speed for about 45 minutes. After 45 minutes, quickly open the vertical mixer, add in ***250 grams of nitronium perchlorate*** of 100 microns, re-seal the vertical mixer, and then continue to blend the mixture for about 1 hour under nitrogen gas purge. After 1 hour, the mixture is ready for use; therefore stop the nitrogen gas purge, open the vertical mixer, and then immediately press the mixture into any desirable rocket motor under a pressure of 8,000 psi, and then cure the rocket motor or mold at 50 Celsius for 3 hours.
**Burn rate:** 2.8 seconds per inch at 8,000 psi; tends to vary based on diameter and density of rocket grain.
**Water resistance:** Poor.

**Stability:** Can be stored for up to 10 years when kept sealed in airtight and watertight containers.
**Sensitivity:** Stable towards friction; stable towards shock; stable to temperatures up to 100 Celsius.
**Flammability (1 to 10, un-contained):** 9.3 solid grain; 9.5 fine grain.
**Ease of ignition (1 to 10 un-contained):** 8.6 solid grain; grain size varies and results in different ignition rates.
**Tendency to cake:** Will not cake.
**Detonation ability:** Can be detonated, but normally requires a blasting cap or detonator.
**Explosive ability (contained; explosive deflagration):** Above moderate.
**Percentage:** ***50% nitronium perchlorate, 25% polybutadiene, 25% metallic lithium***
**Classification:** Deflagrating explosive (classified as propellant).

**Danger of preparation:** Should only be prepared by experienced personnel.
**Use:** Used to propel specialty rockets and missiles for military and commercial applications.

### 04-04-019A: High Performance hydrazine nitrate rocket propellant:

Into a suitable sealable vertical mixer, place ***600 grams of hydrazine nitrate***, followed by ***400 grams of beryllium hydride***, seal the vertical mixer, and then blend the mixture on high speed for about 30 minutes. Thereafter, place 800 grams of this mixture into a clean separate sealable vertical mixer equipped with vacuum port, and then add in ***120 grams of carboxy terminated polybutadiene***, followed by ***40 grams of dioctly azelate***, and then followed by ***40 grams of mineral oil***. Thereafter, seal the vertical mixer, and then blend the mixture for about 1 hour on high speed. After 1 hour, continue to rapidly stir the mixture, but apply a vacuum of 300mm of mercury to de-gas the mixture, and hold said vacuum for about 45 minutes at room temperature. After 45 minutes, remove the vacuum, equalize the pressure, open the vertical mixer, and then place the mixture into a large capacity desiccator, and allow the mixture to stand in the desiccator for 7 days. Note: during this time, the mixture will expand. After 7 days, recover the mixture and then place into a suitable ball mill, filled with 500 grams of Teflon coated steel shot, and then tumble the mixture at 200 RPM for 3 hours. Thereafter, the mixture is ready for use. To use, the tumbled mixture needs to be pressed into any desirable rocket motor or mold under a pressure of 10,000 psi. Use a standard ignition composition for proper burn.
**Burn rate:** 2.98 seconds per inch at 10,000 psi; tends to vary based on diameter and density of rocket grain.
**Water resistance:** Moderate.
**Stability:** Can be stored for up to 15 years when kept sealed in airtight and watertight containers.
**Sensitivity:** Stable towards friction; stable towards shock; stable to temperatures up to 150 Celsius.
**Flammability (1 to 10, un-contained):** 9.1 solid grain; 9.4 fine grain.
**Ease of ignition (1 to 10 un-contained):** 9 solid grain; grain size varies and results in different ignition rates.
**Tendency to cake:** Will not cake.
**Detonation ability:** Can be detonated, but normally requires a blasting cap or detonator.
**Explosive ability (contained; explosive deflagration):** Above moderate.
**Percentage:** ***48% hydrazine nitrate, 32% beryllium hydride, 12% carboxy terminated polybutadiene, 4% dioctyl azelate, 4% mineral oil***
**Classification:** Deflagrating explosive (classified as propellant).

**Danger of preparation:** Should only be prepared by experienced personnel.
**Use:** Used to propel specialty rockets and missiles for military and commercial applications.

### 04-04-020A: High Performance hydrazine nitrate rocket propellant:

Into a suitable sealable vertical mixer, place ***700 grams of hydrazine nitrate***, followed by ***300 grams of sodium borohydride***, and then followed by ***180 grams of red phosphorus***. Thereafter, seal the vertical mixer, and then blend the mixture on high speed for about 1 hour. After 1 hour, open the vertical mixer, add in ***250 grams of potassium nitrate***, and then followed by ***300 grams of epon 815 epoxy binder*** (or any similar liquid binder). Thereafter, re-seal the vertical mixer, and then blend the mixture on moderate speed for about 45 minutes. Thereafter, the mixture is ready for use. To use, the mixture needs to be pressed into any desired rocket motor or mold under a pressure of 7,500 psi, in the usual manner, and then cure the rocket motor or mold in an oven at 60 Celsius for 4 hours. A standard ignition composition can be used for proper burn.
**Burn rate:** 4.5 seconds per inch at 7,500 psi; tends to vary based on diameter and density of rocket grain.
**Water resistance:** Good.
**Stability:** Can be stored for up to 20 years when kept from moisture.
**Sensitivity:** Stable towards friction; stable towards shock; stable to temperatures up to 150 Celsius.
**Flammability (1 to 10, un-contained):** 8.5 solid grain; 9 fine grain.
**Ease of ignition (1 to 10 un-contained):** 8.8 solid grain; grain size varies and results in different ignition rates.
**Tendency to cake:** Will not cake.
**Detonation ability:** Cannot be detonated.
**Explosive ability (contained; explosive deflagration):** Low.

**Percentage:** ***40.46% hydrazine nitrate, 17.34% sodium borohydride, 17.34% epon 815 epoxy binder, 14.45% potassium nitrate, 10.4% red phosphorus, 0.01% residual balance***
**Classification:** Deflagrating explosive (classified as propellant).

**Danger of preparation:** Should only be prepared by experienced personnel.
**Use:** Used to propel specialty rockets and missiles for military and commercial applications.

### 04-04-021A: High Performance potassium perchlorate rocket propellant with nitrocellulose base:

Into a special sealable vertical mixer equipped with vacuum port, place ***200 grams of triacetin***, followed by ***50 grams of butyl phthalyl butyl glycolate***, followed by ***250 grams of nitrocellulose*** (14% nitrogen content), and then followed by ***2.5 grams of lecithin***. Thereafter, blend the mixture on moderate speed for about 15 minutes. After 15 minutes, add in ***480 grams of potassium perchlorate***, and then continue to blend the mixture on high speed for about 30 minutes. Thereafter, seal the vertical mixer, and then de-gas the mixture by applying a vacuum of 75mm of mercury, and hold said vacuum for 45 minutes at room temperature. After 45 minutes, remove the vacuum, equalize the pressure, and open the vertical mixer. Thereafter, the mixture is ready for use. To use, the mixture needs to be pressed into any desired rocket motor or mold under a pressure of 5,000 psi and then cure the rocket motor or mold in an oven at 70 Celsius for 48 hours. A standard ignition composition can be used for proper burn.
**Burn rate:** 2.99 seconds per inch at 5,000 psi; tends to vary based on diameter and density of rocket grain.
**Water resistance:** Good.
**Stability:** Can be stored for decades.
**Sensitivity:** Stable towards friction; stable towards shock; stable to temperatures up to 150 Celsius.
**Flammability (1 to 10, un-contained):** 9.1 solid grain; 9.3 fine grain.
**Ease of ignition (1 to 10 un-contained):** 9.1 solid grain; grain size varies and results in different ignition rates.
**Tendency to cake:** Will not cake.
**Detonation ability:** Can be detonated, but requires a blasting cap or detonator.
**Explosive ability (contained; explosive deflagration):** High.
**Percentage:** ***48.85% potassium perchlorate, 25.44% nitrocellulose, 20.35% triacetin, 5.08% butyl phthalyl butyl glycolate, 0.25% mixed balance***
**Classification:** Deflagrating explosive (classified as propellant).

**Danger of preparation:** Can be prepared by amateurs, but caution should be exercised.
**Use:** Used to propel specialty rockets and missiles for military and commercial applications.

### 04-04-022A: High Performance aluminum hydride propellant with nitrocellulose base:

Into a suitable sealable vertical mixer, place ***200 grams of nitrocellulose*** (14% nitrogen content) followed by ***116 grams of aluminum hydride***, followed by ***56 grams of nitronium perchlorate***, and then followed by ***28 grams of polymeric dichlorostyrene***; thereafter seal the vertical mixer, and then blend the mixture on high speed for about 1 hour. Thereafter, the mixture is ready for use. To use, the mixture needs to be pressed into any desired rocket motor or mold under a pressure of 7,000 psi, in the usual manner, and then cure the rocket motor or mold in an oven at 60 Celsius for 3 days. A standard ignition composition can be used for proper burn.
**Burn rate:** 2.5 seconds per inch at 7,000 psi; tends to vary based on diameter and density of rocket grain.
**Water resistance:** Good.
**Stability:** Can be stored for decades.
**Sensitivity:** Stable towards friction; stable towards shock; stable to temperatures up to 150 Celsius.
**Flammability (1 to 10, un-contained):** 9.3 solid grain; 9.5 fine grain.
**Ease of ignition (1 to 10 un-contained):** 9.8 solid grain; grain size varies and results in different ignition rates.
**Tendency to cake:** Will not cake.
**Detonation ability:** Can be detonated, but requires a blasting cap or detonator.
**Explosive ability (contained; explosive deflagration):** High.
**Percentage:** ***50% nitrocellulose, 29% aluminum hydride, 14% nitronium perchlorate, 7% polymeric dichlorostyrene***
**Classification:** Deflagrating explosive (classified as propellant).

**Danger of preparation:** Can be prepared by amateurs, but caution should be exercised.
**Use:** Used to propel specialty rockets and missiles for military and commercial applications.

**04-04-023A: High Performance potassium perchlorate rocket propellant:**
Into a suitable sealable vertical mixer, place ***100 grams of a silicone elastomer*** (composed of 90 mol percent dimethyl dichlorosilane and 10 mol percent of diphenyl dichlorosilane), followed by ***300 grams of potassium perchlorate***, and then followed by ***2.4 grams of benzoyl peroxide***. Thereafter, seal the vertical mixer, and then blend the mixture on high speed for 45 minutes. Thereafter, the mixture is ready for use. To use, press the mixture into any desired rocket motor under a pressure of 9,000 psi, and then cure the rocket motor in an oven at 120 Celsius for about 2 hours. After 2 hours, raise the temperature to 200 Celsius, and heat at said temperature for about 14 hours. After the heating period, the rocket propellant should be set and hard. Thereafter, the rocket motor is ready to use. The rocket motor needs to be primed with a high temperature ignition composition.
**Burn rate:** 3 seconds per inch at 9,000 psi; tends to vary based on diameter and density of rocket grain.
**Water resistance:** Good.
**Stability:** Can be stored for decades.
**Sensitivity:** Stable towards friction; stable towards shock; stable to temperatures up to 150 Celsius.
**Flammability (1 to 10, un-contained):** 9.1 solid grain; 9.4 fine grain.
**Ease of ignition (1 to 10 un-contained):** 7 solid grain; grain size varies and results in different ignition rates.
**Tendency to cake:** Will not cake.
**Detonation ability:** Can be detonated, but requires a blasting cap or detonator.
**Explosive ability (contained; explosive deflagration):** Moderate.
**Percentage:** ***74.55% potassium perchlorate, 24.85% silicone elastomer, 0.21% benzoyl peroxide, 0.39% mixed impurities***
**Classification:** Deflagrating explosive (classified as propellant).

**Danger of preparation:** Can be prepared by amateurs, but caution should be exercised.
**Use:** Used to propel specialty rockets and missiles for military and commercial applications.

**04-04-023B: High Performance sodium perchlorate rocket propellant:**
Into a suitable sealable vertical mixer, place ***150 grams of dimethyl polysiloxane elastomer*** (having a viscosity of 10,000 centipoises), followed by ***350 grams of sodium perchlorate***, and then followed by ***3 grams of benzoyl peroxide***. Thereafter, seal the vertical mixer, and then blend the mixture on high speed for about 45 minutes. Thereafter, press the mixture into any desired rocket motor under a pressure of 9,000 psi, and then place the rocket motor into an oven and heat to 120 Celsius for about 2 hours. After 2 hours, raise the temperature to 200 Celsius, and heat at said temperature for about 14 hours. After the heating period, the rocket propellant should be set and hard. Thereafter, the rocket motor is ready to use. To use, the motor needs to be primed with a high temperature ignition composition.
**Burn rate:** 3 seconds per inch at 9,000 psi; tends to vary based on diameter and density of rocket grain.
**Water resistance:** Average.
**Stability:** Can be stored for decades when kept from moisture.
**Sensitivity:** Stable towards friction; stable towards shock; stable to temperatures up to 150 Celsius.
**Flammability (1 to 10, un-contained):** 9.1 solid grain; 9.4 fine grain.
**Ease of ignition (1 to 10 un-contained):** 7.1 solid grain; grain size varies and results in different ignition rates.
**Tendency to cake:** Will not cake.
**Detonation ability:** Cannot be detonated.
**Explosive ability (contained; explosive deflagration):** Low.
**Percentage:** ***69.58% sodium perchlorate, 29.82% silicone elastomer, 0.59% benzoyl peroxide, 0.01% mixed impurities***
**Classification:** Deflagrating explosive (classified as propellant).

**Danger of preparation:** Can be prepared by amateurs, but caution should be exercised.
**Use:** Used to propel specialty rockets and missiles for military and commercial applications.

**04-04-023C: High Performance cesium perchlorate rocket propellant with infrared illumination:**
Into a suitable sealable vertical mixer, place ***70 grams of silicone gum*** (having a hydrocarbon groups composed of methyl groups), followed by ***30 grams of powdered aluminum perchlorate*** (anhydrous), followed by ***300 grams of cesium perchlorate*** and then followed by ***2.5 grams of benzoyl peroxide***. Thereafter, seal the vertical mixer, and then blend the mixture on high speed for about 45 minutes. Thereafter, press the mixture into any desired rocket motor under a pressure of 9,000 psi, and then place the rocket motor into an oven and heat to 120 Celsius for about 2 hours. After 2 hours, raise the temperature to 200 Celsius, and heat at said temperature for about 14 hours. After the heating period, the rocket propellant should be set and hard. Thereafter, the rocket motor is ready to use. To use, the motor needs to be primed with a high temperature ignition composition.
**Burn rate:** 3.4 seconds per inch at 9,000 psi; tends to vary based on diameter and density of rocket grain.
**Water resistance:** Average.
**Stability:** Can be stored for decades when kept from moisture.
**Sensitivity:** Stable towards friction; stable towards shock; stable to temperatures up to 150 Celsius.

**Flammability (1 to 10, un-contained):** 9 solid grain; 9.2 fine grain.
**Ease of ignition (1 to 10 un-contained):** 7.6 solid grain; grain size varies and results in different ignition rates.
**Tendency to cake:** Will not cake.
**Detonation ability:** Cannot be detonated.
**Explosive ability (contained; explosive deflagration):** Low.
**Percentage:** ***74.53% cesium perchlorate, 17.39% silicone gum, 7.45% aluminum perchlorate, 0.62% benzoyl peroxide, 0.01% balance***
**Classification:** Deflagrating explosive (classified as propellant).

**Danger of preparation:** Can be prepared by amateurs, but caution should be exercised.
**Use:** Used in specialty rockets and missiles for generating an infrared trail to distract IR guided munitions.

### 04-04-024A: High Performance potassium perchlorate rocket propellant:

Into a standard sealable vertical mixer, place ***125 grams of alpha-caprolactum***, followed by ***2 grams of sodium hydride***, followed by ***75 grams of N-acetyl caprolactam***, and then followed by ***300 grams of potassium perchlorate***. Thereafter, seal the vertical mixer, and then blend the mixture on high speed for 40 minutes. Note; make sure to carryout the operation in the absence of moisture. After blending for 40 minutes, the mixture is ready for use. To use, press the mixture into any desirable rocket motor or mold under a pressure of 10,000 psi, in the usual manner, and then place the rocket motor or mold into an oven and heat to about 160 Celsius; heat at said temperature for about 15 minutes. Thereafter, remove the heat source, and allow the rocket motor or mold to cool to room temperature for 48 hours. Use a standard ignition composition for proper burn.
**Burn rate:** 4.6 seconds per inch at 10,000 psi; tends to vary based on diameter and density of rocket grain.
**Water resistance:** Excellent.
**Stability:** Can be stored for decades.
**Sensitivity:** Stable towards friction; stable towards shock; stable to temperatures up to 250 Celsius.
**Flammability (1 to 10, un-contained):** 8.7 solid grain; 9 fine grain.
**Ease of ignition (1 to 10 un-contained):** 7 solid grain; grain size varies and results in different ignition rates.
**Tendency to cake:** Will not cake.
**Detonation ability:** Cannot be detonated.
**Explosive ability (contained; explosive deflagration):** Low.
**Percentage:** ***59.76% potassium perchlorate, 24.9% alpha-caprolactam, 14.94% N-acetyl caprolactam, 0.39% sodium hydride catalyst, 0.01% residue***
**Classification:** Deflagrating explosive (classified as propellant).

**Danger of preparation:** Can be prepared by amateurs, but caution should be exercised.
**Use:** Used to propel specialty rockets and missiles for military and commercial applications.

### 04-04-025A: High performance thermally stable rocket propellant:

Into a suitable heated vertical mixer, place ***130.4 grams of magnesium dust***, followed by ***385.1 grams of anhydrous barium nitrate***, followed by ***264.5 grams of cyanuric acid***, and then blend the mixture on high speed for about 1 hour at room temperature. Thereafter, add in 250 milliliters of acetone, followed by ***109.5 grams of Viton A copolymer binder***, followed by ***110 grams of Teflon powder***, and then followed by ***500 milligrams of copper chromite***; thereafter, blend the mixture at 60 Celsius until the bulk of the acetone evaporates; once this point is achieved, the mixture is ready for use. To use, the mixture needs to be pressed into any desirable rocket motor or mold under a pressure of 8,000 psi, in the usual manner, and then place the rocket motor or mold into an oven and heat to 140 Celsius for 15 minutes. Thereafter, remove the heat source, and allow the rocket motor or mold to cool to room temperature for 48 hours. Use a standard ignition composition for proper burn.
**Burn rate:** 4.9 seconds per inch at 8,000 psi; tends to vary based on diameter and density of rocket grain.
**Water resistance:** Excellent.
**Stability:** Can be stored for decades.
**Sensitivity:** Stable towards friction; stable towards shock; stable to temperatures up to 250 Celsius.
**Flammability (1 to 10, un-contained):** 8 solid grain; 8.9 fine grain.
**Ease of ignition (1 to 10 un-contained):** 6.8 solid grain; grain size varies and results in different ignition rates.
**Tendency to cake:** Will not cake.
**Detonation ability:** Cannot be detonated.
**Explosive ability (contained; explosive deflagration):** Low.
**Percentage:** ***38.51% barium nitrate, 26.45% cyanuric acid, 26.08% magnesium dust, 11% Teflon, 10.95% Viton A copolymer binder, 0.05% copper chromite burn catalyst***
**Classification:** Deflagrating explosive (classified as propellant).

**Danger of preparation:** Can be prepared by amateurs, but caution should be exercised.
**Use:** Used to propel specialty rockets and missiles for military and commercial applications.

### 04-04-026A: Moderate performance potassium nitrate based rocket propellant:

Into a special heated vertical mixer, place ***375 grams of powdered potassium nitrate***, and then followed by ***125 grams of finely ground asphalt black.*** Thereafter, blend the mixture on high speed for 1 hour at 250 Celsius. Thereafter, the mixture is ready for use. To use, the hot mixture needs to be pressed into any desirable rocket motor or mold under a pressure of 9,000 psi, in the usual manner. Thereafter, the rocket motor or mold needs to set and cool for 3 days. Use a standard ignition composition for proper burn.
**Burn rate:** 3.8 seconds per inch at 9,000 psi; tends to vary based on diameter and density of rocket grain.
**Water resistance:** Excellent.
**Stability:** Can be stored for decades.
**Sensitivity:** Stable towards friction; stable towards shock; stable to temperatures up to 300 Celsius.
**Flammability (1 to 10, un-contained):** 8.6 solid grain; 8.96 fine grain.
**Ease of ignition (1 to 10 un-contained):** 7 solid grain; grain size varies and results in different ignition rates.
**Tendency to cake:** Will not cake.
**Detonation ability:** Cannot be detonated.
**Explosive ability (contained; explosive deflagration):** Low.
**Percentage:** ***75% potassium nitrate, 25% asphalt black***
**Classification:** Deflagrating explosive (classified as propellant).

**Danger of preparation:** Can be prepared by amateurs.
**Use:** Used to propel specialty rockets and missiles for military and commercial applications.

# Chapter 5. Non-Black Powder Gun Propellants

## Section 1: Ammonium Nitrate Containing

***Chemicals used in this section (binders are not included)***

| | | |
|---|---|---|
| **1. Potassium nitrate (see Black Powder)** | | **2. Sulfur (see Black Powder)** |
| **3. Charcoal (see Black Powder)** | | **4. Sugar Carbon (see Modified Black Powder)** |
| **5. Barium Chromate (see Modified Black Powder)** | | **6. Potassium Perchlorate (see Modified Black Powder)** |
| **7. Potassium Dichromate (see Modified Black Powder)** | | **8. Ammonium Bisulfide (see Modified Black Powder)** |
| **9. Potassium chlorate (see Modified Black Powder)** | | **10. Carbon Disulfide (see Modified Black Powder)** |
| **11. Nitrocellulose (see Modified Black Powder)** | | **12. Lead Tetraoxide (see Modified Black Powder)** |
| **13. Diphenylamine (see Modified Black Powder)** | | **14. Titanium Dioxide (see Modified Black Powder)** |
| **15. Manganese Dioxide (see Modified Black Powder)** | | **16. Sodium Benzoate (see Modified Black Powder)** |
| **17. Ammonium Nitrate (see Modified Black Powder)** | | **18. Calcium Carbonate (see Modified Black Powder)** |
| **19. Sodium Nitrate (see Modified Black Powder)** | | **20. Urea (see Modified Black Powder)** |
| **21. Lead Nitrate (see Modified Black Powder)** | | **22. Nitro Starch (see Modified Black Powder)** |
| **23. Ammonium Perchlorate (see Ammonium Perchlorate Rocket Propellants)** | | **24. Aluminum powder (see Ammonium Perchlorate Rocket Propellants)** |
| **25. Magnesium Sulfide (see Ammonium Perchlorate Rocket Propellants)** | | **26. Copper Chromite (see Ammonium Perchlorate Rocket Propellants)** |
| **27. Nitroguanidine (see Ammonium Perchlorate Rocket Propellants)** | | **28. Ammonium Sulfate (see Ammonium Perchlorate Rocket Propellants)** |
| **29. Nitroglycerine (see Ammonium Perchlorate Rocket Propellants)** | | **30. Magnesium Oxide (see Ammonium Perchlorate Rocket Propellants)** |
| **31. Iron-II-oxide (see Ammonium Perchlorate Rocket Propellants)** | | **32. Zirconium Hydride (see Ammonium Perchlorate Rocket Propellants)** |
| **33. Teflon (see Ammonium Perchlorate Rocket Propellants)** | | **34. Zinc Oxide (see Ammonium Perchlorate Rocket Propellants)** |
| **35. Ammonium Dichromate (see Ammonium Perchlorate Rocket Propellants)** | | **36. Lithium Perchlorate (see Ammonium Perchlorate Rocket Propellants)** |
| **37. Magnesium powder (see Ammonium Perchlorate Rocket Propellants)** | | **38. Lithium Aluminum Hydride (see Ammonium Perchlorate Rocket Propellants)** |
| **39. Iron-III-oxide (red iron oxide) (see Ammonium Perchlorate Rocket Propellants)** | | **40. PVC (see Ammonium Perchlorate Rocket Propellants)** |
| **41. Copper Sulfide (see Ammonium Perchlorate Rocket Propellants)** | | **42. Sodium Hydride (see Ammonium Perchlorate Rocket Propellants)** |
| **43. Titanium Hydride (see Ammonium Perchlorate Rocket Propellants)** | | **44. Silicon Nitride (see Ammonium Perchlorate Rocket Propellants)** |
| **45. ADN (see ADN Rocket Propellants)** | | **46. Urea Nitrate (see ADN Rocket Propellants)** |
| **47. Silicon Powder (see ADN Rocket Propellants)** | | **48. Hexamine (see ADN Rocket Propellants)** |
| **49. Boron powder (see ADN Rocket Propellants)** | | **50. Sodium hypophosphite (see ADN Rocket Propellants)** |
| **51. KDN (see ADN Rocket Propellants)** | | **52. Nickel Chloride (see Ammonium Nitrate Rocket Propellants)** |
| **53. TNT (see Ammonium Nitrate Rocket Propellants)** | | **54. Acrylamide (see Miscellaneous Rocket H.P. Rocket Propellants)** |
| **55. Ethylene Glycol (see Miscellaneous Rocket H.P. Rocket Propellants)** | | **56. Magnesium Perchlorate (see Miscellaneous Rocket H.P. Rocket Propellants)** |

| | |
|---|---|
| **57. Metallic Lithium (see Miscellaneous Rocket H.P. Rocket Propellants)** | **58. Beryllium Hydride (see Miscellaneous Rocket H.P. Rocket Propellants)** |
| **59. Red Phosphorus (see Miscellaneous Rocket H.P. Rocket Propellants)** | **60. Sodium Borohydride (see Miscellaneous Rocket H.P. Rocket Propellants)** |
| **61. Sodium chlorate (see Miscellaneous Rocket H.P. Rocket Propellants)** | **62. Lead Dioxide (see Miscellaneous Rocket H.P. Rocket Propellants)** |
| **63. Benzoyl Peroxide (see Miscellaneous Rocket H.P. Rocket Propellants)** | **64. Barium Nitrate (see Miscellaneous Rocket H.P. Rocket Propellants)** |
| **65. Cyanuric Acid (see Miscellaneous Rocket H.P. Rocket Propellants)** | **66. Aluminum Stearate** |
| | $[CH_3(CH_2)_{16}COO]_3Al$ |
| | Forms hard white to crystalline masses, which polymerize to a plastic like mass upon heating. When freshly prepared, the salt is soluble in alcohol, and slightly in water, but reverts to polymer forms upon standing which is insoluble in water and most organic solvents. Is prepared by heating aluminum hydroxide with stearic acid in solvent. |
| **67. Aluminum Hydride** | **68. Magnesium Peroxide** |
| 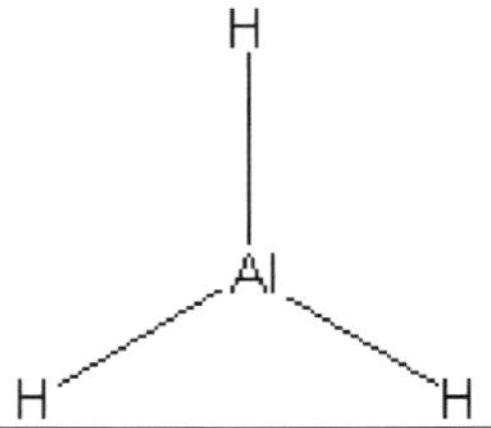 |  |
| Aluminum hydride forms colorless crystals that probably exist as a polymerized solid containing diethyl ether of crystallization that cannot be removed. The hydride is relatively stable at room temperature, but reacts with water with some violence. The salt is prepared by reacting ether solutions of aluminum chloride and lithium hydride. | Magnesium peroxide forms a white crystalline mass or powder. The salt is insoluble in water and most solvents. Magnesium peroxide is relatively unstable and begins to loose oxygen on standing in moist air. |
| **69. Potassium Permanganate** | **70. Calcium Hydride** |
| 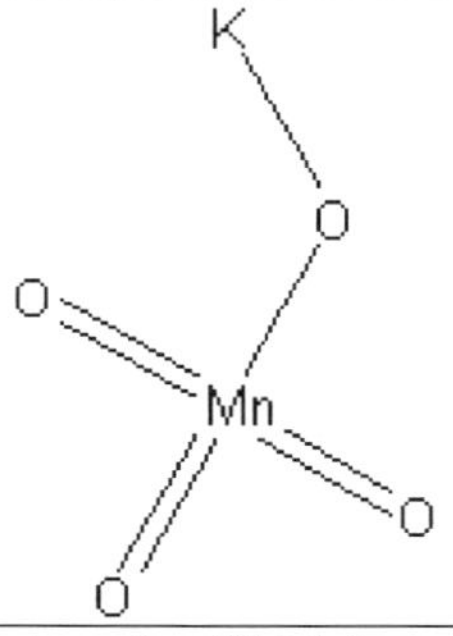 | 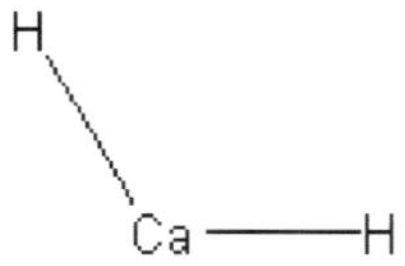 |
| Potassium permanganate forms beautiful dark purple crystals that have a beautiful sheen. The salt is very stable in air, but begins to decompose when heated to 240 Celsius. The salt is soluble in water, and alcohol, but the latter it reacts with. Potassium permanganate is a powerful oxidizing salt, so users should use caution when working with this compound. | White to grayish powder or crystalline mass. The salt decomposes in contact with water and alcohol. The hydride can be stored in the absence of moisture for some time. Best prepared by heating the two elements together. |
| **71. Potassium Tartrate** | **72. Potassium Ferrocyanide** |
| 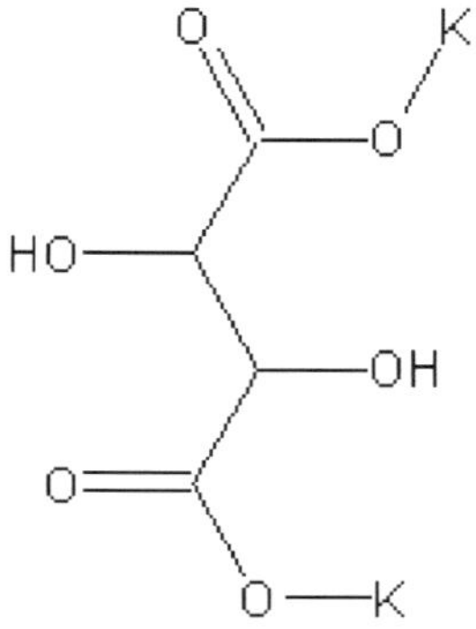 | 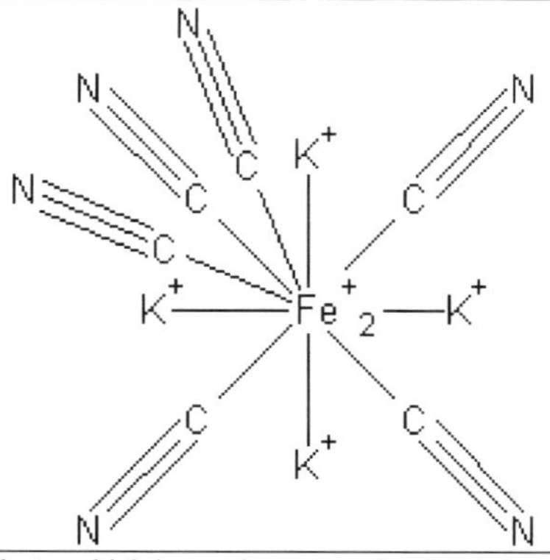 |
| Forms white crystals, granules, or powder. The crystals readily absorb moisture forming a hemihydrate. The salt is soluble in water, but insoluble in alcohol and other common solvents. | Forms a trihydrate, which looses its water when heated to 100 Celsius. The salt is soluble in water, but relatively insoluble in the usual organic solvents. Keep out of contact with concentrated sulfuric acid to avoid the formation of the deadly hydrogen cyanide gas. |

| **73. Sodium Azide** | **74. Sodium Chloride** |
|---|---|
| 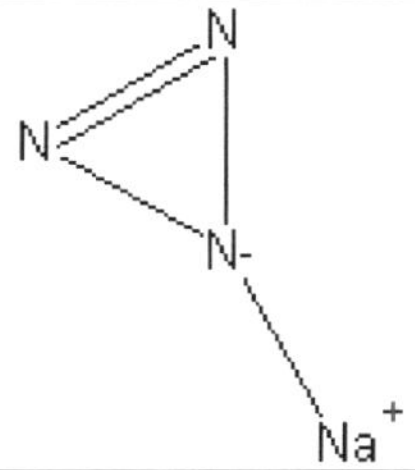 |  |
| Forms colorless crystals or a white powder or granules. The crystals break down into nitrogen and metallic sodium. The crystals are readily soluble in water, but relatively insoluble in most common solvents. Aqueous solutions should not be stored for long periods of time due to formation of hydrazoic acid. Sodium azide is highly toxic, so users should use caution. The salt is prepared by heating nitrous oxide with sodamide | Forms cubic crystals, or white granules or powder. Is capable of forming large beautiful cubic like crystals, some as big as basketballs. Sodium chloride is one of the most common salts in human civilization, and is a major source of chlorine and sodium salts. Its melting point is 804 Celsius. |

***- High Performance Ammonium Nitrate Gun Propellants in this section -***

| | |
|---|---|
| **1. 05-01-001A: High Performance ammonium nitrate gunpowder:** *60% ammonium nitrate, 17.8% charcoal, 10% sulfur, 10% sodium nitrate, 1.1% calcium carbonate, 0.90% urea, 0.150% calcium sulfate, 0.050% impurities* | **2. 05-01-002A: High Performance ammonium nitrate gunpowder:** *70% ammonium nitrate, 15% aluminum, 15% aluminum stearate* |
| **3. 05-01-003A: High Performance ammonium nitrate gunpowder:** *73.55% ammonium nitrate, 12.5% methyl iodide, 9.48% 1,3-butadiene, 3.16% 2-vinylpyridine, 0.62% sodium soap, 0.62% mercaptan blend, 0.037% potassium persulfate catalyst, 0.033% rounded balance* | **4. 05-01-004A: High Performance ammonium nitrate gunpowder:** *81.4% ammonium nitrate, 9.96% copolymer binder, 3.6% ammonium perchlorate, 1.97% triacetin, 1.76% melamine, 0.5% magnesium oxide, 0.3% mixed impurities, 0.3% zinc oxide, 0.21% GMF* |
| **5. 05-01-005A: High performance gun propellant with catalyst accelerant:** *52.2% ammonium nitrate, 33.3% magnesium (percentage can vary), 14.3% magnesium peroxide (percentage can vary), 0.2% potassium permanganate* | **6. 05-01-006A: High performance gun propellant with catalyst accelerant:** *82% ammonium nitrate, 10% potassium nitrate, 7% binder composition, 1% ammonium dichromate* |
| **7. 05-01-007A: High performance ammonium nitrate gun propellant:** *69.1% ammonium nitrate, 29.61% nitrocellulose, 1.28% diphenylamine, 0.01% mixed balance* | **8. 05-01-007B: High performance ammonium nitrate gun propellant (modified):** *55.8% ammonium nitrate, 40.93% nitrocellulose, 1.6% aluminum powder, 1.17% diphenylamine, 0.5% camphor* |
| **9. 05-01-008A: Ammonium nitrate smokeless gun propellant:** *65.23% ammonium nitrate, 20.07% magnesium powder, 9.83% nitrocellulose,4.41% aluminum powder, 0.44% diphenylamine, 0.02% graphite and mixed balance* | **10. 05-01-009A: Standard high performance ammonium nitrate gun propellant:** *46.3% ammonium nitrate, 19.7% nitroguanidine, 16.25% calcium hydride, 9.85% Teflon, 6.4% nitrocellulose, 1.47% diphenylamine, 0.03% mixed residual balance* |
| **11. 05-01-010A: High performance ammonium nitrate smokeless powder:** *71.42% ammonium nitrate, 19.04% sodium oxalate, 4.76% nitrocellulose, 4.76% calcium tartrate, 0.02% residue* | **12. 05-01-011A: High performance nitrocellulose gun propellant:** *59.7% ammonium nitrate, 19.9% nitrocellulose, 9.95% corn starch, 9.95% dinitrotoluene, 0.49% diphenylamine, 0.01% graphite and mixed residues* |
| **13. 05-01-012A: Ammonium nitrate high performance gun propellant:** *44.3% ammonium nitrate, 20.25% sodium benzoate, 12.65% potassium ferrocyanide, 9.49% barium nitrate, 6.32% potassium chlorate, 6.32% nitrocellulose, 0.63% lamp black, 0.04% graphite and mixed residues* | **14. 05-01-013A: High performance ammonium nitrate based gun propellant:** *51.36% ammonium nitrate, 27.39% nitrocellulose, 13.01% wheat flour, 6.16% sulfur, 2.05% sodium azide, 0.03% graphite and impurities* |
| **15. 05-01-014A: High performance nitrate based gun propellant:** *55.55% ammonium nitrate, 22.22% starch, 11.11% charcoal, 11.11% barium nitrate, 0.01% graphite and mixed impurities* | **16. 05-01-015A: High performance ammonium nitrate gun propellant with potassium picrate booster:** *52% ammonium nitrate, 25% potassium picrate, 20% sulfur, 3% soft wood charcoal* |
| **17. 05-01-016A: High performance ammonium nitrate gun propellant:** *49.8% ammonium nitrate, 17.24% nitrocellulose, 15.32% potassium nitrate, 10.53% graphite, 7.08% sodium hypophosphite, 0.03% mixed balance* | **18. 05-01-017A: High performance ammonium nitrate gun propellant:** *54.86% ammonium nitrate, 20% nitrocellulose, 19% dinitrotoluene, 4.4% dibutylphthalate, 1.5% diphenylamine, 0.24% graphite and mixed residue* |
| | |

| | |
|---|---|
| **19. 05-01-018A: High performance ammonium nitrate gun propellant:** *72.81% ammonium nitrate, 14.56% sulfur, 9.7% soft wood charcoal, 2.91% sodium chloride, 0.02% graphite and residual balance* | **20. 05-01-018B: High performance ammonium nitrate gun propellant:** *63.55% ammonium nitrate, 12.71% sulfur, 8.47% soft wood charcoal, 8.47% nitrocellulose, 4.23% barium nitrate, 2.54% sodium chloride, 0.03% mixed residual balance* |

**05-01-001A: High Performance ammonium nitrate gunpowder:**
Into a suitable vertical mixer, add ***600 grams of powdered ammonium nitrate***, ***100 grams of pure sulfur***, ***178 grams of charcoal***, ***100 grams of sodium nitrate***, ***11 grams of calcium carbonate***, ***9 grams of urea***, and then ***1.5 grams calcium sulfate***. Thereafter, pour in 350 milliliters of kerosene or naphtha, and then thoroughly blend the mixture for 1 hour at room temperature. After the blending operation, the mixture is ready for extruding; to do so, extrude the mixture through any extruding device into "spaghetti-like" strands ranging from 3 to 4mm in diameter, by 300mm in length, and then cut the strands into pieces ranging from 1 to 3mm length, and then cure the granules by placing them on a shallow pan or tray, and allow to thoroughly air dry for 7 days. To use the grains, simply pack them loosely into any desirable cartridge casing. Use standard gun casing primers for initiation.
**Chamber Pressure (standard 9mm casing; weight of propellant charge/grain size on average):** 34,098 psi
**Chamber Pressure (standard 5.56mm rifle casing; weight of propellant charge/grain size on average):** 40,098 psi
**Chamber Pressure (standard 105mm artillery casing; weight of propellant charge/grain size on average):** 27,099 psi
**Mortar tube Pressure (standard 81mm mortar propellant charge; weight of propellant charge/grain size on average):** 6,734 psi
**Burn rate (in the open):** 2 seconds per inch; tends to vary based on diameter of grain.
**Water resistance:** Fair.
**Stability:** Can be stored for decades when kept from moisture.
**Sensitivity:** Stable towards friction; stable towards shock; stable to temperatures up to 300 Celsius.
**Flammability (1 to 10, un-contained):** 9.4 solid grain; 9.7 fine grain.
**Tendency to cake:** Will not cake.
**Detonation ability:** Can be detonated but must be primed with HE charge.
**Percentage:** ***60% ammonium nitrate, 17.8% charcoal, 10% sulfur, 10% sodium nitrate, 1.1% calcium carbonate, 0.90% urea, 0.150% calcium sulfate, 0.050% impurities***
**Classification:** Deflagrating explosive (classified as propellant).

**Danger of preparation:** Can be prepared by amateurs.
**Use:** Substitute for black powder in firearms; used in mortar propellants, artillery gun propellants; used in non-firearms weapons to propel explosive devices.

**Common extruding machines**

These extruding machines are identical to the machines used to make spaghetti and other types of pasta; the idea and fashion of use is the same whether extruding pasta, or a gun propellant. Many types of extruding machines and models for use in extruding gun propellants are technically modeled after machines used by the food industry, and all are readily available.

| Extruding machine die castings |
| --- |
| 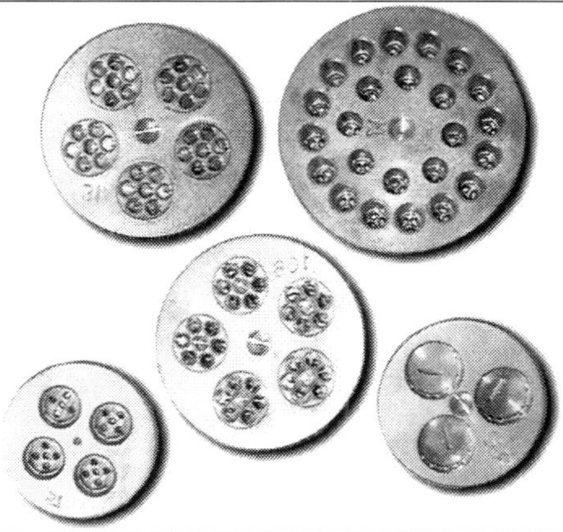  |
| Extruding machines can extrude material in many lengths, sizes, thickness, and body types. |

| Anti-personnel "bouncing betty" land mine | Description: |
| --- | --- |
| 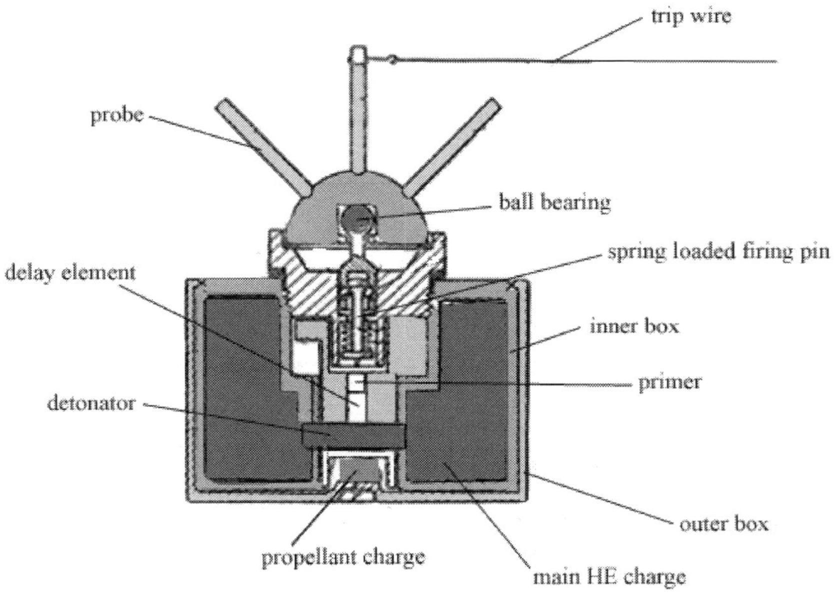  | Bouncing betty type land mines are designed to propel into the air where they detonate. Because they detonate in the air, the fragmenting radius, and hence, casualty producing radius, is increased. |

### 05-01-002A: High Performance ammonium nitrate gunpowder:

Into a suitable vertical mixer, place ***210 grams of powdered ammonium nitrate***, followed by ***45 grams of finely powdered aluminum***, followed by ***45 grams of aluminum stearate***, and then followed by 200 milliliters of hexane; thereafter, blend the mixture on high speed for about 45 minutes. After the blending operation, the mixture is ready for extruding; to do so, extrude the mixture through any extruding device into "spaghetti-like" strands ranging from 3 to 4mm in diameter, by 300mm in length, and then cut the strands into pieces ranging from 1 to 3mm length, and then cure the granules by placing them on a shallow pan or tray, and allow to thoroughly air dry for 7 days. To use the grains, simply pack them loosely into any desirable cartridge casing. Use standard gun casing primers for initiation.

**Chamber Pressure (standard 9mm casing; weight of propellant charge/grain size on average):** 34,704 psi
**Chamber Pressure (standard 5.56mm rifle casing; weight of propellant charge/grain size on average):** 39,984 psi
**Chamber Pressure (standard 105mm artillery casing; weight of propellant charge/grain size on average):** 25,900 psi
**Mortar tube Pressure (standard 81mm mortar propellant charge; weight of propellant charge/grain size on average):** 6,779 psi
**Burn rate (in the open):** 2.2 seconds per inch; tends to vary based on diameter of grain.
**Water resistance:** Fair.
**Stability:** Can be stored for decades when kept from moisture.
**Sensitivity:** Stable towards friction; stable towards shock; stable to temperatures up to 300 Celsius.
**Flammability (1 to 10, un-contained):** 9.3 solid grain; 9.7 fine grain.
**Tendency to cake:** Will not cake.
**Detonation ability:** Can be detonated but must be primed with HE charge.
**Percentage:** ***70% ammonium nitrate, 15% aluminum, 15% aluminum stearate***
**Classification:** Deflagrating explosive (classified as propellant).

**Danger of preparation:** Can be prepared by amateurs.
**Use:** Substitute for black powder in firearms; used in mortar propellants, artillery gun propellants; used in non-firearms weapons to propel explosive devices.

### 05-01-003A: High Performance ammonium nitrate gunpowder:

Into a suitable vertical mixer, place ***25.8 grams of 1,3-butadiene***, followed by ***8.6 grams of 2-vinylpyridine,*** followed by ***1.7 grams of pulverized sodium soap flakes,*** followed by ***62 grams of distilled water,*** followed by ***103 milligrams of potassium persulfate,*** followed by ***1.7 grams of a mercaptan blend*** (composed of C12, C14, and C16 aliphatic mercaptans), followed by ***200 grams of powdered ammonium nitrate,*** and then followed by ***34 grams of methyl iodide***. Thereafter, rapidly blend the mixture on high speed for 45 minutes. After the blending operation, the mixture is ready for extruding; to do so, extrude the mixture through any extruding device into "spaghetti-like" strands ranging from 3 to 4mm in diameter, by 300mm in length, and then cut the strands into pieces ranging from 1 to 3mm length, and then cure the granules by placing them on a shallow pan or tray, and place said pan or tray into an oven, and cure the granules at 125 Celsius for 10 minutes. To use the grains, simply pack them loosely into any desirable cartridge casing. Use standard gun casing primers for initiation.
**Chamber Pressure (standard 9mm casing; weight of propellant charge/grain size on average):** 35,092 psi
**Chamber Pressure (standard 5.56mm rifle casing; weight of propellant charge/grain size on average):** 40,376 psi
**Chamber Pressure (standard 105mm artillery casing; weight of propellant charge/grain size on average):** 24,922 psi
**Mortar tube Pressure (standard 81mm mortar propellant charge; weight of propellant charge/grain size on average):** 7,096 psi
**Burn rate (in the open):** 2.19 seconds per inch; tends to vary based on diameter of grain.
**Water resistance:** Fair.
**Stability:** Can be stored for decades when kept from moisture.
**Sensitivity:** Stable towards friction; stable towards shock; stable to temperatures up to 300 Celsius.
**Flammability (1 to 10, un-contained):** 9.2 solid grain; 9.5 fine grain.
**Tendency to cake:** Will not cake.
**Detonation ability:** Can be detonated but must be primed with HE charge.
**Percentage:** ***73.55% ammonium nitrate, 12.5% methyl iodide, 9.48% 1,3-butadiene, 3.16% 2-vinylpyridine, 0.62% sodium soap, 0.62% mercaptan blend, 0.037% potassium persulfate catalyst, 0.033% rounded balance***
**Classification:** Deflagrating explosive (classified as propellant).

**Danger of preparation:** Can be prepared by amateurs.
**Use:** Substitute for black powder in firearms; used in mortar propellants, artillery gun propellants; used in non-firearms weapons to propel explosive devices.

### 05-01-004A: High Performance ammonium nitrate gunpowder:

Into a suitable vertical mixer, place 250 milliliters of cyclohexane, followed by ***49.8 grams of a copolymer of 1,3-butadiene-2-methyl-5-vinyl pyridine***, followed by ***8.8 grams of melamine***, followed by ***9.85 grams of triacetin***, followed by ***1.5 grams of Flexamine*** (65% diarylamine ketone and 35% N,N-diphenylparaphenylene diamine), followed by ***1.05 grams of GMF (paraquinone dioxime),*** followed by ***2.5 grams of magnesium oxide***, followed by ***1.5 grams of zinc oxide***, followed by ***407 grams of powdered ammonium nitrate***, and then followed by ***18 grams of ammonium perchlorate***. Thereafter, blend the mixture on high speed for about 50 minutes. After the blending operation, the mixture is ready for extruding; to do so, extrude the mixture through any extruding device into "spaghetti-like" strands ranging from 2 to 4mm in diameter, by 300mm in length, and then cut the strands into pieces ranging from 1 to 3mm length, and then cure the granules by placing them on a shallow pan or tray, and place said pan or tray into an oven, and cure the granules at 85 Celsius for 12 hours. To use the grains, simply pack them loosely into any desirable cartridge casing. Use standard gun casing primers for initiation.
**Chamber Pressure (standard 9mm casing; weight of propellant charge/grain size on average):** 33,903 psi
**Chamber Pressure (standard 5.56mm rifle casing; weight of propellant charge/grain size on average):** 37,998 psi
**Chamber Pressure (standard 105mm artillery casing; weight of propellant charge/grain size on average):** 23,092 psi
**Mortar tube Pressure (standard 81mm mortar propellant charge; weight of propellant charge/grain size on average):** 6,550 psi
**Burn rate (in the open):** 2.56 seconds per inch; tends to vary based on diameter of grain.
**Water resistance:** Fair.
**Stability:** Can be stored for decades when kept from moisture.
**Sensitivity:** Stable towards friction; stable towards shock; stable to temperatures up to 300 Celsius.
**Flammability (1 to 10, un-contained):** 9 solid grain; 9.3 fine grain.
**Tendency to cake:** Will not cake.
**Detonation ability:** Can be detonated but must be primed with HE charge.
**Percentage:** ***81.4% ammonium nitrate, 9.96% copolymer binder, 3.6% ammonium perchlorate, 1.97% triacetin, 1.76% melamine, 0.5% magnesium oxide, 0.3% mixed impurities, 0.3% zinc oxide, 0.21% GMF***
**Classification:** Deflagrating explosive (classified as propellant).

**Danger of preparation:** Can be prepared by amateurs.
**Use:** Substitute for black powder in firearms; used in mortar propellants, artillery gun propellants; used in non-firearms weapons to propel explosive devices.

### 05-01-005A: High performance gun propellant with catalyst accelerant:

Into a suitable vertical mixer, place 500 milliliters of warm water, and then add and dissolve 75 grams of potassium permanganate. Thereafter, add in ***365 grams of finely powdered magnesium metal***, and then stir the mixture for about 10 minutes. Thereafter, filter-off the magnesium, and then allow it to thoroughly dry on a shallow tray. Now, into a clean suitable vertical mixer equipped with cooling means, place 400 milliliters of a 30% hydrogen peroxide solution, and then chill this solution to 0 Celsius. When the temperature reaches 0 Celsius, add in, in small portions, the dried coated magnesium particles, and then stir the mixture for about 5 minutes. Thereafter, immediately filter-off the insoluble particles, and then air-dry them. Finally, into a clean vertical mixer, place the dried filtered-off particles, and then add in ***400 grams of powdered ammonium nitrate***, followed by 190 milliliters of hexane, and then blend the mixture for 45 minutes After the blending operation, the mixture is ready for extruding; to do so, extrude the mixture through any extruding device into "spaghetti-like" strands ranging from 2 to 6mm in diameter, by 300mm in length, and then cut the strands into pieces ranging from 1 to 6mm length, and then cure the granules by placing them on a shallow pan or tray, and place said pan or tray into an oven, and cure the granules at 85 Celsius for 12 hours. To use the grains, simply pack them loosely into any desirable cartridge casing. Use standard gun casing primers for initiation.
**Chamber Pressure (standard 9mm casing; weight of propellant charge/grain size on average):** 33,927 psi
**Chamber Pressure (standard 5.56mm rifle casing; weight of propellant charge/grain size on average):** 37,995 psi
**Chamber Pressure (standard 105mm artillery casing; weight of propellant charge/grain size on average):** 24,703 psi
**Mortar tube Pressure (standard 81mm mortar propellant charge; weight of propellant charge/grain size on average):** 7,099 psi
**Burn rate (in the open):** 2.42 seconds per inch; tends to vary based on diameter of grain.
**Water resistance:** Fair.
**Stability:** Can be stored for decades when kept from moisture.
**Sensitivity:** Stable towards friction; stable towards shock; stable to temperatures up to 300 Celsius.
**Flammability (1 to 10, un-contained):** 9 solid grain; 9.5 fine grain.
**Tendency to cake:** Will not cake.
**Detonation ability:** Can be detonated but must be primed with HE charge.
**Percentage:** ***52.2% ammonium nitrate, 33.3% magnesium (percentage can vary), 14.3% magnesium peroxide (percentage can vary), 0.2% potassium permanganate***
**Classification:** Deflagrating explosive (classified as propellant).

**Danger of preparation:** Can be prepared by amateurs.
**Use:** Substitute for black powder in firearms; used in mortar propellants, artillery gun propellants; used in non-firearms weapons to propel explosive devices.

### 05-01-006A: High performance gun propellant with catalyst accelerant:

Into a suitable vertical mixer equipped with cooling means, place 180 grams of 1,3-butadiene, followed by 33 grams of furnace black, followed by 8 grams of finely divided selenium metal, and then followed by 20 grams of 2-methyl-5-vinylpyeridine. Thereafter, chill the mixture to about 5 Celsius, and then blend the mixture for about 1 hour at said temperature. Now, into a clean separate vertical mixer equipped with heating means, place ***330 grams of powdered ammonium nitrate***, followed by ***40 grams of potassium nitrate***, and then followed by 150 milliliters of hexane. Thereafter, rapidly blend the mixture at 65 Celsius for about 4 hours. After 4 hours, add in ***28 grams of the 1,3-butadiene polymer compound*** previously prepared, and then add in ***4 grams of ammonium dichromate*** burn catalyst, and then continue to blend the mixture for about 1 hour on high speed. After the blending operation, the mixture is ready for extruding; to do so, extrude the mixture through any extruding device into "spaghetti-like" strands ranging from 2 to 7mm in diameter, by 300mm in length, and then cut the strands into pieces ranging from 1 to 8mm length, and then cure the granules by placing them on a shallow pan or tray, and place said pan or tray into an oven, and cure the granules at 85 Celsius for 12 hours. To use the grains, simply pack them loosely into any desirable cartridge casing. Use standard gun casing primers for initiation.
**Chamber Pressure (standard 9mm casing; weight of propellant charge/grain size on average):** 32,994 psi
**Chamber Pressure (standard 5.56mm rifle casing; weight of propellant charge/grain size on average):** 36,665 psi
**Chamber Pressure (standard 105mm artillery casing; weight of propellant charge/grain size on average):** 23,093 psi
**Mortar tube Pressure (standard 81mm mortar propellant charge; weight of propellant charge/grain size on average):** 6,779 psi
**Burn rate (in the open):** 3.1 seconds per inch; tends to vary based on diameter of grain.
**Water resistance:** Fair.
**Stability:** Can be stored for decades when kept from moisture.
**Sensitivity:** Stable towards friction; stable towards shock; stable to temperatures up to 300 Celsius.
**Flammability (1 to 10, un-contained):** 8.7 solid grain; 9 fine grain.
**Tendency to cake:** Will not cake.

**Detonation ability:** Can be detonated but must be primed with HE charge.
**Percentage:** ***82% ammonium nitrate, 10% potassium nitrate, 7% binder composition, 1% ammonium dichromate***
**Classification:** Deflagrating explosive (classified as propellant).

**Danger of preparation:** Can be prepared by amateurs.
**Use:** Substitute for black powder in firearms; used in mortar propellants, artillery gun propellants; used in non-firearms weapons to propel explosive devices.

### 05-01-007A: High performance ammonium nitrate gun propellant:

Into a suitable vertical mixer, place ***350 grams of powdered ammonium nitrate***, followed by ***150 grams of nitrocellulose*** (14% nitrogen content), followed by 200 milliliters of hexane, and then followed by 18 milliliters of 95% ethyl alcohol. Thereafter, blend the mixture on high speed for 15 minutes. Thereafter add in ***6.5 grams of diphenylamine***, and then continue to blend the mixture on high speed for about 45 minutes. After the blending operation, the mixture is ready for extruding; to do so, extrude the mixture through any extruding device into "spaghetti-like" strands ranging from 2 to 7mm in diameter, by 300mm in length, and then cut the strands into pieces ranging from 1 to 7mm length, and then cure the granules by placing them on a shallow pan or tray, and place said pan or tray into an oven, and cure the granules at 60 Celsius for 12 hours. To use the grains, simply pack them loosely into any desirable cartridge casing. Use standard gun casing primers for initiation.
**Chamber Pressure (standard 9mm casing; weight of propellant charge/grain size on average):** 36,993 psi
**Chamber Pressure (standard 5.56mm rifle casing; weight of propellant charge/grain size on average):** 37,884 psi
**Chamber Pressure (standard 105mm artillery casing; weight of propellant charge/grain size on average):** 25,902 psi
**Mortar tube Pressure (standard 81mm mortar propellant charge; weight of propellant charge/grain size on average):** 8,025 psi
**Burn rate (in the open):** 2 seconds per inch; tends to vary based on diameter of grain.
**Water resistance:** Fair.
**Stability:** Can be stored for decades when kept from moisture.
**Sensitivity:** Stable towards friction; stable towards shock; stable to temperatures up to 300 Celsius.
**Flammability (1 to 10, un-contained):** 9.3 solid grain; 9.8 fine grain.
**Tendency to cake:** Will not cake.
**Detonation ability:** Can be detonated but must be primed with HE charge.
**Percentage:** ***69.1% ammonium nitrate, 29.61% nitrocellulose, 1.28% diphenylamine, 0.01% mixed balance***
**Classification:** Deflagrating explosive (classified as propellant).

**Danger of preparation:** Can be prepared by amateurs, but caution should be exercised.
**Use:** Substitute for black powder in firearms; used in mortar propellants, artillery gun propellants; used in non-firearms weapons to propel explosive devices.

### 05-01-007B: High performance ammonium nitrate gun propellant (modified):

Into a suitable vertical mixer, place ***558 grams of powdered ammonium nitrate***, followed by ***409.3 grams of nitrocellulose*** (14% nitrogen content), followed by 250 milliliters of hexane, and then followed by 35 milliliters of 95% ethyl alcohol. Thereafter, blend the mixture on high speed for about 15 minutes. Thereafter add in ***11.7 grams of diphenylamine***, followed by ***16 grams of aluminum powder***, and then followed by ***5 grams of camphor***, and then continue to blend the mixture on high speed for about 35 minutes. After the blending operation, the mixture is ready for extruding; to do so, extrude the mixture through any extruding device into "spaghetti-like" strands ranging from 2 to 10mm in diameter, by 300mm in length, and then cut the strands into pieces ranging from 1 to 10mm length, and then cure the granules by placing them on a shallow pan or tray, and place said pan or tray into an oven, and cure the granules at 60 Celsius for 12 hours. To use the grains, simply pack them loosely into any desirable cartridge casing. Use standard gun casing primers for initiation.
**Chamber Pressure (standard 9mm casing; weight of propellant charge/grain size on average):** 38,915 psi
**Chamber Pressure (standard 5.56mm rifle casing; weight of propellant charge/grain size on average):** 39,009 psi
**Chamber Pressure (standard 105mm artillery casing; weight of propellant charge/grain size on average):** 27,998 psi
**Mortar tube Pressure (standard 81mm mortar propellant charge; weight of propellant charge/grain size on average):** 8,976 psi
**Burn rate (in the open):** 1.9 seconds per inch; tends to vary based on diameter of grain.
**Water resistance:** Fair.
**Stability:** Can be stored for decades when kept from moisture.
**Sensitivity:** Stable towards friction; stable towards shock; stable to temperatures up to 300 Celsius.
**Flammability (1 to 10, un-contained):** 9.5 solid grain; 9.9 fine grain.
**Tendency to cake:** Will not cake.
**Detonation ability:** Can be detonated but must be primed with HE charge.
**Percentage:** ***55.8% ammonium nitrate, 40.93% nitrocellulose, 1.6% aluminum powder, 1.17% diphenylamine, 0.5% camphor.***
**Classification:** Deflagrating explosive (classified as propellant).

**Danger of preparation:** Can be prepared by amateurs, but caution should be exercised.
**Use:** Substitute for black powder in firearms; used in mortar propellants, artillery gun propellants; used in non-firearms weapons to propel explosive devices.

### 05-01-008A: Ammonium nitrate smokeless gun propellant:

Into a suitable vertical mixer, place ***325 grams of powdered ammonium nitrate***, followed by ***100 grams of magnesium powder***, followed by 250 milliliters of 95% ethyl alcohol or denatured alcohol, and then blend the mixture on high speed for 10 minutes. Thereafter, add in ***49 grams of nitrocellulose*** (14% nitrogen content), followed by ***2.2 grams of diphenylamine***, followed by ***22 grams of aluminum powder***, and then continue to blend the mixture for 30 minutes on high speed. After the blending operation, the mixture is ready for extruding; to do so, extrude the mixture through any extruding device into "spaghetti-like" strands ranging from 2 to 10mm in diameter, by 300mm in length, and then cut the strands into pieces ranging from 1 to 10mm length, and then cure the granules by placing them on a shallow pan or tray, and place said pan or tray into an oven, and cure the granules at 60 Celsius for 12 hours. To use the grains, first rotate the grains in a slanted open ball bill with graphite powder for about 15 minutes to coat them, and then simply remove the coated granules from the slanted ball mill, and then pack them loosely into any desirable cartridge casing. Use standard gun casing primers for initiation.
**Chamber Pressure (standard 9mm casing; weight of propellant charge/grain size on average):** 39,881 psi
**Chamber Pressure (standard 5.56mm rifle casing; weight of propellant charge/grain size on average):** 39,994 psi
**Chamber Pressure (standard 105mm artillery casing; weight of propellant charge/grain size on average):** 28,056 psi
**Mortar tube Pressure (standard 81mm mortar propellant charge; weight of propellant charge/grain size on average):** 9,061 psi
**Burn rate (in the open):** 1.88 seconds per inch; tends to vary based on diameter of grain.
**Water resistance:** Fair.
**Stability:** Can be stored for decades when kept from moisture.
**Sensitivity:** Stable towards friction; stable towards shock; stable to temperatures up to 300 Celsius.
**Flammability (1 to 10, un-contained):** 9.6 solid grain; 9.9 fine grain.
**Tendency to cake:** Will not cake.
**Detonation ability:** Can be detonated but must be primed with HE charge.
**Percentage:** ***65.23% ammonium nitrate, 20.07% magnesium powder, 9.83% nitrocellulose,4.41% aluminum powder, 0.44% diphenylamine, 0.02% graphite and mixed balance***
**Classification:** Deflagrating explosive (classified as propellant).

**Danger of preparation:** Can be prepared by amateurs, but caution should be exercised.
**Use:** Substitute for black powder in firearms; used in mortar propellants, artillery gun propellants; used in non-firearms weapons to propel explosive devices.

### 05-01-009A: Standard high performance ammonium nitrate gun propellant:

Into a suitable vertical mixer, place ***470 grams of powdered ammonium nitrate***, followed by ***200 grams of nitroguanidine***, followed by ***165 grams of calcium hydride***, followed by ***15 grams of diphenylamine***, followed by 350 milliliters of hexane, followed by ***100 grams of finely divided Teflon,*** and then followed by ***65 grams of nitrocellulose.*** Thereafter, blend the mixture for about 2 hours at room temperature. After the blending operation, the mixture is ready for extruding; to do so, extrude the mixture through any extruding device into "spaghetti-like" strands ranging from 2 to 7mm in diameter, by 300mm in length, and then cut the strands into pieces ranging from 1 to 9mm length, and then cure the granules by placing them on a shallow pan or tray, and place said pan or tray into an oven, and cure the granules at 60 Celsius for 12 hours. To use the grains, simply pack them loosely into any desirable cartridge casing. Use standard gun casing primers for initiation.
**Chamber Pressure (standard 9mm casing; weight of propellant charge/grain size on average):** 38,902 psi
**Chamber Pressure (standard 5.56mm rifle casing; weight of propellant charge/grain size on average):** 39,980 psi
**Chamber Pressure (standard 105mm artillery casing; weight of propellant charge/grain size on average):** 28,036 psi
**Mortar tube Pressure (standard 81mm mortar propellant charge; weight of propellant charge/grain size on average):** 9,008 psi
**Burn rate (in the open):** 1.89 seconds per inch; tends to vary based on diameter of grain.
**Water resistance:** Fair.
**Stability:** Can be stored for decades when kept from moisture.
**Sensitivity:** Stable towards friction; stable towards shock; stable to temperatures up to 300 Celsius.
**Flammability (1 to 10, un-contained):** 9.59 solid grain; 9.9 fine grain.
**Tendency to cake:** Will not cake.
**Detonation ability:** Can be detonated but must be primed with HE charge.
**Percentage:** ***46.3% ammonium nitrate, 19.7% nitroguanidine, 16.25% calcium hydride, 9.85% Teflon, 6.4% nitrocellulose, 1.47% diphenylamine, 0.03% mixed residual balance***
**Classification:** Deflagrating explosive (classified as propellant).

**Danger of preparation:** Can be prepared by amateurs, but caution should be exercised.
**Use:** Substitute for black powder in firearms; used in mortar propellants, artillery gun propellants; used in non-firearms weapons to propel explosive devices.

### 05-01-010A: High performance ammonium nitrate smokeless powder:

Into a suitable vertical mixer, place ***375 grams of powdered ammonium nitrate***, followed by ***100 grams of sodium oxalate,*** followed by ***25 grams of nitrocellulose*** (14% nitrogen content), followed by 210 milliliters of hexane, and then followed by ***25 grams of calcium tartrate***. Thereafter rapidly blend the mixture for about 1 hour at room temperature After the blending operation, the mixture is ready for extruding; to do so, extrude the mixture through any extruding device into "spaghetti-like" strands ranging from 2 to 7mm in diameter, by 300mm in length, and then cut the strands into pieces ranging from 1 to 8mm length, and then cure the granules by placing them on a shallow pan or tray, and place said pan or tray into an oven, and cure the granules at 60 Celsius for 12 hours. To use the grains, simply pack them loosely into any desirable cartridge casing. Use standard gun casing primers for initiation.
**Chamber Pressure (standard 9mm casing; weight of propellant charge/grain size on average):** 39,074 psi
**Chamber Pressure (standard 5.56mm rifle casing; weight of propellant charge/grain size on average):** 39,999 psi
**Chamber Pressure (standard 105mm artillery casing; weight of propellant charge/grain size on average):** 29,100 psi
**Mortar tube Pressure (standard 81mm mortar propellant charge; weight of propellant charge/grain size on average):** 8,923 psi
**Burn rate (in the open):** 1.99 seconds per inch; tends to vary based on diameter of grain.
**Water resistance:** Fair.
**Stability:** Can be stored for decades when kept from moisture.
**Sensitivity:** Stable towards friction; stable towards shock; stable to temperatures up to 300 Celsius.
**Flammability (1 to 10, un-contained):** 9.4 solid grain; 9.81 fine grain.
**Tendency to cake:** Will not cake.
**Detonation ability:** Can be detonated but must be primed with HE charge.
**Percentage:** ***71.42% ammonium nitrate, 19.04% sodium oxalate, 4.76% nitrocellulose, 4.76% calcium tartrate, 0.02% residue***
**Classification:** Deflagrating explosive (classified as propellant).

**Danger of preparation:** Can be prepared by amateurs, but caution should be exercised.
**Use:** Substitute for black powder in firearms; used in mortar propellants, artillery gun propellants; used in non-firearms weapons to propel explosive devices.

### 05-01-011A: High performance nitrocellulose gun propellant:

Into a suitable vertical mixer, place ***300 grams of ammonium nitrate,*** followed by ***100 grams of nitrocellulose*** (14% nitrogen content), followed by 175 milliliters of 95% ethyl alcohol, followed by ***50 grams of cornstarch***, followed by ***50 grams of dinitrotoluene*** (any suitable isomer), and then followed by ***2.5 grams of diphenylamine***. Thereafter, rapidly blend the mixture for 45 minutes at room temperature After the blending operation, the mixture is ready for extruding; to do so, extrude the mixture through any extruding device into "spaghetti-like" strands ranging from 2 to 7mm in diameter, by 300mm in length, and then cut the strands into pieces ranging from 1 to 8mm length, and then cure the granules by placing them on a shallow pan or tray, and place said pan or tray into an oven, and cure the granules at 80 Celsius for 12 hours. To use the grains, simply pack them loosely into any desirable cartridge casing. Use standard gun casing primers for initiation.
**Chamber Pressure (standard 9mm casing; weight of propellant charge/grain size on average):** 39,990 psi
**Chamber Pressure (standard 5.56mm rifle casing; weight of propellant charge/grain size on average):** 39,561 psi
**Chamber Pressure (standard 105mm artillery casing; weight of propellant charge/grain size on average):** 29,984 psi
**Mortar tube Pressure (standard 81mm mortar propellant charge; weight of propellant charge/grain size on average):** 8,700 psi
**Burn rate (in the open):** 2 seconds per inch; tends to vary based on diameter of grain.
**Water resistance:** Fair.
**Stability:** Can be stored for decades when kept from moisture.
**Sensitivity:** Stable towards friction; stable towards shock; stable to temperatures up to 300 Celsius.
**Flammability (1 to 10, un-contained):** 9.2 solid grain; 9.7 fine grain.
**Tendency to cake:** Will not cake.
**Detonation ability:** Can be detonated but must be primed with HE charge.
**Percentage:** ***59.7% ammonium nitrate, 19.9% nitrocellulose, 9.95% corn starch, 9.95% dinitrotoluene, 0.49% diphenylamine, 0.01% graphite and mixed residues***
**Classification:** Deflagrating explosive (classified as propellant).

**Danger of preparation:** Can be prepared by amateurs, but it is advisable that only experienced personnel perform this procedure. Please keep in mind the toxicity of dinitrotoluene.
**Use:** Substitute for black powder in firearms; used in mortar propellants, artillery gun propellants; used in non-firearms weapons to propel explosive devices.

## 05-01-012A: Ammonium nitrate high performance gun propellant:

[omitted]

**Chamber Pressure (standard 9mm casing; weight of propellant charge/grain size on average):** 39,022 psi
**Chamber Pressure (standard 5.56mm rifle casing; weight of propellant charge/grain size on average):** 38,501 psi
**Chamber Pressure (standard 105mm artillery casing; weight of propellant charge/grain size on average):** 28,776 psi
**Mortar tube Pressure (standard 81mm mortar propellant charge; weight of propellant charge/grain size on average):** 8,448 psi
**Burn rate (in the open):** 2.2 seconds per inch; tends to vary based on diameter of grain.
**Water resistance:** Fair.
**Stability:** Can be stored for decades when kept from moisture.
**Sensitivity:** Stable towards friction; stable towards shock; stable to temperatures up to 300 Celsius.
**Flammability (1 to 10, un-contained):** 9.1 solid grain; 9.4 fine grain.
**Tendency to cake:** Will not cake.
**Detonation ability:** Can be detonated but must be primed with HE charge.
**Percentage:** [omitted]
**Classification:** Deflagrating explosive (classified as propellant).

**Danger of preparation:** Can be prepared by amateurs, but caution should be exercised.
**Use:** Substitute for black powder in firearms; used in mortar propellants, artillery gun propellants; used in non-firearms weapons to propel explosive devices.

## 05-01-013A: High performance ammonium nitrate based gun propellant:

[omitted]

**Chamber Pressure (standard 9mm casing; weight of propellant charge/grain size on average):** 39,993 psi
**Chamber Pressure (standard 5.56mm rifle casing; weight of propellant charge/grain size on average):** 38,998 psi
**Chamber Pressure (standard 105mm artillery casing; weight of propellant charge/grain size on average):** 29,981 psi
**Mortar tube Pressure (standard 81mm mortar propellant charge; weight of propellant charge/grain size on average):** 9,216 psi
**Burn rate (in the open):** 2 seconds per inch; tends to vary based on diameter of grain.
**Water resistance:** Fair.
**Stability:** Can be stored for decades when kept from moisture.
**Sensitivity:** Stable towards friction; stable towards shock; stable to temperatures up to 180 Celsius.
**Flammability (1 to 10, un-contained):** 9.3 solid grain; 9.6 fine grain.
**Tendency to cake:** Will not cake.
**Detonation ability:** Can be detonated but must be primed with HE charge.
**Percentage:** [omitted]
**Classification:** Deflagrating explosive (classified as propellant).

**Danger of preparation:** Can be prepared by amateurs, but caution should be exercised.
**Use:** Substitute for black powder in firearms; used in mortar propellants, artillery gun propellants; used in non-firearms weapons to propel explosive devices.

### 05-01-014A: High performance nitrate based gun propellant:

Into a suitable vertical mixer, place 250 milliliters of hexane, followed by ***200 grams of starch***, followed by ***100 grams of powdered soft wood charcoal***, followed by ***100 grams of barium nitrate***, and then followed by ***500 grams of powdered ammonium nitrate***. Thereafter, blend the mixture on high speed for about 1 hour. After the blending operation, the mixture is ready for extruding; to do so, extrude the mixture through any extruding device into "spaghetti-like" strands ranging from 2 to 8mm in diameter, by 300mm in length, and then cut the strands into pieces ranging from 1 to 10mm length, and then cure the granules by placing them on a shallow pan or tray, and place said pan or tray into an oven, and cure the granules at 80 Celsius for 12 hours. To use the grains, simply pack them loosely into any desirable cartridge casing. Use standard gun casing primers for initiation.
**Chamber Pressure (standard 9mm casing; weight of propellant charge/grain size on average):** 38,084 psi
**Chamber Pressure (standard 5.56mm rifle casing; weight of propellant charge/grain size on average):** 37,158 psi
**Chamber Pressure (standard 105mm artillery casing; weight of propellant charge/grain size on average):** 28,099 psi
**Mortar tube Pressure (standard 81mm mortar propellant charge; weight of propellant charge/grain size on average):** 8,377 psi
**Burn rate (in the open):** 2.9 seconds per inch; tends to vary based on diameter of grain.
**Water resistance:** Fair.
**Stability:** Can be stored for decades when kept from moisture.
**Sensitivity:** Stable towards friction; stable towards shock; stable to temperatures up to 180 Celsius.
**Flammability (1 to 10, un-contained):** 9 solid grain; 9.2 fine grain.
**Tendency to cake:** Will not cake.
**Detonation ability:** Can be detonated but must be primed with HE charge.
**Percentage:** ***55.55% ammonium nitrate, 22.22% starch, 11.11% charcoal, 11.11% barium nitrate, 0.01% graphite and mixed impurities***
**Classification:** Deflagrating explosive (classified as propellant).

**Danger of preparation:** Can be prepared by amateurs, but caution should be exercised.
**Use:** Substitute for black powder in firearms; used in mortar propellants, artillery gun propellants; used in non-firearms weapons to propel explosive devices.

### 05-01-015A: High performance ammonium nitrate gun propellant with potassium picrate booster:

Into a suitable vertical mixer, place ***125 grams of potassium picrate***, followed by ***260 grams of powdered ammonium*** nitrate, followed by ***100 grams of flours of sulfur,*** followed by 125 milliliters of hexane, and then followed by ***15 grams of soft wood charcoal.*** Thereafter, blend the mixture on moderate speed for about 45 minutes. After the blending operation, the mixture is ready for extruding; to do so, extrude the mixture through any extruding device into "spaghetti-like" strands ranging from 2 to 8mm in diameter, by 300mm in length, and then cut the strands into pieces ranging from 1 to 10mm length, and then cure the granules by placing them on a shallow pan or tray, and place said pan or tray into an oven, and cure the granules at 80 Celsius for 12 hours. To use the grains, simply pack them loosely into any desirable cartridge casing. Use standard gun casing primers for initiation.
**Chamber Pressure (standard 12 guage shotgun shell casing; weight of propellant charge/grain size on average):** 39,801 psi
**Chamber Pressure (standard 210mm artillery casing; weight of propellant charge/grain size on average):** 29,714 psi
**Burn rate (in the open):** 2 seconds per inch; tends to vary based on diameter of grain.
**Water resistance:** Fair.
**Stability:** Can be stored for decades when kept from moisture.
**Sensitivity:** Stable towards friction; stable towards shock; stable to temperatures up to 150 Celsius.
**Flammability (1 to 10, un-contained):** 9.3 solid grain; 9.5 fine grain.
**Tendency to cake:** Will not cake.
**Detonation ability:** Can be detonated but must be primed with HE charge.
**Percentage:** ***52% ammonium nitrate, 25% potassium picrate, 20% sulfur, 3% soft wood charcoal***
**Classification:** Deflagrating explosive (classified as class A explosive).

**Danger of preparation:** Should only be prepared by experienced personnel. Potassium pircrate is a primary explosive, so caution should be exercised.

**Use:** Used in shotgun propellants, but mainly used in large bore guns such as artillery guns, and battleship deck guns.

### 05-01-016A: High performance ammonium nitrate gun propellant:

Into a suitable vertical mixer, place ***520 grams of powdered ammonium nitrate***, followed by ***110 grams of finely divided graphite***, followed by ***74 grams of sodium hypophosphite,*** followed by 210 milliliters of hexane, followed by ***160 grams of potassium nitrate***, and then followed by ***180 grams of nitrocellulose*** (14% nitrogen content). Thereafter, blend the mixture on high speed at room temperature for about 45 minutes. After the blending operation, the mixture is ready for extruding; to do so, extrude the mixture through any extruding device into "spaghetti-like" strands ranging from 2 to 8mm in diameter, by 300mm in length, and then cut the strands into pieces ranging from 1 to 10mm length, and then cure the granules by placing them on a shallow pan or tray, and place said pan or tray into an oven, and cure the granules at 80 Celsius for 12 hours. To use the grains, simply pack them loosely into any desirable cartridge casing. Use standard gun casing primers for initiation.

**Chamber Pressure (standard 9mm casing; weight of propellant charge/grain size on average):** 38,084 psi
**Chamber Pressure (standard 5.56mm rifle casing; weight of propellant charge/grain size on average):** 37,158 psi
**Chamber Pressure (standard 105mm artillery casing; weight of propellant charge/grain size on average):** 28,099 psi
**Mortar tube Pressure (standard 81mm mortar propellant charge; weight of propellant charge/grain size on average):** 8,377 psi
**Burn rate (in the open):** 2.5 seconds per inch; tends to vary based on diameter of grain.
**Water resistance:** Fair.
**Stability:** Can be stored for decades when kept from moisture.
**Sensitivity:** Stable towards friction; stable towards shock; stable to temperatures up to 180 Celsius.
**Flammability (1 to 10, un-contained):** 9 solid grain; 9.19 fine grain.
**Tendency to cake:** Will not cake.
**Detonation ability:** Can be detonated but must be primed with HE charge.
**Percentage:** ***49.8% ammonium nitrate, 17.24% nitrocellulose, 15.32% potassium nitrate, 10.53% graphite, 7.08% sodium hypophosphite, 0.03% mixed balance***
**Classification:** Deflagrating explosive (classified as propellant).

**Danger of preparation:** Can be prepared by amateurs, but caution should be exercised.
**Use:** Substitute for black powder in firearms; used in mortar propellants, artillery gun propellants; used in non-firearms weapons to propel explosive devices.

### 05-01-017A: High performance ammonium nitrate gun propellant:

Into a suitable vertical mixer, place ***200 grams of nitrocellulose*** (14% nitrogen content), followed by 290 milliliters of hexane, followed by ***190 gram of dinitrotoluene***, followed by ***45 grams of dibutylphthalate***, followed by ***15 grams of diphenylamine***, and then followed by ***547 grams of powdered ammonium nitrate***. Thereafter, blend the mixture on high speed for about 45 minutes at room temperature. After the blending operation, the mixture is ready for extruding; to do so, extrude the mixture through any extruding device into "spaghetti-like" strands ranging from 2 to 8mm in diameter, by 300mm in length, and then cut the strands into pieces ranging from 1 to 10mm length, and then cure the granules by placing them on a shallow pan or tray, and place said pan or tray into an oven, and cure the granules at 70 Celsius for 12 hours. To use the grains, simply pack them loosely into any desirable cartridge casing. Use standard gun casing primers for initiation.

**Chamber Pressure (standard 9mm casing; weight of propellant charge/grain size on average):** 38,997 psi
**Chamber Pressure (standard 5.56mm rifle casing; weight of propellant charge/grain size on average):** 38,086 psi
**Chamber Pressure (standard 105mm artillery casing; weight of propellant charge/grain size on average):** 28,915 psi
**Mortar tube Pressure (standard 81mm mortar propellant charge; weight of propellant charge/grain size on average):** 8,967 psi
**Burn rate (in the open):** 2.3 seconds per inch; tends to vary based on diameter of grain.
**Water resistance:** Fair.
**Stability:** Can be stored for decades when kept from moisture.
**Sensitivity:** Stable towards friction; stable towards shock; stable to temperatures up to 180 Celsius.
**Flammability (1 to 10, un-contained):** 9 solid grain; 9.1 fine grain.
**Tendency to cake:** Will not cake.
**Detonation ability:** Can be detonated but must be primed with HE charge.
**Percentage:** ***54.86% ammonium nitrate, 20% nitrocellulose, 19% dinitrotoluene, 4.4% dibutylphthalate, 1.5% diphenylamine, 0.24% graphite and mixed residua***
**Classification:** Deflagrating explosive (classified as propellant).

**Danger of preparation:** Can be prepared by amateurs, but it is advisable that only experienced personnel perform this procedure. Please keep in mind the toxicity of dinitrotoluene.
**Use:** Substitute for black powder in firearms; used in mortar propellants, artillery gun propellants; used in non-firearms weapons to propel explosive devices.

**05-01-018A: High performance ammonium nitrate gun propellant:**
Into suitable vertical mixer, place ***375 grams of powdered ammonium nitrate***, followed by ***75 grams of flours of sulfur***, followed by ***50 grams of soft wood charcoal***, followed by 190 milliliters of hexane, and then followed by ***15 grams of sodium chloride***. Thereafter, blend the mixture on high speed for about 45 minutes. After the blending operation, the mixture is ready for extruding; to do so, extrude the mixture through any extruding device into "spaghetti-like" strands ranging from 2 to 8mm in diameter, by 300mm in length, and then cut the strands into pieces ranging from 1 to 10mm length, and then cure the granules by placing them on a shallow pan or tray, and place said pan or tray into an oven, and cure the granules at 70 Celsius for 12 hours. To use the grains, simply pack them loosely into any desirable cartridge casing. Use standard gun casing primers for initiation.
**Chamber Pressure (standard 9mm casing; weight of propellant charge/grain size on average):** 37,099 psi
**Chamber Pressure (standard 5.56mm rifle casing; weight of propellant charge/grain size on average):** 37,334 psi
**Chamber Pressure (standard 105mm artillery casing; weight of propellant charge/grain size on average):** 27,093 psi
**Mortar tube Pressure (standard 81mm mortar propellant charge; weight of propellant charge/grain size on average):** 8,003 psi
**Burn rate (in the open):** 2.7 seconds per inch; tends to vary based on diameter of grain.
**Water resistance:** Fair.
**Stability:** Can be stored for decades when kept from moisture.
**Sensitivity:** Stable towards friction; stable towards shock; stable to temperatures up to 180 Celsius.
**Flammability (1 to 10, un-contained):** 8.9 solid grain; 9.1 fine grain.
**Tendency to cake:** Will not cake.
**Detonation ability:** Can be detonated but must be primed with HE charge.
**Percentage:** ***72.81% ammonium nitrate, 14.56% sulfur, 9.7% soft wood charcoal, 2.91% sodium chloride, 0.02% graphite and residual balance***
**Classification:** Deflagrating explosive (classified as propellant).

**Danger of preparation:** Can be prepared by amateurs.
**Use:** Substitute for black powder in firearms; used in mortar propellants.

**05-01-018B: High performance ammonium nitrate gun propellant:**
Into suitable vertical mixer, place ***375 grams of powdered ammonium nitrate***, followed by ***75 grams of flours of sulfur***, followed by ***50 grams of soft wood charcoal,*** followed by ***50 grams of nitrocellulose*** (14% nitrogen content), followed by ***25 grams of barium nitrate***, and then followed by ***15 grams of sodium chloride***. Thereafter, blend the mixture on high speed for about 45 minutes. After the blending operation, the mixture is ready for extruding; to do so, extrude the mixture through any extruding device into "spaghetti-like" strands ranging from 2 to 8mm in diameter, by 300mm in length, and then cut the strands into pieces ranging from 1 to 10mm length, and then cure the granules by placing them on a shallow pan or tray, and place said pan or tray into an oven, and cure the granules at 70 Celsius for 12 hours. To use the grains, simply pack them loosely into any desirable cartridge casing. Use standard gun casing primers for initiation.
**Chamber Pressure (standard 9mm casing; weight of propellant charge/grain size on average):** 37,099 psi
**Chamber Pressure (standard 5.56mm rifle casing; weight of propellant charge/grain size on average):** 37,334 psi
**Chamber Pressure (standard 105mm artillery casing; weight of propellant charge/grain size on average):** 27,093 psi
**Mortar tube Pressure (standard 81mm mortar propellant charge; weight of propellant charge/grain size on average):** 8,003 psi
**Burn rate (in the open):** 2.7 seconds per inch; tends to vary based on diameter of grain.
**Water resistance:** Fair.
**Stability:** Can be stored for decades when kept from moisture.
**Sensitivity:** Stable towards friction; stable towards shock; stable to temperatures up to 180 Celsius.
**Flammability (1 to 10, un-contained):** 8.9 solid grain; 9.1 fine grain.
**Tendency to cake:** Will not cake.
**Detonation ability:** Can be detonated but must be primed with HE charge.
**Percentage:** ***63.55% ammonium nitrate, 12.71% sulfur, 8.47% soft wood charcoal, 8.47% nitrocellulose, 4.23% barium nitrate, 2.54% sodium chloride, 0.03% graphite and mixed residual balance***
**Classification:** Deflagrating explosive (classified as propellant).

**Danger of preparation:** Can be prepared by amateurs.
**Use:** Substitute for black powder in firearms; used in mortar propellants.

# Section 2: Nitrocellulose containing

***Chemicals used in this section (binders are not included)***

| | | |
|---|---|---|
| 1. Potassium nitrate (see Black Powder) | | 2. Sulfur (see Black Powder) |
| 3. Charcoal (see Black Powder) | | 4. Sugar Carbon (see Modified Black Powder) |
| 5. Barium Chromate (see Modified Black Powder) | | 6. Potassium Perchlorate (see Modified Black Powder) |
| 7. Potassium Dichromate (see Modified Black Powder) | | 8. Ammonium Bisulfide (see Modified Black Powder) |
| 9. Potassium chlorate (see Modified Black Powder) | | 10. Carbon Disulfide (see Modified Black Powder) |
| 11. Nitrocellulose (see Modified Black Powder) | | 12. Lead Tetraoxide (see Modified Black Powder) |
| 13. Diphenylamine (see Modified Black Powder) | | 14. Titanium Dioxide (see Modified Black Powder) |
| 15. Manganese Dioxide (see Modified Black Powder) | | 16. Sodium Benzoate (see Modified Black Powder) |
| 17. Ammonium Nitrate (see Modified Black Powder) | | 18. Calcium Carbonate (see Modified Black Powder) |
| 19. Sodium Nitrate (see Modified Black Powder) | | 20. Urea (see Modified Black Powder) |
| 21. Lead Nitrate (see Modified Black Powder) | | 22. Nitro Starch (see Modified Black Powder) |
| 23. Ammonium Perchlorate (see Ammonium Perchlorate Rocket Propellants) | | 24. Aluminum powder (see Ammonium Perchlorate Rocket Propellants) |
| 25. Magnesium Sulfide (see Ammonium Perchlorate Rocket Propellants) | | 26. Copper Chromite (see Ammonium Perchlorate Rocket Propellants) |
| 27. Nitroguanidine (see Ammonium Perchlorate Rocket Propellants) | | 28. Ammonium Sulfate (see Ammonium Perchlorate Rocket Propellants) |
| 29. Nitroglycerine (see Ammonium Perchlorate Rocket Propellants) | | 30. Magnesium Oxide (see Ammonium Perchlorate Rocket Propellants) |
| 31. Iron-II-oxide (see Ammonium Perchlorate Rocket Propellants) | | 32. Zirconium Hydride (see Ammonium Perchlorate Rocket Propellants) |
| 33. Teflon (see Ammonium Perchlorate Rocket Propellants) | | 34. Zinc Oxide (see Ammonium Perchlorate Rocket Propellants) |
| 35. Ammonium Dichromate (see Ammonium Perchlorate Rocket Propellants) | | 36. Lithium Perchlorate (see Ammonium Perchlorate Rocket Propellants) |
| 37. Magnesium powder (see Ammonium Perchlorate Rocket Propellants) | | 38. Lithium Aluminum Hydride (see Ammonium Perchlorate Rocket Propellants) |
| 39. Iron-III-oxide (red iron oxide) (see Ammonium Perchlorate Rocket Propellants) | | 40. PVC (see Ammonium Perchlorate Rocket Propellants) |
| 41. Copper Sulfide (see Ammonium Perchlorate Rocket Propellants) | | 42. Sodium Hydride (see Ammonium Perchlorate Rocket Propellants) |
| 43. Titanium Hydride (see Ammonium Perchlorate Rocket Propellants) | | 44. Silicon Nitride (see Ammonium Perchlorate Rocket Propellants) |
| 45. ADN (see ADN Rocket Propellants) | | 46. Urea Nitrate (see ADN Rocket Propellants) |
| 47. Silicon Powder (see ADN Rocket Propellants) | | 48. Hexamine (see ADN Rocket Propellants) |
| 49. Boron powder (see ADN Rocket Propellants) | | 50. Sodium hypophosphite (see ADN Rocket Propellants) |
| 51. KDN (see ADN Rocket Propellants) | | 52. Nickel Chloride (see Ammonium Nitrate Rocket Propellants) |
| 53. TNT (see Ammonium Nitrate Rocket Propellants) | | 54. Acrylamide (see Miscellaneous Rocket H.P. Rocket Propellants) |
| 55. Ethylene Glycol (see Miscellaneous Rocket H.P. Rocket Propellants) | | 56. Magnesium Perchlorate (see Miscellaneous Rocket H.P. Rocket Propellants) |
| 57. Metallic Lithium (see Miscellaneous Rocket H.P. Rocket Propellants) | | 58. Beryllium Hydride (see Miscellaneous Rocket H.P. Rocket Propellants) |
| 59. Red Phosphorus (see Miscellaneous Rocket H.P. Rocket Propellants) | | 60. Sodium Borohydride (see Miscellaneous Rocket H.P. Rocket Propellants) |
| 61. Sodium chlorate (see Miscellaneous Rocket H.P. Rocket Propellants) | | 62. Lead Dioxide (see Miscellaneous Rocket H.P. Rocket Propellants) |
| 63. Benzoyl Peroxide (see Miscellaneous Rocket H.P. Rocket Propellants) | | 64. Barium Nitrate (see Miscellaneous Rocket H.P. Rocket Propellants) |

| | |
|---|---|
| **65. Cyanuric Acid (see Miscellaneous Rocket H.P. Rocket Propellants)** | **66. Aluminum Stearate (see Ammonium Nitrate Gun Propellants)** |
| **67. Aluminum Hydride (see Ammonium Nitrate Gun Propellants)** | **68. Magnesium Peroxide (see Ammonium Nitrate Gun Propellants)** |
| **69. Potassium Permanganate (see Ammonium Nitrate Gun Propellants)** | **70. Calcium Hydride (see Ammonium Nitrate Gun Propellants)** |
| **71. Potassium Tartrate (see Ammonium Nitrate Gun Propellants)** | **72. Potassium Ferrocyanide (see Ammonium Nitrate Gun Propellants)** |
| **73. Sodium Azide (see Ammonium Nitrate Gun Propellants)** | **74. Sodium Chloride (see Ammonium Nitrate Gun Propellants)** |
| **75. Potassium Sulfate** | **76. Lead Stearate** |
| 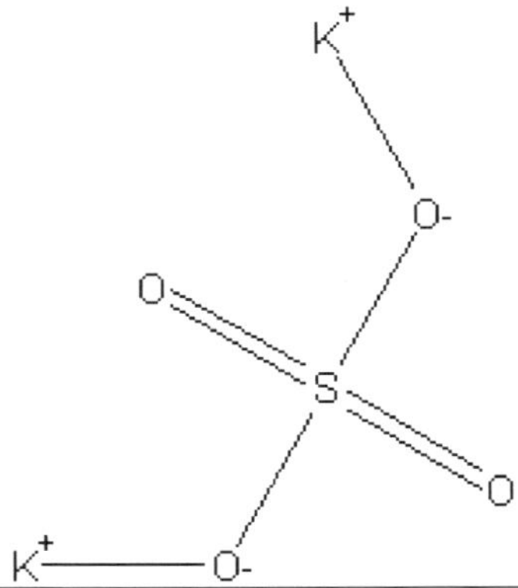 | $[CH_3(CH_2)_{16}COO]_2Pb$ |
| Forms colorless to white crystals, granules, or powder. The sulfate is relatively insoluble in water, alcohol, and other solvents. | Lead stearate forms a white powder or granular mass. The melting point of the salt is 125 Celsius. Lead stearate is toxic, so users should use caution and wear proper gloves. |
| **77. Triethylene Glycol** | |
| 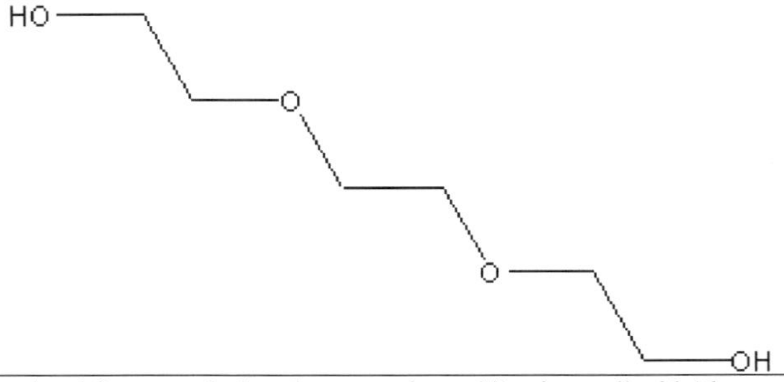 | |
| The glycol forms a colorless, hygroscopic, mobile viscous liquid. The melting point of the glycol is –7.2 Celsius, and its boiling point at STP is 285 Celsius. Triethylene glycol is an important plasticizer for use in explosives and pyrotechnics. | |

***- High Performance Nitrocellulose Gun Propellants in this section -***

| | |
|---|---|
| **1. 05-02-001A: Nitrocellulose high performance gun propellant (military grade gunpowder; double based):** *56.83% nitroglycerine, 27.14% nitrocellulose, 6.33% potassium nitrate, 5.97% 4,4'-diphenylmethane diisocyanate plasticizer, 2.98% glue, 0.67% diethyldiphenylurea catalyst, 0.08% residue* | **2. 05-02-002A: Nitrocellulose high performance gun propellant (military grade gunpowder; double based):** *63.3% nitrocellulose, 17% potassium nitrate, 17% KDN, 1.5% diethyldiphenylurea, 1% potassium sulfate smoke reducer, 0.20% impurities* |
| **3. 05-02-003A: Nitrocellulose "Smokeless powder":** *85.83% nitrocellulose, 12.87% potassium nitrate, 1.28% diphenylamine catalyst, 0.02% impurities* | **4. 05-02-004A: Nitrocellulose and cellulose azidonitrate high performance gun propellant:** *49% nitrocellulose, 49% cellulose azidonitrate, 1.4% diphenylamine, 0.28% graphite, 0.320% residue* |
| **5. 05-02-005A: Nitrocellulose high performance gun propellant with increased ballistic properties:** *84.9% nitrocellulose, 9.7% nitroguanidine, 4.3% barium nitrate, 0.43% diphenylamine, 0.29% residue, 0.24% graphite, 0.14% TNT* | **6. 05-02-006A: Nitrocellulose high performance gun propellant for small arms weapons:** *94.9% nitrocellulose, 1.8% potassium nitrate, 1% methyl centralite, 1% diphenylamine, 0.63% potassium aluminum fluoride, 0.37% graphite, 0.30% moisture* |
| | |

| | |
|---|---|
| **7. 05-02-007A: Nitrocellulose gun propellant utilizing the dispersion/flotation method:** *84.4% nitrocellulose, 9% lead stearate, 4.6% dioctylphthalate, 1.8% 2-nitrodiphenylamine, 0.20% animal glue* | **8. 05-02-008A: Nitrocellulose smokeless gun propellant:** *85.3% nitrocellulose, 9.8% nitroguanidine, 4.4% barium nitrate, 0.44% diphenylamine, 0.06% graphite* |
| **9. 05-02-009A: Improved nitrocellulose smokeless gun propellant:** *80% nitrocellulose, 14% potassium nitrate, 3% barium nitrate, 3% charcoal* | **10. 05-02-010A: Nitrocellulose smokeless gun propellant specifically for shotguns:** *72% nitrocellulose, 18.1% nitroglycerine, 9% methylcellulose, 0.69% diethyldiphenylamine, 0.21% combined balance including, 0.05% graphite* |
| **11. 05-02-010B: Nitrocellulose smokeless gun propellant specifically for shotguns (increased chamber pressure via coal dust):** *75.4% nitrocellulose, 19% nitroglycerine, 4.7% coal dust, 0.73% diethyldiphenylamine, 0.17% combined balance including, 0.05% graphite* | **12. 05-02-011A: Nitrocellulose gun propellant (US ARMY M6 propellant):** *86% nitrocellulose, 10% dinitrotoluene, 3% dibutylphthalate, 1% diphenylamine* |
| **13. 05-02-011B: Nitrocellulose gun propellant (US ARMY M6 propellant), with flash suppression addictive:** *85% nitrocellulose, 10% dinitrotoluene, 3% dibutylphthalate, 1% diphenylamine, 1% potassium sulfate* | **14. 05-02-011C: Nitrocellulose gun propellant (former US Navy propellant—pyro propellant):** *99% nitrocellulose, 1% diphenylamine* |
| **15. 05-02-012A: Standard high performance nitrocellulose gun propellant:** *66% nitrocellulose, 16% nitroglycerine, 16% polyvinyl nitrate, 1% 2-nitrodiphenyl, 1% impurities* | **16. 05-02-013A: Standard high performance nitrocellulose "double base" smokeless powder with reduced flash:** *50.6% nitrocellulose, 30% nitroglycerine, 10% ethyl abietate, 6% barium nitrate, 2% potassium nitrate, 0.76% diphenylamine, 0.64% residue* |
| **17. 05-02-014A: Standard high performance nitrocellulose (high nitroglycerine content) smokeless powder with reduced flash:** *70% nitroglycerine, 23% nitrocellulose, 5% lycopodium, 2% urea* | **18. 05-02-014B: Standard high performance nitrocellulose (high nitroglycerine content) smokeless powder with reduced flash (modified using dinitrobenzene):** *70% nitroglycerine, 23% nitrocellulose, 5% lycopodium, 2% dinitrobenzene* |
| **19. 05-02-015A: High performance nitrocellulose (high nitroglycerine content) smokeless powder with reduced flash:** *55% nitroglycerine, 30% nitrocellulose, 10% Vaseline, 5% camphor* | **20. 05-02-016A: High performance nitrocellulose smokeless powder with reduced flash:** *60% nitrocellulose, 30% nitroglycerine, 5% diamyl phthalate, 5% mineral jelly* |
| **21. 05-02-017A: High performance nitrocellulose smokeless powder**: *52.6% nitrocellulose, 42.1% nitroglycerine, 2.6% Vaseline, 2.6% olive oil, 0.1% impurities* | **22. 05-02-017B: High performance nitrocellulose smokeless powder (fortified with barium tartrate burn enhancer):** *50% nitrocellulose, 40% nitroglycerine, 4% barium tartrate, 2.5Vaseline, 2.6% olive oil, 0.90% potassium tartrate, 0.1% mixed balance* |
| **23. 05-02-018A: High performance nitrocellulose smokeless powder containing diethyldiphenylcarbamid burn enhancer:** **Percentage:** *71.42% nitrocellulose, 23.8% nitroglycerine, 4.76% diethyldiphenhylcarbamide, 0.02% residue* | **24. 05-02-018B: High performance nitrocellulose smokeless powder with potassium lactate stabilizer:** *69.88% nitrocellulose, 23.12% nitroglycerine, 5% Vaseline, 2% potassium lactate* |
| **25. 05-02-019A: High performance nitrocellulose smokeless powder (utilizing the "Schultze" process):** *72.22% nitrocellulose, 16.66% barium nitrate, 4.16% potassium nitrate, 4.16% paraffin oil, 2.77% starch, 0.03% mixed residues* | **26. 05-02-020A: High performance nitrocellulose smokeless powder:** *89.28% nitrocellulose, 8.92% black powder, 1.78% dinitrotoluene binder, 0.02 mixed balance* |
| **27. 05-02-021A: High performance nitrocellulose smokeless powder (military grade):** *40.1% nitroguanidine, 34.09% nitroglycerine, 24.73% nitrocellulose, 1.06% methyl diphenylurea, 0.02% mixed residue* | **28. 05-02-021B: High performance nitrocellulose smokeless powder (military grade):** *52.47% nitrocellulose, 35.64% nitroglycerine, 9.9% nitroguanidine, 1.98% diethyl diphenylamine, 0.01% residues* |
| **29. 05-02-022A: High performance nitrocellulose gun propellant (potassium nitrate fortified):** *56.43% nitrocellulose, 33.86% potassium nitrate, 8.46% dibutyl phthalate, 0.677% dinitrotoluene, 0.22% calcium carbonate, 0.168% ethyl cellulose, 0.168% diphenylamine, 0.017%* | **30. 05-02-023A: High performance nitrocellulose gun propellant (with TNT modifier):** *71.2% nitrocellulose, 20% nitroglycerine, 6.6% TNT, 1.2% dinitrotoluene* |

| | |
|---|---|
| *mixed balance* | |
| **31. 05-02-024A: High performance nitrocellulose gun propellant:** *58.82% nitrocellulose, 37.43% nitroglycerine, 3.74% diamyl phthalate, 0.01% residue* | **32. 05-02-025A: High performance nitrocellulose gun propellant:** *77.98% nitrocellulose, 20.01% nitroglycerine, 2% dinitrotoluene, 0.01% mixed residues* |
| **33. 05-02-026A: High performance nitrocellulose gun propellant:** *79.6% nitrocellulose, 9.95% corn starch, 9.95% dinitrotoluene, 0.49% diphenylamine, 0.01% graphite* | **34. 05-02-027A: High performance nitrocellulose gun propellant:** *74.5% nitrocellulose, 20% dinitrobenzene, 5% mononitroxylene, 0.50% diphenylamine* |
| **35. 05-02-028A: High performance "flashless" nitrocellulose gun propellant:** *85% nitrocellulose, 10% dinitrotoluene, 2.5% triethylene glycol, 2.5% caprylic acid* | **36. 05-02-028B: High performance "flashless" nitrocellulose gun propellant:** *80% nitrocellulose, 15% dinitrotoluene, 5% triethylene glycol caprylate* |
| **37. 05-02-029A: High performance "clean burning" nitrocellulose smokeless gun propellant:** *85% nitrocellulose, 6.18% butyl stearate, 5.67% animal glue, 0.98% ethyl centralite, 0.98% basic lead carbonate, 0.98% potassium sulfate, 0.21% impurities* | |

## *Formulating Nitrocellulose Compositions:*

| Nitrocellulose | Nitroglycerine | Combustible (reducing agent | Addictive (oxidizer) | Filler (binder) |
|---|---|---|---|---|
| 10 parts | 2 parts | 1 part | 1 part | ½ part |
| 20 parts | 5 parts | 2 parts | 1 part | 1 part |
| 30 parts | 11 parts | 3 parts | 2 parts | 2 parts |
| 40 parts (low smoke) | 23 parts | 5 parts | 3 parts | 2.5 parts |
| 60 parts (high power) | 30 parts | 5 parts | 1 part | 2 parts |
| 70 parts (high power 2) | 30 parts | 7 parts | 1 part | 4 parts |
| 90 parts (high power 3: direct fire guns) | 40 parts | 10 parts | 5 parts | 6 parts |
| 100 parts (artillery grade) | 40 parts | 12 parts | 6 parts | 10 parts |
| 120 parts (lower burn) | 30 parts | 20 parts | 10 parts | 10 parts |
| 140 parts (artillery grade: reduced flash) | 60 parts | 30 parts | 20 parts | 20 parts |
| 160 parts | 60 parts | 60 parts | 20 parts | 20 parts |
| 180 parts | 80 parts | 50 parts | 40 parts | 20 parts |

**Warning! Use great care when handling nitroglycerine, and make sure to wear proper protective clothing, gloves, ect., and ALWAYS make sure to wear a respirator or protective mask to avoid inhalation of nitroglycerine vapors. Inhalation of nitroglycerine vapors WILL lead to severe headaches/migraines that can hang around for WEEKS! Protecting yourself from skin absorption, and inhalation of nitroglycerine super precedes safety from its explosion hazards.**

## *Extruding grain sizes for various arms (by length):*

| | | |
|---|---|---|
| **0.22 caliber to 41 caliber handgun** | **60mm to 81mm mortar** | **30mm to 90mm tank gun** |
| 0.5mm fine grain | 6mm to 12mm flat grain | 5mm to 9mm cyndrical grain |
| **10mm to 45 caliber Magnum** | **90mm to 130mm mortar** | **100mm to 125mm tank gun** |
| 0.7mm fine grain | 6mm to 12 flat grain | 6.5mm to 19mm cyndrical grain |
| **20 gauge to 10 gauge shotgun** | **150mm to 210mm mortar** | **130mm to 160mm tank gun** |
| 1mm to 2mm flat grain | 6.6mm to 12.5mm flat grain | 10mm to 18mm cyndrical grain |
| **5.56mm to 8mm rifle** | **90mm to 105mm recoilless rifle** | |
| 1mm to 2mm fine grain | 4.5mm to 11mm fine grain | |
| **9mm to 0.45 caliber sub-machine gun** | **30mm to 90mm artillery** | |
| 1.55mm to 2.1mm fine grain | 5mm to 10mm granular grain | |
| **5.56mm to 7.62mm machine gun** | **105 to 155mm artillery** | |
| 2mm to 3.1mm cyndrical grain | 5mm to 12mm cyndrical grain | |
| **7.62mm to 12.7mm heavy machine gun** | **160mm to 210mm artillery** | |
| 3mm to 10mm cyndrical grain | 10mm to 15mm cyndrical grain | |

| 12.7mm up to 18.9mm heavy rifles | 4" to 6" battleship deck gun | |
|---|---|---|
| 5mm to 10mm flat grain | 11mm to 18mm flat grain | |
| **20mm to 30mm cannon** | **9" to 12" battleship deck gun** | |
| 6mm to 12mm granulated grain | 12mm to 25mm cyndrical/elongated grain | |

### 05-02-001A: Nitrocellulose high performance gun propellant (military grade gunpowder; double based):

Into a suitable sealable vertical mixer, place ***60 grams of nitrocellulose*** (14% nitrogen content), followed by 100 milliliters of hexane, followed by ***125.6 grams of nitroglycerine***, followed by ***1.5 grams of diethyldiphenylurea***, followed by ***14 grams of potassium nitrate***, followed by ***13.2 grams of 4,4'-diphenylmethane diisocyanate***, and then followed by ***6.6 grams of standard glue***. Thereafter, seal the vertical mixer, and then rapidly blend the mixture on high speed for about 45 minutes. After the blending operation, the mixture is ready for extruding; to do so, extrude the mixture through any extruding device under high pressure into "spaghetti-like" strands ranging from 0.5 to 12mm in diameter, by 300mm in length, and then cut the strands into pieces ranging from 0.5 to 18mm length, and then cure the granules by placing them on a shallow pan or tray, and place said pan or tray into an oven, and cure the granules at 70 Celsius for 12 hours. To use the grains, simply pack them loosely into any desirable cartridge casing. Use standard gun casing primers for initiation.

**Chamber Pressure (standard 9mm casing; weight of propellant charge/grain size on average):** 40,008 psi
**Chamber Pressure (standard 12.7mm rifle casing; weight of propellant charge/grain size on average):** 85,904 psi
**Chamber Pressure (standard 105mm artillery casing; weight of propellant charge/grain size on average):** 30,995 psi
**Burn rate (in the open):** 1.5 seconds per inch; tends to vary based on diameter of grain.
**Water resistance:** Fair.
**Stability:** Can be stored for decades when kept from moisture.
**Sensitivity:** Stable towards friction; stable towards shock; stable to temperatures up to 150 Celsius.
**Flammability (1 to 10, un-contained):** 9.6 solid grain; 9.8 fine grain.
**Tendency to cake:** Will not cake.
**Detonation ability:** Can be detonated using a blasting cap or detonator.
**Percentage:** ***56.83% nitroglycerine, 27.14% nitrocellulose, 6.33% potassium nitrate, 5.97% 4,4'-diphenylmethane diisocyanate plasticizer, 2.98% glue, 0.67% diethyldiphenylurea catalyst, 0.08% residue***
**Classification:** Deflagrating explosive (classified as class A explosive).

**Danger of preparation:** Should only be prepared by experienced personnel.
**Use:** Used in military ammunition for pistols, shotguns, rifles, machine guns, tank fired munitions, artillery guns, and ship deck guns.

**Note:** Inexperienced personnel should not attempt this procedure, as there are hazards involved. Make sure to thoroughly stir the mixture to avoid lumps or clumps, which can lead to excessive heat build-up. This procedure is perfectly safe for experienced persons. To gain experience, preparers should use only small quantities to begin with.

### 05-02-002A: Nitrocellulose high performance gun propellant (military grade gunpowder; double based):

Into a suitable sealable vertical mixer, place ***126 grams of nitrocellulose*** (14% nitrogen content), followed by ***2 grams of potassium sulfate,*** followed by ***34 grams of potassium nitrate***, followed by ***34 grams of KDN***, followed by 100 milliliters of hexane, and then followed by ***3 grams of diethyldiphenylurea***. Thereafter, seal the vertical mixer, and then rapidly blend the mixture for about 45 minutes at room temperature. After the blending operation, the mixture is ready for extruding; to do so, extrude the mixture through any extruding device into "spaghetti-like" strands ranging from 0.5 to 10mm in diameter, by 300mm in length, and then cut the strands into pieces ranging from 0.5 to 18mm length, and then cure the granules by placing them on a shallow pan or tray, and place said pan or tray into an oven, and cure the granules at 70 Celsius for 12 hours. To use the grains, simply pack them loosely into any desirable cartridge casing. Use standard gun casing primers for initiation.

**Chamber Pressure (standard 9mm casing; weight of propellant charge/grain size on average):** 38,098 psi
**Chamber Pressure (standard 12.7mm rifle casing; weight of propellant charge/grain size on average):** 80,776 psi
**Chamber Pressure (standard 105mm artillery casing; weight of propellant charge/grain size on average):** 29,006 psi
**Mortar tube Pressure (standard 81mm mortar propellant charge; weight of propellant charge/grain size on average):** 9,091 psi
**Burn rate (in the open):** 1.8 seconds per inch; tends to vary based on diameter of grain.
**Water resistance:** Fair.
**Stability:** Can be stored for decades when kept from moisture.
**Sensitivity:** Stable towards friction; stable towards shock; stable to temperatures up to 200 Celsius.
**Flammability (1 to 10, un-contained):** 9.2 solid grain; 9.7 fine grain.
**Tendency to cake:** Will not cake.
**Detonation ability:** Can be detonated using a blasting cap or detonator.
**Percentage:** ***63.3% nitrocellulose, 17% potassium nitrate, 17% KDN, 1.5% diethyldiphenylurea, 1% potassium sulfate smoke reducer, 0.20% impurities***
**Classification:** Deflagrating explosive (classified as propellant).

**Danger of preparation:** Should only be prepared by experienced personnel.
**Use:** Used in military ammunition for pistols, shotguns, rifles, machine guns, mortars, tank fired munitions, artillery guns, and ship deck guns.

**Note:** Inexperienced personnel should not attempt this procedure, as there are hazards involved. Make sure to thoroughly stir the mixture to avoid lumps or clumps, which can lead to excessive heat build-up. This procedure is perfectly safe for experienced persons. To gain experience, preparers should use only small quantities to begin with.

### 05-02-003A: Nitrocellulose "Smokeless powder":

Into a standard vertical mixer, place ***200 grams of nitrocellulose*** (14% nitrogen content), followed by 90 milliliters of hexane, followed by ***30 grams of potassium nitrate***, and then followed by ***3 grams of diphenylamine***. Thereafter, blend the mixture on high speed for about 45 minutes. After the blending operation, the mixture is ready for extruding; to do so, extrude the mixture through any extruding device into "spaghetti-like" strands ranging from 0.5 to 7mm in diameter, by 300mm in length, and then cut the strands into pieces ranging from 0.5 to 18mm length, and then cure the granules by placing them on a shallow pan or tray, and place said pan or tray into an oven, and cure the granules at 70 Celsius for 12 hours. To use the grains, simply pack them loosely into any desirable cartridge casing. Use standard gun casing primers for initiation.
**Chamber Pressure (standard 9mm casing; weight of propellant charge/grain size on average):** 37,889 psi
**Chamber Pressure (standard 5.56mm rifle casing; weight of propellant charge/grain size on average):** 34,094 psi
**Chamber Pressure (standard 105mm artillery casing; weight of propellant charge/grain size on average):** 28,091 psi
**Mortar tube Pressure (standard 81mm mortar propellant charge; weight of propellant charge/grain size on average):** 8,551 psi
**Burn rate (in the open):** 2.1 seconds per inch; tends to vary based on diameter of grain.
**Water resistance:** Fair.
**Stability:** Can be stored for decades when kept from moisture.
**Sensitivity:** Stable towards friction; stable towards shock; stable to temperatures up to 250 Celsius.
**Flammability (1 to 10, un-contained):** 9.1 solid grain; 9.4 fine grain.
**Tendency to cake:** Will not cake.
**Detonation ability:** Can be detonated using a blasting cap or detonator.
**Percentage:** ***85.83% nitrocellulose, 12.87% potassium nitrate, 1.28% diphenylamine catalyst, 0.02% impurities***
**Classification:** Deflagrating explosive (classified as propellant).

**Danger of preparation:** Can be prepared by amateurs, but caution should be exercised.
**Use:** Used in military ammunition for pistols, shotguns, rifles, machine guns, mortars, tank fired munitions, artillery guns, and ship deck guns.

### 05-02-004A: Nitrocellulose and cellulose azidonitrate high performance gun propellant:

**Step 1:** preparation of cellulose azidonitrate:
Into a suitable beaker, place 625 milliliters of dimethylsulfoxide (DMSO), and then add and dissolve 17.5 grams of nitrocellulose (14% nitrogen content), followed by 12.5 grams of sodium azide. Thereafter, heat the mixture to about 65 Celsius and stir the mixture at this temperature for about 66 hours. Thereafter, remove the heat source, and allow the mixture to cool to room temperature. Thereafter, pour the entire mixture into 625 milliliters of ice-cold water (contained in a suitable beaker), and then stir the entire mixture for about 30 minutes. Then, filter-off the insoluble gelatinous precipitate, and then allow it to air-dry. The result will be about 8.75 grams of cellulose azidonitrate product.
**Step 2:** Preparation of gun propellant:
Into a standard vertical mixer, place ***8.75 grams of cellulose azidonitrate*** product obtained in step1, followed by ***8.75 grams of nitrocellulose*** (14% nitrogen content), followed by 100 milliliters of 95% ethyl alcohol, and then followed by ***250 milligrams of diphenylamine***. Thereafter blend the mixture on high speed for 30 minutes. After the blending operation, the mixture is ready for extruding; to do so, extrude the mixture through any extruding device into "spaghetti-like" strands ranging from 0.5 to 8mm in diameter, by 300mm in length, and then cut the strands into pieces ranging from 0.5 to 16mm length, and then cure the granules by placing them on a shallow pan or tray, and place said pan or tray into an oven, and cure the granules at 80 Celsius for 12 hours. To use the grains, simply pack them loosely into any desirable cartridge casing. Use standard gun casing primers for initiation.
**Chamber Pressure (standard 9mm casing; weight of propellant charge/grain size on average):** 38,915 psi
**Chamber Pressure (standard 5.56mm rifle casing; weight of propellant charge/grain size on average):** 35,907 psi
**Chamber Pressure (standard 105mm artillery casing; weight of propellant charge/grain size on average):** 29,887 psi
**Mortar tube Pressure (standard 81mm mortar propellant charge; weight of propellant charge/grain size on average):** 9,017 psi
**Burn rate (in the open):** 2 seconds per inch; tends to vary based on diameter of grain.
**Water resistance:** Fair.

**Stability:** Can be stored for decades when kept from moisture.
**Sensitivity:** Stable towards friction; stable towards shock; stable to temperatures up to 180 Celsius.
**Flammability (1 to 10, un-contained):** 9.19 solid grain; 9.5 fine grain.
**Tendency to cake:** Will not cake.
**Detonation ability:** Can be detonated using a blasting cap or detonator.
**Percentage:** ***49% nitrocellulose, 49% cellulose azidonitrate, 1.4% diphenylamine, 0.28% graphite, 0.320% residue***
**Classification:** Deflagrating explosive (classified as propellant).

**Danger of preparation:** Can be prepared by amateurs, but caution should be exercised.
**Use:** Used in military ammunition for pistols, shotguns, rifles, machine guns, mortars, tank fired munitions, artillery guns, and ship deck guns.

### 05-02-005A: Nitrocellulose high performance gun propellant with increased ballistic properties:

Into a suitable vertical mixer, place ***426 grams of nitrocellulose*** (14% nitrogen content), followed by 375 milliliters of hexane, followed by ***49 grams of nitroguanidine***, followed by ***22 grams of barium nitrate***, followed by ***2.2 grams of diphenylamine***, followed by ***1.5 grams of finely powdered TNT***, and then followed by ***2.5 grams of graphite powder***. Thereafter, rapidly blend the mixture for 45 minutes. After the blending operation, the mixture is ready for extruding; to do so, extrude the mixture through any extruding device into "spaghetti-like" strands ranging from 0.5 to 8mm in diameter, by 300mm in length, and then cut the strands into pieces ranging from 0.5 to 16mm length, and then cure the granules by placing them on a shallow pan or tray, and place said pan or tray into an oven, and cure the granules at 80 Celsius for 12 hours. To use the grains, simply pack them loosely into any desirable cartridge casing. Use standard gun casing primers for initiation.
**Chamber Pressure (standard 9mm casing; weight of propellant charge/grain size on average):** 39,993 psi
**Chamber Pressure (standard 5.56mm rifle casing; weight of propellant charge/grain size on average):** 36,892 psi
**Chamber Pressure (standard 105mm artillery casing; weight of propellant charge/grain size on average):** 30,990 psi
**Mortar tube Pressure (standard 81mm mortar propellant charge; weight of propellant charge/grain size on average):** 9,990 psi
**Burn rate (in the open):** 2 seconds per inch; tends to vary based on diameter of grain.
**Water resistance:** Fair.
**Stability:** Can be stored for decades when kept from moisture.
**Sensitivity:** Stable towards friction; stable towards shock; stable to temperatures up to 180 Celsius.
**Flammability (1 to 10, un-contained):** 9.3 solid grain; 9.59 fine grain.
**Tendency to cake:** Will not cake.
**Detonation ability:** Can be detonated using a blasting cap or detonator.
**Percentage:** ***84.9% nitrocellulose, 9.7% nitroguanidine, 4.3% barium nitrate, 0.43% diphenylamine, 0.29% residue, 0.24% graphite, 0.14% TNT***
**Classification:** Deflagrating explosive (classified as propellant).

**Danger of preparation:** Can be prepared by amateurs, but caution should be exercised.
**Use:** Used in military ammunition for pistols, shotguns, rifles, machine guns, mortars, tank fired munitions, artillery guns, and ship deck guns.

### 05-02-006A: Nitrocellulose high performance gun propellant for small arms weapons:

Into a suitable vertical mixer, place ***450 grams of nitrocellulose*** (14% nitrogen content), followed by 300 milliliters of hexane, followed by ***9 grams of potassium nitrate***, followed by ***5 grams of diphenylamine***, followed by ***3 grams of potassium aluminum fluoride***, followed by ***5 grams of methyl centralite***, and then followed by ***1.8 grams of finely powdered graphite***. Thereafter, rapidly blend the mixture for 45 minutes. After the blending operation, the mixture is ready for extruding; to do so, extrude the mixture through any extruding device into "spaghetti-like" strands ranging from 0.5 to 9mm in diameter, by 300mm in length, and then cut the strands into pieces ranging from 0.5 to 18mm length, and then cure the granules by placing them on a shallow pan or tray, and place said pan or tray into an oven, and cure the granules at 80 Celsius for 12 hours. To use the grains, simply pack them loosely into any desirable cartridge casing. Use standard gun casing primers for initiation.
**Chamber Pressure (standard 9mm casing; weight of propellant charge/grain size on average):** 39,999 psi
**Chamber Pressure (standard 5.56mm rifle casing; weight of propellant charge/grain size on average):** 36,995 psi
**Chamber Pressure (standard 105mm artillery casing; weight of propellant charge/grain size on average):** 31,099 psi
**Mortar tube Pressure (standard 81mm mortar propellant charge; weight of propellant charge/grain size on average):** 10,008 psi
**Burn rate (in the open):** 2 seconds per inch; tends to vary based on diameter of grain.
**Water resistance:** Fair.
**Stability:** Can be stored for decades when kept from moisture.
**Sensitivity:** Stable towards friction; stable towards shock; stable to temperatures up to 180 Celsius.
**Flammability (1 to 10, un-contained):** 9.39 solid grain; 9.599 fine grain.
**Tendency to cake:** Will not cake.

**Detonation ability:** Can be detonated using a blasting cap or detonator.
**Percentage:** ***94.9% nitrocellulose, 1.8% potassium nitrate, 1% methyl centralite, 1% diphenylamine, 0.63% potassium aluminum fluoride, 0.37% graphite, 0.30% moisture***
**Classification:** Deflagrating explosive (classified as propellant).

**Danger of preparation:** Can be prepared by amateurs, but caution should be exercised.
**Use:** Used in military ammunition for pistols, shotguns, rifles, machine guns, mortars, tank fired munitions, artillery guns, and ship deck guns.

### 05-02-007A: Nitrocellulose gun propellant utilizing the dispersion/flotation method:

Into a large suitable vertical mixer, place ***325 grams of nitrocellulose*** (14% nitrogen content), followed by 4500 milliliters of hexane, and then allow the mixture to stand for 30 minutes. After 30 minutes, add in ***35 grams of lead stearate,*** followed by ***18 grams of dioctylphthalate,*** followed by ***7 grams of 2-nitrodiphenylamine,*** and then blend the mixture on high speed for about 30 minutes to form a homogenous mixture. Now, into a clean glass beaker, prepare a glue solution by adding and dissolving 175 grams of standard animal glue into 400 milliliters of water, and then blend this mixture for about 10 minutes. After 10 minutes, add this glue solution to the nitrocellulose mixture, and then continue to blend the combined mixture on high for another 30 minutes. Thereafter pour the mixture onto a large shallow pan or tray and allow the water and hexane to evaporate. Once the odor of solvent is gone, and the after the water has evaporated, recover the left over mixture, and then place into a ball mill filled with 500 grams of steel shot of 10mm diameter, and then tumble the mixture at 250 RPM for 1 hour. After 1 hour, remove the tumbled mixture, separate it from the steel shot, and then place this tumbled mixture into a clean vertical mixer, add in a solvent mixture composed of 150 milliliters of ethyl acetate and 50 milliliters of isopropyl alcohol, and then rapidly blend the mixture for about 45 minutes. After the blending operation, the mixture is ready for extruding; to do so, extrude the mixture through any extruding device into "spaghetti-like" strands ranging from 0.5 to 12mm in diameter, by 300mm in length, and then cut the strands into pieces ranging from 0.5 to 18mm length, and then cure the granules by placing them on a shallow pan or tray, and place said pan or tray into an oven, and cure the granules at 105 Celsius for 24 hours. To use the grains, simply pack them loosely into any desirable cartridge casing. Use standard gun casing primers for initiation.
**Chamber Pressure (standard 9mm casing; weight of propellant charge/grain size on average):** 39,109 psi
**Chamber Pressure (standard 5.56mm rifle casing; weight of propellant charge/grain size on average):** 36,189 psi
**Chamber Pressure (standard 105mm artillery casing; weight of propellant charge/grain size on average):** 30,026 psi
**Mortar tube Pressure (standard 81mm mortar propellant charge; weight of propellant charge/grain size on average):** 9,448 psi
**Burn rate (in the open):** 2.2 seconds per inch; tends to vary based on diameter of grain.
**Water resistance:** Fair.
**Stability:** Can be stored for decades when kept from moisture.
**Sensitivity:** Stable towards friction; stable towards shock; stable to temperatures up to 180 Celsius.
**Flammability (1 to 10, un-contained):** 9 solid grain; 9.2 fine grain.
**Tendency to cake:** Will not cake.
**Detonation ability:** Can be detonated using a blasting cap or detonator.
**Percentage:** ***84.4% nitrocellulose, 9% lead stearate, 4.6% dioctylphthalate, 1.8% 2-nitrodiphenylamine, 0.20% animal glue***
**Classification:** Deflagrating explosive (classified as propellant).

**Danger of preparation:** Can be prepared by amateurs, but caution should be exercised.
**Use:** Used in military ammunition for pistols, shotguns, rifles, machine guns, mortars, tank fired munitions, artillery guns, and ship deck guns. Also used in low pressure/low velocity gun systems such as 40-millimeter grenade launchers and bazookas.

### 05-02-008A: Nitrocellulose smokeless gun propellant:

Into a suitable vertical mixer, place ***425 grams of nitrocellulose*** (14% nitrogen content), followed by 250 milliliters of hexane, followed by ***49 grams of nitroguanidine***, followed by ***22 grams of barium nitrate***, followed by ***2.2 grams of diphenylamine***. Thereafter, blend the mixture on high speed for about 45 minutes. After the blending operation, the mixture is ready for extruding; to do so, extrude the mixture through any extruding device into "spaghetti-like" strands ranging from 0.5 to 12mm in diameter, by 300mm in length, and then cut the strands into pieces ranging from 0.5 to 18mm length, and then cure the granules by placing them on a shallow pan or tray, and place said pan or tray into an oven, and cure the granules at 80 Celsius for 24 hours. To use the grains, they first need to be placed into a plastic bag filled with graphite powder, shaken for about 10 minutes to coat the grains, and then simply pack the grains loosely into any desirable cartridge casing. Use standard gun casing primers for initiation.
**Chamber Pressure (standard 9mm casing; weight of propellant charge/grain size on average):** 39,887 psi
**Chamber Pressure (standard 5.56mm rifle casing; weight of propellant charge/grain size on average):** 36,991 psi
**Chamber Pressure (standard 105mm artillery casing; weight of propellant charge/grain size on average):** 30,896 psi
**Mortar tube Pressure (standard 81mm mortar propellant charge; weight of propellant charge/grain size on average):** 9,992 psi
**Burn rate (in the open):** 2.29 seconds per inch; tends to vary based on diameter of grain.
**Water resistance:** Fair.

**Stability:** Can be stored for decades when kept from moisture.
**Sensitivity:** Stable towards friction; stable towards shock; stable to temperatures up to 180 Celsius.
**Flammability (1 to 10, un-contained):** 9 solid grain; 9.17 fine grain.
**Tendency to cake:** Will not cake.
**Detonation ability:** Can be detonated using a blasting cap or detonator.
**Percentage:** ***85.3% nitrocellulose, 9.8% nitroguanidine, 4.4% barium nitrate, 0.44% diphenylamine, 0.06% graphite***
**Classification:** Deflagrating explosive (classified as propellant).

**Danger of preparation:** Can be prepared by amateurs, but caution should be exercised.
**Use:** Used in military ammunition for pistols, shotguns, rifles, machine guns, mortars, tank fired munitions, artillery guns, and ship deck guns.

### 05-02-009A: Improved nitrocellulose smokeless gun propellant:

Into a suitable vertical mixer, place ***240 grams of nitrocellulose*** (14% nitrogen content), followed by 150 milliliters of hexane, and then followed by ***42 grams of potassium nitrate,*** followed by ***9 grams of barium nitrate,*** and then followed by ***9 grams of standard wood charcoal***. Thereafter, blend the mixture on high speed for about 45 minutes. After the blending operation, the mixture is ready for extruding; to do so, extrude the mixture through any extruding device into "spaghetti-like" strands ranging from 0.5 to 12mm in diameter, by 300mm in length, and then cut the strands into pieces ranging from 0.5 to 18mm length, and then cure the granules by placing them on a shallow pan or tray, and place said pan or tray into an oven, and cure the granules at 80 Celsius for 12 hours. To use the grains, simply pack the grains loosely into any desirable cartridge casing. Use standard gun casing primers for initiation.
**Chamber Pressure (standard 9mm casing; weight of propellant charge/grain size on average):** 39,004 psi
**Chamber Pressure (standard 5.56mm rifle casing; weight of propellant charge/grain size on average):** 36,013 psi
**Chamber Pressure (standard 105mm artillery casing; weight of propellant charge/grain size on average):** 30,005 psi
**Mortar tube Pressure (standard 81mm mortar propellant charge; weight of propellant charge/grain size on average):** 9,015 psi
**Burn rate (in the open):** 2.3 seconds per inch; tends to vary based on diameter of grain.
**Water resistance:** Fair.
**Stability:** Can be stored for decades when kept from moisture.
**Sensitivity:** Stable towards friction; stable towards shock; stable to temperatures up to 180 Celsius.
**Flammability (1 to 10, un-contained):** 9 solid grain; 9.1 fine grain.
**Tendency to cake:** Will not cake.
**Detonation ability:** Can be detonated using a blasting cap or detonator.
**Percentage:** ***80% nitrocellulose, 14% potassium nitrate, 3% barium nitrate, 3% charcoal***
**Classification:** Deflagrating explosive (classified as propellant).

**Danger of preparation:** Can be prepared by amateurs, but caution should be exercised.
**Use:** Used in military ammunition for pistols, shotguns, rifles, machine guns, mortars, tank fired munitions, artillery guns, and ship deck guns.

**Gun fired 105mm rocket projectile**

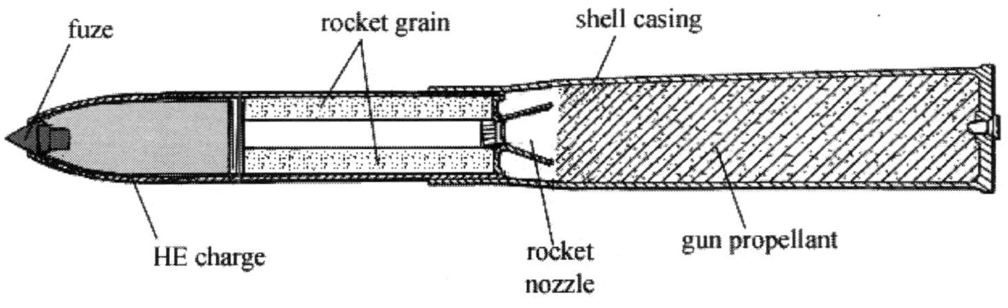

**05-02-010A: Nitrocellulose smokeless gun propellant specifically for shotguns:**
Into a suitable sealable vertical mixer, place 150 milliliters of ethyl acetate, followed by ***60 grams of nitroglycerine***, followed by ***2.3 grams of diethyldiphenylamine***, followed by ***237.6 grams of nitrocellulose*** (14% nitrogen content), and then followed by ***29.9 grams of methylcellulose***. Thereafter, seal the vertical mixer, and then blend the mixture on moderate speed for about 1 hour. After the blending operation, the mixture is ready for extruding; to do so, extrude the mixture through any extruding device into "spaghetti-like" strands ranging from 0.5 to 12mm in diameter, by 300mm in length, and then cut the strands into pieces ranging from 0.5 to 18mm length, and then cure the granules by placing them on a shallow pan or tray, and place said pan or tray into an oven, and cure the granules at 80 Celsius for 12 hours. To use the grains, simply pack the grains loosely into any desirable cartridge casing. Use standard gun casing primers for initiation.
**Chamber Pressure (standard 9mm casing; weight of propellant charge/grain size on average):** 39,981 psi
**Chamber Pressure (standard 12.7mm rifle casing; weight of propellant charge/grain size on average):** 85,093 psi
**Chamber Pressure (standard 105mm artillery casing; weight of propellant charge/grain size on average):** 32,902 psi
**Chamber Pressure (standard 12 gauge shot shell casing; weight of propellant charge/grain size on average):** 14,994 psi
**Burn rate (in the open):** 1.8 seconds per inch; tends to vary based on diameter of grain.
**Water resistance:** Fair.
**Stability:** Can be stored for decades.
**Sensitivity:** Stable towards friction; stable towards shock; stable to temperatures up to 100 Celsius.
**Flammability (1 to 10, un-contained):** 9.2 solid grain; 9.5 fine grain.
**Tendency to cake:** Will not cake.
**Detonation ability:** Can be detonated using a blasting cap or detonator.
**Percentage:** ***72% nitrocellulose, 18.1% nitroglycerine, 9% methylcellulose, 0.69% diethyldiphenylamine, 0.21% combined balance including, 0.05% graphite***
**Classification:** Deflagrating explosive (classified as class A explosive).

**Danger of preparation:** Should only be prepared by experienced personnel.
**Use:** Used in military ammunition for pistols, shotguns, rifles, machine guns, tank fired munitions, artillery guns, and ship deck guns.

**Note:** Inexperienced personnel should not attempt this procedure, as there are hazards involved. Make sure to thoroughly stir the mixture to avoid lumps or clumps, which can lead to excessive heat build-up. This procedure is perfectly safe for experienced persons. To gain experience, preparers should use only small quantities to begin with.

**Rocket assisted 12 gauge 3" Shot gun slug**

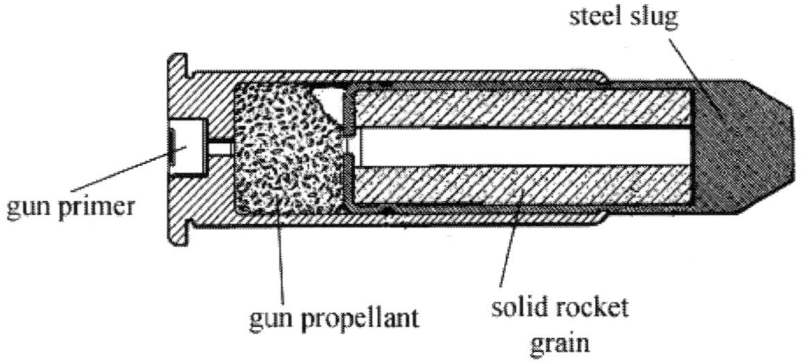


**05-02-010B: Nitrocellulose smokeless gun propellant specifically for shotguns (increased chamber pressure via coal dust):**
Into a suitable sealable vertical mixer, place 150 milliliters of ethyl acetate, followed by ***60 grams of nitroglycerine***, followed by ***2.3 grams of diethyldiphenylamine***, followed by ***237.6 grams of nitrocellulose*** (14% nitrogen content), and then followed by ***15 grams of coal dust***. Thereafter, seal the vertical mixer, and then blend the mixture on moderate speed for about 1 hour. After the blending operation, the mixture is ready for extruding; to do so, extrude the mixture through any extruding device into "spaghetti-like" strands ranging from 0.5 to 12mm in diameter, by 300mm in length, and then cut the strands into pieces ranging from 0.5 to 18mm length, and then cure the granules by placing them on a shallow pan or tray, and place said pan or tray into an oven, and cure the granules at 80 Celsius for 12 hours. To use the grains, simply pack the grains loosely into any desirable cartridge casing. Use standard gun casing primers for initiation.
**Chamber Pressure (standard 9mm casing; weight of propellant charge/grain size on average):** 39,998 psi
**Chamber Pressure (standard 12.7mm rifle casing; weight of propellant charge/grain size on average):** 85,109 psi
**Chamber Pressure (standard 105mm artillery casing; weight of propellant charge/grain size on average):** 33,015 psi
**Chamber Pressure (standard 12 gauge shot shell casing; weight of propellant charge/grain size on average):** 15,903 psi
**Burn rate (in the open):** 1.71 seconds per inch; tends to vary based on diameter of grain.
**Water resistance:** Fair.
**Stability:** Can be stored for decades.
**Sensitivity:** Stable towards friction; stable towards shock; stable to temperatures up to 100 Celsius.

**Flammability (1 to 10, un-contained):** 9.16 solid grain; 9.46 fine grain.
**Tendency to cake:** Will not cake.
**Detonation ability:** Can be detonated using a blasting cap or detonator.
**Percentage:** ***75.4% nitrocellulose, 19% nitroglycerine, 4.7% coal dust, 0.73% diethyldiphenylamine, 0.17% combined balance including, 0.05% graphite***
**Classification:** Deflagrating explosive (classified as class A explosive).

**Danger of preparation:** Should only be prepared by experienced personnel.
**Use:** Used in military ammunition for pistols, shotguns, rifles, machine guns, tank fired munitions, artillery guns, and ship deck guns.

**Note:** Inexperienced personnel should not attempt this procedure, as there are hazards involved. Make sure to thoroughly stir the mixture to avoid lumps or clumps, which can lead to excessive heat build-up. This procedure is perfectly safe for experienced persons. To gain experience, preparers should use only small quantities to begin with.

### 05-02-011A: Nitrocellulose gun propellant (US ARMY M6 propellant):

Into a suitable vertical mixer, place 200 milliliters of hexane, followed by ***413 grams of nitrocellulose*** (14% nitrogen content), followed by ***47.5 grams of dinitrotoluene***, followed by ***14.3 grams of dibutylphthalate***, and then followed by ***4.8 grams of diphenylamine***. Thereafter, blend the mixture on high speed for about 1 hour to form a uniform mass. After the blending operation, the mixture is ready for extruding; to do so, extrude the mixture through any extruding device into "spaghetti-like" strands ranging from 0.5 to 12mm in diameter, by 300mm in length, and then cut the strands into pieces ranging from 0.5 to 18mm length, and then cure the granules by placing them on a shallow pan or tray, and place said pan or tray into an oven, and cure the granules at 80 Celsius for 12 hours. To use the grains, simply pack the grains loosely into any desirable cartridge casing. Use standard gun casing primers for initiation.
**Chamber Pressure (standard 9mm casing; weight of propellant charge/grain size on average):** 38,901 psi
**Chamber Pressure (standard 7.62mm rifle casing; weight of propellant charge/grain size on average):** 40,027 psi
**Chamber Pressure (standard 105mm artillery casing; weight of propellant charge/grain size on average):** 34,904 psi
**Mortar tube Pressure (standard 81mm mortar propellant charge; weight of propellant charge/grain size on average):** 9,993 psi
**Burn rate (in the open):** 2 seconds per inch; tends to vary based on diameter of grain.
**Water resistance:** Fair.
**Stability:** Can be stored for decades.
**Sensitivity:** Stable towards friction; stable towards shock; stable to temperatures up to 180 Celsius.
**Flammability (1 to 10, un-contained):** 9.2 solid grain; 9.4 fine grain.
**Tendency to cake:** Will not cake.
**Detonation ability:** Can be detonated using a blasting cap or detonator.
**Percentage:** ***86% nitrocellulose, 10% dinitrotoluene, 3% dibutylphthalate, 1% diphenylamine***
**Classification:** Deflagrating explosive (classified as propellant).

**Danger of preparation:** Can be prepared by amateurs, but caution should be exercised. Dinitrotoluene is highly toxic, use care.
**Use:** Used in military ammunition for pistols, shotguns, rifles, machine guns, mortars, tank fired munitions, artillery guns, and ship deck guns.

### 05-02-011B: Nitrocellulose gun propellant (US ARMY M6 propellant), with flash suppression addictive:

Into a suitable vertical mixer, place 200 milliliters of hexane, followed by ***413 grams of nitrocellulose of*** (14% nitrogen content), followed by ***47.5 grams of dinitrotoluene***, followed by ***14.3 grams of dibutylphthalate***, followed by ***4.8 grams of diphenylamine***, and then followed by ***4.8 grams of anhydrous potassium sulfate***. Thereafter, blend the mixture on high speed for about 1 hour to form a uniform mass. After the blending operation, the mixture is ready for extruding; to do so, extrude the mixture through any extruding device into "spaghetti-like" strands ranging from 0.5 to 12mm in diameter, by 300mm in length, and then cut the strands into pieces ranging from 0.5 to 18mm length, and then cure the granules by placing them on a shallow pan or tray, and place said pan or tray into an oven, and cure the granules at 80 Celsius for 12 hours. To use the grains, simply pack the grains loosely into any desirable cartridge casing. Use standard gun casing primers for initiation.
**Chamber Pressure (standard 9mm casing; weight of propellant charge/grain size on average):** 38,099 psi
**Chamber Pressure (standard 7.62mm rifle casing; weight of propellant charge/grain size on average):** 39,096 psi
**Chamber Pressure (standard 105mm artillery casing; weight of propellant charge/grain size on average):** 34,002 psi
**Mortar tube Pressure (standard 81mm mortar propellant charge; weight of propellant charge/grain size on average):** 9,558 psi
**Burn rate (in the open):** 2 seconds per inch; tends to vary based on diameter of grain.
**Water resistance:** Fair.
**Stability:** Can be stored for decades.
**Sensitivity:** Stable towards friction; stable towards shock; stable to temperatures up to 180 Celsius.

**Flammability (1 to 10, un-contained):** 9.2 solid grain; 9.4 fine grain.
**Tendency to cake:** Will not cake.
**Detonation ability:** Can be detonated using a blasting cap or detonator.
**Percentage:** ***85% nitrocellulose, 10% dinitrotoluene, 3% dibutylphthalate, 1% diphenylamine, 1% potassium sulfate***
**Classification:** Deflagrating explosive (classified as propellant).

**Danger of preparation:** Can be prepared by amateurs, but caution should be exercised. Dinitrotoluene is highly toxic, use care.
**Use:** Used in military ammunition for pistols, shotguns, rifles, machine guns, mortars, tank fired munitions, artillery guns, and ship deck guns.

### 05-02-011C: Nitrocellulose gun propellant (former US Navy propellant—pyro propellant):

Into a suitable vertical mixer, place 200 milliliters of 95% ethyl alcohol, followed by ***455 grams of nitrocellulose*** (14% nitrogen content), and then followed by ***5 grams of diphenylamine***. Thereafter, simply blend the mixture on high speed for about 1 hour to form a uniform mass. After the blending operation, the mixture is ready for extruding; to do so, extrude the mixture through any extruding device into "spaghetti-like" strands ranging from 0.5 to 12mm in diameter, by 300mm in length, and then cut the strands into pieces ranging from 0.5 to 18mm length, and then cure the granules by placing them on a shallow pan or tray, and place said pan or tray into an oven, and cure the granules at 80 Celsius for 12 hours. To use the grains, simply pack the grains loosely into any desirable cartridge casing. Use standard gun casing primers for initiation.
**Chamber Pressure (standard 9mm casing; weight of propellant charge/grain size on average):** 38,887 psi
**Chamber Pressure (standard 7.62mm rifle casing; weight of propellant charge/grain size on average):** 39,921 psi
**Chamber Pressure (standard 105mm artillery casing; weight of propellant charge/grain size on average):** 34,779 psi
**Mortar tube Pressure (standard 81mm mortar propellant charge; weight of propellant charge/grain size on average):** 9,899 psi
**Burn rate (in the open):** 2 seconds per inch; tends to vary based on diameter of grain.
**Water resistance:** Fair.
**Stability:** Can be stored for decades.
**Sensitivity:** Stable towards friction; stable towards shock; stable to temperatures up to 200 Celsius.
**Flammability (1 to 10, un-contained):** 9.21 solid grain; 9.49 fine grain.
**Tendency to cake:** Will not cake.
**Detonation ability:** Can be detonated using a blasting cap or detonator.
**Percentage:** ***99% nitrocellulose, 1% diphenylamine***
**Classification:** Deflagrating explosive (classified as propellant).

**Danger of preparation:** Can be prepared by amateurs, but caution should be exercised.
**Use:** Used in military ammunition for pistols, shotguns, rifles, machine guns, mortars, tank fired munitions, artillery guns, and ship deck guns.

### 05-02-012A: Standard high performance nitrocellulose gun propellant:

Into a suitable vertical mixer, place ***670 grams of nitrocellulose*** (14% nitrogen content), followed by a nitroglycerine solution prepared by adding and dissolving ***165 grams of nitroglycerine*** into 210 milliliters of acetone, followed by ***15 grams of 2-nitrodiphenylamine***, followed by 100 milliliters of 95% ethyl alcohol and 175 milliliters of hexane. Thereafter blend the mixture for about 30 minutes on moderate speed. Thereafter add in ***165 grams of polyvinyl nitrate***, in 20-gram portions over a period of about 30 minutes with rapid blending of the mixture. After the addition of the polyvinyl nitrate, continue to blend the mixture on moderate speed for about 5 hours at room temperature. After the blending operation, the mixture is ready for extruding; to do so, extrude the mixture through any extruding device into "spaghetti-like" strands ranging from 0.5 to 12mm in diameter, by 300mm in length, and then cut the strands into pieces ranging from 0.5 to 18mm length, and then cure the granules by placing them on a shallow pan or tray, and place said pan or tray into an oven, and cure the granules at 80 Celsius for 12 hours. To use the grains, simply pack the grains loosely into any desirable cartridge casing. Use standard gun casing primers for initiation.
**Chamber Pressure (standard 9mm casing; weight of propellant charge/grain size on average):** 39,099 psi
**Chamber Pressure (standard 7.62mm rifle casing; weight of propellant charge/grain size on average):** 40,081 psi
**Chamber Pressure (standard 105mm artillery casing; weight of propellant charge/grain size on average):** 35,331 psi
**Mortar tube Pressure (standard 81mm mortar propellant charge; weight of propellant charge/grain size on average):** 10,095 psi
**Burn rate (in the open):** 1.8 seconds per inch; tends to vary based on diameter of grain.
**Water resistance:** Fair.
**Stability:** Can be stored for decades.
**Sensitivity:** Stable towards friction; stable towards shock; stable to temperatures up to 100 Celsius.
**Flammability (1 to 10, un-contained):** 9.3 solid grain; 9.59 fine grain.
**Tendency to cake:** Will not cake.
**Detonation ability:** Can be detonated using a blasting cap or detonator.

**Percentage:** ***66% nitrocellulose, 16% nitroglycerine, 16% polyvinyl nitrate, 1% 2-nitrodiphenyl, 1% impurities***
**Classification:** Deflagrating explosive (classified as class A explosive).

**Danger of preparation:** Should only be prepared by experienced personnel.
**Use:** Used in military ammunition for pistols, shotguns, rifles, machine guns, tank fired munitions, artillery guns, and ship deck guns.

**Note:** Inexperienced personnel should not attempt this procedure, as there are hazards involved. Make sure to thoroughly stir the mixture to avoid lumps or clumps, which can lead to excessive heat build-up. This procedure is perfectly safe for experienced persons. To gain experience, preparers should use only small quantities to begin with.

## 05-02-013A: Standard high performance nitrocellulose "double base" smokeless powder with reduced flash:

Into a suitable sealable vertical mixer, place ***250 grams of nitrocellulose*** (14% nitrogen content), followed by ***150 grams of nitroglycerine,*** followed by 150 milliliters of hexane, followed by ***30 grams of barium nitrate***, followed by ***50 grams of ethyl abietate,*** followed by ***10 grams of potassium nitrate***, and then followed by ***3.75 grams diphenylamine***. Thereafter blend the mixture for about 45 minutes on high speed. After the blending operation, the mixture is ready for extruding; to do so, extrude the mixture through any extruding device into "spaghetti-like" strands ranging from 0.5 to 12mm in diameter, by 300mm in length, and then cut the strands into pieces ranging from 0.5 to 18mm length, and then cure the granules by placing them on a shallow pan or tray, and place said pan or tray into an oven, and cure the granules at 80 Celsius for 12 hours. To use the grains, simply pack the grains loosely into any desirable cartridge casing. Use standard gun casing primers for initiation.
**Chamber Pressure (standard 9mm casing; weight of propellant charge/grain size on average):** 40,109 psi
**Chamber Pressure (standard 7.62mm rifle casing; weight of propellant charge/grain size on average):** 40,799 psi
**Chamber Pressure (standard 105mm artillery casing; weight of propellant charge/grain size on average):** 35,998 psi
**Mortar tube Pressure (standard 81mm mortar propellant charge; weight of propellant charge/grain size on average):** 10,401 psi
**Burn rate (in the open):** 1.7 seconds per inch; tends to vary based on diameter of grain.
**Water resistance:** Fair.
**Stability:** Can be stored for decades.
**Sensitivity:** Stable towards friction; stable towards shock; stable to temperatures up to 100 Celsius.
**Flammability (1 to 10, un-contained):** 9.3 solid grain; 9.6 fine grain.
**Tendency to cake:** Will not cake.
**Detonation ability:** Can be detonated using a blasting cap or detonator.
**Percentage:** ***50.6% nitrocellulose, 30% nitroglycerine, 10% ethyl abietate, 6% barium nitrate, 2% potassium nitrate, 0.76% diphenylamine, 0.64% residue***
**Classification:** Deflagrating explosive (classified as class A explosive).

**Danger of preparation:** Should only be prepared by experienced professionals.
**Use:** Used in military ammunition for pistols, shotguns, rifles, machine guns, tank fired munitions, artillery guns, and ship deck guns.

**Note:** Inexperienced personnel should not attempt this procedure, as there are hazards involved. Make sure to thoroughly stir the mixture to avoid lumps or clumps, which can lead to excessive heat build-up. This procedure is perfectly safe for experienced persons. To gain experience, preparers should use only small quantities to begin with.

**Standard 9mm bullet design**

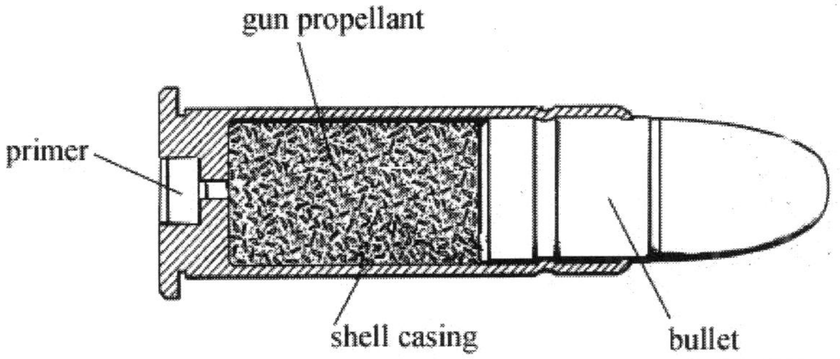

### 05-02-014A: Standard high performance nitrocellulose (high nitroglycerine content) smokeless powder with reduced flash:

Into a suitable vertical mixer, place ***750 grams of nitroglycerine***, followed by 500 milliliters of hexane, followed by ***250 grams of nitrocellulose*** (14% nitrogen content), followed by ***50 grams of lycopodium***, and then followed by ***20 grams of urea***. Thereafter blend the mixture for 2 hours on high speed. After the blending operation, the mixture is ready for extruding; to do so, extrude the mixture through any extruding device into "spaghetti-like" strands ranging from 0.5 to 12mm in diameter, by 300mm in length, and then cut the strands into pieces ranging from 0.5 to 18mm length, and then cure the granules by placing them on a shallow pan or tray, and place said pan or tray into an oven, and cure the granules at 80 Celsius for 12 hours. To use the grains, simply pack the grains loosely into any desirable cartridge casing. Use standard gun casing primers for initiation.
**Chamber Pressure (standard 0.45 caliber casing; weight of propellant charge/grain size on average):** 40,990 psi
**Chamber Pressure (standard 12.7mm rifle casing; weight of propellant charge/grain size on average):** 84,903 psi
**Chamber Pressure (standard 210mm artillery casing; weight of propellant charge/grain size on average):** 36,009 psi
**Burn rate (in the open):** 1.5 seconds per inch; tends to vary based on diameter of grain.
**Water resistance:** Fair.
**Stability:** Can be stored for decades.
**Sensitivity:** Stable towards friction; stable towards shock; stable to temperatures up to 100 Celsius.
**Flammability (1 to 10, un-contained):** 9.6 solid grain; 9.99 fine grain.
**Tendency to cake:** Will not cake.
**Detonation ability:** Can be detonated using a blasting cap or detonator, or when rapidly heated with a strong flame.
**Percentage:** ***70% nitroglycerine, 23% nitrocellulose, 5% lycopodium, 2% urea***
**Classification:** Deflagrating explosive (classified as class A explosive).

**Danger of preparation:** Should only be prepared by experienced professionals.
**Use:** Used in military ammunition for pistols, shotguns, rifles, machine guns, tank fired munitions, artillery guns, and ship deck guns.

**Note:** Inexperienced personnel should not attempt this procedure, as there are hazards involved. Make sure to thoroughly stir the mixture to avoid lumps or clumps, which can lead to excessive heat build-up. This procedure is perfectly safe for experienced persons. To gain experience, preparers should use only small quantities to begin with.

### 05-02-014B: Standard high performance nitrocellulose (high nitroglycerine content) smokeless powder with reduced flash (modified using dinitrobenzene):

Into a suitable vertical mixer, place ***750 grams of nitroglycerine***, followed by 500 milliliters of hexane, followed by ***250 grams of nitrocellulose*** (14% nitrogen content), followed by ***50 grams of lycopodium***, and then followed by ***20 grams of dinitrobenzene***. Thereafter blend the mixture for 2 hours on high speed. After the blending operation, the mixture is ready for extruding; to do so, extrude the mixture through any extruding device into "spaghetti-like" strands ranging from 0.5 to 12mm in diameter, by 300mm in length, and then cut the strands into pieces ranging from 0.5 to 18mm length, and then cure the granules by placing them on a shallow pan or tray, and place said pan or tray into an oven, and cure the granules at 80 Celsius for 12 hours. To use the grains, simply pack the grains loosely into any desirable cartridge casing. Use standard gun casing primers for initiation.
**Chamber Pressure (standard 0.45 caliber casing; weight of propellant charge/grain size on average):** 41,009 psi
**Chamber Pressure (standard 12.7mm rifle casing; weight of propellant charge/grain size on average):** 85,003 psi
**Chamber Pressure (standard 210mm artillery casing; weight of propellant charge/grain size on average):** 36,879 psi
**Burn rate (in the open):** 1.45 seconds per inch; tends to vary based on diameter of grain.
**Water resistance:** Fair.
**Stability:** Can be stored for decades.
**Sensitivity:** Stable towards friction; stable towards shock; stable to temperatures up to 100 Celsius.
**Flammability (1 to 10, un-contained):** 9.6 solid grain; 9.99 fine grain.
**Tendency to cake:** Will not cake.
**Detonation ability:** Can be detonated using a blasting cap or detonator, or when rapidly heated with a strong flame.
**Percentage:** ***70% nitroglycerine, 23% nitrocellulose, 5% lycopodium, 2% dinitrobenzene***
**Classification:** Deflagrating explosive (classified as class A explosive).

**Danger of preparation:** Should only be prepared by experienced professionals.
**Use:** Used in military ammunition for pistols, shotguns, rifles, machine guns, tank fired munitions, artillery guns, and ship deck guns.

**Note:** Inexperienced personnel should not attempt this procedure, as there are hazards involved. Make sure to thoroughly stir the mixture to avoid lumps or clumps, which can lead to excessive heat build-up. This procedure is perfectly safe for experienced persons. To gain experience, preparers should use only small quantities to begin with.

### 05-02-015A: High performance nitrocellulose (high nitroglycerine content) smokeless powder with reduced flash:

Into a suitable vertical mixer, place ***275 grams of nitroglycerine***, followed by 250 milliliters of ethyl acetate, followed by ***150 grams of nitrocellulose*** (14% nitrogen content), followed by ***50 grams of Vaseline***, and then followed by ***25 grams of camphor***. Thereafter blend the mixture for 2 hours on high speed. After the blending operation, the mixture is ready for extruding; to do so, extrude the mixture through any extruding device into "spaghetti-like" strands ranging from 0.5 to 12mm in diameter, by 300mm in length, and then cut the strands into pieces ranging from 0.5 to 18mm length, and then cure the granules by placing them on a shallow pan or tray, and place said pan or tray into an oven, and cure the granules at 90 Celsius for 12 hours. To use the grains, simply pack the grains loosely into any desirable cartridge casing. Use standard gun casing primers for initiation.
**Chamber Pressure (standard 0.45 caliber casing; weight of propellant charge/grain size on average):** 40,014 psi
**Chamber Pressure (standard 12.7mm rifle casing; weight of propellant charge/grain size on average):** 84,332 psi
**Chamber Pressure (standard 210mm artillery casing; weight of propellant charge/grain size on average):** 35,887 psi
**Burn rate (in the open):** 1.9 seconds per inch; tends to vary based on diameter of grain.
**Water resistance:** Fair.
**Stability:** Can be stored for decades.
**Sensitivity:** Stable towards friction; stable towards shock; stable to temperatures up to 100 Celsius.
**Flammability (1 to 10, un-contained):** 9.1 solid grain; 9.4 fine grain.
**Tendency to cake:** Will not cake.
**Detonation ability:** Can be detonated using a blasting cap or detonator, or when rapidly heated with a strong flame.
**Percentage:** ***55% nitroglycerine, 30% nitrocellulose, 10% Vaseline, 5% camphor***
**Classification:** Deflagrating explosive (classified as class A explosive).

**Danger of preparation:** Should only be prepared by experienced professionals.
**Use:** Used in military ammunition for pistols, shotguns, rifles, machine guns, tank fired munitions, artillery guns, and ship deck guns.

**Note:** Inexperienced personnel should not attempt this procedure, as there are hazards involved. Make sure to thoroughly stir the mixture to avoid lumps or clumps, which can lead to excessive heat build-up. This procedure is perfectly safe for experienced persons. To gain experience, preparers should use only small quantities to begin with.

### 05-02-016A: High performance nitrocellulose smokeless powder with reduced flash:

Into a suitable vertical mixer, place ***150 grams of nitroglycerine***, followed by 175 milliliters of ethyl acetate, followed by ***300 grams of nitrocellulose*** (14% nitrogen content), followed by ***25 grams of diamyl phthalate***, and then followed by ***25 grams of mineral jelly***. Thereafter blend the mixture for 2 hours on high speed. After the blending operation, the mixture is ready for extruding; to do so, extrude the mixture through any extruding device into "spaghetti-like" strands ranging from 0.5 to 12mm in diameter, by 300mm in length, and then cut the strands into pieces ranging from 0.5 to 18mm length, and then cure the granules by placing them on a shallow pan or tray, and place said pan or tray into an oven, and cure the granules at 90 Celsius for 12 hours. To use the grains, simply pack the grains loosely into any desirable cartridge casing. Use standard gun casing primers for initiation.
**Chamber Pressure (standard 0.45 caliber casing; weight of propellant charge/grain size on average):** 49,887 psi
**Chamber Pressure (standard 12.7mm rifle casing; weight of propellant charge/grain size on average):** 84,010 psi
**Chamber Pressure (standard 210mm artillery casing; weight of propellant charge/grain size on average):** 35,550 psi
**Burn rate (in the open):** 1.89 seconds per inch; tends to vary based on diameter of grain.
**Water resistance:** Fair.
**Stability:** Can be stored for decades.
**Sensitivity:** Stable towards friction; stable towards shock; stable to temperatures up to 100 Celsius.
**Flammability (1 to 10, un-contained):** 9.1 solid grain; 9.4 fine grain.
**Tendency to cake:** Will not cake.
**Detonation ability:** Can be detonated using a blasting cap or detonator, or when rapidly heated with a strong flame.
**Percentage:** ***60% nitrocellulose, 30% nitroglycerine, 5% diamyl phthalate, 5% mineral jelly***
**Classification:** Deflagrating explosive (classified as class A explosive).

**Danger of preparation:** Should only be prepared by experienced professionals.
**Use:** Used in military ammunition for pistols, shotguns, rifles, machine guns, tank fired munitions, artillery guns, and ship deck guns.

**Note:** Inexperienced personnel should not attempt this procedure, as there are hazards involved. Make sure to thoroughly stir the mixture to avoid lumps or clumps, which can lead to excessive heat build-up. This procedure is perfectly safe for experienced persons. To gain experience, preparers should use only small quantities to begin with.

### 05-02-017A: High performance nitrocellulose smokeless powder:

Into a suitable vertical mixer, place ***200 grams of nitroglycerine***, followed by 150 milliliters of ethyl acetate, followed by ***250 grams of nitrocellulose*** (14% nitrogen content), followed by ***25 grams of Vaseline***, and then followed by ***12.5 grams of olive oil***. Thereafter blend the mixture for 3 hours on high speed. After the blending operation, the mixture is ready for extruding; to do so, extrude the mixture through any extruding device into "spaghetti-like" strands ranging from 0.5 to 12mm in diameter, by 300mm in length, and then cut the strands into pieces ranging from 0.5 to 18mm length, and then cure the granules by placing them on a shallow pan or tray, and place said pan or tray into an oven, and cure the granules at 90 Celsius for 12 hours. To use the grains, simply pack the grains loosely into any desirable cartridge casing. Use standard gun casing primers for initiation.
**Chamber Pressure (standard 0.45 caliber casing; weight of propellant charge/grain size on average):** 49,990 psi
**Chamber Pressure (standard 12.7mm rifle casing; weight of propellant charge/grain size on average):** 84,338 psi
**Chamber Pressure (standard 210mm artillery casing; weight of propellant charge/grain size on average):** 35,994 psi
**Burn rate (in the open):** 1.7 seconds per inch; tends to vary based on diameter of grain.
**Water resistance:** Fair.
**Stability:** Can be stored for decades.
**Sensitivity:** Stable towards friction; stable towards shock; stable to temperatures up to 100 Celsius.
**Flammability (1 to 10, un-contained):** 9.19 solid grain; 9.49 fine grain.
**Tendency to cake:** Will not cake.
**Detonation ability:** Can be detonated using a blasting cap or detonator, or when rapidly heated with a strong flame.
**Percentage:** ***52.6% nitrocellulose, 42.1% nitroglycerine, 2.6% Vaseline, 2.6% olive oil, 0.1% impurities***
**Classification:** Deflagrating explosive (classified as class A explosive).

**Danger of preparation:** Should only be prepared by experienced professionals.
**Use:** Used in military ammunition for pistols, shotguns, rifles, machine guns, tank fired munitions, artillery guns, and ship deck guns.

**Note:** Inexperienced personnel should not attempt this procedure, as there are hazards involved. Make sure to thoroughly stir the mixture to avoid lumps or clumps, which can lead to excessive heat build-up. This procedure is perfectly safe for experienced persons. To gain experience, preparers should use only small quantities to begin with.

### 05-02-017B: High performance nitrocellulose smokeless powder (fortified with barium tartrate burn enhancer):

Into a suitable vertical mixer, place ***200 grams of nitroglycerine***, followed by 200 milliliters of ethyl acetate, followed by ***250 grams of nitrocellulose*** (14% nitrogen content), followed by ***12 grams of Vaseline***, followed by ***12.5 grams of olive oil***, Followed by ***4.5 grams of potassium tartrate***, and then followed by ***20 grams of barium tartrate***. Thereafter blend the mixture for 1 hour on high speed. After the blending operation, the mixture is ready for extruding; to do so, extrude the mixture through any extruding device into "spaghetti-like" strands ranging from 0.5 to 12mm in diameter, by 300mm in length, and then cut the strands into pieces ranging from 0.5 to 18mm length, and then cure the granules by placing them on a shallow pan or tray, and place said pan or tray into an oven, and cure the granules at 90 Celsius for 12 hours. To use the grains, simply pack the grains loosely into any desirable cartridge casing. Use standard gun casing primers for initiation.
**Chamber Pressure (standard 0.45 caliber casing; weight of propellant charge/grain size on average):** 49,559 psi
**Chamber Pressure (standard 12.7mm rifle casing; weight of propellant charge/grain size on average):** 84,299 psi
**Chamber Pressure (standard 210mm artillery casing; weight of propellant charge/grain size on average):** 35,339 psi
**Burn rate (in the open):** 1.76 seconds per inch; tends to vary based on diameter of grain.
**Water resistance:** Fair.
**Stability:** Can be stored for decades.
**Sensitivity:** Stable towards friction; stable towards shock; stable to temperatures up to 100 Celsius.
**Flammability (1 to 10, un-contained):** 9.2 solid grain; 9.4 fine grain.
**Tendency to cake:** Will not cake.
**Detonation ability:** Can be detonated using a blasting cap or detonator, or when rapidly heated with a strong flame.
**Percentage:** ***50% nitrocellulose, 40% nitroglycerine, 4% barium tartrate, 2.5Vaseline, 2.6% olive oil, 0.90% potassium tartrate, 0.1% mixed balance***
**Classification:** Deflagrating explosive (classified as class A explosive).

**Danger of preparation:** Should only be prepared by experienced professionals.
**Use:** Used in military ammunition for pistols, shotguns, rifles, machine guns, tank fired munitions, artillery guns, and ship deck guns.

**Note:** Inexperienced personnel should not attempt this procedure, as there are hazards involved. Make sure to thoroughly stir the mixture to avoid lumps or clumps, which can lead to excessive heat build-up. This procedure is perfectly safe for experienced persons. To gain experience, preparers should use only small quantities to begin with.

**05-02-018A: High performance nitrocellulose smokeless powder containing diethyldiphenylcarbamide burn enhancer:**
Into a suitable vertical mixer, place ***125 grams of nitroglycerine***, followed by 200 milliliters of ethyl acetate, followed by ***375 grams of nitrocellulose*** (14% nitrogen content), and then followed by ***25 grams of diethyldiphenhylcarbamide***. Thereafter blend the mixture for 1 hour on high speed. After the blending operation, the mixture is ready for extruding; to do so, extrude the mixture through any extruding device into "spaghetti-like" strands ranging from 0.5 to 12mm in diameter, by 300mm in length, and then cut the strands into pieces ranging from 0.5 to 18mm length, and then cure the granules by placing them on a shallow pan or tray, and place said pan or tray into an oven, and cure the granules at 90 Celsius for 12 hours. To use the grains, simply pack the grains loosely into any desirable cartridge casing. Use standard gun casing primers for initiation.
**Chamber Pressure (standard 0.45 caliber casing; weight of propellant charge/grain size on average):** 49,339 psi
**Chamber Pressure (standard 12.7mm rifle casing; weight of propellant charge/grain size on average):** 84,100 psi
**Chamber Pressure (standard 210mm artillery casing; weight of propellant charge/grain size on average):** 35,200 psi
**Burn rate (in the open):** 1.8 seconds per inch; tends to vary based on diameter of grain.
**Water resistance:** Fair.
**Stability:** Can be stored for decades.
**Sensitivity:** Stable towards friction; stable towards shock; stable to temperatures up to 100 Celsius.
**Flammability (1 to 10, un-contained):** 9.29 solid grain; 9.49 fine grain.
**Tendency to cake:** Will not cake.
**Detonation ability:** Can be detonated using a blasting cap or detonator, or when rapidly heated with a strong flame.
**Percentage:** ***71.42% nitrocellulose, 23.8% nitroglycerine, 4.76% diethyldiphenhylcarbamide, 0.02% residue***
**Classification:** Deflagrating explosive (classified as class A explosive).

**Danger of preparation:** Should only be prepared by experienced professionals.
**Use:** Used in military ammunition for pistols, shotguns, rifles, machine guns, tank fired munitions, artillery guns, and ship deck guns.

**Note:** Inexperienced personnel should not attempt this procedure, as there are hazards involved. Make sure to thoroughly stir the mixture to avoid lumps or clumps, which can lead to excessive heat build-up. This procedure is perfectly safe for experienced persons. To gain experience, preparers should use only small quantities to begin with.

**05-02-018B: High performance nitrocellulose smokeless powder with potassium lactate stabilizer:**
Into a suitable vertical mixer, place ***115.6 grams of nitroglycerine***, followed by 200 milliliters of ethyl acetate, followed by ***349.4 grams of nitrocellulose*** (14% nitrogen content), followed by ***25 grams of Vaseline***, and then followed by ***10 grams of potassium lactate***. Thereafter blend the mixture for 1 hour on high speed. After the blending operation, the mixture is ready for extruding; to do so, extrude the mixture through any extruding device into "spaghetti-like" strands ranging from 0.5 to 12mm in diameter, by 300mm in length, and then cut the strands into pieces ranging from 0.5 to 18mm length, and then cure the granules by placing them on a shallow pan or tray, and place said pan or tray into an oven, and cure the granules at 90 Celsius for 12 hours. To use the grains, simply pack the grains loosely into any desirable cartridge casing. Use standard gun casing primers for initiation.
**Chamber Pressure (standard 0.45 caliber casing; weight of propellant charge/grain size on average):** 49,760 psi
**Chamber Pressure (standard 12.7mm rifle casing; weight of propellant charge/grain size on average):** 84,551 psi
**Chamber Pressure (standard 210mm artillery casing; weight of propellant charge/grain size on average):** 35,399 psi
**Burn rate (in the open):** 1.6 seconds per inch; tends to vary based on diameter of grain.
**Water resistance:** Fair.
**Stability:** Can be stored for decades.
**Sensitivity:** Stable towards friction; stable towards shock; stable to temperatures up to 100 Celsius.
**Flammability (1 to 10, un-contained):** 9.3 solid grain; 9.5 fine grain.
**Tendency to cake:** Will not cake.
**Detonation ability:** Can be detonated using a blasting cap or detonator, or when rapidly heated with a strong flame.
**Percentage:** ***69.88% nitrocellulose, 23.12% nitroglycerine, 5% Vaseline, 2% potassium lactate***
**Classification:** Deflagrating explosive (classified as class A explosive).

**Danger of preparation:** Should only be prepared by experienced professionals.
**Use:** Used in military ammunition for pistols, shotguns, rifles, machine guns, tank fired munitions, artillery guns, and ship deck guns.

**Note:** Inexperienced personnel should not attempt this procedure, as there are hazards involved. Make sure to thoroughly stir the mixture to avoid lumps or clumps, which can lead to excessive heat build-up. This procedure is perfectly safe for experienced persons. To gain experience, preparers should use only small quantities to begin with.

## 05-02-019A: High performance nitrocellulose smokeless powder (utilizing the "Schultze" process):

Into a suitable sealable vertical mixer, place ***208 grams of nitrocellulose*** (14% nitrogen content), followed by 1500 milliliters of ethyl acetate, followed by ***8 grams of starch***, followed by ***12 grams of paraffin oil***, followed by ***48 grams of barium nitrate***, and then followed by ***12 grams of potassium nitrate***. Thereafter, seal the vertical mixer, and then rapidly blend the mixture on high speed for about 3 hours. After the 3-hour blending operation, open the vertical mixer, and then filter the mixture using a Buchner funnel under mild suction, and then suction off the bulk of the ethyl acetate. Thereafter, the mixture is ready for extruding; to do so, recover the filtered mixture from the Buchner funnel, and then extrude the mixture through any extruding device into "spaghetti-like" strands ranging from 0.5 to 12mm in diameter, by 300mm in length, and then cut the strands into pieces ranging from 0.5 to 18mm length, and then cure the granules by placing them on a shallow pan or tray, and place said pan or tray into an oven, and cure the granules at 90 Celsius for 12 hours. To use the grains, simply pack the grains loosely into any desirable cartridge casing. Use standard gun casing primers for initiation.
**Chamber Pressure (standard 9mm casing; weight of propellant charge/grain size on average):** 39,889 psi
**Chamber Pressure (standard 5.56mm rifle casing; weight of propellant charge/grain size on average):** 36,701 psi
**Chamber Pressure (standard 105mm artillery casing; weight of propellant charge/grain size on average):** 31,902 psi
**Mortar tube Pressure (standard 81mm mortar propellant charge; weight of propellant charge/grain size on average):** 9,778 psi
**Burn rate (in the open):** 2 seconds per inch; tends to vary based on diameter of grain.
**Water resistance:** Fair.
**Stability:** Can be stored for decades.
**Sensitivity:** Stable towards friction; stable towards shock; stable to temperatures up to 100 Celsius.
**Flammability (1 to 10, un-contained):** 9 solid grain; 9.2 fine grain.
**Tendency to cake:** Will not cake.
**Detonation ability:** Can be detonated using a blasting cap or detonator.
**Percentage:** ***72.22% nitrocellulose, 16.66% barium nitrate, 4.16% potassium nitrate, 4.16% paraffin oil, 2.77% starch, 0.03% mixed residues***
**Classification:** Deflagrating explosive (classified as propellant).

**Danger of preparation:** Can be made by amateurs, but caution should be exercised.
**Use:** Used in military ammunition for pistols, shotguns, rifles, machine guns, tank fired munitions, artillery guns, and ship deck guns.

## 05-02-020A: High performance nitrocellulose smokeless powder:

Into a suitable vertical mixer, place ***300 grams of nitrocellulose*** (14% nitrogen content), followed by 175 milliliters of hexane, followed by ***6 grams of dinitrotoluene***, and then followed by ***30 grams of black powder***. Thereafter, blend the mixture on high speed for 2 hours. After the blending operation, the mixture is ready for extruding; to do so, extrude the mixture through any extruding device into "spaghetti-like" strands ranging from 0.5 to 12mm in diameter, by 300mm in length, and then cut the strands into pieces ranging from 0.5 to 18mm length, and then cure the granules by placing them on a shallow pan or tray, and place said pan or tray into an oven, and cure the granules at 80 Celsius for 12 hours. To use the grains, simply pack the grains loosely into any desirable cartridge casing. Use standard gun casing primers for initiation.
**Chamber Pressure (standard 9mm casing; weight of propellant charge/grain size on average):** 35,008 psi
**Chamber Pressure (standard 5.56mm rifle casing; weight of propellant charge/grain size on average):** 35,881 psi
**Chamber Pressure (standard 105mm artillery casing; weight of propellant charge/grain size on average):** 30,990 psi
**Mortar tube Pressure (standard 81mm mortar propellant charge; weight of propellant charge/grain size on average):** 9,558 psi
**Burn rate (in the open):** 2.2 seconds per inch; tends to vary based on diameter of grain.
**Water resistance:** Good.
**Stability:** Can be stored for decades.
**Sensitivity:** Stable towards friction; stable towards shock; stable to temperatures up to 180 Celsius.
**Flammability (1 to 10, un-contained):** 9.1 solid grain; 9.29 fine grain.
**Tendency to cake:** Will not cake.
**Detonation ability:** Can be detonated using a blasting cap or detonator.
**Percentage:** ***89.28% nitrocellulose, 8.92% black powder, 1.78% dinitrotoluene binder, 0.02% mixed balance***
**Classification:** Deflagrating explosive (classified as propellant).

**Danger of preparation:** Can be made by amateurs, but caution should be exercised. Dinitrotoluene is highly toxic so use care.
**Use:** Used in military ammunition for pistols, shotguns, rifles, machine guns, mortars, tank fired munitions, artillery guns, and ship deck guns.

**05-02-021A: High performance nitrocellulose smokeless powder (military grade):**
Into a suitable vertical mixer, place ***185 grams of nitrocellulose*** (14% nitrogen content), followed by 300 millimeters of hexane, followed by ***255 grams of nitroglycerine***, followed by ***300 grams of nitroguanidine***, and then followed by ***8 grams of methyl diphenylurea***. Thereafter, blend the mixture on high speed for about 1 hour. After the blending operation, the mixture is ready for extruding; to do so, extrude the mixture through any extruding device into "spaghetti-like" strands ranging from 0.5 to 12mm in diameter, by 300mm in length, and then cut the strands into pieces ranging from 0.5 to 18mm length, and then cure the granules by placing them on a shallow pan or tray, and place said pan or tray into an oven, and cure the granules at 80 Celsius for 12 hours. To use the grains, simply pack the grains loosely into any desirable cartridge casing. Use standard gun casing primers for initiation.
**Chamber Pressure (standard 9mm casing; weight of propellant charge/grain size on average):** 36,115 psi
**Chamber Pressure (standard 5.56mm rifle casing; weight of propellant charge/grain size on average):** 36,094 psi
**Chamber Pressure (standard 105mm artillery casing; weight of propellant charge/grain size on average):** 31,924 psi
**Mortar tube Pressure (standard 81mm mortar propellant charge; weight of propellant charge/grain size on average):** 10,446 psi
**Burn rate (in the open):** 1.8 seconds per inch; tends to vary based on diameter of grain.
**Water resistance:** Good.
**Stability:** Can be stored for decades.
**Sensitivity:** Stable towards friction; stable towards shock; stable to temperatures up to 100 Celsius.
**Flammability (1 to 10, un-contained):** 9.39 solid grain; 9.4 fine grain.
**Tendency to cake:** Will not cake.
**Detonation ability:** Can be detonated using a blasting cap or detonator.
**Percentage:** ***40.1% nitroguanidine, 34.09% nitroglycerine, 24.73% nitrocellulose, 1.06% methyl diphenylurea, 0.02% mixed residue***
**Classification:** Deflagrating explosive (classified as class A explosive).

**Danger of preparation:** Should only be prepared by experienced professionals.
**Use:** Used in military ammunition for pistols, shotguns, rifles, machine guns, mortars, tank fired munitions, artillery guns, and ship deck guns.

**Note:** Inexperienced personnel should not attempt this procedure, as there are hazards involved. Make sure to thoroughly stir the mixture to avoid lumps or clumps, which can lead to excessive heat build-up. This procedure is perfectly safe for experienced persons. To gain experience, preparers should use only small quantities to begin with.

**05-02-021B: High performance nitrocellulose smokeless powder (military grade):**
Into a suitable vertical mixer, place ***265 grams of nitrocellulose*** (14% nitrogen content), followed by 125 milliliters of hexane, followed by ***180 grams of nitroglycerine***, followed by ***50 grams of nitroguanidine***, followed by ***5 grams of dimethyl diphenylurea***, and then followed by ***5 grams of diethyl dipehnylurea***. Thereafter, rapidly blend the mixture for about 45 minutes on high speed. After the blending operation, the mixture is ready for extruding; to do so, extrude the mixture through any extruding device into "spaghetti-like" strands ranging from 0.5 to 12mm in diameter, by 300mm in length, and then cut the strands into pieces ranging from 0.5 to 18mm length, and then cure the granules by placing them on a shallow pan or tray, and place said pan or tray into an oven, and cure the granules at 80 Celsius for 12 hours. To use the grains, simply pack the grains loosely into any desirable cartridge casing. Use standard gun casing primers for initiation.
**Chamber Pressure (standard 9mm casing; weight of propellant charge/grain size on average):** 37,109 psi
**Chamber Pressure (standard 5.56mm rifle casing; weight of propellant charge/grain size on average):** 36,994 psi
**Chamber Pressure (standard 105mm artillery casing; weight of propellant charge/grain size on average):** 32,910 psi
**Burn rate (in the open):** 1.59 seconds per inch; tends to vary based on diameter of grain.
**Water resistance:** Good.
**Stability:** Can be stored for decades.
**Sensitivity:** Stable towards friction; stable towards shock; stable to temperatures up to 100 Celsius.
**Flammability (1 to 10, un-contained):** 9.49 solid grain; 9.6 fine grain.
**Tendency to cake:** Will not cake.
**Detonation ability:** Can be detonated using a blasting cap or detonator.
**Percentage:** ***52.47% nitrocellulose, 35.64% nitroglycerine, 9.9% nitroguanidine, 1.98% diethyl diphenylamine, 0.01% residues***
**Classification:** Deflagrating explosive (classified as class A explosive).

**Danger of preparation:** Should only be prepared by experienced professionals.
**Use:** Used in military ammunition for pistols, shotguns, rifles, machine guns, tank fired munitions, artillery guns, and ship deck guns.

**Note:** Inexperienced personnel should not attempt this procedure, as there are hazards involved. Make sure to thoroughly stir the mixture to avoid lumps or clumps, which can lead to excessive heat build-up. This procedure is perfectly safe for experienced persons. To gain experience, preparers should use only small quantities to begin with.

### 05-02-022A: High performance nitrocellulose gun propellant (potassium nitrate fortified):

Into a suitable vertical mixer, place ***300 grams of dibutyl phthalate***, followed by ***6 grams of ethyl cellulose***, followed by ***6 grams of diphenylamine***, followed by 1250 milliliters of hexane, followed by ***8 grams of calcium carbonate***, followed by ***1200 grams of potassium nitrate***, followed by ***2000 grams of nitrocellulose*** (14% nitrogen content), and then followed by ***24 grams of dinitrotoluene***. Thereafter, blend the mixture on high speed for about 2 hours. After the blending operation, the mixture is ready for extruding; to do so, extrude the mixture through any extruding device into "spaghetti-like" strands ranging from 0.5 to 12mm in diameter, by 300mm in length, and then cut the strands into pieces ranging from 0.5 to 18mm length, and then cure the granules by placing them on a shallow pan or tray, and place said pan or tray into an oven, and cure the granules at 80 Celsius for 12 hours. To use the grains, simply pack the grains loosely into any desirable cartridge casing. Use standard gun casing primers for initiation.
**Chamber Pressure (standard 9mm casing; weight of propellant charge/grain size on average):** 35,443 psi
**Chamber Pressure (standard 5.56mm rifle casing; weight of propellant charge/grain size on average):** 35,091 psi
**Chamber Pressure (standard 105mm artillery casing; weight of propellant charge/grain size on average):** 31,101 psi
**Mortar tube Pressure (standard 81mm mortar propellant charge; weight of propellant charge/grain size on average):** 9,301 psi
**Burn rate (in the open):** 2 seconds per inch; tends to vary based on diameter of grain.
**Water resistance:** Good.
**Stability:** Can be stored for decades.
**Sensitivity:** Stable towards friction; stable towards shock; stable to temperatures up to 250 Celsius.
**Flammability (1 to 10, un-contained):** 9 solid grain; 9.3 fine grain.
**Tendency to cake:** Will not cake.
**Detonation ability:** Can be detonated using a blasting cap or detonator.
**Percentage:** ***56.43% nitrocellulose, 33.86% potassium nitrate, 8.46% dibutyl phthalate, 0.677% dinitrotoluene, 0.22% calcium carbonate, 0.168% ethyl cellulose, 0.168% diphenylamine, 0.017% mixed balance***
**Classification:** Deflagrating explosive (classified as propellant).

**Danger of preparation:** Can be made by amateurs, but caution should be exercised. Dinitrotoluene is highly toxic so use care.
**Use:** Used in military ammunition for pistols, shotguns, rifles, machine guns, mortars, tank fired munitions, artillery guns, and ship deck guns.

### 05-02-023A: High performance nitrocellulose gun propellant (with TNT modifier):

Into a suitable vertical mixer, place 200 milliliters of hexane, followed by ***356 grams of nitrocellulose*** (14% nitrogen content), followed by ***33 grams of TNT***, followed by ***6 grams of dinitrotoluene***, followed by ***5 grams of diphenylamine***, and then followed by ***100 grams of nitroglycerine***. Thereafter, blend the mixture on high speed for about 3 hours. After the blending operation, the mixture is ready for extruding; to do so, extrude the mixture through any extruding device into "spaghetti-like" strands ranging from 0.5 to 12mm in diameter, by 300mm in length, and then cut the strands into pieces ranging from 0.5 to 18mm length, and then cure the granules by placing them on a shallow pan or tray, and place said pan or tray into an oven, and cure the granules at 80 Celsius for 12 hours. To use the grains, simply pack the grains loosely into any desirable cartridge casing. Use standard gun casing primers for initiation.
**Chamber Pressure (standard 9mm casing; weight of propellant charge/grain size on average):** 36,901 psi
**Chamber Pressure (standard 5.56mm rifle casing; weight of propellant charge/grain size on average):** 36,953 psi
**Chamber Pressure (standard 105mm artillery casing; weight of propellant charge/grain size on average):** 32,590 psi
**Mortar tube Pressure (standard 81mm mortar propellant charge; weight of propellant charge/grain size on average):** 10,036 psi
**Burn rate (in the open):** 1.5 seconds per inch; tends to vary based on diameter of grain.
**Water resistance:** Good.
**Stability:** Can be stored for decades.
**Sensitivity:** Stable towards friction; stable towards shock; stable to temperatures up to 100 Celsius.
**Flammability (1 to 10, un-contained):** 9.4 solid grain; 9.7 fine grain.
**Tendency to cake:** Will not cake.
**Detonation ability:** Can be detonated using a blasting cap or detonator.
**Percentage:** ***71.2% nitrocellulose, 20% nitroglycerine, 6.6% TNT, 1.2% dinitrotoluene***
**Classification:** Deflagrating explosive (classified as class A explosive).

**Danger of preparation:** Should only be prepared by experienced professionals.
**Use:** Used in military ammunition for pistols, shotguns, rifles, machine guns, mortars, tank fired munitions, artillery guns, and ship deck guns.

**Note:** Inexperienced personnel should not attempt this procedure, as there are hazards involved. Make sure to thoroughly stir the mixture to avoid lumps or clumps, which can lead to excessive heat build-up. This procedure is perfectly safe for experienced persons. To gain experience, preparers should use only small quantities to begin with.

### 05-02-024A: High performance nitrocellulose gun propellant:
Into a suitable vertical mixer, place ***550 grams of nitrocellulose*** (14% nitrogen content), followed by 300 milliliters of ethyl acetate, followed by ***350 grams of nitroglycerine***, and then followed by ***35 grams of diamyl phthalate***. Thereafter, blend the mixture on high speed for about 1 hour. After the blending operation, the mixture is ready for extruding; to do so, extrude the mixture through any extruding device into "spaghetti-like" strands ranging from 0.5 to 12mm in diameter, by 300mm in length, and then cut the strands into pieces ranging from 0.5 to 18mm length, and then cure the granules by placing them on a shallow pan or tray, and place said pan or tray into an oven, and cure the granules at 80 Celsius for 12 hours. To use the grains, simply pack the grains loosely into any desirable cartridge casing. Use standard gun casing primers for initiation.
**Chamber Pressure (standard 0.45 caliber casing; weight of propellant charge/grain size on average):** 37,910 psi
**Chamber Pressure (standard 12.7mm rifle casing; weight of propellant charge/grain size on average):** 85,090 psi
**Chamber Pressure (standard 210mm artillery casing; weight of propellant charge/grain size on average):** 33,992 psi
**Burn rate (in the open):** 1.59 seconds per inch; tends to vary based on diameter of grain.
**Water resistance:** Good.
**Stability:** Can be stored for decades.
**Sensitivity:** Stable towards friction; stable towards shock; stable to temperatures up to 100 Celsius.
**Flammability (1 to 10, un-contained):** 9.4 solid grain; 9.7 fine grain.
**Tendency to cake:** Will not cake.
**Detonation ability:** Can be detonated using a blasting cap or detonator.
**Percentage:** ***58.82% nitrocellulose, 37.43% nitroglycerine, 3.74% diamyl phthalate, 0.01% residue***
**Classification:** Deflagrating explosive (classified as class A explosive).

**Danger of preparation:** Should only be prepared by experienced professionals.
**Use:** Used in military ammunition for pistols, shotguns, rifles, machine guns, tank fired munitions, artillery guns, and ship deck guns.

**Note:** Inexperienced personnel should not attempt this procedure, as there are hazards involved. Make sure to thoroughly stir the mixture to avoid lumps or clumps, which can lead to excessive heat build-up. This procedure is perfectly safe for experienced persons. To gain experience, preparers should use only small quantities to begin with.

### 05-02-025A: High performance nitrocellulose gun propellant:
Into a suitable vertical mixer, place ***150 grams of nitrocellulose*** (14% nitrogen content), followed by 90 milliliters of hexane, followed by ***38.5 grams of nitroglycerine***, and then followed by ***3.85 grams of dinitrotoluene***. Thereafter, blend the mixture on high speed for about 50 minutes. After the blending operation, the mixture is ready for extruding; to do so, extrude the mixture through any extruding device into "spaghetti-like" strands ranging from 0.5 to 12mm in diameter, by 300mm in length, and then cut the strands into pieces ranging from 0.5 to 18mm length, and then cure the granules by placing them on a shallow pan or tray, and place said pan or tray into an oven, and cure the granules at 80 Celsius for 12 hours. To use the grains, simply pack the grains loosely into any desirable cartridge casing. Use standard gun casing primers for initiation.
**Chamber Pressure (standard 9mm casing; weight of propellant charge/grain size on average):** 36,889 psi
**Chamber Pressure (standard 5.56mm rifle casing; weight of propellant charge/grain size on average):** 36,501 psi
**Chamber Pressure (standard 105mm artillery casing; weight of propellant charge/grain size on average):** 32,339 psi
**Mortar tube Pressure (standard 81mm mortar propellant charge; weight of propellant charge/grain size on average):** 10,003 psi
**Burn rate (in the open):** 1.59 seconds per inch; tends to vary based on diameter of grain.
**Water resistance:** Good.
**Stability:** Can be stored for decades.
**Sensitivity:** Stable towards friction; stable towards shock; stable to temperatures up to 100 Celsius.
**Flammability (1 to 10, un-contained):** 9.4 solid grain; 9.7 fine grain.
**Tendency to cake:** Will not cake.
**Detonation ability:** Can be detonated using a blasting cap or detonator.
**Percentage:** ***77.98% nitrocellulose, 20.01% nitroglycerine, 2% dinitrotoluene, 0.01% mixed residues***
**Classification:** Deflagrating explosive (classified as class A explosive).

**Danger of preparation:** Should only be prepared by experienced professionals.
**Use:** Used in military ammunition for pistols, shotguns, rifles, machine guns, mortars, tank fired munitions, artillery guns, and ship deck guns.

**Note:** Inexperienced personnel should not attempt this procedure, as there are hazards involved. Make sure to thoroughly stir the mixture to avoid lumps or clumps, which can lead to excessive heat build-up. This procedure is perfectly safe for experienced persons. To gain experience, preparers should use only small quantities to begin with.

**05-02-026A: High performance nitrocellulose gun propellant:**
Into a suitable heated vertical mixer, place 250 milliliters of hexane, followed by ***400 grams of nitrocellulose*** (14% nitrogen content), followed by ***50 grams of finely ground cornstarch***, followed by ***50 grams of dinitrotoluene*** (any suitable isomer), and then followed by ***2.5 grams of diphenylamine***. Thereafter, heat the mixture to about 60 Celsius, and then blend the mixture at this temperature on high speed for about 1 hour. After the blending operation, the mixture is ready for extruding; to do so, extrude the mixture through any extruding device into "spaghetti-like" strands ranging from 0.5 to 12mm in diameter, by 300mm in length, and then cut the strands into pieces ranging from 0.5 to 18mm length, and then cure the granules by placing them on a shallow pan or tray, and place said pan or tray into an oven, and cure the granules at 80 Celsius for 12 hours. After 12 hours, the grains need to be dusted with graphite powder. To use the grains, simply pack the grains loosely into any desirable cartridge casing. Use standard gun casing primers for initiation.
**Chamber Pressure (standard 9mm casing; weight of propellant charge/grain size on average):** 36,005 psi
**Chamber Pressure (standard 5.56mm rifle casing; weight of propellant charge/grain size on average):** 36,031 psi
**Chamber Pressure (standard 105mm artillery casing; weight of propellant charge/grain size on average):** 32,077 psi
**Mortar tube Pressure (standard 81mm mortar propellant charge; weight of propellant charge/grain size on average):** 9,336 psi
**Burn rate (in the open):** 2 seconds per inch; tends to vary based on diameter of grain.
**Water resistance:** Good.
**Stability:** Can be stored for decades.
**Sensitivity:** Stable towards friction; stable towards shock; stable to temperatures up to 180 Celsius.
**Flammability (1 to 10, un-contained):** 9.1 solid grain; 9.3 fine grain.
**Tendency to cake:** Will not cake.
**Detonation ability:** Can be detonated using a blasting cap or detonator.
**Percentage:** ***79.6% nitrocellulose, 9.95% corn starch, 9.95% dinitrotoluene, 0.49% diphenylamine, 0.01% graphite***
**Classification:** Deflagrating explosive (classified as propellant).

**Danger of preparation:** Can be made by amateurs, but caution should be exercised. Dinitrotoluene is highly toxic so use care.
**Use:** Used in military ammunition for pistols, shotguns, rifles, machine guns, mortars, tank fired munitions, artillery guns, and ship deck guns.

**05-02-027A: High performance nitrocellulose gun propellant:**
Into a suitable heated vertical mixer, place 210 milliliters of hexane, followed by ***372.5 grams of nitrocellulose*** (14% nitrogen content), followed by ***25 grams of mononitroxylene*** (any isomer will do), followed by ***100 grams of dinitrobenzene***, and then followed by ***2.5 grams of diphenylamine***. Thereafter, heat the mixture to about 60 Celsius, and then blend the mixture on high speed for about 1 hour. After the blending operation, the mixture is ready for extruding; to do so, extrude the mixture through any extruding device into "spaghetti-like" strands ranging from 0.5 to 12mm in diameter, by 300mm in length, and then cut the strands into pieces ranging from 0.5 to 18mm length, and then cure the granules by placing them on a shallow pan or tray, and place said pan or tray into an oven, and cure the granules at 80 Celsius for 12 hours. To use the grains, simply pack the grains loosely into any desirable cartridge casing. Use standard gun casing primers for initiation.
**Chamber Pressure (standard 9mm casing; weight of propellant charge/grain size on average):** 35,880 psi
**Chamber Pressure (standard 5.56mm rifle casing; weight of propellant charge/grain size on average):** 35,992 psi
**Chamber Pressure (standard 105mm artillery casing; weight of propellant charge/grain size on average):** 31,891 psi
**Mortar tube Pressure (standard 81mm mortar propellant charge; weight of propellant charge/grain size on average):** 9,010 psi
**Burn rate (in the open):** 2 seconds per inch; tends to vary based on diameter of grain.
**Water resistance:** Good.
**Stability:** Can be stored for decades.
**Sensitivity:** Stable towards friction; stable towards shock; stable to temperatures up to 180 Celsius.
**Flammability (1 to 10, un-contained):** 9.1 solid grain; 9.3 fine grain.
**Tendency to cake:** Will not cake.
**Detonation ability:** Can be detonated using a blasting cap or detonator.
**Percentage:** ***74.5% nitrocellulose, 20% dinitrobenzene, 5% mononitroxylene, 0.50% diphenylamine***
**Classification:** Deflagrating explosive (classified as propellant).

**Danger of preparation:** Can be made by amateurs, but caution should be exercised. Dinitrobenzene is highly toxic so use care.
**Use:** Used in military ammunition for pistols, shotguns, rifles, machine guns, mortars, tank fired munitions, artillery guns, and ship deck guns.

## 05-02-028A: High performance "flashless" nitrocellulose gun propellant:

Into a suitable heated vertical mixer, place 200 milliliters of hexane, followed by ***425 grams of nitrocellulose*** (14% nitrogen content), followed by ***50 grams of dinitrotoluene***, followed by ***12.5 grams of triethylene glycol***, and then followed by ***12.5 grams of caprylic acid***. Thereafter, heat the mixture to about 60 Celsius, and then blend the mixture on high speed for about 1 hour. After the blending operation, the mixture is ready for extruding; to do so, extrude the mixture through any extruding device into "spaghetti-like" strands ranging from 0.5 to 12mm in diameter, by 300mm in length, and then cut the strands into pieces ranging from 0.5 to 18mm length, and then cure the granules by placing them on a shallow pan or tray, and place said pan or tray into an oven, and cure the granules at 80 Celsius for 12 hours. To use the grains, simply pack the grains loosely into any desirable cartridge casing. Use standard gun casing primers for initiation.
**Chamber Pressure (standard 9mm casing; weight of propellant charge/grain size on average):** 35,992 psi
**Chamber Pressure (standard 5.56mm rifle casing; weight of propellant charge/grain size on average):** 36,008 psi
**Chamber Pressure (standard 105mm artillery casing; weight of propellant charge/grain size on average):** 31,992 psi
**Mortar tube Pressure (standard 81mm mortar propellant charge; weight of propellant charge/grain size on average):** 9,300 psi
**Burn rate (in the open):** 2 seconds per inch; tends to vary based on diameter of grain.
**Water resistance:** Good.
**Stability:** Can be stored for decades.
**Sensitivity:** Stable towards friction; stable towards shock; stable to temperatures up to 180 Celsius.
**Flammability (1 to 10, un-contained):** 9.1 solid grain; 9.3 fine grain.
**Tendency to cake:** Will not cake.
**Detonation ability:** Can be detonated using a blasting cap or detonator.
**Percentage:** ***85% nitrocellulose, 10% dinitrotoluene, 2.5% triethylene glycol, 2.5% caprylic acid***
**Classification:** Deflagrating explosive (classified as propellant).

**Danger of preparation:** Can be made by amateurs, but caution should be exercised. Dinitrotoluene is highly toxic so use care.
**Use:** Used in military ammunition for pistols, shotguns, rifles, machine guns, mortars, tank fired munitions, artillery guns, and ship deck guns.

## 05-02-028B: High performance "flashless" nitrocellulose gun propellant:

Into a suitable heated vertical mixer, place 200 milliliters of hexane, followed by ***400 grams of nitrocellulose*** (14% nitrogen content), followed by ***75 grams of dinitrotoluene***, and then followed by ***25 grams of propylene glycol caprylate.*** Thereafter, heat the mixture to 60 Celsius, and then blend the mixture on high speed for about 1 hour. After the blending operation, the mixture is ready for extruding; to do so, extrude the mixture through any extruding device into "spaghetti-like" strands ranging from 0.5 to 12mm in diameter, by 300mm in length, and then cut the strands into pieces ranging from 0.5 to 18mm length, and then cure the granules by placing them on a shallow pan or tray, and place said pan or tray into an oven, and cure the granules at 80 Celsius for 12 hours. To use the grains, simply pack the grains loosely into any desirable cartridge casing. Use standard gun casing primers for initiation.
**Chamber Pressure (standard 9mm casing; weight of propellant charge/grain size on average):** 35,887 psi
**Chamber Pressure (standard 5.56mm rifle casing; weight of propellant charge/grain size on average):** 35,992 psi
**Chamber Pressure (standard 105mm artillery casing; weight of propellant charge/grain size on average):** 31,599 psi
**Mortar tube Pressure (standard 81mm mortar propellant charge; weight of propellant charge/grain size on average):** 9,109 psi
**Burn rate (in the open):** 2.1 seconds per inch; tends to vary based on diameter of grain.
**Water resistance:** Good.
**Stability:** Can be stored for decades.
**Sensitivity:** Stable towards friction; stable towards shock; stable to temperatures up to 180 Celsius.
**Flammability (1 to 10, un-contained):** 9.09 solid grain; 9.25 fine grain.
**Tendency to cake:** Will not cake.
**Detonation ability:** Can be detonated using a blasting cap or detonator.
**Percentage:** ***80% nitrocellulose, 15% dinitrotoluene, 5% triethylene glycol caprylate***
**Classification:** Deflagrating explosive (classified as propellant).

**Danger of preparation:** Can be made by amateurs, but caution should be exercised. Dinitrotoluene is highly toxic so use care.
**Use:** Used in military ammunition for pistols, shotguns, rifles, machine guns, mortars, tank fired munitions, artillery guns, and ship deck guns.

**05-02-029A: High performance "clean burning" nitrocellulose smokeless gun propellant:**
Into a suitable heated vertical mixer, place 200 milliliters of hexane, and then add in ***433.5 grams of nitrocellulose*** (14% nitrogen content), followed by ***28.9 grams of standard animal glue***, followed by ***31.5 grams of butyl stearate***, followed by ***5 grams of ethyl centralite***, followed by ***5 grams of basic lead carbonate,*** and then followed by ***5 grams of potassium sulfate***. Thereafter, blend the mixture at 60 Celsius under rapid speed for 45 minutes. After the blending operation, the mixture is ready for extruding; to do so, extrude the mixture through any extruding device into "spaghetti-like" strands ranging from 0.5 to 12mm in diameter, by 300mm in length, and then cut the strands into pieces ranging from 0.5 to 18mm length, and then cure the granules by placing them on a shallow pan or tray, and place said pan or tray into an oven, and cure the granules at 80 Celsius for 12 hours. To use the grains, simply pack the grains loosely into any desirable cartridge casing. Use standard gun casing primers for initiation.
**Chamber Pressure (standard 9mm casing; weight of propellant charge/grain size on average):** 35,993 psi
**Chamber Pressure (standard 5.56mm rifle casing; weight of propellant charge/grain size on average):** 36,090 psi
**Chamber Pressure (standard 105mm artillery casing; weight of propellant charge/grain size on average):** 31,947 psi
**Mortar tube Pressure (standard 81mm mortar propellant charge; weight of propellant charge/grain size on average):** 9,341 psi
**Burn rate (in the open):** 2 seconds per inch; tends to vary based on diameter of grain.
**Water resistance:** Good.
**Stability:** Can be stored for decades.
**Sensitivity:** Stable towards friction; stable towards shock; stable to temperatures up to 180 Celsius.
**Flammability (1 to 10, un-contained):** 9 solid grain; 9.2 fine grain.
**Tendency to cake:** Will not cake.
**Detonation ability:** Can be detonated using a blasting cap or detonator.
**Percentage: *85% nitrocellulose, 6.18% butyl stearate, 5.67% animal glue, 0.98% ethyl centralite, 0.98% basic lead carbonate, 0.98% potassium sulfate, 0.21% impurities***
**Classification:** Deflagrating explosive (classified as propellant).

**Danger of preparation:** Can be made by amateurs, but caution should be exercised.
**Use:** Used in military ammunition for pistols, shotguns, rifles, machine guns, mortars, tank fired munitions, artillery guns, and ship deck guns.

# Section 3: Miscellaneous gun propellants

## *Chemicals used in this section (binders are not included)*

| | | |
|---|---|---|
| **1. Potassium nitrate (see Black Powder)** | | **2. Sulfur (see Black Powder)** |
| **3. Charcoal (see Black Powder)** | | **4. Sugar Carbon (see Modified Black Powder)** |
| **5. Barium Chromate (see Modified Black Powder)** | | **6. Potassium Perchlorate (see Modified Black Powder)** |
| **7. Potassium Dichromate (see Modified Black Powder)** | | **8. Ammonium Bisulfide (see Modified Black Powder)** |
| **9. Potassium chlorate (see Modified Black Powder)** | | **10. Carbon Disulfide (see Modified Black Powder)** |
| **11. Nitrocellulose (see Modified Black Powder)** | | **12. Lead Tetraoxide (see Modified Black Powder)** |
| **13. Diphenylamine (see Modified Black Powder)** | | **14. Titanium Dioxide (see Modified Black Powder)** |
| **15. Manganese Dioxide (see Modified Black Powder)** | | **16. Sodium Benzoate (see Modified Black Powder)** |
| **17. Ammonium Nitrate (see Modified Black Powder)** | | **18. Calcium Carbonate (see Modified Black Powder)** |
| **19. Sodium Nitrate (see Modified Black Powder)** | | **20. Urea (see Modified Black Powder)** |
| **21. Lead Nitrate (see Modified Black Powder)** | | **22. Nitro Starch (see Modified Black Powder)** |
| **23. Ammonium Perchlorate (see Ammonium Perchlorate Rocket Propellants)** | | **24. Aluminum powder (see Ammonium Perchlorate Rocket Propellants)** |
| **25. Magnesium Sulfide (see Ammonium Perchlorate Rocket Propellants)** | | **26. Copper Chromite (see Ammonium Perchlorate Rocket Propellants)** |
| **27. Nitroguanidine (see Ammonium Perchlorate Rocket Propellants)** | | **28. Ammonium Sulfate (see Ammonium Perchlorate Rocket Propellants)** |
| **29. Nitroglycerine (see Ammonium Perchlorate Rocket Propellants)** | | **30. Magnesium Oxide (see Ammonium Perchlorate Rocket Propellants)** |
| **31. Iron-II-oxide (see Ammonium Perchlorate Rocket Propellants)** | | **32. Zirconium Hydride (see Ammonium Perchlorate Rocket Propellants)** |
| **33. Teflon (see Ammonium Perchlorate Rocket Propellants)** | | **34. Zinc Oxide (see Ammonium Perchlorate Rocket Propellants)** |
| **35. Ammonium Dichromate (see Ammonium Perchlorate Rocket Propellants)** | | **36. Lithium Perchlorate (see Ammonium Perchlorate Rocket Propellants)** |

| | | |
|---|---|---|
| **37. Magnesium powder (see Ammonium Perchlorate Rocket Propellants)** | | **38. Lithium Aluminum Hydride (see Ammonium Perchlorate Rocket Propellants)** |
| **39. Iron-III-oxide (red iron oxide) (see Ammonium Perchlorate Rocket Propellants)** | | **40. PVC (see Ammonium Perchlorate Rocket Propellants)** |
| **41. Copper Sulfide (see Ammonium Perchlorate Rocket Propellants)** | | **42. Sodium Hydride (see Ammonium Perchlorate Rocket Propellants)** |
| **43. Titanium Hydride (see Ammonium Perchlorate Rocket Propellants)** | | **44. Silicon Nitride (see Ammonium Perchlorate Rocket Propellants)** |
| **45. ADN (see ADN Rocket Propellants)** | | **46. Urea Nitrate (see ADN Rocket Propellants)** |
| **47. Silicon Powder (see ADN Rocket Propellants)** | | **48. Hexamine (see ADN Rocket Propellants)** |
| **49. Boron powder (see ADN Rocket Propellants)** | | **50. Sodium hypophosphite (see ADN Rocket Propellants)** |
| **51. KDN (see ADN Rocket Propellants)** | | **52. Nickel Chloride (see Ammonium Nitrate Rocket Propellants)** |
| **53. TNT (see Ammonium Nitrate Rocket Propellants)** | | **54. Acrylamide (see Miscellaneous Rocket H.P. Rocket Propellants)** |
| **55. Ethylene Glycol (see Miscellaneous Rocket H.P. Rocket Propellants)** | | **56. Magnesium Perchlorate (see Miscellaneous Rocket H.P. Rocket Propellants)** |
| **57. Metallic Lithium (see Miscellaneous Rocket H.P. Rocket Propellants)** | | **58. Beryllium Hydride (see Miscellaneous Rocket H.P. Rocket Propellants)** |
| **59. Red Phosphorus (see Miscellaneous Rocket H.P. Rocket Propellants)** | | **60. Sodium Borohydride (see Miscellaneous Rocket H.P. Rocket Propellants)** |
| **61. Sodium chlorate (see Miscellaneous Rocket H.P. Rocket Propellants)** | | **62. Lead Dioxide (see Miscellaneous Rocket H.P. Rocket Propellants)** |
| **63. Benzoyl Peroxide (see Miscellaneous Rocket H.P. Rocket Propellants)** | | **64. Barium Nitrate (see Miscellaneous Rocket H.P. Rocket Propellants)** |
| **65. Cyanuric Acid (see Miscellaneous Rocket H.P. Rocket Propellants)** | | **66. Aluminum Stearate (see Ammonium Nitrate Gun Propellants)** |
| **67. Aluminum Hydride (see Ammonium Nitrate Gun Propellants)** | | **68. Magnesium Peroxide (see Ammonium Nitrate Gun Propellants)** |
| **69. Potassium Permanganate (see Ammonium Nitrate Gun Propellants)** | | **70. Calcium Hydride (see Ammonium Nitrate Gun Propellants)** |
| **71. Potassium Tartrate (see Ammonium Nitrate Gun Propellants)** | | **72. Potassium Ferrocyanide (see Ammonium Nitrate Gun Propellants)** |
| **73. Sodium Azide (see Ammonium Nitrate Gun Propellants)** | | **74. Sodium Chloride (see Ammonium Nitrate Gun Propellants)** |
| **75. Potassium Sulfate (see Nitrocellulose Gun Propellants)** | | **76. Lead Stearate (see Nitrocellulose Gun Propellants)** |
| **77. Triethylene Glycol (see Nitrocellulose Gun Propellants)** | | **78. Sugar (sucrose)** |
| | | 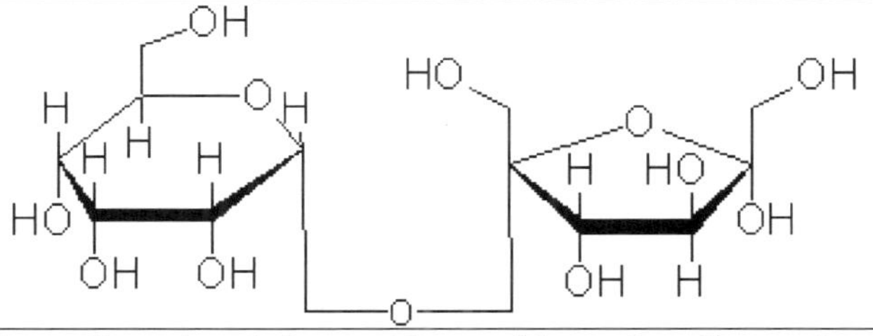 |
| | | Sucrose forms nice colorless crystals, or a white powder, with a sweet taste. Sucrose decomposes when heated to 180 Celsius, but carmelization begins at lower temperatures. The crystals are very soluble in water, but relatively insoluble in most solvents. |
| | | |

**79. Sodium Propionate**

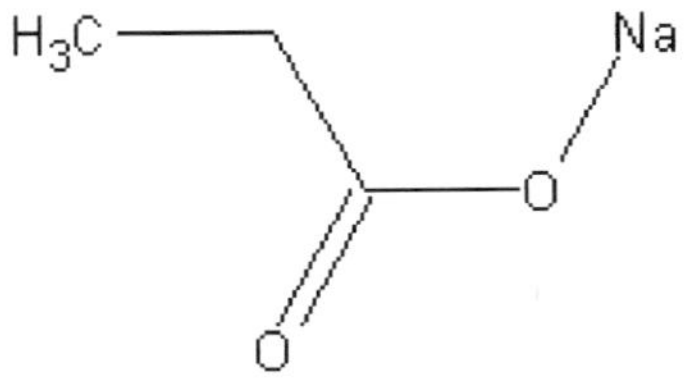

Forms transparent crystals, or granules. The propionate is soluble in water, but relatively insoluble in alcohol and other solvents.

**80. Picric Acid**

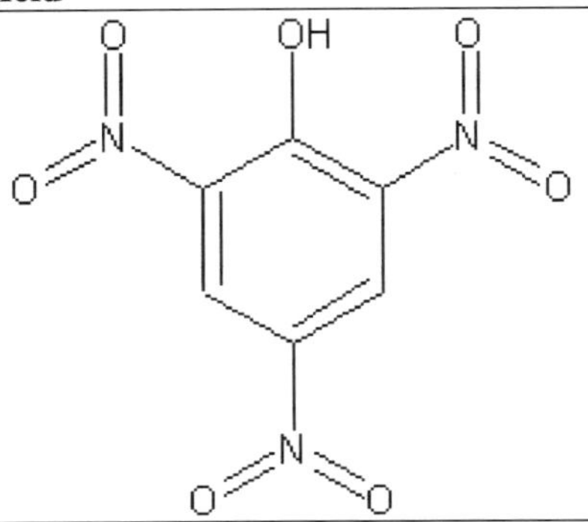

Forms yellowish, to pale-yellowish crystals, granules, or powder. The crystals have a melting point of 123 Celsius, and explode when heated to 300 Celsius. Picric acid is insoluble in the usual organic solvents, and water. The crystals are toxic, so users should wear proper gloves when handling. See The Preparatory Manual of Explosive Fourth Edition for complete preparation.

**81. Copper-II-oxide**

Forms a fine black to brownish-black powder. The powder is insoluble in water and all known solvents. It slowly dissolves in ammonia, and acids, but is soluble in cyanide solutions and ammonium carbonate solution.

**82. Ammonium Picrate**

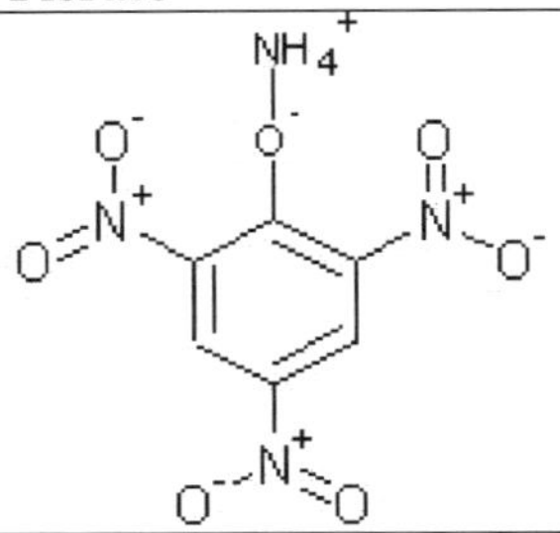

Ammonium picrate forms bright yellow scales, or orthorhombic crystals, which are insoluble in water. The crystals explode easily from heat, shock, friction, fire, and percussion. Ammonium picrate is commonly used in priming compositions for bullets and the like, initiation compositions for blasting caps and detonators, high performance rocket propellants, and fireworks and other pyrotechnic compositions. Ammonium picrate should be stored submerged in kerosene.

**Method of Preparation 1:** Bubble 3 grams of ammonia gas into 400 milliliters of benzene. Then, stir the mixture while slowly adding and dissolving 38 grams of finely divided (powdered) picric acid. Keep the reaction mixture at room temperature (a cold water bath may or may not be needed). While slowly adding the picric acid, bubble another 3 grams ammonia gas into the reaction mixture. After the addition of the ammonia gas and the picric acid, stir the solution for 2 hours at room temperature. The ammonium picrate will begin to separate shortly after the first couple of additions of the picric acid. After 2 hours, filter-off the ammonium picrate, wash with 500 milliliters of water, and then vacuum dry or air-dry the product. Do not use heat to dry the ammonium picrate.

*- High Performance Miscellaneous Gun Propellants in this section -*

| | |
|---|---|
| **1. 05-03-001A: High "heaving" action gun propellant suitable for artillery guns:** *38% potassium chlorate, 33.3% sugar, 28.6% glycerin, 0.10% residue* | **2. 05-03-002B: Low "chamber pressure" gun propellant suitable for grenade launchers:** *52.1% potassium perchlorate, 34.7% sodium propionate, 8.6% sulfur, 4.3% dextrin, 0.30% residue* |
| **3. 05-03-003A: High performance chlorate gun propellant suitable for firearms:** *59.9% potassium chlorate, 39% sugar, 1.1% residue* | **4. 05-03-004A: "Super" high performance gun propellant suitable for propelling projectiles to high altitudes:** *62.4% titanium hydride, 37.5% ammonium azide, 0.10% residue* |
| **5. 05-03-004B: "Super" high performance gun propellant suitable for propelling projectiles to high altitudes (modified, stabilized):** *61.9% titanium hydride, 37.2% ammonium azide, 0.79% diatomaceous earth, 0.11% impurities* | **6. 05-03-005A: Potassium perchlorate high performance gun propellant:** *76% potassium perchlorate, 14.5% epoxy resin, 9.5% aluminum* |
| **7. 05-03-006A: Nitro coated high performance gun propellant containing picric acid:** *43.1% barium nitrate, 34.4% picric acid, 7.1% potassium ferrocyanide, 7.1% wheat flour, 7.1% potassium chlorate, 1% lamp black, 0.20% nitro compounds* | **8. 05-03-007A: High performance sodium azide containing specialty gun propellant:** *66.8% sodium azide, 18.2% iron oxide, 15% copper oxide* |
| **9. 05-03-008A: High performance nitrocellulose gun propellant (triple based):** *60.77% nitroguanidine, 22.22% nitrocellulose, 10% nitroglycerine, 5% dibutyl phthalate, 1.66% 2-nitrodiphenylamine, 0.33% cryolite, 0.02% impurities* | **10. 05-03-008B: High performance nitrocellulose gun propellant (triple based):** *54.86% nitroguanidine, 20.06% nitrocellulose, 19.05% nitroglycerine, 4.51% dibutylphthalate, 1.5% nitrogiphenylamine, 0.02% mixed balance* |
| **11. 05-03-009A: High performance ammonium perchlorate gun propellant:** *46.87% ammonium perchlorate, 31.25% nitrocellulose, 17.96% wheat flour, 3.9% flours of sulfur, 0.02% impurities* | **12. 05-03-009B: High performance perchlorate based gun propellant:** *51.36% potassium perchlorate, 27.39% nitrocellulose, 13.01% wheat flour, 6.16% sulfur, 2.05% sodium azide, 0.03% impurities* |
| **13. 05-03-009C: High performance chlorate based gun propellant:** *66.66% potassium chlorate, 33.33% starch, 0.01% mixed balance* | **14. 05-03-010A: High performance nitroguanidine gun propellant:** *59% nitroguanidine, 21.57% nitrocellulose, 9.7% nitroglycerine, 4.85% dibutylphthalate, 3.23% cryolite, 1.61% 2-nitrodiphenylamine, 0.04% mixed balance* |
| **15. 05-03-010B: High performance nitroguanidine gun propellant:** *54.86% nitroguanidine, 20% nitrocellulose, 19% nitroglycerine, 4.4% dibutylphthalate, 1.5% 2-nitrodiphenylamine, 0.24% mixed residual balance* | **16. 05-03-011A: High performance barium nitrate gun propellant for small arms weapons:** *51% barium nitrate, 17% lead nitrate, 15% nitrocellulose, 9.5% soft wood charcoal, 7.5% sulfur* |
| **17. 05-03-011B: High performance barium nitrate gun propellant for small arms weapons:** *49.8% barium nitrate, 17.24% nitrocellulose, 15.32% potassium nitrate, 10.53% graphite, 7.08% sodium hypophosphite, 0.03% mixed balance* | **18. 05-03-012A: Specialty high performance ammonium picrate gun propellant:** *47% ammonium picrate, 30% barium nitrate, 23% potassium dichromate* |
| **19. 05-03-013A: High performance ammonium picrate gun propellant:** *52% potassium nitrate, 25% ammonium picrate, 10% sulfur, 10% corn starch, 3% soft wood charcoal* | **20. 05-03-014A: High performance barium nitrate gun propellant:** *43.95% barium nitrate, 43.95% pyroxyline, 6.59% nitro-naphthalene, 5.49% potassium nitrate, 0.02% mixed balance* |
| **21. 05-03-014B: High performance barium nitrate gun propellant (sulfur induced):** *61.22% pyroxyline, 24.48% barium nitrate, 10.2% potassium nitrate, 2.04% nitro-naphthalene, 2.04% sulfur, 0.02% mixed balance* | **22. 05-03-015A: High performance "smokeless powder" ammonium picrate gun propellant:** *68% ammonium picrate, 25% potassium dichromate, 7% potassium permanganate* |

**05-03-001A: High "heaving" action gun propellant suitable for artillery guns:**
Into a suitable vertical mixer, place 180 milliliters of hexane, followed by ***170 grams of glycerin***, and then followed by ***198 grams of powdered sugar***. Thereafter, blend the mixture on high speed for about 15 minutes. Thereafter, add in ***226 gram of potassium chlorate***, and then continue to blend the mixture on high speed for about 45 minutes. After the blending operation, the mixture is ready for extruding; to do so, extrude the mixture through any extruding device into "spaghetti-like" strands ranging from 5 to 12mm in diameter, by 300mm in length, and then cut the strands into pieces ranging from 5 to 18mm length, and then cure the granules by placing them on a shallow pan or tray, and place said pan or tray into an oven, and cure the granules at 80 Celsius for 12 hours. To use the grains, simply pack the grains loosely into any desirable cartridge casing. Use standard gun casing primers for initiation.
**Chamber Pressure (standard 105mm artillery casing; weight of propellant charge/grain size on average):** 29,025 psi
**Burn rate (in the open):** 4.1 seconds per inch; tends to vary based on diameter of grain.
**Water resistance:** Good.

**Stability:** Can be stored for decades.
**Sensitivity:** Stable towards friction; stable towards shock; stable to temperatures up to 180 Celsius.
**Flammability (1 to 10, un-contained):** 8 solid grain; 8.2 fine grain.
**Tendency to cake:** Will not cake.
**Detonation ability:** Cannot be detonated.
**Percentage:** ***38% potassium chlorate, 33.3% sugar, 28.6% glycerin, 0.10% residue***
**Classification:** Deflagrating explosive (classified as propellant).

**Danger of preparation:** Can be made by amateurs, but caution should be exercised.
**Use:** Used in military ammunition for artillery guns.

### 05-03-002B: Low "chamber pressure" gun propellant suitable for grenade launchers:

Into a suitable vertical mixer, place ***300 gram of potassium perchlorate***, followed by ***200 grams of sodium propionate***, followed by ***50 grams of flours of sulfur***, followed by ***25 grams of dextrin***, followed by 250 milliliters of hexane, and then blend the mixture on high speed for about 45 minutes. After the blending operation, the mixture is ready for extruding; to do so, extrude the mixture through any extruding device into "spaghetti-like" strands ranging from 0.5 to 4mm in diameter, by 300mm in length, and then cut the strands into pieces ranging from 0.5 to 3mm length, and then cure the granules by placing them on a shallow pan or tray, and place said pan or tray into an oven, and cure the granules at 80 Celsius for 12 hours. To use the grains, simply pack the grains loosely into any desirable cartridge casing. Use standard gun casing primers for initiation.
**Chamber Pressure (standard 40mm grenade launcher casing; weight of propellant charge/grain size on average):** 8,771 psi
**Burn rate (in the open):** 4.6 seconds per inch; tends to vary based on diameter of grain.
**Water resistance:** Good.
**Stability:** Can be stored for decades.
**Sensitivity:** Stable towards friction; stable towards shock; stable to temperatures up to 300 Celsius.
**Flammability (1 to 10, un-contained):** 7.8 solid grain; 8 fine grain.
**Tendency to cake:** Will not cake.
**Detonation ability:** Cannot be detonated.
**Percentage:** ***52.1% potassium perchlorate, 34.7% sodium propionate, 8.6% sulfur, 4.3% dextrin, 0.30% residue***
**Classification:** Deflagrating explosive (classified as propellant).

**Danger of preparation:** Can be made by amateurs, but caution should be exercised.
**Use:** Used in military ammunition for 30mm to 60mm low pressure/low velocity launching systems.

### 05-03-003A: High performance chlorate gun propellant suitable for firearms:

Into a standard vertical mixer, place ***100 grams of powdered sugar,*** followed by 120 milliliters of hexane, and then followed by ***150 grams of potassium chlorate***. Thereafter, rapidly blend the mixture on high speed for about 45 minutes. After the blending operation, the mixture is ready for extruding; to do so, extrude the mixture through any extruding device into "spaghetti-like" strands ranging from 0.5 to 5mm in diameter, by 300mm in length, and then cut the strands into pieces ranging from 0.5 to 4mm length, and then cure the granules by placing them on a shallow pan or tray, and place said pan or tray into an oven, and cure the granules at 80 Celsius for 12 hours. To use the grains, simply pack the grains loosely into any desirable cartridge casing. Use standard gun casing primers for initiation.
**Chamber Pressure (standard 9mm pistol casing; weight of propellant charge/grain size on average):** 33,982 psi
**Chamber Pressure (standard 5.56mm rifle casing; weight of propellant charge/grain size on average):** 34,003 psi
**Chamber Pressure (standard 12 gauge shot shell casing; weight of propellant charge/grain size on average):** 12,441 psi
**Burn rate (in the open):** 2.9 seconds per inch; tends to vary based on diameter of grain.
**Water resistance:** Good.
**Stability:** Can be stored for decades.
**Sensitivity:** Stable towards friction; stable towards shock; stable to temperatures up to 300 Celsius.
**Flammability (1 to 10, un-contained):** 8.7 solid grain; 9 fine grain.
**Tendency to cake:** Will not cake.
**Detonation ability:** Cannot be detonated.
**Percentage:** ***59.9% potassium chlorate, 39% sugar, 1.1% residue***
**Classification:** Deflagrating explosive (classified as propellant).

**Danger of preparation:** Can be made by amateurs, but caution should be exercised.
**Use:** Used in military ammunition for pistols, shotguns, and rifles.

### 05-03-004A: "Super" high performance gun propellant suitable for propelling projectiles to high altitudes:

Into a suitable vertical mixer, place 280 milliliters of hexane, followed by ***312 grams of titanium hydride***, and then followed by ***187.5 grams of ammonium azide***. Thereafter, blend the mixture on high speed for about 45 minutes. After the blending operation, the mixture is ready for extruding; to do so, extrude the mixture through any extruding device into "spaghetti-like" strands ranging from 0.5 to 12mm in diameter, by 300mm in length, and then cut the strands into pieces ranging from 0.5 to 18mm length, and then cure the granules by placing them on a shallow pan or tray, and place said pan or tray into an oven, and cure the granules at 80 Celsius for 12 hours. To use the grains, simply pack the grains loosely into any desirable cartridge casing. Use standard gun casing primers for initiation. **Note: this propellant is not designed for small arms.**
**Chamber Pressure (standard 12.7mm rifle casing; weight of propellant charge/grain size on average):** 90,057 psi
**Chamber Pressure (standard 12.7mm to 18mm anti-aircraft gun casings; weight of propellant charge/grain size on average):** 90,118 psi
**Chamber Pressure (standard 25mm to 90mm anti-aircraft gun casings; weight of propellant charge/grain size on average):** 95,996 psi
**Burn rate (in the open):** 1.2 seconds per inch; tends to vary based on diameter of grain.
**Water resistance:** Poor.
**Stability:** Can be stored for decades when kept sealed from air and moisture.
**Sensitivity:** Stable towards friction; stable towards shock; stable to temperatures up to 120 Celsius.
**Flammability (1 to 10, un-contained):** 9.5 solid grain; 9.9 fine grain.
**Tendency to cake:** Will not cake.
**Detonation ability:** Can be detonated with a blasting cap or detonator; may explode when subjected to rapid strong heating.
**Percentage:** ***62.4% titanium hydride, 37.5% ammonium azide, 0.10% residue***
**Classification:** Deflagrating explosive (classified as class A explosive).

**Danger of preparation:** Should only be prepared by experienced professionals.
**Use:** Used in military ammunition for anti-aircraft systems.

### 05-03-004B: "Super" high performance gun propellant suitable for propelling projectiles to high altitudes (modified, stabilized):

Into a suitable vertical mixer, place 280 milliliters of hexane, followed by ***312 grams of titanium hydride***, followed by ***4 grams of diatomaceous earth*** and then followed by ***187.5 grams of ammonium azide***. Thereafter, blend the mixture on high speed for about 45 minutes. After the blending operation, the mixture is ready for extruding; to do so, extrude the mixture through any extruding device into "spaghetti-like" strands ranging from 0.5 to 12mm in diameter, by 300mm in length, and then cut the strands into pieces ranging from 0.5 to 18mm length, and then cure the granules by placing them on a shallow pan or tray, and place said pan or tray into an oven, and cure the granules at 80 Celsius for 12 hours. To use the grains, simply pack the grains loosely into any desirable cartridge casing. Use standard gun casing primers for initiation. **Note: this propellant is not designed for small arms.**
**Chamber Pressure (standard 12.7mm rifle casing; weight of propellant charge/grain size on average):** 90,000 psi
**Chamber Pressure (standard 12.7mm to 18mm anti-aircraft gun casings; weight of propellant charge/grain size on average):** 90,005 psi
**Chamber Pressure (standard 25mm to 90mm anti-aircraft gun casings; weight of propellant charge/grain size on average):** 95,550 psi
**Burn rate (in the open):** 1.39 seconds per inch; tends to vary based on diameter of grain.
**Water resistance:** Poor.
**Stability:** Can be stored for decades when kept sealed from air and moisture.
**Sensitivity:** Stable towards friction; stable towards shock; stable to temperatures up to 120 Celsius.
**Flammability (1 to 10, un-contained):** 9.4 solid grain; 9.8 fine grain.
**Tendency to cake:** Will not cake.
**Detonation ability:** Can be detonated with a blasting cap or detonator; may explode when subjected to rapid strong heating.
**Percentage:** ***61.9% titanium hydride, 37.2% ammonium azide, 0.79% diatomaceous earth, 0.11% impurities***
**Classification:** Deflagrating explosive (classified as class A explosive).

**Danger of preparation:** Should only be prepared by experienced professionals.
**Use:** Used in military ammunition for anti-aircraft systems.

## 05-03-005A: Potassium perchlorate high performance gun propellant:

Into a suitable sealable vertical mixer, place ***380 grams of potassium perchlorate,*** followed by ***47.5 grams of finely powdered aluminum,*** and then followed by ***72.5 grams of Armstrong epoxy resin E-301-14***. Thereafter, seal the vertical mixer, and then blend the mixture on high speed for about 45 minutes. After the blending operation, the mixture is ready for extruding; to do so, extrude the mixture through any extruding device into "spaghetti-like" strands ranging from 0.5 to 12mm in diameter, by 300mm in length, and then cut the strands into pieces ranging from 0.5 to 18mm length, and then cure the granules by placing them on a shallow pan or tray, and place said pan or tray into an oven, and cure the granules at 80 Celsius for 12 hours. To use the grains, simply pack the grains loosely into any desirable cartridge casing. Use standard gun casing primers for initiation.

**Chamber Pressure (standard 9mm casing; weight of propellant charge/grain size on average):** 34,086 psi
**Chamber Pressure (standard 5.56mm rifle casing; weight of propellant charge/grain size on average):** 35,009 psi
**Chamber Pressure (standard 105mm artillery casing; weight of propellant charge/grain size on average):** 30,026 psi
**Mortar tube Pressure (standard 81mm mortar propellant charge; weight of propellant charge/grain size on average):** 9,221 psi
**Burn rate (in the open):** 2 seconds per inch; tends to vary based on diameter of grain.
**Water resistance:** Good.
**Stability:** Can be stored for decades.
**Sensitivity:** Stable towards friction; stable towards shock; stable to temperatures up to 180 Celsius.
**Flammability (1 to 10, un-contained):** 9 solid grain; 9.2 fine grain.
**Tendency to cake:** Will not cake.
**Detonation ability:** Can be detonated using a blasting cap or detonator.
**Percentage:** ***76% potassium perchlorate, 14.5% epoxy resin, 9.5% aluminum***
**Classification:** Deflagrating explosive (classified as propellant).

**Danger of preparation:** Can be made by amateurs, but caution should be exercised.
**Use:** Used in military ammunition for pistols, shotguns, rifles, machine guns, mortars, tank fired munitions, artillery guns, and ship deck guns.

## 05-03-006A: Nitro coated high performance gun propellant containing picric acid:

Into a large capacity glass beaker equipped with motorized stirrer, place 115 milliliters of 70% nitric acid, followed by 4 liters of refined petroleum ether, and then allow the acid mixture to stand at room temperature for about 40 minutes Thereafter, add in 115 milliliters of concentrated sulfuric acid, and then blend them mixture for about 2 hours. Thereafter, drown the entire mixture into 4 liters of ice water, and then decant-off the upper organic liquid. Thereafter, place this upper collected layer into a suitable vertical mixer, followed by ***200 grams of barium nitrate***, followed by ***33 grams of wheat flour***, followed by ***33 grams of potassium ferrocyanide***, followed by ***160 grams of picric acid***, followed by ***33 grams of potassium chlorate***, followed by 40 grams liquid ammonia, followed by ***5 grams of lamp black***, and then blend the mixture on high speed for about 45 minutes. After 45 minutes, the ammonia would have evaporated-off; thereafter, filter-off the solids by using a Buchner funnel under mild suction, and thereafter, recover the solid material from the Buchner funnel. At this point, the collected material is ready for extruding; to do so, extrude the mixture through any extruding device into "spaghetti-like" strands ranging from 0.5 to 12mm in diameter, by 300mm in length, and then cut the strands into pieces ranging from 0.5 to 18mm length, and then cure the granules by placing them on a shallow pan or tray, and place said pan or tray into an oven, and cure the granules at 60 Celsius for 48 hours. To use the grains, simply pack the grains loosely into any desirable cartridge casing. Use standard gun casing primers for initiation.

**Chamber Pressure (standard 9mm casing; weight of propellant charge/grain size on average):** 34,669 psi
**Chamber Pressure (standard 5.56mm rifle casing; weight of propellant charge/grain size on average):** 35,338 psi
**Chamber Pressure (standard 105mm artillery casing; weight of propellant charge/grain size on average):** 31,332 psi
**Mortar tube Pressure (standard 81mm mortar propellant charge; weight of propellant charge/grain size on average):** 9,089 psi
**Burn rate (in the open):** 2.3 seconds per inch; tends to vary based on diameter of grain.
**Water resistance:** Good.
**Stability:** Can be stored for decades.
**Sensitivity:** Stable towards friction; stable towards shock; stable to temperatures up to 180 Celsius.
**Flammability (1 to 10, un-contained):** 9 solid grain; 9.2 fine grain.
**Tendency to cake:** Will not cake.
**Detonation ability:** Can be detonated using a blasting cap or detonator.
**Percentage:** ***43.1% barium nitrate, 34.4% picric acid, 7.1% potassium ferrocyanide, 7.1% wheat flour, 7.1% potassium chlorate, 1% lamp black, 0.20% nitro compounds***
**Classification:** Deflagrating explosive (classified as propellant).

**Danger of preparation:** Can be made by amateurs, but caution should be exercised.
**Use:** Used in military ammunition for pistols, shotguns, rifles, machine guns, mortars, tank fired munitions, artillery guns, and ship deck guns.

### 05-03-007A: High performance sodium azide containing specialty gun propellant:

Into a suitable vertical mixer, place 210 milliliters of hexane, followed by ***334 grams of sodium azide***, followed by ***91 grams of red iron-III-oxide***, and then followed by ***75 grams of black copper oxide***. Thereafter, blend the mixture on high speed for 1 hour. After the blending operation, the mixture is ready for extruding; to do so, extrude the mixture through any extruding device into "spaghetti-like" strands ranging from 0.5 to 12mm in diameter, by 300mm in length, and then cut the strands into pieces ranging from 0.5 to 18mm length, and then cure the granules by placing them on a shallow pan or tray, and place said pan or tray into an oven, and cure the granules at 80 Celsius for 12 hours. To use the grains, simply pack the grains loosely into any desirable cartridge casing. Use standard gun casing primers for initiation.

**Chamber Pressure (standard 9mm casing; weight of propellant charge/grain size on average):** 33,009 psi
**Chamber Pressure (standard 5.56mm rifle casing; weight of propellant charge/grain size on average):** 34,334 psi
**Chamber Pressure (standard 105mm artillery casing; weight of propellant charge/grain size on average):** 29,100 psi
**Mortar tube Pressure (standard 81mm mortar propellant charge; weight of propellant charge/grain size on average):** 8,795 psi
**Burn rate (in the open):** 2.9 seconds per inch; tends to vary based on diameter of grain.
**Water resistance:** Good.
**Stability:** Can be stored for decades.
**Sensitivity:** Stable towards friction; stable towards shock; stable to temperatures up to 150 Celsius.
**Flammability (1 to 10, un-contained):** 8.8 solid grain; 9.1 fine grain.
**Tendency to cake:** Will not cake.
**Detonation ability:** Can be detonated using a blasting cap or detonator.
**Percentage:** ***66.8% sodium azide, 18.2% iron oxide, 15% copper oxide***
**Classification:** Deflagrating explosive (classified as propellant).

**Danger of preparation:** Can be made by amateurs, but caution should be exercised. Sodium azide is poisonous so use care.
**Use:** Used in military ammunition for pistols, shotguns, rifles, machine guns, mortars, tank fired munitions, artillery guns, and ship deck guns.

**Bazooka fired 66mm projectile with fin stabilization**

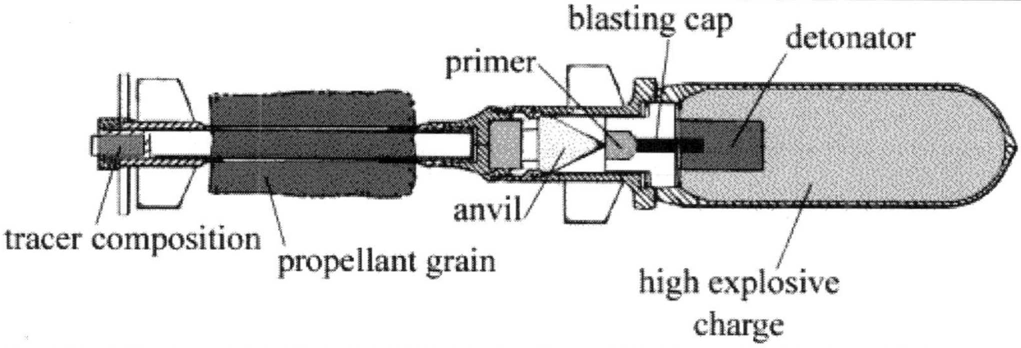


### 05-03-008A: High performance nitrocellulose gun propellant (triple based):

Into a suitable vertical mixer, place ***547 grams of nitroguanidine***, followed by ***200 grams of nitrocellulose*** (14% nitrogen content), followed by 320 milliliters of hexane, followed by ***90 grams of nitroglycerine***, followed by ***3 grams of finely divided cryolite***, followed by ***45 grams of dibutyl phthalate***, and then followed by ***15 grams of 2-nitrodiphenylamine***. Thereafter, blend the mixture for 2 hours on high speed. After the blending operation, the mixture is ready for extruding; to do so, extrude the mixture through any extruding device into "spaghetti-like" strands ranging from 0.5 to 12mm in diameter, by 300mm in length, and then cut the strands into pieces ranging from 0.5 to 18mm length, and then cure the granules by placing them on a shallow pan or tray, and place said pan or tray into an oven, and cure the granules at 80 Celsius for 12 hours. To use the grains, simply pack the grains loosely into any desirable cartridge casing. Use standard gun casing primers for initiation.

**Chamber Pressure (standard 9mm casing; weight of propellant charge/grain size on average):** 34,165 psi
**Chamber Pressure (standard 5.56mm rifle casing; weight of propellant charge/grain size on average):** 35,776 psi
**Chamber Pressure (standard 105mm artillery casing; weight of propellant charge/grain size on average):** 31,109 psi
**Mortar tube Pressure (standard 81mm mortar propellant charge; weight of propellant charge/grain size on average):** 9,887 psi
**Burn rate (in the open):** 2 seconds per inch; tends to vary based on diameter of grain.
**Water resistance:** Good.
**Stability:** Can be stored for decades.
**Sensitivity:** Stable towards friction; stable towards shock; stable to temperatures up to 100 Celsius.
**Flammability (1 to 10, un-contained):** 9.1 solid grain; 9.4 fine grain.
**Tendency to cake:** Will not cake.
**Detonation ability:** Can be detonated using a blasting cap or detonator.
**Percentage:** ***60.77% nitroguanidine, 22.22% nitrocellulose, 10% nitroglycerine, 5% dibutyl phthalate, 1.66% 2-nitrodiphenylamine, 0.33% cryolite, 0.02% impurities***

**Classification:** Deflagrating explosive (classified as class A explosive).

**Danger of preparation:** Should only be prepared by experienced professionals.
**Use:** Used in military ammunition for pistols, shotguns, rifles, machine guns, tank fired munitions, artillery guns, and ship deck guns.

**Note:** Inexperienced personnel should not attempt this procedure, as there are hazards involved. Make sure to thoroughly stir the mixture to avoid lumps or clumps, which can lead to excessive heat build-up. This procedure is perfectly safe for experienced persons. To gain experience, preparers should use only small quantities to begin with.

### 05-03-008B: High performance nitrocellulose gun propellant (triple based):

Into a suitable vertical mixer, place ***200 grams of nitrocellulose*** (14% nitrogen content), followed by 340 milliliters of hexane, followed by ***190 grams of nitroglycerine***, followed by ***45 grams of dibutylphthalate***, followed by ***15 grams of nitrodiphenylamine***, and then followed by ***547 grams of nitroguanidine***. Thereafter, blend the mixture on high speed for about 1 hour. After the blending operation, the mixture is ready for extruding; to do so, extrude the mixture through any extruding device into "spaghetti-like" strands ranging from 0.5 to 12mm in diameter, by 300mm in length, and then cut the strands into pieces ranging from 0.5 to 18mm length, and then cure the granules by placing them on a shallow pan or tray, and place said pan or tray into an oven, and cure the granules at 80 Celsius for 12 hours. To use the grains, simply pack the grains loosely into any desirable cartridge casing. Use standard gun casing primers for initiation.
**Chamber Pressure (standard 9mm casing; weight of propellant charge/grain size on average):** 34,683 psi
**Chamber Pressure (standard 5.56mm rifle casing; weight of propellant charge/grain size on average):** 35,991 psi
**Chamber Pressure (standard 105mm artillery casing; weight of propellant charge/grain size on average):** 31,307 psi
**Mortar tube Pressure (standard 81mm mortar propellant charge; weight of propellant charge/grain size on average):** 10,090 psi
**Burn rate (in the open):** 2 seconds per inch; tends to vary based on diameter of grain.
**Water resistance:** Good.
**Stability:** Can be stored for decades.
**Sensitivity:** Stable towards friction; stable towards shock; stable to temperatures up to 100 Celsius.
**Flammability (1 to 10, un-contained):** 9.1 solid grain; 9.4 fine grain.
**Tendency to cake:** Will not cake.
**Detonation ability:** Can be detonated using a blasting cap or detonator.
**Percentage:** ***54.86% nitroguanidine, 20.06% nitrocellulose, 19.05% nitroglycerine, 4.51% dibutylphthalate, 1.5% nitrogiphenylamine, 0.02% mixed balance***
**Classification:** Deflagrating explosive (classified as class A explosive).

**Danger of preparation:** Should only be prepared by experienced professionals.
**Use:** Used in military ammunition for pistols, shotguns, rifles, machine guns, tank fired munitions, artillery guns, and ship deck guns.

**Note:** Inexperienced personnel should not attempt this procedure, as there are hazards involved. Make sure to thoroughly stir the mixture to avoid lumps or clumps, which can lead to excessive heat build-up. This procedure is perfectly safe for experienced persons. To gain experience, preparers should use only small quantities to begin with.

### 05-03-009A: High performance ammonium perchlorate gun propellant:

Into a suitable vertical mixer, place ***200 grams of nitrocellulose*** (14% nitrogen content), followed by 280 milliliters of hexane, followed by ***25 grams of flours of sulfur***, followed by ***300 grams of ammonium perchlorate***, and then followed by ***115 grams of wheat flour***. Thereafter, blend the mixture on high speed for about 45 minutes. After the blending operation, the mixture is ready for extruding; to do so, extrude the mixture through any extruding device into "spaghetti-like" strands ranging from 0.5 to 12mm in diameter, by 300mm in length, and then cut the strands into pieces ranging from 0.5 to 18mm length, and then cure the granules by placing them on a shallow pan or tray, and place said pan or tray into an oven, and cure the granules at 80 Celsius for 12 hours. To use the grains, simply pack the grains loosely into any desirable cartridge casing. Use standard gun casing primers for initiation.
**Chamber Pressure (standard 9mm casing; weight of propellant charge/grain size on average):** 34,600 psi
**Chamber Pressure (standard 5.56mm rifle casing; weight of propellant charge/grain size on average):** 35,846 psi
**Chamber Pressure (standard 105mm artillery casing; weight of propellant charge/grain size on average):** 31,109 psi
**Mortar tube Pressure (standard 81mm mortar propellant charge; weight of propellant charge/grain size on average):** 9,099 psi
**Burn rate (in the open):** 2 seconds per inch; tends to vary based on diameter of grain.
**Water resistance:** Good.
**Stability:** Can be stored for decades.
**Sensitivity:** Stable towards friction; stable towards shock; stable to temperatures up to 180 Celsius.
**Flammability (1 to 10, un-contained):** 9.1 solid grain; 9.4 fine grain.
**Tendency to cake:** Will not cake.

**Detonation ability:** Can be detonated using a blasting cap or detonator.
**Percentage:** ***46.87% ammonium perchlorate, 31.25% nitrocellulose, 17.96% wheat flour, 3.9% flours of sulfur, 0.02% impurities***
**Classification:** Deflagrating explosive (classified as propellant).

**Danger of preparation:** Can be made by amateurs, but caution should be exercised.
**Use:** Used in military ammunition for pistols, shotguns, rifles, machine guns, mortars, tank fired munitions, artillery guns, and ship deck guns.

## 05-03-009B: High performance perchlorate based gun propellant:

Into a suitable vertical mixer, place ***200 grams of nitrocellulose*** (14% nitrogen content), followed by 260 milliliters of hexane, followed by ***45 grams of flours of sulfur***, followed by ***375 grams of potassium perchlorate***, followed by ***95 grams of wheat flour***, and then followed by ***15 grams of sodium azide***. Thereafter, rapidly blend the mixture on high speed for about 1 hour. After the blending operation, the mixture is ready for extruding; to do so, extrude the mixture through any extruding device into "spaghetti-like" strands ranging from 0.5 to 12mm in diameter, by 300mm in length, and then cut the strands into pieces ranging from 0.5 to 18mm length, and then cure the granules by placing them on a shallow pan or tray, and place said pan or tray into an oven, and cure the granules at 80 Celsius for 12 hours. To use the grains, simply pack the grains loosely into any desirable cartridge casing. Use standard gun casing primers for initiation.
**Chamber Pressure (standard 9mm casing; weight of propellant charge/grain size on average):** 34,558 psi
**Chamber Pressure (standard 5.56mm rifle casing; weight of propellant charge/grain size on average):** 35,701 psi
**Chamber Pressure (standard 105mm artillery casing; weight of propellant charge/grain size on average):** 31,088 psi
**Mortar tube Pressure (standard 81mm mortar propellant charge; weight of propellant charge/grain size on average):** 9,101 psi
**Burn rate (in the open):** 2 seconds per inch; tends to vary based on diameter of grain.
**Water resistance:** Good.
**Stability:** Can be stored for decades.
**Sensitivity:** Stable towards friction; stable towards shock; stable to temperatures up to 100 Celsius.
**Flammability (1 to 10, un-contained):** 9.1 solid grain; 9.4 fine grain.
**Tendency to cake:** Will not cake.
**Detonation ability:** Can be detonated using a blasting cap or detonator.
**Percentage:** ***51.36% potassium perchlorate, 27.39% nitrocellulose, 13.01% wheat flour, 6.16% sulfur, 2.05% sodium azide, 0.03% impurities***
**Classification:** Deflagrating explosive (classified as propellant).

**Danger of preparation:** Can be made by amateurs, but caution should be exercised. Sodium azide is poisonous so use care.
**Use:** Used in military ammunition for pistols, shotguns, rifles, machine guns, mortars, tank fired munitions, artillery guns, and ship deck guns.

## 05-03-009C: High performance chlorate based gun propellant:

Into a suitable vertical mixer, place 260 milliliters of hexane, followed by ***300 grams of starch***, and then followed by ***600 grams of potassium chlorate***. Thereafter, blend the mixture on high speed for 1 hour. After the blending operation, the mixture is ready for extruding; to do so, extrude the mixture through any extruding device into "spaghetti-like" strands ranging from 0.5 to 12mm in diameter, by 300mm in length, and then cut the strands into pieces ranging from 0.5 to 18mm length, and then cure the granules by placing them on a shallow pan or tray, and place said pan or tray into an oven, and cure the granules at 80 Celsius for 12 hours. To use the grains, simply pack the grains loosely into any desirable cartridge casing. Use standard gun casing primers for initiation.
**Chamber Pressure (standard 9mm casing; weight of propellant charge/grain size on average):** 33,090 psi
**Chamber Pressure (standard 5.56mm rifle casing; weight of propellant charge/grain size on average):** 35,002 psi
**Chamber Pressure (standard 105mm artillery casing; weight of propellant charge/grain size on average):** 30,141 psi
**Mortar tube Pressure (standard 81mm mortar propellant charge; weight of propellant charge/grain size on average):** 9,000 psi
**Burn rate (in the open):** 2.9 seconds per inch; tends to vary based on diameter of grain.
**Water resistance:** Good.
**Stability:** Can be stored for decades.
**Sensitivity:** Stable towards friction; stable towards shock; stable to temperatures up to 300 Celsius.
**Flammability (1 to 10, un-contained):** 8.9 solid grain; 9.1 fine grain.
**Tendency to cake:** Will not cake.
**Detonation ability:** Cannot be detonated.
**Percentage:** ***66.66% potassium chlorate, 33.33% starch, 0.01% mixed balance***
**Classification:** Deflagrating explosive (classified as propellant).

**Danger of preparation:** Can be made by amateurs, but caution should be exercised.
**Use:** Used in military ammunition for pistols, shotguns, rifles, machine guns, mortars, tank fired munitions, artillery guns, and ship deck guns.

## 05-03-010A: High performance nitroguanidine gun propellant:

Into a suitable vertical mixer, place ***547 grams of nitroguanidine***, followed by ***200 grams of nitrocellulose*** (14% nitrogen content), followed by ***90 grams of nitroglycerine***, followed by 290 milliliters of hexane, followed by ***30 grams of cryolite***, followed by ***45 grams of dibutylphthalate***, and then followed by ***15 grams of 2-nitrodiphenylamine***. Thereafter, blend the mixture on high speed for about 1 hour. After the blending operation, the mixture is ready for extruding; to do so, extrude the mixture through any extruding device into "spaghetti-like" strands ranging from 0.5 to 12mm in diameter, by 300mm in length, and then cut the strands into pieces ranging from 0.5 to 18mm length, and then cure the granules by placing them on a shallow pan or tray, and place said pan or tray into an oven, and cure the granules at 80 Celsius for 12 hours. To use the grains, simply pack the grains loosely into any desirable cartridge casing. Use standard gun casing primers for initiation.

**Chamber Pressure (standard 9mm casing; weight of propellant charge/grain size on average):** 34,109 psi
**Chamber Pressure (standard 5.56mm rifle casing; weight of propellant charge/grain size on average):** 35,601 psi
**Chamber Pressure (standard 105mm artillery casing; weight of propellant charge/grain size on average):** 31,033 psi
**Mortar tube Pressure (standard 81mm mortar propellant charge; weight of propellant charge/grain size on average):** 9,774 psi
**Burn rate (in the open):** 2.1 seconds per inch; tends to vary based on diameter of grain.
**Water resistance:** Good.
**Stability:** Can be stored for decades.
**Sensitivity:** Stable towards friction; stable towards shock; stable to temperatures up to 100 Celsius.
**Flammability (1 to 10, un-contained):** 9 solid grain; 9.21 fine grain.
**Tendency to cake:** Will not cake.
**Detonation ability:** Can be detonated using a blasting cap or detonator.
**Percentage:** ***59% nitroguanidine, 21.57% nitrocellulose, 9.7% nitroglycerine, 4.85% dibutylphthalate, 3.23% cryolite, 1.61% 2-nitrodiphenylamine, 0.04% mixed balance***
**Classification:** Deflagrating explosive (classified as propellant).

**Danger of preparation:** Can be made by amateurs, but caution should be exercised.
**Use:** Used in military ammunition for pistols, shotguns, rifles, machine guns, mortars, tank fired munitions, artillery guns, and ship deck guns.

## 05-03-010B: High performance nitroguanidine gun propellant:

Into a suitable vertical mixer, place ***200 grams of nitrocellulose*** (14% nitrogen content), followed by 300 milliliters of hexane, followed by ***190 gram of nitroglycerine***, followed by ***45 grams of dibutylphthalate***, followed by ***15 grams of 2-nitrodiphenylamine***, and then followed by ***547 grams of nitroguanidine***. Thereafter, blend the mixture on high speed for 45 minutes. After the blending operation, the mixture is ready for extruding; to do so, extrude the mixture through any extruding device into "spaghetti-like" strands ranging from 0.5 to 12mm in diameter, by 300mm in length, and then cut the strands into pieces ranging from 0.5 to 18mm length, and then cure the granules by placing them on a shallow pan or tray, and place said pan or tray into an oven, and cure the granules at 80 Celsius for 12 hours. To use the grains, simply pack the grains loosely into any desirable cartridge casing. Use standard gun casing primers for initiation.

**Chamber Pressure (standard 9mm casing; weight of propellant charge/grain size on average):** 34,499 psi
**Chamber Pressure (standard 5.56mm rifle casing; weight of propellant charge/grain size on average):** 35,990 psi
**Chamber Pressure (standard 105mm artillery casing; weight of propellant charge/grain size on average):** 31,552 psi
**Mortar tube Pressure (standard 81mm mortar propellant charge; weight of propellant charge/grain size on average):** 9,909 psi
**Burn rate (in the open):** 2.3 seconds per inch; tends to vary based on diameter of grain.
**Water resistance:** Good.
**Stability:** Can be stored for decades.
**Sensitivity:** Stable towards friction; stable towards shock; stable to temperatures up to 100 Celsius.
**Flammability (1 to 10, un-contained):** 9 solid grain; 9.2 fine grain.
**Tendency to cake:** Will not cake.
**Detonation ability:** Can be detonated using a blasting cap or detonator.
**Percentage:** ***54.86% nitroguanidine, 20% nitrocellulose, 19% nitroglycerine, 4.4% dibutylphthalate, 1.5% 2-nitrodiphenylamine, 0.24% mixed residual balance***
**Classification:** Deflagrating explosive (classified as class A explosive).

**Danger of preparation:** Should only be prepared by experienced professionals.
**Use:** Used in military ammunition for pistols, shotguns, rifles, machine guns, tank fired munitions, artillery guns, and ship deck guns.

**Note:** Inexperienced personnel should not attempt this procedure, as there are hazards involved. Make sure to thoroughly stir the mixture to avoid lumps or clumps, which can lead to excessive heat build-up. This procedure is perfectly safe for experienced persons. To gain experience, preparers should use only small quantities to begin with.

### 05-03-011A: High performance barium nitrate gun propellant for small arms weapons:

Into a suitable vertical mixer, place ***255 grams of barium nitrate***, followed by ***47.5 grams of powdered soft wood charcoal***, followed by ***37.5 grams of sulfur powder***, followed by 140 milliliters of hexane, followed by ***85 grams of anhydrous lead nitrate***, and then followed by ***75 grams of nitrocellulose*** (14% nitrogen content). Thereafter, blend the mixture on rapid speed for 1 hour. After the blending operation, the mixture is ready for extruding; to do so, extrude the mixture through any extruding device into "spaghetti-like" strands ranging from 0.5 to 12mm in diameter, by 300mm in length, and then cut the strands into pieces ranging from 0.5 to 18mm length, and then cure the granules by placing them on a shallow pan or tray, and place said pan or tray into an oven, and cure the granules at 80 Celsius for 12 hours. To use the grains, simply pack the grains loosely into any desirable cartridge casing. Use standard gun casing primers for initiation.
**Chamber Pressure (standard 9mm casing; weight of propellant charge/grain size on average):** 33,109 psi
**Chamber Pressure (standard 5.56mm rifle casing; weight of propellant charge/grain size on average):** 34,090 psi
**Chamber Pressure (standard 105mm artillery casing; weight of propellant charge/grain size on average):** 30,443 psi
**Mortar tube Pressure (standard 81mm mortar propellant charge; weight of propellant charge/grain size on average):** 9,221 psi
**Burn rate (in the open):** 2.7 seconds per inch; tends to vary based on diameter of grain.
**Water resistance:** Good.
**Stability:** Can be stored for decades.
**Sensitivity:** Stable towards friction; stable towards shock; stable to temperatures up to 180 Celsius.
**Flammability (1 to 10, un-contained):** 8.3 solid grain; 9.1 fine grain.
**Tendency to cake:** Will not cake.
**Detonation ability:** Can be detonated using a blasting cap or detonator.
**Percentage:** ***51% barium nitrate, 17% lead nitrate, 15% nitrocellulose, 9.5% soft wood charcoal, 7.5% sulfur***
**Classification:** Deflagrating explosive (classified as propellant).

**Danger of preparation:** Can be made by amateurs, but caution should be exercised.
**Use:** Used in military ammunition for pistols, shotguns, rifles, machine guns, mortars, tank fired munitions, artillery guns, and ship deck guns.

### 05-03-011B: High performance barium nitrate gun propellant for small arms weapons:

Into a suitable vertical mixer, place ***260 grams of barium nitrate***, followed by ***55 grams of finely divided graphite***, followed by ***37 grams of sodium hypophosphite***, followed by 120 milliliters of hexane, followed by ***80 grams of potassium nitrate***, and then followed by ***90 grams of nitrocellulose*** (14% nitrogen content). Thereafter, blend the mixture on high speed for about 1 hour. After the blending operation, the mixture is ready for extruding; to do so, extrude the mixture through any extruding device into "spaghetti-like" strands ranging from 0.5 to 12mm in diameter, by 300mm in length, and then cut the strands into pieces ranging from 0.5 to 18mm length, and then cure the granules by placing them on a shallow pan or tray, and place said pan or tray into an oven, and cure the granules at 80 Celsius for 12 hours. To use the grains, simply pack the grains loosely into any desirable cartridge casing. Use standard gun casing primers for initiation.
**Chamber Pressure (standard 9mm casing; weight of propellant charge/grain size on average):** 33,211 psi
**Chamber Pressure (standard 5.56mm rifle casing; weight of propellant charge/grain size on average):** 34,197 psi
**Chamber Pressure (standard 105mm artillery casing; weight of propellant charge/grain size on average):** 30,500 psi
**Mortar tube Pressure (standard 81mm mortar propellant charge; weight of propellant charge/grain size on average):** 9,396 psi
**Burn rate (in the open):** 2.6 seconds per inch; tends to vary based on diameter of grain.
**Water resistance:** Good.
**Stability:** Can be stored for decades.
**Sensitivity:** Stable towards friction; stable towards shock; stable to temperatures up to 100 Celsius.
**Flammability (1 to 10, un-contained):** 8.4 solid grain; 9.1 fine grain.
**Tendency to cake:** Will not cake.
**Detonation ability:** Cannot be detonated.
**Percentage:** ***49.8% barium nitrate, 17.24% nitrocellulose, 15.32% potassium nitrate, 10.53% graphite, 7.08% sodium hypophosphite, 0.03% mixed balance***
**Classification:** Deflagrating explosive (classified as propellant).

**Danger of preparation:** Can be made by amateurs, but caution should be exercised.
**Use:** Used in military ammunition for pistols, shotguns, rifles, machine guns, mortars, tank fired munitions, artillery guns, and ship deck guns.

## 05-03-012A: Specialty high performance ammonium picrate gun propellant:

Into a suitable vertical mixer, place ***235 grams of ammonium picrate***, followed by 200 milliliters of hexane, followed by ***115 grams of potassium dichromate***, and then followed by ***150 grams of barium nitrate***. Thereafter, blend the mixture on moderate speed for about 1 hour. After the blending operation, the mixture is ready for extruding; to do so, extrude the mixture through any extruding device into "spaghetti-like" strands ranging from 0.5 to 12mm in diameter, by 300mm in length, and then cut the strands into pieces ranging from 0.5 to 18mm length, and then cure the granules by placing them on a shallow pan or tray, and place said pan or tray into an oven, and cure the granules at 60 Celsius for 24 hours. To use the grains, simply pack the grains loosely into any desirable cartridge casing. Use standard gun casing primers for initiation.
**Chamber Pressure (standard 12.7mm rifle casing; weight of propellant charge/grain size on average):** 84,093 psi
**Chamber Pressure (standard 210mm artillery casing; weight of propellant charge/grain size on average):** 33,900 psi
**Burn rate (in the open):** 1.9 seconds per inch; tends to vary based on diameter of grain.
**Water resistance:** Good.
**Stability:** Can be stored for decades.
**Sensitivity:** Stable towards friction; stable towards shock; stable to temperatures up to 150 Celsius.
**Flammability (1 to 10, un-contained):** 9.19 solid grain; 9.4 fine grain.
**Tendency to cake:** Will not cake.
**Detonation ability:** Can be detonated using a blasting cap or detonator; can be detonated when rapidly heated with a strong flame.
**Percentage:** ***47% ammonium picrate, 30% barium nitrate, 23% potassium dichromate***
**Classification:** Deflagrating explosive (classified as class A explosive).

**Danger of preparation:** Should only be prepared by experienced professionals. Keep in mind the dangers of working with ammonium picrate.
**Use:** Used in military ammunition for high caliber weapons such as artillery guns, and ship deck guns.

**Note:** Inexperienced personnel should not attempt this procedure, as there are hazards involved. Make sure to thoroughly stir the mixture to avoid lumps or clumps, which can lead to excessive heat build-up. This procedure is perfectly safe for experienced persons. To gain experience, preparers should use only small quantities to begin with.

## 05-03-013A: High performance ammonium picrate gun propellant:

Into a suitable vertical mixer, place ***125 grams of ammonium picrate***, followed by 110 milliliters of hexane, followed by ***260 grams of potassium nitrate***, followed by ***50 grams of flours of sulfur,*** followed by ***15 grams of soft wood charcoal,*** and then followed by ***50 grams of cornstarch.*** Thereafter, blend the mixture on moderate speed for about 1 hour. After the blending operation, the mixture is ready for extruding; to do so, extrude the mixture through any extruding device into "spaghetti-like" strands ranging from 0.5 to 12mm in diameter, by 300mm in length, and then cut the strands into pieces ranging from 0.5 to 18mm length, and then cure the granules by placing them on a shallow pan or tray, and place said pan or tray into an oven, and cure the granules at 60 Celsius for 24 hours. To use the grains, simply pack the grains loosely into any desirable cartridge casing. Use standard gun casing primers for initiation.
**Chamber Pressure (standard 9mm casing; weight of propellant charge/grain size on average):** 34,935 psi
**Chamber Pressure (standard 12.7mm rifle casing; weight of propellant charge/grain size on average):** 82,035 psi
**Chamber Pressure (standard 210mm artillery casing; weight of propellant charge/grain size on average):** 31,995 psi
**Burn rate (in the open):** 2 seconds per inch; tends to vary based on diameter of grain.
**Water resistance:** Good.
**Stability:** Can be stored for decades.
**Sensitivity:** Stable towards friction; stable towards shock; stable to temperatures up to 150 Celsius.
**Flammability (1 to 10, un-contained):** 9.1 solid grain; 9.3 fine grain.
**Tendency to cake:** Will not cake.
**Detonation ability:** Can be detonated using a blasting cap or detonator; can be detonated when rapidly heated with a strong flame.
**Percentage:** ***52% potassium nitrate, 25% ammonium picrate, 10% sulfur, 10% corn starch, 3% soft wood charcoal***
**Classification:** Deflagrating explosive (classified as class A explosive).

**Danger of preparation:** Should only be prepared by experienced professionals. Keep in mind the dangers of working with ammonium picrate.
**Use:** Used in military ammunition for pistols, shotguns, rifles, machine guns, tank fired munitions, artillery guns, and ship deck guns.

**Note:** Inexperienced personnel should not attempt this procedure, as there are hazards involved. Make sure to thoroughly stir the mixture to avoid lumps or clumps, which can lead to excessive heat build-up. This procedure is perfectly safe for experienced persons. To gain experience, preparers should use only small quantities to begin with.

### 05-03-014A: High performance barium nitrate gun propellant:

Into a suitable vertical mixer, place ***60 grams of nitro-naphthalene***, followed by ***400 grams of pyroxyline***, followed by ***400 grams of barium nitrate***, followed by 300 milliliters of hexane, and the followed by ***50 grams of potassium nitrate***. Thereafter, blend the mixture on high speed for about 1 hour. After the blending operation, the mixture is ready for extruding; to do so, extrude the mixture through any extruding device into "spaghetti-like" strands ranging from 0.5 to 12mm in diameter, by 300mm in length, and then cut the strands into pieces ranging from 0.5 to 18mm length, and then cure the granules by placing them on a shallow pan or tray, and place said pan or tray into an oven, and cure the granules at 80 Celsius for 12 hours. To use the grains, simply pack the grains loosely into any desirable cartridge casing. Use standard gun casing primers for initiation.
**Chamber Pressure (standard 9mm casing; weight of propellant charge/grain size on average):** 30,331 psi
**Chamber Pressure (standard 5.56mm rifle casing; weight of propellant charge/grain size on average):** 31,097 psi
**Mortar tube Pressure (standard 81mm mortar propellant charge; weight of propellant charge/grain size on average):** 8,870 psi
**Burn rate (in the open):** 3.9 seconds per inch; tends to vary based on diameter of grain.
**Water resistance:** Good.
**Stability:** Can be stored for decades.
**Sensitivity:** Stable towards friction; stable towards shock; stable to temperatures up to 150 Celsius.
**Flammability (1 to 10, un-contained):** 7.9 solid grain; 8 fine grain.
**Tendency to cake:** Will not cake.
**Detonation ability:** Cannot be detonated.
**Percentage:** ***43.95% barium nitrate, 43.95% pyroxyline, 6.59% nitro-naphthalene, 5.49% potassium nitrate, 0.02% mixed balance***
**Classification:** Deflagrating explosive (classified as propellant).

**Danger of preparation:** Can be made by amateurs.
**Use:** Used in military ammunition for pistols, shotguns, rifles, and mortars.

### 05-03-014B: High performance barium nitrate gun propellant (sulfur induced):

Into a suitable vertical mixer, place ***100 grams of nitro-naphthalene***, followed by ***3000 grams of pyroxyline***, followed by ***1200 grams of barium nitrate***, followed by 1800 milliliters of hexane, followed by ***100 grams of sulfur***, and the followed by ***500 grams of potassium nitrate***. Thereafter, blend the mixture on high speed for about 1 hour. After the blending operation, the mixture is ready for extruding; to do so, extrude the mixture through any extruding device into "spaghetti-like" strands ranging from 0.5 to 12mm in diameter, by 300mm in length, and then cut the strands into pieces ranging from 0.5 to 18mm length, and then cure the granules by placing them on a shallow pan or tray, and place said pan or tray into an oven, and cure the granules at 80 Celsius for 12 hours. To use the grains, simply pack the grains loosely into any desirable cartridge casing. Use standard gun casing primers for initiation.
**Chamber Pressure (standard 9mm casing; weight of propellant charge/grain size on average):** 29,167 psi
**Chamber Pressure (standard 5.56mm rifle casing; weight of propellant charge/grain size on average):** 30,091 psi
**Mortar tube Pressure (standard 81mm mortar propellant charge; weight of propellant charge/grain size on average):** 8,600 psi
**Burn rate (in the open):** 4.1 seconds per inch; tends to vary based on diameter of grain.
**Water resistance:** Good.
**Stability:** Can be stored for decades.
**Sensitivity:** Stable towards friction; stable towards shock; stable to temperatures up to 150 Celsius.
**Flammability (1 to 10, un-contained):** 7.9 solid grain; 8 fine grain.
**Tendency to cake:** Will not cake.
**Detonation ability:** Cannot be detonated.
**Percentage:** ***61.22% pyroxyline, 24.48% barium nitrate, 10.2% potassium nitrate, 2.04% nitro-naphthalene, 2.04% sulfur, 0.02% mixed balance***
**Classification:** Deflagrating explosive (classified as propellant).

**Danger of preparation:** Can be made by amateurs.
**Use:** Used in military ammunition for pistols, shotguns, rifles, and mortars.

### 05-03-015A: High performance "smokeless powder" ammonium picrate gun propellant:

Into a vertical mixer, place ***35 grams of potassium permanganate***, followed by 200 milliliters of hexane, followed by ***125 grams of potassium dichromate***, and then followed by ***340 grams of ammonium picrate***. Thereafter, blend the mixture on moderate speed for about 45 minutes. After the blending operation, the mixture is ready for extruding; to do so, extrude the mixture through any extruding device into "spaghetti-like" strands ranging from 0.5 to 12mm in diameter, by 300mm in length, and then cut the strands into pieces ranging from 0.5 to 18mm length, and then cure the granules by placing them on a shallow pan or tray, and place said pan or tray into an oven, and cure the granules at 60 Celsius for 24 hours. To use the grains, simply pack the grains loosely into any desirable cartridge casing. Use standard gun casing primers for initiation.
**Chamber Pressure (standard 210mm artillery casing; weight of propellant charge/grain size on average):** 34,226 psi
**Burn rate (in the open):** 2 seconds per inch; tends to vary based on diameter of grain.
**Water resistance:** Good.
**Stability:** Can be stored for decades.
**Sensitivity:** Stable towards friction; stable towards shock; stable to temperatures up to 150 Celsius.
**Flammability (1 to 10, un-contained):** 9.1 solid grain; 9.3 fine grain.
**Tendency to cake:** Will not cake.
**Detonation ability:** Can be detonated using a blasting cap or detonator; can be detonated when rapidly heated with a strong flame.
**Percentage:** ***68% ammonium picrate, 25% potassium dichromate, 7% potassium permanganate***
**Classification:** Deflagrating explosive (classified as class A explosive).

**Danger of preparation:** Should only be prepared by experienced professionals. Keep in mind the dangers of working with ammonium picrate.
**Use:** Used in military ammunition for high caliber weapons such as artillery guns, and ship deck guns.

**Note:** Inexperienced personnel should not attempt this procedure, as there are hazards involved. Make sure to thoroughly stir the mixture to avoid lumps or clumps, which can lead to excessive heat build-up. This procedure is perfectly safe for experienced persons. To gain experience, preparers should use only small quantities to begin with.

# Chapter 6. General pyrotechnic compositions I

## Section 1: Bullet Tracer compositions

***Chemicals used in this section (binders are not included)***

| | | |
|---|---|---|
| 1. Potassium nitrate (see Black Powder) | | 2. Sulfur (see Black Powder) |
| 3. Charcoal (see Black Powder) | | 4. Sugar Carbon (see Modified Black Powder) |
| 5. Barium Chromate (see Modified Black Powder) | | 6. Potassium Perchlorate (see Modified Black Powder) |
| 7. Potassium Dichromate (see Modified Black Powder) | | 8. Ammonium Bisulfide (see Modified Black Powder) |
| 9. Potassium chlorate (see Modified Black Powder) | | 10. Carbon Disulfide (see Modified Black Powder) |
| 11. Nitrocellulose (see Modified Black Powder) | | 12. Lead Tetraoxide (see Modified Black Powder) |
| 13. Diphenylamine (see Modified Black Powder) | | 14. Titanium Dioxide (see Modified Black Powder) |
| 15. Manganese Dioxide (see Modified Black Powder) | | 16. Sodium Benzoate (see Modified Black Powder) |
| 17. Ammonium Nitrate (see Modified Black Powder) | | 18. Calcium Carbonate (see Modified Black Powder) |
| 19. Sodium Nitrate (see Modified Black Powder) | | 20. Urea (see Modified Black Powder) |
| 21. Lead Nitrate (see Modified Black Powder) | | 22. Nitro Starch (see Modified Black Powder) |
| 23. Ammonium Perchlorate (see Ammonium Perchlorate Rocket Propellants) | | 24. Aluminum powder (see Ammonium Perchlorate Rocket Propellants) |
| 25. Magnesium Sulfide (see Ammonium Perchlorate Rocket Propellants) | | 26. Copper Chromite (see Ammonium Perchlorate Rocket Propellants) |
| 27. Nitroguanidine (see Ammonium Perchlorate Rocket Propellants) | | 28. Ammonium Sulfate (see Ammonium Perchlorate Rocket Propellants) |
| 29. Nitroglycerine (see Ammonium Perchlorate Rocket Propellants) | | 30. Magnesium Oxide (see Ammonium Perchlorate Rocket Propellants) |
| 31. Iron-II-oxide (see Ammonium Perchlorate Rocket Propellants) | | 32. Zirconium Hydride (see Ammonium Perchlorate Rocket Propellants) |
| 33. Teflon (see Ammonium Perchlorate Rocket Propellants) | | 34. Zinc Oxide (see Ammonium Perchlorate Rocket Propellants) |
| 35. Ammonium Dichromate (see Ammonium Perchlorate Rocket Propellants) | | 36. Lithium Perchlorate (see Ammonium Perchlorate Rocket Propellants) |
| 37. Magnesium powder (see Ammonium Perchlorate Rocket Propellants) | | 38. Lithium Aluminum Hydride (see Ammonium Perchlorate Rocket Propellants) |
| 39. Iron-III-oxide (red iron oxide) (see Ammonium Perchlorate Rocket Propellants) | | 40. PVC (see Ammonium Perchlorate Rocket Propellants) |
| 41. Copper Sulfide (see Ammonium Perchlorate Rocket Propellants) | | 42. Sodium Hydride (see Ammonium Perchlorate Rocket Propellants) |
| 43. Titanium Hydride (see Ammonium Perchlorate Rocket Propellants) | | 44. Silicon Nitride (see Ammonium Perchlorate Rocket Propellants) |
| 45. ADN (see ADN Rocket Propellants) | | 46. Urea Nitrate (see ADN Rocket Propellants) |
| 47. Silicon Powder (see ADN Rocket Propellants) | | 48. Hexamine (see ADN Rocket Propellants) |
| 49. Boron powder (see ADN Rocket Propellants) | | 50. Sodium hypophosphite (see ADN Rocket Propellants) |
| 51. KDN (see ADN Rocket Propellants) | | |

| | |
|---|---|
| | **52. Nickel Chloride (see Ammonium Nitrate Rocket Propellants)** |
| **53. TNT (see Ammonium Nitrate Rocket Propellants)** | **54. Acrylamide (see Miscellaneous Rocket H.P. Rocket Propellants)** |
| **55. Ethylene Glycol (see Miscellaneous Rocket H.P. Rocket Propellants)** | **56. Magnesium Perchlorate (see Miscellaneous Rocket H.P. Rocket Propellants)** |
| **57. Metallic Lithium (see Miscellaneous Rocket H.P. Rocket Propellants)** | **58. Beryllium Hydride (see Miscellaneous Rocket H.P. Rocket Propellants)** |
| **59. Red Phosphorus (see Miscellaneous Rocket H.P. Rocket Propellants)** | **60. Sodium Borohydride (see Miscellaneous Rocket H.P. Rocket Propellants)** |
| **61. Sodium chlorate (see Miscellaneous Rocket H.P. Rocket Propellants)** | **62. Lead Dioxide (see Miscellaneous Rocket H.P. Rocket Propellants)** |
| **63. Benzoyl Peroxide (see Miscellaneous Rocket H.P. Rocket Propellants)** | **64. Barium Nitrate (see Miscellaneous Rocket H.P. Rocket Propellants)** |
| **65. Cyanuric Acid (see Miscellaneous Rocket H.P. Rocket Propellants)** | **66. Aluminum Stearate (see Ammonium Nitrate Gun Propellants)** |
| **67. Aluminum Hydride (see Ammonium Nitrate Gun Propellants)** | **68. Magnesium Peroxide (see Ammonium Nitrate Gun Propellants)** |
| **69. Potassium Permanganate (see Ammonium Nitrate Gun Propellants)** | **70. Calcium Hydride (see Ammonium Nitrate Gun Propellants)** |
| **71. Potassium Tartrate (see Ammonium Nitrate Gun Propellants)** | **72. Potassium Ferrocyanide (see Ammonium Nitrate Gun Propellants)** |
| **73. Sodium Azide (see Ammonium Nitrate Gun Propellants)** | **74. Sodium Chloride (see Ammonium Nitrate Gun Propellants)** |
| **75. Potassium Sulfate (see Nitrocellulose Gun Propellants)** | **76. Lead Stearate (see Nitrocellulose Gun Propellants)** |
| **77. Triethylene Glycol (see Nitrocellulose Gun Propellants)** | **78. Sugar (sucrose) (see Miscellaneous Gun Propellants)** |
| **79. Sodium Propionate (see Miscellaneous Gun Propellants)** | **80. Picric Acid (see Miscellaneous Gun Propellants)** |
| **81. Copper-II-oxide (see Miscellaneous Gun Propellants)** | **82. Ammonium Picrate (see Miscellaneous Gun Propellants)** |
| **83. Barium Peroxide** | **84. Magnesium Carbonate** |
| 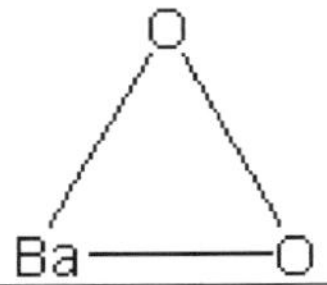 | 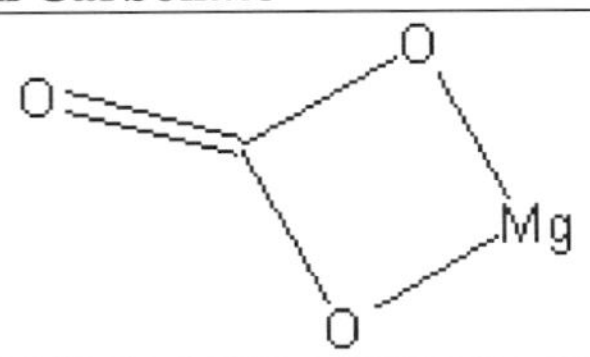 |
| Forms a grayish-white to white powder. The powder is insoluble in water, but is capable of forming an octahydrate. The salt slowly decomposes in moist air, or in contact with water. Barium peroxide decomposes rapidly in the presence of acid. | Forms a white odorless solid, granules, or powder. The commercial grade contains a carbonate hydroxide. The solid breaks down rapidly in the presence of acids. The carbonate is insoluble in water and all organic solvents. |
| **85. Strontium Peroxide** | **86. Strontium Nitrate** |
| 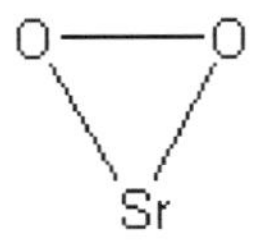 | 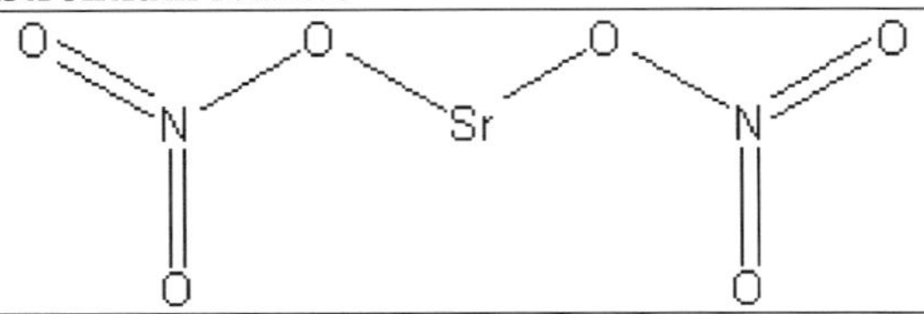 |
| Strontium peroxide forms a white odorless powder. The powder is insoluble in water, but slowly reacts with it forming oxygen. The salt decomposes in the presence of acids. The solid should be stored in airtight bottles away from moist air. | Forms white granules or powder. The powder has a melting point of 570 Celsius. The nitrate is soluble in water, and slowly forms a tetrahdyrate there with. |

## 87. Cupric chloride

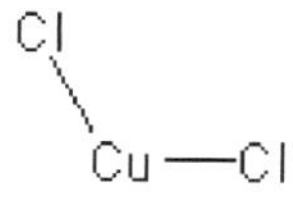


Forms a yellowish to yellowish-brown crystalline solid, which readily forms hydrates with water. The salt slowly decomposes when heated above 300 Celsius, but melts at 498 Celsius. The solid is readily soluble in water, alcohol, and acetone. Can be prepared by reacting dry chlorine with copper scrap, or via electrochemical methods.

## 88. Hexachlorobenzene

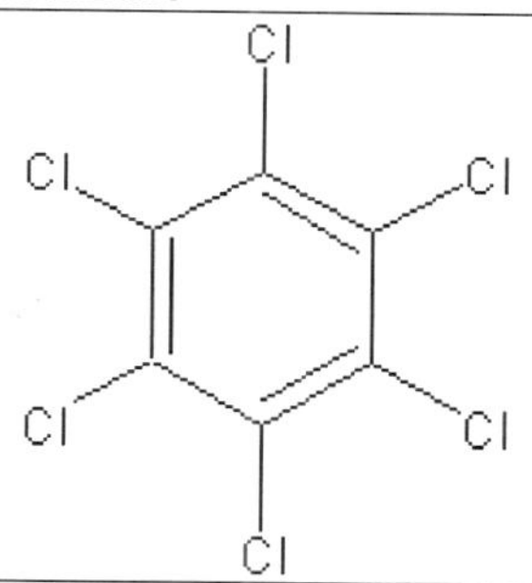


Forms colorless crystalline needles with a melting point of 231 Celsius. The crystals are sublimable and may be volatile with steam. The crystals are insoluble in water and alcohol, but soluble in common solvents.

## 89. Strontium oxalate

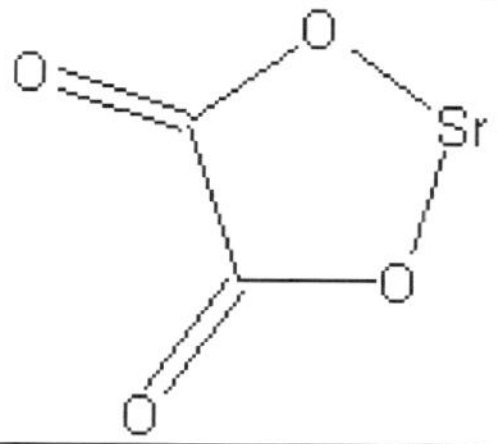


Forms a monohydrate, which forms colorless to white crystals, granules, or powder. The crystals are insoluble in water and the usual solvents, but are capable of dissolving in glacial acetic acid.

## 90. Mercury-I-Chloride

Forms a white heavy powder, which slowly decomposes in air. The salt cannot be melted, as the crystals sublime when heated to 400 Celsius. Mercury chloride is insoluble in water and most organic solvents. A strong reducing agent. Keep bottles well closed and avoid skin contact.

## 91. Zinc Oxalate

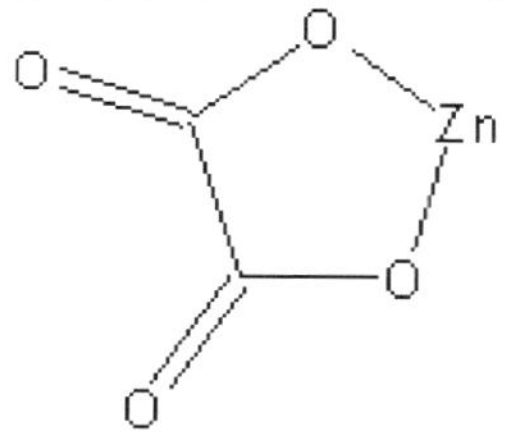


Forms a dihydrate. Slightly soluble in water and the usual solvents. Soluble in acids.

## 92. Zinc Chloride

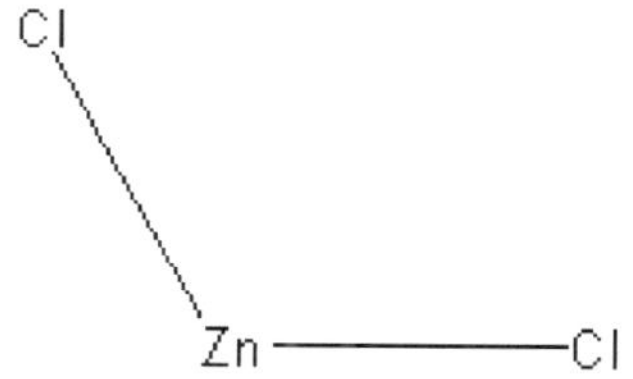


Forms colorless to white crystals, granules, or powder. The crystals are extremely hygroscopic and can absorb moisture from the air until a self-solution is produced. Readily soluble in alcohol and the usual solvents. Zinc chloride has a melting point of 290 Celsius. The crystals are also freely soluble in acetone. Can be made by reacting zinc metal with hydrochloric acid, or by neutralizing zinc hydroxide, oxide, or carbonate with hydrochloric acid.

## 93. Uranium

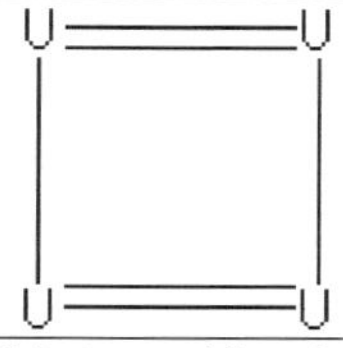


Uranium forms a silvery-white to grayish to dark gray radioactive metal. Uranium is stable at room temperature, but rapidly tarnishes forming an oxide layer. Powdered Uranium is spontaneously combustible. Users should use caution.

## 94. Zirconium nitrate

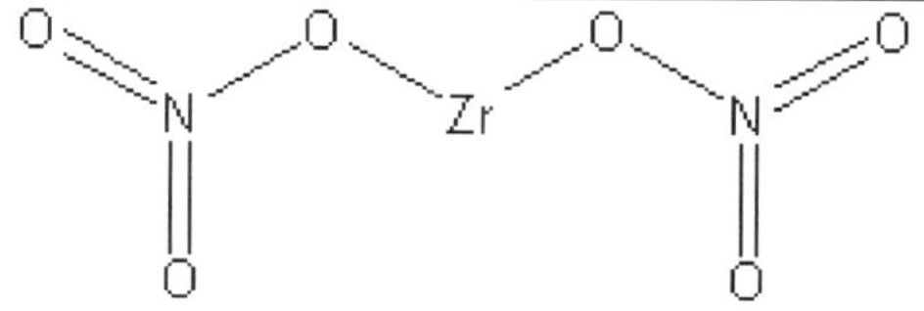


Forms a pentahydrate, which forms white to colorless crystals, granules, or powder. The crystals are very soluble in water and alcohol.

### 95. Yttrium Nitrate

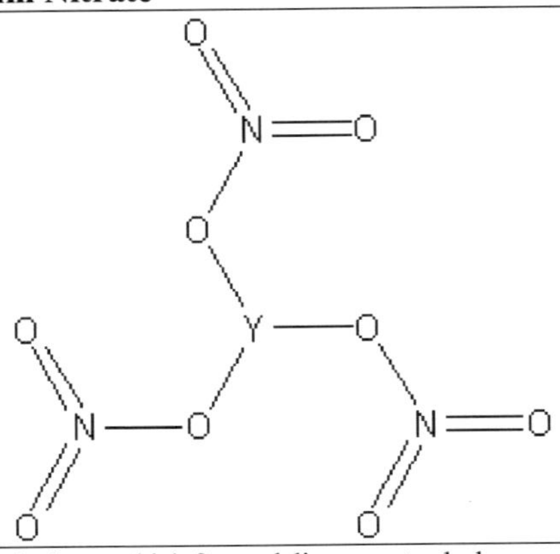

Forms a hexahydrate, which forms deliquescent colorless crystals, or white granules, or powder. The powder is somewhat unstable and begin to breaks down at room temperature upon standing in moist air. Decomposes when heated.

### 96. Yttrium Oxide

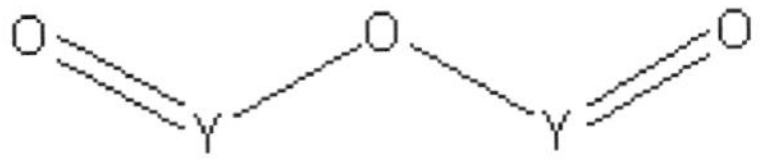

Yttrium oxide is a white powder or lumps. The powder is stable at ordinary temperatures. The powder has a strong tendency for ammonia, and will displace ammonia from ammonium salts. Can be prepared by roasting yttrium salts.

### 97. Zirconium Oxide

O=Zr=O

Forms a white powder or granular masses. Zirconium oxide is insoluble in water and the usual solvents, but soluble in acids.

### 98. Cerium Oxide

O=Ce=O

Ceric oxide forms a heavy white powder, which is often colored yellow due to impurities. The oxide is insoluble in water and the usual solvents. Some forms of the oxide are resistant to acids.

### 99. Hexachloroethane

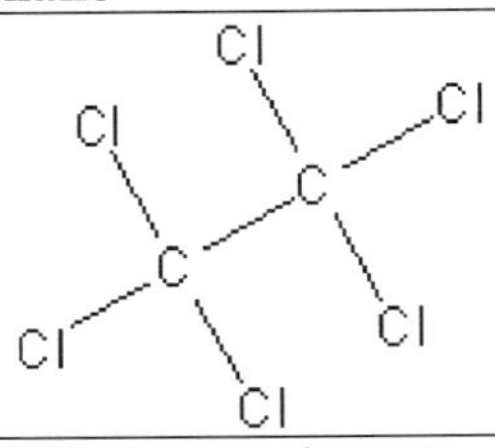

Forms colorless to white crystals, granules, or powder that have a distinct camphor like odor. The crystals are insoluble in water, but soluble in most common solvents.

### 100. Antimony trisulfide

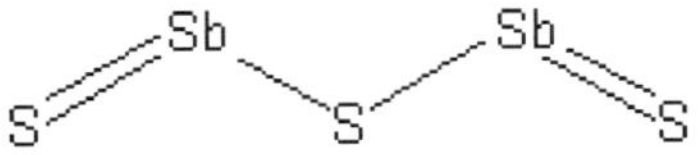

Forms a grayish powder or grayish-black crystalline powder. The powder has a melting point of 550 Celsius. Antimony trisulfide is insoluble in water and all solvents, and reacts with acids forming hydrogen sulfide. Antimony trisulfide is stable, but may ignite upon strong friction or percussion.

### 101. Anthracene

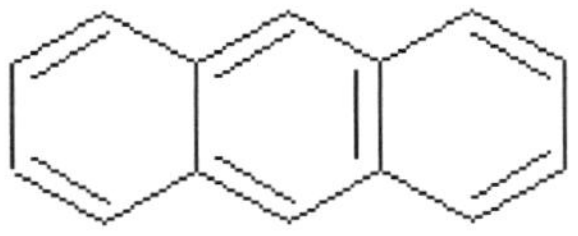

Anthracene forms monoclinic plates, which are colorless when pure, but usually colored yellow with green fluorescence. Anthracene is insoluble in all known solvents. Often a constituent of smoke produced by burning wood.

*- Bullet Tracer Compositions in this section -*

| | |
|---|---|
| **1. 06-01-001A: Infrared Tracer composition for shells:** *32.87% strontium peroxide, 32.87% barium peroxide, 10.95% magnesium carbonate, 10.95% calcium resinate, 10.95% barium nitrate, 1.36% silicon powder, 0.05% impurities* | **2. 06-01-002A: "Invisible" tracer composition visible only by shooters wearing or using night vision goggles, infrared viewers, or thermal imaging devices:** *37.5% strontium peroxide, 37.5% barium peroxide, 15% magnesium carbonate, 10% calcium resinate* |
| **3. 06-01-003A: High performance tracer composition 1:** *46.4% strontium nitrate, 42.5% magnesium, 5% magnesium carbonate, 4% Kynar binder, 2% PVC, 0.10% moisture* | **4. 06-01-004A: High performance white tracer composition with reduced base drag:** *54% magnesium, 30% Teflon binder, 16% Viton A binder* |
| **5. 06-01-005A: High performance pinkish-red tracer composition with reduced base drag:** *45% magnesium, 23% Teflon binder, 17% strontium nitrate, 15% Viton A binder* | **6. 06-01-005B: High performance red tracer composition with reduced base drag:** *45% magnesium, 20% Teflon binder, 20% strontium nitrate, 12% Viton A binder, 3% carbon black* |
| **7. 06-01-006A: Blue tracer composition with gray smoke trail:** *38.5% barium nitrate, 15.4% anhydrous cupric chloride, 15.4% potassium perchlorate, 15.3% hexachlorobenzene, 7.7% sulfur, 7.7% magnesium* | **8. 06-01-007A: High visibility tracer composition:** *49.5% strontium nitrate, 16.8% magnesium, 13.8% calcium resinate, 5.9% aluminum, 4.9% strontium oxalate, 4.9% mercurous chloride, 3.9% sodium nitrate, 0.30% residue* |
| **9. 06-01-007B: High visibility tracer composition (modified):** *49.5% strontium nitrate, 16.8% magnesium, 13.8% calcium resinate, 5.9% aluminum, 4.9% zinc oxalate, 4.9% zinc chloride, 3.9% sodium nitrate, 0.30% residue* | **10. 06-01-007C: High visibility tracer composition with reduced illumination for anti-aircraft guns:** *78% strontium peroxide, 7% calcium resinate, 7% calcium silicide, 4% lead dioxide, 4% barium peroxide* |
| **11. 06-01-008A: Long burning tracer composition containing depleted uranium for use in long range weapons such as sniper rifles, heavy machine guns, mortars, and artillery shells:** *73.2% depleted uranium, 12.2% aluminum/magnesium alloy, 12.2% potassium perchlorate, 2.3% epoxy resin, 0.10% residue* | **12. 06-01-009A: High intensity illuminous tracer composition for use in rifles, heavy machine guns, and flares:** *30% yttrium nitrate, 20% magnesium, 20% barium peroxide, 10% zirconium nitrate, 10% calcium resinate, 10% lead sulpho-cyanate* |
| **13. 06-01-009B: High intensity illuminous tracer composition for use in sniper rifles, heavy machine guns, and flares (modified):** *28.8% sodium nitrate, 24% strontium nitrate, 19.2% calcium resinate, 19.2% magnesium, 4.8% yttrium oxide, 3.8% zirconium oxide, 0.20% residue* | **14. 06-01-010A: High intensity illuminous tracer composition for use in rifles, heavy machine guns, rockets, grenade projectiles, and similar projectiles:** *36.1% strontium nitrate, 24% magnesium, 24% cerium nitrate, 15% calcium resinate, 0.66% thorium nitrate, 0.24% impurities* |
| **15. 06-01-010B: High intensity illuminous tracer composition for use in rifles, heavy machine guns, rockets, grenade projectiles, and similar projectiles (modified):** *31.5% sodium nitrate, 26.3% strontium nitrate, 21% magnesium, 10.5% calcium oxide, 5.2% cerium oxide, 5.2% thorium oxide. 0.30% residue* | **16. 06-01-011A: Tracer composition for use in high velocity projectiles fired by tanks, guns, and other firing weapons whereby the firer/soldier is utilizing laser optic sighting devices:** *40% magnesium, 20% strontium nitrate, 15% potassium nitrate, 9% iron-III-oxide, 5% sodium nitrate, 5% manganese dioxide, 5% Viton A binder, 1% ethyl cellulose* |
| **17. 06-01-011B: Tracer composition for use in high velocity projectiles fired by tanks, guns, and other firing weapons whereby the firer/soldier is utilizing laser optic sighting devices (modified):** *40% magnesium, 20% strontium nitrate, 15% potassium perchlorate, 10% barium peroxide, 5% sodium nitrate, 5% polyvinyl chloride, 5% Viton A binder* | **18. 06-01-012A: Tracer composition with increased red luminosity:** *33.2% strontium nitrate, 33.2% magnesium, 23.9% strontium tartrate, 4.3% hexachlorobenzene, 4.3% stearic acid, 0.78% wood charcoal, 0.32% combined impurities* |
| **19. 06-01-013A: Green tracer composition with increased luminosity:** *41.1% barium nitrate, 29.4% hexachloroethane, 27.4% magnesium, 1.9% linseed oil, 0.2% impurities* | **20. 06-01-013B: Red tracer composition with increased luminosity:** *40% strontium nitrate, 33% magnesium, 25% hexachloroethane, 2% linseed oil* |
| **21. 06-01-013C: Bluish-green tracer composition with high luminosity:** *42% barium nitrate, 40% hexachloroethane, 17% magnesium, 1% copper* | **22. 06-01-014A: Tracer composition with reduced base drag:** *70.9% strontium peroxide, 10% gelatin, 8.2% calcium resinate, 7.3% magnesium, 3.6% carbon black* |
| **23. 06-01-014B: Tracer composition with reduced base drag and increased burn range:** *49.3% strontium nitrate, 25.4% magnesium, 15.3% PVC, 10% gelatin* | **24. 06-01-015A: White light tracer composition:** *57.14% magnesium powder, 38.09% sodium nitrate, 4.76% binder, 0.01% residual balance* |

| | |
|---|---|
| **25. 06-01-016A: White light tracer composition with sparkling effect:** *33.44% aluminum powder, 33.44% sodium chlorate, 16.72% coarse magnesium grains, 6.02% zinc oxide, 5.35% binder, 5.01% potassium dichromate, 0.02% residual balance* | **26. 06-01-017A: Specialty depleted uranium tracer composition for large caliber ammunition:** *50% depleted uranium metal, 25% aluminum magnesium alloy, 25% potassium perchlorate* |
| **27. 06-01-018A: Multi purpose tracer composition for various applications:** *76% barium peroxide, 20% magnesium powder, 4% aluminum stearate* | **28. 06-01-018B: Multi purpose tracer composition for various applications:** *73.5% barium peroxide, 24% magnesium powder, 2.5% magnesium stearate* |
| **29. 06-01-018C: Multi purpose tracer composition for various applications:** *73.07% barium peroxide, 19.23% magnesium powder, 7.69% zinc stearate, 0.01% residual balance* | **30. 06-01-019A: Tracer composition for high caliber projectiles:** *60% barium peroxide, 35% antimony trisulfide, 4% asphaltum, 1% graphite powder* |
| **31. 06-01-020A: Specialty red tracer composition with red smoke trail for use in daytime operations:** *70% strontium peroxide, 15% 1-methylaminoanthraquinone dye agent, 10% calcium resinate, 5% magnesium powder* | **32. 06-01-020B: Specialty tracer composition with red smoke trail for use in daytime operations only:** ***70%*** *strontium peroxide, 10% 1-methylaminoanthraquinone dye agent, 10% calcium resinate, 10% ortho-dihydroxy benzene* |
| **33. 06-01-020C: Specialty tracer composition with yellow smoke trail for use in daytime operations only:** *80% strontium peroxide, 10% auramine dye, 10% calcium resinate* | **34. 06-01-020D: Specialty tracer composition with black smoke trail for use in daytime operations:** *70% strontium peroxide, 15% anthracene, 10% calcium resinate, 5% magnesium powder* |

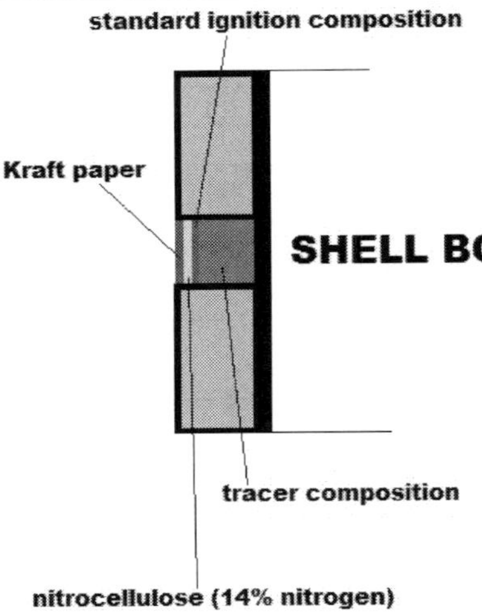


**Standard setup for proper tracer burn**

**06-01-001A: Infrared Tracer composition for shells:**

Into a standard ball mill, filled with 500 grams of steel shot of 10mm diameter, place ***144 grams of strontium peroxide***, followed by ***144 grams of barium peroxide***, followed by ***48 grams of barium nitrate***, followed by ***48 grams of magnesium carbonate***, followed by ***48 grams of calcium resinate***, and finally followed by ***6 grams of finely ground elemental (powdered) silicon***. Thereafter, rotate the mixture at 190 RPM for 6 hours. After 6 hours, remove the mixture from the ball mill, and then separate the mixture from the steel shot in the usual manner. Thereafter, the mixture is ready for pressing. To do so, press the mixture into the tracer compartment of the given projectile under a pressure of 10,000 psi.

**Burn rate (tracer burn out, standard 40mm grenade round on average):** 55 meters
**Burn rate (tracer burn out, standard 7.62mm rifle bullet on average):** 350 meters
**Burn rate (tracer burn out, standard 7.62mm machine gun round on average):** 600 meters
**Burn rate (tracer burn out, standard 12.7mm rifle bullet on average):** 1000 meters
**Burn rate (tracer burn out, standard 25mm cannon round on average):** 1500 meters
**Burn rate (tracer burn out, standard 105mm tank fired projectile on average):** 1800 meters
**Water resistance:** Good.
**Stability:** Can be stored for many years.
**Flammability (1 to 10):** 4 (pressed at 10,000 psi); 4.8 powdered.
**Ease of ignition (1 to 10):** 3 (pressed at 10,000 psi); 5 powdered.
**Tendency to cake:** None.
**Explosive ability:** Cannot be detonated.
**Percentage:** ***32.87% strontium peroxide, 32.87% barium peroxide, 10.95% magnesium carbonate, 10.95% calcium resinate, 10.95% barium nitrate, 1.36% silicon powder, 0.05% impurities***

**Classification:** Deflagrating mixture (classified as 1.3G pyrotechnic composition).

**Danger of preparation:** Can be made by amateurs.
**Use:** Used in military ammunition for rifles, machine guns, low pressure/low velocity projectiles, shoulder fired anti-tank munitions, and tank fired munitions.

<table>
<tr><th>Exploding .50 caliber projectile w/tracer</th><th>Details</th></tr>
<tr><td>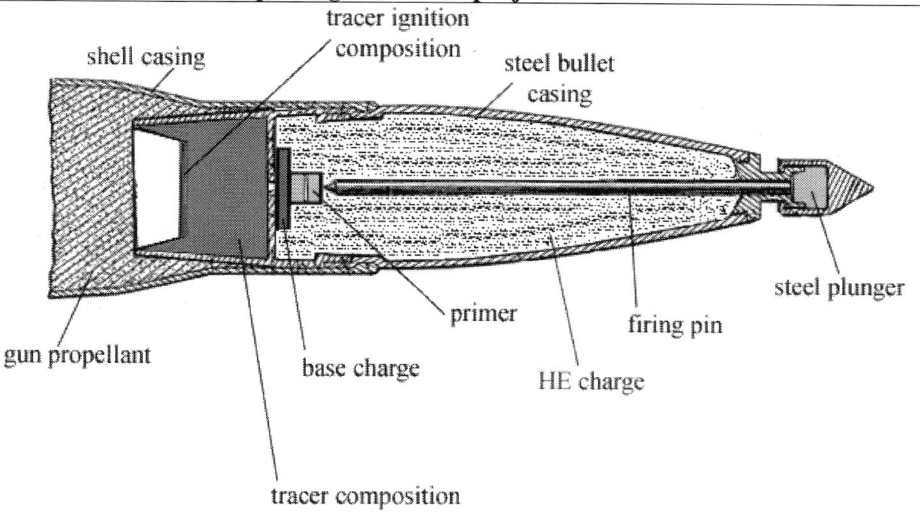
</td><td>In this illustration, a .50 caliber projectile with exploding tip and tracer is shown. The projectile operates using compression, where by the nose of the projectile impacts the target, and the firing pin is compressed inwards, whereby it strikes a primer. The primer sets-off the base charge, which then detonates the HE charge.</td></tr>
</table>

**06-01-002A: "Invisible" tracer composition visible only by shooters wearing or using night vision goggles, infrared viewers, or thermal imaging devices:**
Into a standard ball mill, filled with 500 grams of steel shot of 10mm diameter, place ***150 grams of strontium peroxide***, followed by ***150 grams of barium peroxide***, followed by ***60 grams of magnesium carbonate***, and then followed by ***40 grams of calcium resinate***. Thereafter, rotate the mixture at 180 RPM for 6 hours. After 6 hours, remove the mixture from the ball mill, and then separate the mixture from the steel shot in the usual manner. Thereafter, the mixture is ready for pressing. To do so, press the mixture into the tracer compartment of the given projectile under a pressure of 10,000 psi.
**Burn rate (tracer burn out, standard 40mm grenade round on average):** 45 meters
**Burn rate (tracer burn out, standard 7.62mm rifle bullet on average):** 300 meters
**Burn rate (tracer burn out, standard 7.62mm machine gun round on average):** 600 meters
**Burn rate (tracer burn out, standard 12.7mm rifle bullet on average):** 900 meters
**Burn rate (tracer burn out, standard 25mm cannon round on average):** 1200 meters
**Burn rate (tracer burn out, standard 105mm tank fired projectile on average):** 1850 meters
**Water resistance:** Good.
**Stability:** Can be stored for many years.
**Flammability (1 to 10):** 3.9 (pressed at 10,000 psi); 4.9 powdered.
**Ease of ignition (1 to 10):** 3 (pressed at 10,000 psi); 5 powdered.
**Tendency to cake:** None.
**Explosive ability:** Cannot be detonated.
**Percentage:** ***37.5% strontium peroxide, 37.5% barium peroxide, 15% magnesium carbonate, 10% calcium resinate***
**Classification:** Deflagrating mixture (classified as 1.3G pyrotechnic composition).

**Danger of preparation:** Can be made by amateurs.
**Use:** Used in military ammunition for rifles, machine guns, low pressure/low velocity projectiles, shoulder fired anti-tank munitions, and tank fired munitions.

### 06-01-003A: High performance tracer composition 1:

Into a standard ball mill, filled with 500 grams of steel shot of 10mm diameter, place ***20 grams of Kynar 9301 trademark*** (from Penwalt Co.; terpolymer of vinylidene fluoride, hexafluoropropylene and polytetrafluoroethylene), and then add in ***25 grams of magnesium carbonate,*** followed by ***10 grams of powdered PVC,*** followed by ***232 grams of strontium nitrate***, and then followed by ***212.5 grams of magnesium powder***. Thereafter, rotate the mixture at 180 RPM for 6 hours. After 6 hours, remove the mixture from the ball mill, and then separate the mixture from the steel shot in the usual manner. Thereafter, the mixture is ready for pressing. To do so, press the mixture into the tracer compartment of the given projectile under a pressure of 10,000 psi.

**Burn rate (tracer burn out, standard 40mm grenade round on average):** 55 meters
**Burn rate (tracer burn out, standard 7.62mm rifle bullet on average):** 310 meters
**Burn rate (tracer burn out, standard 7.62mm machine gun round on average):** 610 meters
**Burn rate (tracer burn out, standard 12.7mm rifle bullet on average):** 950 meters
**Burn rate (tracer burn out, standard 25mm cannon round on average):** 1250 meters
**Burn rate (tracer burn out, standard 105mm tank fired projectile on average):** 1950 meters
**Water resistance:** Good.
**Stability:** Can be stored for many years.
**Flammability (1 to 10):** 4.9 (pressed at 10,000 psi); 5.5 powdered.
**Ease of ignition (1 to 10):** 3.9 (pressed at 10,000 psi); 7 powdered.
**Tendency to cake:** None.
**Explosive ability:** Cannot be detonated.
**Percentage:** ***46.4% strontium nitrate, 42.5% magnesium, 5% magnesium carbonate, 4% Kynar binder, 2% PVC, 0.10% moisture***
**Classification:** Deflagrating mixture (classified as 1.3G pyrotechnic composition).

**Danger of preparation:** Can be made by amateurs.
**Use:** Used in military ammunition for rifles, machine guns, low pressure/low velocity projectiles, shoulder fired anti-tank munitions, and tank fired munitions.

### 06-01-004A: High performance white tracer composition with reduced base drag:

Into a standard ball mill, filled with 500 grams of steel shot of 10mm diameter, place ***80 grams of Viton A*** (fluorine-containing copolymer), followed by ***150 grams of Teflon No. 7,*** and then followed by ***270 grams of magnesium powder***. Thereafter, rotate the mixture at 200 RPM for 6 hours. After 6 hours, remove the mixture from the ball mill, and then separate the mixture from the steel shot in the usual manner. Thereafter, the mixture is ready for pressing. To do so, press the mixture into the tracer compartment of the given projectile under a pressure of 10,000 psi.

**Burn rate (tracer burn out, standard 40mm grenade round on average):** 50 meters
**Burn rate (tracer burn out, standard 7.62mm rifle bullet on average):** 300 meters
**Burn rate (tracer burn out, standard 7.62mm machine gun round on average):** 600 meters
**Burn rate (tracer burn out, standard 12.7mm rifle bullet on average):** 1050 meters
**Burn rate (tracer burn out, standard 25mm cannon round on average):** 1200 meters
**Burn rate (tracer burn out, standard 105mm tank fired projectile on average):** 1950 meters
**Water resistance:** Good.
**Stability:** Can be stored for many years.
**Flammability (1 to 10):** 4.7 (pressed at 10,000 psi); 5 powdered.
**Ease of ignition (1 to 10):** 4 (pressed at 10,000 psi); 7.2 powdered.
**Tendency to cake:** None.
**Explosive ability:** Cannot be detonated.
**Percentage:** ***54% magnesium, 30% Teflon binder, 16% Viton A binder***
**Classification:** Deflagrating mixture (classified as 1.3G pyrotechnic composition).

**Danger of preparation:** Can be made by amateurs.
**Use:** Used in military ammunition for rifles, machine guns, low pressure/low velocity projectiles, shoulder fired anti-tank munitions, and tank fired munitions.

**Tank fired 30mm Anti-tank projectile w/tracer**

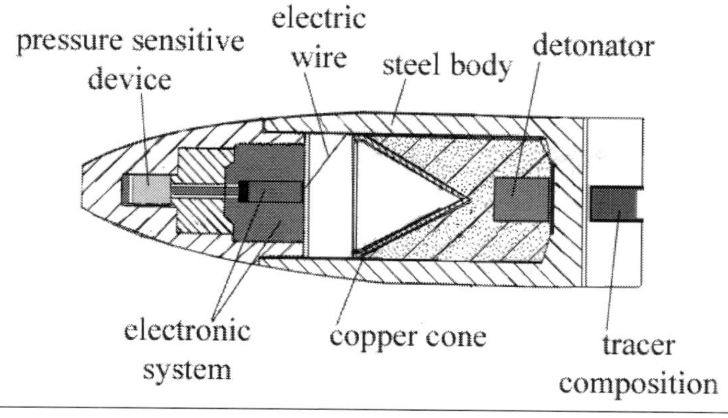


### 06-01-005A: High performance pinkish-red tracer composition with reduced base drag:

Into a standard ball mill, filled with 500 grams of steel shot of 10mm diameter, place ***75 grams of Viton A*** (fluorine-containing copolymer), followed by ***115 grams of Teflon No. 7,*** followed by ***225 grams of magnesium powder***, and then followed by ***85 grams of strontium nitrate***. Thereafter, rotate the mixture at 150 RPM for 6 hours. After 6 hours, remove the mixture from the ball mill, and then separate the mixture from the steel shot in the usual manner. Thereafter, the mixture is ready for pressing. To do so, press the mixture into the tracer compartment of the given projectile under a pressure of 10,000 psi.

**Burn rate (tracer burn out, standard 40mm grenade round on average):** 60 meters
**Burn rate (tracer burn out, standard 7.62mm rifle bullet on average):** 310 meters
**Burn rate (tracer burn out, standard 7.62mm machine gun round on average):** 620 meters
**Burn rate (tracer burn out, standard 12.7mm rifle bullet on average):** 1090 meters
**Burn rate (tracer burn out, standard 25mm cannon round on average):** 1250 meters
**Burn rate (tracer burn out, standard 105mm tank fired projectile on average):** 1960 meters
**Water resistance:** Good.
**Stability:** Can be stored for many years.
**Flammability (1 to 10):** 4.9 (pressed at 10,000 psi); 5.5 powdered.
**Ease of ignition (1 to 10):** 4.6 (pressed at 10,000 psi); 7.9 powdered.
**Tendency to cake:** None.
**Explosive ability:** Cannot be detonated.
**Percentage:** ***45% magnesium, 23% Teflon binder, 17% strontium nitrate, 15% Viton A binder***
**Classification:** Deflagrating mixture (classified as 1.3G pyrotechnic composition).

**Danger of preparation:** Can be made by amateurs.
**Use:** Used in military ammunition for rifles, machine guns, low pressure/low velocity projectiles, shoulder fired anti-tank munitions, and tank fired munitions.

### 06-01-005B: High performance red tracer composition with reduced base drag:

Into a standard ball mill, filled with 500 grams of steel shot of 10mm diameter, place ***60 grams of Viton A*** (fluorine-containing copolymer), followed by ***100 grams of Teflon No. 7,*** followed by ***225 grams of magnesium powder***, followed by ***100 grams of strontium nitrate,*** and then followed by ***15 grams of carbon black***. Thereafter, rotate the mixture at 150 RPM for 6 hours. After 6 hours, remove the mixture from the ball mill, and then separate the mixture from the steel shot in the usual manner. Thereafter, the mixture is ready for pressing. To do so, press the mixture into the tracer compartment of the given projectile under a pressure of 10,000 psi.

**Burn rate (tracer burn out, standard 40mm grenade round on average):** 55 meters
**Burn rate (tracer burn out, standard 7.62mm rifle bullet on average):** 325 meters
**Burn rate (tracer burn out, standard 7.62mm machine gun round on average):** 600 meters
**Burn rate (tracer burn out, standard 12.7mm rifle bullet on average):** 1100 meters
**Burn rate (tracer burn out, standard 25mm cannon round on average):** 1200 meters
**Burn rate (tracer burn out, standard 105mm tank fired projectile on average):** 1900 meters
**Water resistance:** Good.
**Stability:** Can be stored for many years.
**Flammability (1 to 10):** 4.9 (pressed at 10,000 psi); 5.5 powdered.
**Ease of ignition (1 to 10):** 4.9 (pressed at 10,000 psi); 8 powdered.
**Tendency to cake:** None.
**Explosive ability:** Cannot be detonated.
**Percentage:** ***45% magnesium, 20% Teflon binder, 20% strontium nitrate, 12% Viton A binder, 3% carbon black***
**Classification:** Deflagrating mixture (classified as 1.3G pyrotechnic composition).

**Danger of preparation:** Can be made by amateurs.
**Use:** Used in military ammunition for rifles, machine guns, low pressure/low velocity projectiles, shoulder fired anti-tank munitions, and tank fired munitions.

**120mm Tank-fired anti-tank projectile (HEAT)**

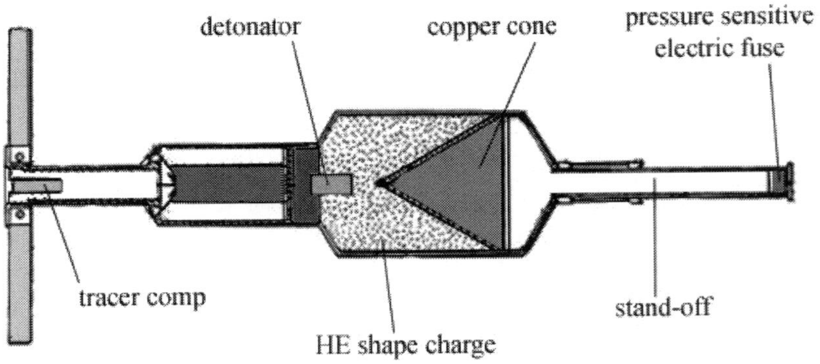


### 06-01-006A: Blue tracer composition with gray smoke trail:

Into a standard ball mill, filled with 500 grams of Teflon coated steel shot of 10mm diameter, place ***76.5 grams of hexachlorobenzene,*** followed by ***77 grams of anhydrous cupric chloride,*** followed by ***38.5 grams of sulfur powder,*** followed by ***192.5 grams of barium nitrate,*** followed by ***38.5 grams of potassium perchlorate,*** and then followed by ***77 grams of magnesium powder***. Thereafter, rotate the mixture at 150 RPM for 6 hours. After 6 hours, remove the mixture from the ball mill, and then separate the mixture from the steel shot in the usual manner. Thereafter, the mixture is ready for pressing. To do so, press the mixture into the tracer compartment of the given projectile under a pressure of 10,000 psi.
**Burn rate (tracer burn out, standard 40mm grenade round on average):** 50 meters
**Burn rate (tracer burn out, standard 7.62mm rifle bullet on average):** 320 meters
**Burn rate (tracer burn out, standard 7.62mm machine gun round on average):** 600 meters
**Burn rate (tracer burn out, standard 12.7mm rifle bullet on average):** 1200 meters
**Burn rate (tracer burn out, standard 25mm cannon round on average):** 1300 meters
**Burn rate (tracer burn out, standard 105mm tank fired projectile on average):** 1950 meters
**Water resistance:** Good.
**Stability:** Can be stored for many years.
**Flammability (1 to 10):** 4.9 (pressed at 10,000 psi); 5.5 powdered.
**Ease of ignition (1 to 10):** 4.9 (pressed at 10,000 psi); 8.1 powdered.
**Tendency to cake:** None.
**Explosive ability:** Cannot be detonated.
**Percentage:** ***38.5% barium nitrate, 15.4% anhydrous cupric chloride, 15.4% potassium perchlorate, 15.3% hexachlorobenzene, 7.7% sulfur, 7.7% magnesium powder***
**Classification:** Deflagrating mixture (classified as 1.3G pyrotechnic composition).

**Danger of preparation:** Can be made by amateurs.
**Use:** Used in military ammunition for rifles, machine guns, low pressure/low velocity projectiles, shoulder fired anti-tank munitions, and tank fired munitions.

**Tandem "binary" 210mm HE anti-ship projectile**

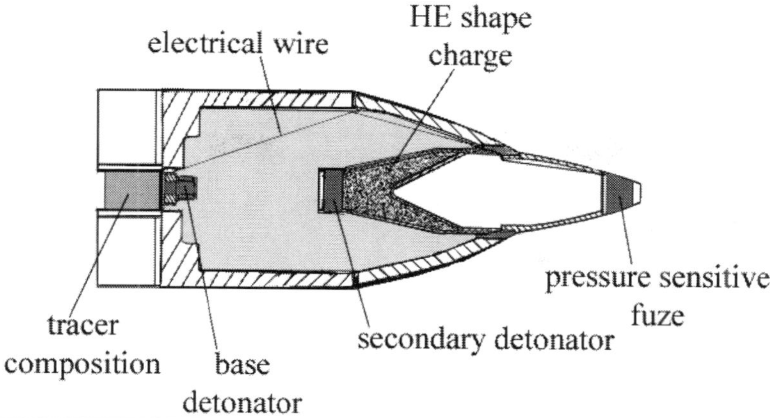


### 06-01-007A: High visibility tracer composition:

Into a standard ball mill, filled with 500 grams of Teflon coated steel shot of 10mm diameter, place ***140 grams of calcium resinate,*** followed by ***60 grams of standard powdered aluminum,*** followed by ***170 grams of standard powdered magnesium,*** followed by ***50 grams of strontium oxalate,*** followed by ***500 grams of strontium nitrate,*** followed by ***40 grams of sodium nitrate,*** followed by ***50 grams of mercurous chloride***. Thereafter, rotate the mixture at 170 RPM for 5 hours. After 5 hours, remove the mixture from the ball mill, and then separate the mixture from the steel shot in the usual manner. Thereafter, the mixture is ready for pressing. To do so, press the mixture into the tracer compartment of the given projectile under a pressure of 10,000 psi.

**Burn rate (tracer burn out, standard 40mm grenade round on average):** 50 meters
**Burn rate (tracer burn out, standard 7.62mm rifle bullet on average):** 300 meters
**Burn rate (tracer burn out, standard 7.62mm machine gun round on average):** 600 meters
**Burn rate (tracer burn out, standard 12.7mm rifle bullet on average):** 1200 meters
**Burn rate (tracer burn out, standard 25mm cannon round on average):** 1050 meters
**Burn rate (tracer burn out, standard 105mm tank fired projectile on average):** 1900 meters
**Water resistance:** Good.
**Stability:** Can be stored for many years.
**Flammability (1 to 10):** 4.2 (pressed at 10,000 psi); 5 powdered.
**Ease of ignition (1 to 10):** 4 (pressed at 10,000 psi); 7.9 powdered.
**Tendency to cake:** None.
**Explosive ability:** Cannot be detonated.
**Percentage:** ***49.5% strontium nitrate, 16.8% magnesium, 13.8% calcium resinate, 5.9% aluminum, 4.9% strontium oxalate, 4.9% mercurous chloride, 3.9% sodium nitrate, 0.30% residue***
**Classification:** Deflagrating mixture (classified as 1.3G pyrotechnic composition).

**Danger of preparation:** Can be made by amateurs.
**Use:** Used in military ammunition for rifles, machine guns, low pressure/low velocity projectiles, shoulder fired anti-tank munitions, and tank fired munitions.

**"Direct fire", close proximity 160mm anti-tank artillery shell w/tracer (tracer burn out, 300 meters)**

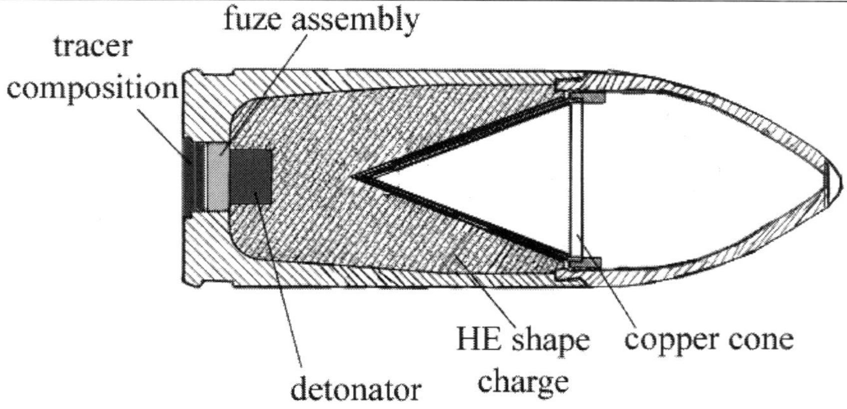

### 06-01-007B: High visibility tracer composition (modified):

Into a suitable ball mill filled with 500 grams of Teflon coated steel shot of 10 millimeters in diameter, place ***140 grams of calcium resinate,*** followed by ***60 grams of standard powdered aluminum,*** followed by ***170 grams of standard powdered magnesium,*** followed by ***50 grams of zinc oxalate,*** followed by ***500 grams of strontium nitrate,*** followed by ***40 grams of sodium nitrate,*** and then followed by ***50 grams of zinc chloride***. Thereafter, rotate the mixture at 170 RPM for 6 hours. After 6 hours, remove the mixture from the ball mill, and then separate the mixture from the steel shot in the usual manner. Thereafter, the mixture is ready for pressing. To do so, press the mixture into the tracer compartment of the given projectile under a pressure of 10,000 psi.

**Burn rate (tracer burn out, standard 40mm grenade round on average):** 60 meters
**Burn rate (tracer burn out, standard 40mm special round (US/NATO MK19) on average):** 350 meters
**Burn rate (tracer burn out, standard 7.62mm rifle bullet on average):** 300 meters
**Burn rate (tracer burn out, standard 7.62mm machine gun round on average):** 600 meters
**Burn rate (tracer burn out, standard 12.7mm rifle bullet on average):** 1200 meters
**Burn rate (tracer burn out, standard 25mm cannon round on average):** 1000 meters
**Burn rate (tracer burn out, standard 105mm tank fired projectile on average):** 1800 meters
**Water resistance:** Good.
**Stability:** Can be stored for many years.
**Flammability (1 to 10):** 4.1 (pressed at 10,000 psi); 4.9 powdered.
**Ease of ignition (1 to 10):** 4 (pressed at 10,000 psi); 7.9 powdered.
**Tendency to cake:** None.
**Explosive ability:** Cannot be detonated.
**Percentage:** ***49.5% strontium nitrate, 16.8% magnesium, 13.8% calcium resinate, 5.9% aluminum, 4.9% zinc oxalate, 4.9% zinc chloride, 3.9% sodium nitrate, 0.30% residue***
**Classification:** Deflagrating mixture (classified as 1.3G pyrotechnic composition).

**Danger of preparation:** Can be made by amateurs.
**Use:** Used in military ammunition for rifles, machine guns, low pressure/low velocity projectiles, shoulder fired anti-tank munitions, and tank fired munitions.

### 06-01-007C: High visibility tracer composition with reduced illumination for anti-aircraft guns:

Into a suitable ball mill filled with 500 grams of Teflon coated steel shot of 10 millimeters in diameter, place ***35 grams of calcium resinate,*** followed by ***35 grams of calcium silicide,*** followed by ***20 grams of lead dioxide,*** followed by ***20 grams of barium peroxide,*** and then followed by ***390 grams of strontium peroxide***. Thereafter, rotate the mixture at 170 RPM for 6 hours. After 6 hours, remove the mixture from the ball mill, and then separate the mixture from the steel shot in the usual manner. Thereafter, the mixture is ready for pressing. To do so, press the mixture into the tracer compartment of the given projectile under a pressure of 10,000 psi.

**Burn rate (tracer burn out, standard 40mm grenade round on average):** 55 meters
**Burn rate (tracer burn out, standard 40mm special round (US/NATO MK19) on average):** 375 meters
**Burn rate (tracer burn out, standard 7.62mm rifle bullet on average):** 300 meters
**Burn rate (tracer burn out, standard 7.62mm machine gun round on average):** 600 meters
**Burn rate (tracer burn out, standard 12.7mm rifle bullet on average):** 1210 meters
**Burn rate (tracer burn out, standard 25mm cannon round on average):** 1000 meters
**Burn rate (tracer burn out, standard 105mm tank fired projectile on average):** 1700 meters
**Water resistance:** Good.
**Stability:** Can be stored for many years.
**Flammability (1 to 10):** 3.7 (pressed at 10,000 psi); 4.6 powdered.
**Ease of ignition (1 to 10):** 3.1 (pressed at 10,000 psi); 6 powdered.
**Tendency to cake:** None.
**Explosive ability:** Cannot be detonated.
**Percentage:** ***78% strontium peroxide, 7% calcium resinate, 7% calcium silicide, 4% lead dioxide, 4% barium peroxide***
**Classification:** Deflagrating mixture (classified as 1.3G pyrotechnic composition).

**Danger of preparation:** Can be made by amateurs.
**Use:** Used in military ammunition for rifles, machine guns, low pressure/low velocity projectiles, shoulder fired anti-tank munitions, and tank fired munitions.

### 06-01-008A: Long burning tracer composition containing depleted uranium for use in long range weapons such as sniper rifles, heavy machine guns, mortars, and artillery shells:

Into a suitable ball mill filled with 500 grams of Teflon coated steel shot of 10 millimeters in diameter, place ***375 grams of depleted uranium*** of 20 to 40 mesh, followed by ***62.5 grams of finely powdered aluminum/magnesium alloy***, containing 50% aluminum and 50% magnesium, followed by ***62.5 grams of potassium perchlorate***, and then followed by ***12 grams of any standard epoxy resin***. Thereafter, rotate the mixture at 200 RPM for 6 hours. After 6 hours, remove the mixture from the ball mill, and then separate the mixture from the steel shot in the usual manner. Thereafter, the mixture is ready for pressing. To do so, press the mixture into the tracer compartment of the given projectile under a pressure of 10,000 psi.
**Burn rate (tracer burn out, standard 7.62mm rifle bullet on average):** 450 meters
**Burn rate (tracer burn out, standard 7.62mm machine gun round on average):** 900 meters
**Burn rate (tracer burn out, standard 12.7mm rifle bullet on average):** 1500 meters
**Burn rate (tracer burn out, standard 12.7mm machine gun round on average):** 2000 meters
**Burn rate (tracer burn out, standard 120mm tank fired projectile on average):** 1900 meters
**Burn rate (tracer burn out, standard 155mm artillery shell practice round on average):** 15,000 meters
**Water resistance:** Good.
**Stability:** Can be stored for many years.
**Flammability (1 to 10):** 3.3 (pressed at 10,000 psi); 4.3 powdered.
**Ease of ignition (1 to 10):** 3 (pressed at 10,000 psi); 5 powdered.
**Tendency to cake:** None.
**Explosive ability:** Cannot be detonated.
**Percentage:** ***73.2% depleted uranium, 12.2% aluminum/magnesium alloy, 12.2% potassium perchlorate, 2.3% epoxy resin, and 0.10% residue***
**Classification:** Deflagrating mixture (classified as 1.3G pyrotechnic composition; radiation hazard class E).

**Danger of preparation:** Can be made by amateurs, but caution should be exercised as depleted uranium is moderately radioactive.
**Use:** Used in sniper rifles of up to .50 caliber, machines guns ranging form 7.62 millimeters up to 18 millimeters, and in training rounds for artillery practice.

### 06-01-009A: High intensity illuminous tracer composition for use in rifles, heavy machine guns, and flares:

Into a suitable ball mill filled with 500 grams of Teflon coated steel shot of 10 millimeters in diameter, place ***50 grams of standard powdered magnesium***, followed by ***75 grams of yttrium nitrate***, followed by ***25 grams of zirconium nitrate***, followed by ***25 grams of calcium resinate***, followed by ***25 grams of lead sulpho-cyanate,*** and then followed by ***50 grams of barium peroxide***. Thereafter, rotate the mixture at 180 RPM for 6 hours. After 6 hours, remove the mixture from the ball mill, and then separate the mixture from the steel shot in the usual manner. Thereafter, the mixture is ready for pressing. To do so, press the mixture into the tracer compartment of the given projectile under a pressure of 10,000 psi.
**Burn rate (tracer burn out, standard 40mm grenade round on average):** 50 meters
**Burn rate (tracer burn out, standard 40mm special round (US/NATO MK19) on average):** 400 meters
**Burn rate (tracer burn out, standard 7.62mm rifle bullet on average):** 325 meters
**Burn rate (tracer burn out, standard 7.62mm machine gun round on average):** 600 meters
**Burn rate (tracer burn out, standard 12.7mm rifle bullet on average):** 1200 meters
**Burn rate (tracer burn out, standard 12.7mm machine gun round on average):** 1500 meters
**Burn rate (tracer burn out, standard 25mm cannon round on average):** 900 meters
**Burn rate (tracer burn out, standard 120mm tank fired projectile on average):** 1750 meters
**Water resistance:** Good.
**Stability:** Can be stored for many years.
**Flammability (1 to 10):** 3.5 (pressed at 10,000 psi); 4.7 powdered.
**Ease of ignition (1 to 10):** 3.2 (pressed at 10,000 psi); 6 powdered.
**Tendency to cake:** None.
**Explosive ability:** Cannot be detonated.
**Percentage:** ***30% yttrium nitrate, 20% magnesium, 20% barium peroxide, 10% zirconium nitrate, 10% calcium resinate, 10% lead sulpho-cyanate***
**Classification:** Deflagrating mixture (classified as 1.3G pyrotechnic composition).

**Danger of preparation:** Can be made by amateurs.
**Use:** Used in military ammunition for rifles, machine guns, low pressure/low velocity projectiles, shoulder fired anti-tank munitions, and tank fired munitions.

### 06-01-009B: High intensity illuminous tracer composition for use in sniper rifles, heavy machine guns, and flares (modified):

Into a suitable ball mill filled with 500 grams of steel shot of 10 millimeters in diameter, place ***200 grams of standard powdered magnesium***, followed by ***50 grams of yttrium oxide ($Y_2O_3$)***, followed by ***40 grams of zirconium oxide***, followed by ***200 grams of calcium resinate***, followed by ***300 grams of sodium nitrate,*** and then followed by ***250 grams of strontium nitrate***. Thereafter, rotate the mixture at 190 RPM for 6 hours. After 6 hours, remove the mixture from the ball mill, and then separate the mixture from the steel shot in the usual manner. Thereafter, the mixture is ready for pressing. To do so, press the mixture into the tracer compartment of the given projectile under a pressure of 10,000 psi.
**Burn rate (tracer burn out, standard 40mm grenade round on average):** 55 meters
**Burn rate (tracer burn out, standard 40mm special round (US/NATO MK19) on average):** 410 meters
**Burn rate (tracer burn out, standard 7.62mm rifle bullet on average):** 300 meters
**Burn rate (tracer burn out, standard 7.62mm machine gun round on average):** 600 meters
**Burn rate (tracer burn out, standard 12.7mm rifle bullet on average):** 1210 meters
**Burn rate (tracer burn out, standard 12.7mm machine gun round on average):** 1550 meters
**Burn rate (tracer burn out, standard 25mm cannon round on average):** 900 meters
**Burn rate (tracer burn out, standard 120mm tank fired projectile on average):** 1700 meters
**Water resistance:** Good.
**Stability:** Can be stored for many years.
**Flammability (1 to 10):** 3.4 (pressed at 10,000 psi); 5 powdered.
**Ease of ignition (1 to 10):** 3 (pressed at 10,000 psi); 5.8 powdered.
**Tendency to cake:** None.
**Explosive ability:** Cannot be detonated.
**Percentage:** ***28.8% sodium nitrate, 24% strontium nitrate, 19.2% calcium resinate, 19.2% magnesium, 4.8% yttrium oxide, 3.8% zirconium oxide, 0.20% residue***
**Classification:** Deflagrating mixture (classified as 1.3G pyrotechnic composition).

**Danger of preparation:** Can be made by amateurs.
**Use:** Used in military ammunition for rifles, machine guns, low pressure/low velocity projectiles, shoulder fired anti-tank munitions, and tank fired munitions.

### 06-01-010A: High intensity illuminous tracer composition for use in rifles, heavy machine guns, rockets, grenade projectiles, and similar projectiles:

Into a suitable ball mill filled with 500 grams of steel shot of 10 millimeters in diameter, place ***300 grams of strontium nitrate***, followed by ***200 grams of cerium nitrate***, followed by ***5.5 grams of thorium nitrate***, followed by ***200 grams of standard magnesium powder***, and then followed by ***125 grams of calcium resinate***. Thereafter, rotate the mixture at 190 RPM for 6 hours. After 6 hours, remove the mixture from the ball mill, and then separate the mixture from the steel shot in the usual manner. Thereafter, the mixture is ready for pressing. To do so, press the mixture into the tracer compartment of the given projectile under a pressure of 10,000 psi.
**Burn rate (tracer burn out, standard 40mm grenade round on average):** 50 meters
**Burn rate (tracer burn out, standard 40mm special round (US/NATO MK19) on average):** 400 meters
**Burn rate (tracer burn out, standard 7.62mm rifle bullet on average):** 300 meters
**Burn rate (tracer burn out, standard 7.62mm machine gun round on average):** 610 meters
**Burn rate (tracer burn out, standard 12.7mm rifle bullet on average):** 1200 meters
**Burn rate (tracer burn out, standard 12.7mm machine gun round on average):** 1590 meters
**Burn rate (tracer burn out, standard 25mm cannon round on average):** 910 meters
**Burn rate (tracer burn out, standard 120mm tank fired projectile on average):** 1750 meters
**Water resistance:** Good.
**Stability:** Can be stored for many years.
**Flammability (1 to 10):** 3.3 (pressed at 10,000 psi); 5.1 powdered.
**Ease of ignition (1 to 10):** 3.1 (pressed at 10,000 psi); 5.9 powdered.
**Tendency to cake:** None.
**Explosive ability:** Cannot be detonated.
**Percentage:** ***36.1% strontium nitrate, 24%magnesium, 24% cerium nitrate, 15% calcium resinate, 0.66% thorium nitrate, 0.24% impurities***
**Classification:** Deflagrating mixture (classified as 1.3G pyrotechnic composition).

**Danger of preparation:** Can be made by amateurs.

**Use:** Used in military ammunition for rifles, machine guns, low pressure/low velocity projectiles, shoulder fired anti-tank munitions, and tank fired munitions.

### 06-01-010B: High intensity illuminous tracer composition for use in rifles, heavy machine guns, rockets, grenade projectiles, and similar projectiles (modified):

Into a suitable ball mill filled with 500 grams of steel shot of 10 millimeters in diameter, place ***250 grams of strontium nitrate***, followed by ***50 grams of cerium oxide***, followed by ***50 grams of thorium oxide***, followed by ***200 grams of standard magnesium powder***, followed by ***100 grams of calcium oxide***, and then followed by ***300 grams of sodium nitrate***. Thereafter, rotate the mixture at 190 RPM for 6 hours. After 6 hours, remove the mixture from the ball mill, and then separate the mixture from the steel shot in the usual manner. Thereafter, the mixture is ready for pressing. To do so, press the mixture into the tracer compartment of the given projectile under a pressure of 10,000 psi.
**Burn rate (tracer burn out, standard 40mm grenade round on average):** 50 meters
**Burn rate (tracer burn out, standard 40mm special round (US/NATO MK19) on average):** 410 meters
**Burn rate (tracer burn out, standard 7.62mm rifle bullet on average):** 300 meters
**Burn rate (tracer burn out, standard 7.62mm machine gun round on average):** 620 meters
**Burn rate (tracer burn out, standard 12.7mm rifle bullet on average):** 1200 meters
**Burn rate (tracer burn out, standard 12.7mm machine gun round on average):** 1600 meters
**Burn rate (tracer burn out, standard 25mm cannon round on average):** 990 meters
**Burn rate (tracer burn out, standard 120mm tank fired projectile on average):** 1800 meters
**Water resistance:** Good.
**Stability:** Can be stored for many years.
**Flammability (1 to 10):** 3.2 (pressed at 10,000 psi); 4.2 powdered.
**Ease of ignition (1 to 10):** 3 (pressed at 10,000 psi); 5.5 powdered.
**Tendency to cake:** None.
**Explosive ability:** Cannot be detonated.
**Percentage:** ***31.5% sodium nitrate, 26.3% strontium nitrate, 21% magnesium, 10.5% calcium oxide, 5.2% cerium oxide, 5.2% thorium oxide. 0.30% residue***
**Classification:** Deflagrating mixture (classified as 1.3G pyrotechnic composition).

**Danger of preparation:** Can be made by amateurs.
**Use:** Used in military ammunition for rifles, machine guns, low pressure/low velocity projectiles, shoulder fired anti-tank munitions, and tank fired munitions.

### 06-01-011A: Tracer composition for use in high velocity projectiles fired by tanks, guns, and other firing weapons whereby the firer/soldier is utilizing laser optic sighting devices:

Into a suitable ball mill filled with 500 grams of steel shot of 10 millimeters in diameter, place ***200 grams of standard powdered magnesium***, followed by ***100 grams of strontium nitrate***, followed by ***25 grams of sodium nitrate***, followed by ***25 grams of manganese dioxide***, followed by ***45 grams of finely divided red iron-III-oxide***, followed by ***75 grams of potassium nitrate***, followed by ***5 grams of ethyl cellulose***, and then followed by ***25 grams of Viton A binder***. Thereafter, rotate the mixture at 200 RPM for 4 hours. After 4 hours, remove the mixture from the ball mill, and then separate the mixture from the steel shot in the usual manner. Thereafter, the mixture is ready for pressing. To do so, press the mixture into the tracer compartment of the given projectile under a pressure of 10,000 psi.
**Burn rate (tracer burn out, standard 40mm grenade round on average):** 50 meters
**Burn rate (tracer burn out, standard 40mm special round (US/NATO MK19) on average):** 400 meters
**Burn rate (tracer burn out, standard 7.62mm rifle bullet on average):** 300 meters
**Burn rate (tracer burn out, standard 7.62mm machine gun round on average):** 600 meters
**Burn rate (tracer burn out, standard 12.7mm rifle bullet on average):** 1200 meters
**Burn rate (tracer burn out, standard 12.7mm machine gun round on average):** 1650 meters
**Burn rate (tracer burn out, standard 25mm cannon round on average):** 980 meters
**Burn rate (tracer burn out, standard 120mm tank fired projectile on average):** 1800 meters
**Water resistance:** Good.
**Stability:** Can be stored for many years.
**Flammability (1 to 10):** 3 (pressed at 10,000 psi); 4.1 powdered.
**Ease of ignition (1 to 10):** 3 (pressed at 10,000 psi); 4.7 powdered.
**Tendency to cake:** None.
**Explosive ability:** Cannot be detonated.
**Percentage:** ***40% magnesium, 20% strontium nitrate, 15% potassium nitrate, 9% iron-III-oxide, 5% sodium nitrate, 5% manganese dioxide, 5% Viton A binder, 1% ethyl cellulose***
**Classification:** Deflagrating mixture (classified as 1.3G pyrotechnic composition).

**Danger of preparation:** Can be made by amateurs.
**Use:** Used in military ammunition for rifles, machine guns, low pressure/low velocity projectiles, shoulder fired anti-tank munitions, and tank fired munitions.

### 06-01-011B: Tracer composition for use in high velocity projectiles fired by tanks, guns, and other firing weapons whereby the firer/soldier is utilizing laser optic sighting devices (modified):

Into a suitable ball mill filled with 500 grams of steel shot of 10 millimeters in diameter, place ***200 grams of standard powdered magnesium***, followed by ***100 grams of strontium nitrate***, followed by ***25 grams of sodium nitrate***, followed by ***25 grams of polyvinyl chloride***, followed by ***50 grams of barium peroxide***, followed by ***75 grams of potassium perchlorate***, and then followed by ***25 grams of Viton A binder***. Thereafter, rotate the mixture at 200 RPM for 3 hours. After 3 hours, remove the mixture from the ball mill, and then separate the mixture from the steel shot in the usual manner. Thereafter, the mixture is ready for pressing. To do so, press the mixture into the tracer compartment of the given projectile under a pressure of 10,000 psi.
**Burn rate (tracer burn out, standard 40mm grenade round on average):** 55 meters
**Burn rate (tracer burn out, standard 40mm special round (US/NATO MK19) on average):** 410 meters
**Burn rate (tracer burn out, standard 7.62mm rifle bullet on average):** 320 meters
**Burn rate (tracer burn out, standard 7.62mm machine gun round on average):** 610 meters
**Burn rate (tracer burn out, standard 12.7mm rifle bullet on average):** 1210 meters
**Burn rate (tracer burn out, standard 12.7mm machine gun round on average):** 1660 meters
**Burn rate (tracer burn out, standard 25mm cannon round on average):** 1000 meters
**Burn rate (tracer burn out, standard 120mm tank fired projectile on average):** 1850 meters
**Water resistance:** Good.
**Stability:** Can be stored for many years.
**Flammability (1 to 10):** 3.5 (pressed at 10,000 psi); 4.8 powdered.
**Ease of ignition (1 to 10):** 3.1 (pressed at 10,000 psi); 5 powdered.
**Tendency to cake:** None.
**Explosive ability:** Cannot be detonated.
**Percentage:** ***40% magnesium, 20% strontium nitrate, 15% potassium perchlorate, 10% barium peroxide, 5% sodium nitrate, 5% polyvinyl chloride, 5% Viton A binder***
**Classification:** Deflagrating mixture (classified as 1.3G pyrotechnic composition).

**Danger of preparation:** Can be made by amateurs.
**Use:** Used in military ammunition for rifles, machine guns, low pressure/low velocity projectiles, shoulder fired anti-tank munitions, and tank fired munitions.

### 06-01-012A: Tracer composition with increased red luminosity:

Into a suitable ball mill filled with 500 grams of steel shot of 10 millimeters in diameter, place ***190 grams of finely powdered magnesium***, followed by ***190 grams of strontium nitrate***, followed by ***136.5 grams of strontium tartrate***, followed by ***25 grams of hexachlorobenzene***, followed by ***4.5 grams of standard soft wood charcoal***, and then followed by ***25 grams of stearic acid***. Thereafter, rotate the mixture at 170 RPM for 6 hours. After 6 hours, remove the mixture from the ball mill, and then separate the mixture from the steel shot in the usual manner. Thereafter, the mixture is ready for pressing. To do so, press the mixture into the tracer compartment of the given projectile under a pressure of 10,000 psi.
**Burn rate (tracer burn out, standard 40mm grenade round on average):** 50 meters
**Burn rate (tracer burn out, standard 40mm special round (US/NATO MK19) on average):** 400 meters
**Burn rate (tracer burn out, standard 7.62mm rifle bullet on average):** 310 meters
**Burn rate (tracer burn out, standard 7.62mm machine gun round on average):** 600 meters
**Burn rate (tracer burn out, standard 12.7mm rifle bullet on average):** 1200 meters
**Burn rate (tracer burn out, standard 12.7mm machine gun round on average):** 1640 meters
**Burn rate (tracer burn out, standard 25mm cannon round on average):** 980 meters
**Burn rate (tracer burn out, standard 120mm tank fired projectile on average):** 1800 meters
**Water resistance:** Good.
**Stability:** Can be stored for many years.
**Flammability (1 to 10):** 3.3 (pressed at 10,000 psi); 4.5 powdered.
**Ease of ignition (1 to 10):** 3 (pressed at 10,000 psi); 4.9 powdered.
**Tendency to cake:** None.
**Explosive ability:** Cannot be detonated.

**Percentage:** ***33.2% strontium nitrate, 33.2% magnesium, 23.9% strontium tartrate, 4.3% hexachlorobenzene, 4.3% stearic acid, 0.78% wood charcoal, 0.32% combined impurities***
**Classification:** Deflagrating mixture (classified as 1.3G pyrotechnic composition).

**Danger of preparation:** Can be made by amateurs.
**Use:** Used in military ammunition for rifles, machine guns, low pressure/low velocity projectiles, shoulder fired anti-tank munitions, and tank fired munitions.

### 06-01-013A: Green tracer composition with increased luminosity:

Into a suitable ball mill filled with 500 grams of steel shot of 10 millimeters in diameter, place ***126 grams of barium nitrate***, followed by ***84 grams of magnesium powder,*** followed by ***90 grams of hexachloroethane***, and then followed by ***6 grams of linseed oil***. Thereafter, rotate the mixture at 150 RPM for 5 hours. After 5 hours, remove the mixture from the ball mill, and then separate the mixture from the steel shot in the usual manner. Thereafter, the mixture is ready for pressing. To do so, press the mixture into the tracer compartment of the given projectile under a pressure of 10,000 psi.
**Burn rate (tracer burn out, standard 40mm grenade round on average):** 60 meters
**Burn rate (tracer burn out, standard 40mm special round (US/NATO MK19) on average):** 410 meters
**Burn rate (tracer burn out, standard 7.62mm rifle bullet on average):** 330 meters
**Burn rate (tracer burn out, standard 7.62mm machine gun round on average):** 650 meters
**Burn rate (tracer burn out, standard 12.7mm rifle bullet on average):** 1250 meters
**Burn rate (tracer burn out, standard 12.7mm machine gun round on average):** 1670 meters
**Burn rate (tracer burn out, standard 25mm cannon round on average):** 1000 meters
**Burn rate (tracer burn out, standard 120mm tank fired projectile on average):** 1890 meters
**Water resistance:** Good.
**Stability:** Can be stored for many years.
**Flammability (1 to 10):** 3.9 (pressed at 10,000 psi); 5 powdered.
**Ease of ignition (1 to 10):** 3.3 (pressed at 10,000 psi); 5 powdered.
**Tendency to cake:** None.
**Explosive ability:** Cannot be detonated.
**Percentage:** ***41.1% barium nitrate, 29.4% hexachloroethane, 27.4% magnesium, 1.9% linseed oil, 0.2% impurities***
**Classification:** Deflagrating mixture (classified as 1.3G pyrotechnic composition).

**Danger of preparation:** Can be made by amateurs.
**Use:** Used in military ammunition for rifles, machine guns, low pressure/low velocity projectiles, shoulder fired anti-tank munitions, and tank fired munitions.

### 06-01-013B: Red tracer composition with increased luminosity:

Into a suitable ball mill filled with 500 grams of steel shot of 10 millimeters in diameter, place ***120 grams of strontium nitrate***, followed by ***99 grams of magnesium powder,*** followed by ***75 grams of hexachloroethane***, and then followed by ***6 grams of linseed oil***. Thereafter, rotate the mixture at 150 RPM for 5 hours. After 5 hours, remove the mixture from the ball mill, and then separate the mixture from the steel shot in the usual manner. Thereafter, the mixture is ready for pressing. To do so, press the mixture into the tracer compartment of the given projectile under a pressure of 10,000 psi.
**Burn rate (tracer burn out, standard 40mm grenade round on average):** 55 meters
**Burn rate (tracer burn out, standard 40mm special round (US/NATO MK19) on average):** 400 meters
**Burn rate (tracer burn out, standard 7.62mm rifle bullet on average):** 320 meters
**Burn rate (tracer burn out, standard 7.62mm machine gun round on average):** 630 meters
**Burn rate (tracer burn out, standard 12.7mm rifle bullet on average):** 1220 meters
**Burn rate (tracer burn out, standard 12.7mm machine gun round on average):** 1630 meters
**Burn rate (tracer burn out, standard 25mm cannon round on average):** 1000 meters
**Burn rate (tracer burn out, standard 120mm tank fired projectile on average):** 1800 meters
**Water resistance:** Good.
**Stability:** Can be stored for many years.
**Flammability (1 to 10):** 3.98 (pressed at 10,000 psi); 5.1 powdered.
**Ease of ignition (1 to 10):** 3.4 (pressed at 10,000 psi); 5.1 powdered.
**Tendency to cake:** None.
**Explosive ability:** Cannot be detonated.
**Percentage:** ***40% strontium nitrate, 33% magnesium, 25% hexachloroethane, 2% linseed oil***
**Classification:** Deflagrating mixture (classified as 1.3G pyrotechnic composition).

**Danger of preparation:** Can be made by amateurs.
**Use:** Used in military ammunition for rifles, machine guns, low pressure/low velocity projectiles, shoulder fired anti-tank munitions, and tank fired munitions.

### 06-01-013C: Bluish-green tracer composition with high luminosity:

Into a suitable ball mill filled with 500 grams of steel shot of 10 millimeters in diameter, place ***126 grams of barium nitrate***, followed by ***51 grams of magnesium powder,*** followed by ***120 grams of hexachloroethane***, and then followed by ***3 grams of finely divided copper metal***. Thereafter, rotate the mixture at 200 RPM for 4 hours. After 4 hours, remove the mixture from the ball mill, and then separate the mixture from the steel shot in the usual manner. Thereafter, the mixture is ready for pressing. To do so, press the mixture into the tracer compartment of the given projectile under a pressure of 10,000 psi.
**Burn rate (tracer burn out, standard 40mm grenade round on average):** 50 meters
**Burn rate (tracer burn out, standard 40mm special round (US/NATO MK19) on average):** 450 meters
**Burn rate (tracer burn out, standard 7.62mm rifle bullet on average):** 300 meters
**Burn rate (tracer burn out, standard 7.62mm machine gun round on average):** 600 meters
**Burn rate (tracer burn out, standard 12.7mm rifle bullet on average):** 1240 meters
**Burn rate (tracer burn out, standard 12.7mm machine gun round on average):** 1650 meters
**Burn rate (tracer burn out, standard 25mm cannon round on average):** 1050 meters
**Burn rate (tracer burn out, standard 120mm tank fired projectile on average):** 1810 meters
**Water resistance:** Good.
**Stability:** Can be stored for many years.
**Flammability (1 to 10):** 4 (pressed at 10,000 psi); 5.5 powdered.
**Ease of ignition (1 to 10):** 3.3 (pressed at 10,000 psi); 5 powdered.
**Tendency to cake:** None.
**Explosive ability:** Cannot be detonated.
**Percentage:** ***42% barium nitrate, 40% hexachloroethane, 17% magnesium, 1% copper***
**Classification:** Deflagrating mixture (classified as 1.3G pyrotechnic composition).

**Danger of preparation:** Can be made by amateurs.
**Use:** Used in military ammunition for rifles, machine guns, low pressure/low velocity projectiles, shoulder fired anti-tank munitions, and tank fired munitions.

### 06-01-014A: Tracer composition with reduced base drag:

Into a suitable ball mill filled with 500 grams of steel shot of 10 millimeters in diameter, place ***36.5 grams of finely powdered magnesium of average mesh***, followed by ***41 grams of calcium resinate***, followed by ***50 grams of gelatin***, followed by ***354.5 grams of strontium peroxide***, and then followed by ***18 grams of finely powdered carbon black***. Thereafter, rotate the mixture at 180 RPM for 5 hours. After 5 hours, remove the mixture from the ball mill, and then separate the mixture from the steel shot in the usual manner. Thereafter, the mixture is ready for pressing. To do so, press the mixture into the tracer compartment of the given projectile under a pressure of 10,000 psi.
**Burn rate (tracer burn out, standard 40mm grenade round on average):** 55 meters
**Burn rate (tracer burn out, standard 40mm special round (US/NATO MK19) on average):** 460 meters
**Burn rate (tracer burn out, standard 7.62mm rifle bullet on average):** 310 meters
**Burn rate (tracer burn out, standard 7.62mm machine gun round on average):** 620 meters
**Burn rate (tracer burn out, standard 12.7mm rifle bullet on average):** 1250 meters
**Burn rate (tracer burn out, standard 12.7mm machine gun round on average):** 1670 meters
**Burn rate (tracer burn out, standard 25mm cannon round on average):** 1060 meters
**Burn rate (tracer burn out, standard 120mm tank fired projectile on average):** 1820 meters
**Water resistance:** Good.
**Stability:** Can be stored for many years.
**Flammability (1 to 10):** 3.9 (pressed at 10,000 psi); 5 powdered.
**Ease of ignition (1 to 10):** 3 (pressed at 10,000 psi); 5 powdered.
**Tendency to cake:** None.
**Explosive ability:** Cannot be detonated.
**Percentage:** ***70.9% strontium peroxide, 10% gelatin, 8.2% calcium resinate, 7.3% magnesium, 3.6% carbon black***
**Classification:** Deflagrating mixture (classified as 1.3G pyrotechnic composition).

**Danger of preparation:** Can be made by amateurs.
**Use:** Used in military ammunition for rifles, machine guns, low pressure/low velocity projectiles, shoulder fired anti-tank munitions, and tank fired munitions.

## 06-01-014B: Tracer composition with reduced base drag and increased burn range:

Into a suitable ball mill filled with 500 grams of steel shot of 10 millimeters in diameter, place ***127 grams of finely powdered magnesium of average mesh***, followed by ***246.5 grams of strontium nitrate,*** followed by ***50 grams of gelatin***, and then followed by ***76.5 grams of PVC***. Thereafter, rotate the mixture at 190 RPM for 5 hours. After 5 hours, remove the mixture from the ball mill, and then separate the mixture from the steel shot in the usual manner. Thereafter, the mixture is ready for pressing. To do so, press the mixture into the tracer compartment of the given projectile under a pressure of 10,000 psi.
**Burn rate (tracer burn out, standard 40mm grenade round on average):** 60 meters
**Burn rate (tracer burn out, standard 40mm special round (US/NATO MK19) on average):** 500 meters
**Burn rate (tracer burn out, standard 7.62mm rifle bullet on average):** 340 meters
**Burn rate (tracer burn out, standard 7.62mm machine gun round on average):** 650 meters
**Burn rate (tracer burn out, standard 12.7mm rifle bullet on average):** 1300 meters
**Burn rate (tracer burn out, standard 12.7mm machine gun round on average):** 1690 meters
**Burn rate (tracer burn out, standard 25mm cannon round on average):** 1080 meters
**Burn rate (tracer burn out, standard 120mm tank fired projectile on average):** 1850 meters
**Water resistance:** Good.
**Stability:** Can be stored for many years.
**Flammability (1 to 10):** 3.7 (pressed at 10,000 psi); 5 powdered.
**Ease of ignition (1 to 10):** 3.1 (pressed at 10,000 psi); 5.1 powdered.
**Tendency to cake:** None.
**Explosive ability:** Cannot be detonated.
**Percentage:** ***49.3% strontium nitrate, 25.4%magnesium, 15.3% PVC, 10% gelatin***
**Classification:** Deflagrating mixture (classified as 1.3G pyrotechnic composition).

**Danger of preparation:** Can be made by amateurs.
**Use:** Used in military ammunition for rifles, machine guns, low pressure/low velocity projectiles, shoulder fired anti-tank munitions, and tank fired munitions.

## 06-01-015A: White light tracer composition:

Into a suitable ball mill filled with 500 grams of steel shot of 10 millimeters in diameter, place ***600 grams of atomized magnesium powder***, and then followed by ***400 grams of dry sodium nitrate***. Thereafter, rotate the mixture at 150 RPM for about 2 hours. Thereafter, add in ***50 grams of "Laminac 4116 binder",*** and then rotate the mixture at 300 RPM for 45 minutes. After 45 minutes, remove the mixture from the ball mill, and then separate the mixture from the steel shot in the usual manner. Thereafter, the mixture is ready for pressing. To do so, press the mixture into the tracer compartment of the given projectile under a pressure of 10,000 psi, and then cure the projectile in an oven at 60 Celsius for 6 hours.
**Burn rate (tracer burn out, standard 40mm grenade round on average):** 50 meters
**Burn rate (tracer burn out, standard 40mm special round (US/NATO MK19) on average):** 450 meters
**Burn rate (tracer burn out, standard 7.62mm rifle bullet on average):** 300 meters
**Burn rate (tracer burn out, standard 7.62mm machine gun round on average):** 600 meters
**Burn rate (tracer burn out, standard 12.7mm rifle bullet on average):** 1210 meters
**Burn rate (tracer burn out, standard 12.7mm machine gun round on average):** 1600 meters
**Burn rate (tracer burn out, standard 25mm cannon round on average):** 1000 meters
**Burn rate (tracer burn out, standard 120mm tank fired projectile on average):** 1800 meters
**Water resistance:** Good.
**Stability:** Can be stored for many years.
**Flammability (1 to 10):** 3 (pressed at 10,000 psi); 4 powdered.
**Ease of ignition (1 to 10):** 2.9 (pressed at 10,000 psi); 4 powdered.
**Tendency to cake:** None.
**Explosive ability:** Cannot be detonated.
**Percentage:** ***57.14% magnesium powder, 38.09% sodium nitrate, 4.76% binder, 0.01% residual balance***
**Classification:** Deflagrating mixture (classified as 1.3G pyrotechnic composition).

**Danger of preparation:** Can be made by amateurs.
**Use:** Used in military ammunition for rifles, machine guns, low pressure/low velocity projectiles, shoulder fired anti-tank munitions, and tank fired munitions.

### 06-01-016A: White light tracer composition with sparkling effect:

Into a suitable ball mill filled with 500 grams of steel shot of 10 millimeters in diameter, place ***500 grams of atomized aluminum powder***, followed by ***250 grams of coarse magnesium grains*** (0.5 to 1 millimeter in diameter), followed by ***500 grams of dry sodium chlorate,*** followed by ***75 grams of potassium dichromate***, and then followed by ***90 grams of zinc oxide***. Thereafter, rotate the mixture at 175 RPM for about 2 hours. Thereafter, add in ***80 grams of "Laminac 4116 binder",*** and then continue to rotate the mixture at 300 RPM for 45 minutes. After 45 minutes, remove the mixture from the ball mill, and then separate the mixture from the steel shot in the usual manner. Thereafter, the mixture is ready for pressing. To do so, press the mixture into the tracer compartment of the given projectile under a pressure of 10,000 psi, and then cure the projectile in an oven at 60 Celsius for 6 hours.
**Burn rate (tracer burn out, standard 40mm grenade round on average):** 45 meters
**Burn rate (tracer burn out, standard 40mm special round (US/NATO MK19) on average):** 400 meters
**Burn rate (tracer burn out, standard 7.62mm rifle bullet on average):** 300 meters
**Burn rate (tracer burn out, standard 7.62mm machine gun round on average):** 600 meters
**Burn rate (tracer burn out, standard 12.7mm rifle bullet on average):** 1200 meters
**Burn rate (tracer burn out, standard 12.7mm machine gun round on average):** 1600 meters
**Burn rate (tracer burn out, standard 25mm cannon round on average):** 1000 meters
**Burn rate (tracer burn out, standard 120mm tank fired projectile on average):** 1800 meters
**Water resistance:** Good, when kept from exposure to air and moisture.
**Stability:** Can be stored for many years.
**Flammability (1 to 10):** 3.2 (pressed at 10,000 psi); 4.2 powdered.
**Ease of ignition (1 to 10):** 3 (pressed at 10,000 psi); 4.2 powdered.
**Tendency to cake:** None.
**Explosive ability:** Cannot be detonated.
**Percentage:** ***33.44% aluminum powder, 33.44% sodium chlorate, 16.72% coarse magnesium grains, 6.02% zinc oxide, 5.35% binder, 5.01% potassium dichromate, 0.02% residual balance***
**Classification:** Deflagrating mixture (classified as 1.3G pyrotechnic composition).

**Danger of preparation:** Can be made by amateurs.
**Use:** Used in military ammunition for rifles, machine guns, low pressure/low velocity projectiles, shoulder fired anti-tank munitions, and tank fired munitions.

### 06-01-017A: Specialty depleted uranium tracer composition for large caliber ammunition:

Into a suitable ball mill filled with 500 grams of steel shot of 10 millimeters in diameter, place ***250 grams of finely powdered depleted uranium metal***, followed by ***125 grams of finely ground 50/50 aluminum magnesium alloy***, and then followed by ***125 grams of potassium perchlorate***. Thereafter, rotate the mixture at 175 RPM for about 3 hours. After 3 hours, remove the mixture from the ball mill, and then separate the mixture from the steel shot in the usual manner. Thereafter, the mixture is ready for pressing. To do so, press the mixture into the tracer compartment of the given projectile under a pressure of 10,000 psi.
**Burn rate (tracer burn out, standard 40mm grenade round on average):** 45 meters
**Burn rate (tracer burn out, standard 40mm special round (US/NATO MK19) on average):** 400 meters
**Burn rate (tracer burn out, standard 7.62mm rifle bullet on average):** 300 meters
**Burn rate (tracer burn out, standard 7.62mm machine gun round on average):** 590 meters
**Burn rate (tracer burn out, standard 12.7mm rifle bullet on average):** 1200 meters
**Burn rate (tracer burn out, standard 12.7mm machine gun round on average):** 1600 meters
**Burn rate (tracer burn out, standard 25mm cannon round on average):** 1000 meters
**Burn rate (tracer burn out, standard 120mm tank fired projectile on average):** 1800 meters
**Water resistance:** Good.
**Stability:** Can be stored for many years.
**Flammability (1 to 10):** 2.9 (pressed at 10,000 psi); 3 powdered.
**Ease of ignition (1 to 10):** 2.8 (pressed at 10,000 psi); 3 powdered.
**Tendency to cake:** None.
**Explosive ability:** Cannot be detonated.
**Percentage:** ***50% depleted uranium metal, 25% aluminum magnesium alloy, 25% potassium perchlorate***
**Classification:** Deflagrating mixture (classified as 1.3G pyrotechnic composition; radiation hazard class E).

**Danger of preparation:** Can be made by amateurs, but caution should be exercised as depleted uranium is moderately radioactive.
**Use:** Used in military ammunition for rifles, machine guns, low pressure/low velocity projectiles, shoulder fired anti-tank munitions, and tank fired munitions.

### 06-01-018A: Multi purpose tracer composition for various applications:

Into a suitable ball mill filled with 500 grams of steel shot of 10 millimeters in diameter, place ***100 grams of magnesium powder***, followed ***380 grams of barium peroxide***, and then followed by ***20 grams of aluminum stearate***. Thereafter, rotate the mixture at 170 RPM for about 3 hours. After 3 hours, remove the mixture from the ball mill, and then separate the mixture from the steel shot in the usual manner. Thereafter, the mixture is ready for pressing. To do so, press the mixture into the tracer compartment of the given projectile under a pressure of 10,000 psi.
**Burn rate (tracer burn out, standard 40mm grenade round on average):** 60 meters
**Burn rate (tracer burn out, standard 40mm special round (US/NATO MK19) on average):** 490 meters
**Burn rate (tracer burn out, standard 7.62mm rifle bullet on average):** 350 meters
**Burn rate (tracer burn out, standard 7.62mm machine gun round on average):** 640 meters
**Burn rate (tracer burn out, standard 12.7mm rifle bullet on average):** 1250 meters
**Burn rate (tracer burn out, standard 12.7mm machine gun round on average):** 1690 meters
**Burn rate (tracer burn out, standard 25mm cannon round on average):** 1050 meters
**Burn rate (tracer burn out, standard 120mm tank fired projectile on average):** 1900 meters
**Water resistance:** Good.
**Stability:** Can be stored for many years.
**Flammability (1 to 10):** 3.4 (pressed at 10,000 psi); 4.9 powdered.
**Ease of ignition (1 to 10):** 3.1 (pressed at 10,000 psi); 5 powdered.
**Tendency to cake:** None.
**Explosive ability:** Cannot be detonated.
**Percentage:** ***76% barium peroxide, 20% magnesium powder, 4% aluminum stearate***
**Classification:** Deflagrating mixture (classified as 1.3G pyrotechnic composition).

**Danger of preparation:** Can be made by amateurs.
**Use:** Used in military ammunition for rifles, machine guns, low pressure/low velocity projectiles, shoulder fired anti-tank munitions, and tank fired munitions.

### 06-01-018B: Multi purpose tracer composition for various applications:

Into a suitable ball mill filled with 500 grams of steel shot of 10 millimeters in diameter, place ***120 grams of magnesium powder***, followed by ***367.5 grams of barium peroxide***, and then followed by ***12.5 grams of magnesium stearate***. Thereafter, rotate the mixture at 180 RPM for about 3 hours. After 3 hours, remove the mixture from the ball mill, and then separate the mixture from the steel shot in the usual manner. Thereafter, the mixture is ready for pressing. To do so, press the mixture into the tracer compartment of the given projectile under a pressure of 10,000 psi.
**Burn rate (tracer burn out, standard 40mm grenade round on average):** 50 meters
**Burn rate (tracer burn out, standard 40mm special round (US/NATO MK19) on average):** 450 meters
**Burn rate (tracer burn out, standard 7.62mm rifle bullet on average):** 310 meters
**Burn rate (tracer burn out, standard 7.62mm machine gun round on average):** 610 meters
**Burn rate (tracer burn out, standard 12.7mm rifle bullet on average):** 1210 meters
**Burn rate (tracer burn out, standard 12.7mm machine gun round on average):** 1650 meters
**Burn rate (tracer burn out, standard 25mm cannon round on average):** 1010 meters
**Burn rate (tracer burn out, standard 120mm tank fired projectile on average):** 1870 meters
**Water resistance:** Good.
**Stability:** Can be stored for many years.
**Flammability (1 to 10):** 3.3 (pressed at 10,000 psi); 4.7 powdered.
**Ease of ignition (1 to 10):** 3.2 (pressed at 10,000 psi); 5 powdered.
**Tendency to cake:** None.
**Explosive ability:** Cannot be detonated.
**Percentage:** ***73.5% barium peroxide, 24% magnesium powder, 2.5% magnesium stearate***
**Classification:** Deflagrating mixture (classified as 1.3G pyrotechnic composition).

**Danger of preparation:** Can be made by amateurs.
**Use:** Used in military ammunition for rifles, machine guns, low pressure/low velocity projectiles, shoulder fired anti-tank munitions, and tank fired munitions.

### 06-01-018C: Multi purpose tracer composition for various applications:

Into a suitable ball mill filled with 500 grams of steel shot of 10 millimeters in diameter, place ***125 grams of magnesium powder***, followed by ***475 grams of barium peroxide***, and then followed by ***50 grams of zinc stearate***. Thereafter, rotate the mixture at 190 RPM for about 3 hours. After 3 hours, remove the mixture from the ball mill, and then separate the mixture from the steel shot in the usual manner. Thereafter, the mixture is ready for pressing. To do so, press the mixture into the tracer compartment of the given projectile under a pressure of 10,000 psi.
**Burn rate (tracer burn out, standard 40mm grenade round on average):** 40 meters
**Burn rate (tracer burn out, standard 40mm special round (US/NATO MK19) on average):** 400 meters
**Burn rate (tracer burn out, standard 7.62mm rifle bullet on average):** 300 meters
**Burn rate (tracer burn out, standard 7.62mm machine gun round on average):** 600 meters
**Burn rate (tracer burn out, standard 12.7mm rifle bullet on average):** 1200 meters
**Burn rate (tracer burn out, standard 12.7mm machine gun round on average):** 1620 meters
**Burn rate (tracer burn out, standard 25mm cannon round on average):** 1000 meters
**Burn rate (tracer burn out, standard 120mm tank fired projectile on average):** 1850 meters
**Water resistance:** Good.
**Stability:** Can be stored for many years.
**Flammability (1 to 10):** 4 (pressed at 10,000 psi); 5.1 powdered.
**Ease of ignition (1 to 10):** 4 (pressed at 10,000 psi); 6 powdered.
**Tendency to cake:** None.
**Explosive ability:** Cannot be detonated.
**Percentage:** ***73.07% barium peroxide, 19.23% magnesium powder, 7.69% zinc stearate, 0.01% residual balance***
**Classification:** Deflagrating mixture (classified as 1.3G pyrotechnic composition).

**Danger of preparation:** Can be made by amateurs.
**Use:** Used in military ammunition for rifles, machine guns, low pressure/low velocity projectiles, shoulder fired anti-tank munitions, and tank fired munitions.

### 06-01-019A: Tracer composition for high caliber projectiles:

Into a suitable ball mill filled with 500 grams of steel shot of 10 millimeters in diameter, place ***300 grams of barium peroxide,*** followed by ***175 grams of antimony trisulfide,*** followed by ***20 grams of asphaltum,*** and then followed by ***5 grams of graphite powder***. Thereafter, rotate the mixture at 50 RPM for about 6 hours. After 6 hours, remove the mixture from the ball mill, and then separate the mixture from the steel shot in the usual manner. Thereafter, the mixture is ready for pressing. To do so, press the mixture into the tracer compartment of the given projectile under a pressure of 10,000 psi.
**Burn rate (tracer burn out, standard 40mm grenade round on average):** 55 meters
**Burn rate (tracer burn out, standard 40mm special round (US/NATO MK19) on average):** 460 meters
**Burn rate (tracer burn out, standard 7.62mm machine gun round on average):** 660 meters
**Burn rate (tracer burn out, standard 12.7mm machine gun round on average):** 1690 meters
**Burn rate (tracer burn out, standard 25mm cannon round on average):** 1100 meters
**Burn rate (tracer burn out, standard 120mm tank fired projectile on average):** 1900 meters
**Water resistance:** Good.
**Stability:** Can be stored for many years.
**Flammability (1 to 10):** 4 (pressed at 10,000 psi); 5 powdered.
**Ease of ignition (1 to 10):** 3.9 (pressed at 10,000 psi); 5.8 powdered.
**Tendency to cake:** None.
**Explosive ability:** Cannot be detonated.
**Percentage:** ***60% barium peroxide, 35% antimony trisulfide, 4% asphaltum, 1% graphite powder***
**Classification:** Deflagrating mixture (classified as 1.3G pyrotechnic composition).

**Danger of preparation:** Can be made by amateurs.
**Use:** Used in military ammunition for machine guns, low pressure/low velocity projectiles, shoulder fired anti-tank munitions, and tank fired munitions.

### 06-01-020A: Specialty red tracer composition with red smoke trail for use in daytime operations:
Into a suitable ball mill filled with 500 grams of steel shot of 10 millimeters in diameter, place ***350 grams of strontium peroxide,*** followed by ***50 grams of calcium resinate,*** followed by ***75 grams of 1-methylaminoanthraquinone,*** and then followed by ***25 grams of magnesium powder***. Thereafter, rotate the mixture at 170 RPM for about 3 hours. After 3 hours, remove the mixture from the ball mill, and then separate the mixture from the steel shot in the usual manner. Thereafter, the mixture is ready for pressing. To do so, press the mixture into the tracer compartment of the given projectile under a pressure of 10,000 psi.
**Burn rate (tracer burn out, standard 40mm grenade round on average):** 55 meters
**Burn rate (tracer burn out, standard 40mm special round (US/NATO MK19) on average):** 420 meters
**Burn rate (tracer burn out, standard 7.62mm rifle bullet on average):** 310 meters
**Burn rate (tracer burn out, standard 7.62mm machine gun round on average):** 615 meters
**Burn rate (tracer burn out, standard 12.7mm rifle bullet on average):** 1200 meters
**Burn rate (tracer burn out, standard 12.7mm machine gun round on average):** 1650 meters
**Burn rate (tracer burn out, standard 25mm cannon round on average):** 1050 meters
**Burn rate (tracer burn out, standard 120mm tank fired projectile on average):** 1900 meters
**Water resistance:** Good.
**Stability:** Can be stored for many years.
**Flammability (1 to 10):** 4.1 (pressed at 10,000 psi); 5.8 powdered.
**Ease of ignition (1 to 10):** 3.9 (pressed at 10,000 psi); 6 powdered.
**Tendency to cake:** None.
**Explosive ability:** Cannot be detonated.
**Percentage:** ***70% strontium peroxide, 15% 1-methylaminoanthraquinone dye agent, 10% calcium resinate, 5% magnesium powder***
**Classification:** Deflagrating mixture (classified as 1.3G pyrotechnic composition).

**Danger of preparation:** Can be made by amateurs.
**Use:** Used in military ammunition for rifles, machine guns, low pressure/low velocity projectiles, shoulder fired anti-tank munitions, and tank fired munitions.

### 06-01-020B: Specialty tracer composition with red smoke trail for use in daytime operations only:
Into a suitable ball mill filled with 500 grams of steel shot of 10 millimeters in diameter, place ***350 grams of strontium peroxide,*** followed by ***50 grams of calcium resinate,*** followed by ***50 grams of 1-methylaminoanthraquinone,*** and then followed by ***50 grams of ortho-dihydroxy benzene***. Thereafter, rotate the mixture at 190 RPM for about 3 hours. After 3 hours, remove the mixture from the ball mill, and then separate the mixture from the steel shot in the usual manner. Thereafter, the mixture is ready for pressing. To do so, press the mixture into the tracer compartment of the given projectile under a pressure of 10,000 psi.
**Burn rate (tracer burn out, standard 40mm grenade round on average):** 60 meters
**Burn rate (tracer burn out, standard 40mm special round (US/NATO MK19) on average):** 460 meters
**Burn rate (tracer burn out, standard 7.62mm rifle bullet on average):** 350 meters
**Burn rate (tracer burn out, standard 7.62mm machine gun round on average):** 625 meters
**Burn rate (tracer burn out, standard 12.7mm rifle bullet on average):** 1240 meters
**Burn rate (tracer burn out, standard 12.7mm machine gun round on average):** 1690 meters
**Burn rate (tracer burn out, standard 25mm cannon round on average):** 1099 meters
**Burn rate (tracer burn out, standard 120mm tank fired projectile on average):** 1950 meters
**Water resistance:** Good.
**Stability:** Can be stored for many years.
**Flammability (1 to 10):** 4.6 (pressed at 10,000 psi); 6 powdered.
**Ease of ignition (1 to 10):** 4.3 (pressed at 10,000 psi); 6.9 powdered.
**Tendency to cake:** None.
**Explosive ability:** Cannot be detonated.
**Percentage:** ***70% strontium peroxide, 10% 1-methylaminoanthraquinone dye agent, 10% calcium resinate, 10% ortho-dihydroxy benzene***
**Classification:** Deflagrating mixture (classified as 1.3G pyrotechnic composition).

**Danger of preparation:** Can be made by amateurs.
**Use:** Used in military ammunition for rifles, machine guns, low pressure/low velocity projectiles, shoulder fired anti-tank munitions, and tank fired munitions.

## 06-01-020C: Specialty tracer composition with yellow smoke trail for use in daytime operations only:

Into a suitable ball mill filled with 500 grams of steel shot of 10 millimeters in diameter, place ***400 grams of strontium peroxide,*** followed by ***50 grams of calcium resinate,*** and then followed by ***50 grams of bis(para-dimethylaminophenyl)-methylenimine hydrochloride (auramine dye).*** Thereafter, rotate the mixture at 190 RPM for about 3 hours. After 3 hours, remove the mixture from the ball mill, and then separate the mixture from the steel shot in the usual manner. Thereafter, the mixture is ready for pressing. To do so, press the mixture into the tracer compartment of the given projectile under a pressure of 10,000 psi.

**Burn rate (tracer burn out, standard 40mm grenade round on average):** 50 meters
**Burn rate (tracer burn out, standard 40mm special round (US/NATO MK19) on average):** 455 meters
**Burn rate (tracer burn out, standard 7.62mm rifle bullet on average):** 340 meters
**Burn rate (tracer burn out, standard 7.62mm machine gun round on average):** 600 meters
**Burn rate (tracer burn out, standard 12.7mm rifle bullet on average):** 1210 meters
**Burn rate (tracer burn out, standard 12.7mm machine gun round on average):** 1650 meters
**Burn rate (tracer burn out, standard 25mm cannon round on average):** 1020 meters
**Burn rate (tracer burn out, standard 120mm tank fired projectile on average):** 1920 meters
**Water resistance:** Good.
**Stability:** Can be stored for many years.
**Flammability (1 to 10):** 4.7 (pressed at 10,000 psi); 6.2 powdered.
**Ease of ignition (1 to 10):** 4 (pressed at 10,000 psi); 5.5 powdered.
**Tendency to cake:** None.
**Explosive ability:** Cannot be detonated.
**Percentage:** ***80% strontium peroxide, 10% auramine dye, 10% calcium resinate***
**Classification:** Deflagrating mixture (classified as 1.3G pyrotechnic composition).

**Danger of preparation:** Can be made by amateurs.
**Use:** Used in military ammunition for rifles, machine guns, low pressure/low velocity projectiles, shoulder fired anti-tank munitions, and tank fired munitions.

## 06-01-020D: Specialty tracer composition with black smoke trail for use in daytime operations:

Into a suitable ball mill filled with 500 grams of steel shot of 10 millimeters in diameter, place ***350 grams of strontium peroxide,*** followed by ***50 grams of calcium resinate,*** followed by ***75 grams of powdered anthracene,*** and then followed by ***25 grams of magnesium powder.*** Thereafter, rotate the mixture at 190 RPM for about 3 hours. After 3 hours, remove the mixture from the ball mill, and then separate the mixture from the steel shot in the usual manner. Thereafter, the mixture is ready for pressing. To do so, press the mixture into the tracer compartment of the given projectile under a pressure of 10,000 psi.

**Burn rate (tracer burn out, standard 40mm grenade round on average):** 55 meters
**Burn rate (tracer burn out, standard 40mm special round (US/NATO MK19) on average):** 450 meters
**Burn rate (tracer burn out, standard 7.62mm rifle bullet on average):** 345 meters
**Burn rate (tracer burn out, standard 7.62mm machine gun round on average):** 610 meters
**Burn rate (tracer burn out, standard 12.7mm rifle bullet on average):** 1250 meters
**Burn rate (tracer burn out, standard 12.7mm machine gun round on average):** 1640 meters
**Burn rate (tracer burn out, standard 25mm cannon round on average):** 1030 meters
**Burn rate (tracer burn out, standard 120mm tank fired projectile on average):** 1950 meters
**Water resistance:** Good.
**Stability:** Can be stored for many years.
**Flammability (1 to 10):** 5.1 (pressed at 10,000 psi); 7 powdered.
**Ease of ignition (1 to 10):** 4.9 (pressed at 10,000 psi); 7 powdered.
**Tendency to cake:** None.
**Explosive ability:** Cannot be detonated.
**Percentage:** ***70% strontium peroxide, 15% anthracene, 10% calcium resinate, 5% magnesium powder***
**Classification:** Deflagrating mixture (classified as 1.3G pyrotechnic composition).

**Danger of preparation:** Can be made by amateurs.
**Use:** Used in military ammunition for rifles, machine guns, low pressure/low velocity projectiles, shoulder fired anti-tank munitions, and tank fired munitions.

# Section 2: Match-head compositions

***Chemicals used in this section (binders are not included)***

| | |
|---|---|
| 1. Potassium nitrate (see Black Powder) | 2. Sulfur (see Black Powder) |
| 3. Charcoal (see Black Powder) | 4. Sugar Carbon (see Modified Black Powder) |
| 5. Barium Chromate (see Modified Black Powder) | 6. Potassium Perchlorate (see Modified Black Powder) |
| 7. Potassium Dichromate (see Modified Black Powder) | 8. Ammonium Bisulfide (see Modified Black Powder) |
| 9. Potassium chlorate (see Modified Black Powder) | 10. Carbon Disulfide (see Modified Black Powder) |
| 11. Nitrocellulose (see Modified Black Powder) | 12. Lead Tetraoxide (see Modified Black Powder) |
| 13. Diphenylamine (see Modified Black Powder) | 14. Titanium Dioxide (see Modified Black Powder) |
| 15. Manganese Dioxide (see Modified Black Powder) | 16. Sodium Benzoate (see Modified Black Powder) |
| 17. Ammonium Nitrate (see Modified Black Powder) | 18. Calcium Carbonate (see Modified Black Powder) |
| 19. Sodium Nitrate (see Modified Black Powder) | 20. Urea (see Modified Black Powder) |
| 21. Lead Nitrate (see Modified Black Powder) | 22. Nitro Starch (see Modified Black Powder) |
| 23. Ammonium Perchlorate (see Ammonium Perchlorate Rocket Propellants) | 24. Aluminum powder (see Ammonium Perchlorate Rocket Propellants) |
| 25. Magnesium Sulfide (see Ammonium Perchlorate Rocket Propellants) | 26. Copper Chromite (see Ammonium Perchlorate Rocket Propellants) |
| 27. Nitroguanidine (see Ammonium Perchlorate Rocket Propellants) | 28. Ammonium Sulfate (see Ammonium Perchlorate Rocket Propellants) |
| 29. Nitroglycerine (see Ammonium Perchlorate Rocket Propellants) | 30. Magnesium Oxide (see Ammonium Perchlorate Rocket Propellants) |
| 31. Iron-II-oxide (see Ammonium Perchlorate Rocket Propellants) | 32. Zirconium Hydride (see Ammonium Perchlorate Rocket Propellants) |
| 33. Teflon (see Ammonium Perchlorate Rocket Propellants) | 34. Zinc Oxide (see Ammonium Perchlorate Rocket Propellants) |
| 35. Ammonium Dichromate (see Ammonium Perchlorate Rocket Propellants) | 36. Lithium Perchlorate (see Ammonium Perchlorate Rocket Propellants) |
| 37. Magnesium powder (see Ammonium Perchlorate Rocket Propellants) | 38. Lithium Aluminum Hydride (see Ammonium Perchlorate Rocket Propellants) |
| 39. Iron-III-oxide (red iron oxide) (see Ammonium Perchlorate Rocket Propellants) | 40. PVC (see Ammonium Perchlorate Rocket Propellants) |
| 41. Copper Sulfide (see Ammonium Perchlorate Rocket Propellants) | 42. Sodium Hydride (see Ammonium Perchlorate Rocket Propellants) |
| 43. Titanium Hydride (see Ammonium Perchlorate Rocket Propellants) | 44. Silicon Nitride (see Ammonium Perchlorate Rocket Propellants) |
| 45. ADN (see ADN Rocket Propellants) | 46. Urea Nitrate (see ADN Rocket Propellants) |
| 47. Silicon Powder (see ADN Rocket Propellants) | 48. Hexamine (see ADN Rocket Propellants) |
| 49. Boron powder (see ADN Rocket Propellants) | 50. Sodium hypophosphite (see ADN Rocket Propellants) |
| 51. KDN (see ADN Rocket Propellants) | 52. Nickel Chloride (see Ammonium Nitrate Rocket Propellants) |
| 53. TNT (see Ammonium Nitrate Rocket Propellants) | 54. Acrylamide (see Miscellaneous Rocket H.P. Rocket Propellants) |
| 55. Ethylene Glycol (see Miscellaneous Rocket H.P. Rocket Propellants) | 56. Magnesium Perchlorate (see Miscellaneous Rocket H.P. Rocket Propellants) |
| 57. Metallic Lithium (see Miscellaneous Rocket H.P. Rocket Propellants) | 58. Beryllium Hydride (see Miscellaneous Rocket H.P. Rocket Propellants) |
| 59. Red Phosphorus (see Miscellaneous Rocket H.P. Rocket Propellants) | 60. Sodium Borohydride (see Miscellaneous Rocket H.P. Rocket Propellants) |
| 61. Sodium chlorate (see Miscellaneous Rocket H.P. Rocket Propellants) | 62. Lead Dioxide (see Miscellaneous Rocket H.P. Rocket Propellants) |
| 63. Benzoyl Peroxide (see Miscellaneous Rocket H.P. Rocket Propellants) | 64. Barium Nitrate (see Miscellaneous Rocket H.P. Rocket Propellants) |
| 65. Cyanuric Acid (see Miscellaneous Rocket H.P. Rocket Propellants) | 66. Aluminum Stearate (see Ammonium Nitrate Gun Propellants) |

| | |
|---|---|
| **67. Aluminum Hydride (see Ammonium Nitrate Gun Propellants)** | **68. Magnesium Peroxide (see Ammonium Nitrate Gun Propellants)** |
| **69. Potassium Permanganate (see Ammonium Nitrate Gun Propellants)** | **70. Calcium Hydride (see Ammonium Nitrate Gun Propellants)** |
| **71. Potassium Tartrate (see Ammonium Nitrate Gun Propellants)** | **72. Potassium Ferrocyanide (see Ammonium Nitrate Gun Propellants)** |
| **73. Sodium Azide (see Ammonium Nitrate Gun Propellants)** | **74. Sodium Chloride (see Ammonium Nitrate Gun Propellants)** |
| **75. Potassium Sulfate (see Nitrocellulose Gun Propellants)** | **76. Lead Stearate (see Nitrocellulose Gun Propellants)** |
| **77. Triethylene Glycol (see Nitrocellulose Gun Propellants)** | **78. Sugar (sucrose) (see Miscellaneous Gun Propellants)** |
| **79. Sodium Propionate (see Miscellaneous Gun Propellants)** | **80. Picric Acid (see Miscellaneous Gun Propellants)** |
| **81. Copper-II-oxide (see Miscellaneous Gun Propellants)** | **82. Ammonium Picrate (see Miscellaneous Gun Propellants)** |
| **83. Barium Peroxide (see Bullet Tracer Compositions)** | **84. Magnesium Carbonate (see Bullet Tracer Compositions)** |
| **85. Strontium Peroxide (see Bullet Tracer Compositions)** | **86. Strontium Nitrate (see Bullet Tracer Compositions)** |
| **87. Cupric chloride (see Bullet Tracer Compositions)** | **88. Hexachlorobenzene (see Bullet Tracer Compositions)** |
| **89. Strontium oxalate (see Bullet Tracer Compositions)** | **90. Mercury-I-Chloride (see Bullet Tracer Compositions)** |
| **91. Zinc Oxalate (see Bullet Tracer Compositions)** | **92. Zinc Chloride (see Bullet Tracer Compositions)** |
| **93. Uranium (see Bullet Tracer Compositions)** | **94. Zirconium nitrate (see Bullet Tracer Compositions)** |
| **95. Yttrium Nitrate (see Bullet Tracer Compositions)** | **96. Yttrium Oxide (see Bullet Tracer Compositions)** |
| **97. Zirconium Oxide (see Bullet Tracer Compositions)** | **98. Cerium Oxide (see Bullet Tracer Compositions)** |
| **99. Hexachloroethane (see Bullet Tracer Compositions)** | **100. Antimony trisulfide (see Bullet Tracer Compositions)** |
| **101. Anthracene (see Bullet Tracer Compositions)** | **102. Phosphorus Sesquisulphide** |
| | 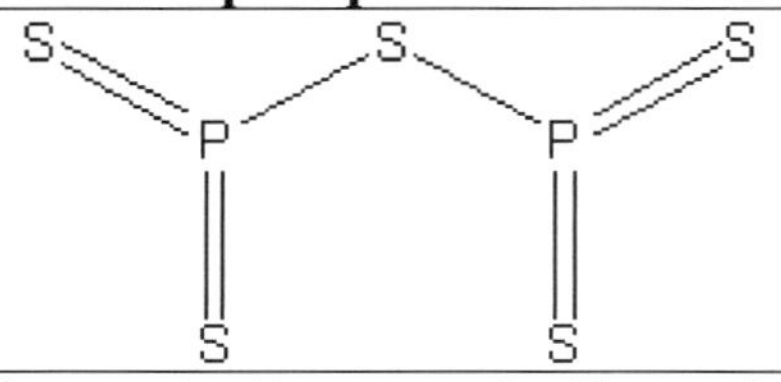 |
| | Forms light yellow crystals with a strange odor. The crystals have melting point of 290 Celsius, and a boiling point of 515 Celsius. The crystals are insoluble in water, and are decomposed by it forming phosphoric acid and hydrogen sulfide. The crystals are soluble in carbon disulfide. |
| **103. Boric acid** | **104. Aluminum Hydroxide** |
| 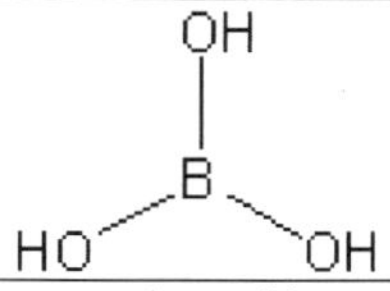 | 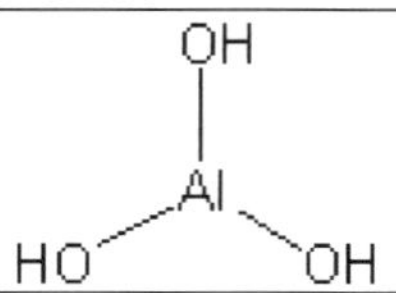 |
| Boric acid forms colorless crystals or white granules or powder. The melting point of boric acid is about 171 Celsius, but transfer to boric oxide begins at higher temperatures. The solid is slightly soluble in water, and alcohol. | Forms a white bulky powder, which is insoluble in water and all known solvents. The solid breaks down into aluminum oxide when heated to high temperatures. Aluminum hydroxide is soluble in acids. |
| **105. Antimony Pentasulfide** | **106. Glucose** |
| 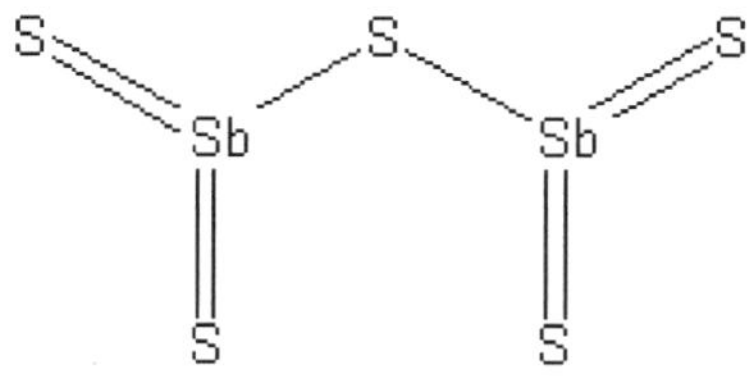 | 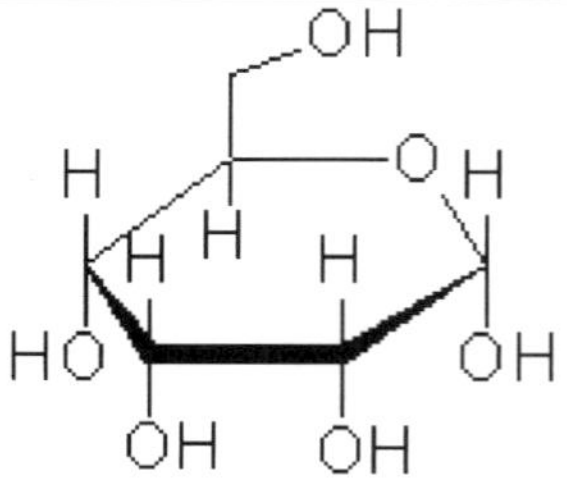 |
| | |

Forms an orange-yellow solid. The solid is stable at room temperature and does not react with water. The powder is insoluble in water and most organic solvents. Antimony pentasulfide reacts with acids forming toxic hydrogen sulfide gas.

Glucose forms colorless, transparent crystals, or white granules, or powder. Glucose exists in several forms, but the main form has a melting point of 83 Celsius. The crystals are very soluble in water, but only slightly soluble in alcohol. Glucose is the main source of energy for all plants. Glucose can be prepared by inverting table sugar with dilute acids.

**107. Sodium Hydroxide**

Sodium hydroxide forms white granules, lumps, or powder. The melting point is 318 Celsius. Sodium hydroxide is very soluble in water, and its dissolving in water generates much heat, so use caution. It is also readily soluble in alcohol. Sodium hydroxide is prepared by electrolysizing a solution of table salt in a divided cell.

**108. Lead Hypophosphite**

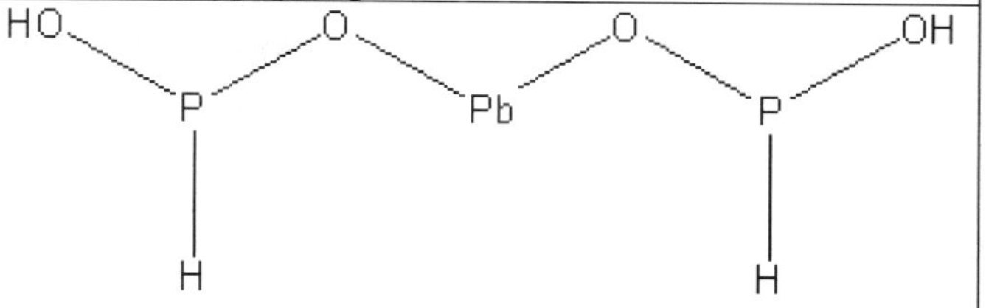


Forms a hygroscopic crystalline powder or granules. The crystals are toxic, so users should use caution. The salt is insoluble in water and the usual solvents.

**109. Calcium Sulfate**

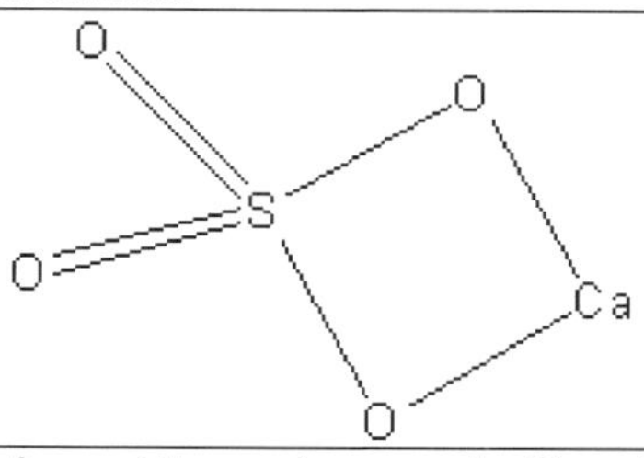


Calcium sulfate forms white granules or powder. The anhydrous form is a white powder that reacts with water forming a series of hydrates. The hydrates are capable of setting to hard masses, such as plaster of Paris.

***- Match Head Compositions in this section -***

| | |
|---|---|
| **1. 06-02-001A: Safety-match composition 1:** *50% potassium chlorate, 12.5% standard animal glue binder, 12% siliceous microspheres, 9% manganese dioxide, 5.5% sulfur, 5.5 diatomaceous earth combustible, 2.5% ferric oxide coloring agent, 1% zinc oxide filler, 1% hydroxyethyl cellulose curative, 1% potassium dichromate catalyst* | **2. 06-02-001B: Safety-match (modified) composition 2:** *49.5% potassium chlorate, 30% siliceous microspheres, 12.5% standard animal glue binder, 5.5% sulfur, 1.1% hydroxyethyl cellulose curative, 1.1% potassium dichromate catalyst, 0.30% impurities* |
| **3. 06-02-002A: "Strike anywhere" match composition 1:** *35% potassium chlorate, 19% siliceous microspheres, 17% standard animal glue binder, 15.5% plaster of Paris, 7% Phosphorus Sesquisulphide, 6% zinc oxide, 0.50% dye* | **4. 06-02-002B: "Strike anywhere" (modified) match composition 2:** *40.4% potassium chlorate, 18.7% phenolic resin microspheres, 17.2% standard animal glue binder, 8.8% plaster of Paris, 8.2% Phosphorus Sesquisulphide, 6% zinc oxide, 0.32% dye, 0.380% impurities* |
| **5. 06-02-003A: Match-book match composition 1:** *55% potassium chlorate, 10% standard animal glue binder, 7.5% siliceous microspheres, 7.5% flour glass, 6.2% diatomaceous earth, 5% zinc oxide, 5% sulfur, 2.2% starch filler, 0.75% potassium dichromate, 0.50% gum Arabic, 0.25% dye, 0.10% impurities* | **6. 06-02-004A:"Double dip" friction resistant match stick composition 1:** *41.8% potassium chlorate, 13.3% standard animal glue binder, 13.35% flour glass, 8% sulfur, 7.8% plaster of Paris, 3.7% clay, 2.9% calcium resinate, 2.9% dye, 2.6% zinc oxide, 2.2% red phosphorus, 0.75% gun Arabic, 0.10% potassium dichromate, 0.60% impurities* |
| **7. 06-02-005A: Clean burning match composition 1:** *46% potassium chlorate, 23.5% glass powder, 13% standard animal glue, 5.8% diatomaceous earth, 3.5% graphite, 3.5% nitrocellulose, 2.3% dextrin, 1.2% dye, 0.70% boric acid* | **8. 06-02-006A: Quick-lite "easy light" weather proof match composition 1:** *48.5% potassium chlorate, 24.9% glass powder, 14.6% standard animal glue, 3.2% diatomaceous earth, 2.6% titanium dioxide, 1.3% wood pulp, 1.3% starch, 1.3% aluminum hydroxide, 1% moisture, 0.65% turpentine, 0.65% shellac,* |
| | |

| | |
|---|---|
| **9. 06-02-007A: "Strike anywhere" match head composition 3:** *49.7% potassium chlorate, 21.3% feldspar, 12.7% gelatin, 7.1% limestone, 2.8% red phosphorus, 2.7% starch, 2.7% titanium dioxide, 0.35% animal glue, 0.2% sodium dithionite, 0.450% residue* | **10. 06-02-007B: Safety match composition 3:** *54.4% potassium chlorate, 20.4% felspar, 6.8% iron oxide, 6.8% limestone, 4% gelatin, 4% starch, 2.7% animal glue, 0.5% red phosphorus, 0.09% Arylan PWS foaming agent, 0.31% residue* |
| **11. 06-02-008A: Match striking composition for flares and similar devices:** *31.7% condensation product of cresylic acid and paraformaldehyde with hexamethylenetetramine, 19.1% slate dust, 17.2% red phosphorus, 12.7% emery, 8.2% antimony sulfide, 6.3% glass powder, 4.4% chalk, 0.4% mixed impurities* | **12. 06-02-009A: Safety match composition:** *48.8% potassium chlorate, 23.9% lead hypophosphate, 13.6% lead nitrate, 12.6% animal glue, 0.79% ground glass, 0.18% gum benzoin, 0.13% mixed residues* |
| **13. 06-02-010A: High heat producing match composition:** *32.7% glass powder, 27.2% potassium chlorate, 16.3% magnetic iron oxide, 14.5% animal glue, 5.4% aluminum, 3.6% iron oxide, 0.30% impurities* | **14. 06-02-011A: Standard match composition:** *26.6% potassium chlorate, 18.6% glass powder, 14.6% ferrous peroxide, 14.6% zinc white, 13.3% animal glue, 12% phosphorus sesquisulfide, 0.30% balance* |
| **15. 06-02-012A: Classic match composition:** *27.5% potassium chlorate, 20.6% ground glass, 13.7% iron-III-oxide, 13.7% phosphorus trisulfide, 13.3% animal glue, 11% zinc oxide, 0.20% balance* | **16. 06-02-013A: Classic match composition with zinc hypothiophosphate burn increaser:** *27.9% potassium chlorate, 20.9% ground glass, 20.9% zinc hypothiophosphate, 13.4% animal glue, 8.3% iron-III-oxide, 8.3% zinc oxide, 0.30% balance* |
| **17. 06-02-014A: Classic match composition with gum ammoniac binder:** *33.8% potassium chlorate, 16.9% glass, 16.9% zinc oxide, 15.2% phosphorus sesquisulfide, 11.8% animal glue, 5% gum ammoniac, 0.40% mixed impurities* | **18. 06-02-015A: Classic match composition:** *45% potassium chlorate, 15% glass, 15% flours of sulfur, 12% diazobenzenesulfonic acid, 10% animal glue, 3% antimony sulfide* |
| **19. 06-02-016A: Classic match composition (circa 1870):** *59% potassium chlorate, 22% gelatin, 18% pumice, 1% iron protoxide* | **20. 06-02-017A: Friction resistant match composition:** *77% potassium chlorate, 15% ammonium sulfate, 8% glucose* |
| **21. 06-02-018A: Classic waterproof match composition:** *49% potassium chlorate, 29% powdered glass, 21% resorcin, 1% sodium hydroxide* | **22. 06-02-019A: Friction sensitive composition for use in matches:** *59% red phosphorus, 18% antimony sulfide, 11.5% gum acacia, 11.5% carbon black* |
| **23. 06-02-020A: Friction sensitive composition for use in matches:** *52.63% potassium chlorate, 14.03% calcium plumbate, 14.03% gum acacia, 7.01% powdered glass, 5.26% red phosphorus, 5.26% gum Arabic, 1.75% sulfur, 0.03% mixed residual balance* | **24. 06-02-021A: Friction sensitive composition for use in matches:** *42.34% potassium chlorate, 28.09% lead hypophosphite, 13.98% dye/pigment, 5.85% gum Arabic, 5.59% potassium dichromate, 4.12% red phosphorus, 0.03% residual balance* |
| **25. 06-02-022A: Friction sensitive composition for use in "strike anywhere" matches:** *39.21% potassium chlorate, 19.6% powdered glass, 12.74% glue, 11.76% whiting, 9.8% red phosphorus, 6.86% plaster of Paris, 0.03% mixed balance* | **26. 06-02-023A: Friction sensitive composition for use in matches:** *34.48% potassium chlorate, 17.24% antimony trisulfide, 17.24% lead nitrate, 8.62% vegetable charcoal, 8.62% red phosphorus, 8.62% potassium dichromate, 5.17% animal glue, 0.01% residual balance* |
| **27. 06-02-024A: Friction sensitive composition for use in matches:** *35.08% potassium chlorate, 22.8% glass powder, 12.28% animal glue, 12.28% zinc oxide, 7.01% red phosphorus, 5.26% lead hypophosphite, 3.5% gum benzoin, 1.75% sulfur, 0.04% mixed balance* | **28. 06-02-025A: Friction sensitive composition for use in matches:** *42.85% potassium chlorate, 20% powdered glass, 15.23% animal glue, 9.52% red phosphorus, 3.8% plaster of Paris, 3.8% zinc oxide, 2.85% sulfur, 1.9% whiting, 0.05% residue* |
| **29. 06-02-026A: Classic match-head composition:** *33.33% sand, 29.16% animal glue, 20.83% potassium perchlorate, 12.5% phosphorus sesqui-sulfide, 4.16% sulfur, 0.02% residual balance* | **30. 06-02-026B: Classic match-head composition:** *33.84% powdered sand, 30.76% ammonium perchlorate, 26.15% animal glue, 9.23% phosphorus sequi-sulfide, 0.02% residual balance* |
| **31. 06-02-027A: Classic "safety" match-head composition:** *30% potassium perchlorate, 27% powdered sand, 16% animal glue 10% magnesium/silicon alloy, 10% red phosphorus, 7% powdered chalk* | **32. 06-02-028A: Match composition for multiple uses:** *56.41% potassium chlorate, 35.89% composition 1, 7.69% composition 2, 0.01% mixed balance* |
| **33. 06-02-029A: Weatherproof match composition:** *38.09% potassium chlorate, 15.87% powdered glass, 15.87% powdered pumice stone, 9.52% sodium nitrate, 7.93% corn starch, 6.34% antimony trisulfide, 3.17% cascarilla bark, 1.58% animal glue, 1.58% lamp black, 0.05% mixed balance* | |

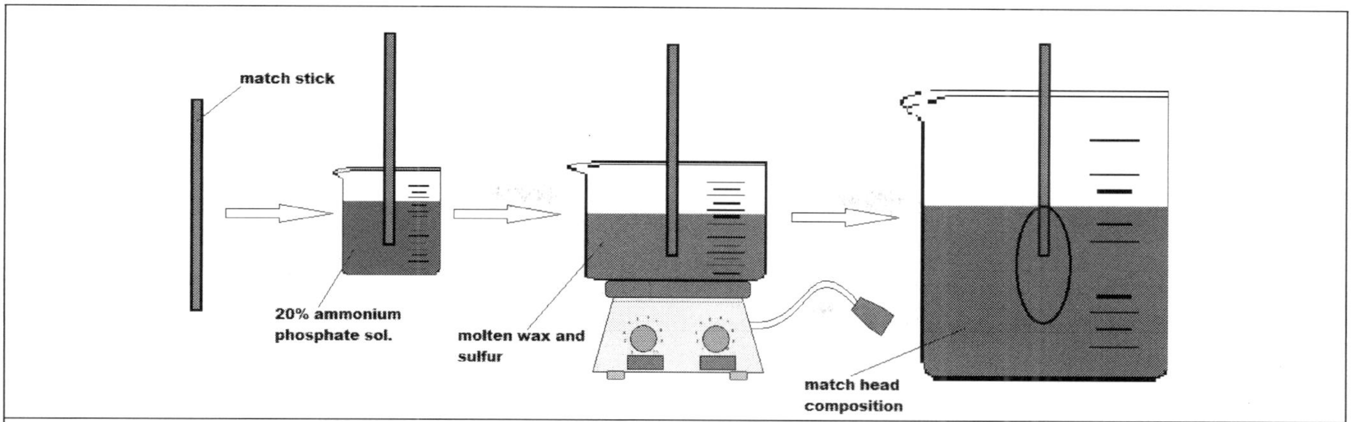


First, take wooden sticks, ranging from 40mm to 200mm in length by 1mm to 5mm diameter, and dip the tips of the wooden sticks into a glass container, filled with a 20% ammonium phosphate [$(NH_4)_3PO_4$] solution. Note: only the exact "tips" of the wood-be wooden match sticks (where the chemicals will be applied) should be dipped into the ammonium phosphate solution. Thereafter, place the dipped wooden sticks on a shallow pan or tray and allow them to air-dry for 48 hours. After 48 hours, take the dry wooden sticks, and dip the tips (the same tips used for the ammonium phosphate) into a glass beaker containing molten paraffin wax mixed with molten sulfur (20 parts paraffin wax, 3 parts sulfur); the dipping process should be very rapid, and only a very thin layer of paraffin wax/sulfur layer is desired (0.09mm to 0.97mm thickness); then allow the sticks to dry for 15 minutes. **Note: Paper "sticks" can be used if desired. Second note: the above illustration is not up to scale.**

**06-02-001A Safety-match composition 1:**

Into a vertical mixer, equipped with flexible Teflon stir blades, place ***10 grams of ferric oxide***, followed by ***4 grams of zinc oxide***, followed by ***4 grams of hydroxyethyl cellulose***, followed by ***22 grams of flours of sulfur***, followed by ***22 grams of diatomaceous earth***, followed by ***48 grams of siliceous microspheres***, followed by ***36 grams of manganese dioxide***, followed by ***4 grams of potassium dichromate***, and then followed by ***200 grams of potassium chlorate***. Thereafter, dry blend the mixture on low speed for 30 minutes. After 30 minutes, add in ***50 grams of standard animal glue***, and then continue to blend the mixture on low speed for 10 minutes. Thereafter, the mixture is ready to be casted ("dipped"). To cast the match head composition, the wood or paper "match sticks" should be dipped and re-dipped into the semi-fluidized mixture to form a uniform thickness ranging from 2mm to 5mm in width (depends on discretion, and the desired thickness of the match head composition can vary). Thereafter, the wood/paper matchsticks can be dipped into a colored dye solution if desired; in either case, the matches should then be cured in an oven at 40 Celsius for 12 hours. The matches can be ignited by "striking" them on a suitable surface.

**Ignition temperature (when match is heated externally):** 560 Celsius
**Ignition temperature (exact point when match ignites):** 990 Celsius
**Ease of ignition ("striking" on suitable surface):** 7
**Water resistance:** Very good.
**Weather resistance (match):** wind resistance below normal; moist and wet conditions below normal.
**Stability:** Can be stored for decades.
**Sensitivity:** Very stable towards shock and percussion; friction safe on soft materials.
**Explosive ability:** Cannot be detonated.
**Percentage:** ***50% potassium chlorate, 12.5% standard animal glue binder, 12% siliceous microspheres, 9% manganese dioxide, 5.5% sulfur, 5.5 diatomaceous earth combustible, 2.5% ferric oxide coloring agent, 1% zinc oxide filler, 1% hydroxyethyl cellulose curative, 1% potassium dichromate catalyst***
**Classification:** Deflagrating mixture (classified as 1.3G pyrotechnic composition).

**Danger of preparation:** Can be made by amateurs.
**Use:** Matches; electric match heads.

### 06-02-001B: Safety-match (modified) composition 2:

Into a vertical mixer, equipped with flexible Teflon stir blades, place, ***4.4 grams of hydroxyethyl cellulose***, followed by ***22 grams of flours of sulfur***, followed by ***120 grams of siliceous microspheres***, followed by ***4.4 grams of potassium dichromate***, and then followed by ***197 grams of potassium chlorate***. Thereafter, dry blend the mixture on low speed for 30 minutes. After 30 minutes, add in ***49.7 grams of standard animal glue***, and then continue to blend the mixture on low speed for 10 minutes. Thereafter, the mixture is ready to be casted ("dipped"). To cast the match head composition, the wood or paper "match sticks" should be dipped and re-dipped into the semi-fluidized mixture to form a uniform thickness ranging from 2mm to 5mm in width (depends on discretion, and the desired thickness of the match head composition can vary). Thereafter, the wood/paper matchsticks can be dipped into a colored dye solution if desired; in either case, the matches should then be cured in an oven at 40 Celsius for 12 hours. The matches can be ignited by "striking" them on a suitable surface.
**Ignition temperature (when match is heated externally):** 540 Celsius
**Ignition temperature (exact point when match ignites):** 970 Celsius
**Ease of ignition ("striking" on suitable surface):** 6.8
**Water resistance:** Very good.
**Weather resistance (match):** wind resistance below normal; moist and wet conditions below normal.
**Stability:** Can be stored for decades.
**Sensitivity:** Very stable towards shock and percussion; friction safe on soft materials.
**Explosive ability:** Cannot be detonated.
**Percentage:** ***49.5% potassium chlorate, 30% siliceous microspheres, 12.5% standard animal glue binder, 5.5% sulfur, 1.1% hydroxyethyl cellulose curative, 1.1% potassium dichromate catalyst, 0.30% impurities***
**Classification:** Deflagrating mixture (classified as 1.3G pyrotechnic composition).

**Danger of preparation:** Can be made by amateurs.
**Use:** Matches; electric match heads.

### 06-02-002A: "Strike anywhere" match composition 1:

Into a vertical mixer, equipped with flexible Teflon stir blades, place ***24 grams of zinc oxide***, followed by ***2 grams of any desired dye*** (use the color of your choice), followed by ***62 grams of plaster of Paris***, followed by ***28 grams of Phosphorus Sesquisulphide***, followed by ***76 grams of siliceous microspheres***, and then followed by ***140 grams of potassium chlorate***. Thereafter, dry blend the mixture on low speed for 30 minutes. After 30 minutes, add in ***68 grams of standard animal glue***, and then continue to blend the mixture on low speed for 10 minutes. Thereafter, the mixture is ready to be casted ("dipped"). To cast the match head composition, the wood or paper "match sticks" should be dipped and re-dipped into the semi-fluidized mixture to form a uniform thickness ranging from 2mm to 5mm in width (depends on discretion, and the desired thickness of the match head composition can vary). Thereafter, the wood/paper matchsticks can be dipped into a colored dye solution if desired; in either case, the matches should then be cured in an oven at 40 Celsius for 12 hours. The matches can be ignited by "striking" them on a suitable surface.
**Ignition temperature (when match is heated externally):** 590 Celsius
**Ignition temperature (exact point when match ignites):** 985 Celsius
**Ease of ignition ("striking" on suitable surface):** 7.9
**Water resistance:** Very good.
**Weather resistance (match):** wind resistance below normal; moist and wet conditions below normal.
**Stability:** Can be stored for decades.
**Sensitivity:** Very stable towards shock and percussion; friction safe on soft materials.
**Explosive ability:** Cannot be detonated.
**Percentage:** ***35% potassium chlorate, 19% siliceous microspheres, 17% standard animal glue binder, 15.5% plaster of Paris, 7% Phosphorus Sesquisulphide, 6% zinc oxide, 0.50% dye***
**Classification:** Deflagrating mixture (classified as 1.3G pyrotechnic composition).

**Danger of preparation:** Can be made by amateurs.
**Use:** Matches; electric match heads.

### 06-02-002B: "Strike anywhere" (modified) match composition 2:

Into a vertical mixer, equipped with flexible Teflon stir blades, place ***24 grams of zinc oxide,*** followed by ***1.3 grams of any desired dye*** (use the color of your choice), followed by ***35 grams of plaster of Paris***, followed by ***33 grams of Phosphorus Sesquisulphide***, followed by ***74.8 grams of phenolic resin microspheres***, and then followed by ***161 grams of potassium chlorate***. Thereafter, dry blend the mixture on low speed for 30 minutes. After 30 minutes, add in ***68.8 grams of standard animal glue***, and then continue to blend the mixture on low speed for 10 minutes. Thereafter, the mixture is ready to be casted ("dipped"). To cast the match head composition, the wood or paper "match sticks" should be dipped and re-dipped into the semi-fluidized mixture to form a uniform thickness ranging from 2mm to 5mm in width (depends on discretion, and the desired thickness of the match head composition can vary). Thereafter, the wood/paper matchsticks can be dipped into a colored dye solution if desired; in either case, the matches should then be cured in an oven at 40 Celsius for 12 hours. The matches can be ignited by "striking" them on a suitable surface.
**Ignition temperature (when match is heated externally):** 510 Celsius
**Ignition temperature (exact point when match ignites):** 995 Celsius
**Ease of ignition ("striking" on suitable surface):** 7.8
**Water resistance:** Very good.
**Weather resistance (match):** wind resistance below normal; moist and wet conditions below normal.
**Stability:** Can be stored for decades.
**Sensitivity:** Very stable towards shock and percussion; friction safe on soft materials.
**Explosive ability:** Cannot be detonated.
**Percentage:** ***40.4% potassium chlorate, 18.7% phenolic resin microspheres, 17.2% standard animal glue binder, 8.8% plaster of Paris, 8.2% Phosphorus Sesquisulphide, 6% zinc oxide, 0.32% dye, 0.380% impurities***
**Classification:** Deflagrating mixture (classified as 1.3G pyrotechnic composition).

**Danger of preparation:** Can be made by amateurs.
**Use:** Matches; electric match heads.

### 06-02-003A: Matchbook match composition 1:

Into a vertical mixer, equipped with flexible Teflon stir blades, place ***20 grams of zinc oxide,*** followed by ***1 grams of any desired dye*** (use the color of your choice), followed by ***25 grams of diatomaceous earth***, followed by ***2 grams of gum Arabic,*** followed by ***30 grams of flour glass***, followed by ***30 grams of siliceous microspheres***, followed by ***20 grams of flours of sulfur,*** followed by ***9 grams of powdered starch,*** followed by ***3 grams of potassium dichromate,*** and then followed by ***220 grams of potassium chlorate***. Thereafter, dry blend the mixture on low speed for 30 minutes. After 30 minutes, add in ***40 grams of standard animal glue***, and then continue to blend the mixture on low speed for 10 minutes. Thereafter, the mixture is ready to be casted ("dipped"). To cast the match head composition, the wood or paper "match sticks" should be dipped and re-dipped into the semi-fluidized mixture to form a uniform thickness ranging from 2mm to 5mm in width (depends on discretion, and the desired thickness of the match head composition can vary). Thereafter, the wood/paper matchsticks can be dipped into a colored dye solution if desired; in either case, the matches should then be cured in an oven at 40 Celsius for 12 hours. The matches can be ignited by "striking" them on a suitable surface.
**Ignition temperature (when match is heated externally):** 515+ Celsius
**Ignition temperature (exact point when match ignites):** 1030 Celsius
**Ease of ignition ("striking" on suitable surface):** 7.9
**Water resistance:** Very good.
**Weather resistance (match):** wind resistance below normal; moist and wet conditions below normal.
**Stability:** Can be stored for decades.
**Sensitivity:** Very stable towards shock and percussion; friction safe on soft materials.
**Explosive ability:** Cannot be detonated.
**Percentage:** ***55% potassium chlorate, 10% standard animal glue binder, 7.5% siliceous microspheres, 7.5% flour glass, 6.2% diatomaceous earth, 5% zinc oxide, 5% sulfur, 2.2% starch filler, 0.75% potassium dichromate, 0.50% gum Arabic, 0.25% dye, 0.10% impurities***
**Classification:** Deflagrating mixture (classified as 1.3G pyrotechnic composition).

**Danger of preparation:** Can be made by amateurs, but caution should be exercised.
**Use:** Matches; electric match heads.

### 06-02-004A: "Double dip" friction resistant matchstick composition 1:

Into a vertical mixer, equipped with flexible Teflon stir blades, place ***10.4 grams of zinc oxide,*** followed by ***11.8 grams of any desired dye*** (use the color of your choice), followed by ***11.8 grams of calcium resinate***, followed by ***3 grams of gum Arabic,*** followed by ***14.8 grams of powdered clay,*** followed by ***32 grams of sulfur,*** followed by ***53 grams of flour glass***, followed by ***31 grams of fresh finely divided plaster of Paris (non moistened)***, followed by ***9 grams of red phosphorus,*** followed by ***400 milligrams of potassium dichromate,*** and then followed by ***166 grams of potassium chlorate***. Thereafter, dry blend the mixture on low speed for 30 minutes. After 30 minutes, add in ***53 grams of standard animal glue***, and then continue to blend the mixture on low speed for 10 minutes. Thereafter, the mixture is ready to be casted ("dipped"). To cast the match head composition, the wood or paper "match sticks" should be dipped and re-dipped into the semi-fluidized mixture to form a uniform thickness ranging from 2mm to 5mm in width (depends on discretion, and the desired thickness of the match head composition can vary). Thereafter, the wood/paper matchsticks can be dipped into a colored dye solution if desired; in either case, the matches should then be cured in an oven at 40 Celsius for 12 hours.

**Ignition temperature (when match is heated externally):** N/A
**Ignition temperature (exact point when match ignites):** 970 Celsius
**Ease of ignition ("striking" on suitable surface):** N/A
**Water resistance:** Very good.
**Weather resistance (match):** wind resistance normal; moist and wet conditions normal.
**Stability:** Can be stored for decades.
**Sensitivity:** Very stable towards shock and percussion; friction safe.
**Explosive ability:** Cannot be detonated.
**Percentage:** ***41.8% potassium chlorate, 13.3% standard animal glue binder, 13.35% flour glass, 8% sulfur, 7.8% plaster of Paris, 3.7% clay, 2.9% calcium resinate, 2.9% dye, 2.6% zinc oxide, 2.2% red phosphorus, 0.75% gun Arabic, 0.10% potassium dichromate, 0.60% impurities***
**Classification:** Deflagrating mixture (classified as 1.3G pyrotechnic composition).

**Danger of preparation:** Can be made by amateurs.
**Use:** Electric match heads, ignition match for igniting other pyrotechnic material.

### 06-02-005A: Clean burning match composition 1:

Into a vertical mixer, equipped with flexible Teflon stir blades, place ***2.8 grams of boric acid,*** followed by ***4.8 grams of any desired dye*** (use the color of your choice), followed by ***23 grams of diatomaceous earth***, followed by ***9 grams of dextrin,*** followed by ***14 grams of powdered graphite,*** followed by ***92 grams of glass powder,*** followed by ***180 grams of potassium chlorate,*** and then followed by ***14 grams of nitrocellulose*** (14% nitrogen content). Thereafter, dry blend the mixture on low speed for 30 minutes. After 30 minutes, add in ***50.8 grams of standard animal glue***, and then continue to blend the mixture on low speed for 10 minutes. Thereafter, the mixture is ready to be casted ("dipped"). To cast the match head composition, the wood or paper "match sticks" should be dipped and re-dipped into the semi-fluidized mixture to form a uniform thickness ranging from 2mm to 5mm in width (depends on discretion, and the desired thickness of the match head composition can vary). Thereafter, the wood/paper matchsticks can be dipped into a colored dye solution if desired; in either case, the matches should then be cured in an oven at 40 Celsius for 12 hours.

**Ignition temperature (when match is heated externally):** 550 Celsius
**Ignition temperature (exact point when match ignites):** 995 Celsius
**Ease of ignition ("striking" on suitable surface):** 6.2
**Water resistance:** Very good.
**Weather resistance (match):** wind resistance normal; moist and wet conditions normal.
**Stability:** Can be stored for decades.
**Sensitivity:** Very stable towards shock and percussion; friction safe on soft materials.
**Explosive ability:** Cannot be detonated.
**Percentage:** ***46% potassium chlorate, 23.5% glass powder, 13% standard animal glue, 5.8% diatomaceous earth, 3.5% graphite, 3.5% nitrocellulose, 2.3% dextrin, 1.2% dye, 0.70% boric acid***
**Classification:** Deflagrating mixture (classified as 1.3G pyrotechnic composition).

**Danger of preparation:** Can be made by amateurs.
**Use:** Matches; electric match heads.

### 06-02-006A: Quick-lite "easy light" weather proof match composition 1:

Into a vertical mixer, equipped with flexible Teflon stir blades, place ***370 grams of potassium chlorate,*** followed by ***5 grams of shellac,*** followed by ***5 grams of turpentine,*** followed by ***10 grams of powdered wood pulp,*** followed by ***10 grams of starch,*** followed by ***190 grams of glass powder,*** followed by ***20 grams of titanium dioxide,*** followed by ***25 grams of diatomaceous earth,*** and then followed by ***10 grams of aluminum hydroxide***. Thereafter, dry blend the mixture on low speed for 30 minutes. After 30 minutes, add in ***112 grams of standard animal glue***, and then continue to blend the mixture on low speed for 10 minutes. Thereafter, the mixture is ready to be casted ("dipped"). To cast the match head composition, the wood or paper "match sticks" should be dipped and re-dipped into the semi-fluidized mixture to form a uniform thickness ranging from 2mm to 5mm in width (depends on discretion, and the desired thickness of the match head composition can vary). Thereafter, the wood/paper matchsticks can be dipped into a colored dye solution if desired; in either case, the matches should then be cured in an oven at 40 Celsius for 24 hours.
**Ignition temperature (when match is heated externally):** 545 Celsius
**Ignition temperature (exact point when match ignites):** 1025 Celsius
**Ease of ignition ("striking" on suitable surface):** 6
**Water resistance:** Very good.
**Weather resistance (match):** wind resistance above normal; moist and wet conditions above normal.
**Stability:** Can be stored for decades.
**Sensitivity:** Very stable towards shock and percussion; friction safe on soft materials.
**Explosive ability:** Cannot be detonated.
**Percentage:** ***48.5% potassium chlorate, 24.9% glass powder, 14.6% standard animal glue, 3.2% diatomaceous earth, 2.6% titanium dioxide, 1.3% wood pulp, 1.3% starch, 1.3% aluminum hydroxide, 1% moisture, 0.65% turpentine, 0.65% shellac***
**Classification:** Deflagrating mixture (classified as 1.3G pyrotechnic composition).

**Danger of preparation:** Can be made by amateurs.
**Use:** Matches; electric match heads.

### 06-02-007A: "Strike anywhere" match head composition 3:

Into a vertical mixer, equipped with flexible Teflon stir blades, place ***62 grams of gelatin***, followed by ***13.5 grams of food grade starch***, followed by ***103.5 grams of finely divided feldspar***, followed by ***1.7 grams of animal glue,*** followed by ***13.5 grams of titanium dioxide***, followed by ***34.5 grams of finely ground limestone***, followed by ***1 gram of sodium dithionite***, followed by ***241.5 grams of potassium chlorate***, and then followed by ***13.8 grams of red phosphorus***. Thereafter, blend the mixture on low speed for 30 minutes. After 30 minutes, the mixture is ready to be casted ("dipped"). To cast the match head composition, the wood or paper "match sticks" should be dipped and re-dipped into the semi-fluidized mixture to form a uniform thickness ranging from 2mm to 5mm in width (depends on discretion, and the desired thickness of the match head composition can vary). Thereafter, the wood/paper matchsticks can be dipped into a colored dye solution if desired; in either case, the matches should then be cured in an oven at 40 Celsius for 12 hours.
**Ignition temperature (when match is heated externally):** 555 Celsius
**Ignition temperature (exact point when match ignites):** 1005 Celsius
**Ease of ignition ("striking" on suitable surface):** 7.9
**Water resistance:** Very good.
**Weather resistance (match):** wind resistance normal; moist and wet conditions normal.
**Stability:** Can be stored for decades.
**Sensitivity:** Very stable towards shock and percussion; friction safe on soft materials.
**Explosive ability:** Cannot be detonated.
**Percentage:** ***49.7% potassium chlorate, 21.3% feldspar, 12.7% gelatin, 7.1% limestone, 2.8% red phosphorus, 2.7% starch, 2.7% titanium dioxide, 0.35% animal glue, 0.2% sodium dithionite, 0.450% residue***
**Classification:** Deflagrating mixture (classified as 1.3G pyrotechnic composition).

**Danger of preparation:** Can be made by amateurs.
**Use:** Matches; electric match heads.

### 06-02-007B: Safety match composition 3:

Into a vertical mixer, equipped with flexible Teflon stir blades, place ***20 grams of gelatin***, followed by ***20 grams of food grade starch***, followed by ***101.5 grams of finely divided feldspar***, followed by ***13.5 grams of animal glue,*** followed by ***33.8 grams of iron oxide***, followed by ***450 milligrams of Arylan PWS***, followed by ***33.8 grams of finely ground limestone***, followed by ***270 grams of potassium chlorate***, and then followed by ***2.5 grams of red phosphorus***. Thereafter, blend the mixture on low speed for 30 minutes. After 30 minutes, the mixture is ready to be casted ("dipped"). To cast the match head composition, the wood or paper "match sticks" should be dipped and re-dipped into the semi-fluidized mixture to form a uniform thickness ranging from 2mm to 5mm in width (depends on discretion, and the desired thickness of the match head composition can vary). Thereafter, the wood/paper matchsticks can be dipped into a colored dye solution if desired; in either case, the matches should then be cured in an oven at 40 Celsius for 12 hours.

**Ignition temperature (when match is heated externally):** 500 Celsius
**Ignition temperature (exact point when match ignites):** 985 Celsius
**Ease of ignition ("striking" on suitable surface):** 6
**Water resistance:** Very good.
**Weather resistance (match):** wind resistance normal; moist and wet conditions normal.
**Stability:** Can be stored for decades.
**Sensitivity:** Very stable towards shock and percussion; friction safe on soft materials.
**Explosive ability:** Cannot be detonated.
**Percentage:** ***54.4% potassium chlorate, 20.4% feldspar, 6.8% iron oxide, 6.8% limestone, 4% gelatin, 4% starch, 2.7% animal glue, 0.5% red phosphorus, 0.09% Arylan PWS foaming agent, 0.31% residue***
**Classification:** Deflagrating mixture (classified as 1.3G pyrotechnic composition).

**Danger of preparation:** Can be made by amateurs.
**Use:** Matches; electric match heads.

## 06-02-008A: Match striking composition for flares and similar devices:

Into a vertical mixer, equipped with flexible Teflon stir blades, place ***600 milliliters of cresylic acid***, followed by ***81 grams of hexamethylenetetramine***, and then blend the mixture on high speed for about 30 minutes to form a uniform mix. Thereafter, add in ***272 grams of paraformaldehyde***, followed by ***489 grams of red phosphorus***, followed by ***235 grams of black antimony sulfide***, followed by ***544 grams of slate dust***, followed by ***362 grams of finely divided emery***, followed by ***181 grams of glass powder***, and then followed by ***127 grams of finely divided chalk***. Thereafter, blend the mixture on low speed for 30 minutes. After 30 minutes, the mixture is ready to be casted. To cast, the match composition should be "applied" to the road flare where needed, and then allowed to cure for 12 hours; for smoke grenades, the mixture should be pressed into the desired shape under low pressure and then allowed to cure in an oven at 50 Celsius for 6 hours; for other munitions, the amount of mixture to be applied is based on the munition itself, so discretion must be observed; use your own discretion based on the outline and design of your munition. In any case, the munition(s) should then be cured in an oven at 60 Celsius for 12 hours.
**Ignition temperature (when match is heated externally):** 525 Celsius
**Ignition temperature (exact point when match ignites):** 1095 Celsius
**Ease of ignition ("striking" on suitable surface):** 8
**Water resistance:** Very good.
**Weather resistance (match):** wind resistance normal; moist and wet conditions normal.
**Stability:** Can be stored for decades.
**Sensitivity:** Very stable towards shock and percussion; friction safe on soft materials.
**Explosive ability:** Cannot be detonated.
**Percentage:** ***31.7% condensation product of cresylic acid and paraformaldehyde with hexamethylenetetramine, 19.1% slate dust, 17.2% red phosphorus, 12.7% emery, 8.2% antimony sulfide, 6.3% glass powder, 4.4% chalk, 0.4% mixed impurities***
**Classification:** Deflagrating mixture (classified as 1.3G pyrotechnic composition).

**Danger of preparation:** Can be made by amateurs.
**Use:** Used in road flares, smoke grenades, and other devices that are initiated by "striking" the device/composition on a suitable surface.

## 06-02-009A: Safety match composition:

Into a vertical mixer, equipped with flexible Teflon stir blades, place ***73.6 grams of lead nitrate***, followed by ***1 gram of gum benzoin***, followed by ***263 grams of potassium chlorate***, followed by ***129 grams of lead hypophosphate***, and then followed by ***4.3 grams of ground glass***. Thereafter, dry blend the mixture on low speed for 30 minutes. After 30 minutes, add in ***68 grams of standard animal glue***, and then continue to blend the mixture on low speed for 10 minutes. Thereafter, the mixture is ready to be casted ("dipped"). To cast the match head composition, the wood or paper "match sticks" should be dipped and re-dipped into the semi-fluidized mixture to form a uniform thickness ranging from 2mm to 5mm in width (depends on discretion, and the desired thickness of the match head composition can vary). Thereafter, the wood/paper matchsticks can be dipped into a colored dye solution if desired; in either case, the matches should then be cured in an oven at 60 Celsius for 24 hours.
**Ignition temperature (when match is heated externally):** 500 Celsius
**Ignition temperature (exact point when match ignites):** 995 Celsius
**Ease of ignition ("striking" on suitable surface):** 4
**Water resistance:** Very good.
**Weather resistance (match):** wind resistance normal; moist and wet conditions normal.
**Stability:** Can be stored for decades.
**Sensitivity:** Very stable towards shock and percussion; friction safe on soft/hard materials.
**Explosive ability:** Cannot be detonated.

**Percentage:** ***48.8% potassium chlorate, 23.9% lead hypophosphate, 13.6% lead nitrate, 12.6% animal glue, 0.79% ground glass, 0.18% gum benzoin, 0.13% mixed residues***
**Classification:** Deflagrating mixture (classified as 1.3G pyrotechnic composition).

**Danger of preparation:** Can be made by amateurs.
**Use:** Used to make matches that are hard to ignite; only ignite on proper surfaces such as flint or "sand paper" type surfaces.

### 06-02-010A: High heat producing match composition:

Into a vertical mixer, equipped with flexible Teflon stir blades, place ***15 grams of finely powdered aluminum*** of average mesh, followed by ***10 grams of finely powdered red iron-III-oxide***, followed by ***75 grams of potassium chlorate***, followed by ***45 grams of magnetic iron oxide***, followed by ***90 grams of glass powder***. Thereafter, blend the mixture on low speed for 30 minutes. After 30 minutes, add in ***108 grams of standard animal glue***, and then continue to blend the mixture on low speed for 10 minutes. Thereafter, the mixture is ready to be casted ("dipped"). To cast the match head composition, the wood or paper "match sticks" should be dipped and re-dipped into the semi-fluidized mixture to form a uniform thickness ranging from 2mm to 5mm in width (depends on discretion, and the desired thickness of the match head composition can vary). Thereafter, the wood/paper matchsticks can be dipped into a colored dye solution if desired; in either case, the matches should then be cured in an oven at 60 Celsius for 24 hours. Note: Instead of using matchsticks, the composition can be pressed into pellets, discs, tablets, or any other desirable shape under a pressure of 2,000 psi, and cured at 60 Celsius for 12 hours.
**Ignition temperature (when match is heated externally):** 690 Celsius
**Ignition temperature (exact point when match ignites):** 1,390 Celsius
**Ease of ignition ("striking" on suitable surface):** 6
**Water resistance:** Very good.
**Weather resistance (match):** wind resistance above normal; moist and wet conditions above normal.
**Stability:** Can be stored for decades.
**Sensitivity:** Very stable towards shock and percussion; friction safe on soft materials.
**Explosive ability:** Cannot be detonated.
**Percentage:** ***32.7% glass powder, 27.2% potassium chlorate, 16.3% magnetic iron oxide, 14.5% animal glue, 5.4% aluminum, 3.6% iron oxide, 0.30% impurities***
**Classification:** Deflagrating mixture (classified as 1.3G pyrotechnic composition).

**Danger of preparation:** Can be made by amateurs.
**Use:** Used to make specialty matches for igniting wet or hard to ignite materials.

### 06-02-011A: Standard match composition:

Into a vertical mixer, equipped with flexible Teflon stir blades, place ***108 grams of phosphorus sesquisulfide,*** followed by ***240 grams of potassium chlorate***, followed by ***132 grams of ferrous peroxide***, followed by ***132 grams of finely divided zinc white***, and then followed by ***168 grams of glass powder***. Thereafter, blend the mixture on low speed for 30 minutes. After 30 minutes, add in ***120 grams of standard animal glue***, and then continue to blend the mixture on low speed for 10 minutes. Thereafter, the mixture is ready to be casted ("dipped"). To cast the match head composition, the wood or paper "match sticks" should be dipped and re-dipped into the semi-fluidized mixture to form a uniform thickness ranging from 2mm to 5mm in width (depends on discretion, and the desired thickness of the match head composition can vary). Thereafter, the wood/paper matchsticks can be dipped into a colored dye solution if desired; in either case, the matches should then be cured in an oven at 60 Celsius for 24 hours.
**Ignition temperature (when match is heated externally):** 455 Celsius
**Ignition temperature (exact point when match ignites):** 905 Celsius
**Ease of ignition ("striking" on suitable surface):** 6.9
**Water resistance:** Very good.
**Weather resistance (match):** wind resistance normal; moist and wet conditions normal.
**Stability:** Can be stored for decades.
**Sensitivity:** Very stable towards shock and percussion; friction safe on soft materials.
**Explosive ability:** Cannot be detonated.
**Percentage:** ***26.6% potassium chlorate, 18.6% glass powder, 14.6% ferrous peroxide, 14.6% zinc white, 13.3% animal glue, 12% phosphorus sesquisulfide, 0.30% balance***
**Classification:** Deflagrating mixture (classified as 1.3G pyrotechnic composition).

**Danger of preparation:** Can be made by amateurs.
**Use:** Matches; electric match heads.

### 06-02-012A: Classic match composition:

Into a vertical mixer, equipped with flexible Teflon stir blades, place ***120 grams of phosphorus trisulfide (P2S3),*** followed by ***96 grams of powdered zinc oxide***, followed by ***240 grams of potassium chlorate***, followed by ***120 grams of red iron-III-oxide***, and then followed by ***180 grams of ground glass***. Thereafter, blend the mixture on low speed for 30 minutes. After 30 minutes, add in ***116 grams of standard animal glue***, and then continue to blend the mixture on low speed for 10 minutes. Thereafter, the mixture is ready to be casted ("dipped"). To cast the match head composition, the wood or paper "match sticks" should be dipped and re-dipped into the semi-fluidized mixture to form a uniform thickness ranging from 2mm to 5mm in width (depends on discretion, and the desired thickness of the match head composition can vary). Thereafter, the wood/paper matchsticks can be dipped into a colored dye solution if desired; in either case, the matches should then be cured in an oven at 60 Celsius for 24 hours.
**Ignition temperature (when match is heated externally):** 600 Celsius
**Ignition temperature (exact point when match ignites):** 900 Celsius
**Ease of ignition ("striking" on suitable surface):** 5.1
**Water resistance:** Very good.
**Weather resistance (match):** wind resistance normal; moist and wet conditions normal.
**Stability:** Can be stored for decades.
**Sensitivity:** Very stable towards shock and percussion; friction safe on soft materials.
**Explosive ability:** Cannot be detonated.
**Percentage:** ***27.5% potassium chlorate, 20.6% ground glass, 13.7% iron-III-oxide, 13.7% phosphorus trisulfide, 13.3% animal glue, 11% zinc oxide, 0.20% balance***
**Classification:** Deflagrating mixture (classified as 1.3G pyrotechnic composition).

**Danger of preparation:** Can be made by amateurs.
**Use:** Matches; electric match heads.

### 06-02-013A: Classic match composition with zinc hypothiophosphate burn increaser:

Into a vertical mixer, equipped with flexible Teflon stir blades, place ***180 grams of zinc hypothiophosphate,*** followed by ***240 grams of potassium chlorate***, followed by ***72 grams of red iron-III-oxide***, followed by ***72 grams of zinc oxide***, and then followed by ***180 grams of ground glass***. Thereafter, blend the mixture on low speed for 30 minutes. After 30 minutes, add in ***116 grams of standard animal glue***, and then continue to blend the mixture on low speed for 10 minutes. Thereafter, the mixture is ready to be casted ("dipped"). To cast the match head composition, the wood or paper "match sticks" should be dipped and re-dipped into the semi-fluidized mixture to form a uniform thickness ranging from 2mm to 5mm in width (depends on discretion, and the desired thickness of the match head composition can vary). Thereafter, the wood/paper matchsticks can be dipped into a colored dye solution if desired; in either case, the matches should then be cured in an oven at 60 Celsius for 24 hours.
**Ignition temperature (when match is heated externally):** 540 Celsius
**Ignition temperature (exact point when match ignites):** 1,090 Celsius
**Ease of ignition ("striking" on suitable surface):** 5
**Water resistance:** Very good.
**Weather resistance (match):** wind resistance above normal; moist and wet conditions above normal.
**Stability:** Can be stored for decades.
**Sensitivity:** Very stable towards shock and percussion; friction safe on soft materials.
**Explosive ability:** Cannot be detonated.
**Percentage:** ***27.9% potassium chlorate, 20.9% ground glass, 20.9% zinc hypothiophosphate, 13.4% animal glue, 8.3% iron-III-oxide, 8.3% zinc oxide, 0.30% balance***
**Classification:** Deflagrating mixture (classified as 1.3G pyrotechnic composition).

**Danger of preparation:** Can be made by amateurs.
**Use:** Matches; electric match heads.

### 06-02-014A: Classic match composition with gum ammoniac binder:

Into a vertical mixer, equipped with flexible Teflon stir blades, place ***27.2 grams of gum ammoniac***, followed by ***63.5 grams of standard animal glue***, followed by ***181 grams of potassium chlorate***, followed by ***90.7 grams of zinc oxide***, followed by ***81.6 grams of phosphorus sesquisulfide***, and then followed by ***90.7 grams of powdered glass***. Thereafter, blend the mixture on low speed for 30 minutes. After 30 minutes, the mixture is ready to be casted ("dipped"). To cast the match head composition, the wood or paper "match sticks" should be dipped and re-dipped into the semi-fluidized mixture to form a uniform thickness ranging from 2mm to 5mm in width (depends on discretion, and the desired thickness of the match head composition can vary). Thereafter, the wood/paper matchsticks can be dipped into a colored dye solution if desired; in either case, the matches should then be cured in an oven at 60 Celsius for 24 hours.
**Ignition temperature (when match is heated externally):** 670 Celsius
**Ignition temperature (exact point when match ignites):** 980 Celsius
**Ease of ignition ("striking" on suitable surface):** 5.9
**Water resistance:** Very good.
**Weather resistance (match):** wind resistance normal; moist and wet conditions normal.
**Stability:** Can be stored for decades.
**Sensitivity:** Very stable towards shock and percussion; friction safe on soft materials.
**Explosive ability:** Cannot be detonated.
**Percentage:** ***33.8% potassium chlorate, 16.9% glass, 16.9% zinc oxide, 15.2% phosphorus sesquisulfide, 11.8% animal glue, 5% gum ammoniac, 0.40% mixed impurities***
**Classification:** Deflagrating mixture (classified as 1.3G pyrotechnic composition).

**Danger of preparation:** Can be made by amateurs.
**Use:** Matches; electric match heads.

### 06-02-015A: Classic match composition:

Into a vertical mixer, equipped with flexible Teflon stir blades, place ***45 grams of fours of sulfur***, followed by ***9 grams of antimony sulfide***, followed by ***36 grams of anhydrous diazobenzenesulfonic acid***, followed by ***45 grams of powdered glass***, and then followed by ***135 grams of potassium chlorate***. Thereafter, blend the mixture on low speed for 30 minutes. After 30 minutes, add in ***30 grams of standard animal glue***, and then continue to blend the mixture on low speed for 10 minutes. Thereafter, the mixture is ready to be casted ("dipped"). To cast the match head composition, the wood or paper "match sticks" should be dipped and re-dipped into the semi-fluidized mixture to form a uniform thickness ranging from 2mm to 5mm in width (depends on discretion, and the desired thickness of the match head composition can vary). Thereafter, the wood/paper matchsticks can be dipped into a colored dye solution if desired; in either case, the matches should then be cured in an oven at 60 Celsius for 24 hours.
**Ignition temperature (when match is heated externally):** 600 Celsius
**Ignition temperature (exact point when match ignites):** 995 Celsius
**Ease of ignition ("striking" on suitable surface):** 5.8
**Water resistance:** Very good.
**Weather resistance (match):** wind resistance above normal; moist and wet conditions normal.
**Stability:** Can be stored for decades.
**Sensitivity:** Very stable towards shock and percussion; friction safe on soft materials.
**Explosive ability:** Cannot be detonated.
**Percentage:** ***45% potassium chlorate, 15% glass, 15% flours of sulfur, 12% diazobenzenesulfonic acid, 10% animal glue, 3% antimony sulfide***
**Classification:** Deflagrating mixture (classified as 1.3G pyrotechnic composition).

**Danger of preparation:** Can be made by amateurs.
**Use:** Matches; electric match heads.

### 06-02-016A: Classic match composition (circa 1870):

First, the wooden matchsticks should be lightly coated with paraffin wax (by simply dipping them into molten paraffin wax). Thereafter, prepare the match composition by placing into a suitable mixing bowl, ***110 grams of gelatin***, followed by 325 milliliters of a 10% acetic acid solution, followed by ***6.5 grams of iron protoxide***. Thereafter, manually blend the mixture for about 10 minutes to dissolve the gelatin. Thereafter, add in ***90 grams of ground pumice***, followed by ***300 grams of potassium chlorate***, and then continue to manually blend the mixture for about 1 hour to form a uniform paste. Thereafter, the paste is ready for use. To use, the paraffin coated match sticks simply need to be dipped and re-dipped as many times as needed, and then dried each time in an oven at 60 Celsius.
**Ignition temperature (when match is heated externally):** 400+ Celsius (decomposes into perchlorate at 450 Celsius, and evolving oxygen).
**Ignition temperature (exact point when match ignites):** 975 Celsius

**Ease of ignition ("striking" on suitable surface):** 6.2
**Water resistance:** Very good.
**Weather resistance (match):** wind resistance above normal; moist and wet conditions normal.
**Stability:** Can be stored for decades.
**Sensitivity:** Very stable towards shock and percussion; friction safe on soft materials.
**Explosive ability:** Cannot be detonated.
**Percentage:** ***59% potassium chlorate, 22% gelatin, 18% pumice, 1% iron protoxide***
**Classification:** Deflagrating mixture (classified as 1.3G pyrotechnic composition).

**Danger of preparation:** Can be made by amateurs.
**Use:** Matches; electric match heads.

### 06-02-017A: Friction resistant match composition:

Into a vertical mixer, equipped with flexible Teflon stir blades, place 150 milliliters of hexane, followed by ***10 grams of finely ground glucose***, followed by ***20 grams of ammonium sulfate***, and then followed by ***100 grams of potassium chlorate***. Thereafter, blend the mixture on low speed for 30 minutes. Thereafter, the mixture is ready to be casted ("dipped"). To cast the match head composition, the wood or paper "match sticks" should be dipped, dried, re-dipped, dried, ect., into the semi-fluidized mixture to form a uniform thickness ranging from 2mm to 5mm in width (depends on discretion, and the desired thickness of the match head composition can vary). Thereafter, the wood/paper matchsticks can be dipped into a colored dye solution if desired; in either case, the matches should then be cured in an oven at 80 Celsius for 12 hours. Note: this match will not ignite upon friction, and must be primed with a suitable ignition composition if desired.
**Ignition temperature (when match is heated externally):** 545 Celsius
**Ignition temperature (exact point when match ignites):** 990 Celsius
**Ease of ignition ("striking" on suitable surface):** 1
**Water resistance:** Very good.
**Weather resistance (match):** wind resistance above normal; moist and wet conditions normal.
**Stability:** Can be stored for decades.
**Sensitivity:** Very stable towards shock and percussion; friction safe.
**Explosive ability:** Cannot be detonated.
**Percentage:** ***77% potassium chlorate, 15% ammonium sulfate, 8% glucose***
**Classification:** Deflagrating mixture (classified as 1.3G pyrotechnic composition).

**Danger of preparation:** Can be made by amateurs.
**Use:** Electric match heads; used as an igniter in smoke grenades and other pyrotechnic devices.

### 06-02-018A: Classic waterproof match composition:

Into a vertical mixer, equipped with flexible Teflon stir blades, place ***75 grams of resorcin***, followed by ***75 grams of a 38 to 40% formaldehyde solution,*** followed by ***70 grams of a 20% sodium hydroxide solution***, followed by ***175 grams of potassium chlorate***, and then followed by ***105 grams of finely powdered glass***. Thereafter, blend the mixture on low speed for 30 minutes. Thereafter, the mixture is ready to be casted ("dipped"). To cast the match head composition, the wood or paper "match sticks" should be dipped, dried, re-dipped, dried, ect., into the semi-fluidized mixture to form a uniform thickness ranging from 2mm to 5mm in width (depends on discretion, and the desired thickness of the match head composition can vary). Thereafter, the wood/paper matchsticks can be dipped into a colored dye solution if desired; in either case, the matches should then be cured in an oven at 60 Celsius for 12 hours.
**Ignition temperature (when match is heated externally):** 560 Celsius
**Ignition temperature (exact point when match ignites):** 995 Celsius
**Ease of ignition ("striking" on suitable surface):** 7.2
**Water resistance:** Very good.
**Weather resistance (match):** wind resistance above normal; moist and wet conditions above normal.
**Stability:** Can be stored for decades.
**Sensitivity:** Very stable towards shock and percussion; friction safe on soft materials.
**Explosive ability:** Cannot be detonated.
**Percentage:** ***49% potassium chlorate, 29% powdered glass, 21% resorcin, 1% sodium hydroxide***
**Classification:** Deflagrating mixture (classified as 1.3G pyrotechnic composition).

**Danger of preparation:** Can be made by amateurs.
**Use:** Used to make weatherproof matches.

### 06-02-019A: Friction sensitive composition for use in matches:

Into a vertical mixer, equipped with flexible Teflon stir blades, place ***500 grams of red phosphorus***, followed by ***150 grams of antimony sulfide***, followed by ***100 grams of gum acacia***, followed by 200 milliliters of hexane, and then followed by ***100 grams of carbon black***. Thereafter, blend the mixture on low speed for 30 minutes. Thereafter, the mixture is ready to be casted ("dipped"). To cast the match head composition, the wood or paper "match sticks" should be dipped, dried, re-dipped, dried, ect., into the semi-fluidized mixture to form a uniform thickness ranging from 2mm to 5mm in width (depends on discretion, and the desired thickness of the match head composition can vary). Thereafter, the wood/paper matchsticks can be dipped into a colored dye solution if desired; in either case, the matches should then be cured in an oven at 80 Celsius for 12 hours.
**Ignition temperature (when match is heated externally):** 510 Celsius
**Ignition temperature (exact point when match ignites):** 990 Celsius
**Ease of ignition ("striking" on suitable surface):** 8.4
**Water resistance:** Very good.
**Weather resistance (match):** wind resistance above normal; moist and wet conditions above normal.
**Stability:** Can be stored for decades.
**Sensitivity:** Very stable towards shock and percussion.
**Explosive ability:** Cannot be detonated.
**Percentage:** ***59% red phosphorus, 18% antimony sulfide, 11.5% gum acacia, 11.5% carbon black***
**Classification:** Deflagrating mixture (classified as 1.3G pyrotechnic composition).

**Danger of preparation:** Can be made by amateurs.
**Use:** Used to make weatherproof matches.

### 06-02-020A: Friction sensitive composition for use in matches:

Into a vertical mixer, equipped with flexible Teflon stir blades, place ***25 grams flours of sulfur***, followed by ***200 grams of calcium plumbate***, followed by ***200 grams of gum acacia***, followed by ***75 grams of red phosphorus***, followed by 350 milliliters of hexane, followed by ***75 grams of gum Arabic***, followed by ***750 grams of potassium chlorate***, and then followed by ***100 grams of powdered glass***. Thereafter, blend the mixture on low speed for 30 minutes. Thereafter, the mixture is ready to be casted ("dipped"). To cast the match head composition, the wood or paper "match sticks" should be dipped, dried, re-dipped, dried, ect., into the semi-fluidized mixture to form a uniform thickness ranging from 2mm to 5mm in width (depends on discretion, and the desired thickness of the match head composition can vary). Thereafter, the wood/paper matchsticks can be dipped into a colored dye solution if desired; in either case, the matches should then be cured in an oven at 80 Celsius for 12 hours.
**Ignition temperature (when match is heated externally):** 590 Celsius
**Ignition temperature (exact point when match ignites):** 1005 Celsius
**Ease of ignition ("striking" on suitable surface):** 8.5
**Water resistance:** Very good.
**Weather resistance (match):** wind resistance above normal; moist and wet conditions above normal.
**Stability:** Can be stored for decades.
**Sensitivity:** Very stable towards shock and percussion.
**Explosive ability:** Cannot be detonated.
**Percentage:** ***52.63% potassium chlorate, 14.03% calcium plumbate, 14.03% gum acacia, 7.01% powdered glass, 5.26%red phosphorus, 5.26% gum Arabic, 1.75% sulfur, 0.03% mixed residual balance***
**Classification:** Deflagrating mixture (classified as 1.3G pyrotechnic composition).

**Danger of preparation:** Can be made by amateurs.
**Use:** Used to make weatherproof matches.

### 06-02-021A: Friction sensitive composition for use in matches:

Into a vertical mixer, equipped with flexible Teflon stir blades, place ***22 grams of gum Arabic***, followed by ***21 grams of potassium dichromate***, followed by ***15.5 grams of red phosphorus***, followed by ***105.5 grams of lead hypophosphite***, followed by ***52.5 grams of any desired color dye or pigment***, followed by 120 milliliters of hexane, and then followed by ***159 grams of potassium chlorate***. Thereafter, blend the mixture on low speed for 30 minutes. Thereafter, the mixture is ready to be casted ("dipped"). To cast the match head composition, the wood or paper "match sticks" should be dipped, dried, re-dipped, dried, ect., into the semi-fluidized mixture to form a uniform thickness ranging from 2mm to 5mm in width (depends on discretion, and the desired thickness of the match head composition can vary). Thereafter,

the wood/paper matchsticks can be dipped into a colored dye solution if desired; in either case, the matches should then be cured in an oven at 80 Celsius for 12 hours.
**Ignition temperature (when match is heated externally):** 575 Celsius
**Ignition temperature (exact point when match ignites):** 1010 Celsius
**Ease of ignition ("striking" on suitable surface):** 8.4
**Water resistance:** Very good.
**Weather resistance (match):** wind resistance above normal; moist and wet conditions above normal.
**Stability:** Can be stored for decades.
**Sensitivity:** Very stable towards shock and percussion.
**Explosive ability:** Cannot be detonated.
**Percentage:** ***42.34% potassium chlorate, 28.09% lead hypophosphite, 13.98% dye/pigment, 5.85% gum Arabic, 5.59% potassium dichromate, 4.12% red phosphorus, 0.03% residual balance***
**Classification:** Deflagrating mixture (classified as 1.3G pyrotechnic composition).

**Danger of preparation:** Can be made by amateurs.
**Use:** Used to make weatherproof matches.

## 06-02-022A: Friction sensitive composition for use in "strike anywhere" matches:

Into a vertical mixer, equipped with flexible Teflon stir blades, place ***400 grams of potassium chlorate***, followed by ***200 grams of powdered glass***, followed by ***120 grams of "whiting",*** followed by ***70 grams of plaster of Paris powder***, and then followed by ***100 grams of red phosphorus***. Thereafter, blend the mixture on low speed for 30 minutes. After 30 minutes, add in ***130 grams of standard animal glue***, and then continue to blend the mixture on low speed for 10 minutes. Thereafter, the mixture is ready to be casted ("dipped"). To cast the match head composition, the wood or paper "match sticks" should be dipped and re-dipped into the semi-fluidized mixture to form a uniform thickness ranging from 2mm to 5mm in width (depends on discretion, and the desired thickness of the match head composition can vary). Thereafter, the wood/paper matchsticks can be dipped into a colored dye solution if desired; in either case, the matches should then be cured in an oven at 60 Celsius for 24 hours.
**Ignition temperature (when match is heated externally):** 600 Celsius
**Ignition temperature (exact point when match ignites):** 995 Celsius
**Ease of ignition ("striking" on suitable surface):** 6.9
**Water resistance:** Very good.
**Weather resistance (match):** wind resistance normal; moist and wet conditions normal.
**Stability:** Can be stored for decades.
**Sensitivity:** Very stable towards shock and percussion.
**Explosive ability:** Cannot be detonated.
**Percentage:** ***39.21% potassium chlorate, 19.6% powdered glass, 12.74% animal glue, 11.76% whiting, 9.8% red phosphorus, 6.86% plaster of Paris, 0.03% mixed balance***
**Classification:** Deflagrating mixture (classified as 1.3G pyrotechnic composition).

**Danger of preparation:** Can be made by amateurs.
**Use:** Used to make weatherproof matches.

## 06-02-023A: Friction sensitive composition for use in matches:

Into a vertical mixer, equipped with flexible Teflon stir blades, place ***100 grams of antimony trisulfide***, followed by ***100 grams of lead nitrate***, followed by ***50 grams of red phosphorus,*** followed by ***50 grams of vegetable charcoal***, followed by ***50 grams of potassium dichromate***, and then followed by ***200 grams of potassium chlorate***. Thereafter, blend the mixture on low speed for 30 minutes. After 30 minutes, add in ***30 grams of standard animal glue***, and then continue to blend the mixture on low speed for 10 minutes. Thereafter, the mixture is ready to be casted ("dipped"). To cast the match head composition, the wood or paper "match sticks" should be dipped and re-dipped into the semi-fluidized mixture to form a uniform thickness ranging from 2mm to 5mm in width (depends on discretion, and the desired thickness of the match head composition can vary). Thereafter, the wood/paper matchsticks can be dipped into a colored dye solution if desired; in either case, the matches should then be cured in an oven at 60 Celsius for 24 hours.
**Ignition temperature (when match is heated externally):** 510 Celsius
**Ignition temperature (exact point when match ignites):** 1090 Celsius
**Ease of ignition ("striking" on suitable surface):** 7.1
**Water resistance:** Very good.
**Weather resistance (match):** wind resistance normal; moist and wet conditions normal.
**Stability:** Can be stored for decades.
**Sensitivity:** Very stable towards shock and percussion.
**Explosive ability:** Cannot be detonated.

**Percentage:** ***34.48% potassium chlorate, 17.24% antimony trisulfide, 17.24% lead nitrate, 8.62% vegetable charcoal, 8.62% red phosphorus, 8.62% potassium dichromate, 5.17% animal glue, 0.01% residual balance***
**Classification:** Deflagrating mixture (classified as 1.3G pyrotechnic composition).

**Danger of preparation:** Can be made by amateurs.
**Use:** Used to make weatherproof matches.

### 06-02-024A: Friction sensitive composition for use in matches:

Into a vertical mixer, equipped with flexible Teflon stir blades, place ***175 grams of zinc oxide***, followed by ***325 grams of ground glass powder***, followed by ***50 grams of gum benzoin***, followed by ***100 grams of red phosphorus***, followed by ***25 grams of flours of sulfur***, followed by ***75 grams of lead hypophosphite***, and then followed by ***500 grams of potassium chlorate.*** Thereafter, blend the mixture on low speed for 30 minutes. After 30 minutes, add in ***175 grams of standard animal glue***, and then continue to blend the mixture on low speed for 10 minutes. Thereafter, the mixture is ready to be casted ("dipped"). To cast the match head composition, the wood or paper "match sticks" should be dipped and re-dipped into the semi-fluidized mixture to form a uniform thickness ranging from 2mm to 5mm in width (depends on discretion, and the desired thickness of the match head composition can vary). Thereafter, the wood/paper matchsticks can be dipped into a colored dye solution if desired; in either case, the matches should then be cured in an oven at 60 Celsius for 12 hours.
**Ignition temperature (when match is heated externally):** 580 Celsius
**Ignition temperature (exact point when match ignites):** 998 Celsius
**Ease of ignition ("striking" on suitable surface):** 7.4
**Water resistance:** Very good.
**Weather resistance (match):** wind resistance normal; moist and wet conditions normal.
**Stability:** Can be stored for decades.
**Sensitivity:** Very stable towards shock and percussion.
**Explosive ability:** Cannot be detonated.
**Percentage:** ***35.08% potassium chlorate, 22.8% glass powder, 12.28% animal glue, 12.28% zinc oxide, 7.01% red phosphorus, 5.26% lead hypophosphite, 3.5% gum benzoin, 1.75% sulfur, 0.04% mixed balance***
**Classification:** Deflagrating mixture (classified as 1.3G pyrotechnic composition).

**Danger of preparation:** Can be made by amateurs.
**Use:** Used to make weatherproof matches.

### 06-02-025A: Friction sensitive composition for use in matches:

Into a vertical mixer, equipped with flexible Teflon stir blades, place ***450 grams of potassium chlorate***, followed by ***210 grams of powdered glass***, followed by ***40 grams of plaster of Paris powder***, followed by ***40 grams of zinc oxide***, followed by ***20 grams of" whiting"***, followed by ***30 grams flours of sulfur***, and then followed by ***100 grams of red phosphorus***. Thereafter, blend the mixture on low speed for 30 minutes. After 30 minutes, add in ***160 grams of animal glue***, and then continue to blend the mixture on low speed for 10 minutes. Thereafter, the mixture is ready to be casted ("dipped"). To cast the match head composition, the wood or paper "match sticks" should be dipped and re-dipped into the semi-fluidized mixture to form a uniform thickness ranging from 2mm to 5mm in width (depends on discretion, and the desired thickness of the match head composition can vary). Thereafter, the wood/paper matchsticks can be dipped into a colored dye solution if desired; in either case, the matches should then be cured in an oven at 60 Celsius for 12 hours.
**Ignition temperature (when match is heated externally):** 559 Celsius
**Ignition temperature (exact point when match ignites):** 980 Celsius
**Ease of ignition ("striking" on suitable surface):** 7.7
**Water resistance:** Very good.
**Weather resistance (match):** wind resistance normal; moist and wet conditions normal.
**Stability:** Can be stored for decades.
**Sensitivity:** Very stable towards shock and percussion.
**Explosive ability:** Cannot be detonated.
**Percentage:** ***42.85% potassium chlorate, 20% powdered glass, 15.23% animal glue, 9.52% red phosphorus, 3.8% plaster of Paris, 3.8% zinc oxide, 2.85% sulfur, 1.9% whiting, 0.05% residue***
**Classification:** Deflagrating mixture (classified as 1.3G pyrotechnic composition).

**Danger of preparation:** Can be made by amateurs.

**Use:** Used to make weatherproof matches.

### 06-02-026A: Classic match-head composition:

Into a vertical mixer, equipped with flexible Teflon stir blades, place ***25 grams of sulfur***, followed by ***75 grams of phosphorus sesqui-sulfide***, followed by ***200 grams of powdered sand***, followed by ***125 grams of potassium perchlorate***, and then followed by ***175 grams of standard animal glue***. Thereafter, blend the mixture on low speed for 30 minutes. After 30 minutes, the mixture is ready to be casted ("dipped"). To cast the match head composition, the wood or paper "match sticks" should be dipped and re-dipped into the semi-fluidized mixture to form a uniform thickness ranging from 2mm to 5mm in width (depends on discretion, and the desired thickness of the match head composition can vary). Thereafter, the wood/paper matchsticks can be dipped into a colored dye solution if desired; in either case, the matches should then be cured in an oven at 60 Celsius for 12 hours.
**Ignition temperature (when match is heated externally):** 890 Celsius
**Ignition temperature (exact point when match ignites):** 885 Celsius
**Ease of ignition ("striking" on suitable surface):** 4
**Water resistance:** Very good.
**Weather resistance (match):** wind resistance normal; moist and wet conditions below normal.
**Stability:** Can be stored for decades.
**Sensitivity:** Very stable towards shock and percussion.
**Explosive ability:** Cannot be detonated.
**Percentage:** ***33.33% sand, 29.16% animal glue, 20.83% potassium perchlorate, 12.5% phosphorus sesqui-sulfide, 4.16% sulfur, 0.02% residual balance***
**Classification:** Deflagrating mixture (classified as 1.3G pyrotechnic composition).

**Danger of preparation:** Can be made by amateurs.
**Use:** Used to make matches.

### 06-02-026B: Classic match-head composition:

Into a vertical mixer, equipped with flexible Teflon stir blades, place ***110 grams of powdered sand***, followed by ***30 grams of phosphorus sesqui-sulfide***, followed by ***100 grams of ammonium perchlorate***, and then followed by ***85 grams of standard animal glue***. Thereafter, blend the mixture on low speed for 30 minutes. After 30 minutes, the mixture is ready to be casted ("dipped"). To cast the match head composition, the wood or paper "match sticks" should be dipped and re-dipped into the semi-fluidized mixture to form a uniform thickness ranging from 2mm to 5mm in width (depends on discretion, and the desired thickness of the match head composition can vary). Thereafter, the wood/paper matchsticks can be dipped into a colored dye solution if desired; in either case, the matches should then be cured in an oven at 60 Celsius for 12 hours.
**Ignition temperature (when match is heated externally):** 910 Celsius
**Ignition temperature (exact point when match ignites):** 875 Celsius
**Ease of ignition ("striking" on suitable surface):** 5.1
**Water resistance:** Very good.
**Weather resistance (match):** wind resistance normal; moist and wet conditions below normal.
**Stability:** Can be stored for decades.
**Sensitivity:** Very stable towards shock and percussion.
**Explosive ability:** Cannot be detonated.
**Percentage:** ***33.84% powdered sand, 30.76% ammonium perchlorate, 26.15% animal glue, 9.23% phosphorus sesqui-sulfide, 0.02% residual balance***
**Classification:** Deflagrating mixture (classified as 1.3G pyrotechnic composition).

**Danger of preparation:** Can be made by amateurs.
**Use:** Used to make matches.

### 06-02-027A: Classic "safety" match-head composition:

Into a vertical mixer, equipped with flexible Teflon stir blades, place ***160 grams of animal glue***, followed by ***70 grams of powdered chalk***, followed by ***100 grams of finely divided magnesium/silicon alloy (50/50),*** followed by ***270 grams of powdered sand***, followed by ***100 grams of red phosphorus***, and then followed by ***300 grams of potassium perchlorate***. Thereafter, blend the mixture on low speed for 30 minutes. After 30 minutes, the mixture is ready to be casted ("dipped"). To cast the match head composition, the wood or paper "match sticks" should be dipped and re-dipped into the semi-fluidized mixture to form a uniform thickness ranging from 2mm to 5mm in width (depends on discretion, and the desired thickness of the match head composition can vary). Thereafter, the wood/paper matchsticks can be dipped into a colored dye solution if desired; in either case, the matches should then be cured in an oven at 60 Celsius for 12 hours.
**Ignition temperature (when match is heated externally):** 950 Celsius

**Ignition temperature (exact point when match ignites):** 910 Celsius
**Ease of ignition ("striking" on suitable surface):** 5
**Water resistance:** Very good.
**Weather resistance (match):** wind resistance normal; moist and wet conditions above normal.
**Stability:** Can be stored for decades.
**Sensitivity:** Very stable towards shock and percussion.
**Explosive ability:** Cannot be detonated.
**Percentage:** ***30% potassium perchlorate, 27% powdered sand, 16% animal glue 10% magnesium/silicon alloy, 10% red phosphorus, 7% powdered chalk***
**Classification:** Deflagrating mixture (classified as 1.3G pyrotechnic composition).

**Danger of preparation:** Can be made by amateurs.
**Use:** Used to make matches.

## 06-02-028A: Match composition for multiple uses:

**Preparation of composition 1:**
Into a vertical mixer, place 77.5 grams of animal glue, and then add in 300 milliliters of water. Thereafter, heat the mixture to about 50 Celsius, and then blend the mixture for about 30 minutes. Thereafter, add in 21.25 grams of gluten, followed by 58.75 grams of finely pulverized red brick powder, followed by 212.5 grams of pulverized wood dust, followed by 43.25 grams of flours of sulfur, followed by 16.25 grams of powdered bituminous coal dust, followed by 4.4 grams of black antimony sulfide, followed by 18.75 grams of calcium hydroxide, followed by 42.5 grams of powdered coke, and then finally followed by 11 grams of pulverized soft wood charcoal. Thereafter, blend the mixture on high speed for about 1 hour. Thereafter, place the mixture aside for the moment.
**Preparation of composition 2:**
Into a vertical mixer, place 120 grams of pulverized coke, followed by 22.5 grams of black antimony sulfide, followed by 102.5 grams of powdered glass, followed by 160 grams of powdered red brick dust, and then followed by 95 grams of potassium dichromate. Thereafter, add in about 100 milliliters of warm water, and then blend the mixture on high speed for 1 hour. Thereafter, place composition 1, and composition 2 on separate shallow trays, and allow both compositions to thoroughly dry. Once they have both dried. Manually pulverize both compositions.
**Preparation of final match composition:**
Into a vertical mixer, equipped with flexible Teflon stir blades, place ***220 grams of potassium chlorate***, followed by ***140 grams of composition 1,*** followed by 110 milliliters of hexane, and then followed by ***30 grams of composition 2***. Thereafter, blend the mixture on low speed for 30 minutes. After 30 minutes, the mixture is ready to be casted ("dipped"). To cast the match head composition, the wood or paper "match sticks" should be dipped and re-dipped into the semi-fluidized mixture to form a uniform thickness ranging from 2mm to 5mm in width (depends on discretion, and the desired thickness of the match head composition can vary). Thereafter, the wood/paper matchsticks can be dipped into a colored dye solution if desired; in either case, the matches should then be cured in an oven at 80 Celsius for 12 hours.
**Ignition temperature (when match is heated externally):** 450 Celsius
**Ignition temperature (exact point when match ignites):** 990 Celsius
**Ease of ignition ("striking" on suitable surface):** 7.2
**Water resistance:** Very good.
**Weather resistance (match):** wind resistance normal; moist and wet conditions above normal.
**Stability:** Can be stored for decades.
**Sensitivity:** Very stable towards shock and percussion; friction resistant to soft materials.
**Explosive ability:** Cannot be detonated.
**Percentage:** ***56.41% potassium chlorate, 35.89% composition 1, 7.69% composition 2, 0.01% mixed balance***
**Classification:** Deflagrating mixture (classified as 1.3G pyrotechnic composition).

**Danger of preparation:** Can be made by amateurs.
**Use:** Used to make matches; electric matches; and ignition compositions.

## 06-02-029A: Weatherproof match composition:

Into a vertical mixer, equipped with flexible Teflon stir blades, place ***20 grams of standard animal glue***, followed by ***100 grams of regular corn starch***, followed by ***200 grams of powdered glass***, followed by ***480 grams of potassium chlorate***, followed by ***200 grams of powdered pumice stone***, followed by ***80 grams of antimony trisulfide***, followed by ***120 grams of sodium nitrate***, followed by ***40 grams of pulverized cascarilla bark***, and then followed by ***20 grams of lamp black***. Thereafter, blend the mixture on low speed for 30 minutes. After 30 minutes, the mixture is ready to be casted ("dipped"). To cast the match head composition, the wood or paper "match sticks" should be dipped and re-dipped into the semi-fluidized mixture to form a uniform thickness ranging from 2mm to 5mm in width (depends on discretion, and the desired thickness of the match head composition can vary). Thereafter, the wood/paper matchsticks can be dipped into a colored dye solution if desired; in either case, the matches should then be cured in an oven at 60 Celsius for 12 hours.

**Ignition temperature (when match is heated externally):** 550 Celsius
**Ignition temperature (exact point when match ignites):** 780 Celsius
**Ease of ignition ("striking" on suitable surface):** 4.9
**Water resistance:** Very good.
**Weather resistance (match):** wind resistance normal; moist and wet conditions above normal.
**Stability:** Can be stored for decades.
**Sensitivity:** Very stable towards shock and percussion; friction resistant to soft materials.
**Explosive ability:** Cannot be detonated.
**Percentage:** ***38.09% potassium chlorate, 15.87% powdered glass, 15.87% powdered pumice stone, 9.52% sodium nitrate, 7.93% corn starch, 6.34% antimony trisulfide, 3.17% cascarilla bark, 1.58% animal glue, 1.58% lamp black, 0.05% mixed balance***
**Classification:** Deflagrating mixture (classified as 1.3G pyrotechnic composition).

**Danger of preparation:** Can be made by amateurs.
**Use:** Used to make weatherproof matches.

# Section 3: Smoke generating compositions

## *Chemicals used in this section (binders are not included)*

| | | |
|---|---|---|
| **1. Potassium nitrate (see Black Powder)** | | **2. Sulfur (see Black Powder)** |
| **3. Charcoal (see Black Powder)** | | **4. Sugar Carbon (see Modified Black Powder)** |
| **5. Barium Chromate (see Modified Black Powder)** | | **6. Potassium Perchlorate (see Modified Black Powder)** |
| **7. Potassium Dichromate (see Modified Black Powder)** | | **8. Ammonium Bisulfide (see Modified Black Powder)** |
| **9. Potassium chlorate (see Modified Black Powder)** | | **10. Carbon Disulfide (see Modified Black Powder)** |
| **11. Nitrocellulose (see Modified Black Powder)** | | **12. Lead Tetraoxide (see Modified Black Powder)** |
| **13. Diphenylamine (see Modified Black Powder)** | | **14. Titanium Dioxide (see Modified Black Powder)** |
| **15. Manganese Dioxide (see Modified Black Powder)** | | **16. Sodium Benzoate (see Modified Black Powder)** |
| **17. Ammonium Nitrate (see Modified Black Powder)** | | **18. Calcium Carbonate (see Modified Black Powder)** |
| **19. Sodium Nitrate (see Modified Black Powder)** | | **20. Urea (see Modified Black Powder)** |
| **21. Lead Nitrate (see Modified Black Powder)** | | **22. Nitro Starch (see Modified Black Powder)** |
| **23. Ammonium Perchlorate (see Ammonium Perchlorate Rocket Propellants)** | | **24. Aluminum powder (see Ammonium Perchlorate Rocket Propellants)** |
| **25. Magnesium Sulfide (see Ammonium Perchlorate Rocket Propellants)** | | **26. Copper Chromite (see Ammonium Perchlorate Rocket Propellants)** |
| **27. Nitroguanidine (see Ammonium Perchlorate Rocket Propellants)** | | **28. Ammonium Sulfate (see Ammonium Perchlorate Rocket Propellants)** |
| **29. Nitroglycerine (see Ammonium Perchlorate Rocket Propellants)** | | **30. Magnesium Oxide (see Ammonium Perchlorate Rocket Propellants)** |
| **31. Iron-II-oxide (see Ammonium Perchlorate Rocket Propellants)** | | **32. Zirconium Hydride (see Ammonium Perchlorate Rocket Propellants)** |
| **33. Teflon (see Ammonium Perchlorate Rocket Propellants)** | | **34. Zinc Oxide (see Ammonium Perchlorate Rocket Propellants)** |
| **35. Ammonium Dichromate (see Ammonium Perchlorate Rocket Propellants)** | | **36. Lithium Perchlorate (see Ammonium Perchlorate Rocket Propellants)** |
| **37. Magnesium powder (see Ammonium Perchlorate Rocket Propellants)** | | **38. Lithium Aluminum Hydride (see Ammonium Perchlorate Rocket Propellants)** |
| **39. Iron-III-oxide (red iron oxide) (see Ammonium Perchlorate Rocket Propellants)** | | **40. PVC (see Ammonium Perchlorate Rocket Propellants)** |
| **41. Copper Sulfide (see Ammonium Perchlorate Rocket Propellants)** | | **42. Sodium Hydride (see Ammonium Perchlorate Rocket Propellants)** |
| **43. Titanium Hydride (see Ammonium Perchlorate Rocket Propellants)** | | **44. Silicon Nitride (see Ammonium Perchlorate Rocket Propellants)** |
| **45. ADN (see ADN Rocket Propellants)** | | **46. Urea Nitrate (see ADN Rocket Propellants)** |
| **47. Silicon Powder (see ADN Rocket Propellants)** | | **48. Hexamine (see ADN Rocket Propellants)** |
| **49. Boron powder (see ADN Rocket Propellants)** | | **50. Sodium hypophosphite (see ADN Rocket Propellants)** |

| | |
|---|---|
| 51. KDN (see ADN Rocket Propellants) | 52. Nickel Chloride (see Ammonium Nitrate Rocket Propellants) |
| 53. TNT (see Ammonium Nitrate Rocket Propellants) | 54. Acrylamide (see Miscellaneous Rocket H.P. Rocket Propellants) |
| 55. Ethylene Glycol (see Miscellaneous Rocket H.P. Rocket Propellants) | 56. Magnesium Perchlorate (see Miscellaneous Rocket H.P. Rocket Propellants) |
| 57. Metallic Lithium (see Miscellaneous Rocket H.P. Rocket Propellants) | 58. Beryllium Hydride (see Miscellaneous Rocket H.P. Rocket Propellants) |
| 59. Red Phosphorus (see Miscellaneous Rocket H.P. Rocket Propellants) | 60. Sodium Borohydride (see Miscellaneous Rocket H.P. Rocket Propellants) |
| 61. Sodium chlorate (see Miscellaneous Rocket H.P. Rocket Propellants) | 62. Lead Dioxide (see Miscellaneous Rocket H.P. Rocket Propellants) |
| 63. Benzoyl Peroxide (see Miscellaneous Rocket H.P. Rocket Propellants) | 64. Barium Nitrate (see Miscellaneous Rocket H.P. Rocket Propellants) |
| 65. Cyanuric Acid (see Miscellaneous Rocket H.P. Rocket Propellants) | 66. Aluminum Stearate (see Ammonium Nitrate Gun Propellants) |
| 67. Aluminum Hydride (see Ammonium Nitrate Gun Propellants) | 68. Magnesium Peroxide (see Ammonium Nitrate Gun Propellants) |
| 69. Potassium Permanganate (see Ammonium Nitrate Gun Propellants) | 70. Calcium Hydride (see Ammonium Nitrate Gun Propellants) |
| 71. Potassium Tartrate (see Ammonium Nitrate Gun Propellants) | 72. Potassium Ferrocyanide (see Ammonium Nitrate Gun Propellants) |
| 73. Sodium Azide (see Ammonium Nitrate Gun Propellants) | 74. Sodium Chloride (see Ammonium Nitrate Gun Propellants) |
| 75. Potassium Sulfate (see Nitrocellulose Gun Propellants) | 76. Lead Stearate (see Nitrocellulose Gun Propellants) |
| 77. Triethylene Glycol (see Nitrocellulose Gun Propellants) | 78. Sugar (sucrose) (see Miscellaneous Gun Propellants) |
| 79. Sodium Propionate (see Miscellaneous Gun Propellants) | 80. Picric Acid (see Miscellaneous Gun Propellants) |
| 81. Copper-II-oxide (see Miscellaneous Gun Propellants) | 82. Ammonium Picrate (see Miscellaneous Gun Propellants) |
| 83. Barium Peroxide (see Bullet Tracer Compositions) | 84. Magnesium Carbonate (see Bullet Tracer Compositions) |
| 85. Strontium Peroxide (see Bullet Tracer Compositions) | 86. Strontium Nitrate (see Bullet Tracer Compositions) |
| 87. Cupric chloride (see Bullet Tracer Compositions) | 88. Hexachlorobenzene (see Bullet Tracer Compositions) |
| 89. Strontium oxalate (see Bullet Tracer Compositions) | 90. Mercury-I-Chloride (see Bullet Tracer Compositions) |
| 91. Zinc Oxalate (see Bullet Tracer Compositions) | 92. Zinc Chloride (see Bullet Tracer Compositions) |
| 93. Uranium (see Bullet Tracer Compositions) | 94. Zirconium nitrate (see Bullet Tracer Compositions) |
| 95. Yttrium Nitrate (see Bullet Tracer Compositions) | 96. Yttrium Oxide (see Bullet Tracer Compositions) |
| 97. Zirconium Oxide (see Bullet Tracer Compositions) | 98. Cerium Oxide (see Bullet Tracer Compositions) |
| 99. Hexachloroethane (see Bullet Tracer Compositions) | 100. Antimony trisulfide (see Bullet Tracer Compositions) |
| 101. Anthracene (see Bullet Tracer Compositions) | 102. Phosphorus Sesquisulphide (see Match Compositions) |
| 103. Boric acid (see Match Compositions) | 104. Aluminum Hydroxide (see Match Compositions) |
| 105. Antimony Pentasulfide (see Match Compositions) | 106. Glucose (see Match Compositions) |
| 107. Sodium Hydroxide (see Match Compositions) | 108. Lead Hypophosphite (see Match Compositions) |
| 109. Calcium Sulfate (see Match Compositions) | 110. Ammonium Chloride |

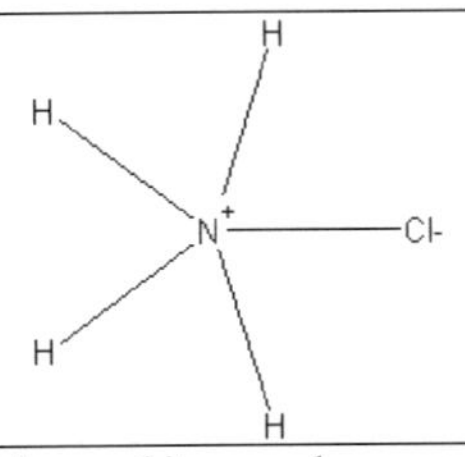


Forms colorless crystals, or white granules or powder. The crystals sublime when heated, and hence cannot be melted. The crystals are soluble in water and alcohol, but insoluble in other common solvents. Ammonium chloride is easily prepared by dissolving ammonia into dilute hydrochloric acid, followed by recrytsallization.

## 111. Manganese

Manganese forms brilliant steel gray crystals. The crystalline powder is stable at room temperature, but will slowly react with water. The powder is a strong reducing agent, and can reduce many different metal oxides. Manganese reacts with the halogens upon heating. Manganese burns with an intense light.

## 112. Lactose

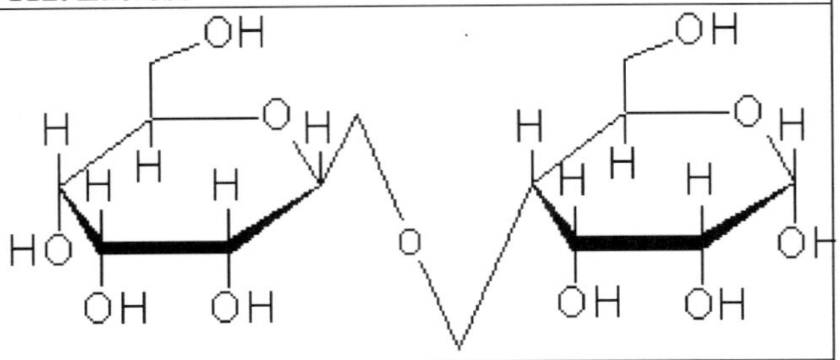

Forms a monohydrate, which is an alpha form. Is the typical milk sugar, and melts when rapidly heated to 202 Celsius. Lactose is stable at room temperature, but tends to absorb odors upon standing. The crystals are soluble in water, but relatively insoluble in other solvents.

## 113. Propylene Glycol

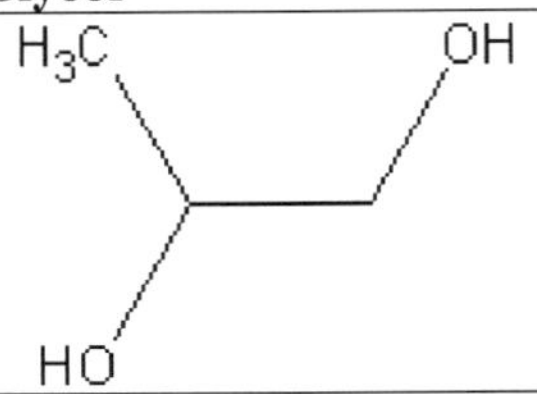


Forms a very hygroscopic viscous liquid. The melting point of the viscous liquid is –59 Celsius, with a boiling point of 188 Celsius at STP. The liquid is miscible with water, acetone, and chloroform, but less soluble in ether.

## 114. Glycerol

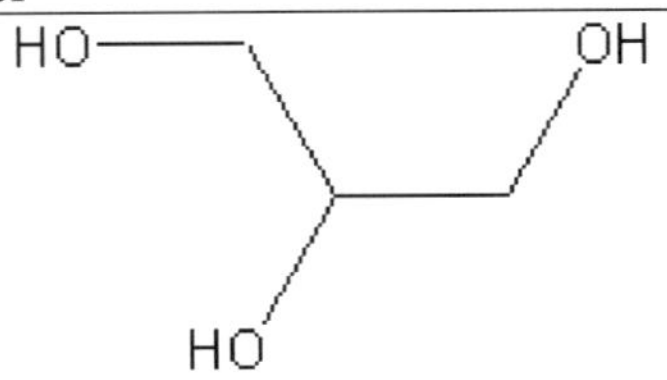


Forms a non-toxic syrupy viscous liquid. The liquid has a melting point of 17 Celsius, with a boiling point of 290 Celsius. The liquid is miscible with water and alcohol, but is relatively insoluble in common organic solvents. Glycerol is a major by-product in the soap manufacturing process, and is the least toxic of all known alcohols, including grain alcohol.

## 115. Potassium Chloride

Forms white to colorless crystals, granules, or powder. The crystals have a melting point of 773 Celsius. The crystals are moderately soluble in water, but insoluble in alcohol and most common solvents.

## 116. Potassium Bicarbonate

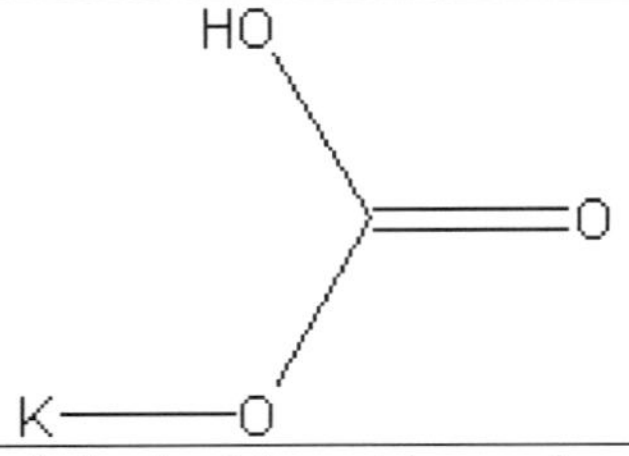


Forms the usual colorless to white crystals, granules, or powder. The crystals are soluble in water, but insoluble in the usual solvents. It reacts vigorously with acids. Potassium bicarbonate is a common ingredient in baking powders.

## 117. Dicyanodiamide

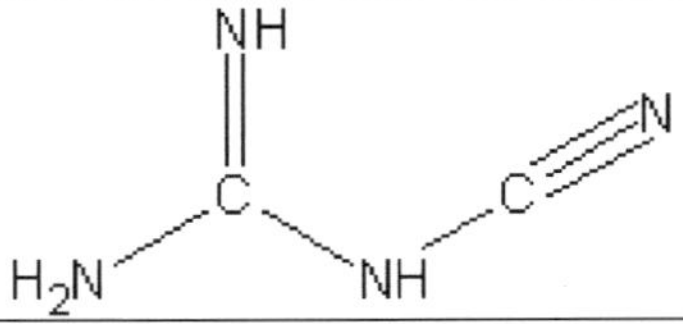


Forms monoclinic crystals with a melting point of 209 Celsius. The crystals are relatively insoluble in water, alcohol, and other common solvents. Can be prepared by polymerization of cyanamide.

## 118. Naphthalene

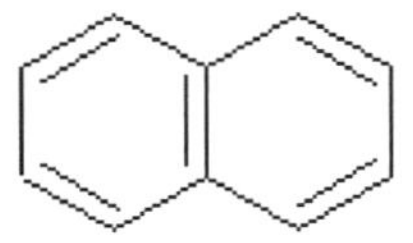

Forms colorless to white granules, crystals, or powder forming monoclinic prismatic crystals. The crystals have a melting point of 80 Celsius. The crystals have the odor of mothballs, which are primarily naphthalene. The crystals tend to sublime when heated to higher temperatures. Naphthalene makes up about 11% of coal tar.

**119. Thiourea**

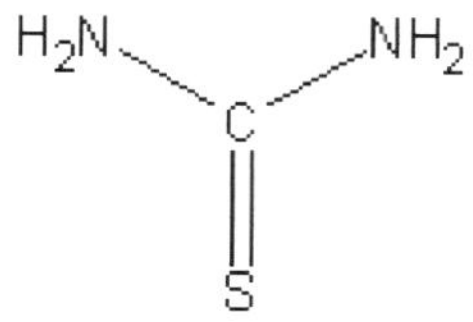

Thiourea forms transparent crystals with a melting point of 178 Celsius. The crystals are not very soluble in water or other common solvents. Thiourea tends to form addition products with metal salts.

**120. Phthalic Anhydride**

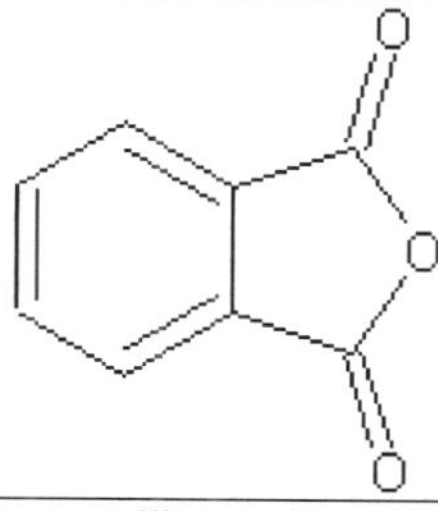

Forms colorless to white crystalline needles with a melting point of 131 Celsius, and a boiling point of 295 Celsius. The crystals tend to partially sublime upon heating. The anhydride breaks down when treated with water.

**122. Cadmium powder**

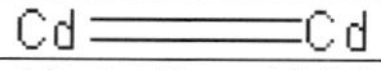

Cadmium forms a silvery-white to silvery-bluish metal. The solid metal can easily be cut with a knife at room temperature. The solid metal has a melting point of 321 Celsius, but the powder readily forms an oxide when heated in air. The powder is usually colored gray due to some oxidation. Cadmium metal is a cancer suspect agent, so users should be careful.

**123. Cadmium Sulfide**

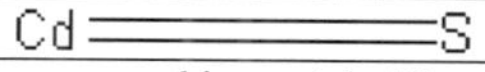

Forms light yellow to orange cubic crystals. The crystals are relatively insoluble in water and most organic solvents. Decomposes when treated with acids.

**124. Melamine**

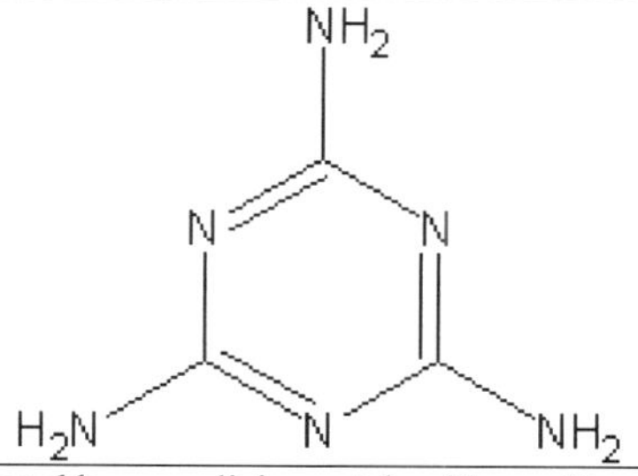

Forms slightly unstable monoclinic crystals with a melting point of 250 Celsius. The crystals tend to sublime at higher temperatures. Melamine is soluble in water, but insoluble in alcohol and most organic solvents. The solid tends to polymerize on standing or when treated with aldehydes.

**125. Malic Acid**

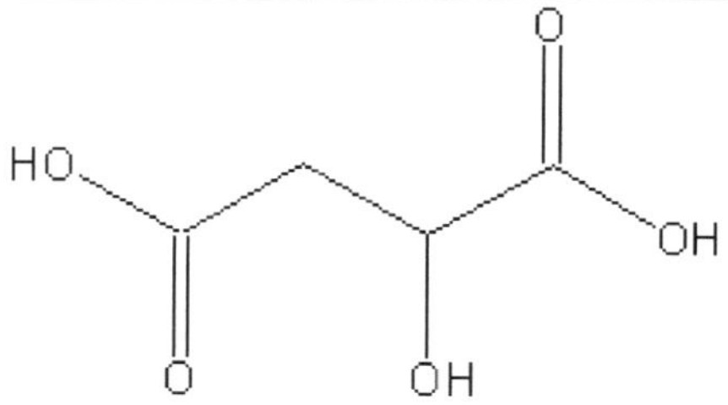

Has three distinct enantiomers. The L form is the most common isomer, and forms colorless crystals with a melting point of 100 Celsius. The crystals tend to break down when heated to 140 Celsius. The compound is readily soluble in water, and soluble in many organic solvents. The L form is found in apple juice.

**126. Calcium Lactate**

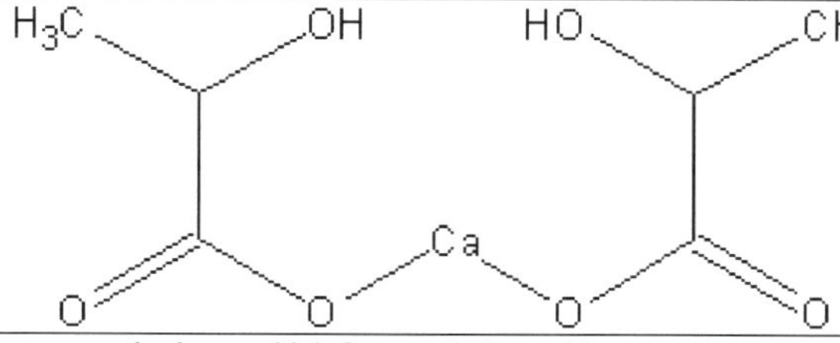

Forms a pentahydrate, which forms odorless white crystals, granules, or powder. The crystals loose their water of hydration when heated to 120 Celsius. The product is soluble in water, but insoluble in alcohol and other solvents.

**127. Metallic Sodium**

Sodium forms a silvery white metallic solid with a beautiful luster. The solid has a melting point of 98 Celsius, and boiling point of 881 Celsius. The solid is very reactive, and tarnishes rapidly when exposed to air. Sodium reacts violently with water and many other chemicals, so caution should be used.

**128. Bismuth Tetraoxide**

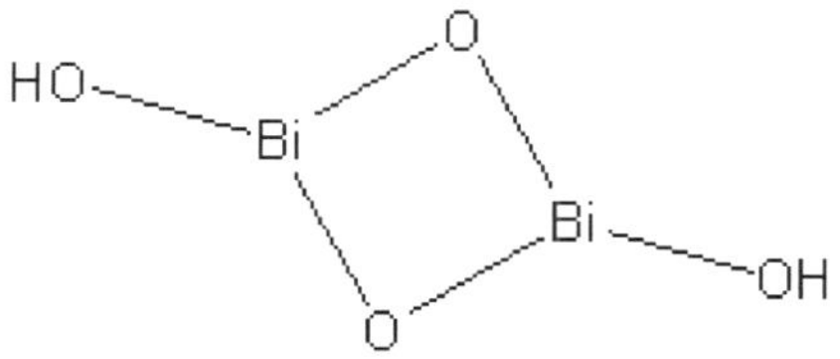

Forms an orange red to yellowish brown solid. The solid is slowly decomposed by water, but rapidly by acids.

**129. Bismuth Subnitrate**

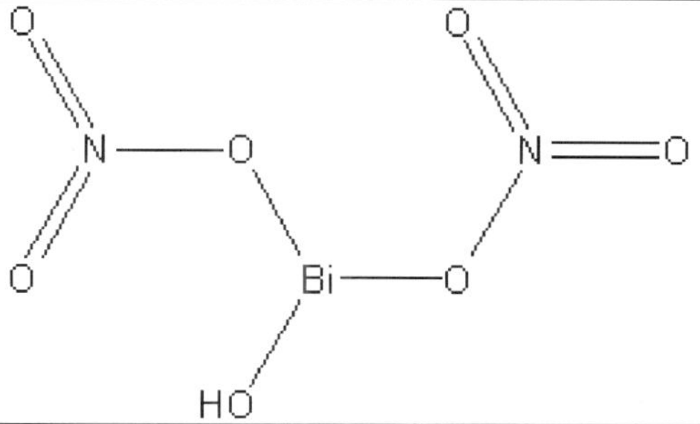

The nitrate forms a heavy hygroscopic microcrystalline solid. The solid is insoluble in water and alcohol and other solvents, but the solid is soluble in acids. The solid decomposes when heated to red heat.

**130. Calcium Iodate**

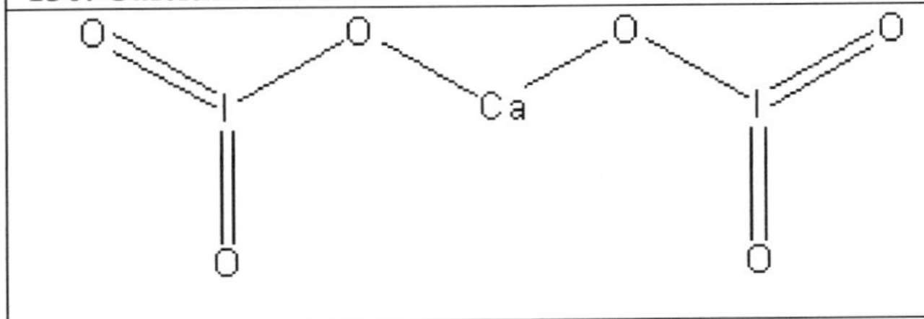


The iodate forms nonhygroscopic prismatic crystals, which are relatively insoluble in water and alcohol. The iodate forms a monohydrate and a hexahydrate.

**131. Potassium Iodate**

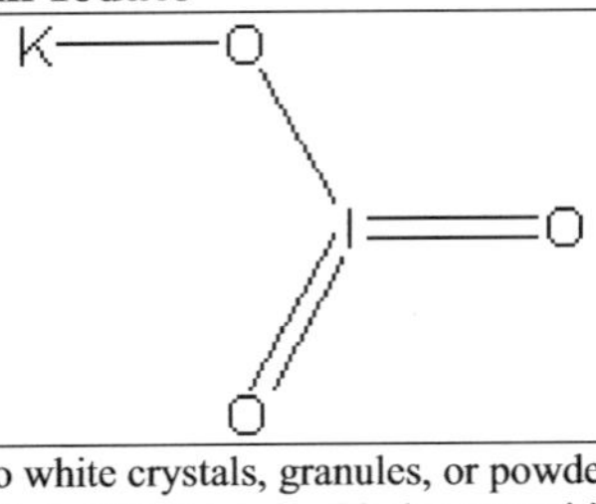


Forms colorless to white crystals, granules, or powder. The melting point of the crystals is 560 Celsius, with decomposition starting at such temperature. The iodate is relatively insoluble in water and the usual organic solvents.

**132. Magnesium Chloride**

$MgCl_2$

Forms colorless to white crystals, granules, or powder The anhydrous form has a melting point of 712 Celsius. The anhydrous crystals dissolve in water with the evolution of heat, and the formation of a hexahydrate. Anhydrous magnesium chloride reacts with alcohols forming addition compounds.

**133. Para-Nitroaniline**

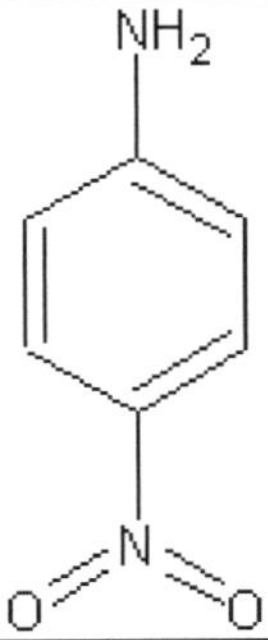


Forms a beautiful bright yellow powder with a melting point of 146 Celsius. The powder is insoluble in water and the usual solvents, but is soluble in benzene and methanol. Para nitro aniline is toxic, so users should avoid skin contact.

**134. Iodine**

I——I

Iodine forms beautiful purple to purple black crystal, plates, or scales that have a brilliant metallic luster. The crystals have an irritating odor and tend to volatize forming a mild purple vapor at STP. The crystals have a melting point of 113 Celsius, and a boiling point of 185 Celsius; however, the crystals will volatize on gentle heating. Iodine is insoluble in water, but soluble in alkaline earth iodide solutions (tinctures), but soluble in chloroform, glacial acetic acid, and glycerol oils.

**Method of Preparation 1:** Into a beaker or other suitable container add 50 grams of potassium iodide or 45 grams of sodium iodide, followed by 150 milliliters of cold tap water. Thereafter, stir the entire mixture for several minutes to dissolve all solids. Thereafter, add in 500 milliliters of regular household bleach (Clorox bleach). Make sure the bleach is just regular household bleach and has no fancy addictives, colors, or the like. The rate of addition of the household bleach to the potassium or sodium iodide solution should be slow, but should only take about 15 to 30 minutes. During the addition of the household bleach to the potassium or sodium iodide solution, casually stir the potassium or sodium iodide solution. After the addition of the household bleach, continue to stir the mixture for about 30 minutes, and then do one of two things: A. filter the mixture to recover the brownish black precipitate formed (crude iodine) using either gravity filtration or vacuum filtration, or B. extract the mixture with methylene chloride (three 75-millilter portions). If filtering the reaction mixture, allow the brownish precipitate of iodine to collect on the filter paper, and thereafter, allow the filter paper to stand over night and partially dry, or if using vacuum filtration, vacuum dry the brownish precipitate. Thereafter, gently scrape-off the brownish precipitate from the filter paper, into a clean beaker or other suitable container, and then add in 500 milliliters of warm tap water. Then gently mix the iodine mixture for about 10 minutes, and then once again, filter-off the insoluble iodine using gravity filtration (this is called washing the iodine with water). Note: if using vacuum filtration, after vacuum drying the initial iodine after filtration, simply wash it several times with three 150-milliliter portions, of warm tap water, and then vacuum dry the iodine. If using gravity filtration, after washing the iodine, and then collecting the iodine by filtration, allow the filter paper containing the iodine to stand over night to partially dry. The next day, gently scrape-off the iodine from the filter paper onto a clean piece of glass or porcelain dish (no metal), and then set this dish aside (no sunlight) for several days to allow the iodine to dry. When the iodine has

**Method of Preparation 2:** Into a suitable beaker or container, place 250 milliliters of warm tap water, followed by 70 grams of bleaching powder. Thereafter, stir the entire mixture for about 1 hour, and then filter-off the precipitated solids (composed of calcium hydroxide). Note: vacuum filtration works best, but if all you have is gravity filtration, place a small amount of clean sand in the bottom of the filter paper before filtering. This will help aid in the filtration process, which can be a slow one. After the filtration process, you will have a yellowish relatively odorless liquid. This yellow liquid will be composed of hypochlorous acid. Now, prepare a potassium or sodium iodide solution by adding and dissolving 50 grams of potassium iodide, or 45 grams of sodium iodide into 200 milliliters of tap water, and then slowly add this iodide solution to your yellowish hypochlorous acid mixture over a period of about 10 to 15 minutes. After 10 to 15 minutes, stir the mixture for about 30 minutes, and then filter-off the precipitated brownish solids (composed of the crude iodine). Once the brown solids have been filtered-off, wash them with several portions of cold water (use the same technique as in method 1 for washing), and then vacuum dry or air-dry these brownish solids. Thereafter, place this dried iodine into a sublimation apparatus, as illustrated in method 1, and then sublime the iodine to purify it. This sublimation process should be repeated once more to produce what is called "double sublimed reagent grade iodine", which will have a purity of no less then 98% iodine.

**Method of Preparation 3:** Now, into a suitable beaker, flask, or container, place 50 grams of potassium iodide or 45 grams of sodium iodide, followed by 150 milliliters of tap water. Thereafter, stir the entire mixture for about 30 minutes to dissolve all solids. Into a chlorine generator apparatus, as illustrated below, place into the reaction flask, 15 grams of manganese dioxide, or 240 grams of a 5% sodium hypochlorite solution (Clorox bleach), or 13 grams of been dried, either by using vacuum filtration, or by using the slow process of air-drying, place the dry iodine into a sublimation apparatus, as illustrated below, and sublime the iodine to purify it. This sublimation process should be repeated once more to produce what is called "double sublimed reagent grade iodine", which will have a purity of no less then 98% iodine

If you extracted the reaction mixture with methylene chloride, combine all methylene chloride portions once finished (if not already done so), and then dry this combined methylene chloride portion by adding to it, 15 grams of anhydrous magnums sulfate. Then stir the entire mixture for about 10 minutes (to absorb moisture), and then filter-off the magnesium sulfate. Once the magnesium sulfate has been filtered-off, place this methylene chloride into a standard atmospheric distillation apparatus, and gently distill-off the methylene chloride at 40 Celsius until no more methylene chloride distills over. When this is the result, recycle the methylene chloride, and then recover the left over remaining brownish residue of iodine. Then place this recovered iodine into a sublimation apparatus, as illustrated below, and sublime the iodine to purify it. This sublimation process should be repeated once more to produce what is called "double sublimed reagent grade iodine", which will have a purity of no less then 98% iodine

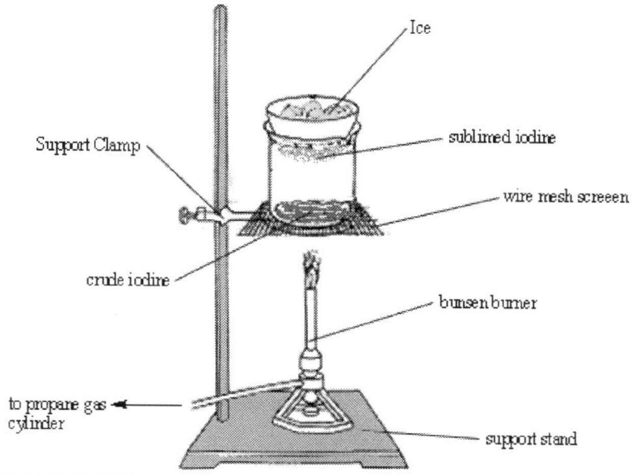


potassium permanganate, or 19 grams of calcium hypochlorite, bleaching powder (65% available chlorine). Then place into the addition funnel, a hydrochloric acid solution prepared by adding and dissolving 30 grams of 35 to 38% hydrochloric acid (muriatic acid of 31% will work) into 40 milliliters of tap water. If you are using an electrochemical cell, fill your cell with 40 milliliters of concentrated hydrochloric acid, followed by 200 milliliters of cold tap water.

Once your chlorine generator is setup (if using the hydrochloric acid drip method), begin dripping the hydrochloric acid solution, onto the oxidizer contained in the reaction flask. If using an electrochemical cell, begin the electrolysis by plugging in your battery charger. In either case, the chlorine gas that is evolved should then be bubbled into your potassium or sodium iodide solution (previously prepared). The chlorine gas will react with the potassium or sodium iodide to form iodine, which will precipitate as a brownish-black solid. When no more chlorine gas passes over, in either case, the reaction is complete. Thereafter, filter-off the precipitated brownish-black iodine solids, as usual, using either vacuum filtration or gravity filtration, and then wash them with cold water, and then vacuum dry or air-dry the solids. Once the iodine is completely dry, sublime it as illustrated in method 1. Repeat the sublimation process one more time to yield high purity reagent grade iodine.

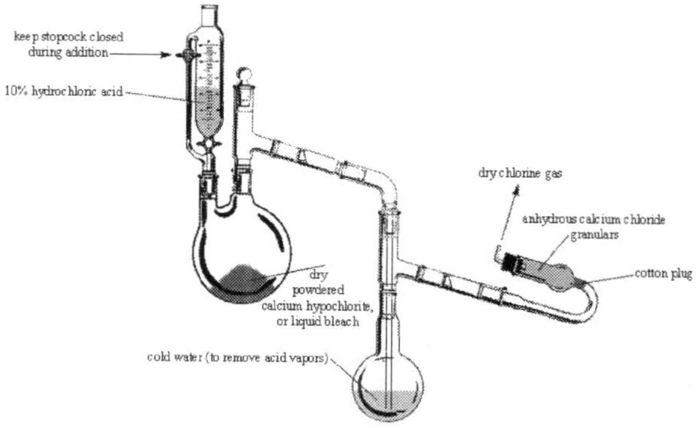


***- Smoke Generating Compositions in this section -***

| | |
|---|---|
| **1. 06-03-001A: White smoke generating composition 1:** *38.6% Terephthalic acid, 23% potassium chlorate, 17.8% pentaerythritol, 10.4% sugar, 4.4% magnesium carbonate, 3.4% polymerized sucrose binder containing some water, 2% stearic acid, 0.40% moisture* | **2. 06-03-001B: Yellow smoke generating composition 1:** *41.6% yellow dye, 22.5% potassium chlorate, 20.4% magnesium carbonate, 11.4% sugar, 3.8% polymerized sucrose binder, 0.3% moisture* |
| **3. 06-03-001C: Green smoke generating composition 1:** *29.4% green dye, 24.5% potassium chlorate, 16.9% magnesium carbonate, 12.5% yellow dye, 12.3% sugar, 4.1% polymerized sucrose binder, 0.3% moisture* | **4. 06-03-002A: Yellow smoke generating composition 2** *35.6% yellow dye "organol", 21.3% potassium chlorate, 17.3% guanidine nitrate, 16.2% glycidyl ether 100 epoxy resin, 5% ethandiol plasticizer, 3% methyl carbazate binder, 1% ethanolamine hardener, 0.20% ferric oxide catalyst, 0.40% moisture* |
| **5. 06-03-002B: Red smoke generating composition 1:** *37.9% red dye "organol", 23.9% guanidine nitrate, 14.9% glycidyl ether 100 epoxy resin, 13.9% potassium chlorate, 4.9% methyl carbazate, 3.9% propanediol plasticizer, 0.199% ferric oxide catalyst, 0.401% moisture* | **6. 06-03-003A: White smoke generating composition 2:** *31.2% zinc oxide, 30% ammonium perchlorate, 15% polychloroisoprene plasticizer, 14.2% ammonium chloride, 9.3 di-otyl-phthalate binder, 0.3% moisture* |
| **7. 06-03-004A: White smoke generating composition 3:** *38% red phosphorus, 30% manganese, 11% magnesium, 8.6% polyglycol resin, 5.3% epoxy resin, 4% lead dioxide, 3% zinc oxide, 0.10% residue* | **8. 06-03-005A: Red smoke generating composition 2:** *34.2% red dye No. 1, 17.5% sucrose, 16.5% potassium chlorate, 14.5% 1,4-benzenedicarboxylic acid, 10.5% magnesium carbonate, 6.8% red dye No. 11* |
| **9. 06-03-006A: High yielding dense red smoke generating composition with long range view ability:** *42.9% red dye No. 1, 28% lactose, 23.1% potassium chlorate, 5.9% red dye No 2* | **10. 06-03-007A: Dense green smoke generating composition with long range view ability:** *25% di(1,4,-amylamino anthraquinone, 25% quinoline yellow base, 20% lactose, 20% potassium chlorate, 10% diatomaceous* |

| | |
|---|---|
| | *earth* |
| **11. 06-03-007B: Dense yellow smoke generating composition with long range view ability:** *50% quinoline yellow base, 25% lactose, 25% potassium chlorate* | **12. 06-03-008A: Excellent yellow smoke generating composition with long duration:** *33% organol yellow dye, 21.5% potassium chlorate, 19.5% glycidil ether 100 epoxy resin, 16.5% guanidine nitrate, 7% propylene glycol, 2.5% ethanolamine hardener* |
| **13. 06-03-008B: Excellent green smoke generating composition with long duration:** *28% organol green thermoplastic dye, 18% guanidine nitrate, 17.6% glycidil ether 100 epoxy resin, 17% potassium chlorate, 8% organol yellow dye, 7% glycerol, 2.4% ethanolamine hardener, 0.8% manganese dioxide catalyst, 0.6% iron oxide catalyst, 0.3% copper oxide catalyst, 0.3% nickel oxide catalyst* | **14. 06-03-008C: Excellent red smoke generating composition with long duration:** *31.8% organol red J dye, 25% guanidine nitrate, 17.3% potassium chlorate, 17.2% glycidil ether 100 epoxy resin, 6.8% glycerol, 1.9% ethanolamine hardener* |
| **15. 06-03-009A: Standard military white smoke generating composition:** *51% red phosphorus, 35% pyrolusite, 8% magnesium, 3% zinc oxide, 3% epoxy binder* | **16. 06-03-010A: Military grade white smoke generating composition for camouflage purposes utilization the "wet mixing" process:** *50% red phosphorus, 44% calcium sulfate, 3% Viton A rubber binder, 3% boron* |
| **17. 06-03-011A: Environmentally safe white smoke generating composition:** *20% magnesium, 20% potassium nitrate, 15% potassium chloride, 15% sodium chloride, 12% potassium bicarbonate, 10% dicyandiamide, 8% potassium perchlorate* | **18. 06-03-012A: Opaque smoke generating composition for interfering with and blocking infrared targeting:** *71.4% hexachlorobenzene, 17.8% magnesium, 8.9% vinylidine polyfluoride binder, 1.7% naphthalene, 0.2% residue* |
| **19. 06-03-012B: Opaque smoke generating composition for interfering with and blocking infrared targeting (modified):** *41% hexachloroethane, 20% naphthalene, 13.3% vinylidine polyfluoride, 13.3% chlorinated paraffin, 12.3% magnesium, 0.1% residue* | **20. 06-03-013A: Improved "HC" white smoke composition (standard US military smoke composition):** *47.2% zinc oxide, 47.2% hexachloroethane, 5.5% aluminum, 0.1% residue* |
| **21. 06-03-014A: Standard white smoke composition utilizing guanidine nitrate:** *36% guanidine nitrate, 34% hexachloroethane, 22% zinc, 8% zinc oxide* | **22. 06-03-014B: Standard white smoke composition utilizing guanidine nitrate (modified):** *56% red phosphorus, 40% guanidine nitrate, 4% hexachloroethane* |
| **23. 06-03-015A: Opaque smoke generating composition for interfering with infrared targeting systems (modified):** *31% hexachloroethane, 31% zinc, 16% potassium perchlorate, 12% zinc oxide, 10% neoprene binder* | **24. 06-03-015B: Opaque smoke generating composition for interfering with infrared targeting systems (modified):** *66.6% hexachlorobenzene, 19% magnesium, 9.5% naphthalene, 4.7% neoprene binder, 0.20% residue* |
| **25. 06-03-016A: Standard white smoke generating composition for general use:** *36.3% ammonium perchlorate, 16.5% zinc oxide, 13.2% thiourea, 12% rubber binder, 11% PVC, 11% ammonium chloride* | **26. 06-03-017A: Standard blue smoke generating composition for general use:** *33.4% potassium nitrate, 25.1% amorphous boron, 12.6% magnesium, 11% wood charcoal, 9.4% Prussian blue dye, 8.4% chloroparaffin, 0.1% moisture* |
| **27. 06-03-018A: White spotting smoke composition for artillery and mortar training and similar use:** *40% zinc, 20% potassium nitrate, 20% aluminum, 20% potassium perchlorate* | **28. 06-03-019A: Orange smoke producing composition:** *42.9% potassium chlorate, 23.9% sucrose, 17.1% 1,8-diaminonapthalen coloring agent, 15.9% phthalic anhydride, 0.2% residue* |
| **29. 06-03-019B: Orange smoke producing composition (modified):** *39.9% potassium chlorate, 20.9% 1,8-diaminonapthalene, 19.4% phthalic anhydride, 13.9% glycerol, 5.9% polyethylene glycol* | **30. 06-03-019C: Orange smoke producing composition (modified 2):** *31% potassium chlorate, 17% 1,8-diaminonapthalene, 12% N-hexyl acid phthalate, 11% phthalic anhydride, 11% Epon 828 epoxy resin, 10% glycerol, 8% sodium barbiturate* |
| **31. 06-03-019D: Yellow smoke producing composition:** *44% potassium chlorate, 34% para-phenylene diamine dye, 22% sucrose* | **32. 06-03-019E: Violet smoke producing composition:** *33% potassium chlorate, 30% 1,8-diaminonapthalene dye, 30% pyromellitic dianhydride dye, 7% sucrose* |
| **33. 06-03-019F: Violet smoke producing composition (modified):** *31% potassium chlorate, 17% 1,8-diaminonapthalene dye, 12% N-hexyl acid phthalate, 11% Epon 828 epoxy resin, 11% phthalic anhydride, 10%* | **34. 06-03-019G: Red smoke producing composition:** *47.3% tetrabromo phthalic anhydride, 32.6% potassium chlorate, 15.9% 1,8-diaminonapthalene, 4.2% sucrose* |

| | |
|---|---|
| *glycerol, 8% sodium barbiturate* | |
| **35. 06-03-019H: Green smoke producing composition:** *49.7% para-dimethylaminobenzaldehyde dye, 32.4% potassium chlorate, 17.8% sucrose, 0.10% residue* | **36. 06-03-020A: Intensely red producing smoke composition for spotting and signaling:** *40% 1-methylaminoanthraquinone dye, 30% potassium chlorate, 12.9% aromatic disulfide ether, 10.1% benzyldimethylamine, 3.5% glycerine glycidyl ether, 3.5% poly sulfide* |
| **37. 06-03-021A: Safe to use, flameless yellow-orange smoke composition for use in highly flammable environments:** *43.2% ammonium iodate, 29.2% hydroxyl terminated polybutadiene binder, 20.7% leucoquinizarin, 6.6% hexamethylene diisocyanate curative agent, 0.121% dibutyl dilaurate curing catalyst, 0.179% residue* | **38. 06-03-022A: Smoke composition for use at night producing a brownish-black smoke, which looks orange from a distance:** *50% cadmium metal, 46% potassium nitrate, 3.9% epoxy resin, 0.10% residue* |
| **39. 06-03-022B: Smoke composition for use at night producing a brownish-black smoke, which looks orange from a distance (modified):** *46.6% black powder, 38.8% cadmium sulfide, 8.7% cadmium metal, 5.8% strontium nitrate, 0.10% mixed impurities* | **40. 06-03-023A: Non organic-chlorine containing white smoke composition for screening:** *49.2% potassium ferric chloride, 37.8% zinc oxide, 10.1% aluminum, 2.7% sodium nitrate, 0.20% moisture* |
| **41. 06-03-024A: Non-toxic smoke generating composition for use in various training operations:** *71.7% trans-cinnamic acid, 12.3% potassium chlorate, 5.2% sucrose, 4.9% nitrocellulose, 3.8% sodium bicarbonate, 1.9% diatomite silica, 0.2% impurities* | **42. 06-03-025A: Pyrotechnic composition for producing a mixed smoke:** *44.19% sodium chlorate, 22.09% maltose, 13.25% melamine freebase, 9.94% malic acid, 7.18% calcium lactate, 2.76% manganese dioxide, 0.55% copper chromite burn rate catalyst, 0.04% impurities* |
| **43. 06-03-026A: Red phosphorus smoke composition:** *67.56% red phosphorus, 27.02% unleaded gasoline, 5.4% butylmethacrylate, 0.02% mixed balance* | **44. 06-03-026B: Red phosphorus smoke composition (modified component):** *40.76% red phosphorus, 21.73% sodium nitrate, 19.02% unleaded gasoline, 13.58% metallic sodium, 4.89% butylmethacrylate, 0.02% impurities* |
| **45. 06-03-027A: Purple smoke composition:** *33.33% ammonium iodide, 25% potassium chlorate, 20.83% iodine, 12.5% lactose, 8.33% magnesium, 0.01% mixed balance* | **46. 06-03-028A: Yellow smoke composition:** *66% potassium dichromate, 20% bismuth tetraoxide, 14% magnesium powder* |
| **47. 06-03-028B: Yellow smoke composition:** *65% potassium dichromate, 20% bismuth subnitrate, 15% magnesium powder* | **48. 06-03-029A: Brown smoke composition:** *50% copper-II-oxide, 35% lead dioxide, 15% magnesium powder* |
| **49. 06-03-030A: Orange smoke composition:** *50% lead dioxide, 35% potassium dichromate, 15% magnesium powder* | **50. 06-03-031A: Pink smoke composition:** *54% potassium iodate, 37% calcium silicide, 9% potassium chromate* |
| **51. 06-03-031B: Pink smoke composition:** *65% calcium iodate, 25% magnesium powder, 10% potassium chromate* | **52. 06-03-032A: Classic HC smoke composition:** *35.82% zinc oxide, 34.82% hexachloroethane, 16.91% ammonium perchlorate, 11.94% magnesium powder, 0.49% linseed oil or mineral oil, 0.02% residual balance* |
| **53. 06-03-032B: Classic smoke composition modified with magnesium chloride:** *29.85% ammonium perchlorate, 24.87% anhydrous magnesium chloride, 24.87% magnesium powder, 19.9% zinc oxide, 0.49% linseed or mineral oil, 0.02% mixed balance* | **54. 06-03-033A: Modified HC smoke composition for generating a hydroscopic smoke for use in humid environments:** *39.5% potassium perchlorate, 20.66% hexachloroethane, 20.5% aluminum powder, 19.34% lithium carbonate* |
| **55. 06-03-034A: Simplified reddish smoke composition for bursting type munitions:** *77.5% lead iodate, 22.5% aluminum powder* | **56. 06-03-034B: Simplified reddish-purple smoke composition for bursting type munitions:** *75% lead iodate, 25% zirconium hydride* |
| **57. 06-03-035A: Dense reddish smoke composition:** *40.9% para-nitroaniline, 22.72% potassium chlorate, 22.72% sublimed iodine, 13.63% lactose, 0.03% residual balance* | **58. 06-03-036A: Opaque smoke composition for marine use:** *40% sodium nitrate, 22.5% ammonium chloride, 20% antimony trisulfide, 12.5% sulfur, 5% coal pitch* |
| **59. 06-03-036B: Opaque smoke composition for marine use:** *30% sodium chlorate, 22.5% ammonium chloride, 20% red phosphorus, 12.5% powdered zinc, 10% manganese dioxide,* | **60. 06-03-036C: Opaque smoke composition for marine use:** *38.46% sodium nitrate, 25.64% ammonium chloride, 20.51% antimony trisulfide, 12.82% sulfur, 2.56% coal* |

| | |
|---|---|
| *5% bituminous coal* | *pitch, 0.01% mixed balance* |
| **61. 06-03-037A: Under-water launched green smoke composition for marine use:** *34.44% solvent green dye, 25.24% potassium chlorate, 20.73% sugar, 6.85% vat yellow dye, 4.59% benzanthrone dye, 4.5% diatomaceous earth, 3.6% baking soda, 0.05% mixed residual balance* | **62. 06-03-037B: Under-water launched yellow smoke composition for marine use:** *27.47% benzanthrone dye, 21.92% potassium chlorate, 19.89% vat yellow dye No. 4, 14.03% powdered sugar, 9.64% diatomaceous earth, 7.01% baking soda, 0.04% mixed residual balance* |

### 06-03-001A: White smoke generating composition 1:

**Step 1: Preparation of special binder:**
Into a suitable flask, place 10 grams of para-toluene sulfonic acid, followed by 15 milliliters of water, followed by 100 grams of table sugar. Thereafter, heat the mixture to 100 Celsius for about 3 hours under heavy agitation. A motorized stirrer should be used as the agitation source. After heating and stirring for about 3 hours, remove the heat source, and allow the reaction mixture to cool to room temperature. Thereafter, pour the cooled reaction mixture mass onto a shallow pan, and allow it to air-dry thoroughly until no trace of water exists. When all the water has been removed, take the remaining mass, and mix it thoroughly into 110 milliliters of water to form a uniform mixture—thereafter, filter the mixture to remove any insoluble materials (if any). After the mixture (the special binder) has been filtered (if needed), your ready to form the actual smoke composition.

**Step 2: Preparation of smoke composition:**
Into a standard ball mill, filled with 500 grams of steel shot of 10mm diameter, place ***52 grams of Terephthalic Acid***, followed by ***24 grams of pentaerythritol***, followed by ***6 grams of magnesium carbonate***, followed by ***2.7 grams of stearic acid***, followed by ***31 grams of potassium chlorate***, and then followed by ***14 grams of table sugar***. Thereafter, tumble the mixture at 150 RPM for 45 minutes. After 45 minutes, add in ***4.7 grams of the special binder mixture*** (prepared in step 1), and then continue tumble the mixture at room temperature at 100 RPM for 30 minutes. Thereafter, the mixture should be removed from the ball mill, and the steel shot separated; thereafter, the mixture is ready for pressing. To do so, press the mixture into any suitable delivery container, such as a grenade, metal tube, plastic tube, cardboard tube, ceramic or clay ball, ect., under a pressure of 5,000 psi. In some cases, the smoke mixture needs to be pressed into blocks, discs, or tablets under a pressure of 6,000 psi for use in some types of smoke grenades. Note: an appropriate igniter composition to begin the burn of this smoke composition should be used.

**Burn rate:** 21 seconds per inch at 6,000 psi; tends to vary based on diameter and density of smoke grain.
**Smoke Volume on average (500 gram solid grain sample enclosed area):** 1500 sq. ft.
**Smoke Volume on average (500 gram solid grain sample open area):** 6000 sq. ft.
**Smoke visibility naked eye (sunny/dry conditions day time; 500 gram solid grain sample):** 5 km
**Smoke visibility naked eye (wet/rainy conditions day time; 500 gram solid grain sample):** 3.2 km
**Smoke visibility naked eye (night time, average environmental conditions; 500 gram solid grain sample):** 600 meters.
**Water resistance:** Very good.
**Stability:** Can be stored for decades.
**Flammability (solid grain):** 4.4
**Flammability (powdered grain):** 6
**Ease of ignition (solid grain):** 4.5 (requires high temperature ignition composition for proper burn).
**Tendency to cake:** Will not cake.
**Explosive ability:** Cannot be detonated.
**Percentage:** ***38.6% Terephthalic acid, 23% potassium chlorate, 17.8% pentaerythritol, 10.4% sugar, 4.4% magnesium carbonate, 3.4% polymerized sucrose binder containing some water, 2% stearic acid, 0.40% moisture***
**Classification:** Deflagrating mixture (classified as 1.3G pyrotechnic composition).

**Danger of preparation:** Can be made by amateurs.
**Use:** Used in professional industrial/commercial smoke munitions for determining gas leaks, spotting, signaling, and in military operations.

**Classic smoke grenade**

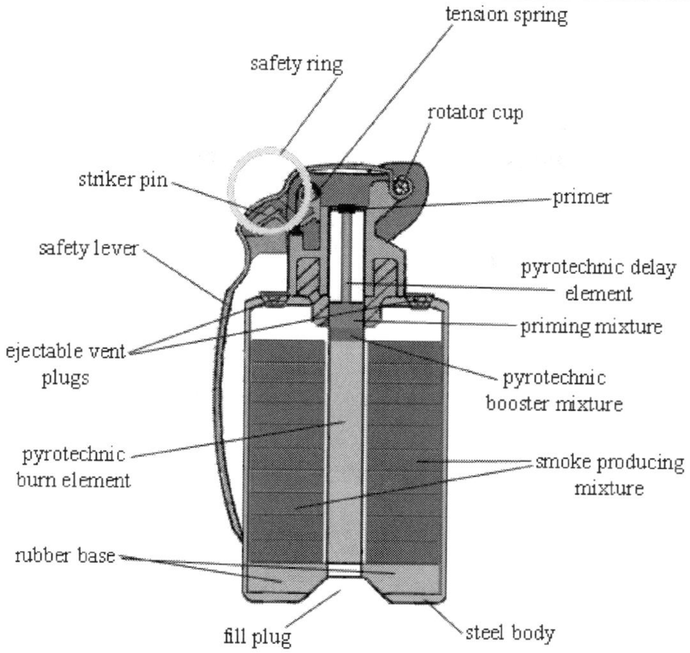


## 06-03-001B: Yellow smoke generating composition 1:

**Step 1: Preparation of special binder:**
Into a suitable flask, place 10 grams of para-toluene sulfonic acid, followed by 15 milliliters of water, followed by 100 grams of table sugar. Thereafter, heat the mixture to 100 Celsius for about 3 hours under heavy agitation. A motorized stirrer should be used as the agitation source. After heating and stirring for about 3 hours, remove the heat source, and allow the reaction mixture to cool to room temperature. Thereafter, pour the cooled reaction mixture mass onto a shallow pan, and allow it to air-dry thoroughly until no trace of water exists. When all the water has been removed, take the remaining mass, and mix it thoroughly into 110 milliliters of water to form a uniform mixture—thereafter, filter the mixture to remove any insoluble materials (if any). After the mixture (the special binder) has been filtered (if needed), your ready to form the actual smoke composition.

**Step 2: Preparation of smoke composition:**
Into a standard ball mill, filled with 500 grams of steel shot of 10mm diameter, place ***56.6 grams of yellow dye No. 33 or equivalent***, followed by ***27.8 grams of magnesium carbonate***, followed by ***30.6 grams of potassium chlorate***, and then followed by ***15.6 grams of table sugar***; thereafter, tumble the mixture at 100 RPM for 15 minutes. Thereafter, add in ***5.2 grams of the special binder mixture*** (prepared in step 1), and then continue to tumble the mixture at room temperature at 100 RPM for 45 minutes. Thereafter, the mixture should be removed from the ball mill, and the steel shot separated; thereafter, the mixture is ready for pressing. To do so, press the mixture into any suitable delivery container, such as a grenade, metal tube, plastic tube, cardboard tube, ceramic or clay ball, ect., under a pressure of 5,000 psi. In some cases, the smoke mixture needs to be pressed into blocks, discs, or tablets under a pressure of 6,000 psi for use in some types of smoke grenades. Note: an appropriate igniter composition to begin the burn of this smoke composition should be used.

**Burn rate:** 26 seconds per inch at 6,000 psi; tends to vary based on diameter and density of smoke grain.
**Smoke Volume on average (500 gram solid grain sample enclosed area):** 1590 sq. ft.
**Smoke Volume on average (500 gram solid grain sample open area):** 6500 sq. ft.
**Smoke visibility naked eye (sunny/dry conditions day time; 500 gram solid grain sample):** 7.1 km
**Smoke visibility naked eye (wet/rainy conditions day time; 500 gram solid grain sample):** 5 km
**Smoke visibility naked eye (night time, average environmental conditions; 500 gram solid grain sample):** 1500 meters.
**Water resistance:** Very good.
**Stability:** Can be stored for decades.
**Flammability (solid grain):** 4.3
**Flammability (powdered grain):** 5.9
**Ease of ignition (solid grain):** 4.2 (requires high temperature ignition composition for proper burn).
**Tendency to cake:** Will not cake.
**Explosive ability:** Cannot be detonated.
**Percentage:** ***41.6% yellow dye, 22.5% potassium chlorate, 20.4% magnesium carbonate, 11.4% sugar, 3.8% polymerized sucrose binder, 0.3% moisture***
**Classification:** Deflagrating mixture (classified as 1.3G pyrotechnic composition).

**Danger of preparation:** Can be made by amateurs.

**Use:** Used in professional industrial/commercial smoke munitions for determining gas leaks, spotting, signaling, and in military operations.

| Bursting type, tube launched smoke grenade | Description |
|---|---|
| 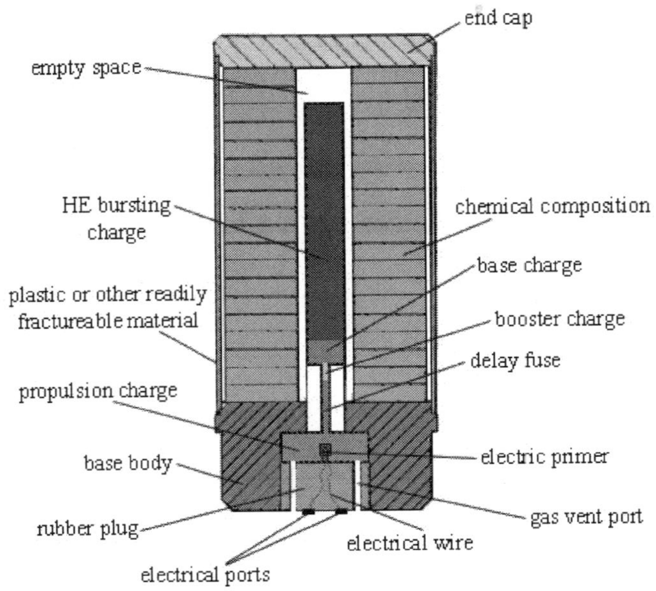  | Bursting type smoke grenades are often found on vehicles and tanks. When they are initiated, they shoot a predetermined distance from the vehicle, where upon they explode, producing a cloud of smoke. |

### 06-03-001C: Green smoke generating composition 1:

**Step 1: Preparation of special binder:**
Into a suitable flask, place 10 grams of para-toluene sulfonic acid, followed by 15 milliliters of water, followed by 100 grams of table sugar. Thereafter, heat the mixture to 100 Celsius for about 3 hours under heavy agitation. A motorized stirrer should be used as the agitation source. After heating and stirring for about 3 hours, remove the heat source, and allow the reaction mixture to cool to room temperature. Thereafter, pour the cooled reaction mixture mass onto a shallow pan, and allow it to air-dry thoroughly until no trace of water exists. When all the water has been removed, take the remaining mass, and mix it thoroughly into 110 milliliters of water to form a uniform mixture—thereafter, filter the mixture to remove any insoluble materials (if any). After the mixture (the special binder) has been filtered (if needed), your ready to form the actual smoke composition.
**Step 2: Preparation of smoke composition:**
Into a standard ball mill, filled with 500 grams of steel shot of 10mm diameter, place ***17 grams of yellow dye No. 33 or equivalent***, followed by ***40 grams of green dye No. 3 or equivalent***, followed by ***23 grams of magnesium carbonate***, followed by ***33.3 grams of potassium chlorate***, and then followed by ***16.7 grams of table sugar***. Thereafter, tumble the mixture at room temperature at 110 RPM for about 30 minutes. After 30 minutes, add in ***5.6 grams of the special binder mixture*** (prepared in step 1), and then continue to tumble the mixture at 110 RPM for about 15 minutes. Thereafter, the mixture should be removed from the ball mill, and the steel shot separated; thereafter, the mixture is ready for pressing. To do so, press the mixture into any suitable delivery container, such as a grenade, metal tube, plastic tube, cardboard tube, ceramic or clay ball, ect., under a pressure of 5,000 psi. In some cases, the smoke mixture needs to be pressed into blocks, discs, or tablets under a pressure of 6,000 psi for use in some types of smoke grenades. Note: an appropriate igniter composition to begin the burn of this smoke composition should be used.
**Burn rate:** 22 seconds per inch at 6,000 psi; tends to vary based on diameter and density of smoke grain.
**Smoke Volume on average (500 gram solid grain sample enclosed area):** 1600 sq. ft.
**Smoke Volume on average (500 gram solid grain sample open area):** 6700 sq. ft.
**Smoke visibility naked eye (sunny/dry conditions day time; 500 gram solid grain sample):** 7.3 km
**Smoke visibility naked eye (wet/rainy conditions day time; 500 gram solid grain sample):** 5.8 km
**Smoke visibility naked eye (night time, average environmental conditions; 500 gram solid grain sample):** 2000 meters.
**Water resistance:** Very good.
**Stability:** Can be stored for decades.
**Flammability (solid grain):** 4.2
**Flammability (powdered grain):** 5.98
**Ease of ignition (solid grain):** 4.1 (requires high temperature ignition composition for proper burn).
**Tendency to cake:** Will not cake.
**Explosive ability:** Cannot be detonated.
**Percentage:** ***29.4% green dye, 24.5% potassium chlorate, 16.9% magnesium carbonate, 12.5% yellow dye, 12.3% sugar, 4.1% polymerized sucrose binder, 0.3% moisture***
**Classification:** Deflagrating mixture (classified as 1.3G pyrotechnic composition).

**Danger of preparation:** Can be made by amateurs.
**Use:** Used in professional industrial/commercial smoke munitions for determining gas leaks, spotting, signaling, and in military operations.

**Classic Smoke Grenade**

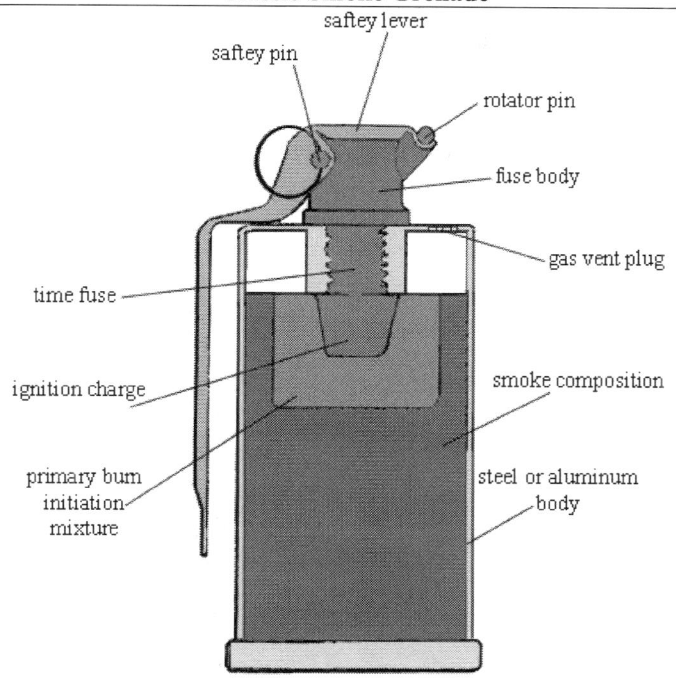


### 06-03-002A: Yellow smoke generating composition 2

**Step 1: Preparation of the binder:**
Into a suitable beaker, or container, place ***25 grams of ethanediol plasticizer,*** followed by ***15 grams of methyl carbazate,*** and then heat the mixture to about 60 Celsius with rapid stirring for about 30 minutes. After 30 minutes add in ***5 grams of ethanolamine***, and then continue to rapidly stir and heat the mixture at 60 Celsius for about 30 minutes. After 30 minutes, remove the heat source, and allow the mixture to cool to room temperature. After it does, mix in ***80 grams of glycidyl ether 100,*** and then rapidly stir the entire mixture at room temperature for about 1 hour. After 1 hour, the binder is ready to use. Note: this mixture should be continuously stirred at room temperature until use.
**Step 2: Preparation of the smoke composition:**
Into a standard ball mill, filled with 500 grams of steel shot of 10mm diameter, place ***1 gram of ferric oxide***, followed by ***175 grams of yellow "organol"dye***, followed by ***85 grams of guanidine nitrate***, and then followed by ***105 grams of potassium chlorate***. Thereafter, tumble the mixture at room temperature at 125 RPM for about 45 minutes. After 45 minutes, remove the mixture from the ball mill, separate the steel shot, and then place the mixture into a suitable vertical mixer; thereafter, add in the binder mixture from step 1, and then rapidly blend the mixture for about 30 minutes. Thereafter, the mixture is ready for pressing. To do so, press the mixture into any suitable delivery container, such as a grenade, metal tube, plastic tube, cardboard tube, ceramic or clay ball, ect., under a pressure of 5,000 psi. In some cases, the smoke mixture needs to be pressed into blocks, discs, or tablets under a pressure of 6,000 psi for use in some types of smoke grenades. Note: an appropriate igniter composition to begin the burn of this smoke composition should be used.
**Burn rate:** 22 seconds per inch at 6,000 psi; tends to vary based on diameter and density of smoke grain.
**Smoke Volume on average (500 gram solid grain sample enclosed area):** 1579 sq. ft.
**Smoke Volume on average (500 gram solid grain sample open area):** 6650 sq. ft.
**Smoke visibility naked eye (sunny/dry conditions day time; 500 gram solid grain sample):** 7.4 km
**Smoke visibility naked eye (wet/rainy conditions day time; 500 gram solid grain sample):** 5.9 km
**Smoke visibility naked eye (night time, average environmental conditions; 500 gram solid grain sample):** 2050 meters.
**Water resistance:** Very good.
**Stability:** Can be stored for decades.
**Flammability (solid grain):** 4.3
**Flammability (powdered grain):** 5.99
**Ease of ignition (solid grain):** 4.3 (requires high temperature ignition composition for proper burn).
**Tendency to cake:** Will not cake.
**Explosive ability:** Cannot be detonated.
**Percentage:** ***35.6% yellow dye "organol", 21.3% potassium chlorate, 17.3% guanidine nitrate, 16.2% glycidyl ether 100 epoxy resin, 5% ethandiol plasticizer, 3% methyl carbazate binder, 1% ethanolamine hardener, 0.20% ferric oxide catalyst, 0.40% moisture***

**Classification:** Deflagrating mixture (classified as 1.3G pyrotechnic composition).

**Danger of preparation:** Can be made by amateurs.
**Use:** Used in professional industrial/commercial smoke munitions for determining gas leaks, spotting, signaling, and in military operations.

## 06-03-002B: Red smoke generating composition 1:

**Step 1: Preparation of the binder:**
Into a suitable beaker, or container, place ***20 grams of propanediol plasticizer,*** followed by ***25 grams of methyl carbazate,*** and then heat the mixture to about 60 Celsius with rapid stirring for about 10 minutes. Afterwards, remove the heat source, and allow the mixture to cool to room temperature. After it does, mix in ***75 grams of glycidyl ether 100,*** and then rapidly stir the entire mixture at room temperature for about 10 minutes. Afterwards, the binder is ready to use. Note: this mixture should be continuously stirred at room temperature until use.
**Step 2: Preparation of the smoke composition:**
Into a standard ball mill, filled with 500 grams of steel shot of 10mm diameter, place ***1 gram of ferric oxide***, followed by ***190 grams of red "organol" dye (methylaminoanthraquinone)***, followed by ***120 grams of guanidine nitrate***, and then followed by ***70 grams of potassium chlorate***. Thereafter, tumble the mixture at room temperature at 125 RPM for about 45 minutes. After 45 minutes, remove the mixture from the ball mill, separate the steel shot, and then place the mixture into a suitable vertical mixer; thereafter, add in the binder mixture from step 1, and then rapidly blend the mixture for about 30 minutes. Thereafter, the mixture is ready for pressing. To do so, press the mixture into any suitable delivery container, such as a grenade, metal tube, plastic tube, cardboard tube, ceramic or clay ball, ect., under a pressure of 5,000 psi. In some cases, the smoke mixture needs to be pressed into blocks, discs, or tablets under a pressure of 6,000 psi for use in some types of smoke grenades. Note: an appropriate igniter composition to begin the burn of this smoke composition should be used.
**Burn rate:** 24 seconds per inch at 6,000 psi; tends to vary based on diameter and density of smoke grain.
**Smoke Volume on average (500 gram solid grain sample enclosed area):** 1599 sq. ft.
**Smoke Volume on average (500 gram solid grain sample open area):** 7109 sq. ft.
**Smoke visibility naked eye (sunny/dry conditions day time; 500 gram solid grain sample):** 7.3 km
**Smoke visibility naked eye (wet/rainy conditions day time; 500 gram solid grain sample):** 5 km
**Smoke visibility naked eye (night time, average environmental conditions; 500 gram solid grain sample):** 2000 meters.
**Water resistance:** Very good.
**Stability:** Can be stored for decades.
**Flammability (solid grain):** 4
**Flammability (powdered grain):** 5.2
**Ease of ignition (solid grain):** 4.1 (requires high temperature ignition composition for proper burn).
**Tendency to cake:** Will not cake.
**Explosive ability:** Cannot be detonated.
**Percentage:** ***37.9% red dye "organol", 23.9% guanidine nitrate, 14.9% glycidyl ether 100 epoxy resin, 13.9% potassium chlorate, 4.9% methyl carbazate, 3.9% propanediol plasticizer, 0.199% ferric oxide catalyst, 0.401% moisture***
**Classification:** Deflagrating mixture (classified as 1.3G pyrotechnic composition).

**Danger of preparation:** Can be made by amateurs.
**Use:** Used in professional industrial/commercial smoke munitions for determining gas leaks, spotting, signaling, and in military operations.

## 06-03-003A: White smoke generating composition 2:

Into a standard ball mill, filled with 500 grams of steel shot of 10mm diameter, place ***37.5 grams of di-octyl-phthalate***, followed by ***125 grams of zinc oxide***, followed by ***120 grams of ammonium perchlorate***, followed by ***60 grams of polychloroisoprene***, and then followed by ***57 grams of ammonium chloride***. Thereafter, tumble the mixture at room temperature at 130 RPM for about 45 minutes. After 45 minutes, the mixture is ready for pressing. To do so, first remove the steel shot, and then press the mixture into any suitable delivery container, such as a grenade, metal tube, plastic tube, cardboard tube, ceramic or clay ball, ect., under a pressure of 6,000 psi. In some cases, the smoke mixture needs to be pressed into blocks, discs, or tablets under a pressure of 6,000 psi for use in some types of smoke grenades. Note: an appropriate igniter composition to begin the burn of this smoke composition should be used.
**Burn rate:** 23 seconds per inch at 6,000 psi; tends to vary based on diameter and density of smoke grain.
**Smoke Volume on average (500 gram solid grain sample enclosed area):** 1600 sq. ft.
**Smoke Volume on average (500 gram solid grain sample open area):** 7100 sq. ft.
**Smoke visibility naked eye (sunny/dry conditions day time; 500 gram solid grain sample):** 7.44 km
**Smoke visibility naked eye (wet/rainy conditions day time; 500 gram solid grain sample):** 5.7 km
**Smoke visibility naked eye (night time, average environmental conditions; 500 gram solid grain sample):** 2150 meters.
**Water resistance:** Very good.
**Stability:** Can be stored for decades.
**Flammability (solid grain):** 4.1
**Flammability (powdered grain):** 5

**Ease of ignition (solid grain):** 4.1 (requires high temperature ignition composition for proper burn).
**Tendency to cake:** Will not cake.
**Explosive ability:** Cannot be detonated.
**Percentage:** ***31.2% zinc oxide, 30% ammonium perchlorate, 15% polychloroisoprene plasticizer, 14.2% ammonium chloride, 9.3 di-otyl-phthalate binder, 0.3% moisture***
**Classification:** Deflagrating mixture (classified as 1.3G pyrotechnic composition).

**Danger of preparation:** Can be made by amateurs.
**Use:** Used in professional industrial/commercial smoke munitions for determining gas leaks, spotting, signaling, and in military operations.

## 06-03-004A: White smoke generating composition 3:

Into a standard ball mill, filled with 500 grams of steel shot of 10mm diameter, place ***26.6 grams of Epoxy resin (D.E.R. 732)***, followed by ***43 grams of polyglycol resin (QX-3812)***, followed by ***15 grams of zinc oxide***, followed by ***55 grams of magnesium powder,*** followed by ***190 grams of red phosphorus,*** followed by ***150 grams of manganese dioxide,*** and then followed by ***20 grams of lead dioxide***. Thereafter, tumble the mixture at room temperature at 150 RPM for about 45 minutes. After 45 minutes, the mixture is ready for pressing. To do so, first remove the steel shot, and then press the mixture into any suitable delivery container, such as a grenade, metal tube, plastic tube, cardboard tube, ceramic or clay ball, ect., under a pressure of 6,000 psi, and then cure the device in an oven at 40 Celsius for 6 hours. In some cases, the smoke mixture needs to be pressed into blocks, discs, or tablets under a pressure of 6,000 psi for use in some types of smoke grenades, followed by curing at 40 Celsius for 6 hours. Note: an appropriate igniter composition to begin the burn of this smoke composition should be used.
**Burn rate:** 27 seconds per inch at 6,000 psi; tends to vary based on diameter and density of smoke grain.
**Smoke Volume on average (500 gram solid grain sample enclosed area):** 1560 sq. ft.
**Smoke Volume on average (500 gram solid grain sample open area):** 7005 sq. ft.
**Smoke visibility naked eye (sunny/dry conditions day time; 500 gram solid grain sample):** 7.39 km
**Smoke visibility naked eye (wet/rainy conditions day time; 500 gram solid grain sample):** 3.1 km
**Smoke visibility naked eye (night time, average environmental conditions; 500 gram solid grain sample):** 1800 meters.
**Water resistance:** Very good.
**Stability:** Can be stored for decades.
**Flammability (solid grain):** 3.9
**Flammability (powdered grain):** 4.9
**Ease of ignition (solid grain):** 3.9 (requires high temperature ignition composition for proper burn).
**Tendency to cake:** Will not cake.
**Explosive ability:** Cannot be detonated.
**Percentage:** ***38% red phosphorus, 30% manganese dioxide, 11% magnesium, 8.6% polyglycol resin, 5.3% epoxy resin, 4% lead dioxide, 3% zinc oxide, 0.10% residue***
**Classification:** Deflagrating mixture (classified as 1.3G pyrotechnic composition).

**Danger of preparation:** Can be made by amateurs.
**Use:** Used in professional industrial/commercial smoke munitions for determining gas leaks, spotting, signaling, and in military operations.

**Standard smoke grenade**

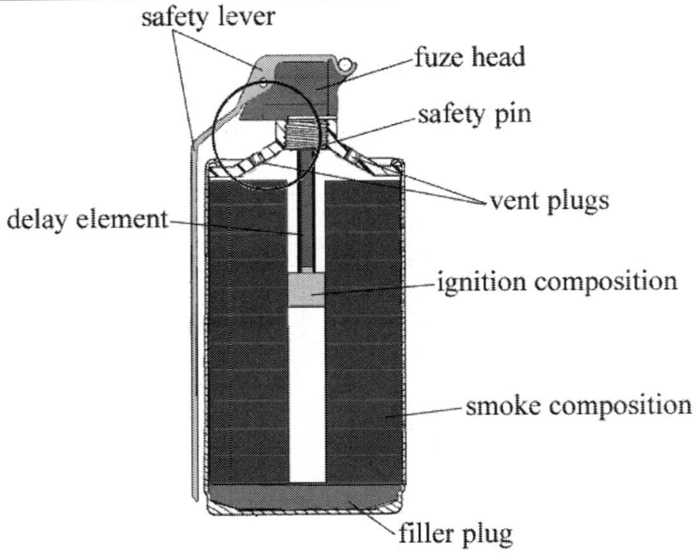


### 06-03-005A: Red smoke generating composition 2:

Into a standard ball mill, filled with 500 grams of steel shot of 10mm diameter, place ***171 grams of solvent red No. 1 dye***, followed by ***34 grams of disperse red No. 11 dye***, followed by ***87.5 grams of sucrose***, followed by ***72.5 grams of 1,4,-benzenedicaroxylic acid,*** followed by ***52.5 grams of magnesium carbonate,*** and then followed by ***82.5 grams of potassium chlorate***. Thereafter, tumble the mixture at room temperature at 150 RPM for about 45 minutes. After 45 minutes, the mixture is ready for pressing. To do so, first remove the steel shot, and then press the mixture into any suitable delivery container, such as a grenade, metal tube, plastic tube, cardboard tube, ceramic or clay ball, ect., under a pressure of 7,000 psi, and then cure the device in an oven at 60 Celsius for 6 hours. In some cases, the smoke mixture needs to be pressed into blocks, discs, or tablets under a pressure of 6,000 psi for use in some types of smoke grenades, followed by curing at 60 Celsius for 6 hours. Note: an appropriate igniter composition to begin the burn of this smoke composition should be used.

**Burn rate:** 26 seconds per inch at 7,000 psi; tends to vary based on diameter and density of smoke grain.
**Smoke Volume on average (500 gram solid grain sample enclosed area):** 1800 sq. ft.
**Smoke Volume on average (500 gram solid grain sample open area):** 7120 sq. ft.
**Smoke visibility naked eye (sunny/dry conditions day time; 500 gram solid grain sample):** 7.5 km
**Smoke visibility naked eye (wet/rainy conditions day time; 500 gram solid grain sample):** 3.9 km
**Smoke visibility naked eye (night time, average environmental conditions; 500 gram solid grain sample):** 3000 meters.
**Water resistance:** Very good.
**Stability:** Can be stored for decades.
**Flammability (solid grain):** 3.8
**Flammability (powdered grain):** 4.8
**Ease of ignition (solid grain):** 3.9 (requires high temperature ignition composition for proper burn).
**Tendency to cake:** Will not cake.
**Explosive ability:** Cannot be detonated.
**Percentage:** ***34.2% red dye No. 1, 17.5% sucrose, 16.5% potassium chlorate, 14.5% 1,4-benzenedicarboxylic acid, 10.5% magnesium carbonate, 6.8% red dye No. 11***
**Classification:** Deflagrating mixture (classified as 1.3G pyrotechnic composition).

**Danger of preparation:** Can be made by amateurs.
**Use:** Used in professional industrial/commercial smoke munitions for determining gas leaks, spotting, signaling, and in military operations.

### 06-03-006A: High yielding dense red smoke generating composition with long-range view ability:

Into a standard ball mill, filled with 200 grams of steel shot of 10mm diameter, place ***65 grams of red dye No. 1***, followed by ***9 grams of red dye No. 2*** (Rhodamine B), followed by ***42.5 grams of lactose***, and then followed by ***35 grams of potassium*** chlorate. Thereafter, tumble the mixture at room temperature at 125 RPM for about 45 minutes. After 45 minutes, the mixture is ready for pressing. To do so, first remove the steel shot, and then press the mixture into any suitable delivery container, such as a grenade, metal tube, plastic tube, cardboard tube, ceramic or clay ball, ect., under a pressure of 7,000 psi. In some cases, the smoke mixture needs to be pressed into blocks, discs, or tablets under a pressure of 7,000 psi for use in some types of smoke grenades. Note: an appropriate igniter composition to begin the burn of this smoke composition should be used.

**Burn rate:** 25 seconds per inch at 7,000 psi; tends to vary based on diameter and density of smoke grain.

**Smoke Volume on average (500 gram solid grain sample enclosed area):** 1890 sq. ft.
**Smoke Volume on average (500 gram solid grain sample open area):** 7200 sq. ft.
**Smoke visibility naked eye (sunny/dry conditions day time; 500 gram solid grain sample):** 8 km
**Smoke visibility naked eye (wet/rainy conditions day time; 500 gram solid grain sample):** 4 km
**Smoke visibility naked eye (night time, average environmental conditions; 500 gram solid grain sample):** 3100 meters.
**Water resistance:** Very good.
**Stability:** Can be stored for decades.
**Flammability (solid grain):** 3.9
**Flammability (powdered grain):** 4.9
**Ease of ignition (solid grain):** 3.91 (requires high temperature ignition composition for proper burn).
**Tendency to cake:** Will not cake.
**Explosive ability:** Cannot be detonated.
**Percentage:** ***42.9% red dye No. 1, 28% lactose, 23.1% potassium chlorate, 5.9% red dye No 2***
**Classification:** Deflagrating mixture (classified as 1.3G pyrotechnic composition).

**Danger of preparation:** Can be made by amateurs.
**Use:** Used in professional industrial/commercial smoke munitions for determining gas leaks, spotting, signaling, and in military operations.

### 06-03-007A: Dense green smoke generating composition with long-range view ability:

Into a standard ball mill, filled with 200 grams of steel shot of 10mm diameter, place ***25 grams of di(1,4-amylamino anthraquinone***, followed by ***25 grams of quinoline yellow base***, followed by ***20 grams of lactose***, followed by ***20 grams of potassium chlorate,*** and then followed by ***10 grams of diatomaceous earth***. Thereafter, tumble the mixture at room temperature at 125 RPM for about 45 minutes. After 45 minutes, the mixture is ready for pressing. To do so, first remove the steel shot, and then press the mixture into any suitable delivery container, such as a grenade, metal tube, plastic tube, cardboard tube, ceramic or clay ball, ect., under a pressure of 7,000 psi. In some cases, the smoke mixture needs to be pressed into blocks, discs, or tablets under a pressure of 7,000 psi for use in some types of smoke grenades. Note: an appropriate igniter composition to begin the burn of this smoke composition should be used.
**Burn rate:** 26 seconds per inch at 7,000 psi; tends to vary based on diameter and density of smoke grain.
**Smoke Volume on average (500 gram solid grain sample enclosed area):** 1880 sq. ft.
**Smoke Volume on average (500 gram solid grain sample open area):** 7205 sq. ft.
**Smoke visibility naked eye (sunny/dry conditions day time; 500 gram solid grain sample):** 8 km
**Smoke visibility naked eye (wet/rainy conditions day time; 500 gram solid grain sample):** 4 km
**Smoke visibility naked eye (night time, average environmental conditions; 500 gram solid grain sample):** 3100 meters.
**Water resistance:** Very good.
**Stability:** Can be stored for decades.
**Flammability (solid grain):** 3.8
**Flammability (powdered grain):** 4.9
**Ease of ignition (solid grain):** 3.9 (requires high temperature ignition composition for proper burn).
**Tendency to cake:** Will not cake.
**Explosive ability:** Cannot be detonated.
**Percentage:** ***25% di-1,4,-amylamino anthraquinone, 25% quinoline yellow base, 20% lactose, 20% potassium chlorate, 10% diatomaceous earth***
**Classification:** Deflagrating mixture (classified as 1.3G pyrotechnic composition).

**Danger of preparation:** Can be made by amateurs.
**Use:** Used in professional industrial/commercial smoke munitions for determining gas leaks, spotting, signaling, and in military operations.

### 06-03-007B: Dense yellow smoke generating composition with long-range view ability:

Into a standard ball mill, filled with 200 grams of steel shot of 10mm diameter, place ***50 grams of quinoline yellow base,*** followed by ***25 grams of lactose***, and then followed by ***25 grams of potassium chlorate***. Thereafter, tumble the mixture at room temperature at 120 RPM for about 45 minutes. After 45 minutes, the mixture is ready for pressing. To do so, first remove the steel shot, and then press the mixture into any suitable delivery container, such as a grenade, metal tube, plastic tube, cardboard tube, ceramic or clay ball, ect., under a pressure of 7,000 psi. In some cases, the smoke mixture needs to be pressed into blocks, discs, or tablets under a pressure of 7,000 psi for use in some types of smoke grenades. Note: an appropriate igniter composition to begin the burn of this smoke composition should be used.
**Burn rate:** 26 seconds per inch at 7,000 psi; tends to vary based on diameter and density of smoke grain.
**Smoke Volume on average (500 gram solid grain sample enclosed area):** 1885 sq. ft.
**Smoke Volume on average (500 gram solid grain sample open area):** 7210 sq. ft.
**Smoke visibility naked eye (sunny/dry conditions day time; 500 gram solid grain sample):** 8.1 km
**Smoke visibility naked eye (wet/rainy conditions day time; 500 gram solid grain sample):** 4.21 km

**Smoke visibility naked eye (night time, average environmental conditions; 500 gram solid grain sample):** 3150 meters.
**Water resistance:** Very good.
**Stability:** Can be stored for decades.
**Flammability (solid grain):** 3.7
**Flammability (powdered grain):** 4.8
**Ease of ignition (solid grain):** 3.8 (requires high temperature ignition composition for proper burn).
**Tendency to cake:** Will not cake.
**Explosive ability:** Cannot be detonated.
**Percentage:** ***50% quinoline yellow base, 25% lactose, 25% potassium chlorate***
**Classification:** Deflagrating mixture (classified as 1.3G pyrotechnic composition).

**Danger of preparation:** Can be made by amateurs.
**Use:** Used in professional industrial/commercial smoke munitions for determining gas leaks, spotting, signaling, and in military operations.

## 06-03-008A: Excellent yellow smoke generating composition with long duration:

Into a standard ball mill, filled with 500 grams of steel shot of 10mm diameter, place ***165 grams of Organol yellow A.D.E. dye*** (sold by Pechiney Ugine Kullman Inc.), followed by ***82.5 grams of guanidine nitrate***, followed by ***107.5 grams of potassium chlorate***, followed by ***35 grams of propylene glycol***, followed by ***97.5 grams of Glycidil ether 100*** (sold by Shell as an epoxyaliphatic resin), and then followed by ***12.5 grams of ethanolamine*** (sold as Hardener HY 938 and aliphatic ethanolamine). Thereafter, tumble the mixture at room temperature at 150 RPM for about 45 minutes. After 45 minutes, the mixture is ready for pressing. To do so, first remove the steel shot, and then press the mixture into any suitable delivery container, such as a grenade, metal tube, plastic tube, cardboard tube, ceramic or clay ball, ect., under a pressure of 7,000 psi, and then cure the device in an oven at 60 Celsius for 6 hours. In some cases, the smoke mixture needs to be pressed into blocks, discs, or tablets under a pressure of 7,000 psi, for use in some types of smoke grenades, followed by curing at 60 Celsius for 6 hours. Note: an appropriate igniter composition to begin the burn of this smoke composition should be used.
**Burn rate:** 29 seconds per inch at 7,000 psi; tends to vary based on diameter and density of smoke grain.
**Smoke Volume on average (500 gram solid grain sample enclosed area):** 1700 sq. ft.
**Smoke Volume on average (500 gram solid grain sample open area):** 7050 sq. ft.
**Smoke visibility naked eye (sunny/dry conditions day time; 500 gram solid grain sample):** 7.3 km
**Smoke visibility naked eye (wet/rainy conditions day time; 500 gram solid grain sample):** 3.2 km
**Smoke visibility naked eye (night time, average environmental conditions; 500 gram solid grain sample):** 2779 meters.
**Water resistance:** Very good.
**Stability:** Can be stored for decades.
**Flammability (solid grain):** 3.5
**Flammability (powdered grain):** 4.6
**Ease of ignition (solid grain):** 3.6 (requires high temperature ignition composition for proper burn).
**Tendency to cake:** Will not cake.
**Explosive ability:** Cannot be detonated.
**Percentage:** ***33% organol yellow dye, 21.5% potassium chlorate, 19.5% glycidyl ether 100 epoxy resin, 16.5% guanidine nitrate, 7% propylene glycol, 2.5% ethanolamine hardener***
**Classification:** Deflagrating mixture (classified as 1.3G pyrotechnic composition).

**Danger of preparation:** Can be made by amateurs.
**Use:** Used in professional industrial/commercial smoke munitions for determining gas leaks, spotting, signaling, and in military operations.

## 06-03-008B: Excellent green smoke generating composition with long duration:

Into a standard ball mill, filled with 500 grams of steel shot of 10mm diameter, place ***3 grams of red iron-III-oxide***, followed by ***4 grams of manganese dioxide***, followed by ***1.5 grams of copper-II-oxide***, followed by ***1.5 grams of nickel oxide***, followed by ***140 grams of Organol green thermoplast dye*** (sold by Pechiney Ugine Kullman Inc.), followed by ***90 grams of guanidine nitrate***, followed by ***40 grams of Organol yellow A.D.E. dye*** (sold by Pechiney Ugine Kullman Inc.), followed by ***85 grams of potassium chlorate***, followed by ***88 grams of Glycidyl ether 100*** (sold by Shell as an epoxy aliphatic resin), and then followed by ***12 grams of ethanolamine*** (sold as Hardener HY 938 and aliphatic ethanolamine). Thereafter, tumble the mixture at room temperature at 150 RPM for about 45 minutes. After 45 minutes, the mixture is ready for pressing. To do so, first remove the steel shot, and then press the mixture into any suitable delivery container, such as a grenade, metal tube, plastic tube, cardboard tube, ceramic or clay ball, ect., under a pressure of 7,000 psi, and then cure the device in an oven at 60 Celsius for 6 hours. In some cases, the smoke mixture needs to be pressed into blocks, discs, or tablets under a pressure of 7,000 psi, for use in some types of smoke grenades, followed by curing at 60 Celsius for 6 hours. Note: an appropriate igniter composition to begin the burn of this smoke composition should be used.
**Burn rate:** 30 seconds per inch at 7,000 psi; tends to vary based on diameter and density of smoke grain.
**Smoke Volume on average (500 gram solid grain sample enclosed area):** 1720 sq. ft.

**Smoke Volume on average (500 gram solid grain sample open area):** 7100 sq. ft.
**Smoke visibility naked eye (sunny/dry conditions day time; 500 gram solid grain sample):** 7.2 km
**Smoke visibility naked eye (wet/rainy conditions day time; 500 gram solid grain sample):** 3.1 km
**Smoke visibility naked eye (night time, average environmental conditions; 500 gram solid grain sample):** 2950 meters.
**Water resistance:** Very good.
**Stability:** Can be stored for decades.
**Flammability (solid grain):** 4
**Flammability (powdered grain):** 4.9
**Ease of ignition (solid grain):** 3.98 (requires high temperature ignition composition for proper burn).
**Tendency to cake:** Will not cake.
**Explosive ability:** Cannot be detonated.
**Percentage:** ***28% organol green thermoplast dye, 18% guanidine nitrate, 17.6% glycidyl ether 100 epoxy resin, 17% potassium chlorate, 8% organol yellow dye, 7% glycerol, 2.4% ethanolamine hardener, 0.8% manganese dioxide catalyst, 0.6% iron oxide catalyst, 0.3% copper oxide catalyst, 0.3% nickel oxide catalyst***
**Classification:** Deflagrating mixture (classified as 1.3G pyrotechnic composition).

**Danger of preparation:** Can be made by amateurs.
**Use:** Used in professional industrial/commercial smoke munitions for determining gas leaks, spotting, signaling, and in military operations.

### 06-03-008C: Excellent red smoke generating composition with long duration:

Into a standard ball mill, filled with 500 grams of steel shot of 10mm diameter, place ***159 grams of Organol red J dye*** (sold by Pechiney Ugine Kullman Inc.), followed by ***125 grams of guanidine nitrate***, followed by ***86.5 grams of potassium chlorate***, followed by ***34 grams of glycerol***, followed by ***86 grams of Glycidyl ether 100*** (sold by Shell as an epoxy aliphatic resin), and then followed by ***9.5 grams of ethanolamine*** (sold as Hardener HY 938 and aliphatic ethanolamine). Thereafter, tumble the mixture at room temperature at 150 RPM for about 45 minutes. After 45 minutes, the mixture is ready for pressing. To do so, first remove the steel shot, and then press the mixture into any suitable delivery container, such as a grenade, metal tube, plastic tube, cardboard tube, ceramic or clay ball, ect., under a pressure of 7,000 psi, and then cure the device in an oven at 60 Celsius for 6 hours. In some cases, the smoke mixture needs to be pressed into blocks, discs, or tablets under a pressure of 7,000 psi, for use in some types of smoke grenades, followed by curing at 60 Celsius for 6 hours. Note: an appropriate igniter composition to begin the burn of this smoke composition should be used.
**Burn rate:** 29 seconds per inch at 7,000 psi; tends to vary based on diameter and density of smoke grain.
**Smoke Volume on average (500 gram solid grain sample enclosed area):** 1700 sq. ft.
**Smoke Volume on average (500 gram solid grain sample open area):** 7150 sq. ft.
**Smoke visibility naked eye (sunny/dry conditions day time; 500 gram solid grain sample):** 7.4 km
**Smoke visibility naked eye (wet/rainy conditions day time; 500 gram solid grain sample):** 3.4 km
**Smoke visibility naked eye (night time, average environmental conditions; 500 gram solid grain sample):** 3000 meters.
**Water resistance:** Very good.
**Stability:** Can be stored for decades.
**Flammability (solid grain):** 4.5
**Flammability (powdered grain):** 5.2
**Ease of ignition (solid grain):** 4.1 (requires high temperature ignition composition for proper burn).
**Tendency to cake:** Will not cake.
**Explosive ability:** Cannot be detonated.
**Percentage:** ***31.8% organol red J dye, 25% guanidine nitrate, 17.3% potassium chlorate, 17.2% glycidyl ether 100 epoxy resin, 6.8% glycerol, 1.9% ethanolamine hardener***
**Classification:** Deflagrating mixture (classified as 1.3G pyrotechnic composition).

**Danger of preparation:** Can be made by amateurs.
**Use:** Used in professional industrial/commercial smoke munitions for determining gas leaks, spotting, signaling, and in military operations.

### 06-03-009A: Standard military white smoke generating composition:

Into a standard ball mill, filled with 500 grams of steel shot of 10mm diameter, place ***255 grams of red amorphous phosphorus***, followed by ***40 grams of standard powdered magnesium***, followed by ***15 grams of zinc oxide***, followed by ***175 grams of finely powdered pyrolusite*** (natural occurring manganese dioxide ore), and then followed by ***15 grams of any desirable epoxy binder***. Thereafter, tumble the mixture at room temperature at 150 RPM for about 45 minutes. After 45 minutes, the mixture is ready for pressing. To do so, first remove the steel shot, and then press the mixture into any suitable delivery container, such as a grenade, metal tube, plastic tube, cardboard tube, ceramic or clay ball, ect., under a pressure of 7,000 psi, and then cure the device at room temperature for 24 hours. In some cases, the smoke mixture needs to be pressed into blocks, discs, or tablets under a pressure of 7,000 psi, for use in some types of smoke grenades, followed by curing at room temperature for 24 hours. Note: an appropriate igniter composition to begin the burn of this smoke composition should be used.

**Burn rate:** 21 seconds per inch at 7,000 psi; tends to vary based on diameter and density of smoke grain.
**Smoke Volume on average (500 gram solid grain sample enclosed area):** 1699 sq. ft.
**Smoke Volume on average (500 gram solid grain sample open area):** 7100 sq. ft.
**Smoke visibility naked eye (sunny/dry conditions day time; 500 gram solid grain sample):** 7 km
**Smoke visibility naked eye (wet/rainy conditions day time; 500 gram solid grain sample):** 3 km
**Smoke visibility naked eye (night time, average environmental conditions; 500 gram solid grain sample):** 2340 meters.
**Water resistance:** Very good.
**Stability:** Can be stored for decades.
**Flammability (solid grain):** 4
**Flammability (powdered grain):** 5
**Ease of ignition (solid grain):** 3.8 (requires high temperature ignition composition for proper burn).
**Tendency to cake:** Will not cake.
**Explosive ability:** Cannot be detonated.
**Percentage:** ***51% red phosphorus, 35% pyrolusite, 8% magnesium, 3% zinc oxide, 3% epoxy binder***
**Classification:** Deflagrating mixture (classified as 1.3G pyrotechnic composition).

**Danger of preparation:** Can be made by amateurs.
**Use:** Used in professional industrial/commercial smoke munitions for determining gas leaks, spotting, signaling, and in military operations.

## 06-03-010A: Military grade white smoke generating composition for camouflage purposes utilizing the "wet mixing" process:

Into a standard vertical mixer, equipped with flexible Teflon blades, place ***15 grams of Viton A fluorocarbon rubber*** (available from E. I. Dupont de Nemours Company), followed by ***250 grams of red phosphorus***, followed by ***15 grams of finely divided boron***, and then followed by ***220 grams of calcium sulfate***. Thereafter, blend the mixture on high speed for about 1 hour at room temperature. After 1 hour, the mixture is ready for pressing. To do so, press the mixture into any suitable delivery container, such as a grenade, metal tube, plastic tube, cardboard tube, ceramic or clay ball, ect., under a pressure of 8,000 psi, and then cure the device at room temperature for 24 hours. In some cases, the smoke mixture needs to be pressed into blocks, discs, or tablets under a pressure of 9,000 psi, for use in some types of smoke grenades, followed by curing at room temperature for 24 hours. Note: an appropriate igniter composition to begin the burn of this smoke composition should be used.
**Burn rate:** 25 seconds per inch at 8,000 psi; tends to vary based on diameter and density of smoke grain.
**Smoke Volume on average (500 gram solid grain sample enclosed area):** 1800 sq. ft.
**Smoke Volume on average (500 gram solid grain sample open area):** 7400 sq. ft.
**Smoke visibility naked eye (sunny/dry conditions day time; 500 gram solid grain sample):** 7.2 km
**Smoke visibility naked eye (wet/rainy conditions day time; 500 gram solid grain sample):** 3.3 km
**Smoke visibility naked eye (night time, average environmental conditions; 500 gram solid grain sample):** 2900 meters.
**Water resistance:** Very good.
**Stability:** Can be stored for decades.
**Flammability (solid grain):** 3.9
**Flammability (powdered grain):** 4
**Ease of ignition (solid grain):** 3.2 (requires high temperature ignition composition for proper burn).
**Tendency to cake:** Will not cake.
**Explosive ability:** Cannot be detonated.
**Percentage:** ***50% red phosphorus, 44% calcium sulfate, 3% Viton A rubber binder, 3% boron***
**Classification:** Deflagrating mixture (classified as 1.3G pyrotechnic composition).

**Danger of preparation:** Can be made by amateurs.
**Use:** Used in military operations for spotting, signaling, concealment, and camouflage operations.

## 06-03-011A: Environmentally safe white smoke generating composition:

Into a standard ball mill, filled with 500 grams of steel shot of 10mm diameter, place ***100 grams of standard magnesium powder***, followed by ***60 grams of potassium bicarbonate***, followed by ***75 grams of potassium chloride***, followed by ***75 grams of sodium chloride***, followed by ***50 grams of dicyandiamide***, followed by ***100 grams of potassium nitrate***, and then followed by ***40 grams potassium perchlorate***. Thereafter, tumble the mixture at 120 RPM for about 45 minutes. After 45 minutes, the mixture is ready for pressing. To do so, first remove the steel shot, and then press the mixture into any suitable delivery container, such as a grenade, metal tube, plastic tube, cardboard tube, ceramic or clay ball, ect., under a pressure of 7,500 psi, and then cure the device at room temperature for 24 hours. In some cases, the smoke mixture needs to be pressed into blocks, discs, or tablets under a pressure of 9,000 psi, for use in some types of smoke grenades, followed by curing at room temperature for 24 hours. Note: an appropriate igniter composition to begin the burn of this smoke composition should be used.

**Burn rate:** 27 seconds per inch at 9,000 psi; tends to vary based on diameter and density of smoke grain.
**Smoke Volume on average (500 gram solid grain sample enclosed area):** 1780 sq. ft.
**Smoke Volume on average (500 gram solid grain sample open area):** 7350 sq. ft.
**Smoke visibility naked eye (sunny/dry conditions day time; 500 gram solid grain sample):** 7 km
**Smoke visibility naked eye (wet/rainy conditions day time; 500 gram solid grain sample):** 3 km
**Smoke visibility naked eye (night time, average environmental conditions; 500 gram solid grain sample):** 2500 meters.
**Water resistance:** Very good.
**Stability:** Can be stored for decades.
**Flammability (solid grain):** 3.7
**Flammability (powdered grain):** 4
**Ease of ignition (solid grain):** 3.3 (requires high temperature ignition composition for proper burn).
**Tendency to cake:** Will not cake.
**Explosive ability:** Cannot be detonated.
**Percentage:** ***20% magnesium, 20% potassium nitrate, 15% potassium chloride, 15% sodium chloride, 12% potassium bicarbonate, 10% dicyandiamide, 8% potassium perchlorate***
**Classification:** Deflagrating mixture (classified as 1.3G pyrotechnic composition).

**Danger of preparation:** Can be made by amateurs.
**Use:** Used in military operations for spotting, signaling, concealment, and camouflage operations.

### 06-03-012A: Opaque smoke generating composition for interfering with and blocking infrared targeting:

Into a standard ball mill, filled with 500 grams of steel shot of 10mm diameter, place ***100 grams of standard magnesium powder***, followed by ***50 grams of vinylidine polyfluoride binder***, followed by ***400 grams of hexachlorobenzene***, and then followed by ***10 grams of naphthalene***. Thereafter, tumble the mixture at 150 RPM for about 90 minutes. After 90 minutes, the mixture is ready for pressing. To do so, first remove the steel shot, and then press the mixture into any suitable delivery container, such as a grenade, metal tube, plastic tube, cardboard tube, ceramic or clay ball, ect., under a pressure of 8,000 psi, and then cure the device at room temperature for 24 hours. In some cases, the smoke mixture needs to be pressed into blocks, discs, or tablets under a pressure of 9,000 psi, for use in some types of smoke devices, followed by curing at room temperature for 24 hours. Note: an appropriate igniter composition to begin the burn of this smoke composition should be used.
**Burn rate:** 25 seconds per inch at 9,000 psi; tends to vary based on diameter and density of smoke grain.
**Smoke Volume on average (500 gram solid grain sample enclosed area):** 1600 sq. ft.
**Smoke Volume on average (500 gram solid grain sample open area):** 7000 sq. ft.
**Smoke visibility naked eye (sunny/dry conditions day time; 500 gram solid grain sample):** 7.1 km
**Smoke visibility naked eye (wet/rainy conditions day time; 500 gram solid grain sample):** 3.1 km
**Smoke visibility naked eye (night time, average environmental conditions; 500 gram solid grain sample):** 2595 meters.
**Max effective IR blocking volume (average environmental conditions; 500 gram solid grain sample):** 341 sq. ft. for 101 seconds.
**Water resistance:** Very good.
**Stability:** Can be stored for decades.
**Flammability (solid grain):** 3.6
**Flammability (powdered grain):** 3.9
**Ease of ignition (solid grain):** 3.1 (requires high temperature ignition composition for proper burn).
**Tendency to cake:** Will not cake.
**Explosive ability:** Cannot be detonated.
**Percentage:** ***71.4% hexachlorobenzene, 17.8% magnesium, 8.9% vinylidine polyfluoride binder, 1.7% naphthalene, 0.2% residue***
**Classification:** Deflagrating mixture (classified as 1.3G pyrotechnic composition).

**Danger of preparation:** Can be made by amateurs.
**Use:** Used in masking vehicles and tanks from infrared-guided weapons.

### 06-03-012B: Opaque smoke generating composition for interfering with and blocking infrared targeting (modified):

Into a standard ball mill, filled with 500 grams of steel shot of 10mm diameter, place ***92.5 grams of standard magnesium powder***, followed by ***100 grams of vinylidine polyfluoride binder*** followed by ***307.5 grams of hexachloroethane***, followed by ***150 grams of naphthalene***, and then followed by ***100 grams of chlorinated paraffin***. Thereafter, tumble the mixture at 150 RPM for about 90 minutes. After 90 minutes, the mixture is ready for pressing. To do so, first remove the steel shot, and then press the mixture into any suitable delivery container, such as a grenade, metal tube, plastic tube, cardboard tube, ceramic or clay ball, ect., under a pressure of 8,000 psi, and then cure the device at room temperature for 24 hours. In some cases, the smoke mixture needs to be pressed into blocks, discs, or tablets under a pressure of 9,000 psi, for use in some types of smoke devices, followed by curing at room temperature for 24 hours. Note: an appropriate igniter composition to begin the burn of this smoke composition should be used.

**Burn rate:** 27 seconds per inch at 9,000 psi; tends to vary based on diameter and density of smoke grain.
**Smoke Volume on average (500 gram solid grain sample enclosed area):** 1590 sq. ft.
**Smoke Volume on average (500 gram solid grain sample open area):** 7005 sq. ft.
**Smoke visibility naked eye (sunny/dry conditions day time; 500 gram solid grain sample):** 7 km
**Smoke visibility naked eye (wet/rainy conditions day time; 500 gram solid grain sample):** 3 km
**Smoke visibility naked eye (night time, average environmental conditions; 500 gram solid grain sample):** 2555 meters.
**Max effective IR blocking volume (average environmental conditions; 500 gram solid grain sample):** 350 sq. ft. for 112 seconds.
**Water resistance:** Very good.
**Stability:** Can be stored for decades.
**Flammability (solid grain):** 3.58
**Flammability (powdered grain):** 3.88
**Ease of ignition (solid grain):** 3 (requires high temperature ignition composition for proper burn).
**Tendency to cake:** Will not cake.
**Explosive ability:** Cannot be detonated.
**Percentage:** ***41% hexachloroethane, 20% naphthalene, 13.3% vinylidine polyfluoride, 13.3% chlorinated paraffin, 12.3% magnesium, 0.1% residue***
**Classification:** Deflagrating mixture (classified as 1.3G pyrotechnic composition).

**Danger of preparation:** Can be made by amateurs.
**Use:** Used in masking vehicles and tanks from infrared-guided weapons.

### 06-03-013A: Improved "HC" white smoke composition (standard US military smoke composition):

Into a standard ball mill, filled with 500 grams of steel shot of 10mm diameter, place ***236 grams of freshly calcined zinc oxide*** (can be prepared by roasting finely divided zinc oxide at 900 Celsius for several hours), followed by ***27.5 grams of standard powdered aluminum***, and then followed by ***236 grams of hexachloroethane***. Thereafter, tumble the mixture at 125 RPM for about 50 minutes. After 50 minutes, the mixture is ready for pressing. To do so, first remove the steel shot, and then press the mixture into any suitable delivery container, such as a grenade, metal tube, plastic tube, cardboard tube, ceramic or clay ball, ect., under a pressure of 8,000 psi, and then cure the device at room temperature for 24 hours. In some cases, the smoke mixture needs to be pressed into blocks, discs, or tablets under a pressure of 9,000 psi, for use in some types of smoke grenades, followed by curing at room temperature for 24 hours. Note: an appropriate igniter composition to begin the burn of this smoke composition should be used.
**Burn rate:** 25 seconds per inch at 9,000 psi; tends to vary based on diameter and density of smoke grain.
**Smoke Volume on average (500 gram solid grain sample enclosed area):** 1805 sq. ft.
**Smoke Volume on average (500 gram solid grain sample open area):** 7440 sq. ft.
**Smoke visibility naked eye (sunny/dry conditions day time; 500 gram solid grain sample):** 7.7 km
**Smoke visibility naked eye (wet/rainy conditions day time; 500 gram solid grain sample):** 3.98 km
**Smoke visibility naked eye (night time, average environmental conditions; 500 gram solid grain sample):** 2980 meters.
**Water resistance:** Very good.
**Stability:** Can be stored for decades.
**Flammability (solid grain):** 3.9
**Flammability (powdered grain):** 4.4
**Ease of ignition (solid grain):** 3.1 (requires high temperature ignition composition for proper burn).
**Tendency to cake:** Will not cake.
**Explosive ability:** Cannot be detonated.
**Percentage:** ***47.2% zinc oxide, 47.2% hexachloroethane, 5.5% aluminum, 0.1% residue***
**Classification:** Deflagrating mixture (classified as 1.3G pyrotechnic composition).

**Danger of preparation:** Can be made by amateurs.
**Use:** Used in military operations for spotting, signaling, concealment, and camouflage operations.

### 06-03-014A: Standard white smoke composition utilizing guanidine nitrate:

Into a standard ball mill, filled with 500 grams of steel shot of 10mm diameter, place ***110 grams of standard powdered zinc***, followed by ***40 grams of zinc oxide***, followed by ***170 grams of hexachloroethane***, and then followed by ***180 grams of guanidine nitrate***. Thereafter, tumble the mixture at 130 RPM for about 45 minutes. After 45 minutes, the mixture is ready for pressing. To do so, first remove the steel shot, and then press the mixture into any suitable delivery container, such as a grenade, metal tube, plastic tube, cardboard tube, ceramic or clay ball, ect., under a pressure of 8,000 psi, and then cure the device at room temperature for 24 hours. In some cases, the smoke mixture needs to be pressed into blocks, discs, or tablets under a pressure of 9,000 psi, for use in some types of smoke grenades, followed by curing at room temperature for 24 hours. Note: an appropriate igniter composition to begin the burn of this smoke composition should be used.
**Burn rate:** 22 seconds per inch at 9,000 psi; tends to vary based on diameter and density of smoke grain.

**Smoke Volume on average (500 gram solid grain sample enclosed area):** 1770 sq. ft.
**Smoke Volume on average (500 gram solid grain sample open area):** 7121 sq. ft.
**Smoke visibility naked eye (sunny/dry conditions day time; 500 gram solid grain sample):** 7.1 km
**Smoke visibility naked eye (wet/rainy conditions day time; 500 gram solid grain sample):** 3.21 km
**Smoke visibility naked eye (night time, average environmental conditions; 500 gram solid grain sample):** 2200 meters.
**Water resistance:** Very good.
**Stability:** Can be stored for decades.
**Flammability (solid grain):** 4.2
**Flammability (powdered grain):** 5.6
**Ease of ignition (solid grain):** 3.7 (requires high temperature ignition composition for proper burn).
**Tendency to cake:** Will not cake.
**Explosive ability:** Cannot be detonated.
**Percentage:** ***36% guanidine nitrate, 34% hexachloroethane, 22% zinc, 8% zinc oxide***
**Classification:** Deflagrating mixture (classified as 1.3G pyrotechnic composition).

**Danger of preparation:** Can be made by amateurs.
**Use:** Used in military operations for spotting, signaling, concealment, and camouflage operations.

### 06-03-014B: Standard white smoke composition utilizing guanidine nitrate (modified):

Into a standard ball mill, filled with 500 grams of steel shot of 10mm diameter, place ***280 grams of red phosphorus***, followed by ***20 grams of hexachloroethane***, and then followed by ***200 grams of guanidine nitrate***. Thereafter, tumble the mixture at 130 RPM for about 45 minutes. After 45 minutes, the mixture is ready for pressing. To do so, first remove the steel shot, and then press the mixture into any suitable delivery container, such as a grenade, metal tube, plastic tube, cardboard tube, ceramic or clay ball, ect., under a pressure of 8,000 psi, and then cure the device at room temperature for 24 hours. In some cases, the smoke mixture needs to be pressed into blocks, discs, or tablets under a pressure of 9,000 psi, for use in some types of smoke grenades, followed by curing at room temperature for 24 hours. Note: an appropriate igniter composition to begin the burn of this smoke composition should be used.
**Burn rate:** 23 seconds per inch at 9,000 psi; tends to vary based on diameter and density of smoke grain.
**Smoke Volume on average (500 gram solid grain sample enclosed area):** 1790 sq. ft.
**Smoke Volume on average (500 gram solid grain sample open area):** 7190 sq. ft.
**Smoke visibility naked eye (sunny/dry conditions day time; 500 gram solid grain sample):** 7.19 km
**Smoke visibility naked eye (wet/rainy conditions day time; 500 gram solid grain sample):** 3.21 km
**Smoke visibility naked eye (night time, average environmental conditions; 500 gram solid grain sample):** 2295 meters.
**Water resistance:** Very good.
**Stability:** Can be stored for decades.
**Flammability (solid grain):** 4.2
**Flammability (powdered grain):** 5.6
**Ease of ignition (solid grain):** 3.6 (requires high temperature ignition composition for proper burn).
**Tendency to cake:** Will not cake.
**Explosive ability:** Cannot be detonated.
**Percentage:** ***56% red phosphorus, 40% guanidine nitrate, 4% hexachloroethane***
**Classification:** Deflagrating mixture (classified as 1.3G pyrotechnic composition).

**Danger of preparation:** Can be made by amateurs.
**Use:** Used in military operations for spotting, signaling, concealment, and camouflage operations.

### 06-03-015A: Opaque smoke generating composition for interfering with infrared targeting systems (modified):

Into a standard ball mill, filled with 500 grams of steel shot of 10mm diameter, place ***50 grams of neoprene binder,*** followed by ***155 grams of hexachloroethane***, followed by ***60 grams of zinc oxide***, followed by ***155 grams of standard zinc powder***, and then followed by ***80 grams of potassium perchlorate***. Thereafter, tumble the mixture at 150 RPM for about 90 minutes. After 90 minutes, the mixture is ready for pressing. To do so, first remove the steel shot, and then press the mixture into any suitable delivery container, such as a grenade, metal tube, plastic tube, cardboard tube, ceramic or clay ball, ect., under a pressure of 8,000 psi, and then cure the device at room temperature for 24 hours. In some cases, the smoke mixture needs to be pressed into blocks, discs, or tablets under a pressure of 9,000 psi, for use in some types of smoke devices, followed by curing at room temperature for 24 hours. Note: an appropriate igniter composition to begin the burn of this smoke composition should be used.
**Burn rate:** 25 seconds per inch at 9,000 psi; tends to vary based on diameter and density of smoke grain.
**Smoke Volume on average (500 gram solid grain sample enclosed area):** 1610 sq. ft.
**Smoke Volume on average (500 gram solid grain sample open area):** 7095 sq. ft.
**Smoke visibility naked eye (sunny/dry conditions day time; 500 gram solid grain sample):** 7.1 km

**Smoke visibility naked eye (wet/rainy conditions day time; 500 gram solid grain sample):** 3.2 km
**Smoke visibility naked eye (night time, average environmental conditions; 500 gram solid grain sample):** 2578 meters.
**Max effective IR blocking volume (average environmental conditions; 500 gram solid grain sample):** 349 sq. ft. for 115 seconds.
**Water resistance:** Very good.
**Stability:** Can be stored for decades.
**Flammability (solid grain):** 3.9
**Flammability (powdered grain):** 4.8
**Ease of ignition (solid grain):** 3.9 (requires high temperature ignition composition for proper burn).
**Tendency to cake:** Will not cake.
**Explosive ability:** Cannot be detonated.
**Percentage:** ***31% hexachloroethane, 31% zinc, 16% potassium perchlorate, 12% zinc oxide, 10% neoprene binder***
**Classification:** Deflagrating mixture (classified as 1.3G pyrotechnic composition).

**Danger of preparation:** Can be made by amateurs.
**Use:** Used in masking vehicles and tanks from infrared-guided weapons.

## 06-03-015B: Opaque smoke generating composition for interfering with infrared targeting systems (modified):

Into a standard ball mill, filled with 500 grams of steel shot of 10mm diameter, place ***25 grams of neoprene binder,*** followed by ***350 grams of hexachlorobenzene***, followed by ***50 grams of naphthalene,*** and then followed by ***100 grams of standard magnesium powder***. Thereafter, tumble the mixture at 150 RPM for about 90 minutes. After 90 minutes, the mixture is ready for pressing. To do so, first remove the steel shot, and then press the mixture into any suitable delivery container, such as a grenade, metal tube, plastic tube, cardboard tube, ceramic or clay ball, ect., under a pressure of 8,000 psi, and then cure the device at room temperature for 24 hours. In some cases, the smoke mixture needs to be pressed into blocks, discs, or tablets under a pressure of 9,000 psi, for use in some types of smoke devices, followed by curing at room temperature for 24 hours. Note: an appropriate igniter composition to begin the burn of this smoke composition should be used.
**Burn rate:** 26 seconds per inch at 9,000 psi; tends to vary based on diameter and density of smoke grain.
**Smoke Volume on average (500 gram solid grain sample enclosed area):** 1699 sq. ft.
**Smoke Volume on average (500 gram solid grain sample open area):** 7105 sq. ft.
**Smoke visibility naked eye (sunny/dry conditions day time; 500 gram solid grain sample):** 7.19 km
**Smoke visibility naked eye (wet/rainy conditions day time; 500 gram solid grain sample):** 3.5 km
**Smoke visibility naked eye (night time, average environmental conditions; 500 gram solid grain sample):** 2605 meters.
**Max effective IR blocking volume (average environmental conditions; 500 gram solid grain sample):** 340 sq. ft. for 110 seconds.
**Water resistance:** Very good.
**Stability:** Can be stored for decades.
**Flammability (solid grain):** 3.98
**Flammability (powdered grain):** 3.89
**Ease of ignition (solid grain):** 3.5 (requires high temperature ignition composition for proper burn).
**Tendency to cake:** Will not cake.
**Explosive ability:** Cannot be detonated.
**Percentage:** ***66.6% hexachlorobenzene, 19% magnesium, 9.5% naphthalene, 4.7% neoprene binder, 0.20% residue***
**Classification:** Deflagrating mixture (classified as 1.3G pyrotechnic composition).

**Danger of preparation:** Can be made by amateurs.
**Use:** Used in masking vehicles and tanks from infrared-guided weapons.

## 06-03-016A: Standard white smoke generating composition for general use:

Into a standard ball mill, filled with 500 grams of steel shot of 10mm diameter, place ***240 grams of any standard viscous binder such as natural rubber***, followed by ***220 grams of finely divided PVC*** of 150 to 250 mesh, followed by ***330 grams of calcined zinc oxide***, followed by ***220 grams of ammonium chloride***, followed by ***264 grams of thiourea***, and then followed by ***726 grams of ammonium perchlorate***. Thereafter, tumble the mixture at 130 RPM for about 45 minutes. After 45 minutes, the mixture is ready for pressing. To do so, first remove the steel shot, and then press the mixture into any suitable delivery container, such as a grenade, metal tube, plastic tube, cardboard tube, ceramic or clay ball, ect., under a pressure of 8,000 psi, and then cure the device at room temperature for 24 hours. In some cases, the smoke mixture needs to be pressed into blocks, discs, or tablets under a pressure of 9,000 psi, for use in some types of smoke grenades, followed by curing at room temperature for 24 hours. Note: an appropriate igniter composition to begin the burn of this smoke composition should be used.
**Burn rate:** 20 seconds per inch at 9,000 psi; tends to vary based on diameter and density of smoke grain.
**Smoke Volume on average (500 gram solid grain sample enclosed area):** 1800 sq. ft.
**Smoke Volume on average (500 gram solid grain sample open area):** 7209 sq. ft.
**Smoke visibility naked eye (sunny/dry conditions day time; 500 gram solid grain sample):** 7.45 km

**Smoke visibility naked eye (wet/rainy conditions day time; 500 gram solid grain sample):** 3.39 km
**Smoke visibility naked eye (night time, average environmental conditions; 500 gram solid grain sample):** 2650 meters.
**Water resistance:** Very good.
**Stability:** Can be stored for decades.
**Flammability (solid grain):** 4.1
**Flammability (powdered grain):** 5.2
**Ease of ignition (solid grain):** 3.98 (requires high temperature ignition composition for proper burn).
**Tendency to cake:** Will not cake.
**Explosive ability:** Cannot be detonated.
**Percentage:** ***36.3% ammonium perchlorate, 16.5% zinc oxide, 13.2% thiourea, 12% rubber binder, 11% PVC, 11% ammonium chloride***
**Classification:** Deflagrating mixture (classified as 1.3G pyrotechnic composition).

**Danger of preparation:** Can be made by amateurs.
**Use:** Used in military operations for spotting, signaling, concealment, and camouflage operations.

### 06-03-017A: Standard blue smoke generating composition for general use:

Into a standard ball mill, filled with 500 grams of steel shot of 10mm diameter, place ***80 grams of chloroparaffin***, followed by ***120 grams of standard magnesium powder***, followed by ***90 grams of Prussian blue dye***, followed by ***239 grams of amorphous elemental boron powder***, followed by ***318 grams of potassium nitrate,*** and then followed by ***105 grams of finely divided wood charcoal***. Thereafter, tumble the mixture at 130 RPM for about 90 minutes. After 90 minutes, the mixture is ready for pressing. To do so, first remove the steel shot, and then press the mixture into any suitable delivery container, such as a grenade, metal tube, plastic tube, cardboard tube, ceramic or clay ball, ect., under a pressure of 8,000 psi, and then cure the device at room temperature for 24 hours. In some cases, the smoke mixture needs to be pressed into blocks, discs, or tablets under a pressure of 9,000 psi, for use in some types of smoke grenades, followed by curing at room temperature for 24 hours. Note: an appropriate igniter composition to begin the burn of this smoke composition should be used.
**Burn rate:** 23 seconds per inch at 9,000 psi; tends to vary based on diameter and density of smoke grain.
**Smoke Volume on average (500 gram solid grain sample enclosed area):** 1670 sq. ft.
**Smoke Volume on average (500 gram solid grain sample open area):** 7180 sq. ft.
**Smoke visibility naked eye (sunny/dry conditions day time; 500 gram solid grain sample):** 7.3 km
**Smoke visibility naked eye (wet/rainy conditions day time; 500 gram solid grain sample):** 3.41 km
**Smoke visibility naked eye (night time, average environmental conditions; 500 gram solid grain sample):** 2700 meters.
**Water resistance:** Very good.
**Stability:** Can be stored for decades.
**Flammability (solid grain):** 4.3
**Flammability (powdered grain):** 5.4
**Ease of ignition (solid grain):** 3.69 (requires high temperature ignition composition for proper burn).
**Tendency to cake:** Will not cake.
**Explosive ability:** Cannot be detonated.
**Percentage:** ***33.4% potassium nitrate, 25.1% amorphous boron, 12.6% magnesium, 11% wood charcoal, 9.4% Prussian blue dye, 8.4% chloroparaffin, 0.1% moisture***
**Classification:** Deflagrating mixture (classified as 1.3G pyrotechnic composition).

**Danger of preparation:** Can be made by amateurs.
**Use:** Used in military operations for spotting, signaling, concealment, and camouflage operations.

### 06-03-018A: White spotting smoke composition for artillery and mortar training and similar use:

Into a standard ball mill, filled with 500 grams of steel shot of 10mm diameter, place ***200 grams of standard mesh zinc dust***, followed by ***100 grams of potassium nitrate***, followed by ***100 grams of standard atomized aluminum powder***, and then followed by ***100 grams of potassium perchlorate***. Thereafter, tumble the mixture at 145 RPM for about 90 minutes. After 90 minutes, the mixture is ready for pressing. To do so, first remove the steel shot, and then press the mixture into any suitable delivery container, such as a grenade, metal tube, plastic tube, cardboard tube, mortar shell, or artillery shell under a pressure of 8,500 psi, and then cure the device at room temperature for 24 hours. In some cases, the smoke mixture needs to be pressed into blocks, discs, or tablets under a pressure of 9,000 psi, for use in some types of mortar or artillery shells, followed by curing at room temperature for 24 hours. Note: an appropriate igniter composition to begin the burn of this smoke composition should be used.
**Burn rate:** 17 seconds per inch at 9,000 psi; tends to vary based on diameter and density of smoke grain.
**Smoke Volume on average (500 gram solid grain sample enclosed area):** 1500 sq. ft.
**Smoke Volume on average (500 gram solid grain sample open area):** 6,187 sq. ft.
**Smoke visibility naked eye (sunny/dry conditions day time; 500 gram solid grain sample):** 8.1 km
**Smoke visibility naked eye (wet/rainy conditions day time; 500 gram solid grain sample):** 4.5 km

**Smoke visibility naked eye (night time, average environmental conditions; 500 gram solid grain sample):** 3900 meters.
**Water resistance:** Very good.
**Stability:** Can be stored for decades.
**Flammability (solid grain):** 6.2
**Flammability (powdered grain):** 7.4
**Ease of ignition (solid grain):** 5.9
**Tendency to cake:** Will not cake.
**Explosive ability:** Cannot be detonated.
**Percentage:** ***40% zinc, 20% potassium nitrate, 20% aluminum, 20% potassium perchlorate***
**Classification:** Deflagrating mixture (classified as 1.3G pyrotechnic composition).

**Danger of preparation:** Can be made by amateurs.
**Use:** Used in spotting for artillery and mortars.

### 06-03-019A: Orange smoke producing composition:

Into a standard ball mill, filled with 500 grams of steel shot of 10mm diameter, place ***85.5 grams of 1,8-diaminonapthalene***, followed by ***79.5 grams phthalic anhydride***, followed by ***119.5 grams of sucrose***, and then followed by ***214 grams of potassium chlorate***. Thereafter, tumble the mixture at 100 RPM for about 90 minutes. After 90 minutes, the mixture is ready for pressing. To do so, first remove the steel shot, and then press the mixture into any suitable delivery container, such as a grenade, metal tube, plastic tube, cardboard tube, ceramic or clay ball, ect., under a pressure of 8,000 psi, and then cure the device at room temperature for 24 hours. In some cases, the smoke mixture needs to be pressed into blocks, discs, or tablets under a pressure of 9,000 psi, for use in some types of smoke grenades, followed by curing at room temperature for 24 hours. Note: an appropriate igniter composition to begin the burn of this smoke composition should be used.
**Burn rate:** 21 seconds per inch at 9,000 psi; tends to vary based on diameter and density of smoke grain.
**Smoke Volume on average (500 gram solid grain sample enclosed area):** 1699 sq. ft.
**Smoke Volume on average (500 gram solid grain sample open area):** 7020 sq. ft.
**Smoke visibility naked eye (sunny/dry conditions day time; 500 gram solid grain sample):** 7.49 km
**Smoke visibility naked eye (wet/rainy conditions day time; 500 gram solid grain sample):** 3.71 km
**Smoke visibility naked eye (night time, average environmental conditions; 500 gram solid grain sample):** 2800 meters.
**Water resistance:** Very good.
**Stability:** Can be stored for decades.
**Flammability (solid grain):** 4.9
**Flammability (powdered grain):** 5.87
**Ease of ignition (solid grain):** 3.99 (requires high temperature ignition composition for proper burn).
**Tendency to cake:** Will not cake.
**Explosive ability:** Cannot be detonated.
**Percentage:** ***42.9% potassium chlorate, 23.9% sucrose, 17.1% 1,8-diaminonapthalene coloring agent, 15.9% phthalic anhydride, 0.2% residue***
**Classification:** Deflagrating mixture (classified as 1.3G pyrotechnic composition).

**Danger of preparation:** Can be made by amateurs.
**Use:** Used in military operations for spotting, signaling, concealment, and camouflage operations.

### 06-03-019B: Orange smoke producing composition (modified):

Into a standard ball mill, filled with 500 grams of steel shot of 10mm diameter, place ***104.5 grams of 1,8-diaminonapthalene***, followed by ***97 grams phthalic anhydride***, followed by ***69.5 grams of glycerol***, followed by ***29.5 grams of polyethylene glycol*** (molecular weight of at least 100,000), and then followed by ***199.5 grams of potassium chlorate***. Thereafter, tumble the mixture at room temperature at 150 RPM for about 45 minutes. After 45 minutes, the mixture is ready for pressing. To do so, first remove the steel shot, and then press the mixture into any suitable delivery container, such as a grenade, metal tube, plastic tube, cardboard tube, ceramic or clay ball, ect., under a pressure of 7,000 psi, and then cure the device in an oven at 60 Celsius for 6 hours. In some cases, the smoke mixture needs to be pressed into blocks, discs, or tablets under a pressure of 7,000 psi, for use in some types of smoke grenades, followed by curing at 60 Celsius for 6 hours. Note: an appropriate igniter composition to begin the burn of this smoke composition should be used.
**Burn rate:** 24 seconds per inch at 7,000 psi; tends to vary based on diameter and density of smoke grain.
**Smoke Volume on average (500 gram solid grain sample enclosed area):** 1700 sq. ft.
**Smoke Volume on average (500 gram solid grain sample open area):** 7198 sq. ft.
**Smoke visibility naked eye (sunny/dry conditions day time; 500 gram solid grain sample):** 7.48 km
**Smoke visibility naked eye (wet/rainy conditions day time; 500 gram solid grain sample):** 3.49 km
**Smoke visibility naked eye (night time, average environmental conditions; 500 gram solid grain sample):** 3010 meters.
**Water resistance:** Very good.

**Stability:** Can be stored for decades.
**Flammability (solid grain):** 4.3
**Flammability (powdered grain):** 5
**Ease of ignition (solid grain):** 4.19 (requires high temperature ignition composition for proper burn).
**Tendency to cake:** Will not cake.
**Explosive ability:** Cannot be detonated.
**Percentage:** ***39.9% potassium chlorate, 20.9% 1,8-diaminonapthalene, 19.4% phthalic anhydride, 13.9% glycerol, 5.9% polyethylene glycol***
**Classification:** Deflagrating mixture (classified as 1.3G pyrotechnic composition).

**Danger of preparation:** Can be made by amateurs.
**Use:** Used in military operations for spotting, signaling, concealment, and camouflage operations.

### 06-03-019C: Orange smoke producing composition (modified 2):

Into a standard ball mill, filled with 500 grams of steel shot of 10mm diameter, place ***85 grams of 1,8-diaminonapthalene***, followed by ***55 grams phthalic anhydride***, followed by ***50 grams of glycerol***, followed by ***55 grams of Epon 828 epoxy resin***, followed by ***40 grams of sodium barbiturate***, followed by ***60 grams of N-hexyl acid phthalate***, and finally followed by ***155 grams of potassium chlorate***. Thereafter, tumble the mixture at room temperature at 190 RPM for about 45 minutes. After 45 minutes, the mixture is ready for pressing. To do so, first remove the steel shot, and then press the mixture into any suitable delivery container, such as a grenade, metal tube, plastic tube, cardboard tube, ceramic or clay ball, ect., under a pressure of 7,000 psi, and then cure the device in an oven at 60 Celsius for 6 hours. In some cases, the smoke mixture needs to be pressed into blocks, discs, or tablets under a pressure of 7,000 psi, for use in some types of smoke grenades, followed by curing at 60 Celsius for 6 hours. Note: an appropriate igniter composition to begin the burn of this smoke composition should be used.
**Burn rate:** 22 seconds per inch at 7,000 psi; tends to vary based on diameter and density of smoke grain.
**Smoke Volume on average (500 gram solid grain sample enclosed area):** 1695 sq. ft.
**Smoke Volume on average (500 gram solid grain sample open area):** 7160 sq. ft.
**Smoke visibility naked eye (sunny/dry conditions day time; 500 gram solid grain sample):** 7.4 km
**Smoke visibility naked eye (wet/rainy conditions day time; 500 gram solid grain sample):** 3.3 km
**Smoke visibility naked eye (night time, average environmental conditions; 500 gram solid grain sample):** 3000 meters.
**Water resistance:** Very good.
**Stability:** Can be stored for decades.
**Flammability (solid grain):** 4.1
**Flammability (powdered grain):** 4.5
**Ease of ignition (solid grain):** 4 (requires high temperature ignition composition for proper burn).
**Tendency to cake:** Will not cake.
**Explosive ability:** Cannot be detonated.
**Percentage:** ***31% potassium chlorate, 17% 1,8-diaminonapthalene, 12% N-hexyl acid phthalate, 11% phthalic anhydride, 11% Epon 828 epoxy resin, 10% glycerol, 8% sodium barbiturate***
**Classification:** Deflagrating mixture (classified as 1.3G pyrotechnic composition).

**Danger of preparation:** Can be made by amateurs.
**Use:** Used in military operations for spotting, signaling, concealment, and camouflage operations.

### 06-03-019D: Yellow smoke producing composition:

Into a standard ball mill, filled with 500 grams of steel shot of 10mm diameter, place ***170 grams of para-phenylene diamine dye***, followed by ***110 grams of finely powdered sucrose***, and then followed by ***220 grams of potassium chlorate***. Thereafter, tumble the mixture at room temperature at 190 RPM for about 45 minutes. After 45 minutes, the mixture is ready for pressing. To do so, first remove the steel shot, and then press the mixture into any suitable delivery container, such as a grenade, metal tube, plastic tube, cardboard tube, ceramic or clay ball, ect., under a pressure of 7,000 psi, and then cure the device in an oven at 60 Celsius for 6 hours. In some cases, the smoke mixture needs to be pressed into blocks, discs, or tablets under a pressure of 7,000 psi, for use in some types of smoke grenades, followed by curing at 60 Celsius for 6 hours. Note: an appropriate igniter composition to begin the burn of this smoke composition should be used.
**Burn rate:** 21 seconds per inch at 7,000 psi; tends to vary based on diameter and density of smoke grain.
**Smoke Volume on average (500 gram solid grain sample enclosed area):** 1690 sq. ft.
**Smoke Volume on average (500 gram solid grain sample open area):** 7170 sq. ft.
**Smoke visibility naked eye (sunny/dry conditions day time; 500 gram solid grain sample):** 7.3 km
**Smoke visibility naked eye (wet/rainy conditions day time; 500 gram solid grain sample):** 3.2 km
**Smoke visibility naked eye (night time, average environmental conditions; 500 gram solid grain sample):** 2890 meters.
**Water resistance:** Very good.

**Stability:** Can be stored for decades.
**Flammability (solid grain):** 4.2
**Flammability (powdered grain):** 5
**Ease of ignition (solid grain):** 4.4 (requires high temperature ignition composition for proper burn).
**Tendency to cake:** Will not cake.
**Explosive ability:** Cannot be detonated.
**Percentage:** ***44% potassium chlorate, 34% para-phenylene diamine dye, 22% sucrose***
**Classification:** Deflagrating mixture (classified as 1.3G pyrotechnic composition).

**Danger of preparation:** Can be made by amateurs.
**Use:** Used in professional industrial/commercial smoke munitions for determining gas leaks, spotting, signaling, and in military operations.

### 06-03-019E: Violet smoke producing composition:

Into a standard ball mill, filled with 500 grams of steel shot of 10mm diameter, place ***150 grams of 1,8-diaminonapthalene***, followed by ***35 grams of finely powdered sucrose***, followed by ***150 grams of pyromellitic dianhydride***, and then followed by ***165 grams of potassium chlorate***. Thereafter, tumble the mixture at room temperature at 150 RPM for about 45 minutes. After 45 minutes, the mixture is ready for pressing. To do so, first remove the steel shot, and then press the mixture into any suitable delivery container, such as a grenade, metal tube, plastic tube, cardboard tube, ceramic or clay ball, ect., under a pressure of 7,000 psi, and then cure the device in an oven at 40 Celsius for 6 hours. In some cases, the smoke mixture needs to be pressed into blocks, discs, or tablets under a pressure of 7,000 psi, for use in some types of smoke grenades, followed by curing at 40 Celsius for 6 hours. Note: an appropriate igniter composition to begin the burn of this smoke composition should be used.
**Burn rate:** 21 seconds per inch at 7,000 psi; tends to vary based on diameter and density of smoke grain.
**Smoke Volume on average (500 gram solid grain sample enclosed area):** 1700 sq. ft.
**Smoke Volume on average (500 gram solid grain sample open area):** 7165 sq. ft.
**Smoke visibility naked eye (sunny/dry conditions day time; 500 gram solid grain sample):** 7.39 km
**Smoke visibility naked eye (wet/rainy conditions day time; 500 gram solid grain sample):** 3.5 km
**Smoke visibility naked eye (night time, average environmental conditions; 500 gram solid grain sample):** 2900 meters.
**Water resistance:** Very good.
**Stability:** Can be stored for decades.
**Flammability (solid grain):** 4.29
**Flammability (powdered grain):** 5.1
**Ease of ignition (solid grain):** 4.2 (requires high temperature ignition composition for proper burn).
**Tendency to cake:** Will not cake.
**Explosive ability:** Cannot be detonated.
**Percentage:** ***33% potassium chlorate, 30% 1,8-diaminonapthalene dye, 30% pyromellitic dianhydride dye, 7% sucrose***
**Classification:** Deflagrating mixture (classified as 1.3G pyrotechnic composition).

**Danger of preparation:** Can be made by amateurs.
**Use:** Used in professional industrial/commercial smoke munitions for determining gas leaks, spotting, signaling, and in military operations.

### 06-03-019F: Violet smoke producing composition (modified):

Into a standard ball mill, filled with 500 grams of steel shot of 10mm diameter, place ***85 grams of 1,8-diaminonapthalene***, followed by ***55 grams phthalic anhydride***, followed by ***50 grams of glycerol***, followed by ***55 grams of Epon 828 epoxy resin***, followed by ***60 grams of N-hexyl acid phthalate***, followed by ***40 grams of sodium barbiturate***, and then followed by ***155 grams of potassium chlorate***. Thereafter, tumble the mixture at room temperature at 190 RPM for about 50 minutes. After 50 minutes, the mixture is ready for pressing. To do so, first remove the steel shot, and then press the mixture into any suitable delivery container, such as a grenade, metal tube, plastic tube, cardboard tube, ceramic or clay ball, ect., under a pressure of 6,000 psi, and then cure the device in an oven at 60 Celsius for 12 hours. In some cases, the smoke mixture needs to be pressed into blocks, discs, or tablets under a pressure of 7,000 psi, for use in some types of smoke grenades, followed by curing at 60 Celsius for 12 hours. Note: an appropriate igniter composition to begin the burn of this smoke composition should be used.
**Burn rate:** 23 seconds per inch at 7,000 psi; tends to vary based on diameter and density of smoke grain.
**Smoke Volume on average (500 gram solid grain sample enclosed area):** 1680 sq. ft.
**Smoke Volume on average (500 gram solid grain sample open area):** 7150 sq. ft.
**Smoke visibility naked eye (sunny/dry conditions day time; 500 gram solid grain sample):** 7.3 km
**Smoke visibility naked eye (wet/rainy conditions day time; 500 gram solid grain sample):** 3.41 km
**Smoke visibility naked eye (night time, average environmental conditions; 500 gram solid grain sample):** 2890 meters.
**Water resistance:** Very good.
**Stability:** Can be stored for decades.

**Flammability (solid grain):** 4.1
**Flammability (powdered grain):** 5
**Ease of ignition (solid grain):** 4.1 (requires high temperature ignition composition for proper burn).
**Tendency to cake:** Will not cake.
**Explosive ability:** Cannot be detonated.
**Percentage:** ***31% potassium chlorate, 17% 1,8-diaminonapthalene dye, 12% N-hexyl acid phthalate, 11% Epon 828 epoxy resin, 11% phthalic anhydride, 10% glycerol, 8% sodium barbiturate***
**Classification:** Deflagrating mixture (classified as 1.3G pyrotechnic composition).

**Danger of preparation:** Can be made by amateurs.
**Use:** Used in professional industrial/commercial smoke munitions for determining gas leaks, spotting, signaling, and in military operations.

## 06-03-019G: Red smoke producing composition:

Into a standard ball mill, filled with 500 grams of steel shot of 10mm diameter, place ***79.5 grams of 1,8-diaminonapthalene***, followed by ***21 grams of finely powdered sucrose***, followed by ***236.5 grams of tetrabromo phthalic anhydride***, and then followed by ***163 grams of potassium chlorate***. Thereafter, tumble the mixture at room temperature at 150 RPM for about 60 minutes. After 60 minutes, the mixture is ready for pressing. To do so, first remove the steel shot, and then press the mixture into any suitable delivery container, such as a grenade, metal tube, plastic tube, cardboard tube, ceramic or clay ball, ect., under a pressure of 6,000 psi, and then cure the device in an oven at 60 Celsius for 12 hours. In some cases, the smoke mixture needs to be pressed into blocks, discs, or tablets under a pressure of 7,000 psi, for use in some types of smoke grenades, followed by curing at 60 Celsius for 12 hours. Note: an appropriate igniter composition to begin the burn of this smoke composition should be used.
**Burn rate:** 25 seconds per inch at 7,000 psi; tends to vary based on diameter and density of smoke grain.
**Smoke Volume on average (500 gram solid grain sample enclosed area):** 1675 sq. ft.
**Smoke Volume on average (500 gram solid grain sample open area):** 7120 sq. ft.
**Smoke visibility naked eye (sunny/dry conditions day time; 500 gram solid grain sample):** 7.4 km
**Smoke visibility naked eye (wet/rainy conditions day time; 500 gram solid grain sample):** 3.5 km
**Smoke visibility naked eye (night time, average environmental conditions; 500 gram solid grain sample):** 2990 meters.
**Water resistance:** Very good.
**Stability:** Can be stored for decades.
**Flammability (solid grain):** 4
**Flammability (powdered grain):** 5
**Ease of ignition (solid grain):** 4.3 (requires high temperature ignition composition for proper burn).
**Tendency to cake:** Will not cake.
**Explosive ability:** Cannot be detonated.
**Percentage:** ***47.3% tetrabromo phthalic anhydride, 32.6% potassium chlorate, 15.9% 1,8-diaminonapthalene, 4.2% sucrose***
**Classification:** Deflagrating mixture (classified as 1.3G pyrotechnic composition).

**Danger of preparation:** Can be made by amateurs.
**Use:** Used in professional industrial/commercial smoke munitions for determining gas leaks, spotting, signaling, and in military operations.

## 06-03-019H: Green smoke producing composition:

Into a standard ball mill, filled with 500 grams of steel shot of 10mm diameter, place ***250 grams of para-dimethylaminobenzaldehyde***, followed by ***89.5 grams of finely powdered sucrose***, and then followed by ***163 grams of potassium chlorate***. Thereafter, tumble the mixture at room temperature at 110 RPM for about 45 minutes. After 45 minutes, the mixture is ready for pressing. To do so, first remove the steel shot, and then press the mixture into any suitable delivery container, such as a grenade, metal tube, plastic tube, cardboard tube, ceramic or clay ball, ect., under a pressure of 7,000 psi. In some cases, the smoke mixture needs to be pressed into blocks, discs, or tablets under a pressure of 7,000 psi, for use in some types of smoke grenades. Note: an appropriate igniter composition to begin the burn of this smoke composition should be used.
**Burn rate:** 21 seconds per inch at 7,000 psi; tends to vary based on diameter and density of smoke grain.
**Smoke Volume on average (500 gram solid grain sample enclosed area):** 1600 sq. ft.
**Smoke Volume on average (500 gram solid grain sample open area):** 7000 sq. ft.
**Smoke visibility naked eye (sunny/dry conditions day time; 500 gram solid grain sample):** 7.29 km
**Smoke visibility naked eye (wet/rainy conditions day time; 500 gram solid grain sample):** 3.41 km
**Smoke visibility naked eye (night time, average environmental conditions; 500 gram solid grain sample):** 2800 meters.
**Water resistance:** Very good.
**Stability:** Can be stored for decades.
**Flammability (solid grain):** 4.2
**Flammability (powdered grain):** 5.4

**Ease of ignition (solid grain):** 4.9
**Tendency to cake:** Will not cake.
**Explosive ability:** Cannot be detonated.
**Percentage:** ***49.7% para-dimethylaminobenzaldehyde dye, 32.4% potassium chlorate, 17.8% sucrose, 0.10% residue***
**Classification:** Deflagrating mixture (classified as 1.3G pyrotechnic composition).

**Danger of preparation:** Can be made by amateurs.
**Use:** Used in professional industrial/commercial smoke munitions for determining gas leaks, spotting, signaling, and in military operations.

### 06-03-020A: Intensely red producing smoke composition for spotting and signaling:

Into a standard ball mill, filled with 500 grams of steel shot of 10mm diameter, place ***64.5 grams of an aromatic disulfide diglycidyl ether***, followed by ***17.5 grams of glycerine glycidyl ether***, followed by ***17.5 grams of a poly sulfide*** (sold as a Thiokol liquid polymer sold as LP-3), followed by ***50.5 grams of benzyldimethylamine***, followed by ***150 grams of potassium chlorate***, and then followed by ***200 grams 1-methylaminoanthraquinone***. Thereafter, tumble the mixture at room temperature at 150 RPM for about 60 minutes. After 60 minutes, the mixture is ready for pressing. To do so, first remove the steel shot, and then press the mixture into any suitable delivery container, such as a grenade, metal tube, plastic tube, cardboard tube, ceramic or clay ball, ect., under a pressure of 7,000 psi, and then cure the device in an oven at 60 Celsius for 12 hours. In some cases, the smoke mixture needs to be pressed into blocks, discs, or tablets under a pressure of 7,000 psi, for use in some types of smoke grenades, followed by curing at 60 Celsius for 12 hours. Note: an appropriate igniter composition to begin the burn of this smoke composition should be used.
**Burn rate:** 22 seconds per inch at 7,000 psi; tends to vary based on diameter and density of smoke grain.
**Smoke Volume on average (500 gram solid grain sample enclosed area):** 1609 sq. ft.
**Smoke Volume on average (500 gram solid grain sample open area):** 7005 sq. ft.
**Smoke visibility naked eye (sunny/dry conditions day time; 500 gram solid grain sample):** 7.2 km
**Smoke visibility naked eye (wet/rainy conditions day time; 500 gram solid grain sample):** 3.8 km
**Smoke visibility naked eye (night time, average environmental conditions; 500 gram solid grain sample):** 2900 meters.
**Water resistance:** Very good.
**Stability:** Can be stored for decades.
**Flammability (solid grain):** 4.2
**Flammability (powdered grain):** 5.6
**Ease of ignition (solid grain):** 4.1 (requires high temperature ignition composition for proper burn).
**Tendency to cake:** Will not cake.
**Explosive ability:** Cannot be detonated.
**Percentage:** ***40% 1-methylaminoanthraquinone dye, 30% potassium chlorate, 12.9% aromatic disulfide ether, 10.1% benzyldimethylamine, 3.5% glycerine glycidyl ether, 3.5% poly sulfide***
**Classification:** Deflagrating mixture (classified as 1.3G pyrotechnic composition).

**Danger of preparation:** Can be made by amateurs.
**Use:** Used in professional industrial/commercial smoke munitions for spotting, signaling, and in military operations.

### 06-03-021A: Safe to use, flameless yellow-orange smoke composition for use in highly flammable environments:

Into a standard ball mill, filled with 500 grams of steel shot of 10mm diameter, place ***110 grams of hexamethylene diisocyanate curative*** (commercially available as Desmodur N-100), followed by ***480 grams of a hydroxyl terminated polybutadiene binder*** (HTPB), followed by ***710 grams of ammonium iodate***, followed by ***340 grams of leucoquinizarin*** (sold under the chemical name, 1,4,9,10-tetrahydroxyanthracene), and then followed by ***2 grams of dibutyltin dilaurate curing catalyst***. Thereafter, tumble the mixture at room temperature at 150 RPM for about 45 minutes. After 45 minutes, the mixture is ready for pressing. To do so, first remove the steel shot, and then press the mixture into any suitable delivery container, such as a grenade, metal tube, plastic tube, cardboard tube, ceramic or clay ball, ect., under a pressure of 7,000 psi, and then cure the device in an oven at 60 Celsius for 12 hours. In some cases, the smoke mixture needs to be pressed into blocks, discs, or tablets under a pressure of 7,000 psi, for use in some types of smoke grenades, followed by curing at 60 Celsius for 12 hours. Note: an appropriate igniter composition to begin the burn of this smoke composition should be used.
**Burn rate:** 23 seconds per inch at 7,000 psi; tends to vary based on diameter and density of smoke grain.
**Smoke Volume on average (500 gram solid grain sample enclosed area):** 1650 sq. ft.
**Smoke Volume on average (500 gram solid grain sample open area):** 7095 sq. ft.
**Smoke visibility naked eye (sunny/dry conditions day time; 500 gram solid grain sample):** 7.29 km
**Smoke visibility naked eye (wet/rainy conditions day time; 500 gram solid grain sample):** 3.74 km
**Smoke visibility naked eye (night time, average environmental conditions; 500 gram solid grain sample):** 2890 meters.
**Water resistance:** Very good.
**Stability:** Can be stored for decades.
**Flammability (solid grain):** 4.2

**Flammability (powdered grain):** 5.5
**Ease of ignition (solid grain):** 4.1 (requires high temperature ignition composition for proper burn).
**Tendency to cake:** Will not cake.
**Explosive ability:** Cannot be detonated.
**Percentage:** ***43.2% ammonium iodate, 29.2% hydroxyl terminated polybutadiene binder, 20.7% leucoquinizarin, 6.6% hexamethylene diisocyanate curative agent, 0.121% dibutyl dilaurate curing catalyst, 0.179% residue***
**Classification:** Deflagrating mixture (classified as 1.3G pyrotechnic composition).

**Danger of preparation:** Can be made by amateurs.
**Use:** Used for spotting, tracking, signaling, and other applications in environments that contain flammable materials such as chemicals, grass, brush, trees, flammable objects, ect.

### 06-03-022A: Smoke composition for use at night producing a brownish-black smoke, which looks orange from a distance:

Into a standard ball mill, filled with 500 grams of steel shot of 10mm diameter, place ***255 grams of cadmium metal powder*** of average mesh, followed by ***235 grams of potassium nitrate***, and then followed by ***20 grams of any desirable epoxy resin such as epoxy resin 815*** or equivalent. Thereafter, tumble the mixture at room temperature at 160 RPM for about 45 minutes. After 45 minutes, the mixture is ready for pressing. To do so, first remove the steel shot, and then press the mixture into any suitable delivery container, such as a grenade, metal tube, plastic tube, cardboard tube, ceramic or clay ball, ect., under a pressure of 6,000 psi, and then cure the device in an oven at 60 Celsius for 12 hours. In some cases, the smoke mixture needs to be pressed into blocks, discs, or tablets under a pressure of 7,000 psi, for use in some types of smoke grenades, followed by curing at 60 Celsius for 12 hours. Note: an appropriate igniter composition to begin the burn of this smoke composition should be used. Note: the smoke from this composition is carcinogenic so exposed personnel should wear proper breathing apparatus.
**Burn rate:** 20 seconds per inch at 7,000 psi; tends to vary based on diameter and density of smoke grain.
**Smoke Volume on average (500 gram solid grain sample enclosed area):** 1600 sq. ft.
**Smoke Volume on average (500 gram solid grain sample open area):** 6900 sq. ft.
**Smoke visibility naked eye (sunny/dry conditions day time; 500 gram solid grain sample):** 7 km
**Smoke visibility naked eye (wet/rainy conditions day time; 500 gram solid grain sample):** 3.1 km
**Smoke visibility naked eye (night time, average environmental conditions; 500 gram solid grain sample):** 2800 meters.
**Water resistance:** Very good.
**Stability:** Can be stored for decades.
**Flammability (solid grain):** 4.9
**Flammability (powdered grain):** 6
**Ease of ignition (solid grain):** 5
**Tendency to cake:** Will not cake.
**Explosive ability:** Cannot be detonated.
**Percentage:** ***50% cadmium metal, 46% potassium nitrate, 3.9% epoxy resin, 0.10% residue***
**Classification:** Deflagrating mixture (classified as 1.3G pyrotechnic composition).

**Danger of preparation:** Can be made by amateurs, but caution should be exercised.
**Use:** Used to generate a brownish-black smoke (that looks orange from a distance), when used near sunset or in the late evening. This composition may be used to produce visual distortions for concealment and other applications.

### 06-03-022B: Smoke composition for use at night producing a brownish-black smoke, which looks orange from a distance (modified):

Into a standard ball mill, filled with 500 grams of steel shot of 10mm diameter, place ***200 grams of cadmium sulfide powder*** of average mesh, followed by ***45 grams of cadmium metal powder,*** followed by ***240 grams of black powder***, and then followed by ***30 grams of strontium nitrate***. Thereafter, tumble the mixture at room temperature at 115 RPM for about 45 minutes. After 45 minutes, the mixture is ready for pressing. To do so, first remove the steel shot, and then press the mixture into any suitable delivery container, such as a grenade, metal tube, plastic tube, cardboard tube, ceramic or clay ball, ect., under a pressure of 7,000 psi. In some cases, the smoke mixture needs to be pressed into blocks, discs, or tablets under a pressure of 7,000 psi, for use in some types of smoke grenades. Note: an appropriate igniter composition to begin the burn of this smoke composition should be used. Note: the smoke from this composition is carcinogenic so exposed personnel should wear proper breathing apparatus.
**Burn rate:** 20 seconds per inch at 7,000 psi; tends to vary based on diameter and density of smoke grain.
**Smoke Volume on average (500 gram solid grain sample enclosed area):** 1615 sq. ft.
**Smoke Volume on average (500 gram solid grain sample open area):** 6950 sq. ft.
**Smoke visibility naked eye (sunny/dry conditions day time; 500 gram solid grain sample):** 7.1 km

**Smoke visibility naked eye (wet/rainy conditions day time; 500 gram solid grain sample):** 3.5 km
**Smoke visibility naked eye (night time, average environmental conditions; 500 gram solid grain sample):** 2900 meters.
**Water resistance:** Very good.
**Stability:** Can be stored for decades.
**Flammability (solid grain):** 5
**Flammability (powdered grain):** 6.9
**Ease of ignition (solid grain):** 5.3
**Tendency to cake:** Will not cake.
**Explosive ability:** Cannot be detonated.
**Percentage:** ***46.6% black powder, 38.8% cadmium sulfide, 8.7% cadmium metal, 5.8% strontium nitrate, 0.10% mixed impurities***
**Classification:** Deflagrating mixture (classified as 1.3G pyrotechnic composition).

**Danger of preparation:** Can be made by amateurs, but caution should be exercised.
**Use:** Used to generate a brownish-black smoke (that looks orange from a distance), when used near sunset or in the late evening. This composition may be used to produce visual distortions for concealment and other applications.

### 06-03-023A: Non organic-chlorine containing white smoke composition for screening:

First, into a clean crucible, place ***81 grams of potassium chloride***, followed by ***217 grams of hydrated ferric chloride*** ($FeCl_3$-$6H_2O$), and then heat the mixture at 200 Celsius for several hours until the mixed product becomes water free. Note: the exact heating time may vary, but is concluded when no more moisture is evolved. You can check for this by placing a piece of glass over the heated mixture. If the glass fogs up, the reaction is not complete. Once the water has been removed, place the remaining heated product (containing a potassium ferric chloride complex) after it has cooled, into a standard ball mill, filled with 500 grams of steel shot of 10mm diameter, followed by adding in ***40 grams of finely powdered aluminum*** of standard mesh, followed by ***150 grams of finely divided zinc oxide***, and then followed by ***11 grams of sodium nitrate***. Thereafter, tumble the mixture at room temperature at 150 RPM for about 45 minutes. After 45 minutes, the mixture is ready for pressing. To do so, first remove the steel shot, and then press the mixture into any suitable delivery container, such as a grenade, metal tube, plastic tube, cardboard tube, ceramic or clay ball, ect., under a pressure of 7,000 psi. In some cases, the smoke mixture needs to be pressed into blocks, discs, or tablets under a pressure of 7,000 psi, for use in some types of smoke grenades. Note: an appropriate igniter composition to begin the burn of this smoke composition should be used.
**Burn rate:** 26 seconds per inch at 7,000 psi; tends to vary based on diameter and density of smoke grain.
**Smoke Volume on average (500 gram solid grain sample enclosed area):** 1600 sq. ft.
**Smoke Volume on average (500 gram solid grain sample open area):** 6900 sq. ft.
**Smoke visibility naked eye (sunny/dry conditions day time; 500 gram solid grain sample):** 7 km
**Smoke visibility naked eye (wet/rainy conditions day time; 500 gram solid grain sample):** 3 km
**Smoke visibility naked eye (night time, average environmental conditions; 500 gram solid grain sample):** 2500 meters.
**Water resistance:** Very good.
**Stability:** Can be stored for decades.
**Flammability (solid grain):** 3.5
**Flammability (powdered grain):** 4.2
**Ease of ignition (solid grain):** 3 (requires high temperature ignition composition for proper burn).
**Tendency to cake:** Will not cake.
**Explosive ability:** Cannot be detonated.
**Percentage:** ***49.2% potassium ferric chloride, 37.8% zinc oxide, 10.1% aluminum, 2.7% sodium nitrate, 0.20% moisture***
**Classification:** Deflagrating mixture (classified as 1.3G pyrotechnic composition).

**Danger of preparation:** Can be made by amateurs.
**Use:** Used in professional industrial/commercial smoke munitions for spotting, signaling, and in military operations.

### 06-03-024A: Non-toxic smoke generating composition for use in various training operations:

Into a standard ball mill, filled with 200 grams of steel shot of 10mm diameter, place ***377.5 grams of trans-cinnamic acid***, followed by ***27.5 grams of sucrose***, followed by ***65 grams of potassium chlorate***, followed by ***20 grams of sodium bicarbonate***, followed by ***10 grams of diatomite silica***, and the followed by ***26 grams of nitrocellulose*** (14% nitrogen content). Thereafter, tumble the mixture at room temperature at 100 RPM for about 45 minutes. After 45 minutes, the mixture is ready for pressing. To do so, first remove the steel shot, and then press the mixture into any suitable delivery container, such as a grenade, metal tube, plastic tube, cardboard tube, ceramic or clay ball, ect., under a pressure of 7,000 psi. In some cases, the smoke mixture needs to be pressed into blocks, discs, or tablets under a pressure of 7,000 psi, for use in some types of smoke grenades. Note: an appropriate igniter composition to begin the burn of this smoke composition should be used.
**Burn rate:** 25 seconds per inch at 7,000 psi; tends to vary based on diameter and density of smoke grain.
**Smoke Volume on average (500 gram solid grain sample enclosed area):** 1605 sq. ft.
**Smoke Volume on average (500 gram solid grain sample open area):** 6940 sq. ft.

**Smoke visibility naked eye (sunny/dry conditions day time; 500 gram solid grain sample):** 7.19 km
**Smoke visibility naked eye (wet/rainy conditions day time; 500 gram solid grain sample):** 3.1 km
**Smoke visibility naked eye (night time, average environmental conditions; 500 gram solid grain sample):** 2800 meters.
**Water resistance:** Very good.
**Stability:** Can be stored for decades.
**Flammability (solid grain):** 3.7
**Flammability (powdered grain):** 4.2
**Ease of ignition (solid grain):** 3.1 (requires high temperature ignition composition for proper burn).
**Tendency to cake:** Will not cake.
**Explosive ability:** Cannot be detonated.
**Percentage:** ***71.7% trans-cinnamic acid, 12.3% potassium chlorate, 5.2% sucrose, 4.9% nitrocellulose, 3.8% sodium bicarbonate, 1.9% diatomite silica, 0.2% impurities***
**Classification:** Deflagrating mixture (classified as 1.3G pyrotechnic composition).

**Danger of preparation:** Can be made by amateurs.
**Use:** Used in professional industrial/commercial smoke munitions for spotting, signaling, and in military operations.

### 06-03-025A: Pyrotechnic composition for producing a mixed smoke:

Into a standard ball mill, filled with 500 grams of steel shot of 10mm diameter, place ***200 grams of dry maltose powder***, followed by ***65 grams of calcium lactate***, followed by ***90 grams of malic acid***, followed by ***25 grams of manganese dioxide***, followed by ***120 grams of melamine freebase***, followed by ***5 grams of copper chromite***, and then followed by ***400 grams of sodium chlorate***. Thereafter, tumble the mixture at room temperature at 120 RPM for about 45 minutes. After 45 minutes, the mixture is ready for pressing. To do so, first remove the steel shot, and then press the mixture into any suitable delivery container, such as a grenade, metal tube, plastic tube, cardboard tube, ceramic or clay ball, ect., under a pressure of 7,500 psi. In some cases, the smoke mixture needs to be pressed into blocks, discs, or tablets under a pressure of 8,000 psi, for use in some types of smoke grenades. Note: an appropriate igniter composition to begin the burn of this smoke composition should be used.
**Burn rate:** 22 seconds per inch at 8,000 psi; tends to vary based on diameter and density of smoke grain.
**Smoke Volume on average (500 gram solid grain sample enclosed area):** 1655 sq. ft.
**Smoke Volume on average (500 gram solid grain sample open area):** 6995 sq. ft.
**Smoke visibility naked eye (sunny/dry conditions day time; 500 gram solid grain sample):** 7.29 km
**Smoke visibility naked eye (wet/rainy conditions day time; 500 gram solid grain sample):** 3.19 km
**Smoke visibility naked eye (night time, average environmental conditions; 500 gram solid grain sample):** 2890 meters.
**Water resistance:** Very good.
**Stability:** Can be stored for decades.
**Flammability (solid grain):** 5
**Flammability (powdered grain):** 6
**Ease of ignition (solid grain):** 4.9
**Tendency to cake:** Will not cake.
**Explosive ability:** Cannot be detonated.
**Percentage:** ***44.19% sodium chlorate, 22.09% maltose, 13.25% melamine freebase, 9.94% malic acid, 7.18% calcium lactate, 2.76% manganese dioxide, 0.55% copper chromite burn rate catalyst, 0.04% impurities***
**Classification:** Deflagrating mixture (classified as 1.3G pyrotechnic composition).

**Danger of preparation:** Can be made by amateurs.
**Use:** Used in professional industrial/commercial smoke munitions for spotting, signaling, and in military operations.

### 06-03-026A: Red phosphorus smoke composition:

Into a suitable sealable vertical mixer, place ***60 grams of butylmethacrylate***, followed by ***300 grams of unleaded premium gasoline***, and then followed by ***750 grams of red phosphorus***. Thereafter, seal the vertical mixer, and then blend the mixture on high speed in the absence of air for about 45 minutes. Thereafter, the mixture is ready for use. To use, the semi-fluidized mass needs to be placed into any desired "pot", metal container, bomb casing, artillery shell, ect., under standard pressure, and then primed with a suitable igniter composition. Note: this composition can be used as a "napalm" substitute when placed into a suitable delivery bomb and primed with an HE bursting charge.
**Burn rate (when ignited with flame):** 17 seconds per 500 gram sample at STP.
**Burn rate (when ignited by HE bursting charge):** 6 seconds per 500 gram sample at STP.
**Smoke Volume on average (500 gram solid grain sample enclosed area):** 1610 sq. ft.
**Smoke Volume on average (500 gram solid grain sample open area):** 6985 sq. ft.
**Smoke visibility naked eye (sunny/dry conditions day time; 500 gram solid grain sample):** 7 km
**Smoke visibility naked eye (wet/rainy conditions day time; 500 gram solid grain sample):** 3 km

**Smoke visibility naked eye (night time, average environmental conditions; 500 gram solid grain sample):** 2760 meters.
**Water resistance:** Very good.
**Stability:** Can be stored for decades.
**Flammability (ignited open air):** 6.9
**Ease of ignition (ignited with flame):** 7.2
**Ease of ignition (ignited with HE bursting charge):** 9
**Tendency to cake:** Will not cake.
**Explosive ability:** Cannot be detonated.
**Percentage:** ***67.56% red phosphorus, 27.02% unleaded gasoline, 5.4% butylmethacrylate, 0.02% mixed balance***
**Classification:** Highly flammable mixture (classified as Class A flammable composition).

**Danger of preparation:** Can be made by amateurs, but caution should be exercised.
**Use:** Used for generating an unbalanced mixed smoke for spotting, patrolling, and signaling. Also used as a substitute for Napalm.

### 06-03-026B: Red phosphorus smoke composition (modified component):

Into a suitable sealable vertical mixer, place ***90 grams of butylmethacrylate***, followed by ***350 grams of unleaded premium gasoline***, followed by ***400 grams of sodium nitrate***, followed by ***250 grams of metallic sodium***, and then followed by ***750 grams of red phosphorus***. Thereafter, seal the vertical mixer, and then blend the mixture on high speed in the absence of air for about 45 minutes. Thereafter, the mixture is ready for use. To use, the semi-fluidized mass needs to be placed into any desired "pot", metal container, bomb casing, artillery shell, ect., under standard pressure, and then primed with a suitable igniter composition. Note: this composition can be used as a "napalm" substitute when placed into a suitable delivery bomb and primed with an HE bursting charge.
**Burn rate (when ignited with flame):** 15 seconds per 500 gram sample at STP.
**Burn rate (when ignited by HE bursting charge):** 4 seconds per 500 gram sample at STP.
**Smoke Volume on average (500 gram solid grain sample enclosed area):** 1600 sq. ft.
**Smoke Volume on average (500 gram solid grain sample open area):** 6970 sq. ft.
**Smoke visibility naked eye (sunny/dry conditions day time; 500 gram solid grain sample):** 6.9 km
**Smoke visibility naked eye (wet/rainy conditions day time; 500 gram solid grain sample):** 3 km
**Smoke visibility naked eye (night time, average environmental conditions; 500 gram solid grain sample):** 2700 meters.
**Water resistance:** Poor; but good when kept sealed from moisture and air.
**Stability:** Can be stored for years when kept from moisture and air.
**Flammability (ignited open air):** 7.9
**Ease of ignition (ignited with flame):** 9
**Ease of ignition (ignited with HE bursting charge):** 9.8
**Tendency to cake:** Will not cake.
**Explosive ability:** Cannot be detonated.
**Percentage:** ***40.76% red phosphorus, 21.73% sodium nitrate, 19.02% unleaded gasoline, 13.58% metallic sodium, 4.89% butylmethacrylate, 0.02% impurities***
**Classification:** Highly flammable mixture (classified as Class A flammable composition).

**Danger of preparation:** Should only be prepared by experienced personnel; caution should be exercised when working with metallic sodium.
**Use:** Used for generating an unbalanced mixed smoke for spotting, patrolling, and signaling. Also used as a substitute for Napalm.

### 06-03-027A: Purple smoke composition:

Into a standard ball mill, filled with 500 grams of steel shot of 10mm diameter, place ***300 grams of potassium chlorate***, followed by ***100 grams of magnesium powder***, followed by ***250 grams of iodine***, followed by ***400 grams of ammonium iodide***, and then followed by ***150 grams of lactose***. Thereafter, tumble the mixture at room temperature at 100 RPM for about 50 minutes. After 50 minutes, the mixture is ready for pressing. To do so, first remove the steel shot, and then press the mixture into any suitable delivery container, such as a grenade, metal tube, plastic tube, cardboard tube, ceramic or clay ball, ect., under a pressure of 7,500 psi. In some cases, the smoke mixture needs to be pressed into blocks, discs, or tablets under a pressure of 8,000 psi, for use in some types of smoke grenades. Note: an appropriate igniter composition to begin the burn of this smoke composition should be used. Note: This composition produces an irritating vapor, so caution should be used.
**Burn rate:** 23 seconds per inch at 8,000 psi; tends to vary based on diameter and density of smoke grain.
**Smoke Volume on average (500 gram solid grain sample enclosed area):** 1590 sq. ft.
**Smoke Volume on average (500 gram solid grain sample open area):** 6800 sq. ft.
**Smoke visibility naked eye (sunny/dry conditions day time; 500 gram solid grain sample):** 7 km
**Smoke visibility naked eye (wet/rainy conditions day time; 500 gram solid grain sample):** 3 km
**Smoke visibility naked eye (night time, average environmental conditions; 500 gram solid grain sample):** 2700 meters.

**Water resistance:** Very good.
**Stability:** Can be stored for decades.
**Flammability (solid grain):** 4.9
**Flammability (powdered grain):** 5.9
**Ease of ignition (solid grain):** 3.99
**Tendency to cake:** Will not cake.
**Explosive ability:** Cannot be detonated.
**Percentage:** ***33.33% ammonium iodide, 25% potassium chlorate, 20.83% iodine, 12.5% lactose, 8.33% magnesium, 0.01% mixed balance***
**Classification:** Deflagrating mixture (classified as 1.3G pyrotechnic composition).

**Danger of preparation:** Can be made by amateurs.
**Use:** Used in professional industrial/commercial smoke munitions for spotting, signaling, and in military operations. Can be used in riot control operations.

### 06-03-028A: Yellow smoke composition:

Into a standard ball mill, filled with 500 grams of steel shot of 10mm diameter, place ***330 grams of potassium dichromate***, followed by ***100 grams of bismuth tetraoxide***, and then followed by ***70 grams of magnesium powder***. Thereafter, tumble the mixture at room temperature at 120 RPM for about 45 minutes. After 45 minutes, the mixture is ready for pressing. To do so, first remove the steel shot, and then press the mixture into any suitable delivery container, such as a grenade, metal tube, plastic tube, cardboard tube, ceramic or clay ball, ect., under a pressure of 8,000 psi. In some cases, the smoke mixture needs to be pressed into blocks, discs, or tablets under a pressure of 8,000 psi, for use in some types of smoke grenades. Note: an appropriate igniter composition to begin the burn of this smoke composition should be used.
**Burn rate:** 29 seconds per inch at 8,000 psi; tends to vary based on diameter and density of smoke grain.
**Smoke Volume on average (500 gram solid grain sample enclosed area):** 1500 sq. ft.
**Smoke Volume on average (500 gram solid grain sample open area):** 6700 sq. ft.
**Smoke visibility naked eye (sunny/dry conditions day time; 500 gram solid grain sample):** 7 km
**Smoke visibility naked eye (wet/rainy conditions day time; 500 gram solid grain sample):** 3 km
**Smoke visibility naked eye (night time, average environmental conditions; 500 gram solid grain sample):** 2500 meters.
**Water resistance:** Very good.
**Stability:** Can be stored for decades.
**Flammability (solid grain):** 3.1
**Flammability (powdered grain):** 3.2
**Ease of ignition (solid grain):** 3 (requires high temperature ignition composition for proper burn).
**Tendency to cake:** Will not cake.
**Explosive ability:** Cannot be detonated.
**Percentage:** ***66% potassium dichromate, 20% bismuth tetraoxide, 14% magnesium powder***
**Classification:** Deflagrating mixture (classified as 1.3G pyrotechnic composition).

**Danger of preparation:** Can be made by amateurs.
**Use:** Used in professional industrial/commercial smoke munitions for spotting, signaling, and in military operations.

### 06-03-028B: Yellow smoke composition:

Into a standard ball mill, filled with 500 grams of steel shot of 10mm diameter, place ***325 grams of potassium dichromate***, followed by ***100 grams of bismuth subnitrate***, and then followed by ***75 grams of magnesium powder***. Thereafter, tumble the mixture at room temperature at 120 RPM for about 45 minutes. After 45 minutes, the mixture is ready for pressing. To do so, first remove the steel shot, and then press the mixture into any suitable delivery container, such as a grenade, metal tube, plastic tube, cardboard tube, ceramic or clay ball, ect., under a pressure of 8,000 psi. In some cases, the smoke mixture needs to be pressed into blocks, discs, or tablets under a pressure of 8,000 psi, for use in some types of smoke grenades. Note: an appropriate igniter composition to begin the burn of this smoke composition should be used.
**Burn rate:** 29 seconds per inch at 8,000 psi; tends to vary based on diameter and density of smoke grain.
**Smoke Volume on average (500 gram solid grain sample enclosed area):** 1510 sq. ft.
**Smoke Volume on average (500 gram solid grain sample open area):** 6710 sq. ft.
**Smoke visibility naked eye (sunny/dry conditions day time; 500 gram solid grain sample):** 7 km
**Smoke visibility naked eye (wet/rainy conditions day time; 500 gram solid grain sample):** 3 km
**Smoke visibility naked eye (night time, average environmental conditions; 500 gram solid grain sample):** 2500 meters.
**Water resistance:** Very good.
**Stability:** Can be stored for decades.
**Flammability (solid grain):** 3.1
**Flammability (powdered grain):** 3.2
**Ease of ignition (solid grain):** 3 (requires high temperature ignition composition for proper burn).

**Tendency to cake:** Will not cake.
**Explosive ability:** Cannot be detonated.
**Percentage:** ***65% potassium dichromate, 20% bismuth subnitrate, 15% magnesium powder***
**Classification:** Deflagrating mixture (classified as 1.3G pyrotechnic composition).

**Danger of preparation:** Can be made by amateurs.
**Use:** Used in professional industrial/commercial smoke munitions for spotting, signaling, and in military operations.

### 06-03-029A: Brown smoke composition:

Into a standard ball mill, filled with 500 grams of steel shot of 10mm diameter, place ***250 grams of copper-II-oxide***, followed by ***175 grams of lead dioxide***, and then followed by ***75 grams of magnesium powder***. Thereafter, tumble the mixture at room temperature at 150 RPM for about 45 minutes. After 45 minutes, the mixture is ready for pressing. To do so, first remove the steel shot, and then press the mixture into any suitable delivery container, such as a grenade, metal tube, plastic tube, cardboard tube, ceramic or clay ball, ect., under a pressure of 8,000 psi. In some cases, the smoke mixture needs to be pressed into blocks, discs, or tablets under a pressure of 8,000 psi, for use in some types of smoke grenades. Note: an appropriate igniter composition to begin the burn of this smoke composition should be used.
**Burn rate:** 25 seconds per inch at 8,000 psi; tends to vary based on diameter and density of smoke grain.
**Smoke Volume on average (500 gram solid grain sample enclosed area):** 1595 sq. ft.
**Smoke Volume on average (500 gram solid grain sample open area):** 6820 sq. ft.
**Smoke visibility naked eye (sunny/dry conditions day time; 500 gram solid grain sample):** 7.2 km
**Smoke visibility naked eye (wet/rainy conditions day time; 500 gram solid grain sample):** 3.7 km
**Smoke visibility naked eye (night time, average environmental conditions; 500 gram solid grain sample):** 2900 meters.
**Water resistance:** Very good.
**Stability:** Can be stored for decades.
**Flammability (solid grain):** 4.1
**Flammability (powdered grain):** 5
**Ease of ignition (solid grain):** 3.4 (requires high temperature ignition composition for proper burn).
**Tendency to cake:** Will not cake.
**Explosive ability:** Cannot be detonated.
**Percentage:** ***50% copper-II-oxide, 35% lead dioxide, 15% magnesium powder***
**Classification:** Deflagrating mixture (classified as 1.3G pyrotechnic composition).

**Danger of preparation:** Can be made by amateurs.
**Use:** Used in professional industrial/commercial smoke munitions for spotting, signaling, and in military operations.

### 06-03-030A: Orange smoke composition:

Into a standard ball mill, filled with 500 grams of steel shot of 10mm diameter, place ***250 grams of lead dioxide***, followed by ***175 grams of potassium dichromate***, and then followed by ***75 grams of magnesium powder***. Thereafter, tumble the mixture at room temperature at 150 RPM for about 45 minutes. After 45 minutes, the mixture is ready for pressing. To do so, first remove the steel shot, and then press the mixture into any suitable delivery container, such as a grenade, metal tube, plastic tube, cardboard tube, ceramic or clay ball, ect., under a pressure of 8,000 psi. In some cases, the smoke mixture needs to be pressed into blocks, discs, or tablets under a pressure of 8,000 psi, for use in some types of smoke grenades. Note: an appropriate igniter composition to begin the burn of this smoke composition should be used.
**Burn rate:** 29 seconds per inch at 8,000 psi; tends to vary based on diameter and density of smoke grain.
**Smoke Volume on average (500 gram solid grain sample enclosed area):** 1500 sq. ft.
**Smoke Volume on average (500 gram solid grain sample open area):** 6800 sq. ft.
**Smoke visibility naked eye (sunny/dry conditions day time; 500 gram solid grain sample):** 7 km
**Smoke visibility naked eye (wet/rainy conditions day time; 500 gram solid grain sample):** 3.2 km
**Smoke visibility naked eye (night time, average environmental conditions; 500 gram solid grain sample):** 2855 meters.
**Water resistance:** Very good.
**Stability:** Can be stored for decades.
**Flammability (solid grain):** 3.8
**Flammability (powdered grain):** 4.2
**Ease of ignition (solid grain):** 3 (requires high temperature ignition composition for proper burn).
**Tendency to cake:** Will not cake.
**Explosive ability:** Cannot be detonated.
**Percentage:** ***50% lead dioxide, 35% potassium dichromate, 15% magnesium powder***
**Classification:** Deflagrating mixture (classified as 1.3G pyrotechnic composition).

**Danger of preparation:** Can be made by amateurs.
**Use:** Used in professional industrial/commercial smoke munitions for spotting, signaling, and in military operations.

### 06-03-031A: Pink smoke composition:

Into a standard ball mill, filled with 500 grams of steel shot of 10mm diameter, place ***185 grams of calcium silicide***, followed by ***270 grams of potassium iodate***, and then followed by ***45 grams of potassium chromate***. Thereafter, tumble the mixture at room temperature at 150 RPM for about 45 minutes. After 45 minutes, the mixture is ready for pressing. To do so, first remove the steel shot, and then press the mixture into any suitable delivery container, such as a grenade, metal tube, plastic tube, cardboard tube, ceramic or clay ball, ect., under a pressure of 8,000 psi. In some cases, the smoke mixture needs to be pressed into blocks, discs, or tablets under a pressure of 8,000 psi, for use in some types of smoke grenades. Note: an appropriate igniter composition to begin the burn of this smoke composition should be used.
**Burn rate:** 26 seconds per inch at 8,000 psi; tends to vary based on diameter and density of smoke grain.
**Smoke Volume on average (500 gram solid grain sample enclosed area):** 1510 sq. ft.
**Smoke Volume on average (500 gram solid grain sample open area):** 6825 sq. ft.
**Smoke visibility naked eye (sunny/dry conditions day time; 500 gram solid grain sample):** 7 km
**Smoke visibility naked eye (wet/rainy conditions day time; 500 gram solid grain sample):** 3.1 km
**Smoke visibility naked eye (night time, average environmental conditions; 500 gram solid grain sample):** 2800 meters.
**Water resistance:** Very good.
**Stability:** Can be stored for decades.
**Flammability (solid grain):** 3.7
**Flammability (powdered grain):** 3.8
**Ease of ignition (solid grain):** 3 (requires high temperature ignition composition for proper burn).
**Tendency to cake:** Will not cake.
**Explosive ability:** Cannot be detonated.
**Percentage:** ***54% potassium iodate, 37% calcium silicide, 9% potassium chromate***
**Classification:** Deflagrating mixture (classified as 1.3G pyrotechnic composition).

**Danger of preparation:** Can be made by amateurs.
**Use:** Used in professional industrial/commercial smoke munitions for spotting, signaling, and in military operations.

### 06-03-031B: Pink smoke composition:

Into a standard ball mill, filled with 500 grams of steel shot of 10mm diameter, place ***325 grams of calcium iodate***, followed by ***50 grams of potassium chromate***, and then followed by ***125 grams of magnesium powder***. Thereafter, tumble the mixture at room temperature at 150 RPM for about 45 minutes. After 45 minutes, the mixture is ready for pressing. To do so, first remove the steel shot, and then press the mixture into any suitable delivery container, such as a grenade, metal tube, plastic tube, cardboard tube, ceramic or clay ball, ect., under a pressure of 8,000 psi. In some cases, the smoke mixture needs to be pressed into blocks, discs, or tablets under a pressure of 8,000 psi, for use in some types of smoke grenades. Note: an appropriate igniter composition to begin the burn of this smoke composition should be used.
**Burn rate:** 24 seconds per inch at 8,000 psi; tends to vary based on diameter and density of smoke grain.
**Smoke Volume on average (500 gram solid grain sample enclosed area):** 1590 sq. ft.
**Smoke Volume on average (500 gram solid grain sample open area):** 6900 sq. ft.
**Smoke visibility naked eye (sunny/dry conditions day time; 500 gram solid grain sample):** 7 km
**Smoke visibility naked eye (wet/rainy conditions day time; 500 gram solid grain sample):** 3.1 km
**Smoke visibility naked eye (night time, average environmental conditions; 500 gram solid grain sample):** 2800 meters.
**Water resistance:** Very good.
**Stability:** Can be stored for decades.
**Flammability (solid grain):** 3.7
**Flammability (powdered grain):** 5
**Ease of ignition (solid grain):** 3 (requires high temperature ignition composition for proper burn).
**Tendency to cake:** Will not cake.
**Explosive ability:** Cannot be detonated.
**Percentage:** ***65% calcium iodate, 25% magnesium powder, 10% potassium chromate***
**Classification:** Deflagrating mixture (classified as 1.3G pyrotechnic composition).

**Danger of preparation:** Can be made by amateurs.

**Use:** Used in professional industrial/commercial smoke munitions for spotting, signaling, and in military operations.

### 06-03-032A: Classic HC smoke composition:

Into a standard ball mill, filled with 500 grams of steel shot of 10mm diameter, place ***120 grams of magnesium powder of average mesh***, followed by ***5 grams of linseed oil or mineral oil***, followed by ***360 grams of zinc oxide***, followed by ***350 grams of hexachloroethane***, and then followed by ***170 grams of ammonium perchlorate***. Thereafter, tumble the mixture at room temperature at 150 RPM for about 45 minutes. After 45 minutes, the mixture is ready for pressing. To do so, first remove the steel shot, and then press the mixture into any suitable delivery container, such as a grenade, metal tube, plastic tube, cardboard tube, ceramic or clay ball, ect., under a pressure of 7,000 psi. In some cases, the smoke mixture needs to be pressed into blocks, discs, or tablets under a pressure of 8,000 psi, for use in some types of smoke grenades. Note: an appropriate igniter composition to begin the burn of this smoke composition should be used.
**Burn rate:** 20 seconds per inch at 8,000 psi; tends to vary based on diameter and density of smoke grain.
**Smoke Volume on average (500 gram solid grain sample enclosed area):** 1600 sq. ft.
**Smoke Volume on average (500 gram solid grain sample open area):** 6990 sq. ft.
**Smoke visibility naked eye (sunny/dry conditions day time; 500 gram solid grain sample):** 7.1 km
**Smoke visibility naked eye (wet/rainy conditions day time; 500 gram solid grain sample):** 3.3 km
**Smoke visibility naked eye (night time, average environmental conditions; 500 gram solid grain sample):** 2950 meters.
**Water resistance:** Very good.
**Stability:** Can be stored for decades.
**Flammability (solid grain):** 3.8
**Flammability (powdered grain):** 5.1
**Ease of ignition (solid grain):** 3.5 (requires high temperature ignition composition for proper burn).
**Tendency to cake:** Will not cake.
**Explosive ability:** Cannot be detonated.
**Percentage:** ***35.82% zinc oxide, 34.82% hexachloroethane, 16.91% ammonium perchlorate, 11.94% magnesium powder, 0.49% linseed oil or mineral oil, 0.02% residual balance***
**Classification:** Deflagrating mixture (classified as 1.3G pyrotechnic composition).

**Danger of preparation:** Can be made by amateurs.
**Use:** Used in professional industrial/commercial smoke munitions for spotting, signaling, and in military operations.

### 06-03-032B: Classic smoke composition modified with magnesium chloride:

Into a standard ball mill, filled with 500 grams of steel shot of 10mm diameter, place ***250 grams of magnesium powder of average mesh***, followed by ***5 grams of linseed oil or mineral oil***, followed by ***250 grams of anhydrous magnesium chloride***, followed by ***200 grams of zinc oxide***, and then followed by ***300 grams of ammonium perchlorate***. Thereafter, tumble the mixture at room temperature at 150 RPM for about 45 minutes. After 45 minutes, the mixture is ready for pressing. To do so, first remove the steel shot, and then press the mixture into any suitable delivery container, such as a grenade, metal tube, plastic tube, cardboard tube, ceramic or clay ball, ect., under a pressure of 7,000 psi. In some cases, the smoke mixture needs to be pressed into blocks, discs, or tablets under a pressure of 8,000 psi, for use in some types of smoke grenades. Note: an appropriate igniter composition to begin the burn of this smoke composition should be used.
**Burn rate:** 24 seconds per inch at 8,000 psi; tends to vary based on diameter and density of smoke grain.
**Smoke Volume on average (500 gram solid grain sample enclosed area):** 1610 sq. ft.
**Smoke Volume on average (500 gram solid grain sample open area):** 7000 sq. ft.
**Smoke visibility naked eye (sunny/dry conditions day time; 500 gram solid grain sample):** 7.19 km
**Smoke visibility naked eye (wet/rainy conditions day time; 500 gram solid grain sample):** 3.35 km
**Smoke visibility naked eye (night time, average environmental conditions; 500 gram solid grain sample):** 2960 meters.
**Water resistance:** Very good.
**Stability:** Can be stored for decades.
**Flammability (solid grain):** 3.8
**Flammability (powdered grain):** 4.1
**Ease of ignition (solid grain):** 3.3 (requires high temperature ignition composition for proper burn).
**Tendency to cake:** Will not cake.
**Explosive ability:** Cannot be detonated.
**Percentage:** ***29.85% ammonium perchlorate, 24.87% anhydrous magnesium chloride, 24.87% magnesium powder, 19.9% zinc oxide, 0.49% linseed or mineral oil, 0.02% mixed balance***
**Classification:** Deflagrating mixture (classified as 1.3G pyrotechnic composition).

**Danger of preparation:** Can be made by amateurs.
**Use:** Used in professional industrial/commercial smoke munitions for spotting, signaling, and in military operations.

**06-03-033A: Modified HC smoke composition for generating a hydroscopic smoke for use in humid environments:** Into a standard ball mill, filled with 500 grams of steel shot of 10mm diameter, place ***96.7 grams of lithium carbonate***, followed by ***197.5 grams of potassium perchlorate,*** followed by ***102.5 grams of aluminum powder***, and then followed by ***103.3 grams of hexachloroethane***. Thereafter, tumble the mixture at room temperature at 100 RPM for about 45 minutes. After 45 minutes, the mixture is ready for pressing. To do so, first remove the steel shot, and then press the mixture into any suitable delivery container, such as a grenade, metal tube, plastic tube, cardboard tube, ceramic or clay ball, ect., under a pressure of 7,000 psi. In some cases, the smoke mixture needs to be pressed into blocks, discs, or tablets under a pressure of 8,000 psi, for use in some types of smoke grenades. Note: an appropriate igniter composition to begin the burn of this smoke composition should be used.
**Burn rate:** 22 seconds per inch at 8,000 psi; tends to vary based on diameter and density of smoke grain.
**Smoke Volume on average (500 gram solid grain sample enclosed area):** 1690 sq. ft.
**Smoke Volume on average (500 gram solid grain sample open area):** 7060 sq. ft.
**Smoke visibility naked eye (sunny/dry conditions day time; 500 gram solid grain sample):** 7.26 km
**Smoke visibility naked eye (wet/rainy conditions day time; 500 gram solid grain sample):** 4.3 km
**Smoke visibility naked eye (night time, average environmental conditions; 500 gram solid grain sample):** 2980 meters.
**Water resistance:** Very good.
**Stability:** Can be stored for decades.
**Flammability (solid grain):** 4.1
**Flammability (powdered grain):** 4.9
**Ease of ignition (solid grain):** 3.7 (requires high temperature ignition composition for proper burn).
**Tendency to cake:** Will not cake.
**Explosive ability:** Cannot be detonated.
**Percentage:** ***39.5% potassium perchlorate, 20.66% hexachloroethane, 20.5% aluminum powder, 19.34% lithium carbonate***
**Classification:** Deflagrating mixture (classified as 1.3G pyrotechnic composition).

**Danger of preparation:** Can be made by amateurs, but caution should be exercised.
**Use:** Used for generating white smoke with increased output in humid environments.

**06-03-034A: Simplified reddish smoke composition for bursting type munitions:**
Into a standard ball mill, filled with 500 grams of steel shot of 10mm diameter, place ***387.5 grams of lead iodate***, and then followed by ***112.5 grams of aluminum powder***. Thereafter, tumble the mixture at 100 RPM for about 1 hour. After 1 hour, the mixture is ready for pressing. To do so, first remove the steel shot, and then press the mixture into any suitable delivery container, such as a plastic tube, plastic container or cardboard tube, under a pressure of 7,500 psi. In some cases, the smoke mixture needs to be pressed into blocks, discs, or tablets under a pressure of 8,000 psi, for use in some types of delivery containers. Note: a bursting charge is placed in the center of the smoke charge, and upon detonation of the bursting charge, a large cloud of smoke will be instantaneously produced.
**Burn rate:** Not applicable.
**Smoke Volume on average (500 gram solid grain sample open area):** 112 sq. ft.
**Smoke visibility naked eye (sunny/dry conditions day time; 500 gram solid grain sample):** 5.2 km
**Smoke visibility naked eye (wet/rainy conditions day time; 500 gram solid grain sample):** 1.9 km
**Smoke visibility naked eye (night time, average environmental conditions; 500 gram solid grain sample):** 1500 meters.
**Water resistance:** Very good.
**Stability:** Can be stored for decades.
**Flammability (solid grain):** Not applicable.
**Flammability (powdered grain):** 6.9
**Ease of ignition (solid grain):** Not applicable.
**Tendency to cake:** Will not cake.
**Explosive ability:** Cannot be detonated.
**Percentage:** ***77.5% lead iodate, 22.5% aluminum powder***
**Classification:** Deflagrating mixture (classified as 1.4G pyrotechnic composition).

**Danger of preparation:** Can be made by amateurs, but caution should be exercised.
**Use:** Used in smoke charges for producing an instantaneous cloud of smoke; used in tube launched bursting-type smoke grenades for masking tanks and vehicles; used in spotting.

**06-03-034B: Simplified reddish-purple smoke composition for bursting type munitions:**
Into a standard ball mill, filled with 500 grams of steel shot of 10mm diameter, place ***375 grams of lead iodate***, and then followed by ***125 grams of zirconium hydride***. Thereafter, tumble the mixture at 100 RPM for about 1 hour. After 1 hour, the mixture is ready for pressing. To do so, first remove the steel shot, and then press the mixture into any suitable delivery container, such as a plastic tube, plastic container or cardboard tube, under a pressure of 8,000 psi. In some cases, the smoke mixture needs to be pressed into blocks, discs, or tablets under a

pressure of 8,000 psi, for use in some types of delivery containers. Note: a bursting charge is placed in the center of the smoke charge, and upon detonation of the bursting charge, a large cloud of smoke will be instantaneously produced.
**Burn rate:** Not applicable.
**Smoke Volume on average (500 gram solid grain sample open area):** 122 sq. ft.
**Smoke visibility naked eye (sunny/dry conditions day time; 500 gram solid grain sample):** 5.3 km
**Smoke visibility naked eye (wet/rainy conditions day time; 500 gram solid grain sample):** 2.1 km
**Smoke visibility naked eye (night time, average environmental conditions; 500 gram solid grain sample):** 1600 meters.
**Water resistance:** Very good when kept sealed from moisture and air.
**Stability:** Can be stored for decades.
**Flammability (solid grain):** Not applicable.
**Flammability (powdered grain):** 7.1
**Ease of ignition (solid grain):** Not applicable.
**Tendency to cake:** Will not cake.
**Explosive ability:** Cannot be detonated.
**Percentage:** ***75% lead iodate, 25% zirconium hydride***
**Classification:** Deflagrating mixture (classified as 1.4G pyrotechnic composition).

**Danger of preparation:** Can be made by amateurs, but caution should be exercised.
**Use:** Used in smoke charges for producing an instantaneous cloud of smoke; used in tube launched bursting-type smoke grenades for masking tanks and vehicles; used in spotting.

### 06-03-035A: Dense reddish smoke composition:

Into a standard ball mill, filled with 500 grams of steel shot of 10mm diameter, place ***150 grams of potassium chlorate***, followed by ***90 grams of powdered lactose***, followed by ***150 grams of sublimed iodine***, and then followed by ***270 grams of para-nitroaniline***. Thereafter, tumble the mixture at room temperature at 90 RPM for about 60 minutes. After 60 minutes, the mixture is ready for pressing. To do so, first remove the steel shot, and then press the mixture into any suitable delivery container, such as a grenade, metal tube, plastic tube, cardboard tube, ceramic or clay ball, ect., under a pressure of 7,000 psi. In some cases, the smoke mixture needs to be pressed into blocks, discs, or tablets under a pressure of 8,000 psi, for use in some types of smoke grenades. Note: an appropriate igniter composition to begin the burn of this smoke composition should be used.
**Burn rate:** 24 seconds per inch at 8,000 psi; tends to vary based on diameter and density of smoke grain.
**Smoke Volume on average (500 gram solid grain sample enclosed area):** 1695 sq. ft.
**Smoke Volume on average (500 gram solid grain sample open area):** 7065 sq. ft.
**Smoke visibility naked eye (sunny/dry conditions day time; 500 gram solid grain sample):** 7.4 km
**Smoke visibility naked eye (wet/rainy conditions day time; 500 gram solid grain sample):** 3.1 km
**Smoke visibility naked eye (night time, average environmental conditions; 500 gram solid grain sample):** 3000 meters.
**Water resistance:** Very good.
**Stability:** Can be stored for decades.
**Flammability (solid grain):** 4
**Flammability (powdered grain):** 4.5
**Ease of ignition (solid grain):** 3.3 (requires high temperature ignition composition for proper burn).
**Tendency to cake:** Will not cake.
**Explosive ability:** Cannot be detonated.
**Percentage:** ***40.9% para-nitroaniline, 22.72% potassium chlorate, 22.72% sublimed iodine, 13.63% lactose, 0.03% residual balance***
**Classification:** Deflagrating mixture (classified as 1.3G pyrotechnic composition).

**Danger of preparation:** Can be made by amateurs.
**Use:** Used in professional industrial/commercial smoke munitions for spotting, signaling, and in military operations.

### 06-03-036A: Opaque smoke composition for marine use:

Into a standard ball mill, filled with 500 grams of steel shot of 10mm diameter, place ***100 grams of hard coal pitch (pre-pulverized)***, followed by ***400 grams of antimony trisulfide***, followed by ***250 grams of sulfur***, followed by ***800 grams of sodium nitrate***, and then followed by ***450 grams of ammonium chloride***. Thereafter, tumble the mixture at room temperature at 150 RPM for about 45 minutes. After 45 minutes, the mixture is ready for pressing. To do so, first remove the steel shot, and then press the mixture into any suitable delivery container, such as a grenade, plastic tube, cardboard tube, ect., under a pressure of 7,000 psi. In some cases, the smoke mixture needs to be pressed into blocks, discs, or tablets under a pressure of 8,000 psi, for use in some types of smoke grenades. Note: an appropriate igniter composition to begin the burn of this smoke composition should be used.
**Burn rate:** 23 seconds per inch at 8,000 psi; tends to vary based on diameter and density of smoke grain.
**Smoke Volume on average (500 gram solid grain sample enclosed area):** 1800 sq. ft.

**Smoke Volume on average (500 gram solid grain sample open area):** 7669 sq. ft.
**Smoke visibility naked eye (sunny/dry conditions day time; 500 gram solid grain sample):** 7.9 km
**Smoke visibility naked eye (wet/rainy conditions day time; 500 gram solid grain sample):** 3.9 km
**Smoke visibility naked eye (night time, average environmental conditions; 500 gram solid grain sample):** 3000 meters.
**Water resistance:** Very good.
**Stability:** Can be stored for decades.
**Flammability (solid grain):** 4.1
**Flammability (powdered grain):** 4.7
**Ease of ignition (solid grain):** 3.9 (requires high temperature ignition composition for proper burn).
**Tendency to cake:** Will not cake.
**Explosive ability:** Cannot be detonated.
**Percentage:** ***40% sodium nitrate, 22.5% ammonium chloride, 20% antimony trisulfide, 12.5% sulfur, 5% coal pitch***
**Classification:** Deflagrating mixture (classified as 1.3G pyrotechnic composition).

**Danger of preparation:** Can be made by amateurs.
**Use:** Used in professional industrial/commercial smoke munitions for spotting, signaling, and in military operations.

### 06-03-036B: Opaque smoke composition for marine use:

Into a standard ball mill, filled with 500 grams of steel shot of 10mm diameter, place ***100 grams of pulverized bituminous coal***, followed by ***400 grams of red phosphorus***, followed by ***250 grams of powdered zinc***, followed by ***200 grams of manganese dioxide,*** followed by ***600 grams of sodium chlorate***, and then followed by ***450 grams of ammonium chloride***. Thereafter, tumble the mixture at room temperature at 150 RPM for about 45 minutes. After 45 minutes, the mixture is ready for pressing. To do so, first remove the steel shot, and then press the mixture into any suitable delivery container, such as a grenade, plastic tube, cardboard tube, ect., under a pressure of 8,000 psi. In some cases, the smoke mixture needs to be pressed into blocks, discs, or tablets under a pressure of 8,000 psi, for use in some types of smoke grenades. Note: an appropriate igniter composition to begin the burn of this smoke composition should be used.
**Burn rate:** 17 seconds per inch at 8,000 psi; tends to vary based on diameter and density of smoke grain.
**Smoke Volume on average (500 gram solid grain sample enclosed area):** 1710 sq. ft.
**Smoke Volume on average (500 gram solid grain sample open area):** 7400 sq. ft.
**Smoke visibility naked eye (sunny/dry conditions day time; 500 gram solid grain sample):** 7.2 km
**Smoke visibility naked eye (wet/rainy conditions day time; 500 gram solid grain sample):** 3.3 km
**Smoke visibility naked eye (night time, average environmental conditions; 500 gram solid grain sample):** 2998 meters.
**Water resistance:** Very good.
**Stability:** Can be stored for decades.
**Flammability (solid grain):** 6.8
**Flammability (powdered grain):** 8.9
**Ease of ignition (solid grain):** 5.7
**Tendency to cake:** Will not cake.
**Explosive ability:** Cannot be detonated, but can burn violently due to formation of some ammonium chlorate during burn.
**Percentage:** ***30% sodium chlorate, 22.5% ammonium chloride, 20% red phosphors, 12.5% powdered zinc, 10% manganese dioxide, 5% bituminous coal***
**Classification:** Deflagrating mixture (classified as 1.4G pyrotechnic composition).

**Danger of preparation:** Can be made by amateurs, but caution should be excersized
**Use:** Used in professional industrial/commercial smoke munitions for spotting, signaling, and in military operations.

### 06-03-036C: Opaque smoke composition for marine use:

Into a standard ball mill, filled with 500 grams of steel shot of 10mm diameter, place ***100 grams of hard coal pitch (pre-pulverized)***, followed by ***800 grams of antimony trisulfide***, followed by ***500 grams of sulfur***, followed by ***1500 grams of sodium nitrate***, and then followed by ***1000 grams of ammonium chloride***. Thereafter, tumble the mixture at room temperature at 100 RPM for about 45 minutes. After 45 minutes, the mixture is ready for pressing. To do so, first remove the steel shot, and then press the mixture into any suitable delivery container, such as a grenade, plastic tube, cardboard tube, ect., under a pressure of 8,000 psi. In some cases, the smoke mixture needs to be pressed into blocks, discs, or tablets under a pressure of 8,000 psi, for use in some types of smoke grenades. Note: an appropriate igniter composition to begin the burn of this smoke composition should be used.
**Burn rate:** 15 seconds per inch at 8,000 psi; tends to vary based on diameter and density of smoke grain.
**Smoke Volume on average (500 gram solid grain sample enclosed area):** 1721 sq. ft.
**Smoke Volume on average (500 gram solid grain sample open area):** 7435 sq. ft.
**Smoke visibility naked eye (sunny/dry conditions day time; 500 gram solid grain sample):** 7.29 km
**Smoke visibility naked eye (wet/rainy conditions day time; 500 gram solid grain sample):** 3.39 km

**Smoke visibility naked eye (night time, average environmental conditions; 500 gram solid grain sample):** 3000 meters.
**Water resistance:** Very good.
**Stability:** Can be stored for decades.
**Flammability (solid grain):** 6.81
**Flammability (powdered grain):** 8.99
**Ease of ignition (solid grain):** 5.8
**Tendency to cake:** Will not cake.
**Explosive ability:** Cannot be detonated, but can burn violently due to formation of some ammonium chlorate during burn.
**Percentage:** ***38.46% sodium nitrate, 25.64% ammonium chloride, 20.51% antimony trisulfide, 12.82% sulfur, 2.56% coal pitch, 0.01% mixed balance***
**Classification:** Deflagrating mixture (classified as 1.4G pyrotechnic composition).

**Danger of preparation:** Can be made by amateurs, but caution should be excersized
**Use:** Used in professional industrial/commercial smoke munitions for spotting, signaling, and in military operations.

## 06-03-037A: Under-water launched green smoke composition for marine use:

Into a standard ball mill, filled with 500 grams of steel shot of 10mm diameter, place ***191 grams of solvent green dye No. 3***, followed by ***25.5 grams of benzanthrone dye***, followed by ***38 grams of vat yellow No. 4 dye***, followed by ***140 grams of potassium chlorate***, followed by ***115 grams of powdered sugar,*** followed by ***20 grams of baking soda powder,*** and then followed by ***25 grams of diatomaceous earth***. Thereafter, tumble the mixture at room temperature at 120 RPM for about 45 minutes. After 45 minutes, the mixture is ready for pressing. To do so, first remove the steel shot, and then press the mixture into any suitable delivery container, such as a grenade, plastic tube, cardboard tube, ect., under a pressure of 8,000 psi. In some cases, the smoke mixture needs to be pressed into blocks, discs, or tablets under a pressure of 8,000 psi, for use in some types of smoke grenades. Note: an appropriate igniter composition to begin the burn of this smoke composition should be used.
**Burn rate:** 21 seconds per inch at 8,000 psi; tends to vary based on diameter and density of smoke grain.
**Smoke Volume on average (500 gram solid grain sample enclosed area):** 1700 sq. ft.
**Smoke Volume on average (500 gram solid grain sample open area):** 7400 sq. ft.
**Smoke visibility naked eye (sunny/dry conditions day time; 500 gram solid grain sample):** 7.39 km
**Smoke visibility naked eye (wet/rainy conditions day time; 500 gram solid grain sample):** 3.31 km
**Smoke visibility naked eye (night time, average environmental conditions; 500 gram solid grain sample):** 3000 meters.
**Water resistance:** Very good.
**Stability:** Can be stored for decades.
**Flammability (solid grain):** 3.3
**Flammability (powdered grain):** 4
**Ease of ignition (solid grain):** 3.7
**Tendency to cake:** Will not cake.
**Explosive ability:** Cannot be detonated.
**Percentage:** ***34.44% solvent green dye, 25.24% potassium chlorate, 20.73% sugar, 6.85% vat yellow dye, 4.59% benzanthrone dye, 4.5% diatomaceous earth, 3.6% baking soda, 0.05% mixed residual balance***
**Classification:** Deflagrating mixture (classified as 1.3G pyrotechnic composition).

**Danger of preparation:** Can be made by amateurs.
**Use:** Used in professional industrial/commercial smoke munitions for spotting, signaling, and in military operations.

## 06-03-037B: Under-water launched yellow smoke composition for marine use:

Into a standard ball mill, filled with 500 grams of steel shot of 10mm diameter, place ***113.4 grams of vat yellow dye No. 4***, followed by ***156.6 grams of benzanthrone dye***, followed by ***125 grams of potassium chlorate***, followed by ***80 grams of powdered sugar,*** followed by ***40 grams of baking soda powder,*** and then followed by ***55 grams of diatomaceous earth***. Thereafter, tumble the mixture at room temperature at 120 RPM for about 45 minutes. After 45 minutes, the mixture is ready for pressing. To do so, first remove the steel shot, and then press the mixture into any suitable delivery container, such as a grenade, plastic tube, cardboard tube, ect., under a pressure of 8,000 psi. In some cases, the smoke mixture needs to be pressed into blocks, discs, or tablets under a pressure of 8,000 psi, for use in some types of smoke grenades. Note: an appropriate igniter composition to begin the burn of this smoke composition should be used.
**Burn rate:** 23 seconds per inch at 8,000 psi; tends to vary based on diameter and density of smoke grain.
**Smoke Volume on average (500 gram solid grain sample enclosed area):** 1705 sq. ft.
**Smoke Volume on average (500 gram solid grain sample open area):** 7390 sq. ft.
**Smoke visibility naked eye (sunny/dry conditions day time; 500 gram solid grain sample):** 7.3 km
**Smoke visibility naked eye (wet/rainy conditions day time; 500 gram solid grain sample):** 3.3 km
**Smoke visibility naked eye (night time, average environmental conditions; 500 gram solid grain sample):** 2900 meters.

**Water resistance:** Very good.
**Stability:** Can be stored for decades.
**Flammability (solid grain):** 3.3
**Flammability (powdered grain):** 4
**Ease of ignition (solid grain):** 3.6
**Tendency to cake:** Will not cake.
**Explosive ability:** Cannot be detonated.
**Percentage:** ***27.47% benzanthrone dye, 21.92% potassium chlorate, 19.89% vat yellow dye No. 4, 14.03% powdered sugar, 9.64% diatomaceous earth, 7.01% baking soda, 0.04% mixed residual balance***
**Classification:** Deflagrating mixture (classified as 1.3G pyrotechnic composition).

**Danger of preparation:** Can be made by amateurs.
**Use:** Used in professional industrial/commercial smoke munitions for spotting, signaling, and in military operations.

# Section 4: Pyrotechnic igniter (starter) and "primer" compositions

***Chemicals used in this section (binders are not included)***

| | |
|---|---|
| **1. Potassium nitrate (see Black Powder)** | **2. Sulfur (see Black Powder)** |
| **3. Charcoal (see Black Powder)** | **4. Sugar Carbon (see Modified Black Powder)** |
| **5. Barium Chromate (see Modified Black Powder)** | **6. Potassium Perchlorate (see Modified Black Powder)** |
| **7. Potassium Dichromate (see Modified Black Powder)** | **8. Ammonium Bisulfide (see Modified Black Powder)** |
| **9. Potassium chlorate (see Modified Black Powder)** | **10. Carbon Disulfide (see Modified Black Powder)** |
| **11. Nitrocellulose (see Modified Black Powder)** | **12. Lead Tetraoxide (see Modified Black Powder)** |
| **13. Diphenylamine (see Modified Black Powder)** | **14. Titanium Dioxide (see Modified Black Powder)** |
| **15. Manganese Dioxide (see Modified Black Powder)** | **16. Sodium Benzoate (see Modified Black Powder)** |
| **17. Ammonium Nitrate (see Modified Black Powder)** | **18. Calcium Carbonate (see Modified Black Powder)** |
| **19. Sodium Nitrate (see Modified Black Powder)** | **20. Urea (see Modified Black Powder)** |
| **21. Lead Nitrate (see Modified Black Powder)** | **22. Nitro Starch (see Modified Black Powder)** |
| **23. Ammonium Perchlorate (see Ammonium Perchlorate Rocket Propellants)** | **24. Aluminum powder (see Ammonium Perchlorate Rocket Propellants)** |
| **25. Magnesium Sulfide (see Ammonium Perchlorate Rocket Propellants)** | **26. Copper Chromite (see Ammonium Perchlorate Rocket Propellants)** |
| **27. Nitroguanidine (see Ammonium Perchlorate Rocket Propellants)** | **28. Ammonium Sulfate (see Ammonium Perchlorate Rocket Propellants)** |
| **29. Nitroglycerine (see Ammonium Perchlorate Rocket Propellants)** | **30. Magnesium Oxide (see Ammonium Perchlorate Rocket Propellants)** |
| **31. Iron-II-oxide (see Ammonium Perchlorate Rocket Propellants)** | **32. Zirconium Hydride (see Ammonium Perchlorate Rocket Propellants)** |
| **33. Teflon (see Ammonium Perchlorate Rocket Propellants)** | **34. Zinc Oxide (see Ammonium Perchlorate Rocket Propellants)** |
| **35. Ammonium Dichromate (see Ammonium Perchlorate Rocket Propellants)** | **36. Lithium Perchlorate (see Ammonium Perchlorate Rocket Propellants)** |
| **37. Magnesium powder (see Ammonium Perchlorate Rocket Propellants)** | **38. Lithium Aluminum Hydride (see Ammonium Perchlorate Rocket Propellants)** |
| **39. Iron-III-oxide (red iron oxide) (see Ammonium Perchlorate Rocket Propellants)** | **40. PVC (see Ammonium Perchlorate Rocket Propellants)** |
| **41. Copper Sulfide (see Ammonium Perchlorate Rocket Propellants)** | **42. Sodium Hydride (see Ammonium Perchlorate Rocket Propellants)** |
| **43. Titanium Hydride (see Ammonium Perchlorate Rocket Propellants)** | **44. Silicon Nitride (see Ammonium Perchlorate Rocket Propellants)** |
| **45. ADN (see ADN Rocket Propellants)** | **46. Urea Nitrate (see ADN Rocket Propellants)** |
| **47. Silicon Powder (see ADN Rocket Propellants)** | **48. Hexamine (see ADN Rocket Propellants)** |
| **49. Boron powder (see ADN Rocket Propellants)** | **50. Sodium hypophosphite (see ADN Rocket Propellants)** |
| **51. KDN (see ADN Rocket Propellants)** | **52. Nickel Chloride (see Ammonium Nitrate Rocket Propellants)** |

| | | |
|---|---|---|
| **53. TNT (see Ammonium Nitrate Rocket Propellants)** | | **54. Acrylamide (see Miscellaneous Rocket H.P. Rocket Propellants)** |
| **55. Ethylene Glycol (see Miscellaneous Rocket H.P. Rocket Propellants)** | | **56. Magnesium Perchlorate (see Miscellaneous Rocket H.P. Rocket Propellants)** |
| **57. Metallic Lithium (see Miscellaneous Rocket H.P. Rocket Propellants)** | | **58. Beryllium Hydride (see Miscellaneous Rocket H.P. Rocket Propellants)** |
| **59. Red Phosphorus (see Miscellaneous Rocket H.P. Rocket Propellants)** | | **60. Sodium Borohydride (see Miscellaneous Rocket H.P. Rocket Propellants)** |
| **61. Sodium chlorate (see Miscellaneous Rocket H.P. Rocket Propellants)** | | **62. Lead Dioxide (see Miscellaneous Rocket H.P. Rocket Propellants)** |
| **63. Benzoyl Peroxide (see Miscellaneous Rocket H.P. Rocket Propellants)** | | **64. Barium Nitrate (see Miscellaneous Rocket H.P. Rocket Propellants)** |
| **65. Cyanuric Acid (see Miscellaneous Rocket H.P. Rocket Propellants)** | | **66. Aluminum Stearate (see Ammonium Nitrate Gun Propellants)** |
| **67. Aluminum Hydride (see Ammonium Nitrate Gun Propellants)** | | **68. Magnesium Peroxide (see Ammonium Nitrate Gun Propellants)** |
| **69. Potassium Permanganate (see Ammonium Nitrate Gun Propellants)** | | **70. Calcium Hydride (see Ammonium Nitrate Gun Propellants)** |
| **71. Potassium Tartrate (see Ammonium Nitrate Gun Propellants)** | | **72. Potassium Ferrocyanide (see Ammonium Nitrate Gun Propellants)** |
| **73. Sodium Azide (see Ammonium Nitrate Gun Propellants)** | | **74. Sodium Chloride (see Ammonium Nitrate Gun Propellants)** |
| **75. Potassium Sulfate (see Nitrocellulose Gun Propellants)** | | **76. Lead Stearate (see Nitrocellulose Gun Propellants)** |
| **77. Triethylene Glycol (see Nitrocellulose Gun Propellants)** | | **78. Sugar (sucrose) (see Miscellaneous Gun Propellants)** |
| **79. Sodium Propionate (see Miscellaneous Gun Propellants)** | | **80. Picric Acid (see Miscellaneous Gun Propellants)** |
| **81. Copper-II-oxide (see Miscellaneous Gun Propellants)** | | **82. Ammonium Picrate (see Miscellaneous Gun Propellants)** |
| **83. Barium Peroxide (see Bullet Tracer Compositions)** | | **84. Magnesium Carbonate (see Bullet Tracer Compositions)** |
| **85. Strontium Peroxide (see Bullet Tracer Compositions)** | | **86. Strontium Nitrate (see Bullet Tracer Compositions)** |
| **87. Cupric chloride (see Bullet Tracer Compositions)** | | **88. Hexachlorobenzene (see Bullet Tracer Compositions)** |
| **89. Strontium oxalate (see Bullet Tracer Compositions)** | | **90. Mercury-I-Chloride (see Bullet Tracer Compositions)** |
| **91. Zinc Oxalate (see Bullet Tracer Compositions)** | | **92. Zinc Chloride (see Bullet Tracer Compositions)** |
| **93. Uranium (see Bullet Tracer Compositions)** | | **94. Zirconium nitrate (see Bullet Tracer Compositions)** |
| **95. Yttrium Nitrate (see Bullet Tracer Compositions)** | | **96. Yttrium Oxide (see Bullet Tracer Compositions)** |
| **97. Zirconium Oxide (see Bullet Tracer Compositions)** | | **98. Cerium Oxide (see Bullet Tracer Compositions)** |
| **99. Hexachloroethane (see Bullet Tracer Compositions)** | | **100. Antimony trisulfide (see Bullet Tracer Compositions)** |
| **101. Anthracene (see Bullet Tracer Compositions)** | | **102. Phosphorus Sesquisulphide (see Match Compositions)** |
| **103. Boric acid (see Match Compositions)** | | **104. Aluminum Hydroxide (see Match Compositions)** |
| **105. Antimony Pentasulfide (see Match Compositions)** | | **106. Glucose (see Match Compositions)** |
| **107. Sodium Hydroxide (see Match Compositions)** | | **108. Lead Hypophosphite (see Match Compositions)** |
| **109. Calcium Sulfate (see Match Compositions)** | | **110. Ammonium Chloride (see Smoke Generating Compositions)** |
| **111. Manganese (see Smoke Generating Compositions)** | | **112. Lactose (see Smoke Generating Compositions)** |
| **113. Propylene Glycol (see Smoke Generating Compositions)** | | **114. Glycerol (see Smoke Generating Compositions)** |
| **115. Potassium Chloride (see Smoke Generating Compositions)** | | **116. Potassium Bicarbonate (see Smoke Generating Compositions)** |
| **117. Dicyanodiamide (see Smoke Generating Compositions)** | | **118. Naphthalene (see Smoke Generating Compositions)** |

| | |
|---|---|
| **119. Thiourea (see Smoke Generating Compositions)** | **120. Phthalic Anhydride (see Smoke Generating Compositions)** |
| **122. Cadmium powder (see Smoke Generating Compositions)** | **123. Cadmium Sulfide (see Smoke Generating Compositions)** |
| **124. Melamine (see Smoke Generating Compositions)** | **125. Malic Acid (see Smoke Generating Compositions)** |
| **126. Calcium Lactate (see Smoke Generating Compositions)** | **127. Metallic Sodium (see Smoke Generating Compositions)** |
| **128. Bismuth Tetraoxide (see Smoke Generating Compositions)** | **129. Bismuth Subnitrate (see Smoke Generating Compositions)** |
| **130. Calcium Iodate (see Smoke Generating Compositions)** | **131. Potassium Iodate (see Smoke Generating Compositions)** |
| **132. Magnesium Chloride (see Smoke Generating Compositions)** | **133. Para-Nitroaniline (see Smoke Generating Compositions)** |
| **134. Iodine (see Smoke Generating Compositions)** | **135. Potassium Ferriccyanide** |
| | 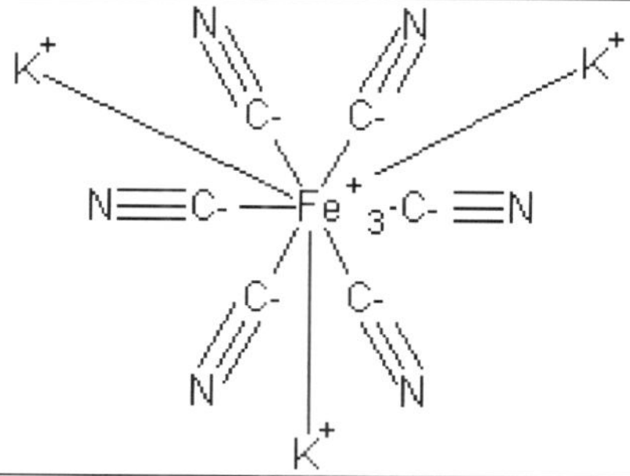 |
| | Forms brilliant ruby red crystals. The crystals are soluble in water, but only slightly soluble in alcohol and other solvents. The crystals react with acids forming the highly toxic hydrogen cyanide gas. Keep out of contact with acids. |
| **136. Potassium hexacyanocobaltate** | **137. Bismuth Trioxide** |
| 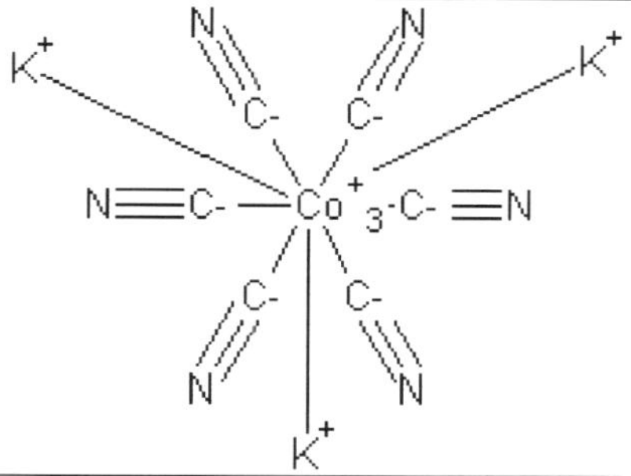 | 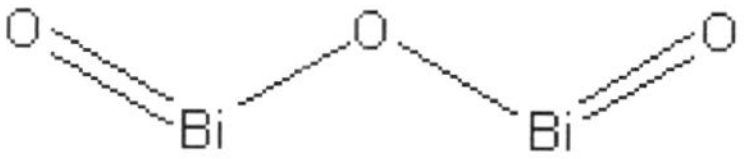 |
| Forms slightly yellow to pale-yellow crystals. The crystals are freely soluble in water and glacial acetic acid. The crystals are insoluble in alcohol and most common solvents. The solid decomposes into hydrogen cyanide when exposed to acids. Hydrogen cyanide also slowly forms on standing. | Forms a yellowish powder, which is stable in air. The crystals are insoluble in water and the usual solvents, but they are soluble in acids. |
| **138. Titanium powder** | **139. Tungsten powder** |
| 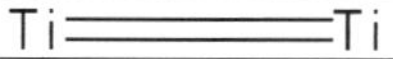 | 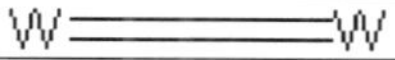 |
| Titanium forms a dark gray powder, which is stable in moist air, but ignites readily upon a flame. The powder is insoluble in water and all known solvents, but reacts with acids in the usual manner. Titanium powder should be stored in a cool dry place. | Powdered tungsten forms a dark gray to bluish gray solid. The solid may ignite spontaneously under certain conditions. The powder readily reacts with acids. |
| **140. Lead Powder** | **141. Lead-II-Oxide (red lead; litharge)** |
| 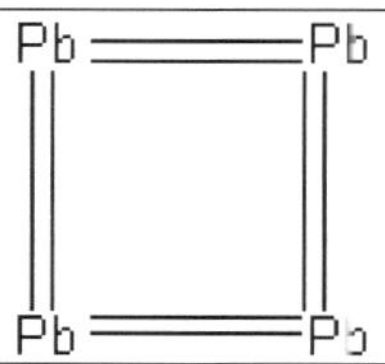 | 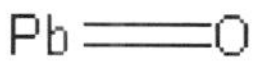 |
| Lead powder is a dark gray solid. Some forms of lead powder may be pyrophoric, and should be stored in air free bottles. Lead powder is insoluble in water and the usual organic solvents. | Exists in two forms, from which the alpha form is the most common, forming a reddish powder. The powder is insoluble in water and the usual solvents. Avoid inhalation of the dust, as the oxide is toxic. |
| | |

### 142. Selenium powder

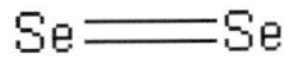

Selenium has several forms, but the powder can range from dark red to reddish-bluish-black, or grayish bluish black in appearance. The powder is insoluble in water and all known solvents. Selenium powder is stable at room temperature and resists chemical attack from a wide range of chemicals.

### 143. Sodium Bicarbonate

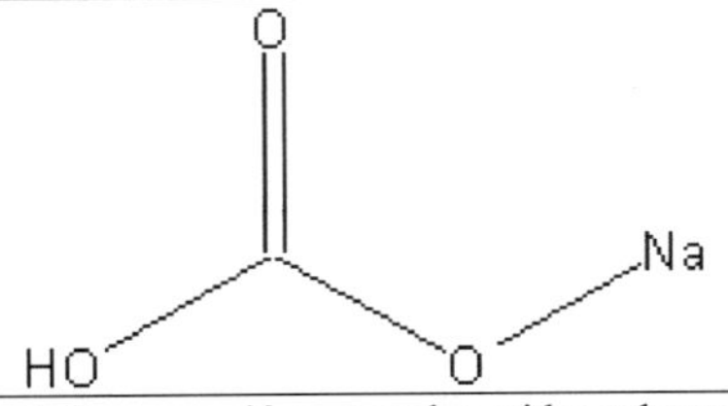

Sodium bicarbonate forms a uniform powder, with a salty sour taste. The powder is ready converted into sodium carbonate when heated to high temperature. The bicarbonate is also known as baking soda. Like many carbonates, sodium bicarbonate reacts vigorously with acids.

### 144. Iron powder

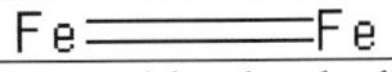

Iron powder forms a distinct grayish colored solid, the powder readily ignites when exposed to flames, and the powder readily converts to the oxide when heated. Iron powder is very reactive and will react with many other elements either on contact or when heated. However, iron powder is stable at room temperature.

### 145. Silicon Dioxide

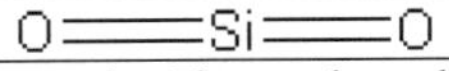

Forms a white amorphous powder. The powder melts to a glass when heated. Silicon dioxide powder is almost insoluble in water and all organic solvents. One of only a few chemicals capable of reacting with it at ordinary temperatures is hydrofluoric acid.

### 146. Lead Thiocyanate

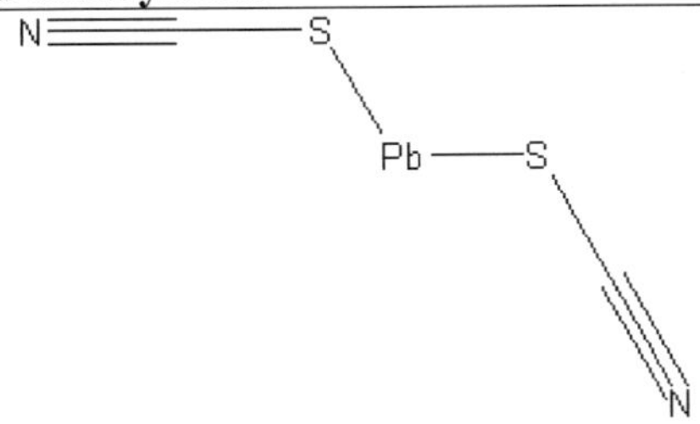

Forms a white odorless powder. The powder is toxic so users should avoid skin contact and inhalation of the dust. The powder is insoluble in water and the usual organic solvents.

### 147. Para-Nitrotoluene

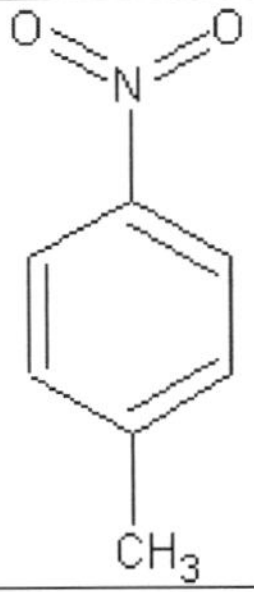

The para isomer forms yellowish crystals with a melting point of 54 Celsius, and a boiling point of 238 Celsius. The crystals are insoluble in water but soluble in the usual organic solvents. Nitrotoluene is a toxic liquid, so users should wear proper gloves when handling.

### 148. Silver powder

**Ag**

Silver powder forms a grayish to grayish white solid. The solid is stable at room temperature and does not tarnish. The powder ready ignites in the presence of flame. The solid is insoluble in water and the usual solvents. Silver is relatively stable to chemical attack.

### 149. Sodium Tungstate

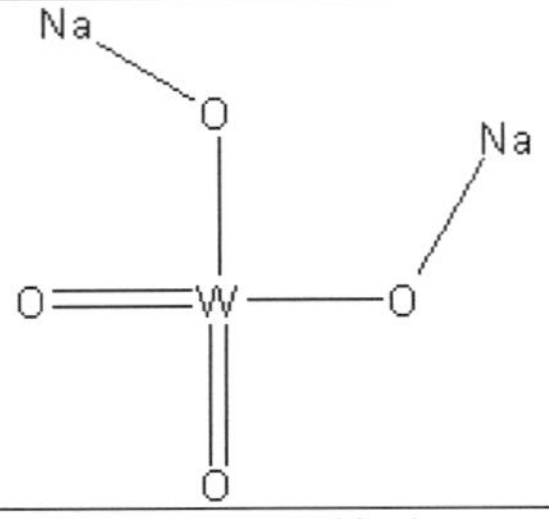

Forms a dihydrate, which is colorless to white in nature. The crystals become anhydrous when heated to 100 Celsius. The solid is soluble in water, but insoluble in the usual solvents.

### 150. Zirconium powder

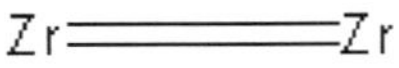

Zirconium metal powder forms a bluish gray to bluish black solid. The powder may be a mixture of the crystalline and amorphous styles. Like most metal powders, it readily burns when ignited, and is much more reactive then the solid metal. Mixtures of the powder with oxidizing agents form explosive mixtures.

### 151. Bismuth powder

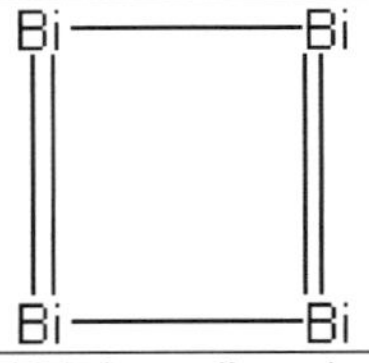

Forms a dull gray powder, which forms dispersion in water. The powder is insoluble in water and the usual solvents. The powder is stable at room temperature but reacts in the usual manner as other powders.

## 152. Copper-I-oxide

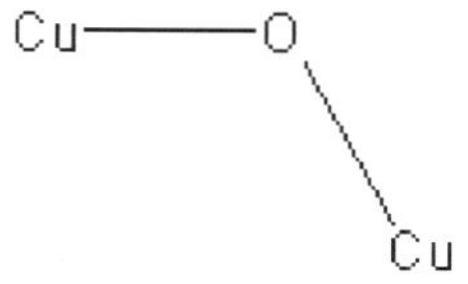


Forms a brick red powder, granules, or solid masses. The oxide slowly oxidizes to the hydroxide on standing in moist air. The oxide is readily converted to the black version when heated. The oxide is insoluble in water and the usual solvents. It can be made by electrolysizing a dilute hydrochloric acid solution between two copper electrodes.

## 153. Lead Styphnate

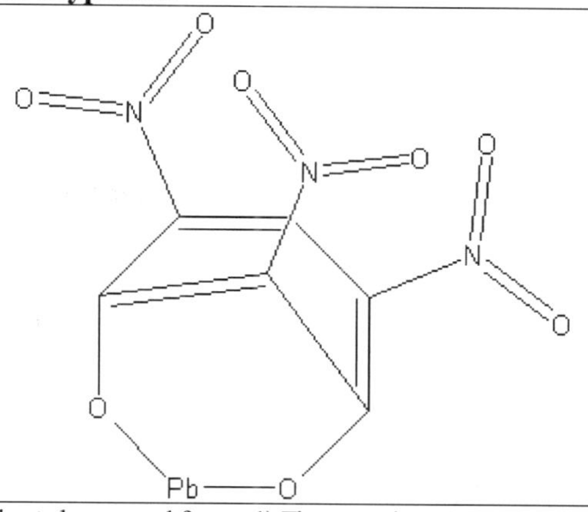


Lead Styphnate has several forms: 1) The amorphous basic lead styphnate forms a yellowish powder, needle-like crystals, or granules, 2) the red basic lead styphnate forms reddish or reddish-brown square prisms, or reddish hexagonal plates, and 3) the normal lead styphnate forms yellowish-orange crystals. All these forms of lead styphnate detonate when heated to about 260 Celsius. All forms are an excellent initiating explosive for use in blasting caps, and detonators. They are also widely used in the preparation of priming mixtures for a variety of applications. All forms of lead styphnate are sensitive to shock, heat, friction, and percussion, but the red basic lead styphnate tends to be the most unstable form. For use in initiation (ignition) compositions, all forms should be desensitized with small amounts of gum Arabic, sulfur, carbon black, dextrose, magnesium sulfate, or starch and mixed with an oxidizer such as barium chromate, lead chromate, copper permanganate, copper perchlorate, or magnesium dichromate.

**Methodof Preapration 1:** Place 1500 milliliters of water into a beaker, and then add and dissolve 24.4 grams of styphnic acid, and then 16 grams of sodium hydroxide. Then, heat this mixture to 100 Celsius with stirring. Then, prepare a solution by dissolving 100 grams of lead nitrate into 2500 milliliters of water, and then add drop-wise the styphnic acid/sodium hydroxide solution to the lead nitrate solution as a steady drip, while stirring the lead nitrate solution. The rate of addition should be as such that the styphnic acid/sodium hydroxide mixtures temperature does not fall below 90 Celsius. Note: Simply place the heated styphnic acid/sodium hydroxide solution into a dropping funnel while it's boiling hot, and then begin the addition. After the addition, stir the reaction mixture for about 20 minutes, and then filter-off the precipitated lead styphnate. After filtering, allow the basic lead styphnate to cool to room temperature. Thereafter, wash with 200 milliliters of cold water, and then vacuum dry or air-dry.

**Method of Preparation 2:** Place 800 milliliters of water into a beaker, and then add, while rapidly stirring the water, 24.4 grams of styphnic acid, and then 16 grams of sodium hydroxide. Then, heat this mixture to 70 Celsius while stirring. Then dissolve 70 grams of lead nitrate into 700 milliliters of water. Thereafter, place the heated styphnic acid/sodium hydroxide into a dropping funnel, and then slowly add, drop-wise, drop-by-drop the styphnic acid/sodium hydroxide mixture into the lead nitrate solution while stirring the lead nitrate. Now, in the beginning, slowly add the styphnic acid/sodium hydroxide mixture very slowly, drop-by-drop until a small amount of an amorphous yellow precipitate forms. When this happens, stop the addition, and allow this amorphous yellow solid to crystallize as reddish-brown diamond shaped crystals. Where upon, add the styphnic acid/sodium hydroxide mixture at a much faster pace, and continue the addition until all the styphnic acid/sodium hydroxide mixture has been added. During the whole operation, stir the lead nitrate solution. After the addition, continue to stir the reaction mixture for 20 minutes, and then filter-off the precipitated lead styphnate. Allow it to cool to room temperature, and then wash with 200 milliliters of cold water, followed by vacuum drying or air-drying.

**Method of preparation 3:**

**Step 1: Preparation of magnesium styphnate**

Add 182 grams of styphnic acid to 466 milliliters of distilled water, and then heat the mixture to 45 Celsius. When the temperature reaches 45 Celsius, add 74 grams magnesium carbonate in small quantities while stirring the mixture. Gas evolution will begin immediately. After the addition of some of the magnesium carbonate, raise the temperature to 55 Celsius and continue adding the magnesium carbonate in small quantities while continuously stirring the mixture. After the addition of the magnesium carbonate, remove the heat source and allow the solution to cool to room temperature. Then add cold water to the reaction mixture until its total volume equals 732 milliliters. After which, filter the reaction mixture to remove any insoluble impurities. After filtration, add 2.2 milliliters of 70% nitric acid while stirring the filtered reaction mixture.

**Step 2: Preparation of lead nitrate solution**

Prepare a lead nitrate solution by adding 248 grams of crystalline lead nitrate to 670 milliliters of distilled water. Then warm the solution to 50 Celsius while stirring. Continue stirring the mixture until all the lead nitrate has dissolved. When all the lead nitrate has dissolved, cool the solution to room temperature and then filter to remove any insoluble impurities.

**Step 3: Preparation of lead styphnate**

Place 132 milliliters of the magnesium styphnate solution (prepared in step 1) into a flask and heat to 80 Celsius while stirring. Then slowly add 114 milliliters of the lead nitrate solution (prepared in step 2) over a period of 40 minutes while maintaining the temperature of the magnesium styphnate solution at 80 Celsius and stirring. After addition of the lead nitrate solution, continue heating the mixture at 80 Celsius for 10 minutes and then remove the heat source, and allow the mixture to cool to room temperature. Afterwards, filter-off the lead styphnate precipitate, and then wash with 1000 milliliters of water. Then vacuum dry or air-dry the product (no heat).

## 154. Tellurium Dioxide

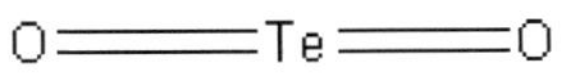


Forms white crystals with a melting point of 733 Celsius. The crystals turn yellow upon heating, forming a yellow liquid at its melting point. The powder is insoluble in water and the usual solvents, but will dissolve in sodium hydroxide solutions, and acids.

## 155. Tetracene

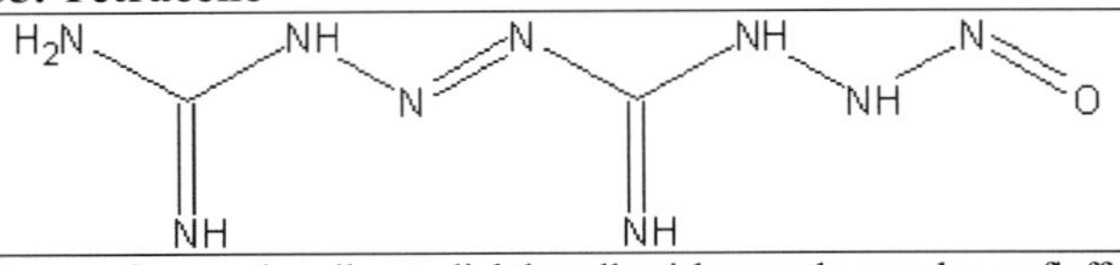


Tetracene forms pale yellow to lightly yellowish crystals, powder, or fluffy flakes which begin to decompose when heated to 150+ Celsius. The crystals may explode if rapidly heated to 161 Celsius producing a flash of mild light and a smoky bang with the evolution of ammonia. Tetracene is decomposed by hot water and alkalies, and should be stored dry in amber glass bottles, and in a cool place away from light. Tetracene is used heavily in the formation of initiating charges and booster charges in combination with lead azide, diazodinitrophenol, lead styphnate, and other common primary explosives. Tetracene has been replaced by many other primary explosives such as sulfur nitride, and salts of triazoles and tetrazoles for use in blasting caps and detonators. However it can still be used in blasting caps and detonators with satisfactory results when compressed using hydraulic presses—tetracene should be packed into blasting caps and detonators under high pressure.

**Method of Preparation 1:** Into a suitable flask, equipped with a thermometer, motorized stirrer and powder funnel, add 30 grams of aminoguanidine hydrochloride, followed by 20 milliliters of glacial acetic acid, followed by 5 milliliters of acetic anhydride. Thereafter, gently heat this mixture to 40 Celsius, and stir for 1 hour. Then slowly add, in small portions, 30 grams of sodium nitrite while stirring the reaction mixture and maintaining its temperature at 40 Celsius. After the addition of the sodium nitrite, continue to stir the reaction mixture at 40 Celsius for 1 hour. Afterwards, remove the heat source, and allow the reaction mixture to cool to room temperature. Then drown the entire reaction mixture into 500 milliliters of ice-cold water, and then stir the entire mixture for 24 hours at room temperature. After 24 hours, filter-off the insoluble crystals, and then wash them with three 250 milliliter portions of cold water. Thereafter, vacuum dry or air-dry the crystals, and then place the crystals into a desiccator and dry for 48 hours. Thereafter, place the crystals into a amber glass bottle and store in a cool place until use.

## 156. Iron Sulfide

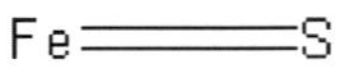


## 157. Zinc Phosphide

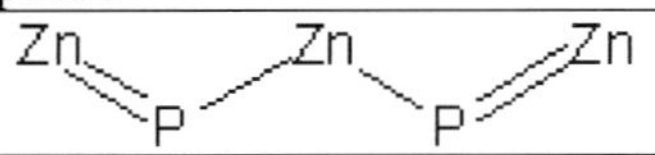

| | |
|---|---|
| The pure material is a colorless to white solid. Usual commercial grades are a grayish color. The melting point is 1194 Celsius. The crystals are oxidized by moist air, and are insoluble in water and all known solvents; however, the powder reacts with acids with the evolution of hydrogen sulfide gas. | Forms a dark gray to grayish lustrous powder, granules, or masses. The solid has a melting point of 420 Celsius, but the solid usually sublimes before the melting point. Zinc phosphide is stable at room temperature in the absence of moisture, but slowly oxidizes upon standing in moist air. The powder reacts with acids with the formation of the spontaneously flammable gas, phosphine. |
| **158. Copper powder** | **159. Hafnium powder** |
|  | 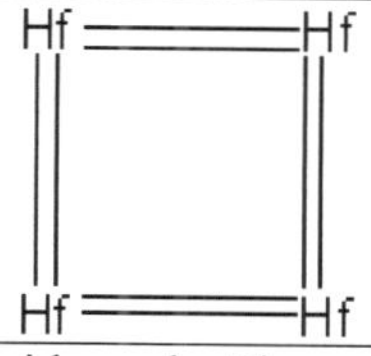 |
| Copper powder forms a reddish ductile solid. The powder readily converts to the oxide upon heating. Copper powder is insoluble in water and all known solvents. | Forms a silvery-white to grayish powder. The powder is highly reactive and may tarnish on exposure to air. |

***- Ignition and Priming Compositions in this section -***

| | |
|---|---|
| **1. 06-04-001A: Shock resistant igniter/primer composition (produces flash and loud report upon ignition):** *61% potassium perchlorate, 39% potassium ferricyanide* | **2. 06-04-001B: Modified shock resistant igniter/primer composition (produces flash and loud report upon ignition):** *61% potassium perchlorate, 39% potassium hexacyanocobaltate* |
| **3. 06-04-002A: Igniter composition (produces no flash or report upon ignition):** *69.4% potassium dichromate, 30.5% potassium ferricyanide, 0.10% moisture* | **4. 06-04-003A: High altitude igniter composition (can be used in supersonic munitions):** *53.8% zirconium, 23% sulfur, 15.3% potassium perchlorate, 7.6% rubber binder, 0.15% potassium dichromate, 0.15% sodium bisulfite* |
| **5. 06-04-004A: "Invisible", dark burning pyrotechnic igniter mixture for tracer compositions:** *63% bismuth trioxide, 35% manganese, 2% graphite* | **6. 06-04-004B: "Invisible", dark burning pyrotechnic igniter mixture for tracer compositions:** *76% barium peroxide, 23% antimony pentasulfide, 1% graphite* |
| **7. 06-04-005A: "Non-violent burning" pyrotechnic igniter element for initiating compositions such as thermite type compositions that have ignition temperatures exceeding 1000 Celsius:** *70% iron oxide, 30% titanium* | **8. 06-04-006A: "Therm 64C" igniter composition for magnesium bombs:** *44% iron oxide, 29% barium nitrate, 25% aluminum, 2% sulfur* |
| **9. 06-04-007A: Igniter composition for low velocity rocket propellants:** *41.5% potassium chlorate, 28.8% sodium bicarbonate, 16.1% sulfur, 9.6% corn starch, 3.8% nitrocellulose, 0.20% residue* | **10. 06-04-008A: Non-luminous igniter composition for various applications:** *65% tungsten, 25% barium chromate, 10% potassium perchlorate* |
| **11. 06-04-009A: Priming composition (ignition composition) for use in electric blasting caps:** *60% lead, 25% selenium, 15% lead-II-oxide* | **12. 06-04-010A: Ignition composition for high performance rocket propellants:** *51.4% iron, 42.6% potassium perchlorate, 3% cellulose acetate, 2.9% aluminum* |
| **13. 06-04-011A: Gasless ignition composition for "thermite" type compositions:** *57% copper oxide, 40% manganese, 3% sulfur* | **14. 06-04-011B: Gasless ignition composition for flares, smoke, and other general pyrotechnic compositions:** *35.2% copper oxide, 35.2% silicon, 17.6% lead dioxide, 11.7% calcium silicide, 0.30% residue* |
| **15. 06-04-012A: Ignition composition for tracer compositions:** *66.6% barium peroxide, 22.3% magnesium, 11.1% lead tetraoxide* | **16. 06-04-013A: Ignition composition for electric blasting caps and detonators:** *59.25% lead, 27.51% selenium, 13.22% lead-II-oxide, 0.02% residual lead oxides* |
| **17. 06-04-014A: Ignition composition for high performance rocket propellants:** *65.6% Dupont Teflon No. 6, 32.3% magnesium, 1.9% fumes silica, 0.2% mixed residues* | **18. 06-04-015A: Ignition composition for incendiary compositions:** *50% silicon dioxide, 40% magnesium, 10% polybutadiene binder* |
| **19. 06-04-016A: Ignition composition for electric primers:** *68% potassium chlorate, 27% titanium, 5% boron* | **20. 06-04-016B: Ignition composition for electric primers with increased sensitivity (modified):** *76.5% potassium chlorate, 20.4% titanium, 2% boron, 1% lead thiocyanate, 0.10% mixed residues* |
| **21. 06-04-017A: Ignition composition for flares, signals, and smoke candles utilizing black powder:** *85% black powder, 10% nitro starch, 5% para-nitrotoluene* | **22. 06-04-018A: Priming composition for blasting caps, and for use in fireworks:** *44.8% potassium chlorate, 25.4% silver, 22.6% antimony sulfide, 4.7% lead dioxide, 0.94% sodium tungstate, 0.94% strontium nitrate, 0.47% copper, 0.15% mixed impurities* |
| | |

| | |
|---|---|
| **23. 06-04-018B: Priming composition for blasting caps, and for use in fireworks with increased shock resistance:** *40% copper ammonium nitrate, 28% aluminum, 25% potassium nitrate, 7% sulfur* | **24. 06-04-019A: Ignition composition for high performance ammonium nitrate/perchlorate rocket propellants:** *53.8% zirconium/nickel alloy, 26.4% potassium perchlorate, 16.5% barium nitrate, 3% ethyl cellulose binder, 0.3% mixed residues* |
| **25. 06-04-019B: Ignition composition for high performance ammonium nitrate/perchlorate rocket propellants (modified):** *16.6% aluminum, 16.1% zirconium/nickel alloy, 14% barium nitrate, 13.6% potassium perchlorate, 13% titanium metal, 11.6% zirconium dichromate, 9.4% magnesium, 3.0% ethyl cellulose binder, 1.8% boron, 0.50% mixed residues, 0.4% zirconium hydride burn catalyst* | **26. 06-04-020A: Gasless ignition composition for high performance rocket propellants:** *52.3% iron, 42.8% potassium perchlorate, 4.7% black powder, 0.2% residues* |
| **27. 06-04-021A: Non-gaseous high temperature ignition composition for incendiary mixtures:** *65.2% iron powder, 21.7% copper-II-oxide, 13% magnesium, 0.1% mixed impurities* | **28. 06-04-022A: Standard ignition composition for tracer mixtures, smoke compositions, and similar mixtures:** *89.5% barium nitrate, 10.5% magnesium* |
| **29. 06-04-022B: Standard ignition composition for tracer mixtures, smoke compositions, and similar mixtures (modified):** *66.6% barium peroxide, 22.3% magnesium, 11.1% lead tetraoxide* | **30. 06-04-023A: Priming composition with heat resistant properties for use in firing lead azide in explosive trains:** *42.8% lead dioxide, 28.5% calcium silicide, 14.2% zirconium, 14.2% sulfur, 0.30% mixed residues* |
| **31. 06-04-023B: Priming composition with heat resistant properties for use in firing lead azide in explosive trains (modified):** *37.5% lead dioxide, 20.8% zirconium, 16.6% calcium silicide, 16.6% manganese dioxide, 8.3% sulfur, 0.20% mixed residues* | **32. 06-04-024A: Ignition composition for use in non-electric blasting caps:** *70% metallic bismuth, 25% selenium metal, 5% potassium chlorate* |
| **33. 06-04-025A: Easily ignitable gasless ignition composition for use in igniting incendiary compositions:** *70% iron-III-oxide, 30% titanium* | **34. 06-04-025B: Easily ignitable gasless ignition composition for use in igniting incendiary compositions:** *70% iron-III-oxide, 30% zirconium metal* |
| **35. 06-04-026A: Easily ignitable gasless ignition composition for use in igniting incendiary compositions:** *47.5% manganese, 40% magnesium, 12.2% titanium/manganese alloy, 0.14% mixed residues, 0.08% titanium, 0.08% zirconium* | **36. 06-04-027A: Priming composition for use in percussion caps:** *25% animal glue binder, 25% oxalic acid, 25% antimony sulfide, 25% potassium chlorate* |
| **37. 06-04-028A: Slow burning, flashless, ignition composition for tracer compositions:** *78.4% barium peroxide, 18.4% antimony pentasulfude, 2.2% magnesium, 1% graphite* | **38. 06-04-029A: Ignition composition for welding specifically for tin producing welding compositions:** *36.1% copper-I-oxide, 30.5% copper-II-oxide, 16.6% metal alloy, 8.3% aluminum, 8.3% red phosphorus, 0.20% combined impurities* |
| **39. 06-04-030A: Gasless ignition composition for use in electric blasting caps and detonators:** *50% zirconium, 50% lead monoxide* | **40. 06-04-031A: Friction sensitive composition for use in primers:** *30% lead styphnate, 28% barium nitrate, 20% glass, 12% lead nitrate, 10% lead hypophosphite* |
| **41. 06-04-031B: Friction sensitive composition for use in primers (tetracene containing):** *40% lead styphnate, 30% lead nitrate, 20% glass, 8% lead sulfocyanate, 2% tetracene* | **42. 06-04-032A: Conductive friction sensitive composition for use in primers (non primary explosive containing):** *38.9% zirconium, 35.5% barium nitrate, 25.4% lead dioxide, 0.2% balance* |
| **43. 06-04-033A: Ignition composition for use in delay trains and for generating heat for other ignition operations:** *60.6% aluminum, 29.9% iron-III-oxide, 6.7% silicon, 1.9% graphite, 0.78% titanium, 0.15% impurities* | **44. 06-04-033B: Primer composition for use in detonators:** *54.5% potassium chlorate, 27.5% aluminum, 13.5% iron-III-oxide, 3% silicon, 0.86% graphite, 0.35% titanium, 0.29% mixed impurities and balance* |
| **45. 06-04-034A: Priming composition for use in electric blasting caps or for use in electric ignition devices:** *57.1% silver metal, 17.1% potassium chlorate, 14.2% mercury metal, 8.5% sulfur, 2.8% red phosphorus, 0.30% balance* | **46. 06-04-035A: Ignition composition for flares and illumination compositions:** *85% barium chromate, 10% boron, 5% magnesium* |
| | |

| | |
|---|---|
| **47. 06-04-036A: Ignition composition for incendiary agents (magnesium ophorite):** *64.3% potassium perchlorate, 33.5% magnesium flake, 1.9% castor oil, 0.3% balance* | **48. 06-04-036B: Ignition composition for incendiary agents (Aluminum ophorite) with increased heat of combustion:** *64.3% potassium perchlorate, 33.5% aluminum flake, 1.9% castor oil, 0.3% balance* |
| **49. 06-04-037A: Heat resistant ignition composition with high thermal stability:** *80% tellurium dioxide, 17% aluminum, 3% carbon black* | **50. 06-04-038A: Ignition composition for use in aircraft ejection seats:** *68.37% strontium nitrate, 17% zirconium metal, 6.8% calcium resinate, 5.1% chlorinated rubber, 1.7% zinc stearate, 0.85% dye, 0.18% balance* |
| **51. 06-04-038B: Ignition composition for use in aircraft ejection seats:** *68.37% lead dioxide, 17% boron, 6.8% strontium resinate, 5.1% chlorinated rubber, 1.7% zinc stearate, 0.85% dye, 0.18% balance* | **52. 06-04-038C: Ignition composition for use in aircraft ejection seats:** *68.37% barium nitrate, 17% magnesium, 6.8% calcium resinate, 5.1% chlorinated rubber, 1.7% zinc stearate, 0.85% toludiene red, 0.18% balance* |
| **53. 06-04-039A: Priming composition for multiple uses (friction sensitive):** *48% potassium chlorate, 36% potassium ferrocyanide, 16% glass powder* | **54. 06-04-040A: Priming composition for use in electric blasting caps with good thermal stability:** *85% cesium decahydrodecaborate, 15% potassium perchlorate* |
| **55. 06-04-040B: Priming composition for use in electric blasting caps with good thermal stability (with increased sensitivity):** *73% cesium decahydrodecaborate, 27% potassium perchlorate* | **56. 06-04-040C: Priming composition for use in electric blasting caps with good thermal stability utilizing potassium permanganate:** *55% potassium permanganate, 45% cesium decahydrodecaborate* |
| **57. 06-04-041A: Igniter composition for use in electric ignition devices:** *67% potassium perchlorate, 33% titanium hydride* | **58. 06-04-042A: Thermally stable, percussion primer composition for use in primers for shell casings:** *59% potassium perchlorate, 41% titanium metal* |
| **59. 06-04-043A: Friction sensitive ignition composition for friction resistant match head compositions:** *40% potassium chlorate, 40% antimony trisulfide, 20% gum Arabic* | **60. 06-04-044A: Ignition composition for smoke grenades:** *57.6% barium nitrate, 17.3% zirconium hydride, 11.5% tetranitrocarbazole, 7.6% silicon, 5.7% epoxy resin, 0.3% mixed impurities* |
| **61. 06-04-045A: Classic highly insensitive ignition composition for use in rapid accelerating projectiles such as tank rounds or artillery rounds:** *60% magnesium, 28% silicon dioxide, 12% polybutadiene* | **62. 06-04-045B: Classic highly insensitive ignition composition for use in moderately accelerating projectiles such as mortar rounds:** *70% magnesium, 15% potassium perchlorate, 7.5% silicon dioxide, 7.5% polybutadiene* |
| **63. 06-04-046A: Special primer composition containing high explosives for use in shot gun primers:** *37% lead styphnate, 25% antimony sulfide 11% barium nitrate, 10% aluminum, 8% PETN, 6% tetracene; 3% polyvinyl alcohol* | **64. 06-04-047A: Ignition composition for use on the tips of matchsticks:** *66.13% potassium chlorate, 13.22% pulverized glass, 9.92% Prussian blue, 4.76% red phosphorus, 3.3% soft wood charcoal, 1.98% animal glue, 0.66% nitrocellulose, 0.03% impurities* |
| **65. 06-04-047B: Ignition composition for use on the tips of matchsticks (modified):** *52.79% potassium chlorate, 15.52% pulverize glass, 14.9% Prussian blue, 7.45% furnace black, 6.21% red phosphorus, 3.1% animal glue, 0.03% impurities* | **66. 06-04-048A: Insensitive ignition composition for rockets and missiles:** *49.9% aluminum powder, 40% sodium azide, 10.1% potassium permanganate* |
| **67. 06-04-048B: Insensitive ignition composition for rockets and missiles:** *60.1% potassium azide, 29.9% boron powder, 10% potassium permanganate* | **68. 06-04-048C: Insensitive ignition composition for rockets and missiles:** *68.26% zirconium powder, 16.56% sodium azide, 15.06% potassium permanganate, 0.12% mixed balance* |
| **69. 06-04-049A: Insensitive priming/ignition composition for multiple uses:** *71.25% boron powder, 23.75% sodium azide, 5% potassium perchlorate* | **70. 06-04-049B: Insensitive priming/ignition composition for multiple uses:** *68.25% aluminum powder, 22.75% sodium azide, 9% barium nitrate* |
| **71. 06-04-050A: Insensitive priming composition for multiple uses:** *52.5% boron powder, 30% potassium chlorate, 17.5% sodium azide* | **72. 06-04-051A: Insensitive ignition composition for multiple uses:** *54.9% sodium azide*, *27% boron powder, 10.1% potassium perchlorate, 8% Viton A copolymer binder* |
| **73. 06-04-051B: Insensitive ignition composition for multiple uses:** *57.3% potassium azide, 28.7% boron powder, 10.7% potassium perchlorate, 3.3% Kel-F elastomer* | **74. 06-04-051C: Insensitive ignition composition for multiple uses:** *38.4% sodium azide, 38.4% aluminum powder, 19.2% Kel-F copolymer binder, 4% Viton A binder* |
| | |

| | |
|---|---|
| **75. 06-04-052A: Friction sensitive ignition composition for electric matches, matchsticks, and friction initiated devices:** *68.08% phosphorus sulfur compound, 14.89% powdered glass 8.5% iron sulfide, 6.38% animal glue, 2.12% lamp black, 0.03% balance,* | **76. 06-04-053A: Non-friction sensitive ignition composition for various operations:** *40% food starch, 35% potassium chlorate, 10% vegetable charcoal, 10% pulverized sawdust, 3% tragacanth gum, 2% potassium dichromate* |
| **77. 06-04-053B: Moderate friction sensitive ignition composition for matchsticks, and other operations:** *45% potassium chlorate, 30% antimony trisulfide, 10% glass powder, 8% gum Arabic, 5% lead dioxide, 2% potassium dichromate* | **78. 06-04-053C: Rough friction producing ignition surface for use in combination with matches and other friction sensitive ignition compositions:** *35% glass powder, 30% red phosphorus, 18% antimony trisulfide, 15% animal glue, 2% potassium dichromate* |
| **79. 06-04-054A: Ignition composition for initiating thermite and other incendiary compositions:** *77.27% potassium dichromate, 13.63% zinc phosphide, 9.09% aluminum powder, 0.01% mixed residual balance* | **80. 06-04-055A: Priming composition for use in blasting caps:** *37.5% potassium dichromate, 35% copper powder, 15% mossy zinc, 12.5% zinc phosphide* |
| **81. 06-04-056A: Priming composition for use in small arms primers:** *50% barium nitrate, 17.85% red phosphorus, 17.85% lead dioxide, 14.28% zirconium metal powder, 0.02% mixed balance* | **82. 06-04-056B: Priming composition for use in small arms primers:** *58.33% barium nitrate, 20.83% red phosphorus, 20.83% antimony trisulfide, 0.01% residual balance* |
| **83. 06-04-057A: Priming composition for use in small arms primers:** *72.99% red phosphorus/lead plumbate addition salt, 14.59% potassium chlorate, 10.94% powdered glass, 0.72% gum tragacanth, 0.72% gum Arabic, 0.04% balance* | **84. 06-04-058A: Priming composition for use in small arms primers:** *37.68% potassium chlorate, 37.18% black antimony sulfide, 20.1% nitrocellulose, 5.02% red phosphorus, 0.02% mixed residual balance* |
| **85. 06-04-059A: Gasless ignition composition for smoke pots and smoke grenades:** *53.3% red lead, 33.7% silicon, 11.2% manganese, 1.8% celluloid* | **86. 06-04-059B: Gasless ignition composition for smoke pots and smoke grenades:** *76.9% litharge, 19.3% silicon, 1.8% celluloid, 1.5% fullers earth, 0.5% graphite* |
| **87. 06-04-059C: Gasless ignition composition for smoke pots and smoke grenades:** *59% lead chromate, 19.6% silicon, 19.6% magnesium powder, 1.8% celluloid* | **88. 06-04-060A: Ignition composition for tracer compositions (with ADN addictive):** *50% barium nitrate, 20% silicon, 15% zirconium hydride, 10% ADN, 5% nitrocellulose* |
| **89. 06-04-061A: Priming composition for use in blasting caps:** *44.44% special metal alloy, 22.22% barium nitrate, 22.22% coarse zirconium metal, 11.11% lead dioxide, 0.01% mixed balance* | **90. 06-04-061B: Priming composition for use in blasting caps:** *50% special metal alloy, 20% barium nitrate, 16% zirconium powder, 10% lead dioxide, 4% antimony trisulfide* |
| **91. 06-04-061C: Priming composition for use in blasting caps:** *46% special metal alloy, 20% barium nitrate, 20% zirconium metal, 10% lead dioxide, 4% antimony trisulfide* | **92. 06-04-061D: Priming composition for use in blasting caps:** *46% special metal alloy, 20% barium nitrate, 20% zirconium metal, 10% lead dioxide, 4% antimony trisulfide* |
| **93. 06-04-062A: Igniter composition for tracer compositions:** *74.86% strontium peroxide, 10.16% calcium silicide, 7.39% calcium resinate, 3.78% barium peroxide, 3.78% lead peroxide. 0.03% residual balance* | **94. 06-04-063A: Specialty high-temperature igniter composition for use in electric "squibs":** *68.6% ground hafnium metal, 26.7% potassium perchlorate, 4.7% nylon binder* |
| **95. 06-04-064A: Classic lead nitrate priming composition for rim-fire ammunition:** *33% lead thiocyanate, 30% potassium chlorate, 25% glass powder, 12% lead nitrate* | **96. 06-04-065A: Priming composition for use in non-electric blasting caps:** *40% nitrocellulose, 30% potassium chlorate, 30% potassium ferrocyanide* |
| **97. 06-04-066A: Specialty priming composition for use in blasting caps and detonators:** *33% lead styphnate, 21% lead calcium hypophosphate/nitrate complex salt, 20% powdered glass, 16% barium nitrate* | **98. 06-04-067A: Starter composition for ignition of smoke compositions:** *79.92% lead tetraoxide, 16.04% ground silicon powder, 4.02% binder, 0.02% mixed balance* |
| **99. 06-04-068A: "Electric match" priming composition/device for initiating rocket motors:** *51.61% potassium chlorate, 16.12% carboxymethylcellulose, 9.67% lamp black, 9.67% magnalium alloy, 6.45% atomized aluminum, 6.45% zirconium metal, 0.03% nitrocellulose* | |

**06-04-001A: Shock resistant igniter/primer composition (produces flash and loud report upon ignition):**
Into a suitable sealable vertical mixer, equipped with flexible Teflon stirrer blades, place 250 milliliters of hexane, followed by ***305 grams of potassium perchlorate***, and then followed by ***195 grams of potassium ferricyanide***. Thereafter, seal the vertical mixer, and then blend the mixture on moderate speed for about 45 minutes. After 45 minutes, the mixture is ready for use. To use in explosive electric bridge wire devices, the metal wire (usually a thin NiChrome wire) should be evenly coated with the moldable mixture coating the exposed portion of the "wire" with a 1 to 3mm thickness of the mixture, and then cured in an oven at 60 Celsius for several hours. To use in blasting caps, the mixture should be pressed into molds under a pressure of 2,000 psi, forming tablets that need to be cured in an oven at 60 Celsius for 12 hours; the diameter and length of these "tablets" is based on the diameter and length of the given blasting cap; use your own discretion.
**Burn rate (when heated externally):** 561 Celsius; explodes.
**Burn rate (when heated directly with a flame or hot NiChrome wire):** Explodes.
**Water resistance:** Very good.
**Sensitivity:** Moderate stability to friction; resistant to above normal shock; heat resistant (external) up to 560 Celsius.
**Stability:** Can be stored for many years.
**Flammability:** Explodes.
**Ease of ignition (when heated by a flame or hot wire):** 9.8
**Tendency to cake:** Will not cake.
**Explosive ability:** Detonates when ignited by a flame; detonation velocity: 1,206 meters/per/second.
**Percentage:** ***61% potassium perchlorate, 39% potassium ferricyanide***
**Classification:** Low explosive (classified as Class B explosive mixture).

**Danger of preparation:** Should only be prepared by experienced pyrotechnicians.
**Use:** Used in explosive electric "bridge" wire devices to initiate detonating cord; used as a priming agent in non-electric and electric blasting caps. Do not use to fire high performance rocket fuels or high performance gun propellants.

**Note:** Inexperienced personnel should not attempt this procedure, as there are hazards involved. Avoid excessive heat or friction. This procedure is perfectly safe for experienced persons. To gain experience, preparers should use only small quantities to begin with. Either way, safety shields should be used at all times, and any possible source of ignition, including static electricity should be avoided—use static resistant tables and floor mats.

**06-04-001B: Modified shock resistant igniter/primer composition (produces flash and loud report upon ignition):**
Into a suitable sealable vertical mixer, equipped with flexible Teflon stirrer blades, place 275 milliliters of hexane, followed by ***305 grams of potassium perchlorate***, and then followed by ***195 grams of potassium hexacyanocobaltate***. Thereafter, seal the vertical mixer, and then blend the mixture on moderate speed for about 45 minutes. After 45 minutes, the mixture is ready for use. To use in explosive electric bridge wire devices, the metal wire (usually a thin NiChrome wire) should be evenly coated with the moldable mixture coating the exposed portion of the "wire" with a 1 to 3mm thickness of the mixture, and then cured in an oven at 60 Celsius for several hours. To use in blasting caps, the mixture should be pressed into molds under a pressure of 2,000 psi, forming tablets that need to be cured in an oven at 60 Celsius for 12 hours; the diameter and length of these "tablets" is based on the diameter and length of the given blasting cap; use your own discretion.
**Burn rate (when heated externally):** 559 Celsius; explodes.
**Burn rate (when heated directly with a flame or hot NiChrome wire):** Explodes.
**Water resistance:** Very good.
**Sensitivity:** Moderate stability to friction; resistant to above normal shock; heat resistant (external) up to 558 Celsius.
**Stability:** Can be stored for many years.
**Flammability:** Explodes.
**Ease of ignition (when heated by a flame or hot wire):** 9.8
**Tendency to cake:** Will not cake.
**Explosive ability:** Detonates when ignited by a flame; detonation velocity: 1,188 meters/per/second.
**Percentage:** ***61% potassium perchlorate, 39% potassium hexacyanocobaltate***
**Classification:** Low explosive (classified as Class B explosive mixture).

**Danger of preparation:** Should only be prepared by experienced pyrotechnicians.
**Use:** Used in explosive electric "bridge" wire devices to initiate detonating cord; used as a priming agent in non-electric and electric blasting caps. Do not use to fire high performance rocket fuels or high performance gun propellants.

**Note:** Inexperienced personnel should not attempt this procedure, as there are hazards involved. Avoid excessive heat or friction. This procedure is perfectly safe for experienced persons. To gain experience, preparers should use only small quantities to begin with. Either way, safety shields should be used at all times, and any possible source of ignition, including static electricity should be avoided—use static resistant tables and floor mats.

### 06-04-002A: Igniter composition (produces no flash or report upon ignition):

Into a large suitable glass beaker, equipped with mechanical stirrer, place 1700 milliliters of water followed immediately by ***164 grams of potassium dichromate***, and then followed by ***72 grams of potassium ferricyanide***. Thereafter, blend the entire mixture rapidly to dissolve all solids. Note: if all solids fail to dissolve, continue adding water in small portions until they do. Once all solids have dissolved, slowly pour in, 1700 milliliters of 99% isopropyl alcohol while rapidly stirring the potassium dichromate/potassium ferricyanide solution. After the addition of all the alcohol, continue to rapidly stir the resulting mixture for about 30 minutes at room temperature. After 30 minutes, stop stirring, and then filter-off the insoluble mass using gravity filtration, or preferably vacuum filtration under rapid suction. In either case, place the filtered-off mass onto a shallow tray and allow it to thoroughly air-dry. Once the mass is thoroughly dried, place it into a suitable vertical mixer, add in 100 milliliters of hexane, and then rapidly blend the mixture for about 45 minutes. After 45 minutes, the mixture is ready for pressing. To do so, press the mixture into discs, tablets, and/or blocks, ranging from 1mm to 10mm thickness under a pressure of 6,000 psi; use your own discretion, and then cure the pressed mixture in an oven at 60 Celsius for 12 hours.
**Burn rate (when heated externally; when pressed at 6,000 psi):** Ignites at 607 Celsius.
**Burn rate (when heated directly with a flame or hot NiChrome wire; when pressed at 6,000 psi; uncontained):** 7.8 seconds per inch; tends to vary based on diameter and density of grain.
**Burn rate (when heated directly with a flame or hot NiChrome wire; when pressed at 6,000 psi; contained):** Tends to explode.
**Burn rate (when heated directly with a flame or hot NiChrome wire; when powdered; uncontained):** 1.2 seconds per inch.
**Burn rate (when heated directly with a flame or hot NiChrome wire; when powdered; contained):** Tends to explode.
**Water resistance:** Very good.
**Sensitivity:** Very stable to friction; resistant to above normal shock; heat resistant (external) up to 606 Celsius.
**Stability:** Can be stored for many years.
**Flammability (ignited by flame or hot wire; uncontained):** 9.8
**Flammability (ignited by flame or hot wire; contained):** Explodes.
**Ease of ignition (when heated by a flame or hot wire; pressed at 6,000 psi):** 9.4
**Ease of ignition (when heated by a flame or hot wire; when powdered):** 9.7
**Tendency to cake:** Will not cake.
**Explosive ability:** Can be detonated when confined; detonation velocity: 1,094 meters/per/second.
**Percentage:** ***69.4% potassium dichromate, 30.5% potassium ferricyanide, 0.10% moisture***
**Classification:** Low explosive (classified as Class C explosive mixture).

**Danger of preparation:** Should only be prepared by experienced pyrotechnicians.
**Use:** Used to fire other pyrotechnic compositions such as smoke mixtures, flares, fireworks, and general pyrotechnic mixtures. Note: this mixture is suitable for use as a primer in fireworks. Do not use to fire high performance rocket fuels or high performance gun propellants.

### 06-04-003A: High altitude igniter composition (can be used in supersonic munitions):

Into a suitable glass beaker, equipped with mechanical stirrer and sitting on top of a hot plate, place 1000 milliliters of water, and then add and dissolve 1.5 grams of sodium dichromate, followed by adding and dissolving 1.5 grams of sodium bisulfate. Thereafter, add in ***350 grams of zirconium powder*** of 325 mesh, and then immediately thereafter heat the mixture at 50 Celsius for 45 minutes. After 45 minutes, remove the heat source and allow the mixture to cool to room temperature. Thereafter, filter-off the coated zirconium powder (using gravity filtration, or preferably vacuum filtration under moderate suction), and then place the collected mass onto a shallow pan and allow the coated zirconium to air-dry. Once thoroughly air dried, place the dried coated zirconium powder into a suitable sized sealable vertical mixer, followed by adding in ***150 grams of flours of sulfur***, followed by ***50 grams of "Thiokol"*** (polysulfide synthetic rubber), and then followed by ***100 grams of potassium perchlorate***. Thereafter, seal the vertical mixer, and then blend the mixture on moderate speed for about 45 minutes. After 45 minutes, the mixture is ready for pressing. To do so, press the mixture into discs, tablets, and/or blocks, ranging from 1mm to 10mm thickness under a pressure of 6,000 psi; use your own discretion, and then cure the pressed mixture in an oven at 60 Celsius for 12 hours.
**Burn rate (when heated externally; when pressed at 6,000 psi):** Ignites at 699 Celsius.
**Burn rate (when heated directly with a flame or hot NiChrome wire; when pressed at 6,000 psi; uncontained):** 17 seconds per inch; tends to vary based on diameter and density of grain.
**Burn rate (when heated directly with a flame or hot NiChrome wire; when pressed at 6,000 psi; contained):** Not applicable.
**Burn rate (when heated directly with a flame or hot NiChrome wire; when powdered; uncontained):** 12 seconds per inch.
**Burn rate (when heated directly with a flame or hot NiChrome wire; when powdered; contained):** Not applicable.
**Water resistance:** Very good.
**Sensitivity:** Very stable to friction; resistant to above normal shock; heat resistant (external) up to 698 Celsius.
**Stability:** Can be stored for many years.
**Flammability (ignited by flame or hot wire; uncontained):** 7.2
**Flammability (ignited by flame or hot wire; contained):** Not applicable.
**Ease of ignition (when heated by a flame or hot wire; pressed at 6,000 psi):** 7.7
**Ease of ignition (when heated by a flame or hot wire; when powdered):** 8.6
**Tendency to cake:** Will not cake.
**Explosive ability:** Cannot be detonated.
**Percentage:** ***53.8% zirconium, 23% sulfur, 15.3% potassium perchlorate, 7.6% rubber binder, 0.15% potassium dichromate, 0.15% sodium bisulfite***

**Classification:** Deflagrating mixture (classified as 1.3G pyrotechnic composition).

**Danger of preparation:** Can be made by amateurs.
**Use:** Used to fire pyrotechnic compositions such as smoke mixtures, flares, illumination, and general pyrotechnic mixtures used in high-speed/high altitude mortar shells, artillery shells, rockets, and missiles.

## 06-04-004A: "Invisible", dark burning pyrotechnic igniter mixture for tracer compositions:

Into a suitable vertical mixer, place 120 milliliters of hexane, followed by ***175 grams of powdered manganese*** of 300 mesh, followed by ***10 grams of graphite powder***, and then followed by ***315 grams of bismuth trioxide*** of 300 mesh. Thereafter, blend the mixture on moderate speed for about 45 minutes. After 45 minutes the mixture is ready for pressing. To do, the mixture needs to be pressed into a thin disc ranging from 1.5mm to 2mm in thickness adhering to the same diameter as the tracer composition used for the given projectile in question and under a pressure of 4,000 psi, and the resulting munition needs to be cured in an oven at 60 Celsius for 6 hours.
**Burn rate:** 1.5mm per second.
**Burn temperature:** 881 Celsius.
**Ignition temperature (average temperature to ignite mixture):** 790 Celsius.
**Water resistance:** Very good.
**Stability:** Can be stored for many years.
**Flammability (pressed grain at 4,000 psi):** 7.2
**Ease of ignition (average temperature to ignite mixture):** 4.2
**Tendency to cake:** None.
**Explosive ability:** Cannot be detonated.
**Percentage:** ***63% bismuth trioxide, 35% manganese, 2% graphite***
**Classification:** Deflagrating mixture (classified as 1.3G pyrotechnic composition).

**Danger of preparation:** Can be made by amateurs.
**Use:** Used to initiate tracer compositions in tracer bullets for military ammunition ranging from riffles to heavy machine guns, to high caliber projectiles.

## 06-04-004B: "Invisible", dark burning pyrotechnic igniter mixture for tracer compositions:

Into a suitable vertical mixer, place 120 milliliters of hexane, followed by ***152 grams of barium peroxide*** followed by ***46 grams of antimony pentasulfide***, and then followed by ***2 grams of powdered graphite***. Thereafter, blend the mixture on moderate speed for about 45 minutes. After 45 minutes the mixture is ready for pressing. To do so, the mixture needs to be pressed into a thin disc ranging from 1.5mm to 2mm in thickness adhering to the same diameter as the tracer composition used for the given projectile in question and under a pressure of 4,000 psi, and the resulting munition needs to be cured in an oven at 60 Celsius for 6 hours.
**Burn rate:** 1.3mm per second.
**Burn temperature:** 879 Celsius.
**Ignition temperature (average temperature to ignite mixture):** 780 Celsius.
**Water resistance:** Very good.
**Stability:** Can be stored for many years.
**Flammability (pressed grain at 4,000 psi):** 7.1
**Ease of ignition (average temperature to ignite mixture):** 4.1
**Tendency to cake:** None.
**Explosive ability:** Cannot be detonated.
**Percentage:** ***76% barium peroxide, 23% antimony pentasulfide, 1% graphite***
**Classification:** Deflagrating mixture (classified as 1.3G pyrotechnic composition).

**Danger of preparation:** Can be made by amateurs.
**Use:** Used to initiate tracer compositions in tracer bullets for military ammunition ranging from riffles to heavy machine guns, to high caliber projectiles.

**06-04-005A: "Non-violent burning" pyrotechnic igniter element for initiating compositions such as thermite type compositions that have ignition temperatures exceeding 1000 Celsius:**
Into a suitable vertical mixer, place 100 milliliters of hexane, followed by ***37.5 grams of titanium powder*** of standard mesh, and then followed by ***87.5 grams of red iron-III-oxide powder***. Thereafter, blend the mixture on high speed for about 45 minutes. After 45 minutes the mixture is ready for pressing. To do so, press the mixture into molds, discs, tablets, cubes, pellets, ect., of diameter and thickness (as dictated by the device/munition in question; use your discretion), under a pressure of 6,000 psi. Thereafter, cure the molds, discs, tablets, cubes, pellets, ect., in an oven at 60 Celsius for 12 hours.
**Burn rate:** 1.1mm per second.
**Burn temperature:** 1120 Celsius.
**Ignition temperature (average temperature to ignite mixture at 6,000 psi):** 708 Celsius.
**Water resistance:** Very good.
**Stability:** Can be stored for many years.
**Flammability (pressed grain at 6,000 psi):** 3.8
**Ease of ignition (average temperature to ignite mixture at 6,000 psi):** 4.1
**Tendency to cake:** None.
**Explosive ability:** Cannot be detonated.
**Percentage:** ***70% iron oxide, 30% titanium***
**Classification:** Deflagrating mixture (classified as 1.3G pyrotechnic composition).

**Danger of preparation:** Can be made by amateurs.
**Use:** Used to initiate incendiary compositions and thermite-type compositions.

**06-04-006A: "Therm 64C" igniter composition for magnesium bombs:**
Into a suitable vertical mixer, place 210 milliliters of hexane, followed by ***80 grams of aluminum powder*** of standard mesh, followed by ***45 grams of standard grained aluminum***, followed by ***220 grams of red iron-III-oxide***, followed by ***10 grams of flours of sulfur***, and then followed by ***145 grams of barium nitrate.*** Thereafter, blend the mixture on high speed for about 45 minutes. After 45 minutes the mixture is ready for pressing. To do so, press the mixture into molds, discs, tablets, cubes, pellets, ect., of diameter and thickness (as dictated by the device/munition in question; use your discretion), under a pressure of 7,000 psi. Thereafter, cure the molds, discs, tablets, cubes, pellets, ect., in an oven at 60 Celsius for 12 hours.
**Burn rate:** 1.6mm per second. Depends on diameter and density of solid grain.
**Burn temperature:** 1099 Celsius.
**Ignition temperature (average temperature to ignite mixture at 7,000 psi):** 712 Celsius.
**Water resistance:** Very good.
**Stability:** Can be stored for many years.
**Flammability (pressed grain at 7,000 psi):** 4.1
**Ease of ignition (average temperature to ignite mixture at 7,000 psi):** 4.3
**Tendency to cake:** None.
**Explosive ability:** Cannot be detonated.
**Percentage:** ***44% iron oxide, 29% barium nitrate, 25% aluminum, 2% sulfur***
**Classification:** Deflagrating mixture (classified as 1.3G pyrotechnic composition).

**Danger of preparation:** Can be made by amateurs.
**Use:** Used to initiate magnesium containing incendiary munitions.

**06-04-007A: Igniter composition for low velocity rocket propellants:**
Into a suitable vertical mixer, place 190 milliliters of hexane, followed by ***150 grams of sodium bicarbonate***, followed by ***50 grams of corn starch***, followed by ***84 grams of flours of sulfur***, followed by ***216 grams of potassium chlorate***, and then followed by ***20 grams of nitrocellulose*** (14% nitrogen content) Thereafter, thoroughly blend the mixture on high speed for about 1 hour at room temperature. After 1 hour, the mixture is read for pressing. To do so, press the mixture into molds, discs, tablets, cubes, pellets, ect., of diameter and thickness (as dictated by the device/munition in question; use your discretion), under a pressure of 7,000 psi. Thereafter, cure the molds, discs, tablets, cubes, pellets, ect., in an oven at 60 Celsius for 12 hours.
**Burn rate:** 2.9mm per second. Depends on diameter and density of solid grain.
**Burn temperature:** 1021 Celsius.
**Ignition temperature (average temperature to ignite mixture at 7,000 psi):** 799 Celsius.
**Water resistance:** Very good.
**Stability:** Can be stored for many years.
**Flammability (pressed grain at 7,000 psi):** 4.4
**Ease of ignition (average temperature to ignite mixture at 7,000 psi):** 4.2

**Tendency to cake:** None.
**Explosive ability:** Cannot be detonated.
**Percentage:** ***41.5% potassium chlorate, 28.8% sodium bicarbonate, 16.1% sulfur, 9.6% corn starch, 3.8% nitrocellulose, 0.20% residue***
**Classification:** Deflagrating mixture (classified as 1.3G pyrotechnic composition).

**Danger of preparation:** Can be made by amateurs.
**Use:** Used to initiate low velocity rocket propellants.

### 06-04-008A: Non-luminous igniter composition for various applications:

Into a suitable vertical mixer, place 195 milliliters of hexane, followed by ***325 grams of standard tungsten powder,*** followed by ***125 grams of barium chromate,*** and then followed by ***50 grams of potassium perchlorate***. Thereafter, thoroughly blend the mixture on high speed for about 1 hour at room temperature. After 1 hour, the mixture is read for pressing. To do so, press the mixture into molds, discs, tablets, cubes, pellets, ect., of diameter and thickness (as dictated by the device/munition in question; use your discretion), under a pressure of 7,000 psi. Thereafter, cure the molds, discs, tablets, cubes, pellets, ect., in an oven at 60 Celsius for 12 hours.
**Burn rate:** 3mm per second. Depends on diameter and density of solid grain.
**Burn temperature:** 1389 Celsius.
**Ignition temperature (average temperature to ignite mixture at 7,000 psi):** 700 Celsius.
**Water resistance:** Very good.
**Stability:** Can be stored for many years.
**Flammability (pressed grain at 7,000 psi):** 5.1
**Ease of ignition (average temperature to ignite mixture at 7,000 psi):** 4.9
**Tendency to cake:** None.
**Explosive ability:** Cannot be detonated.
**Percentage:** ***65% tungsten, 25% barium chromate, 10% potassium perchlorate***
**Classification:** Deflagrating mixture (classified as 1.3G pyrotechnic composition).

**Danger of preparation:** Can be made by amateurs.
**Use:** Used to initiate flares, smoke grenades, tracer compositions, and incendiary grenades.

### 06-04-009A: Priming composition (ignition composition) for use in electric/non-electric blasting caps:

Into a suitable heated vertical mixer, place ***375 grams of standard lead powder***, and then dry blend the powder on high speed while heating to 250 Celsius for 4 hours. After 4 hours, remove the heat, and allow the lead powder to cool to room temperature; continue to dry blend the mixture on high speed during the cool down period. Once cooled add in 195 milliliters of hexane, followed by ***125 grams of standard powdered selenium***, and then continue to rapidly blend the mixture at room temperature for 45 minutes. After 45 minutes, the mixture is ready for use. To use in explosive electric bridge wire devices, the metal wire (usually a thin NiChrome wire) should be evenly coated with the moldable mixture coating the exposed portion of the "wire" with a 1 to 3mm thickness of the mixture, and then cured in an oven at 60 Celsius for several hours. To use in blasting caps, the mixture should be pressed into molds under a pressure of 2,000 psi, forming tablets that need to be cured in an oven at 60 Celsius for 12 hours; the diameter and length of these "tablets" is based on the diameter and length of the given blasting cap; use your own discretion.
**Burn rate (when heated externally):** 600 Celsius. Explodes.
**Burn rate (when heated directly with a flame or hot NiChrome wire):** Explodes.
**Ignition temperature (average temperature to ignite mixture at 2,000 psi):** 731 Celsius.
**Water resistance:** Very good.
**Sensitivity:** Above moderate stability to friction; resistant to above normal shock; heat resistant (external) up to 599 Celsius.
**Stability:** Can be stored for many years.
**Flammability:** Explodes.
**Ease of ignition (when heated by a flame or hot wire):** 9.1
**Tendency to cake:** Will not cake.
**Explosive ability:** Does not detonate, but undergoes extreme explosive combustion.
**Percentage:** ***60% lead, 25% selenium, 15% lead-II-oxide***
**Classification:** Deflagrating mixture (classified as 1.4G pyrotechnic composition).

**Danger of preparation:** Can be made by amateurs, but caution should be exercised.

**Use:** Used in explosive electric "bridge" wire devices to initiate detonating cord; used as a priming agent in non-electric and electric blasting caps. Do not use to fire high performance rocket fuels or high performance gun propellants.

<table>
<tr><th>Standard Non-electric blasting cap</th><th>Description</th></tr>
<tr><td>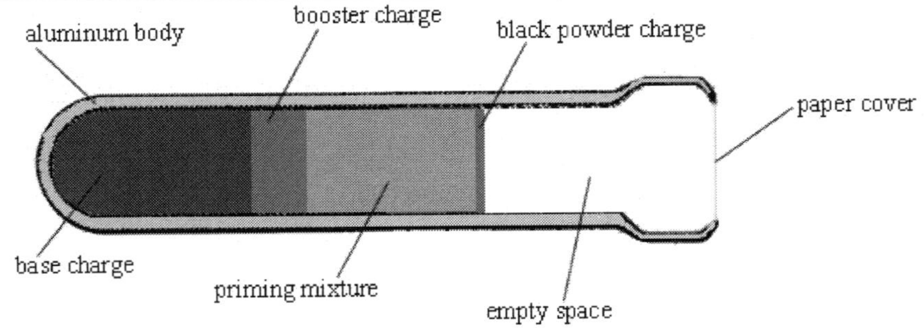
</td><td>Non-electric blasting caps all function in a similar manner. In this case, a black powder time fuse is inserted into the empty space, and this end is then crimped to hold it in place. Thereafter, when the fuse is ignited, the heat reaches the priming mixture. The priming composition is heat sensitive and produces significant energy upon ignition. The priming composition then sets off the booster charge, which in turn sets off the base charge</td></tr>
<tr><th>Standard electric blasting cap</th><th>Description</th></tr>
<tr><td>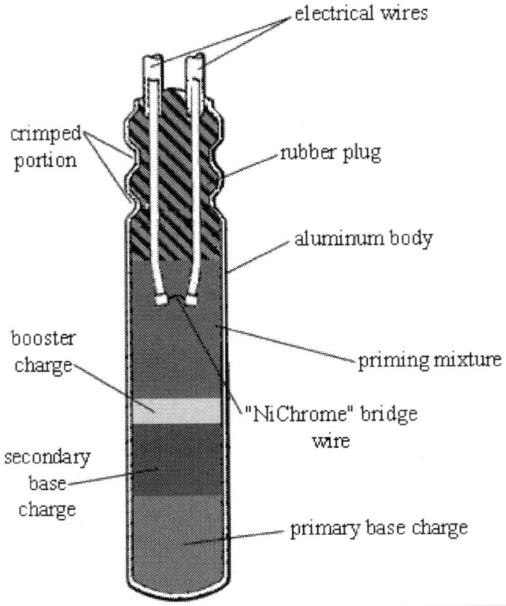
</td><td>Electric blasting caps work in the same manner as non-electric caps, but electrical lead wires replace the black powder time fuse. When an electric current is passed between the Nichrome wire, the wire heats up and sets off the priming mixture.</td></tr>
</table>

### 06-04-010A: Ignition composition for high performance rocket propellants:

Into a suitable vertical mixer, place 195 milliliters of hexane, followed by ***265 grams of standard iron powder***, followed by ***15 grams of standard powdered aluminum,*** followed by ***220 grams of potassium perchlorate,*** and then followed by ***15.5 grams of cellulose acetate***. Thereafter, thoroughly blend the mixture on high speed for about 45 minutes at room temperature. After 45 minutes, the mixture is ready for pressing or for forming electric bridge wire igniters. To do so, first press the mixture into molds, discs, tablets, cubes, pellets, ect., of diameter and thickness (as dictated by the device/munition in question; use your discretion), under a pressure of 8,000 psi. Thereafter, cure the molds, discs, tablets, cubes, pellets, ect., in an oven at 60 Celsius for 12 hours. Second, for forming electric bridge wire igniters, the metal wire (usually a thin NiChrome wire) should be evenly coated with the moldable mixture coating the exposed portion of the "wire" with a 1.5 to 3.5mm thickness of the mixture, and then cured in an oven at 60 Celsius for several hours.

**Burn rate (when heated externally; when pressed at 8,000 psi):** Ignites at 695 Celsius.
**Burn rate (when heated directly with a flame or hot NiChrome wire; when pressed at 8,000 psi; uncontained):** 4.9 seconds per inch; tends to vary based on diameter and density of grain.
**Burn rate (when heated directly with a flame or hot NiChrome wire; when powdered; uncontained):** 1.4 seconds per inch.
**Burn rate (when heated directly with a flame or hot NiChrome wire; electric bridge wire):** Instantaneous ignition.
**Burn temperature:** 1300 Celsius.
**Ignition temperature (average temperature to ignite mixture at 8,000 psi):** 671 Celsius.
**Water resistance:** Very good.
**Sensitivity:** Very stable to friction; resistant to above normal shock; heat resistant (external) up to 694 Celsius.
**Stability:** Can be stored for many years.
**Flammability (ignited by flame or hot wire; uncontained):** 9.1
**Ease of ignition (when heated by a flame or hot wire; pressed at 8,000 psi):** 9.7
**Ease of ignition (when heated by a flame or hot wire; when powdered):** 9.86
**Tendency to cake:** Will not cake.
**Explosive ability:** Cannot be detonated.

**Percentage:** ***51.4% iron, 42.6% potassium perchlorate, 3% cellulose acetate, 2.9% aluminum***
**Classification:** Deflagrating mixture (classified as 1.4G pyrotechnic composition).

**Danger of preparation:** Can be made by amateurs, but caution should be excersized.
**Use:** Used to initiate high performance rocket propellants.

**06-04-011A: Gasless ignition composition for "thermite" type compositions:**
Into a suitable vertical mixer, place 180 milliliters of hexane, followed by ***285 grams of copper-II-oxide***, followed by ***200 grams of finely divided manganese***, and then followed by ***15 grams of powdered sulfur***. Thereafter, blend the mixture on high speed for about 45 minutes. After 45 minutes the mixture is ready for pressing. To do so, press the mixture into molds, discs, tablets, cubes, pellets, ect., of diameter and thickness (as dictated by the device/munition in question; use your discretion), under a pressure of 6,000 psi. Thereafter, cure the molds, discs, tablets, cubes, pellets, ect., in an oven at 60 Celsius for 12 hours.
**Burn rate:** 1.91mm per second. Depends on diameter and density of solid grain.
**Burn temperature:** 1299 Celsius.
**Ignition temperature (average temperature to ignite mixture at 6,000 psi):** 806 Celsius.
**Water resistance:** Very good.
**Stability:** Can be stored for many years.
**Flammability (pressed grain at 6,000 psi):** 4
**Ease of ignition (average temperature to ignite mixture at 6,000 psi):** 4.7
**Tendency to cake:** None.
**Explosive ability:** Cannot be detonated.
**Percentage:** ***57% copper oxide, 40% manganese, 3% sulfur***
**Classification:** Deflagrating mixture (classified as 1.3G pyrotechnic composition).

**Danger of preparation:** Can be made by amateurs.
**Use:** Used to initiate incendiary compositions.

**06-04-011B: Gasless ignition composition for flares, smoke, and other general pyrotechnic compositions:**
Into a suitable vertical mixer, place 180 milliliters of hexane, followed by ***150 grams of copper-II-oxide***, followed by ***150 grams of finely divided silicon powder***, followed by ***50 grams of calcium silicide***, and then followed by ***75 grams of lead dioxide***. Thereafter, blend the mixture on high speed for about 45 minutes. After 45 minutes the mixture is ready for pressing. To do so, press the mixture into molds, discs, tablets, cubes, pellets, ect., of diameter and thickness (as dictated by the device/munition in question; use your discretion), under a pressure of 6,000 psi. Thereafter, cure the molds, discs, tablets, cubes, pellets, ect., in an oven at 60 Celsius for 12 hours.
**Burn rate:** 1.87mm per second. Depends on diameter and density of solid grain.
**Burn temperature:** 1100 Celsius.
**Ignition temperature (average temperature to ignite mixture at 6,000 psi):** 888 Celsius.
**Water resistance:** Very good.
**Stability:** Can be stored for many years.
**Flammability (pressed grain at 6,000 psi):** 3.7
**Ease of ignition (average temperature to ignite mixture at 6,000 psi):** 4.1
**Tendency to cake:** None.
**Explosive ability:** Cannot be detonated.
**Percentage:** ***35.2% copper oxide, 35.2% silicon, 17.6% lead dioxide, 11.7% calcium silicide, 0.30% residue***
**Classification:** Deflagrating mixture (classified as 1.3G pyrotechnic composition).

**Danger of preparation:** Can be made by amateurs.
**Use:** Used to initiate general pyrotechnic compositions.

**06-04-012A: Ignition composition for tracer compositions:**
Into a suitable vertical mixer, place 195 milliliters of hexane, followed by ***111.5 grams of standard magnesium powder***, followed by ***55.5 grams of lead tetraoxide*** ($Pb_3O_4$), and then followed by ***333 grams of barium peroxide***. Thereafter, blend the mixture on moderate speed for about 45 minutes. After 45 minutes the mixture is ready for pressing. To do so, the mixture needs to be pressed into a thin disc ranging from 1.5mm to 2mm in thickness adhering to the same diameter as the tracer composition used for the given projectile in question and under a pressure of 4,000 psi, and the resulting munition needs to be cured in an oven at 60 Celsius for 6 hours.

**Burn rate:** 1.1mm per second.
**Burn temperature:** 809 Celsius.
**Ignition temperature (average temperature to ignite mixture):** 799 Celsius.
**Water resistance:** Very good.
**Stability:** Can be stored for many years.
**Flammability (pressed grain at 4,000 psi):** 7.4
**Ease of ignition (average temperature to ignite mixture):** 4
**Tendency to cake:** None.
**Explosive ability:** Cannot be detonated.
**Percentage:** ***66.6% barium peroxide, 22.3% magnesium, 11.1% lead tetraoxide***
**Classification:** Deflagrating mixture (classified as 1.3G pyrotechnic composition).

**Danger of preparation:** Can be made by amateurs.
**Use:** Used to initiate tracer compositions.

## 06-04-013A: Ignition composition for electric blasting caps and detonators:

Into a suitable heated vertical mixer, place ***411 grams of standard lead powder***, and then dry blend the powder on high speed while heating to 250 Celsius for 4 hours. After 4 hours, remove the heat, and allow the lead powder to cool to room temperature; continue to dry blend the mixture on high speed during the cool down period. Once cooled add in 195 milliliters of hexane, followed by ***156 grams of standard powdered selenium***, and then continue to rapidly blend the mixture at room temperature for 45 minutes. After 45 minutes, the mixture is ready for use. To use in explosive electric bridge wire devices, the metal wire (usually a thin NiChrome wire) should be evenly coated with the moldable mixture coating the exposed portion of the "wire" with a 1 to 3mm thickness of the mixture, and then cured in an oven at 60 Celsius for several hours. To use in blasting caps, the mixture should be pressed into molds under a pressure of 2,000 psi, forming tablets that need to be cured in an oven at 60 Celsius for 12 hours; the diameter and length of these "tablets" is based on the diameter and length of the given blasting cap; use your own discretion.
**Burn rate (when heated externally):** 609 Celsius. Explodes.
**Burn rate (when heated directly with a flame or hot NiChrome wire):** Explodes.
**Ignition temperature (average temperature to ignite mixture at 2,000 psi):** 706 Celsius.
**Water resistance:** Very good.
**Sensitivity:** Above moderate stability to friction; resistant to above normal shock; heat resistant (external) up to 608 Celsius.
**Stability:** Can be stored for many years.
**Flammability:** Explodes.
**Ease of ignition (when heated by a flame or hot wire):** 9.29
**Tendency to cake:** Will not cake.
**Explosive ability:** Does not detonate, but undergoes extreme explosive combustion.
**Percentage:** ***59.25% lead, 27.51% selenium, 13.22% lead-II-oxide, 0.02% residual lead oxides***
**Classification:** Deflagrating mixture (classified as 1.4G pyrotechnic composition).

**Danger of preparation:** Can be made by amateurs, but caution should be exercised.
**Use:** Used in explosive electric "bridge" wire devices to initiate detonating cord; used as a priming agent in non-electric and electric blasting caps. Used in place of lead styphnate/barium chromate priming mixtures. Do not use to fire high performance rocket fuels or high performance gun propellants.

| Standard blasting cap for typical means |
|---|
| 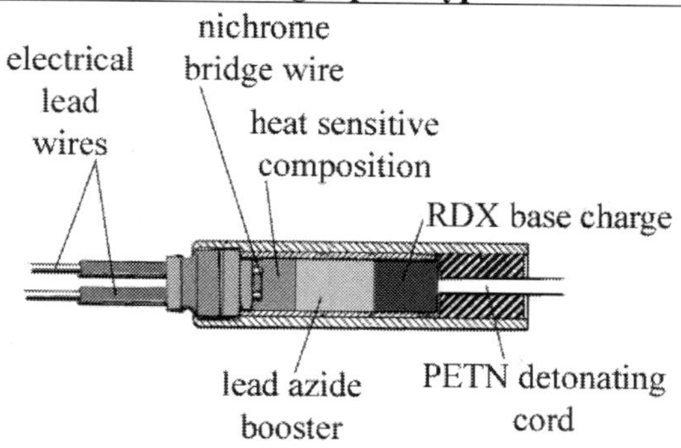  |
| **Details:** |
| Electrically initiated blasting cap, with PETN detonating cord insert. These caps are designed to initiate PETN detonating cords for initiating large masses of explosives tied in series. |

### 06-04-014A: Ignition composition for high performance rocket propellants:

Into a suitable vertical mixer, place 200 milliliters of hexane, followed by ***165 grams of finely powdered magnesium*** of 325 mesh, followed by ***335 grams of finely divided Dupont Teflon No. 6*** product (commercially available), and then followed by ***10 grams of fumed colloidal silica*** (sold as the trade name Cab-O-Sil). Thereafter, thoroughly blend the mixture on high speed for about 45 minutes at room temperature. After 45 minutes, the mixture is ready for pressing or for forming electric bridge wire igniters. To do so, first press the mixture into molds, discs, tablets, cubes, pellets, ect., of diameter and thickness (as dictated by the device/munition in question; use your discretion), under a pressure of 8,000 psi. Thereafter, cure the molds, discs, tablets, cubes, pellets, ect., in an oven at 60 Celsius for 12 hours. Second, for forming electric bridge wire igniters, the metal wire (usually a thin NiChrome wire) should be evenly coated with the moldable mixture coating the exposed portion of the "wire" with a 1.5 to 3.5mm thickness of the mixture, and then cured in an oven at 60 Celsius for several hours.
**Burn rate (when heated externally; when pressed at 8,000 psi):** Ignites at 890 Celsius.
**Burn rate (when heated directly with a flame or hot NiChrome wire; when pressed at 8,000 psi; uncontained):** 5.9 seconds per inch; tends to vary based on diameter and density of grain.
**Burn rate (when heated directly with a flame or hot NiChrome wire; when powdered; uncontained):** 3.9 seconds per inch.
**Burn rate (when heated directly with a flame or hot NiChrome wire; electric bridge wire):** Instantaneous ignition.
**Burn temperature:** 1045 Celsius.
**Ignition temperature (average temperature to ignite mixture at 8,000 psi):** 519 Celsius.
**Water resistance:** Very good.
**Sensitivity:** Very stable to friction; resistant to above normal shock; heat resistant (external) up to 889 Celsius.
**Stability:** Can be stored for many years.
**Flammability (ignited by flame or hot wire; uncontained):** 7.8
**Ease of ignition (when heated by a flame or hot wire; pressed at 8,000 psi):** 9.1
**Ease of ignition (when heated by a flame or hot wire; when powdered):** 9.2
**Tendency to cake:** Will not cake.
**Explosive ability:** Cannot be detonated.
**Percentage:** ***65.6% Dupont Teflon No. 6, 32.3% magnesium, 1.9% fumes silica, 0.2% mixed residues***
**Classification:** Deflagrating mixture (classified as 1.3G pyrotechnic composition).

**Danger of preparation:** Can be made by amateurs.
**Use:** Used to initiate solid rocket propellants in rockets and missiles.

### 06-04-015A: Ignition composition for incendiary compositions:

Into a suitable vertical mixer, place 190 milliliters of hexane, followed by ***200 grams of magnesium powder*** of standard mesh, followed by ***250 grams of powdered silicon dioxide***, and then followed by ***50 grams of finely powdered polybutadiene***. Thereafter, blend the mixture on high speed for about 45 minutes. After 45 minutes the mixture is ready for pressing. To do so, press the mixture into molds, discs, tablets, cubes, pellets, ect., of diameter and thickness (as dictated by the device/munition in question; use your discretion), under a pressure of 7,000 psi. Thereafter, cure the molds, discs, tablets, cubes, pellets, ect., in an oven at 60 Celsius for 12 hours.
**Burn rate:** 1.3mm per second. Depends on diameter and density of solid grain.
**Burn temperature:** 1390 Celsius.
**Ignition temperature (average temperature to ignite mixture at 7,000 psi):** 983 Celsius.
**Water resistance:** Very good.
**Stability:** Can be stored for many years.
**Flammability (pressed grain at 7,000 psi):** 4.4
**Ease of ignition (average temperature to ignite mixture at 7,000 psi):** 4
**Tendency to cake:** None.

**Explosive ability:** Cannot be detonated.
**Percentage:** ***50% silicon dioxide, 40% magnesium, 10% polybutadiene binder***
**Classification:** Deflagrating mixture (classified as 1.3G pyrotechnic composition).

**Danger of preparation:** Can be made by amateurs.
**Use:** Used to initiate incendiary compositions.

### 06-04-016A: Ignition composition for electric primers:

Into a suitable vertical mixer, place 205 milliliters of hexane, followed by ***135 grams of standard titanium powder*** of average commercial mesh, followed by ***340 grams of potassium chlorate***, and then followed by ***25 grams of boron powder*** of average mesh. Thereafter, moderately blend the mixture at room temperature for 60 minutes. After 60 minutes, the mixture is ready for use. To use in explosive electric bridge wire devices, the metal wire (usually a thin NiChrome wire) should be evenly coated with the moldable mixture coating the exposed portion of the "wire" with a 1 to 3mm thickness of the mixture, and then cured in an oven at 60 Celsius for several hours. To use in blasting caps, the mixture should be pressed into molds under a pressure of 2,000 psi, forming tablets that need to be cured in an oven at 60 Celsius for 12 hours; the diameter and length of these "tablets" is based on the diameter and length of the given blasting cap; use your own discretion.
**Burn rate (when heated externally):** 470 Celsius. Explodes.
**Burn rate (when heated directly with a flame or hot NiChrome wire):** Explodes.
**Ignition temperature (average temperature to ignite mixture at 2,000 psi):** 553 Celsius.
**Water resistance:** Very good.
**Sensitivity:** Above moderate stability to friction; resistant to above normal shock; heat resistant (external) up to 469 Celsius.
**Stability:** Can be stored for many years.
**Flammability:** Explodes.
**Ease of ignition (when heated by a flame or hot wire):** 9.39
**Tendency to cake:** Will not cake.
**Explosive ability:** Does not detonate, but undergoes extreme explosive combustion.
**Percentage:** ***68% potassium chlorate, 27% titanium, 5% boron***
**Classification:** Deflagrating mixture (classified as 1.4G pyrotechnic composition).

**Danger of preparation:** Can be made by amateurs, but caution should be exercised.
**Use:** Used in explosive electric "bridge" wire devices to initiate detonating cord; used as a priming agent in non-electric and electric blasting caps. Used in place of lead styphnate/barium chromate priming mixtures. Do not use to fire high performance rocket fuels or high performance gun propellants.

<table>
<tr><th>Classic electric squibb</th><th>Details:</th></tr>
<tr><td>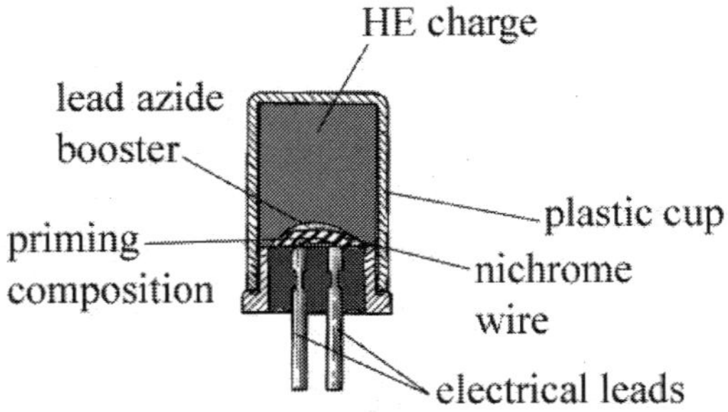
</td><td>Electric squibs are commonly used in commercial and industrial applications, such as movie sets, special effects, and in mining operations; squibs vary in size from very small ones to large ones. The devices work just like electric blasting caps. Medium to large squibs can be dangerous and are able to blow off fingers and even hands.</td></tr>
</table>

### 06-04-016B: Ignition composition for electric primers with increased sensitivity (modified):

Into a suitable vertical mixer, place 225 milliliters of hexane, followed by ***100 grams of standard titanium powder*** of average commercial mesh, followed by ***375 grams of potassium chlorate***, followed by ***10 grams of boron powder*** of average mesh, and then followed by ***5 grams of lead thiocyanate***. Thereafter, moderately blend the mixture at room temperature for 45 minutes. After 45 minutes, the mixture is ready for use. To use in explosive electric bridge wire devices, the metal wire (usually a thin NiChrome wire) should be evenly coated with the moldable mixture coating the exposed portion of the "wire" with a 1 to 3mm thickness of the mixture, and then cured in an oven at 60 Celsius for several hours. To use in blasting caps, the mixture should be pressed into molds under a pressure of 2,000 psi, forming tablets that need to be cured in an oven at 60 Celsius for 12 hours; the diameter and length of these "tablets" is based on the diameter and length of the given blasting cap; use your own discretion.
**Burn rate (when heated externally):** 454 Celsius. Explodes.
**Burn rate (when heated directly with a flame or hot NiChrome wire):** Explodes.

**Ignition temperature (average temperature to ignite mixture at 2,000 psi):** 509 Celsius.
**Water resistance:** Very good.
**Sensitivity:** Moderate stability to friction; resistant to normal shock; heat resistant (external) up to 453 Celsius.
**Stability:** Can be stored for many years.
**Flammability:** Explodes.
**Ease of ignition (when heated by a flame or hot wire):** 9.49
**Tendency to cake:** Will not cake.
**Explosive ability:** Does not detonate, but undergoes extreme explosive combustion.
**Percentage:** ***76.5% potassium chlorate, 20.4% titanium, 2% boron, 1% lead thiocyanate, 0.10% mixed residues***
**Classification:** Deflagrating mixture (classified as 1.4G pyrotechnic composition).

**Danger of preparation:** Can be made by amateurs, but caution should be exercised.
**Use:** Used in explosive electric "bridge" wire devices to initiate detonating cord; used as a priming agent in non-electric and electric blasting caps. Used in place of lead styphnate/barium chromate priming mixtures. Do not use to fire high performance rocket fuels or high performance gun propellants.

| Improvised Barb-wire anti-personnel mine | Details/description: |
|---|---|
| 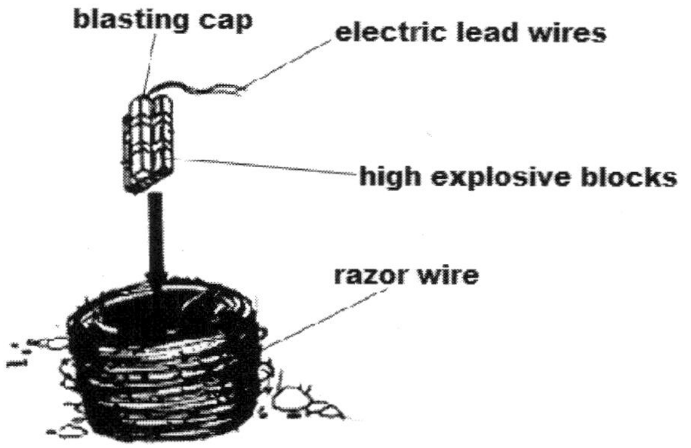  | Barbed wire anti-personnel mine. Note: as the firer, you must maintain at least a 150 meter distance from the mine to prevent self-injury. |

| **Improvised Explosive mixtures utilizing low velocity explosives:** | | |
|---|---|---|
| ***Low velocity explosive*** | ***Priming type and weight*** | ***Fragmentation weight*** |
| 1. Ammonium nitrate (10 parts by weight). | Prime with TNT (2 ½ parts by weight) | Use 5 parts of steel shot, nuts, bolts, or nails |
| 2. Black powder (10 parts by weight). | Prime with TNT, or dynamite (3 ½ parts by weight) | Use 5 parts of steel shot, nuts, bolts, or nails |
| 3. Potassium nitrate (10 parts by weight). | Prime with TNT, dynamite (4 ½ parts by weight) | Use 5 parts of steel shot, nuts, bolts, or nails |
| 4. Potassium chlorate (10 parts by weight). | Prime with 3 ½ parts of TNT, or dynamite | Use 4 ½ parts of steel shot, nuts, bolts, or nails |
| 5. Potassium perchlorate (10 parts by weight). | Prime with 3 ½ parts of TNT, or dynamite | Use 4 ½ parts of steel shot, nuts, bolts, or nails |
| 6. Potassium permanganate (10 parts by weight). | Prime with a mixture of 3 parts of TNT, and 1 part of black powder (mix the black powder thoroughly with powdered TNT, and then press into block. Use this block for direct priming). | Use 4 ½ parts of steel shot, nuts, bolts, or nails. |
| 7. Potassium dichromate (10 parts by weight). | Prime with a mixture of 3 parts of TNT, or PETN, and 1 part of lead-VI-chromate (mix the chromate thoroughly with powdered TNT, and then press into block. Use this block for direct priming). | Use 4 ½ parts of steel shot, nuts, bolts, or nails. |
| 8. Barium nitrate (10 parts by weight). | Prime with a mixture of 3 parts of TNT, or PETN, and 1 part of antimony trisulfide (mix the trisulfide thoroughly with powdered TNT, and then press into block. Use this block for direct priming). | Use 4 parts of steel shot, nuts, bolts, or nails. |
| 9. Ammonium perchlorate (10 parts by weight). | Prime with TNT (3 parts by weight) | Use 5 parts of steel shot, nuts, bolts, or nails. |
| 10. Lead nitrate (anhydrous) (10 parts) | Prime with TNT, PETN, or dynamite (3 parts by weight) | Use 5 to 6 parts steel shot, nuts, bolts, or nails. |

**06-04-017A: Ignition composition for flares, signals, and smoke candles utilizing black powder:**
Into a suitable vertical mixer, place 200 milliliters of hexane, followed by ***425 grams of finely divided black powder***, followed by ***50 grams of nitro starch***, and then followed by ***25 grams of para-nitrotoluene***. Thereafter, blend the mixture on high speed for about 45 minutes. After 45 minutes the mixture is ready for pressing. To do so, press the mixture into molds, discs, tablets, cubes, pellets, ect., of diameter and thickness (as dictated by the device/munition in question; use your discretion), under a pressure of 7,000 psi. Thereafter, cure the molds, discs, tablets, cubes, pellets, ect., in an oven at 60 Celsius for 12 hours.
**Burn rate (average solid grain at 7,000 psi):** 6.9mm per second. Depends on diameter and density of solid grain.
**Burn temperature:** 1093 Celsius.
**Ignition temperature (average temperature to ignite mixture at 7,000 psi):** 668 Celsius.
**Water resistance:** Very good.
**Stability:** Can be stored for many years.
**Flammability (pressed grain at 7,000 psi):** 6.9
**Ease of ignition (average temperature to ignite mixture at 7,000 psi):** 9
**Tendency to cake:** None.
**Explosive ability:** Cannot be detonated.
**Percentage:** ***85% black powder, 10% nitro starch, 5% para-nitrotoluene***
**Classification:** Deflagrating mixture (classified as 1.3G pyrotechnic composition).

**Danger of preparation:** Can be made by amateurs.
**Use:** Used to ignite flares, signal compositions, smoke compositions, and similar pyrotechnics.

**06-04-018A: Priming composition for blasting caps, and for use in fireworks:**
Into a suitable vertical mixer, place 175 milliliters of hexane, followed by ***4 grams of sodium tungstate***, followed by ***2 grams of powdered copper***, followed by ***4 grams of strontium nitrate***, followed by ***96 grams of antimony sulfide,*** followed by ***108 grams of precipitated silver powder***, and then followed by ***20 grams of lead dioxide***. Thereafter, slowly blend the mixture at room temperature for 60 minutes. After 60 minutes, the mixture is ready for use. To use in explosive electric bridge wire devices, the metal wire (usually a thin NiChrome wire) should be evenly coated with the moldable mixture coating the exposed portion of the "wire" with a 1 to 3mm thickness of the mixture, and then cured in an oven at 60 Celsius for several hours. To use in blasting caps, the mixture should be pressed into molds under a pressure of 2,000 psi, forming tablets that need to be cured in an oven at 60 Celsius for 12 hours; the diameter and length of these "tablets" is based on the diameter and length of the given blasting cap; use your own discretion.
**Burn rate (when heated externally):** 490 Celsius. Explodes.
**Burn rate (when heated directly with a flame or hot NiChrome wire):** Explodes.
**Ignition temperature (average temperature to ignite mixture at 2,000 psi):** 599 Celsius.
**Water resistance:** Very good.
**Sensitivity:** Moderate stability to friction; resistant to normal shock; heat resistant (external) up to 489 Celsius.
**Stability:** Can be stored for many years.
**Flammability:** Explodes.
**Ease of ignition (when heated by a flame or hot wire):** 9.8
**Tendency to cake:** Will not cake.
**Explosive ability:** Does not detonate, but undergoes extreme explosive combustion.
**Percentage:** ***44.8% potassium chlorate, 25.4% silver, 22.6% antimony sulfide, 4.7% lead dioxide, 0.94% sodium tungstate, 0.94% strontium nitrate, 0.47% copper, 0.15% mixed impurities***
**Classification:** Deflagrating mixture (classified as 1.4G pyrotechnic composition).

**Danger of preparation:** Should only be prepared by experienced pyrotechnicians.
**Use:** Used in explosive electric "bridge" wire devices to initiate detonating cord; used as a priming agent in non-electric and electric blasting caps. Used for priming aerial shells for use in fireworks. Do not use to fire high performance rocket fuels or high performance gun propellants.

**Note:** Inexperienced personnel should not attempt this procedure, as there are hazards involved. Avoid excessive heat or friction. This procedure is perfectly safe for experienced persons. To gain experience, preparers should use only small quantities to begin with. Either way, safety shields should be used at all times, and any possible source of ignition, including static electricity should be avoided—use static resistant tables and floor mats.

**Electric blasting system**

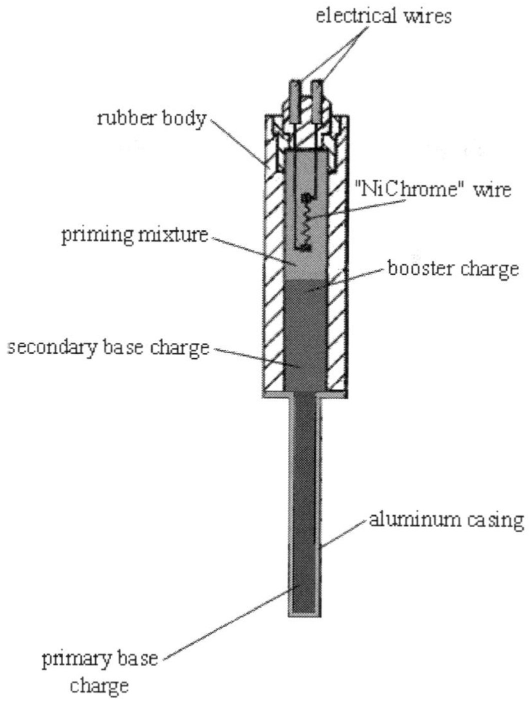


**06-04-018B: Priming composition for blasting caps, and for use in fireworks with increased shock resistance:**
Into a suitable vertical mixer, place 200 milliliters of hexane, followed by ***200 grams of copper ammonium nitrate***, followed by ***125 grams of potassium nitrate***, followed by ***35 grams of flours of sulfur***, and then followed by ***140 grams of finely divided aluminum powder***. Thereafter, slowly blend the mixture at room temperature for 60 minutes. After 60 minutes, the mixture is ready for use. To use in explosive electric bridge wire devices, the metal wire (usually a thin NiChrome wire) should be evenly coated with the moldable mixture coating the exposed portion of the "wire" with a 1 to 3mm thickness of the mixture, and then cured in an oven at 60 Celsius for several hours. To use in blasting caps, the mixture should be pressed into molds under a pressure of 2,500 psi, forming tablets that need to be cured in an oven at 60 Celsius for 12 hours; the diameter and length of these "tablets" is based on the diameter and length of the given blasting cap; use your own discretion.
**Burn rate (when heated externally):** 451 Celsius. Explodes.
**Burn rate (when heated directly with a flame or hot NiChrome wire):** Explodes.
**Ignition temperature (average temperature to ignite mixture at 2,500 psi):** 607 Celsius.
**Water resistance:** Very good.
**Sensitivity:** Moderate stability to friction; resistant to above normal shock; heat resistant (external) up to 450 Celsius.
**Stability:** Can be stored for many years.
**Flammability:** Explodes.
**Ease of ignition (when heated by a flame or hot wire):** 9.8
**Tendency to cake:** Will not cake.
**Explosive ability:** Does not detonate, but undergoes extreme explosive combustion.
**Percentage:** ***40% copper ammonium nitrate, 28% aluminum, 25% potassium nitrate, 7% sulfur***
**Classification:** Deflagrating mixture (classified as 1.4G pyrotechnic composition).

**Danger of preparation:** Should only be prepared by experienced pyrotechnicians.
**Use:** Used in explosive electric "bridge" wire devices to initiate detonating cord; used as a priming agent in non-electric and electric blasting caps. Used for priming aerial shells for use in fireworks. Do not use to fire high performance rocket fuels or high performance gun propellants.

**Note:** Inexperienced personnel should not attempt this procedure, as there are hazards involved. Avoid excessive heat or friction. This procedure is perfectly safe for experienced persons. To gain experience, preparers should use only small quantities to begin with. Either way, safety shields should be used at all times, and any possible source of ignition, including static electricity should be avoided—use static resistant tables and floor mats.

**Standard blasting cap for initiating PETN detonating cord**

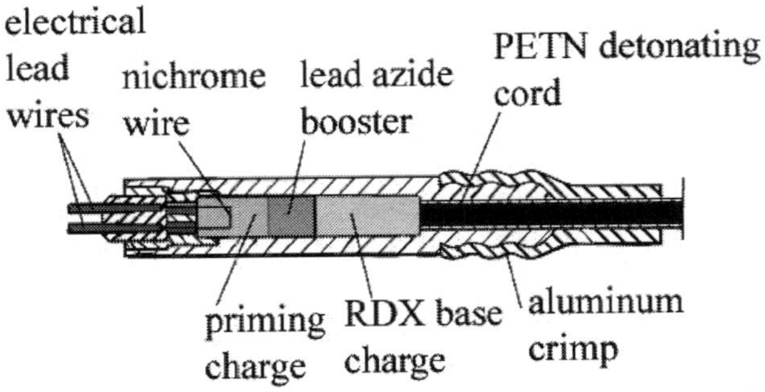


### 06-04-019A: Ignition composition for high performance ammonium nitrate/perchlorate rocket propellants:

Into a suitable vertical mixer, place 275 milliliters of hexane, followed by ***25 grams of ethyl cellulose***, followed by ***219.9 grams of potassium perchlorate***, followed by ***137.7 grams of barium nitrate***, and then followed by ***447.3 grams of finely powdered zirconium/nickel alloy*** containing 25% zirconium and 75% nickel. Thereafter, thoroughly blend the mixture on high speed for about 45 minutes at room temperature. After 45 minutes, the mixture is ready for pressing or for forming electric bridge wire igniters. To do so, first press the mixture into molds, discs, tablets, cubes, pellets, ect., of diameter and thickness (as dictated by the device/munition in question; use your discretion), under a pressure of 8,000 psi. Thereafter, cure the molds, discs, tablets, cubes, pellets, ect., in an oven at 60 Celsius for 12 hours. Second, for forming electric bridge wire igniters, the metal wire (usually a thin NiChrome wire) should be evenly coated with the moldable mixture coating the exposed portion of the “wire” with a 1.5 to 3.5mm thickness of the mixture, and then cured in an oven at 60 Celsius for several hours.

**Burn rate (when heated externally; when pressed at 8,000 psi):** Ignites at 800 Celsius.
**Burn rate (when heated directly with a flame or hot NiChrome wire; when pressed at 8,000 psi; uncontained):** 5.99 seconds per inch; tends to vary based on diameter and density of grain.
**Burn rate (when heated directly with a flame or hot NiChrome wire; when powdered; uncontained):** 4.1 seconds per inch.
**Burn rate (when heated directly with a flame or hot NiChrome wire; electric bridge wire):** Instantaneous ignition.
**Burn temperature:** 1169 Celsius.
**Ignition temperature (average temperature to ignite mixture at 8,000 psi):** 601 Celsius.
**Water resistance:** Very good.
**Sensitivity:** Very stable to friction; resistant to above normal shock; heat resistant (external) up to 799 Celsius.
**Stability:** Can be stored for many years.
**Flammability (ignited by flame or hot wire; uncontained):** 7.9
**Ease of ignition (when heated by a flame or hot wire; pressed at 8,000 psi):** 9.198
**Ease of ignition (when heated by a flame or hot wire; when powdered):** 9.38
**Tendency to cake:** Will not cake.
**Explosive ability:** Cannot be detonated.
**Percentage:** ***53.8% zirconium/nickel alloy, 26.4% potassium perchlorate, 16.5% barium nitrate, 3% ethyl cellulose binder, 0.3% mixed residues***
**Classification:** Deflagrating mixture (classified as 1.3G pyrotechnic composition).

**Danger of preparation:** Can be made by amateurs.
**Use:** Used to initiate solid rocket propellants in rockets and missiles.

### 06-04-019B: Ignition composition for high performance ammonium nitrate/perchlorate rocket propellants (modified):

Into a suitable vertical mixer, place 280 milliliters of hexane, followed by ***30 grams of ethyl cellulose***, followed by ***136 grams of potassium perchlorate***, followed by ***140 grams of barium nitrate***, followed by ***160 grams of finely powdered zirconium/nickel alloy***, containing 25% zirconium and 75% nickel, followed by ***116 grams of zirconium dichromate***, followed by ***165 grams of finely powdered aluminum***, followed by ***94 grams of finely powdered magnesium***, followed by ***130 grams of finely powdered titanium metal***, followed by ***4 grams of zirconium hydride,*** and then followed by ***18 grams of finely divided boron***. Thereafter, thoroughly blend the mixture on high speed for about 45 minutes at room temperature. After 45 minutes, the mixture is ready for pressing or for forming electric bridge wire igniters. To do so, first press the mixture into molds, discs, tablets, cubes, pellets, ect., of diameter and thickness (as dictated by the device/munition in question; use your discretion), under a pressure of 8,000 psi. Thereafter, cure the molds, discs, tablets, cubes, pellets, ect., in an oven at 60 Celsius for 12 hours. Second, for forming electric bridge wire igniters, the metal wire (usually a thin NiChrome wire) should be evenly coated with the moldable mixture coating the exposed portion of the “wire” with a 1.5 to 3.5mm thickness of the mixture, and then cured in an oven at 60 Celsius for several hours.

**Burn rate (when heated externally; when pressed at 8,000 psi):** Ignites at 709 Celsius.
**Burn rate (when heated directly with a flame or hot NiChrome wire; when pressed at 8,000 psi; uncontained):** 5.7 seconds per inch; tends to vary based on diameter and density of grain.
**Burn rate (when heated directly with a flame or hot NiChrome wire; when powdered; uncontained):** 4 seconds per inch.

**Burn rate (when heated directly with a flame or hot NiChrome wire; electric bridge wire):** Instantaneous ignition.
**Burn temperature:** 1207 Celsius.
**Ignition temperature (average temperature to ignite mixture at 8,000 psi):** 598 Celsius.
**Water resistance:** Very good.
**Sensitivity:** Very stable to friction; resistant to above normal shock; heat resistant (external) up to 708 Celsius.
**Stability:** Can be stored for many years.
**Flammability (ignited by flame or hot wire; uncontained):** 7.99
**Ease of ignition (when heated by a flame or hot wire; pressed at 8,000 psi):** 9.23
**Ease of ignition (when heated by a flame or hot wire; when powdered):** 9.4
**Tendency to cake:** Will not cake.
**Explosive ability:** Cannot be detonated.
**Percentage:** ***16.6% aluminum, 16.1% zirconium/nickel alloy, 14% barium nitrate, 13.6% potassium perchlorate, 13% titanium metal, 11.6% zirconium dichromate, 9.4% magnesium, 3.0% ethyl cellulose binder, 1.8% boron, 0.50% mixed residues, 0.4% zirconium hydride burn catalyst***
**Classification:** Deflagrating mixture (classified as 1.3G pyrotechnic composition).

**Danger of preparation:** Can be made by amateurs, but caution should be exercised.
**Use:** Used to initiate solid rocket propellants in rockets and missiles.

### 06-04-020A: Gasless ignition composition for high performance rocket propellants:

Into a suitable vertical mixer, place 210 milliliters of hexane, followed by ***225 grams of potassium perchlorate***, followed by ***275 grams of iron powder***, and then followed by ***25 grams of black powder***. Thereafter, blend the mixture on high speed for about 45 minutes. After 45 minutes the mixture is ready for pressing. To do so, press the mixture into molds, discs, tablets, cubes, pellets, ect., of diameter and thickness (as dictated by the device/munition in question; use your discretion), under a pressure of 7,500 psi. Thereafter, cure the molds, discs, tablets, cubes, pellets, ect., in an oven at 60 Celsius for 12 hours.
**Burn rate:** 1.15mm per second. Depends on diameter and density of solid grain.
**Burn temperature:** 1388 Celsius.
**Ignition temperature (average temperature to ignite mixture at 7,500 psi):** 873 Celsius.
**Water resistance:** Very good.
**Stability:** Can be stored for many years.
**Flammability (pressed grain at 7,500 psi):** 4.7
**Ease of ignition (average temperature to ignite mixture at 7,500 psi):** 4.9
**Tendency to cake:** None.
**Explosive ability:** Cannot be detonated.
**Percentage:** ***52.3% iron, 42.8% potassium perchlorate, 4.7% black powder, 0.2% residues***
**Classification:** Deflagrating mixture (classified as 1.3G pyrotechnic composition).

**Danger of preparation:** Can be made by amateurs.
**Use:** Used to initiate incendiary compositions.

### 06-04-021A: Non-gaseous high temperature ignition composition for incendiary mixtures:

Into a suitable vertical mixer, place 210 milliliters of hexane, followed by ***150 grams of iron powder of average mesh***, followed by ***50 grams of finely powdered copper-II-oxide***, and then followed by ***30 grams of finely powdered magnesium***. Thereafter, blend the mixture on high speed for about 45 minutes. After 45 minutes the mixture is ready for pressing. To do so, press the mixture into molds, discs, tablets, cubes, pellets, ect., of diameter and thickness (as dictated by the device/munition in question; use your discretion), under a pressure of 7,500 psi. Thereafter, cure the molds, discs, tablets, cubes, pellets, ect., in an oven at 60 Celsius for 12 hours.
**Burn rate:** 1.27mm per second. Depends on diameter and density of solid grain.
**Burn temperature:** 1399 Celsius.
**Ignition temperature (average temperature to ignite mixture at 7,500 psi):** 900 Celsius.
**Water resistance:** Very good.
**Stability:** Can be stored for many years.
**Flammability (pressed grain at 7,500 psi):** 4.79
**Ease of ignition (average temperature to ignite mixture at 7,500 psi):** 4
**Tendency to cake:** None.
**Explosive ability:** Cannot be detonated.
**Percentage:** ***65.2% iron powder, 21.7% copper-II-oxide, 13% magnesium, 0.1% mixed impurities***
**Classification:** Deflagrating mixture (classified as 1.3G pyrotechnic composition).

**Danger of preparation:** Can be made by amateurs.
**Use:** Used to initiate incendiary compositions.

### 06-04-022A: Standard ignition composition for tracer mixtures, smoke compositions, and similar mixtures:

Into a suitable vertical mixer, place 205 milliliters of hexane, followed by ***447.5 grams of barium nitrate***, and then followed by ***52.5 grams of finely powdered magnesium***. Thereafter, blend the mixture on moderate speed for about 45 minutes. After 45 minutes the mixture is ready for pressing. For use in tracers, the mixture needs to be pressed into a thin disc ranging from 1.5mm to 2mm in thickness adhering to the same diameter as the tracer composition used for the given projectile in question and under a pressure of 4,000 psi, and the resulting munition needs to be cured in an oven at 60 Celsius for 6 hours. For use in smoke compositions, flares, ect., the mixture needs to pressed into pellets, tablets, discs, rods, ect., of diameter and thickness (as dictated by the device/munition in question; use your discretion), under a pressure of 7,000 psi, and then cured in an oven at 60 Celsius for 8 hours.
**Burn rate:** 1mm per second.
**Burn temperature:** 768 Celsius.
**Ignition temperature (average temperature to ignite mixture):** 825 Celsius.
**Water resistance:** Very good.
**Stability:** Can be stored for many years.
**Flammability (pressed grain at 7,000 psi):** 7.1
**Ease of ignition (average temperature to ignite mixture):** 4
**Tendency to cake:** None.
**Explosive ability:** Cannot be detonated.
**Percentage:** ***89.5% barium nitrate, 10.5% magnesium***
**Classification:** Deflagrating mixture (classified as 1.3G pyrotechnic composition).

**Danger of preparation:** Can be made by amateurs.
**Use:** Widely used mixture for the ignition of tracer compositions, smoke compositions, flares, and similar mixtures.

### 06-04-022B: Standard ignition composition for tracer mixtures, smoke compositions, and similar mixtures (modified):

Into a suitable vertical mixer, place 215 milliliters of hexane, followed by ***333 grams of barium peroxide***, followed by ***55.5 grams of lead tetraoxide ($Pb_3O_4$)***, and then followed by ***111.5 grams of finely powdered magnesium***. Thereafter, blend the mixture on moderate speed for about 45 minutes. After 45 minutes the mixture is ready for pressing. For use in tracers, the mixture needs to be pressed into a thin disc ranging from 1.5mm to 2mm in thickness adhering to the same diameter as the tracer composition used for the given projectile in question and under a pressure of 4,000 psi, and the resulting munition needs to be cured in an oven at 60 Celsius for 6 hours. For use in smoke compositions, flares, ect., the mixture needs to pressed into pellets, tablets, discs, rods, ect., of diameter and thickness (as dictated by the device/munition in question; use your discretion), under a pressure of 7,000 psi, and then cured in an oven at 60 Celsius for 8 hours.
**Burn rate:** 1.3mm per second.
**Burn temperature:** 799 Celsius.
**Ignition temperature (average temperature to ignite mixture):** 760 Celsius.
**Water resistance:** Very good.
**Stability:** Can be stored for many years.
**Flammability (pressed grain at 7,000 psi):** 7.3
**Ease of ignition (average temperature to ignite mixture):** 4.9
**Tendency to cake:** None.
**Explosive ability:** Cannot be detonated.
**Percentage:** ***66.6% barium peroxide, 22.3% magnesium, 11.1% lead tetraoxide***
**Classification:** Deflagrating mixture (classified as 1.3G pyrotechnic composition).

**Danger of preparation:** Can be made by amateurs.
**Use:** Widely used mixture for the ignition of tracer compositions, smoke compositions, flares, and similar mixtures.

**06-04-023A: Priming composition with heat resistant properties for use in firing lead azide in explosive trains:**
Into a suitable vertical mixer, place 120 milliliters of hexane, followed by ***36 grams of calcium silicide***, followed by ***18 grams of finely powdered zirconium***, followed by ***18 grams of finely powdered sulfur***, and then followed by ***54 grams of lead dioxide***. Thereafter, slowly blend the mixture at room temperature for 60 minutes. After 60 minutes, the mixture is ready for use. To use in explosive electric bridge wire devices, the metal wire (usually a thin NiChrome wire) should be evenly coated with the moldable mixture coating the exposed portion of the "wire" with a 1 to 3mm thickness of the mixture, and then cured in an oven at 60 Celsius for several hours. To use in blasting caps, the mixture should be pressed into molds under a pressure of 3,000 psi, forming tablets that need to be cured in an oven at 60 Celsius for 12 hours; the diameter and length of these "tablets" is based on the diameter and length of the given blasting cap; use your own discretion.
**Burn rate (when heated externally):** 400 Celsius. Explodes.
**Burn rate (when heated directly with a flame or hot NiChrome wire):** Explodes.
**Ignition temperature (average temperature to ignite mixture at 3,000 psi):** 624 Celsius.
**Water resistance:** Very good.
**Sensitivity:** Moderate stability to friction; resistant to normal shock; heat resistant (external) up to 399 Celsius.
**Stability:** Can be stored for many years.
**Flammability:** Explodes.
**Ease of ignition (when heated by a flame or hot wire):** 9.8
**Tendency to cake:** Will not cake.
**Explosive ability:** Does not detonate, but undergoes extreme explosive combustion.
**Percentage:** ***42.8% lead dioxide, 28.5% calcium silicide, 14.2% zirconium, 14.2% sulfur, 0.30% mixed residues***
**Classification:** Deflagrating mixture (classified as 1.4G pyrotechnic composition).

**Danger of preparation:** Should only be prepared by experienced pyrotechnicians.
**Use:** Used in explosive electric "bridge" wire devices to initiate detonating cord; used as a priming agent in non-electric and electric blasting caps. Used for priming aerial shells for use in fireworks. Do not use to fire high performance rocket fuels or high performance gun propellants.

**Note:** Inexperienced personnel should not attempt this procedure, as there are hazards involved. Avoid excessive heat or friction. This procedure is perfectly safe for experienced persons. To gain experience, preparers should use only small quantities to begin with. Either way, safety shields should be used at all times, and any possible source of ignition, including static electricity should be avoided—use static resistant tables and floor mats.

| Fuze for hard target projectiles | Description: |
|---|---|
| 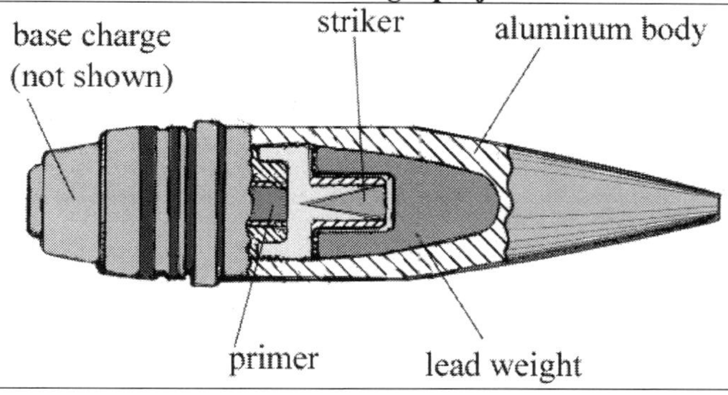 | The fuze illustrated adjacent works upon impact of a hard target. When the projectile hits a hard target, the nose collapses rearwards, causing the striker assembly to impact into a primer. The primer sets of the explosive train, which detonates the main charge (not shown). |

**06-04-023B: Priming composition with heat resistant properties for use in firing lead azide in explosive trains (modified):**
Into a suitable vertical mixer, place 100 milliliters of hexane, followed by ***24 grams of calcium silicide***, followed by ***30 grams of finely powdered zirconium***, followed by ***12 grams of finely powdered sulfur***, followed by ***24 grams of manganese dioxide***, and then followed by ***54 grams of lead dioxide***. Thereafter, slowly blend the mixture at room temperature for 60 minutes. After 60 minutes, the mixture is ready for use. To use in explosive electric bridge wire devices, the metal wire (usually a thin NiChrome wire) should be evenly coated with the moldable mixture coating the exposed portion of the "wire" with a 1 to 3mm thickness of the mixture, and then cured in an oven at 60 Celsius for several hours. To use in blasting caps, the mixture should be pressed into molds under a pressure of 3,000 psi, forming tablets that need to be cured in an oven at 60 Celsius for 12 hours; the diameter and length of these "tablets" is based on the diameter and length of the given blasting cap; use your own discretion.
**Burn rate (when heated externally):** 455 Celsius. Explodes.
**Burn rate (when heated directly with a flame or hot NiChrome wire):** Explodes.
**Ignition temperature (average temperature to ignite mixture at 3,000 psi):** 659 Celsius.
**Water resistance:** Very good.
**Sensitivity:** Moderate stability to friction; below resistance to shock (explodes upon percussion); heat resistant (external) up to 454 Celsius.
**Stability:** Can be stored for many years.

**Flammability:** Explodes.
**Ease of ignition (when heated by a flame or hot wire):** 9.8
**Tendency to cake:** Will not cake.
**Explosive ability:** Does not detonate, but undergoes extreme explosive combustion.
**Percentage:** ***37.5% lead dioxide, 20.8% zirconium, 16.6% calcium silicide, 16.6% manganese dioxide, 8.3% sulfur, 0.20% mixed residues***
**Classification:** Deflagrating mixture (classified as 1.4G pyrotechnic composition).

**Danger of preparation:** Should only be prepared by experienced pyrotechnicians.
**Use:** Used in explosive electric "bridge" wire devices to initiate detonating cord; used as a priming agent in non-electric and electric blasting caps. Used for priming aerial shells for use in fireworks. Do not use to fire high performance rocket fuels or high performance gun propellants.

**Note:** Inexperienced personnel should not attempt this procedure, as there are hazards involved. Avoid excessive heat or friction. This procedure is perfectly safe for experienced persons. To gain experience, preparers should use only small quantities to begin with. Either way, safety shields should be used at all times, and any possible source of ignition, including static electricity should be avoided—use static resistant tables and floor mats.

<table>
<tr><th>Classic fuze for artillery and mortar shells</th><th>Details:</th></tr>
<tr><td>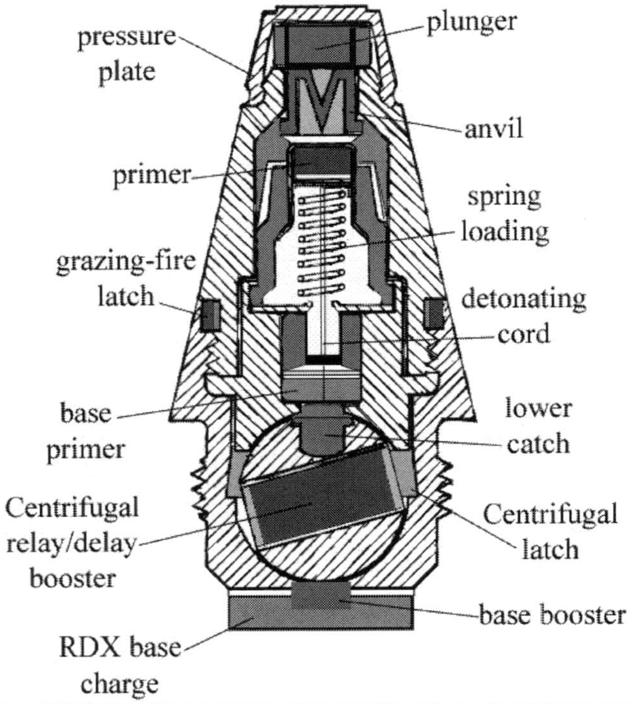
</td><td>In this illustration, a classic impact fuze is shown. The fuze is armed, when the projectile is fired. The arming is caused by the centrifugal force, which forces the relay/delay booster to right it self north to south of the fuze. Thereafter, when the shell impacts with a target, the plunger is forced rearward striking a primer. The primer ignites the detonating cord, which in turn sets-off the explosive train contained in the centrifugal relay/delay booster assembly. The spring loading is carried upon the grazing fire latch, and releases a striker pin, that sets off a base primer assembly. The grazing fire latch is activated when the shell "grazes" the target, rather then hitting it head on.</td></tr>
</table>

### 06-04-024A: Ignition composition for use in non-electric blasting caps:

Into a suitable vertical mixer, place 120 milliliters of hexane, followed by ***140 grams of finely powdered metallic bismuth***, followed by ***50 grams of finely powdered selenium metal***, and then followed by ***10 grams of potassium chlorate***. Thereafter, slowly blend the mixture at room temperature for 60 minutes. After 60 minutes, the mixture is ready for use. To use in blasting caps, the mixture should be pressed into molds under a pressure of 4,000 psi, forming tablets that need to be cured in an oven at 60 Celsius for 12 hours; the diameter and length of these "tablets" is based on the diameter and length of the given blasting cap; use your own discretion.
**Burn rate (when heated externally):** 501 Celsius. Explodes.
**Burn rate (when heated directly with a flame or hot NiChrome wire):** Explodes.
**Ignition temperature (average temperature to ignite mixture at 4,000 psi):** 600 Celsius.
**Water resistance:** Very good.
**Sensitivity:** Moderate stability to friction; above resistance to shock; heat resistant (external) up to 500 Celsius.
**Stability:** Can be stored for many years.
**Flammability:** Explodes.
**Ease of ignition (when heated by a flame or hot wire):** 9.6
**Tendency to cake:** Will not cake.
**Explosive ability:** Does not detonate, but undergoes extreme explosive combustion.
**Percentage:** ***70% metallic bismuth, 25% selenium metal, 5% potassium chlorate***

**Classification:** Deflagrating mixture (classified as 1.4G pyrotechnic composition).

**Danger of preparation:** Should only be prepared by experienced pyrotechnicians.
**Use:** Used as an ignition composition in non-electric blasting caps to initiate lead azide and other primary explosive elements. This ignition composition is ignited by a standard black powder time fuse.

### 06-04-025A: Easily ignitable gasless ignition composition for use in igniting incendiary compositions:

Into a suitable vertical mixer, place 100 milliliters of hexane, followed by ***37.5 grams of finely powdered titanium metal***, and then followed by ***87.5 grams of finely powdered red iron-III-oxide***. Thereafter, blend the mixture on high speed for about 45 minutes. After 45 minutes the mixture is ready for pressing. To do so, press the mixture into molds, discs, tablets, cubes, pellets, ect., of diameter and thickness (as dictated by the device/munition in question; use your discretion), under a pressure of 7,500 psi. Thereafter, cure the molds, discs, tablets, cubes, pellets, ect., in an oven at 60 Celsius for 12 hours.
**Burn rate:** 1.2mm per second. Depends on diameter and density of solid grain.
**Burn temperature:** 1471 Celsius.
**Ignition temperature (average temperature to ignite mixture at 7,500 psi):** 870 Celsius.
**Water resistance:** Very good.
**Stability:** Can be stored for many years.
**Flammability (pressed grain at 7,500 psi):** 4.84
**Ease of ignition (average temperature to ignite mixture at 7,500 psi):** 4.2
**Tendency to cake:** None.
**Explosive ability:** Cannot be detonated.
**Percentage:** ***70% iron-III-oxide, 30% titanium***
**Classification:** Deflagrating mixture (classified as 1.3G pyrotechnic composition).

**Danger of preparation:** Can be made by amateurs.
**Use:** Used to initiate incendiary compositions.

### 06-04-025B: Easily ignitable gasless ignition composition for use in igniting incendiary compositions:

Into a suitable vertical mixer, place 100 milliliters of hexane, followed by ***54 grams of finely powdered zirconium metal***, and then followed by ***66 grams of finely powdered red iron-III-oxide***. Thereafter, blend the mixture on high speed for about 45 minutes. After 45 minutes the mixture is ready for pressing. To do so, press the mixture into molds, discs, tablets, cubes, pellets, ect., of diameter and thickness (as dictated by the device/munition in question; use your discretion), under a pressure of 7,500 psi. Thereafter, cure the molds, discs, tablets, cubes, pellets, ect., in an oven at 60 Celsius for 12 hours.
**Burn rate:** 1.21mm per second. Depends on diameter and density of solid grain.
**Burn temperature:** 1450 Celsius.
**Ignition temperature (average temperature to ignite mixture at 7,500 psi):** 860 Celsius.
**Water resistance:** Very good.
**Stability:** Can be stored for many years.
**Flammability (pressed grain at 7,500 psi):** 4.89
**Ease of ignition (average temperature to ignite mixture at 7,500 psi):** 4.5
**Tendency to cake:** None.
**Explosive ability:** Cannot be detonated.
**Percentage:** ***70% iron-III-oxide, 30% zirconium metal***
**Classification:** Deflagrating mixture (classified as 1.3G pyrotechnic composition).

**Danger of preparation:** Can be made by amateurs.
**Use:** Used to initiate incendiary compositions.

### 06-04-026A: Easily ignitable gasless ignition composition for use in igniting incendiary compositions:

Into a suitable vertical mixer, place 100 milliliters of hexane, followed by ***75 grams of finely powdered magnesium metal***, followed by ***35.2 grams of finely divided manganese metal*** of 200 mesh, followed by ***48 grams of divided manganese metal*** of 325 mesh, followed by ***6 grams of manganese metal*** of 50 mesh, followed by ***23 grams of a titanium/manganese alloy*** containing about 40% titanium and 60% manganese, followed by ***150 milligrams of finely powdered titanium metal***, and then followed by ***150 milligrams of finely powdered zirconium metal***. Thereafter, blend the mixture on high speed for about 45 minutes. After 45 minutes the mixture is ready for pressing. To do so, press the mixture into molds, discs, tablets, cubes, pellets, ect., of diameter and thickness (as dictated by the device/munition in question; use your discretion), under a pressure of 7,500 psi. Thereafter, cure the molds, discs, tablets, cubes, pellets, ect., in an oven at 60 Celsius for 12 hours.

**Burn rate:** 1.69mm per second. Depends on diameter and density of solid grain.
**Burn temperature:** 1300 Celsius.
**Ignition temperature (average temperature to ignite mixture at 7,500 psi):** 900 Celsius.
**Water resistance:** Very good.
**Stability:** Can be stored for many years.
**Flammability (pressed grain at 7,500 psi):** 4
**Ease of ignition (average temperature to ignite mixture at 7,500 psi):** 4.41
**Tendency to cake:** None.
**Explosive ability:** Cannot be detonated.
**Percentage:** ***47.5% manganese, 40% magnesium, 12.2% titanium/manganese alloy, 0.14% mixed residues, 0.08% titanium, 0.08% zirconium***
**Classification:** Deflagrating mixture (classified as 1.3G pyrotechnic composition).

**Danger of preparation:** Can be made by amateurs.
**Use:** Used to initiate incendiary compositions.

### 06-04-027A: Priming composition for use in percussion caps:

Into a suitable sealable vertical mixer, equipped with flexible plastic stir blades, place ***54 grams of oxalic acid***, followed by ***54 grams of antimony pentasulfide***, and then followed by ***54 grams of potassium chlorate***. Thereafter, dry blend the mixture on moderate speed for 50 minutes. After 50 minutes, add in ***54 grams of standard animal glue***, seal the vertical mixer, and then blend the mixture on moderate speed for 15 minutes. After 15 minutes, open the vertical mixer, recover the mixture, and then extrude the mixture through any desired extruding machine into strands ranging from 0.5mm to 3mm in diameter by 10cm in length, then cut the strands into 0.5mm to 3mm pieces and then cure the pieces on a flat tray at room temperature for 48 hours. Note: the size of the cured pieces depends on the gun primer in question. Small arms weapons such as handguns, and rifles should use 0.5 to 1mm sized pieces; whereas larger bore weapons such as 12.7mm up to 25mm casings should use 1.5 to 2mm sized pieces; use your discretion).

**Burn rate (when heated externally):** 499 Celsius. Explodes.
**Burn rate (when heated directly with a flame or hot NiChrome wire):** Explodes.
**Drop Test:** 10cm with 500 gram weight: Explodes.
**Water resistance:** Very good.
**Sensitivity:** Unstable to friction; sensitive to shock and percussion; heat resistant (external) up to 450~ Celsius.
**Stability:** Can be stored for many years.
**Flammability:** Explodes.
**Ease of ignition (when heated by a flame or hot wire):** 9.9
**Tendency to cake:** Will not cake.
**Explosive ability:** Does not detonate, but undergoes extreme explosive combustion.
**Percentage:** ***25% animal glue binder, 25% oxalic acid, 25% antimony sulfide, 25% potassium chlorate***
**Classification:** Deflagrating mixture (classified as 1.4G pyrotechnic composition).

**Danger of preparation:** Should only be prepared by experienced pyrotechnicians.
**Use:** Used in gun casing primers.

### 06-04-028A: Slow burning, flashless, ignition composition for tracer compositions:

Into a suitable vertical mixer, place 180 milliliters of hexane, followed by ***235.2 grams of barium peroxide***, followed by ***55.2 grams of antimony pentasulfide***, followed by ***6.6 grams of finely powdered magnesium*** of 200 mesh, and then followed by ***3 grams of finely powdered graphite***. Thereafter, blend the mixture on moderate speed for about 45 minutes. After 45 minutes the mixture is ready for pressing. To do so, the mixture needs to be pressed into a thin disc ranging from 1.5mm to 2mm in thickness adhering to the same diameter as the tracer composition used for the given projectile in question and under a pressure of 4,000 psi, and the resulting munition needs to be cured in an oven at 60 Celsius for 6 hours.

**Burn rate:** 1.19mm per second.

**Burn temperature:** 781 Celsius.
**Ignition temperature (average temperature to ignite mixture):** 700 Celsius.
**Water resistance:** Very good.
**Stability:** Can be stored for many years.
**Flammability (pressed grain at 4,000 psi):** 7.49
**Ease of ignition (average temperature to ignite mixture):** 4.6
**Tendency to cake:** None.
**Explosive ability:** Cannot be detonated.
**Percentage:** ***78.4% barium peroxide, 18.4% antimony pentasulfude, 2.2% magnesium, 1% graphite***
**Classification:** Deflagrating mixture (classified as 1.3G pyrotechnic composition).

**Danger of preparation:** Can be made by amateurs.
**Use:** Used to initiate tracer compositions.

### 06-04-029A: Ignition composition for welding compositions, specifically for tin producing welding compositions:

Into a suitable vertical mixer, place 200 milliliters of hexane, followed by ***180 grams of copper-I-oxide***, followed by ***152.5 grams of copper-II-oxide***, followed by ***83 grams of a finely powdered metal alloy*** (commercially available; containing 60% copper, and 40% aluminum), followed by ***41.5 grams of finely powdered aluminum*** of average mesh and then followed by ***41.5 grams of finely grained red phosphorus***. Thereafter, blend the mixture on moderate speed for about 45 minutes. After 45 minutes the mixture is ready for pressing. To do so, the mixture needs to be pressed into tablets, pellets, or rods ranging from 2mm to 5mm in thickness by 3mm to 6mm in diameter under a pressure of 7,000 psi, and the resulting tablets, pellets, or rods need to be cured in an oven at 60 Celsius for 12 hours.
**Burn rate:** 1.56mm per second.
**Burn temperature:** 984 Celsius.
**Ignition temperature (average temperature to ignite mixture):** 707~ Celsius.
**Water resistance:** Very good.
**Stability:** Can be stored for many years.
**Flammability (pressed grain at 7,000 psi):** 6.12
**Ease of ignition (average temperature to ignite mixture):** 4.3
**Tendency to cake:** None.
**Explosive ability:** Cannot be detonated.
**Percentage:** ***36.1% copper-I-oxide, 30.5% copper-II-oxide, 16.6% metal alloy, 8.3% aluminum, 8.3% red phosphorus, 0.20% combined impurities***
**Classification:** Deflagrating mixture (classified as 1.3G pyrotechnic composition).

**Danger of preparation:** Can be made by amateurs.
**Use:** Used to initiate welding compositions.

### 06-04-030A: Gasless ignition composition for use in electric blasting caps and detonators:

Into a suitable vertical mixer, place 150 milliliters of hexane, followed by ***100 grams of finely powdered zirconium of average mesh***, and then followed by ***100 grams of lead monoxide***. Thereafter, moderately blend the mixture at room temperature for 45 minutes. After 45 minutes, the mixture is ready for use. To use in explosive electric bridge wire devices, the metal wire (usually a thin NiChrome wire) should be evenly coated with the moldable mixture coating the exposed portion of the "wire" with a 1 to 3mm thickness of the mixture, and then cured in an oven at 60 Celsius for several hours. To use in blasting caps, the mixture should be pressed into molds under a pressure of 9,000 psi, forming tablets that need to be cured in an oven at 60 Celsius for 12 hours; the diameter and length of these "tablets" is based on the diameter and length of the given blasting cap; use your own discretion.
**Burn rate (when heated externally):** 607 Celsius. Explodes.
**Burn rate (when heated directly with a flame or hot NiChrome wire):** Explodes.
**Ignition temperature (average temperature to ignite mixture at 9,000 psi):** 788 Celsius.
**Water resistance:** Very good.
**Sensitivity:** Good stability to friction; good resistance to shock; heat resistant (external) up to 606 Celsius.
**Stability:** Can be stored for many years.
**Flammability:** Explodes.
**Ease of ignition (when heated by a flame or hot wire):** 9.6
**Tendency to cake:** Will not cake.
**Explosive ability:** Does not detonate, but undergoes extreme explosive combustion.
**Percentage:** ***50% zirconium, 50% lead monoxide***
**Classification:** Deflagrating mixture (classified as 1.4G pyrotechnic composition).

**Danger of preparation:** Can be prepared by amateurs, but caution should be exercised.
**Use:** Used as an ignition composition in electric blasting caps to initiate lead azide and other primary explosive elements.

| Classic impact fuze for high velocity projectiles | Details: |
|---|---|
| 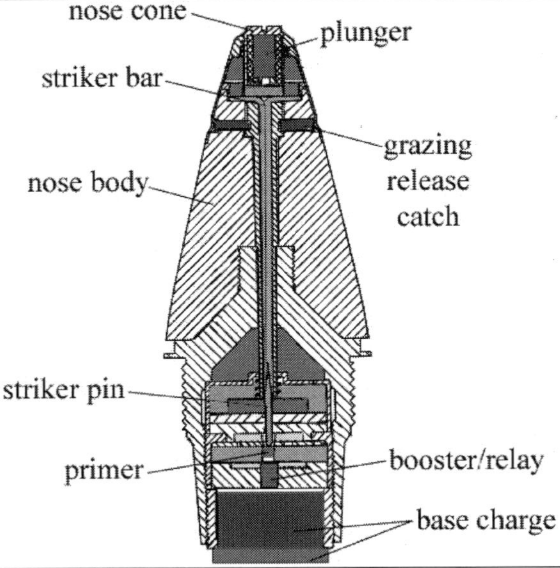  | This impact fuze is armed when the projectile is fired, and is carried about by centrifugal force caused by the spinning and acceleration of the projectile at the time of firing. Upon impacting the target, the striker bar hits a striker pin, which strikes the primer. The primer sets off the booster/relay, which in turn sets off the base charge. |

### 06-04-031A: Friction sensitive composition for use in primers:

Into a suitable sealable vertical mixer, equipped with flexible plastic stir blades, place ***90 grams of lead styphnate***, followed by ***30 grams of lead hypophosphite***, followed by ***36 grams of lead nitrate***, followed by ***84 grams of barium nitrate***, followed by ***30 grams of finely powdered glass***, followed by ***30 grams of ground glass*** of 200 mesh, and then followed by 125 milliliters of dry hexane. Thereafter, carefully and slowly blend the mixture on low speed for 45 minutes. After 45 minutes, open the vertical mixer, recover the mixture, and then extrude the mixture through any desired extruding machine into strands ranging from 0.5mm to 3mm in diameter by 10cm in length, then cut the strands into 0.5mm to 3mm pieces and then cure the pieces on a flat tray at room temperature for 48 hours. Note: the size of the cured pieces depends on the gun primer in question. Small arms weapons such as handguns, and rifles should use 0.5 to 1mm sized pieces; whereas larger bore weapons such as 12.7mm up to 25mm casings should use 1.5 to 2mm sized pieces; use your discretion).
**Burn rate (when heated externally):** 491 Celsius. Explodes.
**Burn rate (when heated directly with a flame or hot NiChrome wire):** Explodes.
**Drop Test:** 8cm with 750 gram weight: Explodes.
**Water resistance:** Acceptable.
**Sensitivity:** Unstable to friction; sensitive to shock and percussion; heat resistant (external) up to 450~ Celsius.
**Stability:** Can be stored for many years.
**Flammability:** Explodes.
**Ease of ignition (when heated by a flame or hot wire):** 7.1
**Tendency to cake:** Will not cake.
**Explosive ability:** Can be detonated.
**Percentage:** ***30% lead styphnate, 28% barium nitrate, 20% glass, 12% lead nitrate, 10% lead hypophosphite***
**Classification:** Explosive mixture (classified as Class A explosive composition).

**Danger of preparation:** Should only be prepared by experienced pyrotechnicians.
**Use:** Used in gun primers. Do not use to fire high performance rocket fuels or high performance gun propellants.

**Note:** Inexperienced personnel should not attempt this procedure, as there are hazards involved. Avoid excessive heat or friction. This procedure is perfectly safe for experienced persons. To gain experience, preparers should use only small quantities to begin with. Either way, safety shields should be used at all times, and any possible source of ignition, including static electricity should be avoided—use static resistant tables and floor mats.

**06-04-031B: Friction sensitive composition for use in primers (tetracene containing):**
Into a suitable sealable vertical mixer, equipped with flexible plastic stir blades, place ***6 grams of tetracene***, followed by ***120 grams of lead styphnate***, followed by ***90 grams of lead nitrate***, followed by ***24 grams of lead sulfocyanate***, followed by ***24 grams of finely powdered glass***, followed by ***36 grams of ground glass*** of 200 mesh, and then followed by 120 milliliters of dry hexane. Thereafter, carefully and slowly blend the mixture on low speed for 45 minutes. After 45 minutes, open the vertical mixer, recover the mixture, and then extrude the mixture through any desired extruding machine into strands ranging from 0.5mm to 3mm in diameter by 10cm in length, then cut the strands into 0.5mm to 3mm pieces and then cure the pieces on a flat tray at room temperature for 48 hours. Note: the size of the cured pieces depends on the gun primer in question. Small arms weapons such as handguns, and rifles should use 0.5 to 1mm sized pieces; whereas larger bore weapons such as 12.7mm up to 25mm casings should use 1.5 to 2mm sized pieces; use your discretion).
**Burn rate (when heated externally):** 195 Celsius. Explodes.
**Burn rate (when heated directly with a flame or hot NiChrome wire):** Explodes.
**Drop Test:** 7.8cm with 500 gram weight: Explodes.
**Water resistance:** Acceptable.
**Sensitivity:** Unstable to friction; sensitive to shock and percussion; heat resistant (external) up to 190~ Celsius.
**Stability:** Can be stored for many years.
**Flammability:** Explodes.
**Ease of ignition (when heated by a flame or hot wire):** 8.3
**Tendency to cake:** Will not cake.
**Explosive ability:** Can be detonated.
**Percentage:** ***40% lead styphnate, 30% lead nitrate, 20% glass, 8% lead sulfocyanate, 2% tetracene***
**Classification:** Explosive mixture (classified as Class A explosive composition).

**Danger of preparation:** Should only be prepared by experienced pyrotechnicians.
**Use:** Used in gun primers. Do not use to fire high performance rocket fuels or high performance gun propellants.

**Note:** Inexperienced personnel should not attempt this procedure, as there are hazards involved. Avoid excessive heat or friction. This procedure is perfectly safe for experienced persons. To gain experience, preparers should use only small quantities to begin with. Either way, safety shields should be used at all times, and any possible source of ignition, including static electricity should be avoided—use static resistant tables and floor mats.

**06-04-032A: Conductive friction sensitive composition for use in primers (non primary explosive containing):**
Into a suitable sealable vertical mixer, equipped with flexible plastic stir blades, place ***50 grams of lead dioxide***, followed by ***120 grams of lead styphnate***, followed by ***15 grams of zirconium with grain size of 5 microns***, followed by ***32.5 grams of coarse zirconium metal of 200 mesh***, followed by ***22.7 grams of zirconium metal of 325 mesh***, followed by ***6.5 grams of zirconium metal of 10 microns***, followed by ***70 grams of barium nitrate,*** and then followed by 135 milliliters of dry hexane. Thereafter, carefully and slowly blend the mixture on low speed for 45 minutes. After 45 minutes, open the vertical mixer, recover the mixture, and then extrude the mixture through any desired extruding machine into strands ranging from 0.5mm to 3mm in diameter by 10cm in length, then cut the strands into 0.5mm to 3mm pieces and then cure the pieces on a flat tray at room temperature for 48 hours. Note: the size of the cured pieces depends on the gun primer in question. Small arms weapons such as handguns, and rifles should use 0.5 to 1mm sized pieces; whereas larger bore weapons such as 12.7mm up to 25mm casings should use 1.5 to 2mm sized pieces; use your discretion).
**Burn rate (when heated externally):** 990 Celsius. Explodes.
**Burn rate (when heated directly with a flame or hot NiChrome wire):** Explodes.
**Drop Test:** 10.5cm with 750 gram weight: Explodes.
**Water resistance:** Acceptable.
**Sensitivity:** Unstable to friction; sensitive to shock and percussion; heat resistant (external) up to 900~ Celsius.
**Stability:** Can be stored for many years.
**Flammability:** Explodes.
**Ease of ignition (when heated by a flame or hot wire):** 7.0
**Tendency to cake:** Will not cake.
**Explosive ability:** Cannot be detonated.
**Percentage:** ***38.9% zirconium, 35.5% barium nitrate, 25.4% lead dioxide, 0.2% balance***
**Classification:** Explosive mixture (classified as Class A explosive composition).

**Danger of preparation:** Should only be prepared by experienced pyrotechnicians.
**Use:** Used in gun primers. Do not use to fire high performance rocket fuels or high performance gun propellants.

**Note:** Inexperienced personnel should not attempt this procedure, as there are hazards involved. Avoid excessive heat or friction. This procedure is perfectly safe for experienced persons. To gain experience, preparers should use only small quantities to begin with. Either way, safety shields should be used at all times, and any possible source of ignition, including static electricity should be avoided—use static resistant tables and floor mats.

| Improvised Shape Charge (ISC) |
|---|
| 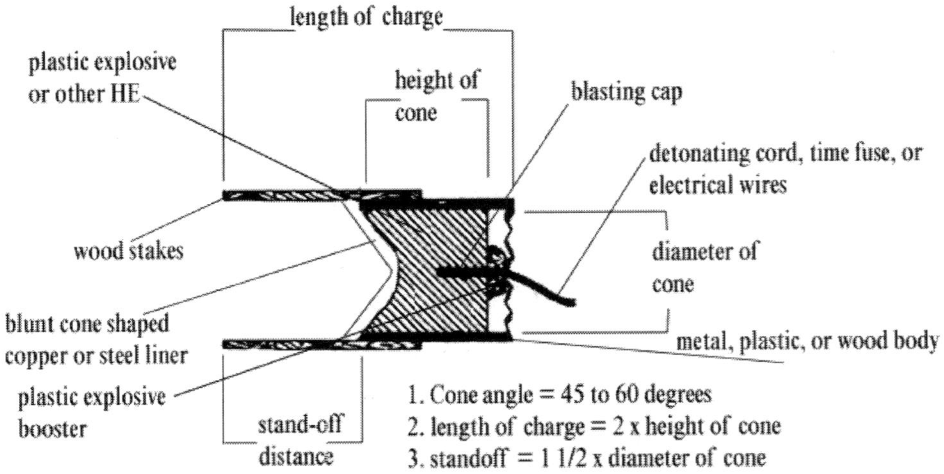 |
| **Details:** |
| Improvised shape charges are used by soldiers in times of war when supplies are limited. They work in identical manners as regular factory built shape charges, but are not as "ideal", and are used only when needed. |



**Note: the following table represents a 3.6 kilogram improvised shape charge. Second note: professionally designed shape charges would have much higher penetrating powers**

| Material | Penetration | Diameter of hole | Standoff |
|---|---|---|---|
| Reinforced concrete | 38.1 centimeters (15 inches) | 3.5 centimeters (1.4 inches) | Average |
| Armor plate | 15.3 centimeters (6 inches) | 1.9 centimeters (0.75 inches) | Average |
| Ground | 91.5 centimeters (36 inches) | 7.6 centimeters (3 inches) | At 15 inches |
| Soil | 1.05 meters (3.5 feet) | 8.9 centimeters (3.5 inches) | At 15 inches |

**06-04-033A: Ignition composition for use in delay trains and for generating heat for other ignition operations:**
Into a suitable ball mill, filled with 150 grams of Teflon coated steel shot of the usual diameter, place ***303 grams of finely powdered aluminum*** of average mesh, followed by ***33.5 grams of finely divided silicon*** of average mesh, followed by ***3.9 grams of finely divided titanium metal***, followed by ***149.5 grams of finely divided red iron-III-oxide***, and then followed by ***9.5 grams of finely divided graphite powder***. Thereafter, tumble the mixture at 150 RPM for 45 minutes. After 45 minutes, separate the mixture from the Teflon coated steel shot in the usual manner, and then depending on the type of operation to be used, the powder should be pressed into the desired diameter and length under a pressure of 6,000 psi.
**Ignition temperature (when heated externally, in powder form):** 1,284 Celsius.
**Ignition temperature (when heated externally, when pressed at 6,000 psi):** 1,389 Celsius.
**Burn rate (when heated directly with a flame or hot NiChrome wire, in powder form):** 7.6 seconds per inch.
**Burn rate (when heated directly with a flame or hot NiChrome wire, when pressed at 6,000 psi):** 10.1 seconds per inch.
**Water resistance:** Acceptable.
**Sensitivity:** Stable to friction, shock and percussion.
**Stability:** Can be stored for many years.
**Ease of ignition (when heated by a flame or hot wire):** 5.98
**Tendency to cake:** Will not cake.
**Explosive ability:** Cannot be detonated.
**Percentage:** ***60.6% aluminum, 29.9% iron-III-oxide, 6.7% silicon, 1.9% graphite, 0.78% titanium, 0.15% impurities***
**Classification:** Deflagrating mixture (classified as 1.3G pyrotechnic composition).

**Danger of preparation:** Can be prepared by amateurs.
**Use:** Used in grenade fuses, and other ignition devices.

**06-04-033B: Primer composition for use in detonators:**
Into a suitable sealable vertical mixer, equipped with flexible plastic stir blades, place ***303 grams of finely powdered aluminum*** of average mesh, followed ***33.5 grams of finely divided silicon*** of average mesh, followed by ***3.9 grams of finely divided titanium metal***, followed by ***149.5 grams of finely divided red iron-III-oxide***, followed by ***9.5 grams of finely divided graphite powder***, followed by ***600 grams of potassium chlorate,*** and then followed by 250 milliliters of dry hexane. Thereafter, slowly blend the mixture on low speed for 45 minutes. After 45 minutes, the mixture is ready for use. To use in explosive electric bridge wire devices, the metal wire (usually a thin NiChrome wire) should be evenly coated with the moldable mixture coating the exposed portion of the "wire" with a 1 to 3mm thickness of the mixture, and then cured in an oven at 60 Celsius for several hours. To use in blasting caps or detonators, the mixture should be pressed into molds under a pressure of 3,000 psi, forming tablets that need to be cured in an oven at 60 Celsius for 12 hours; the diameter and length of these "tablets" is based on the diameter and length of the given blasting cap; use your own discretion.
**Burn rate (when heated externally):** 401 Celsius. Explodes.
**Burn rate (when heated directly with a flame or hot NiChrome wire):** Explodes.
**Drop Test:** 11.9cm with 700 gram weight: Explodes.
**Water resistance:** Acceptable.
**Sensitivity:** Stable to friction; acceptable stability to shock and percussion; heat resistant (external) up to 399~ Celsius.
**Stability:** Can be stored for many years.
**Flammability:** Explodes.
**Ease of ignition (when heated by a flame or hot wire):** 8.4
**Tendency to cake:** Will not cake.
**Explosive ability:** Can be detonated.
**Percentage:** ***54.5% potassium chlorate, 27.5% aluminum, 13.5% iron-III-oxide, 3% silicon, 0.86% graphite, 0.35% titanium, 0.29% mixed impurities and balance***
**Classification:** Deflagrating mixture (classified as 1.4G pyrotechnic composition).

**Danger of preparation:** Can be prepared by amateurs, but caution should be exercised.
**Use:** Used in detonators.

**Detonator assembly**

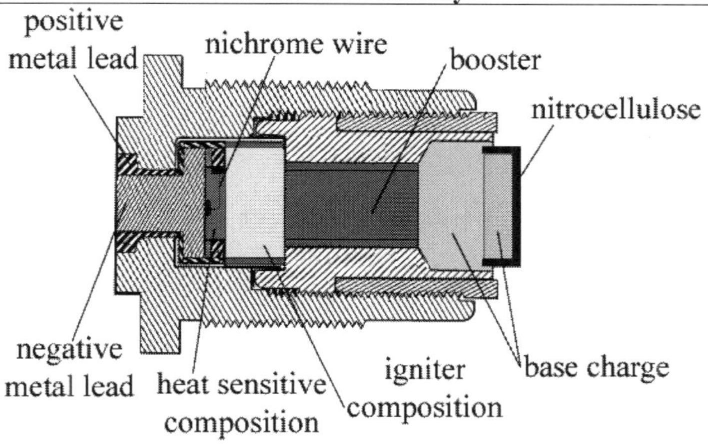


**06-04-034A: Priming composition for use in electric blasting caps or for use in electric ignition devices:**
Into a suitable sealable vertical mixer, equipped with flexible plastic stir blades, place ***6 grams of red phosphorus***, followed ***18 grams of sulfur***, followed by ***120 grams of finely powdered silver metal***, followed by ***30 grams of liquid mercury***, followed by ***36 grams of potassium chlorate***, and then followed by 80 milliliters of dry hexane. Thereafter, slowly blend the mixture on low speed for 45 minutes. After 45 minutes, the mixture is ready for use. To use in explosive electric bridge wire devices, the metal wire (usually a thin NiChrome wire) should be evenly coated with the moldable mixture coating the exposed portion of the "wire" with a 1 to 3mm thickness of the mixture, and then cured in an oven at 60 Celsius for several hours. To use in blasting caps or detonators, the mixture should be pressed into molds under a pressure of 3,000 psi, forming tablets that need to be cured in an oven at 60 Celsius for 12 hours; the diameter and length of these "tablets" is based on the diameter and length of the given blasting cap; use your own discretion.
**Burn rate (when heated externally):** 617 Celsius. Explodes.
**Burn rate (when heated directly with a flame or hot NiChrome wire):** Explodes.
**Drop Test:** 15.9cm with 1000 gram weight: Explodes.
**Water resistance:** Acceptable.
**Sensitivity:** Stable to friction; acceptable stability to shock and percussion; heat resistant (external) up to 600~ Celsius.
**Stability:** Can be stored for many years.
**Flammability:** Explodes.

**Ease of ignition (when heated by a flame or hot wire):** 8.4
**Tendency to cake:** Will not cake.
**Explosive ability:** Cannot be detonated.
**Percentage:** ***57.1% silver metal, 17.1% potassium chlorate, 14.2% mercury metal, 8.5% sulfur, 2.8% red phosphorus, 0.30% balance***
**Classification:** Deflagrating mixture (classified as 1.3G pyrotechnic composition).

**Danger of preparation:** Can be prepared by amateurs, but caution should be exercised.
**Use:** Used in blasting caps, and in electric ignition devices.

## 06-04-035A: Ignition composition for flares and illumination compositions:

Into a suitable vertical mixer, place 200 milliliters of hexane, followed by ***425 grams of barium chromate***, followed by ***50 grams of boron***, and then followed by ***25 grams of magnesium powder*** of about 200 mesh. Thereafter, blend the mixture on high speed for about 45 minutes. After 45 minutes the mixture is ready for pressing. To do so, press the mixture into molds, discs, tablets, cubes, pellets, ect., of diameter and thickness (as dictated by the device/munition in question; use your discretion), under a pressure of 7,300 psi. Thereafter, cure the molds, discs, tablets, cubes, pellets, ect., in an oven at 60 Celsius for 12 hours.
**Burn rate:** 1.95mm per second. Depends on diameter and density of solid grain.
**Burn temperature:** 1277 Celsius.
**Ignition temperature (average temperature to ignite mixture at 7,300 psi):** 977 Celsius.
**Water resistance:** Very good.
**Stability:** Can be stored for many years.
**Flammability (pressed grain at 7,300 psi):** 4.21
**Ease of ignition (average temperature to ignite mixture at 7,300 psi):** 4.18
**Tendency to cake:** None.
**Explosive ability:** Cannot be detonated.
**Percentage:** ***85% barium chromate, 10% boron, 5% magnesium***
**Classification:** Deflagrating mixture (classified as 1.3G pyrotechnic composition).

**Danger of preparation:** Can be made by amateurs.
**Use:** Used to ignite flare and illumination compositions.

## 06-04-036A: Ignition composition for incendiary agents (magnesium ophorite):

Into a suitable ball mill, filled with 150 grams of Teflon coated steel shot of the usual diameter, place ***102.6 grams of fine magnesium flake of average size***, followed by ***197.4 grams of potassium perchlorate***, and then followed by ***6 grams of castor oil***. Thereafter, tumble the mixture at 200 RPM for 45 minutes. After 45 minutes, separate the mixture from the Teflon coated steel shot in the usual manner, and then depending on the type of operation to be used, the powder should be pressed into the desired diameter and length under a pressure of 7,000 psi.
**Ignition temperature (when heated externally, in powder form):** 1,067 Celsius.
**Ignition temperature (when heated externally, when pressed at 7,000 psi):** 1,200~ Celsius.
**Burn rate (when heated directly with a flame or hot NiChrome wire, in powder form):** 1.9 seconds per inch.
**Burn rate (when heated directly with a flame or hot NiChrome wire, when pressed at 7,000 psi):** 3.3 seconds per inch.
**Water resistance:** Acceptable.
**Sensitivity:** Stable to friction, shock and percussion.
**Stability:** Can be stored for many years.
**Ease of ignition (when heated by a flame or hot wire):** 8.99
**Tendency to cake:** Will not cake.
**Explosive ability:** Can be detonated; when in powder form, it explodes when contained and ignited.
**Percentage:** ***64.3% potassium perchlorate, 33.5% magnesium flake, 1.9% castor oil, 0.3% balance***
**Classification:** Deflagrating mixture (classified as 1.4G pyrotechnic composition).

**Danger of preparation:** Can be prepared by amateurs, but caution should be exercised.
**Use:** Used to ignite incendiary agents. Note: can also be used in making firecrackers, as the mixture explodes when confined and then ignited.

**06-04-036B: Ignition composition for incendiary agents (Aluminum ophorite) with increased heat of combustion:**
Into a suitable ball mill, filled with 150 grams of Teflon coated steel shot of the usual diameter, place ***102.6 grams of fine aluminum flake of average size***, followed by ***197.4 grams of potassium perchlorate***, and then followed by ***6 grams of castor oil***. Thereafter, tumble the mixture at 200 RPM for 45 minutes. After 45 minutes, separate the mixture from the Teflon coated steel shot in the usual manner, and then depending on the type of operation to be used, the powder should be pressed into the desired diameter and length under a pressure of 7,250 psi.
**Ignition temperature (when heated externally, in powder form):** 1,000 Celsius.
**Ignition temperature (when heated externally, when pressed at 7,250 psi):** 1,109~ Celsius.
**Burn rate (when heated directly with a flame or hot NiChrome wire, in powder form):** 1.5 seconds per inch.
**Burn rate (when heated directly with a flame or hot NiChrome wire, when pressed at 7,250 psi):** 3.0 seconds per inch.
**Water resistance:** Acceptable.
**Sensitivity:** Stable to friction, shock and percussion.
**Stability:** Can be stored for many years.
**Ease of ignition (when heated by a flame or hot wire):** 9.1
**Tendency to cake:** Will not cake.
**Explosive ability:** Can be detonated; when in powder form, it explodes when contained and ignited.
**Percentage:** ***64.3% potassium perchlorate, 33.5% aluminum flake, 1.9% castor oil, 0.3% balance***
**Classification:** Deflagrating mixture (classified as 1.4G pyrotechnic composition).

**Danger of preparation:** Can be prepared by amateurs, but caution should be exercised.
**Use:** Used to ignite incendiary agents. Note: can also be used in making firecrackers, as the mixture explodes when confined and then ignited.

**06-04-037A: Heat resistant ignition composition with high thermal stability:**
Into a suitable vertical mixer, place 90 milliliters of hexane, followed by ***160 grams of finely divided tellurium dioxide of 4.1 microns***, followed by ***34 grams of finely powdered aluminum of 0.8 microns***, and then followed by ***6 grams of finely divided carbon black of 0.4 microns***. Thereafter, blend the mixture on high speed for about 45 minutes. After 45 minutes the mixture is ready for pressing. To do so, press the mixture into molds, discs, tablets, cubes, pellets, ect., of diameter and thickness (as dictated by the device/munition in question; use your discretion), under a pressure of 8,000 psi. Thereafter, cure the molds, discs, tablets, cubes, pellets, ect., in an oven at 60 Celsius for 12 hours.
**Burn rate:** 2.98mm per second. Depends on diameter and density of solid grain.
**Burn temperature:** 1355 Celsius.
**Ignition temperature (average temperature to ignite mixture at 8,000 psi):** 900 Celsius.
**Water resistance:** Very good.
**Stability:** Can be stored for many years.
**Flammability (pressed grain at 8,000 psi):** 4.00
**Ease of ignition (average temperature to ignite mixture at 8,000 psi):** 400~ Celsius.
**Tendency to cake:** None.
**Explosive ability:** Cannot be detonated.
**Percentage:** ***80% tellurium dioxide, 17% aluminum, 3% carbon black***
**Classification:** Deflagrating mixture (classified as 1.3G pyrotechnic composition).

**Danger of preparation:** Can be made by amateurs.
**Use:** Used in intercontinental missiles for withstanding the strains, stresses, and heat of atmospheric re-entry. Also used to ignite incendiary agents.

**06-04-038A: Ignition composition for use in aircraft ejection seats:**
Into a suitable vertical mixer, place 170 milliliters of hexane, followed by ***100 grams of finely divided zirconium metal***, followed by ***400 grams of strontium nitrate***, followed by ***10 grams of zinc stearate***, followed by ***30 grams of standard chlorinated rubber***, followed by ***5 grams of any desired dye compound***, and then followed by ***40 grams of calcium resinate***. Thereafter, blend the mixture on high speed for about 45 minutes. After 45 minutes the mixture is ready for pressing. To do so, press the mixture into molds, discs, tablets, cubes, pellets, ect., of diameter and thickness (as dictated by the device/ejection seat in question; use your discretion), under a pressure of 8,000 psi. Thereafter, cure the molds, discs, tablets, cubes, pellets, ect., in an oven at 60 Celsius for 12 hours.
**Burn rate:** 2.01mm per second. Depends on diameter and density of solid grain.
**Burn temperature:** 1199 Celsius.
**Ignition temperature (average temperature to ignite mixture at 8,000 psi):** 877 Celsius.
**Water resistance:** Very good.
**Stability:** Can be stored for many years.

**Flammability (pressed grain at 8,000 psi):** 6.4
**Ease of ignition (average temperature to ignite mixture at 8,000 psi):** 499~ Celsius.
**Tendency to cake:** None.
**Explosive ability:** Cannot be detonated.
**Percentage:** ***68.37% strontium nitrate, 17% zirconium metal, 6.8% calcium resinate, 5.1% chlorinated rubber, 1.7% zinc stearate, 0.85% dye, 0.18% balance***
**Classification:** Deflagrating mixture (classified as 1.3G pyrotechnic composition).

**Danger of preparation:** Can be made by amateurs.
**Use:** Used to initiate rocket propellants used in aircraft ejection seats.

### 06-04-038B: Ignition composition for use in aircraft ejection seats:

Into a suitable vertical mixer, place 175 milliliters of hexane, followed by ***100 grams of finely divided boron***, followed by ***400 grams of lead dioxide***, followed by ***10 grams of zinc stearate***, followed by ***30 grams of standard chlorinated rubber***, followed by ***5 grams of any desired dye compound***, and then followed by ***40 grams of strontium resinate***. Thereafter, blend the mixture on high speed for about 45 minutes. After 45 minutes the mixture is ready for pressing. To do so, press the mixture into molds, discs, tablets, cubes, pellets, ect., of diameter and thickness (as dictated by the device/ejection seat in question; use your discretion), under a pressure of 8,100 psi. Thereafter, cure the molds, discs, tablets, cubes, pellets, ect., in an oven at 60 Celsius for 12 hours.
**Burn rate:** 2.45mm per second. Depends on diameter and density of solid grain.
**Burn temperature:** 1197 Celsius.
**Ignition temperature (average temperature to ignite mixture at 8,100 psi):** 801 Celsius.
**Water resistance:** Very good.
**Stability:** Can be stored for many years.
**Flammability (pressed grain at 8,100 psi):** 6.499
**Ease of ignition (average temperature to ignite mixture at 8,000 psi):** 426~ Celsius.
**Tendency to cake:** None.
**Explosive ability:** Cannot be detonated.
**Percentage:** ***68.37% lead dioxide, 17% boron, 6.8% strontium resinate, 5.1% chlorinated rubber, 1.7% zinc stearate, 0.85% dye, 0.18% balance***
**Classification:** Deflagrating mixture (classified as 1.3G pyrotechnic composition).

**Danger of preparation:** Can be made by amateurs.
**Use:** Used to initiate rocket propellants used in aircraft ejection seats.

### 06-04-038C: Ignition composition for use in aircraft ejection seats:

Into a suitable vertical mixer, place 190 milliliters of hexane, followed by ***100 grams of finely divided magnesium metal***, followed by ***400 grams of barium nitrate***, followed by ***10 grams of zinc stearate***, followed by ***30 grams of standard chlorinated rubber***, followed by ***5 grams of toludiene red dye compound***, and then followed by ***40 grams of calcium resinate***. Thereafter, blend the mixture on high speed for about 45 minutes. After 45 minutes the mixture is ready for pressing. To do so, press the mixture into molds, discs, tablets, cubes, pellets, ect., of diameter and thickness (as dictated by the device/ejection seat in question; use your discretion), under a pressure of 7,200 psi. Thereafter, cure the molds, discs, tablets, cubes, pellets, ect., in an oven at 60 Celsius for 12 hours.
**Burn rate:** 2.01mm per second. Depends on diameter and density of solid grain.
**Burn temperature:** 1278 Celsius.
**Ignition temperature (average temperature to ignite mixture at 7,200 psi):** 798 Celsius.
**Water resistance:** Very good.
**Stability:** Can be stored for many years.
**Flammability (pressed grain at 7,200 psi):** 6.99
**Ease of ignition (average temperature to ignite mixture at 8,000 psi):** 400~ Celsius.
**Tendency to cake:** None.
**Explosive ability:** Cannot be detonated.
**Percentage:** ***68.37% barium nitrate, 17% magnesium, 6.8% calcium resinate, 5.1% chlorinated rubber, 1.7% zinc stearate, 0.85% toludiene red, 0.18% balance***
**Classification:** Deflagrating mixture (classified as 1.3G pyrotechnic composition).

**Danger of preparation:** Can be made by amateurs.
**Use:** Used to initiate rocket propellants used in aircraft ejection seats.

### 06-04-039A: Priming composition for multiple uses (friction sensitive):

Into a suitable sealable vertical mixer, equipped with flexible plastic stir blades, place ***240 grams of potassium chlorate***, followed by ***180 grams of potassium ferrocyanide***, followed by ***80 grams of finely powdered glass***, and then followed by 125 milliliters of dry hexane. Thereafter, carefully and slowly blend the mixture on low speed for 45 minutes. After 45 minutes, open the vertical mixer, recover the mixture, and then extrude the mixture through any desired extruding machine into strands ranging from 0.5mm to 3mm in diameter by 10cm in length, then cut the strands into 0.5mm to 3mm pieces and then cure the pieces on a flat tray at room temperature for 48 hours. Note: the size of the cured pieces depends on the gun primer in question. Small arms weapons such as handguns, and rifles should use 0.5 to 1mm sized pieces; whereas larger bore weapons such as 12.7mm up to 25mm casings should use 1.5 to 2mm sized pieces; use your discretion). For use in priming charges for blasting caps, the mixture should be pressed into tablets of any desired diameter and length under a pressure of 4,000 psi, and then cured at room temperature for 72 hours.
**Burn rate (when heated externally):** 400 Celsius. Explodes.
**Burn rate (when heated directly with a flame or hot NiChrome wire):** Explodes.
**Drop Test:** 8.9cm with 750 gram weight: Explodes.
**Water resistance:** Acceptable.
**Sensitivity:** Unstable to friction; sensitive to shock and percussion; heat resistant (external) up to 350~ Celsius.
**Stability:** Can be stored for many years.
**Flammability:** Explodes.
**Ease of ignition (when heated by a flame or hot wire):** 9
**Tendency to cake:** Will not cake.
**Explosive ability:** Can be detonated.
**Percentage:** ***48% potassium chlorate, 36% potassium ferrocyanide, 16% glass powder***
**Classification:** Explosive mixture (classified as Class A explosive composition).

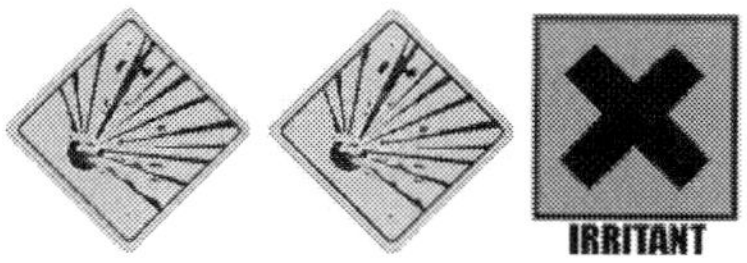


**Danger of preparation:** Should only be prepared by experienced pyrotechnicians.
**Use:** Used to initiate a wide variety of pyrotechnic compositions; used in non-electric blasting caps. Can also be used as a primer for gun ammunition casings. Do not use to fire high performance rocket fuels or high performance gun propellants.

### 06-04-040A: Priming composition for use in electric blasting caps with good thermal stability:

Into a suitable sealable vertical mixer, equipped with flexible plastic stir blades, place ***70 grams of potassium perchlorate***, followed by ***384 grams of cesium decahydrodecaborate***, and then followed by 135 milliliters of dry hexane. Thereafter, carefully and slowly blend the mixture on low speed for 45 minutes. After 45 minutes, open the vertical mixer, recover the mixture, and then the mixture should be pressed into tablets of any desired diameter and length under a pressure of 4,700 psi, and then cured at room temperature for 72 hours.
**Burn rate (when heated externally):** 455+ Celsius. Explodes.
**Burn rate (when heated directly with a flame or hot NiChrome wire):** Explodes.
**Drop Test:** 10cm with 1000 gram weight: Explodes.
**Water resistance:** Acceptable.
**Sensitivity:** Unstable to friction; sensitive to shock and percussion; heat resistant (external) up to 400~ Celsius.
**Stability:** Can be stored for many years.
**Flammability:** Explodes.
**Ease of ignition (when heated by a flame or hot wire):** 8.9
**Tendency to cake:** Will not cake.
**Explosive ability:** Cannot be detonated.
**Percentage:** ***85% cesium decahydrodecaborate, 15% potassium perchlorate***
**Classification:** Explosive mixture (classified as Class A1 explosive composition).

**Danger of preparation:** Can be prepared by amateurs, but caution should be exercised.
**Use:** Used in electric blasting caps. Do not use to fire high performance rocket fuels or high performance gun propellants.

### 06-04-040B: Priming composition for use in electric blasting caps with good thermal stability (with increased sensitivity):

Into a suitable sealable vertical mixer, equipped with flexible plastic stir blades, place ***139 grams of potassium perchlorate***, followed by ***384 grams of cesium decahydrodecaborate***, and then followed by 190 milliliters of dry hexane. Thereafter, carefully and slowly blend the mixture on low speed for 45 minutes. After 45 minutes, open the vertical mixer, recover the mixture, and then the mixture should be pressed into tablets of any desired diameter and length under a pressure of 4,000 psi, and then cured at room temperature for 72 hours.
**Burn rate (when heated externally):** 399 Celsius. Explodes.
**Burn rate (when heated directly with a flame or hot NiChrome wire):** Explodes.
**Drop Test:** 9.2cm with 800 gram weight: Explodes.
**Water resistance:** Acceptable.
**Sensitivity:** Unstable to friction; sensitive to shock and percussion; heat resistant (external) up to 350~ Celsius.
**Stability:** Can be stored for many years.
**Flammability:** Explodes.
**Ease of ignition (when heated by a flame or hot wire):** 8.99
**Tendency to cake:** Will not cake.
**Explosive ability:** Cannot be detonated.
**Percentage:** ***73% cesium decahydrodecaborate, 27% potassium perchlorate***
**Classification:** Explosive mixture (classified as Class A1 explosive composition).

**Danger of preparation:** Can be prepared by amateurs, but caution should be exercised.
**Use:** Used in electric blasting caps. Do not use to fire high performance rocket fuels or high performance gun propellants.

### 06-04-040C: Priming composition for use in electric blasting caps with good thermal stability utilizing potassium permanganate:

Into a suitable sealable vertical mixer, equipped with flexible plastic stir blades, place ***474 grams of potassium permanganate***, followed by ***384 grams of cesium decahydrodecaborate***, and then followed by 220 milliliters of dry hexane. Thereafter, carefully and slowly blend the mixture on low speed for 45 minutes. After 45 minutes, open the vertical mixer, recover the mixture, and then the mixture should be pressed into tablets of any desired diameter and length under a pressure of 4,100 psi, and then cured at room temperature for 72 hours.
**Burn rate (when heated externally):** 402 Celsius. Explodes.
**Burn rate (when heated directly with a flame or hot NiChrome wire):** Explodes.
**Drop Test:** 10.1cm with 1000 gram weight: Explodes.
**Water resistance:** Acceptable.
**Sensitivity:** Unstable to friction; sensitive to shock and percussion; heat resistant (external) up to 350~ Celsius.
**Stability:** Can be stored for many years.
**Flammability:** Explodes.
**Ease of ignition (when heated by a flame or hot wire):** 8.9~
**Tendency to cake:** Will not cake.
**Explosive ability:** Cannot be detonated.
**Percentage:** ***55% potassium permanganate, 45% cesium decahydrodecaborate***
**Classification:** Explosive mixture (classified as Class A1 explosive composition).

**Danger of preparation:** Can be prepared by amateurs, but caution should be exercised.
**Use:** Used in electric blasting caps. Do not use to fire high performance rocket fuels or high performance gun propellants.

### 06-04-041A: Igniter composition for use in electric ignition devices:

Into a suitable sealable vertical mixer, equipped with flexible plastic stir blades, place ***99 grams of finely divided titanium hydride***, followed by ***201 grams of finely divided potassium perchlorate***, and then followed by 100 milliliters of dry hexane. Thereafter, carefully and slowly blend the mixture on low speed for 45 minutes. After 45 minutes, open the vertical mixer, recover the mixture, and then the mixture should be pressed into tablets of any desired diameter and length under a pressure of 2,000 psi, and then cured at room temperature for 72 hours.
**Burn rate (when heated externally):** 377 Celsius. Explodes.
**Burn rate (when heated directly with a flame or hot NiChrome wire):** Explodes.
**Drop Test:** 8cm with 750 gram weight: Explodes.
**Water resistance:** Acceptable.
**Sensitivity:** Stable to friction; sensitive to shock and percussion; heat resistant (external) up to 350~ Celsius.
**Stability:** Can be stored for many years.
**Flammability:** Explodes.
**Ease of ignition (when heated by a flame or hot wire):** 9.1

**Tendency to cake:** Will not cake.
**Explosive ability:** Can be detonated when primed with lead azide, silver fulminate, lead styphnate, or tetracene.
**Percentage:** ***67% potassium perchlorate, 33% titanium hydride***
**Classification:** Explosive mixture (classified as Class A1 explosive composition).

**Danger of preparation:** Can be prepared by amateurs, but special caution should be exercised. Make sure all electrical equipment is grounded, and avoid static electricity.
**Use:** Used in electric activated ignition devices such as blasting caps, detonators, and similar devices. Do not use to fire high performance rocket fuels or high performance gun propellants.

### 06-04-042A: Thermally stable, percussion primer composition for use in primers for shell casing:

Into a suitable sealable vertical mixer, equipped with flexible plastic stir blades, place ***123 grams of finely powdered titanium metal*** of average commercial mesh, followed ***177 grams of potassium perchlorate***, and then followed by 100 milliliters of dry hexane. Thereafter, carefully and slowly blend the mixture on low speed for 45 minutes. After 45 minutes, open the vertical mixer, recover the mixture, and then extrude the mixture through any desired extruding machine into strands ranging from 0.4mm to 3.2mm in diameter by 10cm in length, then cut the strands into 0.1mm to 5mm pieces and then cure the pieces on a flat tray at room temperature for 48 hours. Note: the size of the cured pieces depends on the gun primer in question. Small arms weapons such as handguns, and rifles should use 0.5 to 1mm sized pieces; whereas larger bore weapons such as 12.7mm up to 25mm casings should use 1.5 to 2mm sized pieces; use your discretion).
**Burn rate (when heated externally):** 400~ Celsius. Explodes.
**Burn rate (when heated directly with a flame or hot NiChrome wire):** Explodes.
**Drop Test:** 9.3cm with 750 gram weight: Explodes.
**Water resistance:** Acceptable.
**Sensitivity:** Stable to friction; sensitive to shock and percussion; heat resistant (external) up to 350~ Celsius.
**Stability:** Can be stored for many years.
**Flammability:** Explodes.
**Ease of ignition (when heated by a flame or hot wire):** 9.3
**Tendency to cake:** Will not cake.
**Explosive ability:** Can be detonated when primed with lead azide, silver fulminate, lead styphnate, or tetracene.
**Percentage:** ***59% potassium perchlorate, 41% titanium metal***
**Classification:** Explosive mixture (classified as Class A explosive composition).

**Danger of preparation:** Should only be prepared by experienced pyrotechnicians.
**Use:** Used in gun primers. Do not use to fire high performance rocket fuels or high performance gun propellants.

### 06-04-043A: Friction sensitive ignition composition for friction resistant matchhead compositions:

Into a vertical mixer, equipped with flexible Teflon stir blades, place ***50 grams of gum Arabic***, followed by ***100 grams of potassium chlorate,*** followed by 110 milliliters of hexane, and then followed by ***100 grams of finely divided black antimony trisulfide***. Thereafter, blend the mixture on low speed for 30 minutes. After 30 minutes, the mixture is ready to be casted ("dipped"). To cast the match head composition, the wood or paper "match sticks" should be dipped and re-dipped into the semi-fluidized mixture to form a uniform thickness ranging from 2mm to 5mm in width (depends on discretion, and the desired thickness of the match head composition can vary). Thereafter, the wood/paper matchsticks can be dipped into a colored dye solution if desired; in either case, the matches should then be cured in an oven at 80 Celsius for 12 hours.
**Ignition temperature (when match is heated externally):** 512 Celsius
**Ignition temperature (exact point when match ignites):** 998 Celsius
**Ease of ignition ("striking" on suitable surface):** 7.1
**Water resistance:** Very good.
**Weather resistance (match):** wind resistance normal; moist and wet conditions above normal.
**Stability:** Can be stored for decades.
**Sensitivity:** Very stable towards shock and percussion; non friction resistant to soft materials.
**Explosive ability:** Cannot be detonated.
**Percentage:** ***40% potassium chlorate, 40% antimony trisulfide, 20% gum Arabic***
**Classification:** Deflagrating mixture (classified as 1.3G pyrotechnic composition).

**Danger of preparation:** Can be prepared by amateurs, but caution should be exercised.

**Use:** Used to initiate friction resistant match head compositions and for use in fuses for grenades and similar devices.

### 06-04-044A: Ignition composition for smoke grenades:

Into a standard ball mill, filled with 500 grams of steel shot of 10mm diameter, place ***150 grams of barium nitrate***, followed by ***20 grams of silicon***, followed by ***30 grams of tetranitrocarbazole***, followed by ***45 grams of zirconium hydride***, and then followed by ***15 grams of any standard liquid epoxy resin***. Thereafter, tumble the mixture at room temperature at 250 RPM for about 30 minutes. Thereafter, the mixture should be removed from the ball mill, and the steel shot separated; thereafter, the mixture is ready for pressing. To do so, press the mixture into blocks, discs, or tablets under a pressure of 6,000 psi.
**Burn rate:** 17 seconds per inch at 6,000 psi; tends to vary based on diameter and density of grain.
**Burn temperature:** 1,090 Celsius.
**Ignition temperature (average temperature to ignite mixture):** 652~ Celsius.
**Water resistance:** Very good.
**Stability:** Can be stored for many years.
**Flammability (pressed grain at 6,000 psi):** 6.79
**Ease of ignition (average temperature to ignite mixture):** 6.99
**Tendency to cake:** None.
**Explosive ability:** Cannot be detonated.
**Percentage:** ***57.6% barium nitrate, 17.3% zirconium hydride, 11.5% tetranitrocarbazole, 7.6% silicon, 5.7% epoxy resin, 0.3% mixed impurities***
**Classification:** Deflagrating mixture (classified as 1.3G pyrotechnic composition).

**Danger of preparation:** Can be prepared by amateurs.
**Use:** Used to initiate smoke compositions.

### 06-04-045A: Classic highly insensitive ignition composition for use in rapid accelerating projectiles such as tank rounds or artillery rounds:

Into a standard ball mill, filled with 500 grams of steel shot of 10mm diameter, place ***300 grams of finely divided magnesium powder of the usual mesh***, followed by ***140 grams of silicon dioxide powder***, and then followed by ***60 grams of polybutadiene***. Thereafter, tumble the mixture at room temperature at 200 RPM for about 45 minutes. Thereafter, the mixture should be removed from the ball mill, and the steel shot separated; thereafter, the mixture is ready for pressing. To do so, press the mixture into blocks, discs, or tablets under a pressure of 2,100 psi.
**Burn rate:** 21 seconds per inch at 2,100 psi; tends to vary based on diameter and density of grain.
**Burn temperature:** 1,339 Celsius.
**Ignition temperature (average temperature to ignite mixture):** 900~ Celsius.
**Water resistance:** Very good.
**Stability:** Can be stored for many years.
**Flammability (pressed grain at 2,100 psi):** 5.91
**Ease of ignition (average temperature to ignite mixture):** 6
**Tendency to cake:** None.
**Explosive ability:** Cannot be detonated.
**Percentage:** ***60% magnesium, 28% silicon dioxide, 12% polybutadiene***
**Classification:** Deflagrating mixture (classified as 1.3G pyrotechnic composition).

**Danger of preparation:** Can be prepared by amateurs.
**Use:** Used in tank rounds or artillery shells.

**HE shape charge projectile**

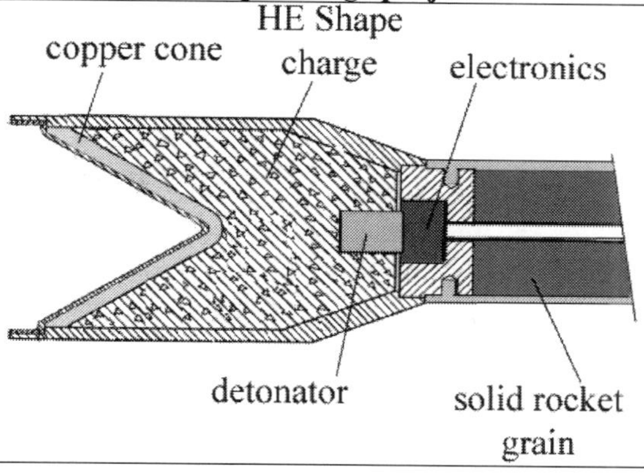


**06-04-045B: Classic highly insensitive ignition composition for use in moderately accelerating projectiles such as mortar rounds:**

Into a standard ball mill, filled with 500 grams of steel shot of 10mm diameter, place ***350 grams of finely ground magnesium powder*** of average mesh, followed by ***75 grams of potassium perchlorate***, followed by ***37.5 grams of silicon dioxide*** and then followed by ***37.5 grams of polybutadiene***. Thereafter, tumble the mixture at room temperature at 260 RPM for about 45 minutes. Thereafter, the mixture should be removed from the ball mill, and the steel shot separated; thereafter, the mixture is ready for pressing. To do so, press the mixture into blocks, discs, or tablets under a pressure of 2,500 psi.

**Burn rate:** 19 seconds per inch at 2,500 psi; tends to vary based on diameter and density of grain.
**Burn temperature:** 1,260 Celsius.
**Ignition temperature (average temperature to ignite mixture):** 955~ Celsius.
**Water resistance:** Very good.
**Stability:** Can be stored for many years.
**Flammability (pressed grain at 2,500 psi):** 5.68
**Ease of ignition (average temperature to ignite mixture):** 6.2
**Tendency to cake:** None.
**Explosive ability:** Cannot be detonated.
**Percentage:** ***70% magnesium, 15% potassium perchlorate, 7.5% silicon dioxide, 7.5% polybutadiene***
**Classification:** Deflagrating mixture (classified as 1.3G pyrotechnic composition).

**Danger of preparation:** Can be prepared by amateurs.
**Use:** Used in mortar rounds, and grenade launchers.

<table>
<tr><th>Classic impact fuze for mortar rounds</th><th>Details:</th></tr>
<tr><td>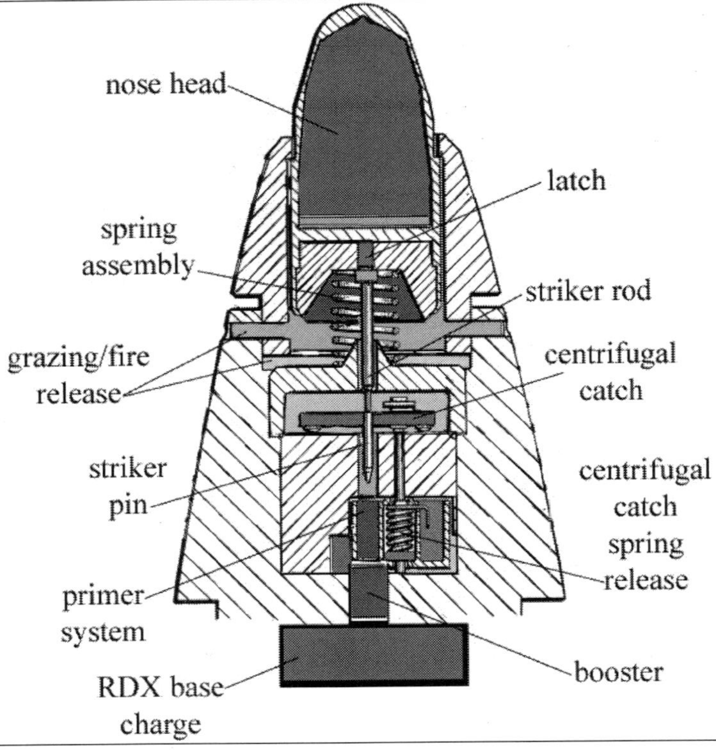
</td><td>Another typical fuze armed by the centrifugal force caused by the firing and spinning of the projectile. Upon impact, the nose head collapse rearward releasing a spring assembly that allows a firing pin to strike a primer, thereby initiating an explosive train in the usual manner.</td></tr>
</table>

**06-04-046A: Special primer composition containing high explosives for use in shot gun primers:**
Into a suitable sealable vertical mixer, equipped with flexible plastic stir blades, place ***163.5 grams of lead styphnate,*** followed by ***112.5 grams of antimony sulfide***, followed by ***51 grams of barium nitrate***, followed by 10 millimeters of a 5% polyvinyl alcohol in water, followed by ***30 grams of tetracene***, followed by ***37.5 grams of PETN***, followed by ***45 grams of finely powdered aluminum*** and then followed by 100 milliliters of dry hexane. Thereafter, carefully and slowly blend the mixture on low speed for 45 minutes. After 45 minutes, open the vertical mixer, recover the mixture, and then extrude the mixture through any desired extruding machine into strands ranging from 0.4mm to 3.2mm in diameter by 10cm in length, then cut the strands into 0.1mm to 5mm pieces and then cure the pieces on a flat tray at room temperature for 48 hours. Note: the size of the cured pieces depends on the gun primer in question. Small arms weapons such as handguns, and rifles should use 0.5 to 1mm sized pieces; whereas larger bore weapons such as 12.7mm up to 25mm casings should use 1.5 to 2mm sized pieces; use your discretion).
**Burn rate (when heated externally):** 288~ Celsius. Explodes.
**Burn rate (when heated directly with a flame or hot NiChrome wire):** Explodes.
**Drop Test:** 7.1cm with 900 gram weight: Explodes.
**Water resistance:** Acceptable.
**Sensitivity:** Stable to friction; sensitive to shock and percussion; heat resistant (external) up to 200~ Celsius.
**Stability:** Can be stored for many years.
**Flammability:** Explodes.
**Ease of ignition (when heated by a flame or hot wire):** 9.19
**Tendency to cake:** Will not cake.
**Explosive ability:** Can be detonated when primed with lead azide, silver fulminate or a blasting cap.
**Percentage:** ***37% lead styphnate, 25% antimony sulfide 11% barium nitrate, 10% aluminum, 8% PETN, 6% tetracene; 3% polyvinyl alcohol***
**Classification:** Explosive mixture (classified as Class A explosive composition).

**Danger of preparation:** Should only be prepared by experienced pyrotechnicians.
**Use:** Used in gun primers. Do not use to fire high performance rocket fuels or high performance gun propellants.

**06-04-047A: Ignition composition for use on the tips of matchsticks:**
Into a vertical mixer, equipped with flexible Teflon stir blades, place ***25 grams of pulverized soft wood charcoal***, followed by ***5 grams of powdered nitrocellulose*** of average nitrogen content, followed by 190 milliliters of hexane, followed by ***75 grams of Prussian blue*** coloring agent, followed by ***36 grams of red phosphorus***, followed by ***500 grams of potassium chlorate***, followed by ***100 grams of pulverized glass powder*** and then followed by ***15 grams of standard animal glue***. Thereafter, blend the mixture on low speed for 30 minutes. After 30 minutes, the mixture is ready to be casted ("dipped"). To cast the match head composition, the wood or paper "match sticks" should be dipped and re-dipped into the semi-fluidized mixture to form a uniform thickness ranging from 2mm to 5mm in width (depends on discretion, and the desired thickness of the match head composition can vary). Thereafter, the wood/paper matchsticks can be dipped into a colored dye solution if desired; in either case, the matches should then be cured in an oven at 80 Celsius for 12 hours.
**Ignition temperature (when match is heated externally):** 299 Celsius
**Ignition temperature (exact point when match ignites):** 1,026 Celsius
**Ease of ignition ("striking" on suitable surface):** 8
**Water resistance:** Very good.
**Weather resistance (match):** wind resistance normal; moist and wet conditions above normal.
**Stability:** Can be stored for decades.
**Sensitivity:** Very stable towards shock and percussion; friction resistant to soft materials.
**Explosive ability:** Cannot be detonated.
**Percentage:** ***66.13% potassium chlorate, 13.22% pulverized glass, 9.92% Prussian blue, 4.76% red phosphorus, 3.3% soft wood charcoal, 1.98% animal glue, 0.66% nitrocellulose, 0.03% impurities***
**Classification:** Deflagrating mixture (classified as 1.3G pyrotechnic composition).

**Danger of preparation:** Can be prepared by amateurs, but caution should be exercised.
**Use:** Used for matches.

**06-04-047B: Ignition composition for use on the tips of matchsticks (modified):**
Into a vertical mixer, equipped with flexible Teflon stir blades, place ***60 grams of furnace black***, followed by ***120 grams of Prussian blue,*** followed by 125 milliliters of hexane, followed by ***50 grams of red phosphorus***, followed by ***425 grams of potassium chlorate***, followed by ***125 grams of pulverized glass powder***, and then followed by ***25 grams of standard animal glue***. Thereafter, blend the mixture on low speed for 30 minutes. After 30 minutes, the mixture is ready to be casted ("dipped"). To cast the match head composition, the wood or paper "match sticks" should be dipped and re-dipped into the semi-fluidized mixture to form a uniform thickness ranging from 2mm to 5mm in width (depends on discretion, and the desired thickness of the match head composition can vary). Thereafter, the wood/paper matchsticks can be dipped into a colored dye solution if desired; in either case, the matches should then be cured in an oven at 80 Celsius for 12 hours.
**Ignition temperature (when match is heated externally):** 299 Celsius
**Ignition temperature (exact point when match ignites):** 1,026 Celsius
**Ease of ignition ("striking" on suitable surface):** 8
**Water resistance:** Very good.
**Weather resistance (match):** wind resistance normal; moist and wet conditions above normal.
**Stability:** Can be stored for decades.
**Sensitivity:** Very stable towards shock and percussion; friction resistant to soft materials.
**Explosive ability:** Cannot be detonated.
**Percentage:** ***52.79% potassium chlorate, 15.52% pulverized glass, 14.9% Prussian blue, 7.45% furnace black, 6.21% red phosphorus, 3.1% animal glue, 0.03% impurities***
**Classification:** Deflagrating mixture (classified as 1.3G pyrotechnic composition).

**Danger of preparation:** Can be prepared by amateurs, but caution should be exercised.
**Use:** Used for matches.

**06-04-048A: Insensitive ignition composition for rockets and missiles:**
Into a suitable vertical mixer, place 150 milliliters of hexane, followed by ***50.5 grams of potassium permanganate***, followed by ***249.5 grams of aluminum powder***, and then followed by ***200 grams of sodium azide***. Thereafter, moderately blend the mixture at room temperature for 45 minutes. After 45 minutes, the mixture is ready for use. To use in ignition bridge wire devices, the metal wire (usually a thin NiChrome wire) should be evenly coated with the moldable mixture coating the exposed portion of the "wire" with a 3 to 5mm thickness of the mixture, and then cured in an oven at 60 Celsius for several hours. To use in pellet form for ignition of rocket grains, the mixture should be pressed into molds under a pressure of 9,000 psi, forming tablets that need to be cured in an oven at 60 Celsius for 12 hours; the diameter and length of these "tablets" is based on the diameter of the given rocket grain and munition casing; use your own discretion.
**Burn rate (when heated externally):** 210 Celsius. Explosive ignition.
**Burn rate (when heated directly with a flame or hot NiChrome wire):** Explosive ignition.

**Ignition temperature (average temperature to ignite mixture at 9,000 psi):** 499 Celsius.
**Ignition Burn temperature:** 1,108 Celsius.
**Water resistance:** Very good.
**Sensitivity:** Good stability to friction; good resistance to shock and percussion; heat resistant (external) up to 200 Celsius.
**Stability:** Can be stored for many years.
**Flammability:** Explosive combustion.
**Ease of ignition (when heated by a flame or hot wire):** 9.6
**Tendency to cake:** Will not cake.
**Explosive ability:** Does not detonate, but undergoes extreme explosive combustion.
**Percentage:** ***49.9% aluminum powder, 40% sodium azide, 10.1% potassium permanganate***
**Classification:** Deflagrating/explosive mixture (classified as Class A pyrotechnic composition).

**Danger of preparation:** Can be prepared by amateurs, but caution should be exercised.
**Use:** Used to ignite rockets and missiles. Can also be used as a gas generating composition.

### 06-04-048B: Insensitive ignition composition for rockets and missiles:

Into a suitable vertical mixer, place 180 milliliters of hexane, followed by ***50 grams of potassium permanganate***, followed by ***149.5 grams of boron powder***, and then followed by ***300.5 grams of potassium azide***. Thereafter, moderately blend the mixture at room temperature for 45 minutes. After 45 minutes, the mixture is ready for use. To use in ignition bridge wire devices, the metal wire (usually a thin NiChrome wire) should be evenly coated with the moldable mixture coating the exposed portion of the "wire" with a 3 to 5mm thickness of the mixture, and then cured in an oven at 60 Celsius for several hours. To use in pellet form for ignition of rocket grains, the mixture should be pressed into molds under a pressure of 8,000 psi, forming tablets that need to be cured in an oven at 60 Celsius for 12 hours; the diameter and length of these "tablets" is based on the diameter of the given rocket grain and munition casing; use your own discretion.
**Burn rate (when heated externally in powder form):** 286 Celsius. Explosive ignition.
**Burn rate (when heated directly with a flame or hot NiChrome wire, "bridge wire"):** Explosive ignition.
**Ignition temperature (average temperature to ignite mixture at 8,000 psi):** 401 Celsius.
**Ignition Burn temperature:** 1,088 Celsius.
**Water resistance:** Very good.
**Sensitivity:** Good stability to friction; good resistance to shock and percussion; heat resistant (external) up to 200 Celsius.
**Stability:** Can be stored for many years.
**Flammability:** Explosive combustion.
**Ease of ignition (when heated by a flame or hot wire):** 9.7
**Tendency to cake:** Will not cake.
**Explosive ability:** Does not detonate, but undergoes extreme explosive combustion.
**Percentage:** ***60.1% potassium azide, 29.9% boron powder, 10% potassium permanganate***
**Classification:** Deflagrating/explosive mixture (classified as Class A pyrotechnic composition).

**Danger of preparation:** Can be prepared by amateurs, but caution should be exercised.
**Use:** Used to ignite rockets and missiles. Can also be used as a gas generating composition.

### 06-04-048C: Insensitive ignition composition for rockets and missiles:

Into a suitable vertical mixer, place 165 milliliters of hexane, followed by ***75.5 grams of potassium permanganate***, followed by ***342.5 grams of zirconium powder***, and then followed by ***83 grams of sodium azide***. Thereafter, moderately blend the mixture at room temperature for 45 minutes. After 45 minutes, the mixture is ready for use. To use in ignition bridge wire devices, the metal wire (usually a thin NiChrome wire) should be evenly coated with the moldable mixture coating the exposed portion of the "wire" with a 3 to 5mm thickness of the mixture, and then cured in an oven at 60 Celsius for several hours. To use in pellet form for ignition of rocket grains, the mixture should be pressed into molds under a pressure of 6,000 psi, forming tablets that need to be cured in an oven at 60 Celsius for 12 hours; the diameter and length of these "tablets" is based on the diameter of the given rocket grain and munition casing; use your own discretion.
**Burn rate (when heated externally):** 401 Celsius. Explosive ignition.
**Burn rate (when heated directly with a flame or hot NiChrome wire):** Explosive ignition.
**Ignition temperature (average temperature to ignite mixture at 6,000 psi):** 510 Celsius.
**Ignition Burn temperature:** 1,449 Celsius.
**Water resistance:** Very good.
**Sensitivity:** Good stability to friction; good resistance to shock and percussion; heat resistant (external) up to 250 Celsius.
**Stability:** Can be stored for many years.
**Flammability:** Explosive combustion.
**Ease of ignition (when heated by a flame or hot wire):** 9.1
**Tendency to cake:** Will not cake.

**Explosive ability:** Does not detonate, but undergoes extreme explosive combustion.
**Percentage:** ***68.26% zirconium powder, 16.56% sodium azide, 15.06% potassium permanganate, 0.12% mixed balance***
**Classification:** Deflagrating/explosive mixture (classified as Class A pyrotechnic composition).

**Danger of preparation:** Can be prepared by amateurs, but caution should be exercised.
**Use:** Used to ignite rockets and missiles. Can also be used as a gas generating composition, or for igniting incendiary compositions and flares.

### 06-04-049A: Insensitive priming/ignition composition for multiple uses

Into a suitable vertical mixer, place 195 milliliters of hexane, followed by ***25 grams of potassium perchlorate***, followed by ***118.75 grams of sodium azide***, and then followed by ***356.25 grams of boron powder***. Thereafter, moderately blend the mixture at room temperature for 45 minutes. After 45 minutes, the mixture is ready for use. To use in ignition bridge wire devices, the metal wire (usually a thin NiChrome wire) should be evenly coated with the moldable mixture coating the exposed portion of the "wire" with a 3 to 5mm thickness of the mixture, and then cured in an oven at 60 Celsius for several hours. To use in pellet form for ignition of rocket grains, the mixture should be pressed into molds under a pressure of 9,000 psi, forming tablets that need to be cured in an oven at 60 Celsius for 12 hours; the diameter and length of these "tablets" is based on the diameter of the given rocket grain and munition casing; use your own discretion.
**Burn rate (when heated externally, in powder form):** 412 Celsius. Explosive ignition.
**Burn rate (when heated directly with a flame or hot NiChrome wire):** Explosive ignition.
**Ignition temperature (average temperature to ignite mixture at 9,000 psi):** 531 Celsius.
**Ignition Burn temperature:** 1,199 Celsius.
**Water resistance:** Very good.
**Sensitivity:** Good stability to friction; good resistance to shock and percussion; heat resistant (external) up to 310 Celsius.
**Stability:** Can be stored for many years.
**Flammability:** Explosive combustion.
**Ease of ignition (when heated by a flame or hot wire):** 9.29
**Tendency to cake:** Will not cake.
**Explosive ability:** Does not detonate, but undergoes extreme explosive combustion.
**Percentage:** ***71.25% boron powder, 23.75% sodium azide, 5% potassium perchlorate***
**Classification:** Deflagrating/explosive mixture (classified as Class A pyrotechnic composition).

**Danger of preparation:** Can be prepared by amateurs, but caution should be exercised.
**Use:** Can be used to ignite rocket and missile propellants, or for use as a priming composition for fireworks, blasting caps, or detonators. May also be used in air-bag inflation devices, but not for use in automobiles.

### 06-04-049B: Insensitive priming/ignition composition for multiple uses

Into a suitable vertical mixer, place 195 milliliters of hexane, followed by ***45 grams of barium nitrate***, followed by ***113.75 grams of sodium azide***, and then followed by ***341.25 grams of aluminum powder***. Thereafter, moderately blend the mixture at room temperature for 45 minutes. After 45 minutes, the mixture is ready for use. To use in ignition bridge wire devices, the metal wire (usually a thin NiChrome wire) should be evenly coated with the moldable mixture coating the exposed portion of the "wire" with a 3 to 5mm thickness of the mixture, and then cured in an oven at 60 Celsius for several hours. To use in pellet form for ignition of rocket grains, the mixture should be pressed into molds under a pressure of 7,000 psi, forming tablets that need to be cured in an oven at 60 Celsius for 12 hours; the diameter and length of these "tablets" is based on the diameter of the given rocket grain and munition casing; use your own discretion.
**Burn rate (when heated externally, in powder form):** 455 Celsius. Explosive ignition.
**Burn rate (when heated directly with a flame or hot NiChrome wire):** Explosive ignition.
**Ignition temperature (average temperature to ignite mixture at 7,000 psi):** 531 Celsius.
**Ignition Burn temperature:** 1,370 Celsius.
**Water resistance:** Very good.
**Sensitivity:** Good stability to friction; good resistance to shock and percussion; heat resistant (external) up to 400 Celsius.
**Stability:** Can be stored for many years.
**Flammability:** Explosive combustion.
**Ease of ignition (when heated by a flame or hot wire):** 9.54
**Tendency to cake:** Will not cake.
**Explosive ability:** Does not detonate, but undergoes extreme explosive combustion.
**Percentage:** ***68.25% aluminum powder, 22.75% sodium azide, 9% barium nitrate***
**Classification:** Deflagrating/explosive mixture (classified as Class A pyrotechnic composition).

**Danger of preparation:** Can be prepared by amateurs, but caution should be exercised.
**Use:** Can be used to ignite rocket and missile propellants, or for use as a priming composition for fireworks, blasting caps, or detonators. May also be used in air-bag inflation devices, but not for use in automobiles.

**Tandem "double" shape charge**

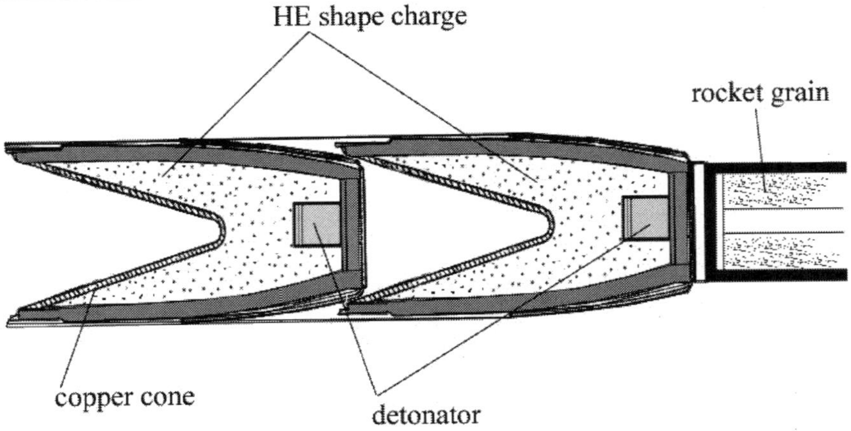


**Details:**

A tandem double shape charge with rocket propulsion. Double shape charges can be used for anti-tank means, or for anti-ship means. In either case, the preceding ignition composition is used to ignite the rocket.

### 06-04-050A: Insensitive priming composition for multiple uses

Into a suitable vertical mixer, place 155 milliliters of hexane, followed by ***45 grams of barium nitrate***, followed by ***113.75 grams of sodium azide***, and then followed by ***341.25 grams of aluminum powder***. Thereafter, moderately blend the mixture at room temperature for 45 minutes. After 45 minutes, the mixture is ready for use. To use in ignition bridge wire devices, the metal wire (usually a thin NiChrome wire) should be evenly coated with the moldable mixture coating the exposed portion of the "wire" with a 3 to 5mm thickness of the mixture, and then cured in an oven at 60 Celsius for several hours. To use in pellet form for use in priming, the mixture should be pressed into molds under a pressure of 7,600 psi, forming tablets that need to be cured in an oven at 60 Celsius for 12 hours; the diameter and length of these "tablets" is based on the diameter of the given munition "casing"; use your own discretion.
**Burn rate (when heated externally):** 401 Celsius. Explodes.
**Burn rate (when heated directly with a flame or hot NiChrome wire):** Explodes.
**Ignition temperature (average temperature to ignite mixture at 7,600 psi):** 419 Celsius.
**Water resistance:** Very good.
**Sensitivity:** Good stability to friction; good resistance to shock and percussion; heat resistant (external) up to 380~ Celsius.
**Stability:** Can be stored for many years.
**Flammability:** Explodes.
**Ease of ignition (when heated by a flame or hot wire):** 9.09
**Tendency to cake:** Will not cake.
**Explosive ability:** Does not detonate, but undergoes extreme explosive combustion.
**Percentage:** ***52.5% boron powder, 30% potassium chlorate, 17.5% sodium azide***
**Classification:** Explosive mixture (classified as Class A explosive composition).

**Danger of preparation:** Should only be prepared by experienced pyrotechnicians.
**Use:** Used for priming fireworks, and for use in blasting caps, or detonators.

| Platter charge primed with blasting cap utilizing priming composition | Details: |
| --- | --- |
| 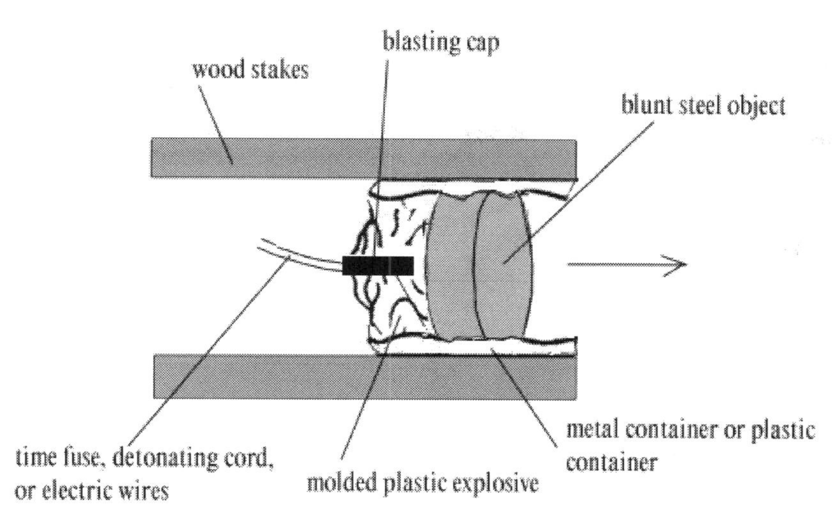 | Platter charges are field expedient munitions utilizing a few basic elements. These elements include a blasting cap, which is inserted into a mass of plastic explosive, which is placed underneath a blunt heavy object such a steel disc or equivalent. Upon detonation of the plastic explosive, the energy released causes the blunt object to accelerate to high speed forming a lethal projectile with significant penetrating power due to its size and weight. |

### 06-04-051A: Insensitive ignition composition for multiple uses:

Into a suitable vertical mixer, place 190 milliliters of hexane, followed by ***40 grams of Viton A copolymer binder***, followed by ***135 grams of boron powder***, followed by ***274.5 grams of sodium azide***, and then followed by ***50.5 grams of potassium perchlorate***. Thereafter, moderately blend the mixture at room temperature for 45 minutes. After 45 minutes, the mixture is ready for use. To use in ignition bridge wire devices, the metal wire (usually a thin NiChrome wire) should be evenly coated with the moldable mixture coating the exposed portion of the "wire" with a 3 to 5mm thickness of the mixture, and then cured in an oven at 60 Celsius for several hours. To use in pellet form for use in ignition, the mixture should be pressed into molds under a pressure of 8,100 psi, forming tablets that need to be cured in an oven at 60 Celsius for 12 hours; the diameter and length of these "tablets" is based on the diameter of the given munition "casing"; use your own discretion.
**Burn rate (when heated externally, in powder form):** 398 Celsius. Explosive ignition.
**Burn rate (when heated directly with a flame or hot NiChrome wire):** Explosive ignition.
**Ignition temperature (average temperature to ignite mixture at 8,100 psi):** 399 Celsius.
**Ignition Burn temperature:** 996 Celsius.
**Water resistance:** Very good.
**Sensitivity:** Good stability to friction; good resistance to shock and percussion; heat resistant (external) up to 350~ Celsius.
**Stability:** Can be stored for many years.
**Flammability:** Explosive combustion.
**Ease of ignition (when heated by a flame or hot wire):** 9
**Tendency to cake:** Will not cake.
**Explosive ability:** Does not detonate, but undergoes extreme explosive combustion.
**Percentage:** ***54.9% sodium azide, 27% boron powder, 10.1% potassium perchlorate, 8% Viton A copolymer binder***
**Classification:** Deflagrating/explosive mixture (classified as Class A pyrotechnic composition).

**Danger of preparation:** Can be prepared by amateurs, but caution should be exercised.
**Use:** Used in grenades and flares.

### 06-04-051B: Insensitive ignition composition for multiple uses:

Into a suitable vertical mixer, place 185 milliliters of hexane, followed by ***16.5 grams of "Kel-F elastomer*** (homopolymer of chlorotrifluoroethylene and vinylidene fluoride), followed by ***286.5 grams of potassium azide***, followed by ***143.5 grams of boron powder***, and then followed by ***53.5 grams of potassium perchlorate***. Thereafter, moderately blend the mixture at room temperature for 45 minutes. After 45 minutes, the mixture is ready for use. To use in ignition bridge wire devices, the metal wire (usually a thin NiChrome wire) should be evenly coated with the moldable mixture coating the exposed portion of the "wire" with a 3 to 5mm thickness of the mixture, and then cured in an oven at 60 Celsius for several hours. To use in pellet form for use in ignition, the mixture should be pressed into molds under a pressure of 8,000 psi, forming tablets that need to be cured in an oven at 60 Celsius for 12 hours; the diameter and length of these "tablets" is based on the diameter of the given munition "casing"; use your own discretion.
**Burn rate (when heated externally, in powder form):** 370 Celsius. Explosive ignition.
**Burn rate (when heated directly with a flame or hot NiChrome wire):** Explosive ignition.
**Ignition temperature (average temperature to ignite mixture at 8,000 psi):** 379 Celsius.
**Ignition Burn temperature:** 1,021 Celsius.
**Water resistance:** Very good.

**Sensitivity:** Good stability to friction; good resistance to shock and percussion; heat resistant (external) up to 350~ Celsius.
**Stability:** Can be stored for many years.
**Flammability:** Explosive combustion.
**Ease of ignition (when heated by a flame or hot wire):** 8.9
**Tendency to cake:** Will not cake.
**Explosive ability:** Does not detonate, but undergoes extreme explosive combustion.
**Percentage:** ***57.3% potassium azide, 28.7% boron powder, 10.7% potassium perchlorate, 3.3% Kel-F elastomer***
**Classification:** Deflagrating/explosive mixture (classified as Class A pyrotechnic composition).

**Danger of preparation:** Can be prepared by amateurs, but caution should be exercised.
**Use:** Used in grenades and flares.

### 06-04-051C: Insensitive ignition composition for multiple uses:

Into a suitable vertical mixer, place 120 milliliters of hexane, followed by ***96 grams of "Kel-F" wax*** (homopolymer of chlorotrifluoroethylene), followed by ***20 grams of Viton A copolymer binder***, followed by ***192 grams of aluminum powder***, and then followed by ***192 grams of sodium azide***. Thereafter, moderately blend the mixture at room temperature for 45 minutes. After 45 minutes, the mixture is ready for use. To use in ignition bridge wire devices, the metal wire (usually a thin NiChrome wire) should be evenly coated with the moldable mixture coating the exposed portion of the "wire" with a 3 to 5mm thickness of the mixture, and then cured in an oven at 60 Celsius for several hours. To use in pellet form for use in ignition, the mixture should be pressed into molds under a pressure of 8,000 psi, forming tablets that need to be cured in an oven at 60 Celsius for 12 hours; the diameter and length of these "tablets" is based on the diameter of the given munition "casing"; use your own discretion.
**Burn rate (when heated externally, in powder form):** 389 Celsius. Explosive ignition.
**Burn rate (when heated directly with a flame or hot NiChrome wire):** Explosive ignition.
**Ignition temperature (average temperature to ignite mixture at 8,000 psi):** 369 Celsius.
**Ignition Burn temperature:** 1,199 Celsius.
**Water resistance:** Very good.
**Sensitivity:** Good stability to friction; good resistance to shock and percussion; heat resistant (external) up to 350~ Celsius.
**Stability:** Can be stored for many years.
**Flammability:** Explosive combustion.
**Ease of ignition (when heated by a flame or hot wire):** 8.71
**Tendency to cake:** Will not cake.
**Explosive ability:** Does not detonate, but undergoes extreme explosive combustion.
**Percentage:** ***38.4% sodium azide, 38.4% aluminum powder, 19.2% Kel-F copolymer binder, 4% Viton A binder***
**Classification:** Deflagrating/explosive mixture (classified as Class A pyrotechnic composition).

**Danger of preparation:** Can be prepared by amateurs, but caution should be exercised.
**Use:** Used in grenades and flares.

### 06-04-052A: Friction sensitive ignition composition for electric matches, matchsticks, and friction initiated devices:

Into a suitable ball mill, filled with 100 grams of steel shot of 5mm diameter, place ***100 gram of red phosphorus***, followed by ***20 grams of iron sulfide***. Thereafter, tumble the mixture at 300 RPM for about 1 hour. Thereafter, into a clean beaker or suitable container, equipped with motorized stirrer, place 80 grams of sulfur monochloride liquid, and then heat this liquid to about 65 Celsius. Thereafter, add in the red phosphorus/lead sulfide mixture, and then blend the entire mixture for about 1 hour at 65 Celsius. Note: a violent reaction will ensue, so use proper ventilation and keep clear of the reaction mixture as it may ignite into flames. After the initial reaction, remove the heat source, and allow the mixture to cool; thereafter, filter the mixture to recover the insoluble materials using a clean Buchner funnel under moderate suction, and then suction dry as best as possible. Thereafter, place this suction dried mixture into a suitable mixing bowl, equipped with motorized stirrer, and then add in 100 milliliters of hexane, followed by ***35 grams of powdered glass***, followed by ***15 grams of animal glue***, and then followed by ***5 grams of lamp black***. Thereafter, moderately blend the mixture at room temperature for 45 minutes. After 45 minutes, the mixture is ready for use. To use in ignition bridge wire devices, the metal wire (usually a thin NiChrome wire) should be evenly coated with the mixture coating the exposed portion of the "wire" with a 1.5 to 4.5mm thickness of the mixture (depending on the desired use), and then cured in an oven at 60 Celsius for several hours. To use on matchsticks, the mixture is ready to be casted ("dipped"). To cast the match head composition, the wood or paper "match sticks" should be dipped and re-dipped into the semi-fluidized mixture to form a uniform thickness ranging from 2mm to 5mm in width (depends on discretion, and the desired thickness of the match head composition can vary). Thereafter, the wood/paper matchsticks can be dipped into a colored dye solution if desired; in either case, the matches should then be cured in an oven at 80 Celsius for 12 hours. For use in "alternative" friction initiated devices, the freshly blended mixture should be pressed in tablets (square, rectangular, or oval shaped, based on the desired munition at hand, use your own discretion) under a pressure of 4,000 psi, and then cured in an oven at 60 Celsius for 12 hours.
**Burn rate (when heated externally):** 499 Celsius. Aggressive deflagration.

**Burn rate (when heated directly with a flame or hot NiChrome wire):** Aggressive deflagration.
**Ignition temperature (average temperature to ignite mixture at 4,000 psi):** 510 Celsius.
**Water resistance:** Very good.
**Sensitivity:** Sensitive to friction; good resistance to shock and percussion; heat resistant (external) up to 450~ Celsius.
**Stability:** Can be stored for many years.
**Flammability:** Aggressive deflagration.
**Ease of ignition (when heated by a flame or hot wire):** 8.91
**Tendency to cake:** Will not cake.
**Explosive ability:** Does not detonate, but undergoes Aggressive deflagration.
**Percentage:** ***68.08% phosphorus sulfur compound, 14.89% powdered glass 8.5% iron sulfide, 6.38% animal glue, 2.12% lamp black, 0.03% balance***
**Classification:** Deflagrating mixture (classified as 1.4G pyrotechnic composition).

**Danger of preparation:** Can be prepared by amateurs, but caution should be exercised.
**Use:** Used in electric matches, and for friction-activated devices.

### 06-04-053A: Non-friction sensitive ignition composition for various operations:

Into a suitable vertical mixer, place 300 milliliters of hexane, followed by ***100 gram of standard vegetable charcoal***, followed by ***400 grams of food grade starch***, followed by ***30 grams of tragacanth gum***, followed by ***100 grams of pulverized sawdust***, followed by ***350 grams of potassium chlorate***, and then followed by ***20 grams of potassium dichromate***. Thereafter, moderately blend the mixture at room temperature for 45 minutes. After 45 minutes, the mixture is ready for use. To use, the mixture should be pressed into molds, tablets, pills, ect., under a pressure of 2,000 psi, and then cured in an oven at 60 Celsius for 12 hours; the diameter and length of these "tablets" is based on the diameter of the given munition "casing"; use your own discretion.
**Burn rate (when heated externally, in powder form):** 23 seconds per inch.
**Burn rate (when heated directly with a flame or hot NiChrome wire at 2,000 psi):** 27 seconds per inch.
**Ignition temperature (average temperature to ignite mixture at 2,000 psi):** 600~ Celsius.
**Ignition Burn temperature:** 899 Celsius.
**Water resistance:** Very good.
**Sensitivity:** Good stability to friction; good resistance to shock and percussion; heat resistant (external) up to 550~ Celsius.
**Stability:** Can be stored for many years.
**Flammability:** Standard deflagration.
**Ease of ignition (when heated by a flame or hot wire):** 6.89
**Tendency to cake:** Will not cake.
**Explosive ability:** Does not detonate.
**Percentage:** ***40% food starch, 35% potassium chlorate, 10% vegetable charcoal, 10% pulverized sawdust, 3% tragacanth gum, 2% potassium dichromate***
**Classification:** Deflagrating mixture (classified as 1.3G pyrotechnic composition).

**Danger of preparation:** Can be prepared by amateurs.
**Use:** Used for non-friction-activated devices.

### 06-04-053B: Moderate friction sensitive ignition composition for matchsticks, and other operations:

Into a suitable vertical mixer, place 310 milliliters of hexane, followed by ***100 gram of glass powder***, followed by ***80 grams of gum Arabic***, followed by ***300 grams of antimony trisulfide***, followed by ***50 grams of lead dioxide***, followed by ***20 grams of potassium dichromate***, and then followed by ***450 grams of potassium chlorate***. Thereafter, moderately blend the mixture at room temperature for 45 minutes. After 45 minutes, the mixture is ready for use. To use, the mixture should be pressed into molds, tablets, pills, ect., under a pressure of 3,500 psi, and then cured in an oven at 60 Celsius for 12 hours; the diameter and length of these "tablets" is based on the diameter of the given munition "casing"; use your own discretion.
**Burn rate (when heated externally, in powder form):** Explosive deflagration.
**Burn rate (when heated directly with a flame or hot NiChrome wire at 3,500 psi):** Explosive deflagration.
**Ignition temperature (average temperature to ignite mixture at 3,500 psi):** 399~ Celsius.
**Ignition Burn temperature:** 1,055 Celsius.
**Water resistance:** Very good.
**Sensitivity:** Moderate stability to friction; good resistance to shock and percussion; heat resistant (external) up to 320~ Celsius.
**Stability:** Can be stored for many years.
**Flammability:** Explosive deflagration.
**Ease of ignition (when heated by a flame or hot wire):** 8.99
**Tendency to cake:** Will not cake.

**Explosive ability:** Does not detonate.
**Percentage:** ***45% potassium chlorate, 30% antimony trisulfide, 10% glass powder, 8% gum Arabic, 5% lead dioxide, 2% potassium dichromate***
**Classification:** Deflagrating/explosive mixture (classified as Class A pyrotechnic composition).

**Danger of preparation:** Can be prepared by amateurs, but caution should be exercised.
**Use:** Can be used for medium tolerance friction-activated devices.

### 06-04-053C: Rough friction producing ignition surface for use in combination with matches and other friction sensitive ignition compositions:

Into a suitable vertical mixer, place 600 milliliters of hexane, followed by ***150 grams of standard animal glue***, followed by ***350 grams of glass powder***, followed by ***300 grams of red phosphorus***, followed by ***180 grams of antimony trisulfide***, and then followed by ***20 grams of potassium dichromate***. Thereafter, moderately blend the mixture at room temperature for 15 minutes. After 15 minutes, strain the mixture to remove excess solvent, and then coat an igniting surface with a 1 to 5mm thickness layer of said mixture, and then cure in an oven at 80 Celsius for 24 hours.
**Burn rate (when heated externally, in powder form):** N/A.
**Burn rate (when heated directly with a flame or hot NiChrome wire):** N/A.
**Ignition temperature (average temperature to ignite mixture):** N/A.
**Ignition Burn temperature:** N/A.
**Water resistance:** Very good.
**Sensitivity:** Excellent stability to friction; excellent resistance to shock and percussion.
**Stability:** Can be stored for many years.
**Flammability:** Standard deflagration.
**Ease of ignition (when heated by a flame or hot wire):** N/A
**Tendency to cake:** Will not cake.
**Explosive ability:** Does not detonate.
**Percentage:** ***35% glass powder, 30% red phosphorus, 18% antimony trisulfide, 15% animal glue, 2% potassium dichromate***
**Classification:** Combustible mixture (classified as 1.1G pyrotechnic composition).

**Danger of preparation:** Can be prepared by amateurs.
**Use:** Used to form a rough ignition surface whereby friction sensitive mixtures can be ignited there from.

### 06-04-054A: Ignition composition for initiating thermite and other incendiary compositions:

Into a suitable vertical mixer, place 300 milliliters of hexane, followed by ***50 grams of aluminum powder***, followed by ***75 grams of zinc phosphide***, and then followed by ***425 grams of potassium dichromate***. Thereafter, moderately blend the mixture at room temperature for 45 minutes. After 45 minutes, the mixture is ready for use. To use, the mixture should be pressed into molds, tablets, pills, ect., under a pressure of 2,500 psi, and then cured in an oven at 60 Celsius for 12 hours; the diameter and length of these "tablets" is based on the diameter of the given munition "casing"; use your own discretion.
**Burn rate (when heated externally, in powder form):** 19 seconds per inch.
**Burn rate (when heated directly with a flame or hot NiChrome wire at 2,500 psi):** 26 seconds per inch.
**Ignition temperature (average temperature to ignite mixture at 2,500 psi):** 588~ Celsius.
**Ignition Burn temperature:** 1,339 Celsius.
**Water resistance:** Very good.
**Sensitivity:** Good stability to friction; good resistance to shock and percussion; heat resistant (external) up to 500~ Celsius.
**Stability:** Can be stored for many years.
**Flammability:** Standard deflagration.
**Ease of ignition (when heated by a flame or hot wire):** 6
**Tendency to cake:** Will not cake.
**Explosive ability:** Does not detonate.
**Percentage:** ***77.27% potassium dichromate, 13.63% zinc phosphide, 9.09% aluminum powder, 0.01% mixed residual balance***
**Classification:** Deflagrating mixture (classified as 1.3G pyrotechnic composition).

**Danger of preparation:** Can be prepared by amateurs.

**Use:** Used to ignite incendiary compositions.

## 06-04-055A: Priming composition for use in blasting caps:

Into a suitable vertical mixer, place 475 milliliters of hexane, followed ***150 grams of finely divided mossy zinc***, followed by ***350 grams of copper powder***, and then followed by ***375 grams of potassium dichromate***. Thereafter, moderately blend the mixture at room temperature for 45 minutes. After 45 minutes, the mixture is ready for use. To use, the mixture should be pressed into molds under a pressure of 4,000 psi, and then cured in an oven at 60 Celsius for 12 hours; the diameter and length of these "molds" is based on the diameter and length of the given munition "casing"; use your own discretion.
**Burn rate (when heated externally, in powder form):** Explosive deflagration.
**Burn rate (when heated directly with a flame or hot NiChrome wire at 4,000 psi):** Explosive deflagration.
**Ignition temperature (average temperature to ignite mixture at 4,000 psi):** 401~ Celsius.
**Ignition Burn temperature:** 1,227 Celsius.
**Water resistance:** Very good.
**Sensitivity:** Good stability to friction; good resistance to shock and percussion; heat resistant (external) up to 360~ Celsius.
**Stability:** Can be stored for many years.
**Flammability:** Explosive deflagration.
**Ease of ignition (when heated by a flame or hot wire):** 8.91
**Tendency to cake:** Will not cake.
**Explosive ability:** Does not detonate.
**Percentage:** ***37.5% potassium dichromate, 35% copper powder, 15% mossy zinc, 12.5% zinc phosphide***
**Classification:** Deflagrating/explosive mixture (classified as Class A pyrotechnic composition).

**Danger of preparation:** Can be prepared by amateurs, but caution should be exercised.
**Use:** Used as a priming composition for use in blasting caps and detonators.

**Commercial Dynamite**

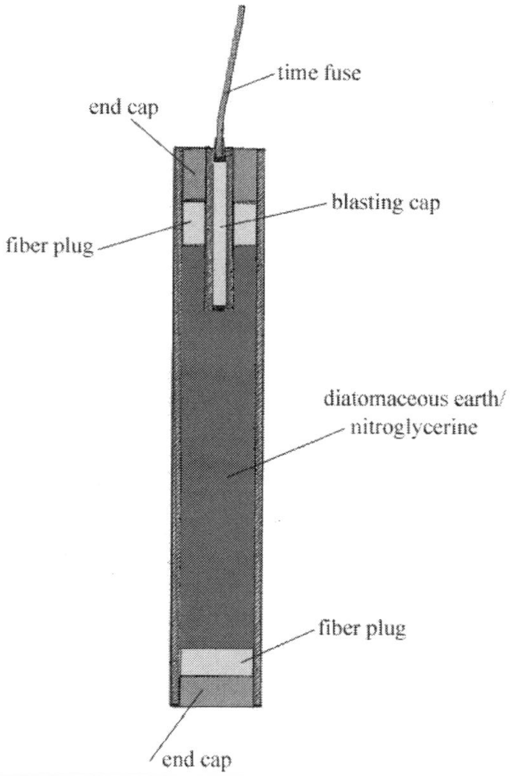


## 06-04-056A: Priming composition for use in small arms primers:

Into a suitable vertical mixer, place 190 milliliters of hexane, followed by ***125 grams of red phosphorus***, followed by ***125 grams of lead dioxide***, followed by ***350 grams of barium nitrate***, and then followed by ***100 grams of zirconium metal powder***. Thereafter, moderately blend the mixture at room temperature for 45 minutes. After 45 minutes, open the vertical mixer, recover the mixture, and then extrude the mixture through any desired extruding machine into strands ranging from 0.4mm to 3.2mm in diameter by 10cm in length, then cut the strands into 0.1mm to 5mm pieces and then cure the pieces on a flat tray at room temperature for 48 hours. Note: the size of the cured pieces

depends on the gun primer in question. Small arms weapons such as handguns, and rifles should use 0.5 to 1mm sized pieces; whereas larger bore weapons such as 12.7mm up to 25mm casings should use 1.5 to 2mm sized pieces; use your discretion).
**Burn rate (when heated externally, in powder form):** 400 Celsius. Explodes.
**Burn rate (when heated directly with a flame or hot NiChrome wire):** Explodes.
**Ignition temperature (average temperature to ignite mixture):** 501 Celsius.
**Water resistance:** Very good.
**Sensitivity:** Good stability to friction; sensitive to shock and percussion; heat resistant (external) up to 400 Celsius.
**Stability:** Can be stored for many years.
**Flammability:** Explodes.
**Ease of ignition (when heated by a flame or hot wire):** 9.67
**Tendency to cake:** Will not cake.
**Explosive ability:** Does not detonate, but undergoes extreme explosive ignition.
**Percentage:** ***50% barium nitrate, 17.85% red phosphorus, 17.85% lead dioxide, 14.28% zirconium metal powder, 0.02% mixed balance***
**Classification:** Deflagrating/explosive mixture (classified as Class A pyrotechnic composition).

**Danger of preparation:** Can be prepared by amateurs, but caution should be exercised.
**Use:** Used as a priming composition for use in small arms ammunition. Should not be used in rim fire casings. Do not use to fire high performance rocket fuels or high performance gun propellants.

**Method of priming an outdated artillery shell to form an IED (US ARMY)**

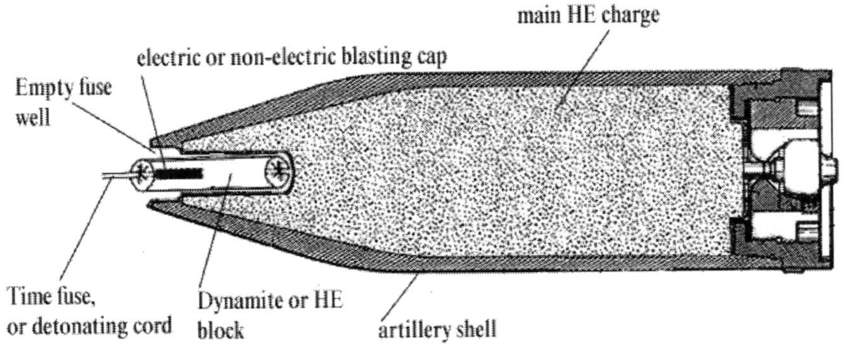


### 06-04-056B: Priming composition for use in small arms primers:

Into a suitable vertical mixer, place 210 milliliters of hexane, followed by ***125 grams of red phosphorus***, followed by ***125 grams of antimony trisulfide***, and then followed by ***350 grams of barium nitrate***. Thereafter, moderately blend the mixture at room temperature for 45 minutes. After 45 minutes, open the vertical mixer, recover the mixture, and then extrude the mixture through any desired extruding machine into strands ranging from 0.4mm to 3.2mm in diameter by 10cm in length, then cut the strands into 0.1mm to 5mm pieces and then cure the pieces on a flat tray at room temperature for 48 hours. Note: the size of the cured pieces depends on the gun primer in question. Small arms weapons such as handguns, and rifles should use 0.5 to 1mm sized pieces; whereas larger bore weapons such as 12.7mm up to 25mm casings should use 1.5 to 2mm sized pieces; use your discretion).
**Burn rate (when heated externally, in powder form):** 455 Celsius. Explodes.
**Burn rate (when heated directly with a flame or hot NiChrome wire):** Explodes.
**Ignition temperature (average temperature to ignite mixture):** 433 Celsius.
**Water resistance:** Very good.
**Sensitivity:** Good stability to friction; sensitive to shock and percussion; heat resistant (external) up to 400~ Celsius.
**Stability:** Can be stored for many years.
**Flammability:** Explodes.
**Ease of ignition (when heated by a flame or hot wire):** 9.4
**Tendency to cake:** Will not cake.
**Explosive ability:** Does not detonate, but undergoes extreme explosive ignition.
**Percentage:** ***58.33% barium nitrate, 20.83% red phosphorus, 20.83% antimony trisulfide, 0.01% residual balance***
**Classification:** Deflagrating/explosive mixture (classified as Class A pyrotechnic composition).

**Danger of preparation:** Can be prepared by amateurs, but caution should be exercised.

**Use:** Used as a priming composition for use in small arms ammunition. Should not be used in rim fire casings. Do not use to fire high performance rocket fuels or high performance gun propellants.

### 06-04-057A: Priming composition for use in small arms primers:

Into a suitable beaker or similar container, place ***100 grams of potassium chlorate***, followed by ***100 grams of red phosphorus***, and then followed by 450 milliliters of hot water (the water should be heated to about 90 Celsius prior to adding). Thereafter, gently blend the mixture for about 10 minutes. Thereafter, add in ***400 grams of lead plumbate***, and then continue to blend the mixture for about 45 minutes at 90 Celsius. Make sure to keep the water mixture heated to around 90 Celsius. Thereafter, remove the heat source, and then pour the entire mixture onto a shallow pan or tray, and allow it to thoroughly air dir. Once it has, the mixture needs to be scraped up, and placed into a suitable vertical mixer, followed by adding in 260 milliliters of hexane, followed by ***5 grams of powdered gum tragacanth***, followed by ***5 grams of gum Arabic***, and then followed by ***75 grams of powdered glass***. Thereafter, moderately blend the mixture at room temperature for 45 minutes. After 45 minutes, open the vertical mixer, recover the mixture, and then extrude the mixture through any desired extruding machine into strands ranging from 0.4mm to 3.2mm in diameter by 10cm in length, then cut the strands into 0.1mm to 5mm pieces and then cure the pieces on a flat tray at room temperature for 48 hours. Note: the size of the cured pieces depends on the gun primer in question. Small arms weapons such as handguns, and rifles should use 0.5 to 1mm sized pieces; whereas larger bore weapons such as 12.7mm up to 25mm casings should use 1.5 to 2mm sized pieces; use your discretion).
**Burn rate (when heated externally, in powder form):** 404 Celsius. Explodes.
**Burn rate (when heated directly with a flame or hot NiChrome wire):** Explodes.
**Ignition temperature (average temperature to ignite mixture):** 490 Celsius.
**Water resistance:** Very good.
**Sensitivity:** Good stability to friction; sensitive to shock and percussion; heat resistant (external) up to 380~ Celsius.
**Stability:** Can be stored for many years.
**Flammability:** Explodes.
**Ease of ignition (when heated by a flame or hot wire):** 9.69
**Tendency to cake:** Will not cake.
**Explosive ability:** Does not detonate, but undergoes extreme explosive ignition.
**Percentage:** ***72.99% red phosphorus/lead plumbate addition salt, 14.59% potassium chlorate, 10.94% powdered glass, 0.72% gum tragacanth, 0.72% gum Arabic, 0.04% balance***
**Classification:** Deflagrating/explosive mixture (classified as Class A pyrotechnic composition).

**Danger of preparation:** Can be prepared by amateurs, but caution should be exercised.
**Use:** Used as a priming composition for use in small arms ammunition, specifically for use in rim fire cartridges. Do not use to fire high performance rocket fuels or high performance gun propellants.

### 06-04-058A: Priming composition for use in small arms primers:

Into a suitable vertical mixer, place 295 milliliters of hexane, followed by ***200 grams of nitrocellulose***, followed by ***50 grams of red phosphorus***, followed by ***375 grams of potassium chlorate***, and then followed by ***370 grams of black antimony sulfide***. Thereafter, moderately blend the mixture at room temperature for 45 minutes. After 45 minutes, open the vertical mixer, recover the mixture, and then extrude the mixture through any desired extruding machine into strands ranging from 0.4mm to 3.2mm in diameter by 10cm in length, then cut the strands into 0.1mm to 5mm pieces and then cure the pieces on a flat tray at room temperature for 48 hours. Note: the size of the cured pieces depends on the gun primer in question. Small arms weapons such as handguns, and rifles should use 0.5 to 1mm sized pieces; whereas larger bore weapons such as 12.7mm up to 25mm casings should use 1.5 to 2mm sized pieces; use your discretion).
**Burn rate (when heated externally, in powder form):** 400 Celsius; 13mm per second. Explosive ignition.
**Burn rate (when heated directly with a flame or hot NiChrome wire):** Explosive ignition.
**Ignition temperature (average temperature to ignite mixture):** 209 Celsius.
**Water resistance:** Very good.
**Sensitivity:** Good stability to friction; sensitive to shock and percussion; heat resistant (external) up to 200~ Celsius.
**Stability:** Can be stored for many years.
**Flammability:** Explosive ignition.
**Ease of ignition (when heated by a flame or hot wire):** 9.8
**Tendency to cake:** Will not cake.
**Explosive ability:** Does not detonate, but undergoes extreme explosive ignition.
**Percentage:** ***37.68% potassium chlorate, 37.18% black antimony sulfide, 20.1% nitrocellulose, 5.02% red phosphorus, 0.02% mixed residual balance***
**Classification:** Deflagrating/explosive mixture (classified as Class A pyrotechnic composition).

**Danger of preparation:** Should only be prepared by experienced pyrotechnicians.

**Use:** Used as a priming composition for use in small arms ammunition. Do not use to fire high performance rocket fuels or high performance gun propellants.

### 06-04-059A: Gasless ignition composition for smoke pots and smoke grenades:

Into a standard ball mill, filled with 900 grams of steel shot of 10mm diameter, place ***18 grams of celluloid product***, followed by ***533 grams of red lead***, followed by ***112 grams of finely divided manganese powder***, and then followed by ***337 grams of finely divided silicon***. Thereafter, tumble the mixture at room temperature at 300 RPM for about 45 minutes. Thereafter, the mixture should be removed from the ball mill, and the steel shot separated; thereafter, the mixture is ready for pressing. To do so, press the mixture into blocks, discs, or tablets under a pressure of 9,000 psi.
**Burn rate:** 32 seconds per inch at 9,000 psi; tends to vary based on diameter and density of grain.
**Burn temperature:** 1,099 Celsius.
**Ignition temperature (average temperature to ignite mixture):** 871 Celsius.
**Water resistance:** Very good.
**Stability:** Can be stored for many years.
**Flammability (pressed grain at 9,000 psi):** 3.87
**Ease of ignition (average temperature to ignite mixture):** 4.1
**Tendency to cake:** None.
**Explosive ability:** Cannot be detonated.
**Percentage:** ***53.3% red lead, 33.7% silicon, 11.2% manganese, 1.8% celluloid***
**Classification:** Deflagrating mixture (classified as 1.1G pyrotechnic composition).

IRRITANT

**Danger of preparation:** Can be prepared by amateurs.
**Use:** Used to initiate smoke pots, smoke grenades, and other smoke generating devices.

### 06-04-059B: Gasless ignition composition for smoke pots and smoke grenades:

Into a standard ball mill, filled with 900 grams of steel shot of 10mm diameter, place ***18 grams of celluloid product***, followed by ***769 grams of litharge***, followed by ***193 grams of finely divided silicon***, followed by ***15 grams of fullers earth***, and then followed by ***5 grams of finely divided graphite***. Thereafter, tumble the mixture at room temperature at 300 RPM for about 45 minutes. Thereafter, the mixture should be removed from the ball mill, and the steel shot separated; thereafter, the mixture is ready for pressing. To do so, press the mixture into blocks, discs, or tablets under a pressure of 9,000 psi.
**Burn rate:** 29 seconds per inch at 9,000 psi; tends to vary based on diameter and density of grain.
**Burn temperature:** 1,055 Celsius.
**Ignition temperature (average temperature to ignite mixture):** 899 Celsius.
**Water resistance:** Very good.
**Stability:** Can be stored for many years.
**Flammability (pressed grain at 9,000 psi):** 3.8
**Ease of ignition (average temperature to ignite mixture):** 4
**Tendency to cake:** None.
**Explosive ability:** Cannot be detonated.
**Percentage:** ***76.9% litharge, 19.3% silicon, 1.8% celluloid, 1.5% fullers earth, 0.5% graphite***
**Classification:** Deflagrating mixture (classified as 1.1G pyrotechnic composition).

IRRITANT

**Danger of preparation:** Can be prepared by amateurs.
**Use:** Used to initiate smoke pots, smoke grenades, and other smoke generating devices.

### 06-04-059C: Gasless ignition composition for smoke pots and smoke grenades:

Into a standard ball mill, filled with 900 grams of steel shot of 10mm diameter, place ***18 grams of celluloid product***, followed by ***590 grams of lead-II-chromate***, followed by ***196 grams of finely divided silicon***, and then followed by ***196 grams of finely divided magnesium powder***. Thereafter, tumble the mixture at room temperature at 150 RPM for about 50 minutes. Thereafter, the mixture should be removed from the ball mill, and the steel shot separated; thereafter, the mixture is ready for pressing. To do so, press the mixture into blocks, discs, or tablets under a pressure of 7,000 psi.
**Burn rate:** 23 seconds per inch at 7,000 psi; tends to vary based on diameter and density of grain.
**Burn temperature:** 1,177 Celsius.
**Ignition temperature (average temperature to ignite mixture):** 877 Celsius.
**Water resistance:** Very good.
**Stability:** Can be stored for many years.
**Flammability (pressed grain at 7,000 psi):** 5.1

**Ease of ignition (average temperature to ignite mixture):** 5
**Tendency to cake:** None.
**Explosive ability:** Cannot be detonated.
**Percentage:** ***59% lead chromate, 19.6% silicon, 19.6% magnesium powder, 1.8% celluloid***
**Classification:** Deflagrating mixture (classified as 1.3G pyrotechnic composition).

**Danger of preparation:** Can be prepared by amateurs.
**Use:** Used to initiate smoke pots, smoke grenades, and other smoke generating devices.

## 06-04-060A: Ignition composition for tracer compositions (with ADN addictive):

Into a suitable vertical mixer, place 210 milliliters of hexane, followed by ***50 grams of nitrocellulose***, followed by ***500 grams of barium nitrate***, followed by ***100 grams of ADN***, followed by ***150 grams of zirconium hydride***, and then followed by ***200 grams of finely divided silicon***. Thereafter, moderately blend the mixture at room temperature for 45 minutes. After 45 minutes, the mixture is ready for use. To use, the mixture should be pressed into molds under a pressure of 5,000 psi, and then cured in an oven at 60 Celsius for 12 hours; the diameter and length of these "molds" is based on the diameter and length of the given munition "housing/cavity"; use your own discretion. ,
**Burn rate:** 1.9 seconds per inch at 5,000 psi; tends to vary based on diameter and density of grain.
**Burn temperature:** 1,079 Celsius.
**Ignition temperature (average temperature to ignite mixture):** 600 Celsius.
**Water resistance:** Acceptable.
**Stability:** Can be stored for many years.
**Flammability (pressed grain at 5,000 psi):** 9+
**Ease of ignition (average temperature to ignite mixture):** 9+
**Tendency to cake:** None.
**Explosive ability:** Can be detonated; can be primed with lead azide, silver azide, mercury fulminate, diazodinitrophenol, or tetracene.
**Percentage:** ***50% barium nitrate, 20% silicon, 15% zirconium hydride, 10% ADN, 5% nitrocellulose***
**Classification:** Deflagrating/explosive mixture (classified as 1.4G pyrotechnic composition).

**Danger of preparation:** Can be prepared by amateurs, but special caution should be exercised.
**Use:** Used to initiate tracer compositions for use in ammunition.

## 06-04-061A: Priming composition for use in blasting caps:

Into a suitable vertical mixer, place 190 milliliters of hexane, followed by ***200 grams of finely divided alloy powder*** (43% titanium, 47% lead, and 10% tin), followed by ***100 grams of barium nitrate***, followed by ***50 grams of lead dioxide***, and then followed by ***100 grams of coarse zirconium metal***. Thereafter, moderately blend the mixture at room temperature for 45 minutes. After 45 minutes, the mixture is ready for use. To use, the mixture should be pressed into molds under a pressure of 3,500 psi, and then cured in an oven at 60 Celsius for 12 hours; the diameter and length of these "molds" is based on the diameter and length of the given munition "casing"; use your own discretion.
**Burn rate (when heated externally, in powder form):** Explosive deflagration.
**Burn rate (when heated directly with a flame or hot NiChrome wire at 3,500 psi):** Explosive deflagration.
**Ignition temperature (average temperature to ignite mixture at 3,500 psi):** 499 Celsius.
**Ignition Burn temperature:** 1,004 Celsius.
**Water resistance:** Very good.
**Sensitivity:** Good stability to friction; good resistance to shock and percussion; heat resistant (external) up to 400~ Celsius.
**Stability:** Can be stored for many years.
**Flammability:** Explosive deflagration.
**Ease of ignition (when heated by a flame or hot wire):** 8.5
**Tendency to cake:** Will not cake.
**Explosive ability:** Does not detonate.
**Percentage:** ***44.44% special metal alloy, 22.22% barium nitrate, 22.22% coarse zirconium metal, 11.11% lead dioxide, 0.01% mixed balance***
**Classification:** Deflagrating/explosive mixture (classified as Class A pyrotechnic composition).

**Danger of preparation:** Can be prepared by amateurs, but caution should be exercised.
**Use:** Used in non-electric blasting caps to be initiated by black powder time fuse. Can also be used as a primer for small arms ammunition.

### 06-04-061B: Priming composition for use in blasting caps:

Into a suitable vertical mixer, place 90 milliliters of hexane, followed by ***250 grams of finely divided alloy powder*** (25% titanium, 25% zirconium, and 50% tin), followed by ***100 grams of barium nitrate***, followed by ***50 grams of lead dioxide***, followed by ***80 grams of zirconium powder***, and then followed by ***20 grams of antimony trisulfide***. Thereafter, moderately blend the mixture at room temperature for 45 minutes. After 45 minutes, the mixture is ready for use. To use, the mixture should be pressed into molds under a pressure of 4,000 psi, and then cured in an oven at 60 Celsius for 12 hours; the diameter and length of these "molds" is based on the diameter and length of the given munition "casing"; use your own discretion.
**Burn rate (when heated externally, in powder form):** Explosive deflagration.
**Burn rate (when heated directly with a flame or hot NiChrome wire at 4,000 psi):** Explosive deflagration.
**Ignition temperature (average temperature to ignite mixture at 4,000 psi):** 600 Celsius.
**Ignition Burn temperature:** 1,109 Celsius.
**Water resistance:** Very good.
**Sensitivity:** Good stability to friction; good resistance to shock and percussion; heat resistant (external) up to 490~ Celsius.
**Stability:** Can be stored for many years.
**Flammability:** Explosive deflagration.
**Ease of ignition (when heated by a flame or hot wire):** 7.9
**Tendency to cake:** Will not cake.
**Explosive ability:** Does not detonate.
**Percentage:** ***50% special metal alloy, 20% barium nitrate, 16% zirconium powder, 10% lead dioxide, 4% antimony trisulfide***
**Classification:** Deflagrating/explosive mixture (classified as Class A pyrotechnic composition).

**Danger of preparation:** Can be prepared by amateurs, but caution should be exercised.
**Use:** Used in non-electric blasting caps to be initiated by black powder time fuse. Can also be used as a primer for small arms ammunition.

### 06-04-061C: Priming composition for use in blasting caps:

Into a suitable vertical mixer, place 145 milliliters of hexane, followed by ***230 grams of special metal alloy*** (35% zirconium, 35% lead, 27% manganese, and 3% antimony), followed by ***100 grams of coarse zirconium metal***, followed by ***50 grams of lead dioxide***, followed by ***20 grams of antimony trisulfide***, and then followed by ***100 grams of barium nitrate***. Thereafter, moderately blend the mixture at room temperature for 45 minutes. After 45 minutes, the mixture is ready for use. To use, the mixture should be pressed into molds under a pressure of 5,500 psi, and then cured in an oven at 60 Celsius for 12 hours; the diameter and length of these "molds" is based on the diameter and length of the given munition "casing"; use your own discretion.
**Burn rate (when heated externally, in powder form):** Explosive deflagration.
**Burn rate (when heated directly with a flame or hot NiChrome wire at 5,500 psi):** Explosive deflagration.
**Ignition temperature (average temperature to ignite mixture at 5,500 psi):** 709 Celsius.
**Ignition Burn temperature:** 1,100 Celsius.
**Water resistance:** Very good.
**Sensitivity:** Good stability to friction; good resistance to shock and percussion; heat resistant (external) up to 455~ Celsius.
**Stability:** Can be stored for many years.
**Flammability:** Explosive deflagration.
**Ease of ignition (when heated by a flame or hot wire):** 7.5
**Tendency to cake:** Will not cake.
**Explosive ability:** Does not detonate.
**Percentage:** ***46% special metal alloy, 20% barium nitrate, 20% zirconium metal, 10% lead dioxide, 4% antimony trisulfide***
**Classification:** Deflagrating/explosive mixture (classified as Class A pyrotechnic composition).

**Danger of preparation:** Can be prepared by amateurs, but caution should be exercised.
**Use:** Used in non-electric blasting caps to be initiated by black powder time fuse. Can also be used as a primer for small arms ammunition.

### 06-04-061D: Priming composition for use in blasting caps:

Into a suitable vertical mixer, place 140 milliliters of hexane, followed by ***230 grams of special metal alloy*** (8% zirconium, 62% manganese, and 30% antimony), followed by ***50 grams of lead dioxide***, followed by ***100 grams of zirconium metal powder***, followed by ***20 grams of antimony trisulfide***, and then followed by ***100 grams of barium nitrate***. Thereafter, moderately blend the mixture at room temperature for 45 minutes. After 45 minutes, the mixture is ready for use. To use, the mixture should be pressed into molds under a pressure of 5,300 psi, and then cured in an oven at 60 Celsius for 12 hours; the diameter and length of these "molds" is based on the diameter and length of the given munition "casing"; use your own discretion.
**Burn rate (when heated externally, in powder form):** Explosive deflagration.

**Burn rate (when heated directly with a flame or hot NiChrome wire at 5,300 psi):** Explosive deflagration.
**Ignition temperature (average temperature to ignite mixture at 5,300 psi):** 722 Celsius.
**Ignition Burn temperature:** 1,192 Celsius.
**Water resistance:** Very good.
**Sensitivity:** Good stability to friction; good resistance to shock and percussion; heat resistant (external) up to 495~ Celsius.
**Stability:** Can be stored for many years.
**Flammability:** Explosive deflagration.
**Ease of ignition (when heated by a flame or hot wire):** 7.9
**Tendency to cake:** Will not cake.
**Explosive ability:** Does not detonate.
**Percentage:** ***46% special metal alloy, 20% barium nitrate, 20% zirconium metal, 10% lead dioxide, 4% antimony trisulfide*** **Classification:** Deflagrating/explosive mixture (classified as Class A pyrotechnic composition).

**Danger of preparation:** Can be prepared by amateurs, but caution should be exercised.
**Use:** Used in non-electric blasting caps to be initiated by black powder time fuse. Can also be used as a primer for small arms ammunition.

### 06-04-062A: Igniter composition for tracer compositions:

Into a suitable vertical mixer, place 200 milliliters of hexane, followed by ***405 grams of strontium peroxide***, followed by ***20.5 grams of barium peroxide***, followed by ***20.5 grams of lead peroxide***, followed by ***40 grams of calcium resinate,*** and then followed by ***55 grams of calcium silicide***. Thereafter, moderately blend the mixture at room temperature for 45 minutes. After 45 minutes, the mixture is ready for use. To use, the mixture should be pressed into molds under a pressure of 6,000 psi, and then cured in an oven at 60 Celsius for 12 hours; the diameter and length of these "molds" is based on the diameter and length of the given munition "housing/cavity"; use your own discretion.
**Burn rate:** 9.2 seconds per inch at 6,000 psi; tends to vary based on diameter and density of grain.
**Burn temperature:** 994 Celsius.
**Ignition temperature (average temperature to ignite mixture):** 690 Celsius.
**Water resistance:** Acceptable.
**Stability:** Can be stored for many years.
**Flammability (pressed grain at 9,000 psi):** 7.8
**Ease of ignition (average temperature to ignite mixture):** 7.67
**Tendency to cake:** None.
**Explosive ability:** Cannot be detonated.
**Percentage:** ***74.86% strontium peroxide, 10.16% calcium silicide, 7.39% calcium resinate, 3.78% barium peroxide, 3.78% lead peroxide. 0.03% residual balance***
**Classification:** Deflagrating mixture (classified as 1.3G pyrotechnic composition).

**Danger of preparation:** Can be prepared by amateurs, but caution should be exercised.
**Use:** Used to initiate tracer compositions for use in ammunition.

**Typical 60mm mortar shell design**

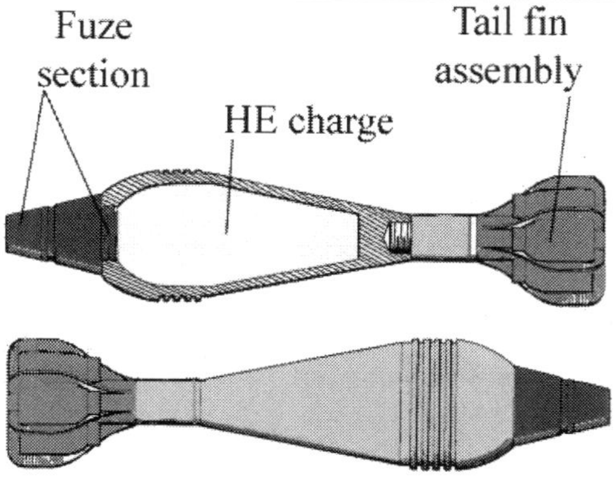

### 06-04-063A: Specialty high-temperature igniter composition for use in electric "squibs":

Into a suitable vertical mixer, place 200 milliliters of hexane, followed by ***343 grams of finely ground hafnium metal***, followed by ***133.5 grams of potassium perchlorate***, and then followed by ***23.5 grams of nylon binder***. Thereafter, moderately blend the mixture at room temperature for 45 minutes. After 45 minutes, the mixture is ready for use. To use in ignition bridge wire devices, the metal wire (usually a thin NiChrome wire) should be evenly coated with the moldable mixture coating the exposed portion of the "wire" with a 3 to 5mm thickness of the mixture, and then cured in an oven at 60 Celsius for several hours. To use in pellet form for use in ignition, the mixture should be pressed into molds under a pressure of 8,400 psi, forming tablets that need to be cured in an oven at 60 Celsius for 12 hours; the diameter and length of these "tablets" is based on the diameter of the given munition "casing/housing"; use your own discretion.
**Burn rate:** 17 seconds per inch at 8,400 psi; tends to vary based on diameter and density of grain.
**Burn temperature:** 1,200 Celsius.
**Ignition temperature (average temperature to ignite mixture):** 563 Celsius.
**Water resistance:** Acceptable.
**Stability:** Can be stored for many years.
**Flammability (pressed grain at 8,400 psi):** 7
**Ease of ignition (average temperature to ignite mixture):** 7.1
**Tendency to cake:** None.
**Explosive ability:** Cannot be detonated.
**Percentage:** ***68.6% ground hafnium metal, 26.7% potassium perchlorate, 4.7% nylon binder***
**Classification:** Deflagrating mixture (classified as 1.3G pyrotechnic composition).

**Danger of preparation:** Can be prepared by amateurs, but caution should be exercised.
**Use:** Used in electric squibs.

### 06-04-064A: Classic lead nitrate priming composition for rim-fire ammunition:

Into a suitable vertical mixer, place 200 milliliters of hexane, followed by ***150 grams of potassium chlorate***, followed by ***60 grams of lead nitrate anhydrous***, followed by ***165 grams of lead thiocyanate***, and then followed by ***125 grams of glass powder***. Thereafter, moderately blend the mixture at room temperature for 45 minutes. After 45 minutes, open the vertical mixer, recover the mixture, and then extrude the mixture through any desired extruding machine into strands ranging from 0.4mm to 3.2mm in diameter by 10cm in length, then cut the strands into 0.1mm to 5mm pieces and then cure the pieces on a flat tray at room temperature for 48 hours. Note: the size of the cured pieces depends on the gun primer in question. Small arms weapons such as handguns, and rifles should use 0.5 to 1mm sized pieces; whereas larger bore weapons such as 12.7mm up to 25mm casings should use 1.5 to 2mm sized pieces; use your discretion).
**Burn rate (when heated externally, in powder form):** 412 Celsius; Explodes.
**Burn rate (when heated directly with a flame or hot NiChrome wire):** Explodes.
**Ignition temperature (average temperature to ignite mixture):** 234 Celsius.
**Water resistance:** Very good.
**Sensitivity:** Good stability to friction; sensitive to shock and percussion; heat resistant (external) up to 190~ Celsius.
**Stability:** Can be stored for many years.
**Flammability:** Explodes.
**Ease of ignition (when heated by a flame or hot wire):** 9+
**Tendency to cake:** Will not cake.
**Explosive ability:** Does not detonate, but undergoes extreme explosive ignition.
**Percentage:** ***33% lead thiocyanate, 30% potassium chlorate, 25% glass powder, 12% lead nitrate***
**Classification:** Explosive mixture (classified as Class A1 explosive composition).

**Danger of preparation:** Should only be prepared by experienced pyrotechnicians.
**Use:** Used as a priming composition for use in small arms ammunition. Do not use to fire high performance rocket fuels or high performance gun propellants.

### 06-04-065A: Priming composition for use in non-electric blasting caps:

Into a suitable vertical mixer, place 195 milliliters of hexane, followed by ***150 grams of potassium chlorate***, followed by ***200 grams of nitrocellulose***, and then followed by ***150 grams of potassium ferrocyanide***. Thereafter, moderately blend the mixture at room temperature for 45 minutes. After 45 minutes, the mixture is ready for use. To use, the mixture should be pressed into molds under a pressure of 3,000 psi, and then cured in an oven at 60 Celsius for 12 hours; the diameter and length of these "molds" is based on the diameter and length of the given munition "casing"; use your own discretion.
**Burn rate (when heated externally, in powder form):** Explosive deflagration.
**Burn rate (when heated directly with a flame or hot NiChrome wire at 3,000 psi):** Explosive deflagration.
**Ignition temperature (average temperature to ignite mixture at 3,000 psi):** 559 Celsius.

**Ignition Burn temperature:** 1,037 Celsius.
**Water resistance:** Very good.
**Sensitivity:** Good stability to friction; good resistance to shock and percussion; heat resistant (external) up to 300~ Celsius.
**Stability:** Can be stored for many years.
**Flammability:** Explosive deflagration.
**Ease of ignition (when heated by a flame or hot wire):** 9+
**Tendency to cake:** Will not cake.
**Explosive ability:** Can be detonated when primed with lead azide, silver azide, mercury fulminate, or lead styphnate.
**Percentage:** ***40% nitrocellulose, 30% potassium chlorate, 30% potassium ferrocyanide***
**Classification:** Deflagrating/explosive mixture (classified as Class A1 explosive composition).

**Danger of preparation:** Should only be prepared by experienced pyrotechnicians.
**Use:** Used in blasting caps and detonators as a first fire ignition charge initiated by a black powder time fuse.

## 06-04-066A: Specialty priming composition for use in blasting caps and detonators:

Into a suitable vertical mixer equipped with plastic stir blades, place 100 milliliters of 95% ethyl alcohol, followed by ***70 grams of calcium hypophosphite***, and then followed by ***140 grams of lead nitrate***. Thereafter, slowly blend the mixture at room temperature for 40 minutes. After 40 minutes, recover the mixture from the vertical mixer, place onto a shallow pan or tray, and allow to thoroughly air dry. Once it has, add it to a clean suitable vertical mixer, followed by adding in 250 milliliters of hexane, followed by ***100 grams of lead sulfocyante***, followed by ***200 grams of powdered glass***, followed by ***160 grams of barium nitrate*** and then followed by ***330 grams of lead styphnate***. Thereafter, moderately blend the mixture at room temperature for 45 minutes. After 45 minutes, the mixture is ready for use. To use, the mixture should be pressed into molds under a pressure of 3,100 psi, and then cured in an oven at 60 Celsius for 12 hours; the diameter and length of these "molds" is based on the diameter and length of the given munition "casing"; use your own discretion.
**Burn rate (when heated externally, in powder form):** Detonates.
**Burn rate (when heated directly with a flame or hot NiChrome wire):** Detonates.
**Ignition temperature (average temperature to initiate mixture at 3,100):** 199 Celsius.
**Water resistance:** Very good.
**Sensitivity:** Good stability to friction; shock and percussion sensitive; heat resistant (external) up to 120~ Celsius.
**Stability:** Can be stored for many years.
**Flammability:** Detonates.
**Ease of ignition (when heated by a flame or hot wire):** 10
**Tendency to cake:** Will not cake.
**Explosive ability:** Can easily be detonated by touching it with a flame or hot wire. Use extreme caution.
**Percentage:** ***33% lead styphnate, 21% lead calcium hypophosphate/nitrate complex salt, 20% powdered glass, 16% barium nitrate, 10% lead sulfocyanate***
**Classification:** Explosive mixture (classified as Class A1 explosive composition).

**Danger of preparation:** Should only be prepared by experienced pyrotechnicians.
**Use:** Used in non-electric and electric blasting caps and detonators.

**Note:** Inexperienced personnel should not attempt this procedure, as there are hazards involved. Make sure to thoroughly stir the mixture to avoid lumps or clumps, which can lead to excessive heat build-up. This procedure is perfectly safe for experienced persons. To gain experience, preparers should use only small quantities to begin with.

## 06-04-067A: Starter composition for ignition of smoke compositions:

Into a standard ball mill, filled with 400 grams of steel shot of 10mm diameter, place ***401 grams of red lead tetraoxide***, followed by ***80.5 grams of finely ground silicon powder***, and then followed by ***20.2 grams of a binder mixture*** composed of 79% vinylidene and 21% hexafluoropropylene. Thereafter, tumble the mixture at room temperature at 300 RPM for about 30 minutes. Thereafter, the mixture should be removed from the ball mill, and the steel shot separated; thereafter, the mixture is ready for pressing. To do so, press the mixture into blocks, discs, or tablets under a pressure of 9,500 psi.
**Burn rate:** 34 seconds per inch at 9,500 psi; tends to vary based on diameter and density of grain.
**Burn temperature:** 1,121 Celsius.
**Ignition temperature (average temperature to ignite mixture):** 776 Celsius.
**Water resistance:** Very good.
**Stability:** Can be stored for many years.
**Flammability (pressed grain at 9,500 psi):** 3.6
**Ease of ignition (average temperature to ignite mixture):** 5.1

**Tendency to cake:** None.
**Explosive ability:** Cannot be detonated.
**Percentage:** ***79.92% lead tetraoxide, 16.04% ground silicon powder, 4.02% binder, 0.02% mixed balance***
**Classification:** Deflagrating mixture (classified as 1.2G pyrotechnic composition).

**Danger of preparation:** Can be prepared by amateurs.
**Use:** Used to initiate smoke pots, smoke grenades, and other smoke generating devices.

### 06-04-068A: "Electric match" priming composition/device for initiating rocket motors:

Into a suitable sealable vertical mixer, equipped with stainless steel blades, place ***100 grams of carboxymethylcellulose,*** followed by ***40 grams of finely divided zirconium metal***, followed by ***40 grams of fine-grained atomized aluminum,*** followed by ***60 grams of 50/50 magnalium alloy***, followed by ***60 grams of lamp black,*** and then followed by ***320 grams of potassium chlorate***. Thereafter, seal the vertical mixer, and then blend the mixture on high speed for about 45 minutes; thereafter, the mixture is ready. To use the mixture, it needs to be wetted with 10% nitrocellulose laqour to form a thick paste; thereafter, this paste needs to be coated onto a NiChrome wire (bridge wire between negative and positive wires), and then allowed to cure in an oven at 70 Celsius for 24 hours. Note: the thickness of the layer is dictated by the thickness of the NiChrome wire. The thickness of the NiChrome wire is entirely dictated by the rocket motor to be initiated.
**Burn rate (when heated externally, in powder form):** Explosive deflagration (no detonation).
**Burn rate (dried layer when heated directly with a flame or hot NiChrome wire):** Explosive deflagration (no detonation).
**Water resistance:** Very good.
**Sensitivity:** Good stability to friction; shock and percussion insensitive; heat resistant (external) up to 105~ Celsius.
**Stability:** Can be stored for many years.
**Flammability:** Explosive deflagration.
**Ease of ignition (dried layer when heated by a flame or hot wire):** 9.87~
**Tendency to cake:** Will not cake.
**Explosive ability:** Does not detonate, but undergoes explosive combustion.
**Percentage:** ***51.61% potassium chlorate, 16.12% carboxymethylcellulose, 9.67% lamp black, 9.67% magnalium alloy, 6.45% atomized aluminum, 6.45% zirconium metal, 0.03% nitrocellulose***
**Classification:** Deflagrating mixture (classified as 1.4G Class A1 pyrotechnic composition).

**Danger of preparation:** Can be made by amateurs, but caution should be exercised.
**Use:** Used to ignite rocket propellants using electric current.

# References


Jared Ledgard, The Preparatory Manual of explosives, 2nd Edition, ADN, and KDN pages 151 to 153

Jared Ledgard, The Preparatory Manual of explosives, 2nd Edition, Nitrocellulose pages 212 to 220

Jared Ledgard, The Preparatory Manual of explosives, 2nd Edition, Ammonium picramate, pages 308 to 309

Jared Ledgard, The Preparatory Manual of explosives, 2nd Edition, AP, pages 310 to 311

Jared Ledgard, The Preparatory Manual of explosives, 2nd Edition, Sodium picramate, pages 313 to 314

Jared Ledgard, The Preparatory Manual of explosives, 2nd Edition, AN, pages 315 to 319

Jared Ledgard, The Preparatory Manual of explosives, 2nd Edition, AC, pages 319 to 320

Jared Ledgard, The Preparatory Manual of explosives, 2nd Edition, APC, pages 320 to 322

Jared Ledgard, The Preparatory Manual of explosives, 2nd Edition, Nitroguanidine, pages 361 to 362

Jared Ledgard, Kings Chemical Survival Guide, pdf version 1.2, Oct. 2003, "electro preparation 3: sodium chlorate, pages 67 to 73

"Black powder composition", Serial number: 438,186, March 4th, 1930 to Benjamin Austin Gillie, Woodbury, NJ; assigned by E.I. Du Pont de Nemours & Company.

"Slow burning black powder composition", Serial number: 288,135, April 7th, 1919, to Joseph Percy Gray, of Newport, Delaware; assigned by E.I. Du Pont De Nemours & Company.

"Thermoplastic composite rocket propellant", application No. 272,859, June 12, 1981, to: Henry C. Allen, Decatur, AL. assigned by: The United States Army

Application No. 903,000, September 2, 1986, by Morton Thiokol, Inc., to Rodney L. Willer, New London, PA, James A. Hartwell, Elkton, MD, and Robert G. Gleeson, Newark, DE

"Granular propellant", Application No. 964,040, November 27, 1978, by Nippon Oil and Fats Co., Ltd. , to Kazuhiro Inoue, Aichi, JP, and Fumio Matsui, Handa, JP

"Propellant formulations based on dinitramide salts and energetic binders", Application No. 052035, April 21, 1993, by Hinshaw; Carol J. (Ogden, UT); Wardle; Robert B. (Logan, UT); Highsmith; Tom K. (North Ogden, UT), assigned by Thiokol Corporation (Ogden, UT)

"Infrared tracer for ammunition", Application No. 926376, September 9, 1997, by Henry, III; Guy H. (Marion, IL); Owens; Michael A. (Johnson City, IL); Terry; Jarrett L. (Carbondale, IL); Tucker; Mark A. (Buncombe, IL); Bone; Frank M. (Carterville, IL), assigned to Primex Technologies, Inc. (Marion, IL)

"Solid propellant composition", Application No. 161364, July 6, 1971, by Conyers; John A. (Fort Worth, TX); Oberth; Adolph E. (Fair Oaks, CA); Santerre; George M. (Sacramento, CA), assigned to: The United States of America as represented by the Secretary of the Navy (Washington, DC)

"Polyurethane propellant", Application No. 212,762, July 24th, 1962, to Peter L. Nichols, Sacramento, CA, assigned to: The United States Army.

"Composite polyurethane propellants with negative pressure exponent of ammonium sulfate", Serial number: 25,606, March 14th, 1970, to Gilbert A. Zimmerman, Sacramento, Ca, and Marshall L. Sherman, Fair Oaks, Ca, assigned to: Aerojet-General Corporation

"Pyrotechnic cloud seeding composition", Application number: 805854, June 13, 1977, to Slusher; Thomas W. (Louisville, CO), assigned by: Nuclei Engineering, Inc. (Louisville, CO)

"Match-head compositions", Application No. 693898, June 8, 1976, to Lanham; Albert Frank (Maidstone, EN); Barned; Philip Sydney (Welling, EN); Potter; Sydney Ralph (London, EN), assigned to: Bryant and May (London, EN)

"Process for preparing a match head composition containing particulate nitrocellulose and no sulfur", Application No. 654346, February 2, 1976, to Nagatugi; Shuzo (Kobe, JA); Miyagawa; Akira (Himeji, JA); Fukuoka; Suguru (Himeji, JA), assigned by: Daicel Ltd. (Osaka, JA)

"Low light level tracer mix", Serial No. 18.320, March 10th, 1970, to Thomas A. Doris, Jr., Palmyra, NJ, Assigned by the United States Army

"Chemical composition for combustible tip used for friction matches and method for manufacturing same", Application No. 280,596, August 14th, 1972, to Hiroshi Hijikata, Shizuoka, Japan, assigned to: Shizuoka match Co.

"Smoke generating compositions and methods of making the same", application No. 912021, July 24, 2001, to Tadros; Raef M. (White Hall, AR), assigned by: The United States of America as represented by the Secretary of the Army (Washington, DC)

"Castable smoke generating pyrotechnic composition and process for its preparation", application No. 512524, July 11, 1983, to: Sedat; Georges (Saint-Raphael, FR), assigned: Etat Francais represente par le Delegue General pour l'Armement (Paris, FR)

"Smoke composition", application No. 175710, August 6, 1980, to: Krone; Uwe (Hamfelde, DE); Moeller; Klaus (Trittau, DE), assigned by: Nico-Pyrotechnik Hanns-Juergen Diederichs KG. (Trittau, DE)

"Castable infrared illuminant compositions", application No. 478511, January 6, 2000, to: Nielson; Daniel B. (Brigham City, UT), assigned: Cordant Technologies Inc. (Salt Lake City, UT)

"Infrared illuminating composition", application No. 150427, November 9, 1993, to: Dillehay; David R. (Marshall, TX); Turner; David W. (Marshall, TX), assigned: Thiokol Corporation (Ogden, UT)

"Flare composition and flare comprising said composition", application No. 113015, October 23, 1987, to: Singer; Stewart M. (1800 Elsinore Rd., Riverside, CA 92506); Fitzgerald; Stanley (7721 San Diego Ave., Rancho Cucamonga, CA 91730)

"Pyrotechnic composition for cutting torch", application No. 440482, November 9, 1982, to: Kennedy; Katherine L. (White Plains, MD); Proctor; Paul W. (White Plains, MD); Dow; Robert L. (LaPlata, MD), assigned by: The United States of America as represented by the Secretary of the Navy (Washington, DC)

"Simultaneous red smoke and bright flame composition containing ammonium iodate", application: 405648, August 5, 1982, to: Webster, III; Henry A. (Ellettsville, IN), assigned by: The United States of America as represented by the Secretary of the Navy (Washington, DC)

"Green flare compositions", application no. 934659, August 17, 1978, to: Webster, III; Henry A. (Ellettsville, IN), assigned by: The United States of America as represented by the Secretary of the Navy (Washington, DC)

"Simultaneous yellow smoke and yellow flame composition containing bismuth subnitrate", application No. 935423, August 21, 1978, to: Tanner, Jr.; John E. (Bloomington, IN); Webster, III; Henry A. (Ellettsville, IN), assigned by: The United States of America as represented by the Secretary of the Navy (Washington, DC)

"Tracer composition and method of producing same", application No. 533412, June 5, 1990, to: Briere; Paul (Neufchatel, CA), assigned by: Minister of National Defence of Her Majesty's Canadian Government (Ontario, CA)

"Tracer and composition", application No. 829995, September 1, 1977, to: Ramnarace; Jawaharlal (Dana Point, CA), assigned by: Ford Aerospace & Communications Corporation (Dearborn, MI)

"Blue-burning tracer mix", application No. 577508, May 14, 1975, to: Mancinelli; Dennis J. (Philadelphia, PA); Puchalski; Walter J. (Southampton, PA), assigned to: The United States of America as represented by the Secretary of the Army (Washington, DC)

"Improvement in the manufacture of gunpowder", Letters patent No. 79,229, June 23rd, 1868, to: William H. Jackson

"Improvement application of grahamite in the manufacture of gun powder and lamp black", Letters patent No. 87,382, March 2nd, 1868, February 13th, 1869, to: P.H. Vander, New York, NY,

"Improved gun powder", Letters patent No. 97,567, December 7th, 1869, to: Thomas Taylor, of Washington D.C.

"Black powder composition", serial No. 671,700, May 18th, 1933, to: Clarence W. Brooks, Jr., Woodbury, NJ, assigned by: E.I. Dupont de Nemours & Company

"Black powder", serial No. 120,120, January 11th, 1937, to: William H. Rinkenbach, and Vernon C. Allison, both of Dover, NJ, assigned by: The United States by the secretary of war

"Process for preparing black powder pellets", application No. 499,248, August 21st, 1974, to: H. William Voigt, Jr., Stanhope, NJ, Lawrence W. Pell, West Orange, NJ, Jean P. Picard, Morristown, NJ, assigned by: The United States Army

"Pyrotechnic composition", serial No. 772,067, October 28th, 1968, to: Duane M. Johnson, and Donald R. Razelton, both of Winslow IN, assigned by: the United States Navy

"Delay powder", serial No. 586,063, January 11th, 1932, George C. Hale, US Army, Dover, NJ

"Signal flare composition", serial No. 244,479, August 30th, 1951, to: John Vinton, San Francisco, CA

"Flare composition comprising dry blend of metal fuel and eutectic mixture of oxidizer salts", serial No. 866,069, October 1st, 1969, to: Howard W. Kruse, Ridgecrest, CA, assigned to: The United States Navy

"Flare composition", serial No. 78,642, October 6th, 1970, to: George A. Lane, Midland, MI, and William Arthur Smith, Westland, MI, assigned by: Dow Chemical Company

"Thermite compositions for use as gas generants", application No. 103768, August 10, 1993, to: Hinshaw; Jerald C. (Ogden, UT); Blau; Reed J. (Richmond, UT), assigned by: Thiokol Corporation (Ogden, UT)

"Igniter composition comprising a perchlorate, and potassium hexacyanocobaltate (III)", application No. 308,965, November 24th, 1972, to: John W. Fronabarger, of Phoenix, AZ, assigned by: UniDynamics/Phoenix, AZ

"First fire and igniter composition", serial No. 717,275, February 24th, 1958, to: David Hart, Dover, NJ, assigned to: The United Sates Army

"Non Gaseous pyrotechnic delay mixture", serial No. 177,544, August 3rd, 1950, to: Raymond H. Comyn, Silver Spring, Maryland

"Dark burning igniter composition", serial No. 130,584, December 1st, 1949, to: Raymond H. Heiskell, Compton, CA

"Igniting composition and method of preparing same", serial No. 228,962, May 29th, 1951, to: Sidney J. Magram, of Baltimore, Maryland, and John J. Blissel, of Kingsville, Maryland

"Advanced solid propellant for anti-tank propulsion subsystems", serial No. 238,704, March 23, 1972, to: David C. Sayles, of Huntsville, Alabama, assigned by: The United States Army.

"Rocket fuel", application No. 900,911, August 26th, 1986, to: Gerald P. Young, Newark, Delaware, assigned by: The United States Army

"Propellant explosive", serial No. 403,132, October 28th, 1929, to: William M. Dehn, Seattle, Wa

"Gun powder and process of making the same", serial No. 760,730, April 12th, 1913, to: Cantella-Dell McDowell, Pontiac, MI

"Pyrotechnic disseminating formulation", application No. 596,008, November 21st, 1966, to: Earl Thomas Niles, Midland, MI, assigned by: The Dow Chemical Company

"White smoke composition containing red phosphorus", application No. 876,118, November 12th, 1969, to: Bernard E. Douda, Bloomfield, IN, assigned by: The United States Navy

"Colored smoke producing smoke composition", application No. 903,630, August 29th, 1986, to: Gene V. Tracey, of Bel Air, Delaware, Joseph A. Domanica, Edgewood, Delaware, and Gerald P. Young, of Newark, NJ, assigned by: The United States Army

"Propellant compositions", application No. 715,221, March, 22nd, 1985, to: Everett E. Gilbert, Morristown, NJ, assigned by: The United States Army

"Pyrotechnic compound", letters patent No. 590,231, September 21st, 1897, to: Ernst Leusmann, Hanover, Germany

"Deflagrating propellant compositions", application No. 598,850, July 24th, 1975, to: Daniel E. Pawlak, Issaquah, Wa, and Michael Levenson, Cleveland heights, OH

"Pyrotechnic formulation with free oxygen consumption", application No. 395,480, September 10th, 1973, to: Brian K. Hamilton, and Eugene F. Garner, of Saugus, CA, assigned by: Allied Chemical Corporation

"Pyrotechnic composition", application No. 847,389, August 4th, 1969, to: Francis R. Taylor, of Mt. Arlington, NJ, Patricia J. Leader, of Rockaway, NJ, and Ramond P. Westerdahl, of Denville, NJ, assigned by: The United State Army

"Deflagrating composition", serial No. 18,220, March 23rd, 1960, to: Frank Abraham Loving, Jr., Wenonah, NJ, assigned by: E.I. Du pont De Nemours and company.

"Non explosive pyrotechnic composition", serial No. 73,768, April 10th, 1936, to: George C. Hale, Dover, NJ

"Pyrotechnic composition", serial No. 756,260, December 6th, 1934, to: George J. Schladt, Dover, NJ

"Signal smoke compositions", serial No. 749,360, May 20th, 1947, to: John Vinton, of San Francisco, CA, and Douglas M. MacMillan, of Berkeley, CA

"Colored Smoke composition", serial No. 494,694, July 14th, 1943, to: John W. Orelup, Short Hills, NJ

"Chlorate powder", serial No. 277,795, September 9th, 1905, to: Charles E. Mathews, of Rio, Wisconsin

"Combustible", serial No. 55,120, September 8th, 1925, to: Sterner St. P. Meek, of the United States Army

"Combustible", serial No. 82,120, January 18th, 1926, to: Sterner St. P. Meek, of the United States Army

"Flash powder", Letters Patent No. 659,331, October 9th, 1900, to: August Weiss, of Strasburg, Germany

"Improved fuse composition", Letters Patent No. 103,931, June 7th, 1870, to: William H. Rogers, of Brooklyn, NY

"Ignition mixture", serial No. 250,949, October 11th, 1951, to: Thomas J. Mulqueeny, of Dodfrey, ILL, assigned by: Olin Mathieson Chemical Corporation

"Igniter compositions", serial No. 329,111, December 31st, 1952, to: Franklin B. Clay, of Westport, Connecticut, Richard A. Sahlin, of Fairfield, Connecticut, assigned by: Remington Arms Company

"Lower burning rate, reduced hazard, high temperature incendiary", application No. 697252, October 27, 2000, to: Gill; Robert C. (White Plains, MD); Gotzmer; Carl (Accokeek, MD); Carpenter; Pamela (Indian Head, MD); Schlegel; Eric (Indian Head, MD), assigned by: The United States of America as represented by the Secretary of the Navy (Washington, DC)

"Low slag, reduced hazard, high temperature incendiary", application No. 697245, October 27, 2000, To: Gill; Robert C. (White Plains, MD); Gotzmer; Carl (Accokeek, MD); Carpenter; Pamela (Indian Head, MD); Schlegel; Eric (Indian Head, MD), assigned by: The United States of America as represented by the Secretary of the Navy (Washington, DC)

"Thermite destructive device", application No. 606514, to: Song; Eugene (Ellicott City, MD), assigned by: The United States of America as represented by the Secretary of the Army (Washington, DC)

"Low temperature gas generator propellant", application No. 500,810, August 26th, 1974, to: Eugene F. Garner, of Saugus, Ca, assigned by: Allied Chemical Corporation

"Solid propellant formulation based on hydroxylamine perchlorates", serial No. 242,941, December 10th, 1962, to: James P. Flynn, and Ernst J. Strayer, both of Midland, MI, assigned by: The Dow Chemical Company.

"Guanidine and lithium perchlorate salts and eutectics", serial No. 207,471, July 2nd, 1962, to: Joseph J. Bryne, of Boston, Mass, assigned to: Monsanto Research Corporation

"Starting mixture", serial No. 510,789, November 18th, 1943, to: Sidney J. Magram, of Baltimore, Maryland

"Pyrotechnic starter composition", serial No. 612, 149, May 18th, 1932, to: George U. Graff, of Washington DC

"Igniting mixture for tracer compositions", serial No. 670,096, October 22nd, 1923, to: Louis Hendler, of Philadelphia, PA, assigned by: The secretary of war of the United States.

"Near infrared illuminating composition", serial No. 255,333, May 22nd, 1972, to: Carl W. Lohkamp, Bloomfield, IN, assigned to: The United States Navy

"Pyrotechnic compound", serial No. 524,513, October 25th, 1909, to: Bernard J. Dever, and Bertram J. Delzeit, Philadelphia, PA

"Match head formulation", application No. 28,260, March 9th, 1993, to: Michael G. C. Cox, Kingsley England, assigned by: Bryant A May Ltd

"Colored Fire", serial No. 294,489, December 24th, 1888, to: John G. Stuttz, of San Diego, CA

"Pyrotechnic composition", serial No. 758,173, August 27th, 1958, to: Donald E. Olander, Florissant, MO, assigned by: Universal Match Corporation

Smokeless ashless signal flare composition containing ammonium perchlorate", serial No. 385,518, July 27th, 1964, to: James B. Mcgriffin, of Linton, IN, and William L. Ripley, of Bedford, IN, assigned by: The United States Navy

"Metal-cutting pyrotechnic composition", application No. 189410, September 22, 1980, to: Mueller; Kurt F. (Oxon Hill, MD); Chang; Marguerite S. (Oxon Hill, MD); Farncomb; Robert F. (Accokeek, MD); Malotky; Lyle O. (Oxon Hill, MD), assigned by: The United Stated of America as represented by the Secretary of the Navy (Washington, DC)

"Floatable incendiary composition", application No. 590370, June 25, 1975, to: Schroeder; Fred (Lakeview Terrace, CA), assigned by: Dow Corning Corporation (Midland, MI)

"Incendiary compositions of magnesium and fluoroalkyl phosphate esters", application No. 577199, May 14, 1975, to: Shaw; Graham C. (Garland, UT); Reed, Jr.; Russell (Ridgecrest, CA), assigned by: Thiokol Corporation (Newtown, PA)

"Pyrotechnic thermite composition", application No. 062500, February 5, 2002, to: Carter, Jr.; Greg (P.O. Box 3355, Joplin, MO 64803); Carter, Sr.; Greg (P.O. Box 3355, Joplin, MO 64803)

"Gas generant composition with coolant", application No. 798640, March 2, 2001, to: Knowlton; Gregory D. (Chandler, AZ); Ludwig; Christopher P. (Fountain Hills, AZ), assigned by: Talley Defense Systems, Inc. (Mesa, AZ)

"Thermally stable nonazide automotive airbag propellants", application No. 851503, May 5, 1997, to: Khandhadia; Paresh S. (Troy, MI); Burns; Sean P. (Ann Arbor, MI), assigned by: Automotive Systems Laboratory, Inc. (Farmington Hills, MI)

"Gas generating compositions", application No. 114196, July 13, 1998, to: Knowlton; Gregory D. (Chandler, AZ); Ludwig; Christopher P. (Fountain Hills, AZ), assigned by: Talley Defense Systems, Inc. (Mesa, AZ)

"High velocity gun propellant", application No. 262452, June 20, 1994, to: Williams; Laurence O. (Colchester, VT), assigned by: Martin Marietta Corporation (Bethesda, MA)

"Pyrotechnic composition for producing radiation-blocking screen", application No. 639685, April 4, 1984, to: Simpson; Geoffrey M. (Nottingham, GB), assigned by: Haley & Weller Limited (Draycott, GB)

"Smoke-producing pyrotechnic composition and its application", application No. 169204, July 15, 1980, to: Sedat; Georges A. (16h rue Jonyh-Prirrugue, 83700 St. Raphael, FR)

"Pyrotechnic disseminating formulations", application No. 853609, August 21, 1969, to: Flynn; James P. (Midland, MI), assigned by: The Dow Chemical Company (Midland, MI)

"Smoke-generating composition", application No. 817896, July 21, 1977, to: Johnson; Duane M. (Loogootee, IN); Webster, III; Henry A. (Ellettsville, IN), assigned by: The United States of America as represented by the Secretary of the Navy (Washington, DC)

"Pyrotechnical mixture for producing a smoke screen", application No. 350737, April 26, 1989, PCT NO:, PCT/EP88/00378, April 26, 1989, to: Krone; Uwe (Hamfelde, DE), assigned by: Rheinmetall GmbH (Dusseldorf, DE)

"Pyrotechnical composition which generates smoke that is opaque to infrared radiance and smoke ammunition as obtained", application No. 517939, July 27, 1983, to: Espagnacq; Andre (Bourges, FR); Sauvestre; Gerard D. (Bourges, FR), assigned by: Etat Francais (Paris, FR)

"Smoke projectile charge and process for its manufacture", application No. 752637, December 8, 1976, to: Huber; Paul (Haus Nr. 9, 8234 Unterjettenberg, DE); Schiessl; Alois (Hochfeldstrasse 26, 8231 Weissbach-Marzoll, DE); Krone; Hartmut (Bertholdstrasse 4, 7844 Neuenburg, DE)

"Pyrotechnic smoke charge containing guanidine", application No. 036423, May 7, 1979, to: Prahauser; Georg (Oberalm, AT); Schiessl; Alois (Marzoll, DE), assigned by: Buck Chemisch-Technische Werke GmbH & Co. (Reichenhall, DE)

"Combustible compositions for generating aerosols, particularly suitable for cloud modification and weather control and aerosolization process", application No. 362680, May 22, 1973, Papee; Henry M. (Rome, IT); Montefinale; Alberto C. (Rome, IT); Petriconi; Gianna L. (Rome, IT); Zawidzky; Tadeusz W. (Ottawa, CA), Consiglio Nazionale delle Richerche (Rome, IT)

"Method for opaquing visible and infrared radiance and smoke-producing ammunition which implements this method", application No. 517940, July 27, 1983, to: Espagnacq; Andre (Bourges, FR); Sauvestre; Gerard D. (Bourges, FR), assigned by: Etat Francais (Paris, FR)

"Pyrotechnic smoke charge with preset breaking points and channel igniter", application No. 293909, August 18, 1981, to: Weber; Manfred (Eisenberg, DE); Hinzmann; Friedmar (Eisenberg, DE), assigned by: Pyrotechnische Fabrik F. Feistel GmbH & Co., KG (Gollheim, DE)

"White smoke spotting composition for training ammunition", application No. 399601, July 19, 1982, to: Knapp; Charles A. (Wayne, NJ), assigned by: The United States of America as represented by the Secretary of the Army (Washington, DC)

"*Pyrotechnic* composition for producing IR-radiation", application No. 339999, January 10, 2003, to: Koch; Ernst-Christian (Kaiserslautern, DE), assigned by: Diehl Munitionssysteme GmbH & Co. KG (Rothenbach, DE)

"Gun powder substituted composition and method", application No. 437,965, November 1st, 1982, to: Earl F. Kurtz, of Longmont, CO

"Pyrotechnic composition containing nickel or aluminum", serial No. 726,548, May 3rd, 1968, to: Horace H. Helms, Jr., of Silver Spring Md, and Alexander G. Rozner, of Bethesda, Md, assigned by: The United States Navy

"Aluminothermic reaction mixture based on copper oxide and iron oxide", application No. 471, 757, May 20th, 1974, to: Theodore Finster, of Hattinfgen Germany, and Horst Schumann, of Mulheim, Germany, assigned by: Th. Goldschmidt AG,

"Welding composition", serial No. 850,105, November 2nd, 1959, to: Rolf E. Rylander, of Cleveland OH, and Lynn J. Ebert, of Euclid, OH, assigned by: Erico Products, Inc.

"Heat producing mixtures", serial No. 3,262, January 19th, 1960, to: Albert Stanley, of John Cross, Birmingham, England, assigned by: Foundry Services International Limited

"Thermite mixture", serial No. 510,372, October 25th, 1921, to: Glenn Walter Merrefeild, of Johnstown, PA

"Thermic mixture", serial No. 301,096, February 14th, 1906, to: Hans Goldschmidt, of Essen-on-the-Ruhr, Germany, assigned by: The Firm of the Goldschmidt

"Long-burning pyrotechnic material containing depleted uranium for spotting rifle projectiles", serial No. 167,449, January 19th, 1962, to: Thomas Stevenson, of Huntingdon Valley, PA, assigned by: The United States Army

"Fire Cracker", serial No. 188,362, August 27th, 1917, to: Thomas G. Hitt, of Seattle, Wa

"Consolidated thermite compositions", application No. 533112, September 25, 1995, to: Taylor; Robert D. (Hyrum, UT), assigned by: Morton International, Inc. (Chicago, IL)

"Composition Causing Combustion When Contacted With Water", application No. 825,059, May 15th, 1969, to: Karl Schneider, of Basel Switzerland, assigned by: Scheidemandel-Motard-Werke AG,

"Pyrotechnic composition for delay elements", application No. 238787, February 27, 1981, to: Schulte; Hans (Siegburg, DE); Voges; Christoph (St. Augustin, DE), assigned by: Dynamit Nobel Aktiengesellschaft (Troisdorf, DE)

"Tracer Composition", serial No. 714,919, May 21st, 1924, to: Sterner St. P. Meek, of Philadelphia, PA

"Tracer Composition", serial No. 715,457, May 23rd, 1924, to: Henry C. Pritham, of Upper Derby, PA

"Incendiary Mixture", serial No. 343,571, December 9th, 1919, to: Walter T. Scheele, of Hackensack, NJ, assigned by: To The United States Navy.

"Ignitable compositions", application No. 073649, September 18th, 1970, to: Robert J. Benge, of Waltham Abbey, Great Britain, and Cecil H. Miller, of Woodford Green, England, assigned by: Ministry of technology in Her Britannic Majesty's Government of the United Kingdom.

"Incendiary Mixture", serial No. 343, 570, December 9th, 1919, to: Walter T. Scheele, of Hackensack, NJ, assigned by: The United States Navy.

"Propellant powder", serial No. 667,359, October 8th, 1923, to: Charles R. Franklin, of Dover NJ

"Granular Propellant Containing KNO3 and Unsaturated Ester", application No. 304,948, November 9th, 1972, to: Herbert Leslie Wood, of West Kilbridge, Scotland, assigned by: Imperial Chemical Industries Limited.

"Ignition Mixture", serial No. 250,949, October 11th, 1951, to: Thomas J. Mulqueeny, of Godfrey, ILL, assigned by: Olin Mathleson Chemical Corporation

"Phosphorus-containing compositions", application No. 356,888, May 3rd, 1973, to: Malcolm George Palmer, and Anthony Donald Bal, both of Coseley, England, assigned by: Albright & Wilson Limited

"Fuze powder composition", serial No. 142,721, February 6th, 1950, to: David Hart, of Dover, NJ

"Flare composition", serial No. 41,746, July 31st, 1948, to: Herbert C. Clauser, and Robert S. Long, both of Westerville, OH, assigned by: Mesne assignments

"Incendiary with high penetrating power", serial No. 452, 079, July 23rd, 1942, to: Albert Grobstein, of Washington DC

"Pyrotechnic composition and device containing such composition", application No. 451150, March 14th, 1974, to: Jean Rene Duguet, of Sarniguet, France, and Marie Louise Sanmartin, of Semeac, France.

"Dry exothermic composition containing fibrous material having oxidizer salt absorbed therein", serial No. 292,715, July 3rd, 1963, to: Edward John Jago, of Cleveland, OH, assigned by: Foseco Trading A.G.

"Pyrotechnic composition for generating lead based smoke", application No. 773,694, October 28th, 1968, to: Duane M. Johnson, of Bicknell, IN, and Donald R. Hazelton, of Winslow, IN, assigned by: The United States Navy

"Ammonium nitrate propellant compositions", application No. 797,697, February 7th, 1969, to: Lionel A. Henderson, of Columbus, Ohio, assigned by: Standard Oil Company, Chicago, Ill.

"Pyrotechnic composition", serial No. 772,067, October 28th, 1968, to: Duane M. Johnson, Bicknell, and Donald R. Razelton, both of Winslow, IN, assigned by: The United States Navy

"Pyrotechnic composition for colored smoke production", serial No. 63,033, August 11th, 1970, to: Frank C. Gunderloy, of Santa Susana, CA, Charles L. Hamermesh, of Tarzana, CA, and Ross I. Wagner, of Woodland Hills, CA, assigned by: North American Rockwell Corporation

"Flare composition", serial No. 78,642, October 6th, 1970, to: George A. Lane, of Midland MI, and William Arthur Smith, of Westland, MI, assigned by: The Dow Chemical Company

"Solid propellant containing cross-linked diamine-terminated polyglycol", serial No. 480,840, August 16th, 1965, to: Don L. Stevens, of Sanford MI, and Harold E. Filter, of Midland, MI, assigned by: The Dow Chemical Company

"Pyrotechnic disseminating composition containing an agent to be disseminated", serial No. 664,919, September 1st, 1967, to: Gary E. Wernette, of Weidman MI, and Erwin M. Jankowiak, of Midland, MI, assigned by: The Dow Chemical Company

"Illuminating flare composition composed of magnesium, sodium nitrate, and an epoxy resin-polyglycol resin binder", serial No. 657,726, July 31st, 1967, to: Bernard E. Douda, of Bloomfeild, IN, assigned by: The United Sates Navy

"Re-ignitable Match", serial No. 439,947, March 15th, 1965, to: John Burton Tigrett, of Jackson, Tenn, and Voorhis F. Wigal, of Jackson, Tenn, assigned by: said Tigrett

"Calcium containing pyrotechnic compositions for high altitudes", serial No. Seymour Lopatin, of Dover, NJ, and David Hart, of Dover, NJ, by Ruth Hart, of Dover, NJ, assigned by: The United States Army

"Solid propellant", serial No. 181,794, March 22nd, 1962, to: Theodore C. Parker, of Sunnyvale CA, and William H. Colburn, Jr., of Irvington, CA, assigned by: General Precision, Inc.

""Polymeric base propellant compositions containing lithium fluoride catalyst", serial No. 374,214, June 10th, 1964, to: Marvin Duane Horton, of Provo Utah, assigned by: The United States Navy

"Solid combustible composition containing epoxy resin", serial No. 151,165, November 9th, 1961, to: Malcolm E. Pruitt, Wallace T. McMichael, and Elmer L. Pendleton, all of Lake Jackson, TX, assigned by: The Dow Chemical Company

"In situ dye smoke", application No. 057888, May 22, 1987, to: Reed, Jr.; Russell (Ridgecrest, CA); Ramnarace; Jawaharlal (Fort Washington, MD), assigned by: The United States of America as represented by the Secretary of the Navy (Washington, DC)

"Solid propellant composition with a polyurethane binder", application No. 791,540, January 10th, 1969, to: Adolf E. Oberth, of Fair Oaks CA, and Rolf S. Bruenner, of Orangevale, CA, assigned by: Aerojet-General Corporation

"Composite modified double base propellant with metal oxide stabilizer", application No. 362,980, May 23rd, 1973, to: Dominic A. Berta, of Wilmington, Del., assigned by: The United Sates Navy

"Modified double base propellants with diisocyante crosslinker", application No. 202,355, June 8th, 1962, to: Donald E. Elrick, assigned by: The United States Navy

"High solids ratio solid rocket motor propellant grains and method of construction thereof", application No. 165641, March 8, 1988, to: Hightower, Jr.; James O. (Huntsville, AL); Sato; Tomio (Huntsville, AL); Hamner; James W. (Huntsville, AL); Rector; Carl M. (Brownsboro, AL), assigned by: Morton Thiokol, Inc. (Chicago, IL)

"High performance, low cost solid propellant compositions producing halogen free exhaust", application No. 531728, June 1, 1990, to: Doll; Daniel W. (N. Ogden, UT); Lund; Gary K. (Ogden, UT), assigned by: Thiokol Corporation (Ogden, UT)

"Polyurethane propellant", application No. 212,762, July 24th, 1962, to: Peter L. Nichols, Jr., of Sacramento, Ca, assigned by: The United Sates Army

"Composite propellant containing molybdenum to reduce electromagnetic radiation interference in combustion products", application No. 657,001, July 31st, 1967, to: Howard G. Cutforth, of Los Altos, CA, Clarence A. Le Febvre, and Melvin E. Steinle, both of San Jose, CA, assigned by: United Aircraft Corporation

"Ignition composition", application No. 870674, May 22, 1986, to: Mackenzie; Gerald L. (Port Tobacco, MD); Petri; Robert O. (La Plata, MD); Rose; James E. (Bryans Road, MD), assigned by: The United States of America as represented by the Secretary of the Navy (Washington, DC)

"Pyrotechnic composition", application No. 323436, November 20, 1981, to: Olander; Donald E. (Huntington Beach, CA); Petersen; Donald W. (Tehachapi, CA), assigned by: Hi-Shear Corporation (Torrance, CA)

"High density impulse solid propellant", application No. 099967, March 31, 1961, to: Kaufman; Martin H. (China Lake, CA); Stang; Peter L. (China Lake, CA), assigned by: The United States of America as represented by the Secretary of the Navy (Washington, DC)

"Compositions for producing flickering signals", application No. 433645, January 16, 1974, to: Cornia; Richard P. (Logan, UT); Reed, Jr.; Russell (Ridgecrest, CA), assigned by: Thiokol Corporation (Newton, PA)

"Propellant pressure burning rate slope modification", application No. 293,258, September 25th, 1972, to: Martin H. Kaufman, of China Lake, CA, assigned by: The United States Navy

"Gas generating compositions", application No. 187797, September 16, 1980, to: Jackson, Jr.; Charles H.

"Manufacture of match-striking compositions", serial No. 89,223, February 18th, 1926, to: Herbert William Robinson, of West Bromwich, England

"Self-hardening pyrotechnic composition", serial No. 121,216, October 13th, 1949, to: David Hart and Hentry J. Eppig, both of Dover, NJ, and William J. Powers, of Wharton, NJ

"Pyrotechnic masses", serial No. 490,063, February, 23rd, 1955, to: Ronald Gibson Hall, of Hemel Hempstead, England, assigned by Brock's Crystals Palace Fireworks Limited

"Propellant powder", serial No. 231,648, June 14th, 1951, to: Don W. Ryker, of Alton, ILL, and Gilbert R. Cox, both of Creve Coeur, MO, assigned by: Olin Mathieson Chemical Corporation

"Solid composite propellant containing polysulfide rubber fuel binder", serial No. 714,127, December 5th, 1946, to: Charles E. Bartley, of Altadena, CA, assigned by: California Institute Research Foundation

"Solid composite propellants containing lithium perchlorate and polyamide polymers", serial No. 719,502, March 6th, 1958, to: Ross M. Hedrick, and Edward H. Mottus, both of Dayton, OH, assigned by: Monsanto Chemical Company

"Solid composite propellants with burning rate catalyst", serial No. 638,176, February 4th, 1957, to: Willard M. St. John, Jr., of Waco, TX, assigned by Phillips Petroleum Company

"Pyrotechnic compound", serial No. 507,306, April 12th, 1894, to: John Agostini, of NY, NY

"Fireworks and method of making same", serial No. 848,979, July 3rd, 1914, to: Aaron Jedel, of New York, NY

"Manufacture of solid propellants", serial No. 475,000, December 13th, 1954, to: Howard, G. Cutforth, of Bartlesville, OK, assigned by: Phillips Petroleum Company

"Composite rubber base propellants and process of preparation employing non-sulfur metallic curing agents", serial No. 622,741, November 19th, 1956, to: Eugene D. Guth, of Bartlesville, OK, assigned by: Phillips Petroleum Company

"Propellant and gas producing compositions of elastic gels containing inorganic oxidizing salts", serial No. 529,342, August 18th, 1955, to: Harry R. Ferguson of Morrisville, PA, assigned by: Thiokol Chemical Corporation

"Solid rocket propellants containing polylactams", serial No. 719,501, March 6th, 1958, to: Ross M. Hedrick and Edward H. Mottus, both of Dayton, OH, assigned by: Monsanto Chemical Company

"Gunpowder", serial No. 560,699, August 27th, 1895, to: Edward Dickson, of Oak Lake, Canada

"Auto ignition system for airbag inflator", application No. 739582, October 30, 1996, to: Seeger; Donald Edwin (Lakeland, FL), assigned by: Breed Automotive Technology, Inc. (Lakeland, FL)

"Ignition compositions for inflator gas generators", application No. 405384, March 15, 1995, to: Khandhadia; Paresh S. (Rochester Hills, MI), assigned by: Automotive Systems Laboratory, Inc. (Farmington Hills, MI)

"Electric primer with intrinsic conductive mix", application No. 348440, May 8, 1989, to: Mei; George C. (Creve Coeur, MO), assigned by: Olin Corporation (Stamford, CT)

"Tracer mixture for use with laser hardened optics", application No. 358290, December 19, 1994, to: Doris; Thomas A. (Sparta, NJ); Martin; Raymond (Tallahassee, FL), assigned by: The United States of America as represented by the Secretary of the Army (Washington, DC)

"Torch fuse-lighter", serial No, 467,130, May 5th, 1921, to: Evelyn Paul Aurand, of Tamaque, PA, assigned by: Atlas Powder Company

"Propellant powder", serial No. 667,359, October 8th, 1923, to: Charles R. Franklin, of Dover, NJ

"Pyrotechnic compound", serial No. 612,141, November 14th, 1896, to: Ernst Leusmann, of Hanover, Germany

"Composition for detonators", serial No. 699,329, December 15th, 1998, to: George Percy Bickford Smith, of Helstone, England

"Detonating composition", serial No. 84,463, December 2nd, 1901, to: Hans Von Dahmen, of Vienna Austria-Hungary

"Smoke producing compound", serial No. 135,943, December 9th, 1916, to: Walter Arthur, of Philadelphia, PA

"Composition for railway signal torpedoes", serial No. 415,264, October 7th, 1920, to: Louis S. Ross, of Newtonville, MA, assigned by: Central Railway Signal Company

"Combustible composition", serial No. 82,120, January 18th, 1926, to: Sterner St. P. Meek, of Philadelphia, PA

"Non explosive pyrotechnic composition", serial No. 73,768, April 10th, 1936, to: George C. Hale, of Dover, NJ

"Production of chromium-bearing reaction mixtures and alloys", serial No. 301,375, October 26th, 1939, to: Marvin J. Udy, of Niagara, NY

"Metal chloride screening smoke mixture", serial No. 558,610, October 13th, 1944, to: John C. Bailar, Jr., and Robert W. Parry, both of Urbana, ILL., assigned by: The United States Navy

"Pelleted igniter composition and method of manufacture", serial No. 592,995, June 21st, 1956, to: Liles G. Herring, of McGregor, TX, assigned by: Phillips Petroleum Company

"Exothermic compositions containing perlite", serial No. 204,633, June 22nd, 1962, to: John P. Walsh and Harold F. Bishop, both of Conneaut, OH, assigned by: Exomet Incorporated

"Gasless ignition composition for solid rocket propellants", serial No. 167,193, January 18th, 1962, to: Barnet R. Adelman, of Atherton, CA, assigned by: United States Army

"Smoke-making compound for signal-rockets", serial No. 292,353, April 24th, 1919, to: Robert Clyde Gowdy, of Cincinnati, OH, assigned by: Government of the United States

"Chemical composition for producing heat", serial No. 139,613, September 21st, 1961, to: Albert T. Camp and Gerald L. MacKenzie, both of Redlands, CA, assigned by: mesne assignments, to Lockheed Aircraft Corporation.

"Solid rocket propellant composition", serial No. 566,103, February 17th, 1956, to: Joe M. Burton, of McGregor, TX, assigned by: Phillips Petroleum Company

"Composite propellants containing activated carbon", serial No. 801,392, March 23rd, 1959, to: Myron L. Dickey, of Whittier, CA, assigned by: Phillips Petroleum Company

"Solid composite rubber base ammonium nitrate propellant cured with metal oxide", serial No. 574,042, March 26th, 1956, to: Barney W. Williams, of McGregor, TX, David R. Smith, of Bartlesville, OK, and Wallace T. McMichael, of Lake Jackson, TX, assigned by: Phillips Petroleum Company.

"Improvement in gunpowder", serial No. 17,321, May 19th, 1857, to: Lammot Du Pont, of Wilmington, Delaware

"Improved composition for friction-matches", serial No. 50,370, October 10th, 1865, to: Louis Lanszweert, of San Francisco, CA

"Gas generating compositions", serial No. 504,117, April 26th, 1955, to: Paul O. Marti, Jr., of Hammond, IN, assigned by: Standard Oil Company

"Flash-lite composition", serial No. 265,806, to: Arthur Eichengrun, of Elberfeld, Germany, assigned by: Farbenfabriken, of Elberfeld, Co.

"Thermic mixture", serial No. 301,096, February 14th, 1906, to: Hans Goldschmidt, of Essen-on-the-ruhr, Germany, assigned by: The firm of the gold Schmidt,

"Manufacture of thermic mixtures", serial No. 294,971, January 6th, 1906, to: Hans Goldschmidt, of Essen-on-the-ruhr, Germany, assigned by: The firm of the goldschmidt

"Combustible mixture for producing actinic light", serial No. 446,060, July 30th, 1908, to: Carl Bethge, of Berlin Germany

"Composition of matter for the generation of heat", serial No. 262,183, May 25th, 1905, to: Frank J. Tone, of Niagara Falls, NY, assigned by: The Carborundum Company

"Composition for match-heads", serial No. 54,310, October 6th, 1915, to: William A. Fairburn, of Short Hills, NJ, assigned by: The Diamond Match Company

"Pyrotechnic composition", serial No. 461,955, April 16th, 1921, to: Robert R.Fulton, of Indian Head, IN

"Thermite mixture", serial No. 510,372, October 25th, 1921, to: Glenn Walter Merefield of Johnstown, PA

"Igniting mixture for tracer compositions", serial No. 670,096, October 22nd, 1923, to: Louise Hendler, of Philadelphia, PA, assigned by: Secretary of War, The United States of America

"Delay powder", serial No. 586,063, January 11th, 1932, to: George C. Hale, of The United States Army, of Dover, NJ

"Primer composition", serial No. 258,724, February 27th, 1939, to: George U. Graff, of Washington, DC

"Delay compositions for electric blasting caps", serial No. 25,366, May 6th, 1948, to: Hugh Mowat Kerr, of Brownsburg, Quebec, Canada, and Cecil R.L. Hall, of Saltcoats, Scotland, assigned by: Canadian Industries Limited

"Ignition composition", serial No. 62,878, December 1st, 1948, to: Charles Philip Spaeth, of Woodbury, NJ, and Charles Pierce Williams, Of Pitman, NJ, assigned by: E.I. du Pont de Nemours & Company

"Igniting composition and method of preparing same" serial No. 228,962, May 29th, 1951, to: Sidney J. Magram, of Baltimore MD, and John J. Blissel, of Kingsville, MD

"Dark burning igniter composition, serial No. 130,584, December 1st, 1949, to: Raymond H. Heiskell, Compton, CA

"Delay composition for delay electric detonators", serial No. 281,029, April 7th, 1952, to: Idwal O. Lewis, of Ardrossan, Scotland, assigned by: Imperial Chemical Industries Limited

"Improvement in percussion-powder", Letters Patent No. 18,016, August 18th, 1857, to: Magnus Kling, of Reading, PA

"Composition for Bengal lights", letters patent No. 323,662, August 4th, 1885, to: Charles Gerhard, of New York, NY

"Solidifying colored fire", serial No. 311,992, May 24th, 1889, to: Hermann O. Frank, of Milwaukee, WI

"Exothermic reaction mixture for producing a molten copper alloy", serial No. 331,634, January 16th, 1953, to: Donald J. Burke, of Cleveland, OH, assigned by: Erico Products, Inc.

"Delay compositions", serial No. 574,663, March 29th, 1956, to: Hugh M. Kerr, of Cartierville, Quebec, Canada, and Gordon Towell, of Brownsburg, Quebec, Canada, assigned by: Canadian Industries Limited

"Hydrogen gas generating propellant compositions", serial No. 394,481, November 25th, 1953, to: John A. Fitzpatrick, of Washington DC, assigned by: Airtronics Inc.

"Delay fuse", serial No. 631,468, December 31st, 1956, to: David T. Zebree, of Kingston, NY, assigned by: Hercules Powder Company

"Improvement in preparation of granulated gunpowder to serve as charges for fire-arms", letters patent No. 33,069, August 20th, 1861, to: John Harcourt Brown, of Abbey Mill House, Romsey, England

"Composition for tracer unit", serial No. 35,960, June 29th, 1948, to: Raymond H. Heiskell, of Compton, CA

"Propellant system", serial No. 719,116, March 4th, 1958, to: Thomas L. Fritzlen, of Henrico County, VA, Assigned by: Reynolds, Metals Company

"Material for welding aluminum and other metals" serial No. 503,569, April 25th, 1955, to: Melvin F. Rejdak, of Cleveland, OH, assigned by: Erico Products, Inc.

"Delay compositions for delay electric detonators", serial No. 147,146,October 24th, 1961, to: Thomas Z. Ball, of New Ringgold, PA, and William D. Trevorrow, of Tamaqua, PA, assigned by: Atlas Chemical Industries, Inc.

"Delay compositions for delay electric detonators", serial No. 147,147, October 24th, 1961, to: Thomas Z. Ball, of New Ringgold, PA, and William D. Trevorrow, of Tamaqua, PA, assigned by: Atlas Chemical Industries, Inc.

"Gasless igniter composition", serial No. 319,890, October 29th, 1963, to: Earl E. Kilmer, of College Park, MD, assigned by: United States Navy

"Flameproof delay composition with a delayed-action time in the milliseconds range", serial No. 395,047, September 8th, 1964. to: Helmut Menke, of Troisdorf, Germany, assigned by: Dynamite Nobel Aktiengesellschaft

"Pyrotechnic composition", serial No. 49,980, June 25th, 1970, to: Horace H. Helms, Jr., of Silver Springs MD, and Alexander G. Rozner, of Bethesda, MD, assigned by: United States Navy

"Propellant powder", serial No. 721,203, June 20th, 1924, to: Arthur S. O'Neil, of Alton, IL, assigned by: Western Cartridge Company

"Progressive burning smokeless powder", serial No. 179,764, December 14th, 1937, to: Lewis C. Weldin, of Kenvil, NJ, assigned by: Hercules Powder Company

"Priming composition", serial No. 403,535, July 22nd, 1941, to: Willi Brun, of Bridgeport, Conn., assigned by: Remington Arms Company

"Conductive priming mixture", serial No. 740,170, June 5th, 1958, to: Thomas Q. Ciccone, of Langhorne, PA, assigned by: The United States Army

"Composite solid propellant containing preshaped (salt – metal – burning rate controller) particles", serial No. 698,389, January 17th, 1968, to: Kiroku Yamazaki, of Tokyo, Japan, and Kazuo Kishi, of Saitama-ken, Japan

"Aluminum-iron oxide incendiary composition containing a (cyclopentadienyl) iron compound", serial No. 600,006, December 6th, 1966, to: Urho Albert Lehikoinen, of Detroit, MI, assigned by: Ethyl corporation

"Detonating firework composition", serial No. 259,377, March 5th, 1928, to: William F. Gehrig, of Berkley Heights, NJ

"Rocket propellant", serial No. 477,294, December 23rd, 1954, James R. Eiszner, of Park Forest, ILL., and William G. Stanley, of Hammond, IN, assigned by: Standard Oil Company

"Improved composition for filling shells", letters patent No. 38,424, May 5th, 1863, to: Levi Short, of Philadelphia, PA

"Fireworks", serial No. 413,579, November 30th, 1891, to: Carl Schmidt, of Berlin, Germany, assigned by: Gustav Gillischewski

"Explosive compound", serial No. 170,469, August 22nd, 1903, to: Charles M. Hall, of Niagara Falls, NY

"Process for making alloys", serial No. 736,343, July 21st, 1934, to: William C. Read, of New Rockelle, NY, assigned by: Electro Metallurgical Company

"Pyrotechnic composition", serial No. 689,600, September 15th, 1933, to: Samuel Wiley, of Metuchen, NJ

"Flash powder and method of preparation", serial No. 350,081, August 2nd, 1940, to Clifford A. Nickle, of Schenectady, NY, assigned by: General Electric Company

"Pyrotechnic compositions", serial No. 224,303, August 11th, 1938, to: George J. Schladt, of Dover, NJ, assigned by: The United States army

"Incendiary material", serial No. 466,658, November 23rd, 1942, to: David L. Woodberry, of The United States Army, William W. Howerton, of Bel Air, MD, and Alver Dunbar, of Teaneck, NJ, assigned by: The United Sates Army

"Metallurgy", serial No. 617,447, September 19th, 1945, to: Marvin J. Udy, of Niagara Falls, NY, assigned by: Guaranty Investment Corporation Limited

"Exothermic alloy addition agent", serial No. 419,096, March 26th, 1954, to: Joseph H. Brennan, of Niagara Falls, NY, and Earle R. Saunders, of Grand Island, NY, assigned by: Union Carbide Corporation

"Fuel comprising a mixture of ammonium nitrate and metal particles", serial No. 714,942, February 13th, 1958, to: Harry A. Toulmin, Jr., of Dayton, OH, assigned by: The Commonwealth Engineering Company

"Exothermic compositions containing zinc and ammonium chloride", serial No. 342,287, February 3rd, 1964, to: John E. Claiborne, of Opelousas, LA

"Solid fuel composition", application No. 515,564, June 14th, 1955, to: Seymour A. Genden, of The United States Army

"Thermoplastic composite rocket propellant", application No. 272859, June 12, 1981, to: Allen; Henry C. (Decatur, AL), assigned by: The United States of America as represented by the Secretary of the Army (Washington, DC)

"Projectile containing pyrotechnic composition for reducing base drag thereof", application No. 424,508, December 13th, 1973, to: Walter J. Puchalski, of Southampton, PA, assigned by: The United States Army

"Improved compound for priming electric fuses", letters patent No. 96,465, November 2nd, 1869, to: George M. Mowbray, of Titusvilled, PA

"Match composition", serial No. 686,357, July, 19th, 1898, to: Henri Sevene and Emile David Cahen, of Paris France

"Igniting composition and method of making same", serial No. 136,470, December 24th, 1902, to: Otto Dieffenbach, and Adolf Schafer, of Griesheim-on-the-main, Germany, assigned by: The Firm of Chemische Fabrik Griedheim Electron

"Match-head or similar composition", serial No. 15,072, March 17th, 1915, to: Henry Staier, of New York, NY

"Detonating firework composition", serial No. 238,467, December 7th, 1927, to: William F. Gehrig, of Berkeley Heights, NJ, assigned by: Essex Specialty Co., Inc.

"Detonating firework composition", serial No. 346,498, March 12th, 1929, to: William F. Gehrig, of Berkeley Heights, NJ, assigned by: Essex Specialty Co., Inc.

"Detonating firework composition", serial No. 259,378, March 5th, 1928, to: William F Gehrig, of Berkeley Heights, NJ

"Pyrotechnic device", serial No. 176,833, March 19th, 1927, to: Hans Reihlen, of Karlsruhe, of Germany, assigned by: To Milton M. Adler

"Percussion detonating composition", serial No. 573,524, November 6th, 1931, to: Andre Blanc, of Salaise-sur-Sanne, France

"Pyrotechnic device", serial No. 58,143, January 8th, 1936, to: Jacob D. Neuss, of Woodbridge, NJ, assigned by: Essex Specialty Co

"Explosives", serial No. 593,971, May 15th, 1945, to: Daniel P. O'Brien, of Stamford, Conn.

"Infra-red tracking flare", application No. 757645, October 21, 1997, to: Hahn; George T. (China Lake, CA); Rivette; Paul G. (Baton Rouge, LA); Weldon; Rodney G. (China Lake, CA), assigned by: The United States of America as represented by the Secretary of the Navy (Washington, DC)

"Igniting composition for matches", serial No. 328,565, July, 31st, 1906, to: Louis Stange, of Aachen, Germany, assigned by: The Chemical Industrial Syndicate Limited

"Aluminum ophorite", serial No. 10,982, February 21st, 1925, Donald B. Bradner, of Philadelphia, PA

"Incendiary composition", serial No. 736,932, May 21st, 1958, to: Thomas Stevenson, of Huntington Valley, PA, assigned by: United States Army

"Solid propellant compositions", serial No. 489,732, February 18th, 1955, to: Horace M. Higgins and Calvin A. Gongwer, of Glendora, CA, assigned by: Aerojet-General Corporation

"High temperature ignition composition of tellurium dioxide, aluminum, and carbon black", serial No. 671,044, to: Hubert G. Timmermans, of Manhattan Beach, CA, assigned by: Space Ordnance Systems, Inc.

"Filler rich powder and method of making", serial No. Richard L. Lou, of Rocklin CA, and Arthur Katzakian, Jr. and Stanley C. Burket, both of Sacramento, CA, assigned by: Aerojet-General Corporation

"Gas generator composition and method", application No. 167,943, August 2nd, 1971, to: Hubert G. Timmerman, of Manhattan Beach, CA, and Vincent O. Catanzarite, of Las Vegas, CA, assigned by: Specialty Products Development Corporation

"Solid propellant", serial No. 181,794, March 22nd, 1962, to: Theodore C. Parker, of Sunnyvale, CA, and William H. Colburn, Jr., of Irvington, CA, assigned by: General Precision, Inc.

"Castable ammonium perchlorate propellant compositions", serial No. 241,506, December 3rd, 1962, to: Hans H. Ender, of Buffalo, NY, assigned by: Union Carbide Corporation

"Solvent process for preparing epoxy base propellants", serial No. 337,107, January 10th, 1964, to: Walter S. Baker, Yucaipa, CA, assigned by: United States of Army

"Composite polysulfide propellants containing addictives for producing extremely fast burning", serial No. 330,365, January 8th, 1953, to: John H. Robson III, of China Lake, CA

"Igniter composition for aircraft ejection apparatus", serial No. 470,302, July 7th, 1965, to: Thomas A. Doris, Jr., Palmyra, NJ, and Thomas Q. Ciccone, of Langhorne, PA, assigned by: The United States Army

"Staple containing solid propellant grain and method of preparation", serial No. 511,027, November 30th, 1965, to: David C. Sayles, of Huntsville, AL, assigned by: The United States Army

"High energy solid propellants containing fluoropolymers and metallic fuels", serial No. 53,566, September 1st, 1960, to: Isadore Shapiro, of Beverly Hills, CA, and Albert Cane, of Los Angeles, CA, assigned by: Hughes Tool Company, Aircraft division

"Composite ammonium nitrate propellants containing boron", serial No. 743,465, June 20th, 1958, to: Charles C. Bice, or Bartlesville, OK, assigned by: Phillips Petroleum Company

"Solid composite propellants containing inorganic oxidizing and polymeric quaternary ammonium compounds", serial No. 284,447, April 25th, 1952, to: William B. Reynolds, and James E. Pritchard, from Bartlesville, OK, assigned by: Phillips Petroleum Company

"Burning rate acceleration catalysts for solid propellant compositions", serial No. 508,314, May 13th, 1955, to: Ralph W. Lawrence, of Glendora, CA, assigned by: Aerojet-General Corporation

"Semi-solid rocket fuel with aluminum naphthenate containing about seven percent combined aluminum", serial No. 824,939 July 6th, 1959, to Charles R. Morton, of Summir, NJ, and Gerald J. Horvitz, of Woodmere, NY, and Nicholas J. Nicholas, of Greenlawn, NY, of assigned by: National Research & Development Corporation

"Composite propellant reinforced with nitrogen compounds", serial No. 703,986, December 19th, 1957, to: Reinhard W. Scharf, of McGregor, TX, assigned by: Phillips Petroleum Company

"Incendiary composition", serial No. 555,765, September 25th, 1944, to: William F. Van Loenen, Los Altos, CA, assigned by: mesne assignments, to Kaiser Aluminum & Chemical Corporation

"Improvement in explosive compounds", Letters patent No., 111,642, February 7th, 1871, to: Joseph Hafenegger, of San Francisco, CA

"Process of making explosives", serial No. 699,943, December 21st, 1898, to: Ernest Auguste, of Paris France

"Explosive compound", serial No. 40,208, December 17th, 1900, to: William A. Gill, of Tarrytown, NY, assigned by: Rendrock Powder Company

"Priming mixtures", serial No. 662,086, March 22nd, 1933, to: Joseph Doerfer, and Jacob Doerfer, both of Littleton, CO

"Ignition compositions comprising boron containing salts", serial No. 166,469, January 12th, 1962, to: Robert K. Armstrong, of Glassboro, NJ, assigned by: E.I. du Pont de Nemours and Company

"Igniter containing titanium hydride and potassium perchlorate", application No. 555754, June 1, 1976, to: Dietzel; Russel W. (Albuquerque, NM); Leslie; William B. (Albuquerque, NM), assigned by: The United States of America as represented by the United States Government (Washington, DC)

"Method of inflating an automobile passenger restraint bag", application No. 407268, October 17, 1973, to: Catanzarite; Vincent Owen (Las Vegas, NV), Specialty Products Development Corporation (Medina, OH)

"Chlorite containing pyrotechnic composition and method of inflating an inflatable automobile safety restraint", application No. 856958, December 5, 1977, to: Garner; Eugene F. (Canyon Country, CA); Hamilton; Brian K. (Canyon Country, CA), assigned by: Allied Chemical Corporation (Morris Township, Morris County, NJ)

"Pyrotechnic composition and method of inflating an inflatable automobile safety restraint", application No. 840692, October 11, 1977, to: Garner; Eugene F. (Canyon County, CA), assigned by: Allied Chemical Corporation (Morris Township, Morris County, NJ)

"Ignition enhancer coating compositions for azide propellant", application No. 905828, May 15, 1978, to: Garner; Eugene F. (Canyon County, CA); Hamilton; Brian K. (Canyon County, CA), assigned by: Allied Chemical Corporation (Morris County, Morris Township, NJ)

"Pyrotechnical delay charge", application No. 377888, May 13, 1982, to: Boberg; Tore (Karlskoga, SE), assigned by: Aktiebolaget Bofors (Bofors, SE)

"Incendiary mixture constituted of metals", application No. 555708, November 28, 1983, to: Klober; Martin (Nuremberg, DE), assigned by: Diehl GmbH & Co. (DE)

"Primer mix, percussion primer and method for initiating combustion", application No. 587344, 8 March, 1984, to: Yates, Jr.; Donald N. (Katy, TX); Oestrich; Michael (Katy, TX), assigned by: GEO Vann, Inc. (Houston, TX)

"Porous propellant grain and method of making same", application No. 062547, June 15, 1987, to: Pietz; John F. (Mesa, AZ), Talley Industries, Inc. (Meza, AZ)

"Improved fuse composition", Letters patent No. 103,931, June 7th, 1870, to: William H. Rogers, of Brooklyn, NY

"Improvement in the manufacture of safety matches", letters patent No. 149,324, April 7th, 1874, to: L. Otto P. Meyer, of Newtown, Conn.

"Inflammable composition for matches", serial No. 322,115, August 27th, 1889, to: Johann Lutz, of Vienna, Austria-Hungary, assigned by: Paul Oltosy & Sohne

"Process for producing damp-proof matches", serial No. 392,799, June 29th, 1920, to: Rene Dubrisay, of Paris, France

"Self lighting cigarette", serial No. 90,012, July 10th, 1936, to: John Biddle, of Whitwick, England, assigned by: Morris Margolis

"Smoke and illumination signal", application No. 512262, October 4, 1974, to: Beatty; Bobby D. (Bloomfield, IN); Daniel; Russell D. (Bloomington, IN); Humerickhouse; Billy J. (Odon, IN); Norris; Gary G. (Burns City, IN), assigned by: The United States of America as represented by the Secretary of the Navy (Washington, DC)

"Cinnamic acid containing pyrotechnic smoke composition", application No. 725606, September 22, 1976, to: Douda; Bernard E. (Bloomfield, IN); Tanner, Jr.; John E. (Bloomington, IN), assigned by: The United States of America as represented by the Secretary of the Navy (Washington, DC)

"Pyrotechnic filled molding powder", application No. 804190, June 6, 1977, to: Hartzel; Lawrence W. (Dayton, OH); Kettling; George E. (Cincinnati, OH), assigned by: The United States of America as represented by the United States (Washington, DC)

"Solid propellant motor", application No. 048621, May 13, 1970, to: Shafer; John I. (Pasadena, CA); Marsh, Jr.; Harold E. (La Canada, CA), assigned by: The United States of America as represented by the Administrator of the (Washington, DC)

"Pyrotechnic composition", application No. 323436, November 20, 1981, to: Olander; Donald E. (Huntington Beach, CA); Petersen; Donald W. (Tehachapi, CA), assigned by: Hi-Shear Corporation (Torrance, CA)

"Process for preparing sterile solid propellants", serial No. 440,033, March 15th, 1965, to: James E. Webb, Administrator at NASA, invented by Lawrence C. Montgomery and Frank A. Morelli

"Purification of combustion catalysts and solid propellant compositions containing the same", application No. 356676, May 2, 1973, to: Rudy; Thomas P. (Saratoga, CA); Nakagawa; Toshio W. (San Jose, CA), assigned by: United Technologies Corporation (E. Hartford, CT)

"Method for making primer constituents", application No. 182362, August 29, 1980, to: Urs; Venkataramaraj S. (Godfrey, IL), assigned by: Olin Corporation (Stamford, CT)

"Granular propellant powder based on nitrocellulose, oily nitrate ester and polyvinyl nitrate, and process", application No. 100646, to: Zeller; Bernard M. (Brunoy, FR); Couturier; Rene J. (Mennecy, FR); Tribot; Robert A. (Hiersac, FR), assigned by: Societe Nationale des Poudres et Explosifs (Paris, FR)

"Method for binary propellant", application No. 183856, September 3, 1980, to: Levenson; Michael K. (Tempe, AZ), assigned by: Pyrodex Corporation (Shawnee Mission, KS)

"Increasing burning rate of solid propellants by electric field effects", application No. 222400, January 7, 1981, to: Sayles; David C. (Huntsville, AL), assigned by: The United States of America as represented by the Secretary of the Army (Washington, DC)

"Smokeless powder", serial No. 281,360, May 28th, 1928, to: Sereno G. Norton, of Brunswick, GA, assigned by: Hercules Powder Company

"Smokeless powder", serial No. 485,417, September 13th, 1893, Mason E. Leonard, of Manchester, NJ, assigned by: The Leonard Smokeless Powder Company

"Combustible compound", serial No. 611,372, November 7th, 1896, to: David Duncan Esson, of Faversham England, assigned by: Esson Motor, Limited

"Smokeless gunpowder", serial No. 96,175, February 28th, 1902, to: Carl Olof Lundholm, of Stevenston, Scotland

"Smokeless powder", serial No. 327,367, July 23rd, 1906, to: Allan Thomas Cocking, of Four Oaks, Birmingham, England

"Smokeless gunpowder", serial No. 329,801, August 9th, 1906, to: Conrad Paul Hubert Claessen, of Wilmersdorf, Berlin Germany

"Gunpowder", serial No. 335,819, September 22nd, 1906, to: Conrad Paul Hubert Claessen, of Berlin Germany

"Propellant mixing", serial No. 64,691, October 25th, 1960, to: Willem Schaafsma, Los Altos, CA assigned by: United Aircraft Corporation

"Nitronium perchlorate propellant composition", serial No. 268,457, March 27th, 1963, to: Thomas N. Scortia, of Los Gatos, and Cornelius J. McCormick, of Palo Alto, CA, assigned by: United Aircraft Corporation

"Solid propellant compositions having epoxy cured, carboxy-terminated rubber binder", application No. 667623, September 6, 1967, to: Elrick; Donald E. (Rawlings, MD); Gilbert; Harry (Cumberland, MD), assigned by: Hercules Incorporated (Wilmington, DE)

"Incendiary Gels", serial No. 508,632, November 1st, 1943 to: Louis F. Fieser, of Belmont, MASS, assigned by: The United States Army

"Process of preparing a Napalm composition containing a finely divided inert carrier", serial No. 250,560, October 9th, 1951, to: Albert O. Herron of Cheshire, Conn, assigned by: Safety Fuel & Chemical Corporation

"Solid propellant with burning rate catalyst", application No. 678523, November 8th, 1977, to: Richard A. Biddle, of Newark, DE, assigned by: Thiokol Corporation

"Inhibited solid propellant composition containing beryllium hydride", application no. 849272, August 5th, 1969, to: T.O. Paine, of Administrator of the National Aeronautics and Space, and Wallace w. Thompson

"Smoke agent", serial no. 562,328, November 7th, 1944, to: John P. Clay and Henry B Elkins, assigned by: The United States Army

"Flash-light composition", serial no. 585,011, March 26th, 1896, to: York Schwartz. Of Hanover, Germany

"Process for manufacturing of fuses and composition therefore", serial no. 6,827, February 6th, 1948, to: Henryk Zenftman and Joseph E. Forlin, both of Salt coats, Scotland, assigned by: Imperial Chemical Industries Limited

"Nitrocellulose compositions and method for making same", serial No. 464,899, October 26th, 1954, to: Lester L. Well, of Alexandria, VA, assigned by: Atlantic Research Corporation

"Nitroguanidine propellant composition and process of preparation", serial no. 83,849, January 19th, 1961, to: Russell L. Trask, or Morris Plains, NJ, Samual Sage, of Bronx, NY, and Isidore G. Nadel, of Little Falls, NJ, assigned by: United States Of America

"Gas generating Material", application number: 547623, June 28th, 1990, to: George W. Goetz, of Rochester Hills, MD, and Thomas H. Vos, of Rochester, MI, assigned by: TRW Vehicle Safety System, Inc.

"Fuse powder for metal delays", Serial No. 609,488, May 5th, 1932, to: Jean Piccard, of Marshalltown, DEL, assigned by: Hercules Powder Company

"Plasticized nitrocellulose propellant composition containing aluminum hydride and nitronium perchlorate", Application 465,252, filed: June 16th, 1965, to: James P. Flynn, George A. Lane, and John J. Plonmer, all of Midland, MI, assigned by: Adamas Carbide Corporation

"Composite Solid Propellant with additive to improve the mechanical properties thereof", Serial No. 105,938, January 12th, 1971, to: Henry C. Allen, of Decatur, Alabama, assigned by: The United States of America as represented by the US Army

"Extrusion aids for extrudable solid compositions", Serial No. 783,646, filed December 19th, 1958, to: Charles C. Bice, of Bartlesville, OK, assigned by: Phillips Petroleum Company

"Smokeless powder and process of making same", Serial No. 12,024, March 4th, 1915 to: James Madison Brown, of Johnson City, TN

"Smokeless powder", Serial No. 169,030, August, 10th, 1903, to: James L. Brown, of Indianhead, Maryland, and David P. Moore, of Washington DC, assigned by: Manuel Antonio Gomez Himalaya, of Arcos-de-val-de-vez Portugal

"Composition for fuses and process of making same", Serial No. 175,485, October 2nd, 1903, to: Wilhelm Theodore Unge, of Stockholm, Sweden

"Ignition composition for matches", Serial No. 13,065, April 26th, 1900, to: John Landin and August Jernander, of Stockholm, Sweden

"NH4NO3 Nitrocellulose coated with nitrocellulose and aluminum", application No: 240,299, filed: March 31st, 1972, to: Louis Leneveu, Frank Pierre Antoine Villey-desmeserets, both from Point de Buis, of France, assigned by: Etat Francais represente par le Ministre d'Etat Charge de la Defense Nationale-Delegation Ministerielle pour l'Armement-Direction des Pourders

"Signal torch", serial No. 758,343, December 19th, 1934, to: Harry E. Sherman, of Alton, ILL, assigned by: The Equitable Powder Manufacturing Company

"Solid gas generating charge", serial No. 128,240, November 18th, 1949, to: Alexander C. Hutchison, of Saltcoats, Scotland, assigned by: Imperial Chemical Industries Limited

"Propellant bulk smokeless powder", Serial No. 344,244, filed December 11th, 1919, to: Richard G. Woodbridge, Jr., of Wilmington, Delaware, assigned by: E.I. Du Pont De Nemours & Company

"Propellant bulk smokeless powder", Serial No. 360,194, filed February 20th, 1920, to: Richard G. Woodbridge, Jr., of Wilmington, Delaware, assigned by: E.I. Du Pont De Nemours & Company

"Novel Polybasic Propellant", serial No. 180,853, filed by September 15th, 1971, to: helmuth Nolte and Paul Linemann, both of waldkraiburg, Germany, and Walter Hahn, of Aschau, Germany, assigned by: Wasagchemie G.m.b.H., of Munich Germany

"Process for preparing rocket fuel containing polymerized olefins and boron", serial No. 179,889, filed August 16th, 1950, to: Edward Whitworth, of West Kilbride, Scotland, assigned by: Imperial Chemical Industries Limited, UK

"Propellant", serial No. 340,663, March 5th, 1953, to: John D. Clark, of Webster Groves, MO, assigned by: Olin Mathieson Chemical Corporation

"Propellant and process for manufacture", serial No. 263,047, March 3rd, 1939, to: George C. Hale, of Dover, NJ, assigned by: The United States Army

"Propellant powder and process of making the same", serial No. 358,708, May 1st, 1929, to: William E. Wagner, of Alton ILL, assigned by: Western Cartridge Company

"Propellant composition", serial No. 612,967, Filed: October 1st, 1956, to: Frank B. Cramer, of Reseda, CA, assigned by: North American Aviation, Inc.

"Solid rocket propellants containing polylactams", serial No. 719,501, Filed: March 6th, 1958, to: Ross M. Hedrick and Edward H. Mottus, both of Dayton, OH, assigned by: Monsanto Chemical Company

"Slow burning propellant composition", serial No. 767,752, Filed: October 16th, 1958, to: William A. Gey, of China Lake, CA, assigned by: United States of America as represented by the US Navy

"Novel pyrotechnics" serial No. 154,909, filed: November 24th, 1961, to: Martin H. Kaufman, of China Lake, CA, and Julian S. Davidson, of Glendale, CA

"Flare composition", serial No. 566,317, Filed: February 17th, 1956, to: Edgar A. Cadwallader, of Rockville, MD, assigned by: United States Navy

"High Temperature Composite propellant system", serial No. 615,345, Filed: February 13th, 1967, to: Joseph T. Hamrick, of Roanoke County, VA, 24014

"Ignition material containing Tellurium Dioxide, Boron and Fluoropolymeric binder", serial No. 619,567, Filed: March 1st, 1967, to: Frank H. Gardner, of Huntington Beach, CA, assigned by: Space Ordnance Systems, Inc.

"Gas generating compositions containing hydroxyl ammonium oxalate coolants and method for their preparation", serial No. 266,497, filed: July 6th, 1965, to: Ernest S. Sutton, Jr., of Newark, Del, assigned by: Thiokol Chemical Corporation

"Solid composite smokeless slow burning, low flame propellant", serial No. 276,703, Filed: March 14th, 1952, to: George W. Batchelder, of Glendora, CA, assigned by: Mesne Assignments, and Aerojet-General Corporation

"Propellant compositions", serial No. 399,596, Filed: December 21st, 1953, to: George W. Batchelder, of Glendora, CA, and Gilbert A. Zimmerman, of Monrovia, CA, assigned by: Aerojet-General Corporation

"Propellant compositions", serial No. 428,791, Filed: May 10th, 1954, to: Ralph W. Lawrence, of Gelndora, CA, assigned by: Aero-jet-General Corporation

"Carboxyl-terminated linear polyester gas-generating composition and method of preparation", serial No. 207,063, Filed: July 2nd, 1962, to: Calvia W. Vriesen, of Brookside, Del, assigned by: Thiokol Chemical Corporation

"Flare composition comprising magnesium, sodium perchlorate, and a ternary organic binder", serial No. 770,380, Filed: October 24th, 1968, to: Yoshiyuki Arikawa, of Destin Fl, and Hal R. Waite, of Walton Beach, Fl, assigned by: Unites States Navy

"Calcium containing pyrotechnic compositions for high altitudes", serial No. 375,596, Filed: June 8th, 1964, to: Seymour Lopatin, of Dover, NJ, and David Hart, of Dover, NJ, assigned by: United States Army

"Ammonium nitrate propellant with low flame temperature exhaust gases", serial No. 210,829, Filed: July 18th, 1962, to: Don E. Kennedy, of Palos Verdes Estates, CA, assigned by: Standard Oil Company

"Nitroguanidine propellant composition and process of preparation", serial No. 83,849, Filed: June 19th, 1961, to: Russell L. Trask, of Morris Plains, NJ, Samuel Sage, of Bronx, NY, and Isidore G. Nadel, of Little Falls, NJ, assigned by: United states Army

"Pyrotechnic composition", serial No. 245,627, Filed: December 18th, 1962, to: Mary M. Williams, of Santa Monica, and Lohr A. Burkardt, of China Lake, CA, assigned by: United states Navy

"Pyrotechnics comprising silver Iodate, Ammonium nitrate, Nitrocellulose, and Nitrate esters, serial No. 609,968, Filed: January, 17th, 1967, to: William G. Finnegan and Lohr A. Burkardt, both of China Lake, CA, assigned by: United states of Navy

"Composite propellants containing activated carbon", serial No. 801,392, Filed: March 23rd, 1959, to Myron L. Dickey, of Whittier, CA, assigned by: Phillips Petroleum Company

"Solid propellant and a process for its preparation", serial No. 725,453, Filed: March 31st, 1958, to: Eugene D. Guth, of Idaho Falls, ID, assigned by: Phillips Petroleum Company

"Self-Hardening pyrotechnic composition", serial No. 121,216, Filed: October 13th, 1949, to: David Hart and Henry J. Eppig, of Dover NJ, and William J. Powers, of Wharton, NJ

"Solid propellant compositions and process for making same", serial No. 183,457, Filed: September 6th, 1950, to: Lester L. Weil, of Alexander, VA, assigned by: Atlantic Research Corporation

"Illuminants", serial No. 749,577, Filed: July 15th, 1958, to: Jean A. M. Brock, of St. Louise, MO, assigned by: Olin Mathieson Chemical Corporation

"Propellant mixing", serial No. 64,691, Filed: October 25th, 1960, to: Willem Schaafsma, of Los Altos, CA, assigned by: United Aircraft Corporation

"Pyrotechnic compound", serial No. 525,611, Filed: October 11th, 1894, to: Charles Gerhard, of Jersey City, NJ

"Pyrotechnic composition", serial No. 343,573, Filed: December 9th, 1919, to: Walter T. Scheele, of Hackensack, NJ, assigned by: United States Navy

"Process of producing match-composition ingredients", serial No. 592,598, Filed: November 16th, 1910, to: Henry Staier, of New York, NY

"Igniting substance", serial No. 307,851, Filed: June 30th, 1919, to: Regino A. Martinez, of Buenos Aires, Argentina, assigned by: Pedro Eugenio Girard

"Pyrotechnic device", serial No. 494,452, Filed: November 8th, 1930, to: Herbert Clauser, of Elkton, MD

"Ignition and priming compound", serial No. 63,555, Filed: February 12th, 1936, to: Mario Palmieri, of South Glastonbury, of Conn, assigned by: one-half to Samuel D. Ehrlich

"Noncorrosive priming compositions", serial No. 194,384, Filed: March 6th, 1938, to: Charles H. Pritham, of Upper Darby, and Ernest R. Rechel, and Thomas Stevenson, both of Philadelphia, PA

"Fuse powder composition", serial No. 535,054, Filed: May 11th, 1944, to: George C. Hale, and David Hart, both of Dover NJ

"Flash powder", serial No. 136,384, Filed: December 11th, 1916, to: John L. Crabtree, of Rochester, NY, assigned by: Eastman Kodak Company

"Pyrotechnic composition", serial No. 456,986, Filed: March 30th, 1921, to: Harry J. Nichols, of Washington, DC

"Delay powder", serial No. 534,289, Filed: May 5th, 1944, to: George C. Hale, of Dover, NJ

"Delay powder", serial No. 632,101, Filed: November 30th, 1945, to: George C. Hale, and David Hart, both of Dover, NJ

"Deflagrating compound", serial No. 228,935, Filed: June 15th, 1880, to: John A. Robinson and Robert H. Dimock, both of New Haven, CONN.

"Match", serial No. 562,426, Filed: June 23rd, 1896, to: Carl Rudolph Adolph George Schwiening, of Bettenhausen, Germany

"Match-paste composition", serial No. 694,216, Filed: October 21st, 1898, to: Gustav Hackel of Vienna Austria, assigned by: Carl Kastner

"Match-composition", serial No. 698,352, Filed: December 5th, 1898, to: Walter G. Cordes, of London England, assigned by: Non-Poisonous "strike anywhere" match syndicate, limited

"Match composition", serial No. 178,891, Filed: October 28th, 1903, to: John A. Wendel, of Milwaukee, WI

"Match composition", serial No. 188,652, Filed: January 11th, 1904, to: Frederick E. Grimm, of Utica, NY

"Match composition", serial No. 206,855, Filed: May 7th, 1904, to: Henry Staier, of Brooklyn, NY

"Pyrotechnic compound", serial No. 524,513, Filed: October 25th, 1909, to: Bernard J. Dever, and Bertram J. Delzeit both of Philadelphia, PA

"Priming composition and method of preparing the same", serial No. 763,071, Filed: April, 23rd, 1913, to: Maurice R. Swope, of Alton, ILL

"Match-head composition", serial No. 298,429, Filed: May 20th, 1919, to: John R. Mardick, of NY, NY

"Gas producing nondetonating composition", serial No. 101,273, Filed: June 24th, 1949, to: James Taylor, and Alexander Cantlay Hutchison, both of Saltscoats, Scotland, assigned by: Imperial Chemical Industries Limited

"Ignition mixture", serial No. 581,538, Filed: March 7th, 1945, to: Joseph, H. McLain, and Theodore A. Ruble, of USA

"Gas producing compositions", serial No. 667,969, Filed: June 25th, 1957, to: Kenneth Urmston Holker, of West Kilbride, Scotland, assigned by: Imperial Chemical Industries Limited, London, England

"Combustible composition", serial No. 638,925, Filed: February 8th, 1957, to: David E. Pearsall, of Avon, CONN, assigned by: Bickford Research Laboratories, INC.

"Fuel and propellant composition", serial No. 721,697, Filed: March 17th, 1958, to: Harry A. Toulmin, Jr., of Dayton, OH, assigned by: The Commonwealth Engineering Company of Ohio

"Tracer mechanism", serial No. 776,655, Filed: November 26th, 1958, to: Nicholas, M. Matusewicz, of Upland, CA, assigned by: United Sates Navy

"Percussion sensitive pyrotechnic or pyrophoric alloy-type priming mixture", serial No. 371,813, Filed: June 1st, 1964, to: Donald S. Foote, of Green Farms, and Richard A. Sahlin, of Fairfield, CONN, assigned by: Remington Arms Company, INC.

"Colored smoke", serial No. 270,576, Filed: January 10th, 1919, to: Edwin C. Weisgerber, of The United states Army

"Composition for matches and process for making same", serial No. 346,111, Filed: December 19th, 1919, to: Manuel Prado P., Santiago, Chile

"Pyrotechnic composition for producing yellow smoke", serial No. 652,430, Filed: January 18th, 1933, to: George U. Graff, of Washington, DC

"Pyrotechnic composition for producing brown smoke", serial No. 652,427, Filed: January 18th, 1933, to: George U. Graff, of Washington DC

"Pyrotechnic composition for producing orange smoke", serial No. 652,431, Filed: January 18th, 1933, to: George U. Graff, of Washington DC

"Pyrotechnic compositions", serial No. 598,539, Filed: March 12th, 1932, to: Joseph B. Decker and Herbert C. Clauser, both of Elkton, MD

"Pyrotechnic composition for producing pink smoke", serial No. 40,796, Filed: September 16th, 1935, to: George U. Graff, of Washington, DC

"Flashlight powder", serial No. 66,751, Filed: March 2nd, 1936, to: Albert Kalber, of Pforzheim, Germany

"Material for use in welding", serial No. 245,954, Filed: December 15th, 1938, to: Charles A. Cadwell, of Cleveland Heights, OH, assigned by: The Electric Railway Improvement Company

"Exothermic mixture", serial No. 338,585, Filed: June 3rd, 1940, to: Walter M. Bruner, of Wilmington, DEL, assigned by: W.G. Shelton Company

"Welding", serial No. 375,099, Filed: January 21st, 1941, to: Charles A. Cadwell, of Cleveland, OH, assigned by: Electric Railway Improvement Company

"Igniter compositions", serial No. 329,111, Filed: December 31st, 1952, to: Franklin B. Clay, of Westport CONN, and Richard A. Sahlin, Fairfield, CONN, assigned by: Remington Arms Company

"Material For Welding Aluminum and other metals, serial No. 554,234, Filed: December 20th, 1955, to: Melvin F. rejdak, of Cleveland, OH, assigned by: Erico Products, INC.

"Pyrotechnic composition", serial No. 729,526, Filed: April 21st, 1958, to: Jean A. M. Brock, of St. Louise, MO, assigned by: Olin Mathieson Chemical Corporation

"Heat producing mixtures", serial No. 3,262, Filed: January 19th, 1960, to: Albert Stanley John Cross, of Birmingham, England, assigned by: Foundry Services International Limited

"Pyrotechnic compositions", serial No. 758,173, Filed: August 27th, 1958, to: Donald E. Olander, of Florissant, MO, assigned by: Universal Match Corporation

"Method of propulsion using solid propellant compositions", serial No. 489,731, Filed: February 18th, 1955, to: Horace M. Higgins, and Calvin A. gongwer, both of Glendora, CA, assigned by: Aerojet-General Corporation

"Metallic composition for production of hygroscopic smoke", serial No. 375,402, Filed: June 15th, 1964, to: Lohr A. Burkardt, and William G. Finnegan, both of China Lake, CA, assigned by: United States Navy

"Pressure compensated pyrotechnic time delay composition", serial No. 147,690, Filed: May 27th, 1971, to: Roswitha Zimmer-Galler, of Niceville, FL, and James E. Rose, of Bryans Road, MD, assigned by: United States Navy

"Pyrotechnic delay compositions containing heavy metal soap", serial No. 236,173, Filed: March 20th, 1972, to: James E. Rose, of Bryans Road, MD, and Roswitha Zimmer-Galler, of Niceville, FL, assigned by: United States Navy

"Thermite composition and method of making", serial No. 234,855, Filed: March 15th, 1972, to: John W. Jones, of Redlands, CA, assigned by: Lockheed Aircraft Corporation

"First fire and igniter composition", serial No. 717,275, Filed: February 24th, 1958, to: David Hart, of Dover, NJ, assigned by: United States Army

"Chemically produced colored smokes", serial No. 767,751, Filed: October 16th, 1958, to: Lohr A. Burkardt, William G. Finnegan, and, Rex L. Smith, all from China Lake, CA

"Hafnium-potassium perchlorate pyrotechnic composition", serial No. 214,147, Filed: August 1st, 1962, to: Robert E. Betts, of Huntsville, Alabama, assigned by: United States Army

"Smokeless ashless signal flare composition containing ammonium perchlorate", serial No. 385,518, Filed: July 27th, 1964, to: James B. McGriffin, of Linton, IND, and William L. Ripley, of Bedford, IND, assigned by: United States Navy

"Zirconium composition with potassium perchlorate and graphite", serial No. 168,555, Filed: January 24th, 1962, to: Charles H. Jackson, of Redondo Beach, CA, assigned by: Hi-Shear Corporation

"Long-burning pyrotechnic material containing depleted uranium for spotting rifle projectiles", serial No. 167,449, Filed: January 19th, 1962, to: Thomas Stevenson, of Huntington Valley, PA, assigned by: United States Army

"Pyrotechnic Compositions", serial No. 155,023, Filed: July 22nd, 1937, to: Charles H. Pritham, of Philadelphia, PA

"Nonluminous pyrotechnic mixture for a projectile", serial No. 35,961, Filed: June 29th, 1948, to: Raymond H. Heiskell, of Compton, CA

"Smoke tracer composition", serial No. 484,599, Filed: January 27th, 1955, to: Thomas Stevenson, of Huntington Valley, PA, and Winton W. Cavell, of Philadelphia, PA, assigned by: United States Army

"Priming mixture", serial No. 489,587, Filed: August 3rd, 1921, to: James E. Burns, of Lowell, MASS

"Propellant powder", serial No. 381,389, Filed: July 26th, 1929, to: Joseph D. McNutt, of New Haven, CONN, assigned by: Mesne Assignments, to Winchester Repeating Arms Company

"Flash Composition", serial No. 609,489, Filed: May 5th, 1932, to: Jean Piccard, of Marshallton, DEL, assigned by: Hercules Powder Company

"Ignition composition", serial No. 623,473, Filed: October 20th, 1945, to: William F. Filbert, of Woodbury, NJ, assigned by: E.I. du Pont de Nemours & Company

"Ammonium nitrate compositions", serial No. 645,641, Filed: March 12th, 1957, to: Aubrey Edward Harper, of Glasgow, Scotland, Kenneth Harrison, of Saltcoats, Scotland, and Edward Graham Cooke, of West Kilbride, Scotland, assigned by: Imperial Chemical Industries Limited, London

"Gas-producing charges suitable for the generation of gas pressure for the operation of mechanical devices and for blasting operations", serial No. 615,195, Filed: September 7th, 1945, to: Alexander Thomson Tyre, of Newton Stewart, Scotland, assigned by: Imperial Chemical Industries Limited

"Process for preparing sterile solid propellants", serial No. 440,033, Filed: March 15th, 1965 to: James E. Webb, Administrator of the National Aeronautics and Space Administration, to Lawrence C. Montgomery and Frank A. Morelli

"Propellant powder to be used in guns and howitzers", serial No. 520,131, Filed: December 5th, 1921, to: John B. Fidlar, of The United States Army

"Propellant powder", serial No. 146,835, Filed: November 6th, 1926, to: Charles H. Keck, of United States Army

"Fleshless propellant powder composition", serial No. 433,717, Filed: March 7th, 1942, to: George Barsky, of NY, NY, assigned by: E.F. Drew & Co., Inc.

"Ignitable Stick", serial No. 358,383, Filed: April 26th, 1929, to: Ferdinand Ringer, of Vienna Austria

"Colored smoke producing composition", serial No. 270,570, Filed: January 10th, 1919, to: Edwin C. Weisgerber, of Washington, D.C

"Process of manufacturing gunpowder", serial No. 209,268, Filed: December 28th, 1917, to: John Buxbaum, of Georgetown, WA

"Slow burning black powder composition", serial No. 288,135, Filed: April 7th, 1919, to: Joseph Percy Gray, of Newport, Delaware, assigned by: E.I. Du Pont de Nemours & Company

"Deflagrating composition", serial No. 534,148, Filed: May 4th, 1944, to: Russel Charrosin Payn, of Watford, England, and Elwyn Jones, of Ardrossan, England, and John Stocks Flanders, of West Kilbride, Scotland, assigned by: Imperial Chemical Industries Limited

"Slow burning powder for fuses, ect.,", serial No. 20,889, Filed: May 10th, 1935, to: David E. Pearsall, of Pittsburgh, PA, assigned by: The Ensign-Bickford Company

"Fuse powder compositions", serial No. 518,242, Filed: January 14th, 1944, to: Harrison H. Holmes, of Woodbury, NJ, and Walter E. Lawson, of Wilmington, DEL, assigned by: E.I. Du Pont De Nemours & Company

"Black powder", serial No. 120,120, Filed: January 11th, 1937, to: William H. Rinkenbach, and Vernon C. Allison, of Dover, NJ, assigned by: The United States Government

"Improvement in compositions for safety blazing fuses", serial No. 153,181, Filed: July 21st, 1874, to: George Carl Julius Schneider, of Erie, PENN

"Colored fire compound", serial No. 309,948, Filed: December 30th, 1884, to: John Herzog, of Baltimore, Maryland

"Composition for Bengal lights", serial No. 363,224, Filed: May 17th, 1887, to: Charles Gerhard, of New York, NY

"Bengal-light compound", serial No. 449,530, Filed: March 31st, 1891, to: Charles Gerhard, of Jersey City, New Jersey

"Gunpowder", serial No. 698,140, Filed: December 3rd, 1898, to: Francis A. Halsey, of San Rafael, CA, assigned by: The Economic Smokeless Powder Company

"Process of preparing pyrotechnics", serial No. 90,816, Filed: January 22nd, 1902, to: Nicholas Del Grande, of Petersburg, Virginia, assigned by: Virginia Fireworks Company

"Semi-smokeless powder", serial No. 305,001, Filed: March 8th, 1906, to: George W. Gentieu, of Peoria, ILL

"Gunpowder", serial No. 222,631, Filed: December 27th, 1886, to: Eduard, Schultze, of Darmstadt, Grand Duchy of Hesse, Germany

"Pyrotechnic compound", serial No. 82,427, Filed: November 15th, 1901, to: Antonio Delgrande, of Petersburg, Virginia, assigned by: Two Thirds, to Robert H. Seabury, and Nicholas Delgrande, both of Petersburg, VA

"Igniter charge for blasting caps", serial No. 582,871, Filed: December 23rd, 1931, to: Lloyd H. Fisher, of Kenvil, NJ, assigned by: Hercules Powder Company

"Improvement in pyrotechnic signals", Filed: July 26th, 1873, to: Alexandre Lamarre, of Paris France, assigned by: Gratiot Washburne

"Material for producing smoke screens", serial No. 271,126, Filed: January 14th, 1919, to: Harry F. French and Raymond C. Benner, both of Fremont, OH, assigned by: National Carbon Company, Inc.

"Chemical compounds", serial No. 161,031, Filed: August 26th, 1937, to: Willi Brun, of Bridgeport, CONN, assigned by: Remington Arms Company, Inc.

"Process of producing an incendiary composition", serial No. 516,313, Filed: December 30th, 1943, to: Morris S. Kharasch, of Chicago, ILL, assigned by: Untied States by the Secretary of War

"Solid propellant containing improved asphalt black", serial No. 244,588, Filed: December 14th, 1962, to: Robert J. Convery, of Wilmington, DEL, assigned by: Sun Oil Company

"Ammonium nitrate composite propellant and method of preparation", serial No. 527,133, Filed: August 8th, 1955, to: Barnet R. Adelman, of Waco, TX, assigned by: Phillips Petroleum Company

"Pyrotechnic disseminating composition", serial No. 667,043, Filed: September 1st, 1967, to: Alan C. Kott, of Mount Pleasant, and Erwin M. Jankowiak, and George A. Lane, both of Midland, Michigan, assigned by: Dow Chemical Company

"Smokeless powder", serial No. 587,846, Filed: April 16th, 1896, to: Francis A. Halsey, of San Rafael, CA

"Cool, nitrocellulose base, non-carbon forming propellant", serial No. 101,020, Filed: April 5th, 1961, to: Beverley W. Lewis, of Hampton, VA, and Carl Boyars, of Silver Spring, MD, assigned by: United States Navy

"Smoke and illumination signal", application No. 512,262, filed: October, 4th, 1974, to: Bobby D. Beatty, of Bloomfield, IN, Russel D. Daniel, of Bloomington, IN, Billy J. Humerickhouse, of Odon, IN, and Gary G. Norris, of Burns City, IN, assigned by: The United States Navy

Printed in Poland
by Amazon Fulfillment
Poland Sp. z o.o., Wrocław

53761536R00210